Progress in Nonlinear Differential Equations and Their Applications
Volume 64

Nonlinear Elliptic and Parabolic Problems

A Special Tribute to the Work of Herbert Amann

Michel Chipot
Joachim Escher
Editors

Birkhäuser
Basel · Boston · Berlin

Editors:

Michel Chipot
Universität Zürich
Angewandte Mathematik
Winterthurerstr. 190
8057 Zürich
Switzerland
e-mail: chipot@amath.unizh.ch

Joachim Escher
Institute of Applied Mathematics
University of Hannover
Welfengarten 1
30167 Hannover
Germany
e-mail: escher@ifam.uni-hannover.de

2000 Mathematics Subject Classification 35Jxx, 35Kxx, 37Gxx, 76xx

A CIP catalogue record for this book is available from the Library of Congress, Washington D.C., USA

Bibliographic information published by Die Deutsche Bibliothek
Die Deutsche Bibliothek lists this publication in the Deutsche Nationalbibliografie; detailed bibliographic data is available in the Internet at <http://dnb.ddb.de>.

ISBN 3-7643-7266-4 Birkhäuser Verlag, Basel – Boston – Berlin

Part of Springer Science+Business Media
Printed on acid-free paper produced of chlorine-free pulp. TCF ∞
Printed in Germany
ISBN 10: 3-7643-7266-4
ISBN 13: 978-3-7643-7266-8
e-ISBN 3-7643-7385-7

9 8 7 6 5 4 3 2 1

www.birkhauser.ch

Contents

Preface

Celebrating the work of a renowned mathematician, it is our pleasure to present this volume containing the proceedings of the conference "Nonlinear Elliptic and Parabolic Problems: A Special Tribute to the Work of Herbert Amann", held in Zurich, June 28–30, 2004.

Herbert Amann had a significant and decisive impact in developing Nonlinear Analysis and one goal of this conference was to reflect his broad scientific interest. It is our hope that this collection of papers gives the reader some idea of the subjects in which Herbert Amann had and still has a deep influence. Of particular importance are fluid dynamics, reaction-diffusion systems, bifurcation theory, maximal regularity, evolution equations, and the theory of function spaces.

The organizers thank the following institutions for provided support for the conference:

- Swiss National Foundation
- Zürcher Hochschulstiftung
- Zürcher Universitätsverein
- Mathematisch-naturwissenschaftliche Fakultät (MNF).

Finally, it is our pleasure to thank all contributors, referees, and Birkhäuser Verlag, particularly T. Hempfling for their help and cooperation in making possible this volume.

The Editors

Progress in Nonlinear Differential Equations
and Their Applications, Vol. 64, 1–15

Bounded Imaginary Powers and H_∞-Calculus of the Stokes Operator in Unbounded Domains

Helmut Abels

Abstract. In the present contribution we study the Stokes operator $A_q = -P_q\Delta$ on $L^q_\sigma(\Omega)$, $1 < q < \infty$, where Ω is a suitable bounded or unbounded domain in $\mathbb{R}^n$, $n \geq 2$, with $C^{1,1}$-boundary. We present some conditions on Ω and the related function spaces and basic equations which guarantee that $c + A_q$ for suitable $c \in \mathbb{R}$ is of positive type and admits a bounded H_∞-calculus. This implies the existence of bounded imaginary powers of $c + A_q$. Most domains studied in the theory of Navier-Stokes like, e.g., bounded, exterior, and aperture domains as well as asymptotically flat layers satisfy the conditions. The proof is done by constructing an approximate resolvent based on the results of [3], which were obtained by applying the calculus of pseudo-differential boundary value problems. Finally, the result is used to proof the existence of a bounded H_∞-calculus of the Stokes operator A_q on an aperture domain.

Mathematics Subject Classification (2000). 35 Q 30, 76 D 07, 47 A 60, 47 F 05.

Keywords. Stokes equations, exterior domains, bounded imaginary powers, H_∞-calculus, aperture domain.

1. Introduction

In this article we consider the Stokes operator $A_q = -P_q\Delta$ on $L^q_\sigma(\Omega)$ with domain

$$\mathcal{D}(A_q) = \{f \in W^2_q(\Omega)^n : f|_{\partial\Omega} = 0\} \cap L^q_\sigma(\Omega)$$

where $P_q\colon L^q(\Omega)^n \to L^q_\sigma(\Omega)$ denotes the Helmholtz projection,

$$L^q_\sigma(\Omega) := \overline{C^\infty_{0,\sigma}(\Omega)}^{\|.\|_q}, \quad C^\infty_{0,\sigma}(\Omega) := \{u \in C^\infty_0(\Omega)^n : \operatorname{div} u = 0\}, \text{ and } \Omega \subseteq \mathbb{R}^n,\ n \geq 2,$$

is a domain specified in Assumption 1.1 below. Properties of the Stokes operator are important for the associated instationary Stokes and Navier-Stokes equations. Since the latter equations arise in mathematical fluid mechanics, many different kinds of bounded and unbounded domains are of interest and have been studied.

The purpose of the present contribution is to present some conditions on Ω and the related function spaces which guarantee that $c + A_q$ for suitable $c \in \mathbb{R}$ is of positive type and admits a bounded H_∞-calculus w.r.t. $\delta \in (0,\pi)$. Here $c + A_q$ is of positive type w.r.t. δ if and only if $\Sigma_\delta \cup \{0\} \subseteq \rho(-c - A_q)$ and

$$\|(\lambda + c + A_q)^{-1}\|_{\mathcal{L}(L^q_\sigma(\Omega))} \le \frac{C_{q,\delta}}{|\lambda|}, \qquad \lambda \in \Sigma_\delta, \tag{1.1}$$

where $\Sigma_\delta := \{z \in \mathbb{C} \setminus \{0\} : |\arg z| < \delta\}$. – Note that, if $\delta > \frac{\pi}{2}$, (1.1) implies that $-c - A_q$ generates a bounded, strongly continuous, analytic semi-group. – Moreover, $A := c + A_q$ is said to admit a *bounded* H_∞*-calculus* w.r.t. δ if and only if

$$h(A) := \frac{1}{2\pi i}\int_\Gamma h(-\lambda)(\lambda + A)^{-1}\,d\lambda \tag{1.2}$$

is a bounded operator satisfying

$$\|h(A)\|_{\mathcal{L}(L^q_\sigma(\Omega))} \le C_{q,\delta}\|h\|_\infty \qquad \text{for all } h \in H_\infty(\delta), \tag{1.3}$$

where $H_\infty(\delta)$ denotes the Banach algebra of all bounded holomorphic functions $h\colon \Sigma_{\pi-\delta} \to \mathbb{C}$, cf. McIntosh [25], and Γ is the negatively orientated boundary of Σ_δ. We note that in order to prove (1.3) for all $h \in H_\infty(\delta)$ it is sufficient to show the estimate for $h \in H(\delta)$, which consists of all $h \in H_\infty(\delta)$ such that

$$|h(z)| \le C\frac{|z|^s}{1+|z|^{2s}} \qquad \text{for all } z \in \Sigma_{\pi-\delta}$$

for some $s > 0$, cf. [8, Lemma 2.1]. For $h \in H(\delta)$ the integral (1.2) is well defined as a Bochner integral.

The property of admitting a bounded H_∞-calculus is a generalization of possessing bounded imaginary powers since $h_y(z) = z^{iy} \in H_\infty(\delta)$ for all $\delta \in (0,\pi)$, which has many important consequences. In particular, (1.3) for $\delta > \frac{\pi}{2}$ yields the *maximal regularity* of $-c - A_q$ by the result of Dore and Venni [13].

The resolvent estimate (1.1) with arbitrary $\delta \in (0,\pi)$, $1 < q < \infty$ and $c = 0$, has been proved for various kinds of domains $\Omega \subset \mathbb{R}^n$, $n \ge 2$, cf. Giga [19] for bounded domains, Borchers and Sohr [9] and Borchers and Varnhorn [9] for exterior domains, Farwig and Sohr [16] for aperture domains, Abels and Wiegner [6] for an infinite layer $\Omega = \mathbb{R}^{n-1} \times (-1,1)$, and Abels [2] for asymptotically flat layers. Moreover, see Farwig and Sohr [15] for a general treatment and Desch, Hieber, and Prüss [12] for the case of a half-space $\mathbb{R}^n_+$ and $\mathbb{R}^n$, where also the borderline cases $q = 1, \infty$ have been studied.

The fact that A_q possesses bounded imaginary powers and admits a bounded H_∞-calculus was proved by Giga [20], Giga and Sohr [21], and Noll and Saal [27] for bounded domains and for exterior domains in $\mathbb{R}^n$, $n \ge 3$, by Giga and Sohr [22] for the half-space $\mathbb{R}^n_+$, see also [12], and by Abels [5, 4, 3] for two-dimensional exterior domains, an infinite layer, and asymptotically flat layers.

In the following we will present an approach, proving (1.1) for large λ and (1.3) for suitable c simultaneously for a class of domains, which includes all previously mentioned cases provided that some auxiliary results are known. More precisely, we make the following assumption.

Assumption 1.1. *Let $1 < q < \infty$ and $\delta \in (0,\pi)$ be fixed. Moreover, let $\Omega \subseteq \mathbb{R}^n$, $n \geq 2$, be a domain satisfying the following conditions:*

(A1) *There is a finite covering of $\overline{\Omega}$ with relatively open sets U_j, $j = 1,\ldots,m$, such that U_j coincides (after rotation) with a relatively open set of $\overline{\mathbb{R}^n_{\gamma_j}}$, where $\mathbb{R}^n_{\gamma_j} := \{(x',x_n) \in \mathbb{R}^n : x_n > \gamma_j(x')\}$, $\gamma_j \in C^{1,1}(\mathbb{R}^{n-1})$. Moreover, suppose that there are cut-off functions $\varphi_j, \psi_j \in C_b^\infty(\overline{\Omega})$, $j = 1,\ldots,m$, such that φ_j, $j = 1,\ldots,m$, is a partition of unity, $\psi_j \equiv 1$ on $\operatorname{supp}\varphi_j$, and $\operatorname{supp}\psi_j \subset U_j$, $j = 1,\ldots,m$.*

(A2) *The Helmholtz decomposition is valid for $L^r(\Omega)^n$ with $r = q, q'$, i.e., for every $f \in L^r(\Omega)^n$ there is a unique decomposition $f = f_0 + \nabla p$ with $f_0 \in L^r_\sigma(\Omega)$ and $p \in \dot{W}^1_r(\Omega)$. Moreover,*

$$L^q_\sigma(\Omega) = \{f \in L^q(\Omega)^n : \operatorname{div} f = 0, \gamma_\nu f = 0\}. \tag{1.4}$$

(A3) *For every $p \in \dot{W}^1_r(\Omega)$, $r = q, q'$, there is a decomposition $p = p_1 + p_2$ such that $p_1 \in W^1_r(\Omega)$, $p_2 \in L^r_{\mathrm{loc}}(\Omega)$ with $\nabla p_2 \in W^1_r(\Omega)$ and $\|(p_1, \nabla p_2)\|_{1,r} \leq C\|\nabla p\|_r$.*

We refer to Section 2 below for the definitions of the function spaces and the normal trace γ_ν.

Remark 1.2. *First of all, we note that* (A1) *can be generalized to the case of a* locally finite *covering U_j, $j \in \mathbb{N}$, if uniform bounds on $\gamma_j, \varphi_j, \psi_j$ in $C^{1,1}$-norm are assumed and if for every $x \in \overline{\Omega}$ the number of sets U_j containing x is bounded by a constant independent of x. Moreover, it is easy to see that* (A1) *is fulfilled for all kinds of domains with $C^{1,1}$-boundary mentioned above. We refer to* [28, 15, 16, 26, 14, 2] *for the validity of the Helmholtz decomposition for these types of domains. The characterization* (1.4) *holds as well in these cases except for an aperture domain if $q > \frac{n}{n-1}$ for which the characterization is different, cf.* [16, Lemma 3.1]. *The identity* (1.4) *is used in Lemma* 3.1 *below. Moreover,* (A3) *is a technical condition needed in the proof of Lemma* 4.2 *below. It is satisfied if the following extension property is valid: For every $p \in \dot{W}^1_q(\Omega)$ there is an extension $\tilde{p} \in \dot{W}^1_q(\mathbb{R}^n)$ such that $\tilde{p}|_\Omega = p$ and $\|\nabla\tilde{p}\|_q \leq C\|\nabla p\|_q$. This is the case for every (ε,∞)-domain, cf.* [10], *in particular, for exterior domains. This extension property does not hold for layer-like domains, cf.* [2, Section 2.4]. *Nevertheless* (A3) *is also valid in layer-like domains, cf.* [2, Lemma 2.4].

The main result is the following:

Theorem 1.3. *Let $\Omega \subset \mathbb{R}^n$, $n \geq 2$, $\delta \in (0,\pi)$, and $1 < q < \infty$ satisfy Assumption* 1.1. *Then there is an $R > 0$ such that $(\lambda + A_q)^{-1}$ exists and*

$$\|(\lambda + A_q)^{-1}\|_q \leq \frac{C_{q,\delta}}{1+|\lambda|}$$

for all $\lambda \in \Sigma_\delta$ *with* $|\lambda| \geq R$. *Moreover,*

$$\left\| \int_{\Gamma_R} h(-\lambda)(\lambda + A_q)^{-1}\, d\lambda \right\|_{\mathcal{L}(L^q_\sigma(\Omega))} \leq C_{q,\delta} \|h\|_\infty \tag{1.5}$$

for every $h \in H(\delta)$, *where* $\Gamma_R = \Gamma \setminus \overline{B_R(0)}$. *In particular, for every* $c \in \mathbb{R}$ *and* $0 < \delta' \leq \delta$ *such that* $c + \Sigma_{\delta'} \subset \rho(-A_q)$ *the shifted Stokes operator* $c + A_q$ *admits a bounded* H_∞*-calculus with respect to* δ'.

Remark: The statement of Theorem 1.3. also holds for the reduced Stokes operator introduced in Section 3 below, which is shown in the proof. We note that the characterization (1.4) is not needed in this case.

The proof of Theorem 1.3 is based on the construction of an approximative resolvent R_λ that coincides with $(\lambda + A_q)^{-1}$ modulo an operator which decays of order $O(|\lambda|^{-1-\varepsilon})$ as $|\lambda| \to \infty$ for some $\varepsilon > 0$. The construction is based on a localization procedure and a suitable result for the reduced Stokes operator on a curved half-space $\mathbb{R}^n_\gamma$, $\gamma \in C^{1,1}(\mathbb{R}^{n-1})$, cf. Theorem 4.1 below. The latter result was basically obtained in [3] and is achieved using the calculus of pseudodifferential boundary value problems developed by Grubb [23] in a non-smooth version, cf. [3, Section 4] and [1].

In particular, if it is known that $\sigma(-A_q) \subseteq (-\infty, 0]$, then Theorem 1.3 implies that $c + A_q$ admits a bounded H_∞-calculus for every $c > 0$. In order to prove that this is also true for $c = 0$ it remains to analyze the resolvent near 0. This is done for the case of an aperture domain in $\mathbb{R}^n$, $n \geq 2$, in Section 5 below, where the following result is proved:

Theorem 1.4. *Let* $1 < q < \infty$, $\delta \in (0, \pi)$, *and let* $\Omega \subset \mathbb{R}^n$, $n \geq 2$, *be an aperture domain with* $C^{2,\mu}$*-boundary,* $\mu > 0$, *as defined in Section* 5 *below. Then the Stokes operator* A_q *admits a bounded* H_∞*-calculus with respect to* δ.

In particular, this yields that the Stokes operator has maximal regularity on $L^q_\sigma(\Omega)$, which was also obtained by Fröhlich [18] in the context of weighted L^q-spaces.

2. Preliminaries

Let $\Omega \subseteq \mathbb{R}^n$, $n \geq 2$, be a domain. Then $C_0^\infty(\Omega)$ denotes the set of all smooth $f\colon \Omega \to \mathbb{C}$ with compact support, and

$$C^\infty_{(0)}(\overline{\Omega}) := \{u = v|_{\overline{\Omega}} : v \in C_0^\infty(\mathbb{R}^n)\}.$$

Moreover, $C_b^\infty(\overline{\Omega})$ denotes the set of all smooth and bounded $f\colon \overline{\Omega} \to \mathbb{C}$ with bounded derivatives and $C^{1,1}(\mathbb{R}^n)$ is the set of all bounded functions $f\colon \mathbb{R}^n \to \mathbb{C}$ with bounded and globally Lipschitz continuous first-order derivatives.

The usual Lebesgue space will be denoted by $L^q(\Omega)$ and $L^q(\partial\Omega)$, $1 \leq q \leq \infty$, normed by $\|.\|_q \equiv \|.\|_{L^q(\Omega)}$ and $\|.\|_{q,\partial\Omega} \equiv \|.\|_{L^q(\partial\Omega)}$, respectively. Furthermore, $L^q_{\mathrm{loc}}(\overline{\Omega})$ consists of all $f\colon \Omega \to \mathbb{C}$ such that $f \in L^q(B \cap \Omega)$ for all balls B with

$B \cap \Omega \neq \emptyset$. The scalar product on $L^2(M)$ is denoted by $(.,.)_M$ for $M = \Omega$ or $M = \partial\Omega$. Moreover, the usual Sobolev-Slobodeckij spaces based on L^q, $1 < q < \infty$, are denoted by $W^s_q(\Omega)$ and $W^s_q(\partial\Omega)$, $s \geq 0$, with norms $\|.\|_{s,q}$ and $\|.\|_{s,q,\partial\Omega}$, resp., cf., e.g., [7]. As usual $W^s_{q,0}(\Omega)$, $s \geq 0$ with $s - \frac{1}{q} \notin \mathbb{N}$, is defined as the closure of $C^\infty_0(\Omega)$ in $W^s_q(\Omega)$, and

$$W^{-s}_q(\Omega) := (W^s_{q',0}(\Omega))', \quad W^{-s}_{q,0}(\Omega) := (W^s_{q'}(\Omega))', \quad W^{-t}_q(\partial\Omega) := (W^t_{q'}(\partial\Omega))'$$

for $s, t > 0$ with $s - \frac{1}{q'} \notin \mathbb{N}$ where $\frac{1}{q} + \frac{1}{q'} = 1$. Finally, the homogeneous Sobolev space of order 1 is defined as

$$\dot{W}^1_q(\Omega) := \{p \in L^q_{\mathrm{loc}}(\overline{\Omega}) : \nabla p \in L^q(\Omega)\}$$

normed by $\|\nabla \cdot \|_q$.

In the following let $\Omega \subseteq \mathbb{R}^n$, $n \geq 2$, be a domain satisfying (A1). Using the cut-off functions φ_j, ψ_j, $j = 1, \ldots, m$, many properties of the Sobolev-Slobodeckij spaces on Ω and $\partial\Omega$ can be reduced to the case of a curved half-space $\mathbb{R}^n_\gamma$, $\gamma \in C^{1,1}(\mathbb{R}^{n-1})$. Then the diffeomorphism

$$F\colon \mathbb{R}^n_+ \to \mathbb{R}^n_\gamma : x \mapsto (x', x_n + \gamma(x')) \tag{2.1}$$

can be used to reduce the statement on $\mathbb{R}^n_\gamma$ to the case of a half-space $\mathbb{R}^n_+ = \{x \in \mathbb{R}^n : x_n > 0\}$. More precisely, if $(F^*f)(x) := f(F(x))$ and $(F^{*,-1}f)(x) := f(F^{-1}(x))$ denote the pull-back and push-forward, resp., of a scalar function f, then

$$F^*\colon W^s_q(\mathbb{R}^n_+) \to W^s_q(\mathbb{R}^n_\gamma), \qquad F^*\colon W^s_{q,0}(\mathbb{R}^n_+) \to W^s_{q,0}(\mathbb{R}^n_\gamma),$$
$$F^*_0\colon W^s_q(\mathbb{R}^{n-1}) \to W^s_q(\partial\mathbb{R}^n_\gamma)$$

are linear isomorphisms for $|s| \leq 2$, where F_0 denotes the restriction of F on $\partial\mathbb{R}^n_+$ and $\partial\mathbb{R}^n_+$ is identified with $\mathbb{R}^{n-1}$. In particular, we note that, if $2 \geq s > j + \frac{1}{q}$, $j = 0, 1$, with $s - \frac{1}{q} \notin \mathbb{N}$, there is a bounded and surjective linear operator

$$\gamma_j\colon W^s_q(\Omega) \to W^{s-\frac{1}{q}}_q(\partial\Omega) \tag{2.2}$$

such that $\gamma_j u = \partial^j_\nu u|_{\partial\Omega}$ for all $f \in C^\infty_{(0)}(\overline{\Omega})$, cf., e.g., [7], where ν denotes the exterior normal. Moreover, if $0 < s < \frac{1}{q}$, by [29, Theorem 2.9.3] $W^s_{q,0}(\Omega) = W^s_q(\Omega)$, where Ω is as in Assumption 1.1. Thus

$$(W^s_q(\Omega))' = W^{-s}_{q'}(\Omega) \qquad \text{for all } s \in \left(-\frac{1}{q'}, \frac{1}{q}\right). \tag{2.3}$$

Furthermore, recall that for $f \in L^q(\Omega)$ with $\operatorname{div} f = 0$ it is possible to define a weak trace of the normal component $\gamma_\nu f \in W^{-\frac{1}{q}}_{q,\nu}(\partial\Omega) := \left(\gamma_0 \dot{W}^1_{q'}(\Omega)\right)'$ by

$$\langle \gamma_\nu f, \gamma_0 v\rangle_{\partial\Omega} := (f, \nabla v)_\Omega \qquad \text{for all } v \in \dot{W}^1_{q'}(\Omega), \tag{2.4}$$

where $\gamma_0 \dot{W}^1_{q'}(\Omega) := \{a \in L^{q'}_{\mathrm{loc}}(\partial\Omega) : a = A|_{\partial\Omega}, A \in \dot{W}^1_{q'}(\Omega)\}$ is equipped with the quotient norm. Of course, if $f \in C^\infty_{(0)}(\overline{\Omega})$ with $\operatorname{div} f = 0$, the definition of $\gamma_\nu f$ by

(2.4) coincides with the usual trace $\nu\cdot f|_{\partial\Omega}$, i.e., $\langle\gamma_\nu f,\gamma_0 v\rangle_{\partial\Omega} = (\nu\cdot f, v)_{\partial\Omega}$ for all $v\in \dot W^1_{q'}(\Omega)$.

Moreover, we note that, if $f = f_0+\nabla p$, $f_0\in L^q_\sigma(\Omega)$, $p\in\dot W^1_q(\Omega)$, is the Helmholtz decomposition of $f\in L^q(\Omega)^n$ and (1.4) is valid, then p is uniquely determined as solution of the weak Neumann problem

$$\Delta p = \operatorname{div} f \qquad \text{in } \Omega, \tag{2.5}$$

$$\partial_\nu p|_{\partial\Omega} = \nu\cdot f|_{\partial\Omega} \quad \text{on } \partial\Omega, \tag{2.6}$$

where (2.5) is understood in the sense of distributions and (2.6) is understood as $\gamma_\nu(f-\nabla p)=0$, cf. [28]. Because of the definition of γ_ν, $p\in\dot W^1_q(\Omega)$ solves (2.5)–(2.6) if and only if

$$(\nabla p,\nabla v)_\Omega = (f,\nabla v)_\Omega \qquad \text{for all } v\in\dot W^1_{q'}(\Omega). \tag{2.7}$$

Moreover, the existence of the (unique) Helmholtz decomposition is equivalent to the existence of a unique solution $p\in\dot W^1_q(\Omega)$ of (2.7) for every $f\in L^q(\Omega)^n$. We note that every $F\in\dot W^{-1}_{q,0}(\Omega) := (\dot W^1_{q'}(\Omega))'$ can be represented as $\langle F, v\rangle_\Omega = (f,\nabla v)_\Omega$ for some $f\in L^q(\Omega)^n$ with $\|f\|_q\le C\|F\|_{\dot W^{-1}_{q,0}}$, which is a consequence of the Hahn-Banach theorem. Finally, let $K_N\colon W^{-\frac1q}_{q,\nu}(\partial\Omega)\to\dot W^1_q(\Omega)$ be defined by

$$(\nabla K_N a,\nabla v)_\Omega = \langle a,\gamma_0 v\rangle_{\partial\Omega} \qquad \text{for all } v\in\dot W^1_{q'}(\Omega). \tag{2.8}$$

By definition of $W^{-\frac1q}_{q,\nu}(\partial\Omega)$, $a=\gamma_\nu A$ for some $A\in L^q_\sigma(\Omega)$. Hence (2.8) is equivalent to $\Delta K_N a = 0$ and $\partial_\nu K_N a|_{\partial\Omega} = a$.

3. The reduced Stokes operator

In the following let $\Omega\subseteq\mathbb{R}^n$, $n\ge 2$, $\delta\in(0,\pi)$, and let $1<q<\infty$ be as in Assumption 1.1.

In order to apply the results from [3] for the construction of an approximate resolvent for large λ, we need the *reduced* Stokes operator $A_{0,q}$ defined as

$$A_{0,q}u \;:=\; (-\Delta+\nabla K_N T)u, \qquad Tu := \gamma_\nu(\Delta-\nabla\operatorname{div})u,$$

for $u\in\mathcal{D}(A_{0,q}) := W^2_q(\Omega)^n\cap W^1_{q,0}(\Omega)^n$, where K_N is defined by (2.8). Since $\operatorname{div}(\Delta-\nabla\operatorname{div})u=0$, we conclude that

$$T\colon\mathcal{D}(A_{0,q})\to W^{-\frac1q}_{q,\nu}(\partial\Omega).$$

Hence $\nabla K_N T\colon\mathcal{D}(A_{0,q})\to L^q(\Omega)^n$ is well defined. – Note that $A_{0,q}$ is a densely defined unbounded operator on $L^q(\Omega)^n$ in contrast to the Stokes operator, which acts on the subspace $L^q_\sigma(\Omega)$. We refer to [2, Section 3] for explanations of the relation between the Stokes and the reduced Stokes operator.

Introducing local coordinates $Tu = \operatorname{div}_\tau \gamma_1 u_\tau$ for every $u \in C^\infty(\overline{\Omega})$ with $u|_{\partial\Omega} = 0$ where div_τ denotes the tangential divergence, cf. [24, Lemma A.1]. Hence by (2.2)

$$T\colon W^{2+s}_q(\Omega)^n \cap W^1_{q,0}(\Omega)^n \to W^{s-\frac{1}{q}}_q(\partial\Omega) \quad \text{for all } 0 \geq s > -1 + \frac{1}{q}. \tag{3.1}$$

We will use the following additional assumption:

(A4) *There is an* $R > 0$ *such that for every* $\lambda \in \Sigma_\delta$ *with* $|\lambda| \geq R$ *there is no non-trivial solution* $g \in W^1_q(\Omega)$ *of*

$$\lambda(g,v)_\Omega + (\nabla g, \nabla v)_\Omega = 0 \qquad \textit{for all } v \in W^1_{q'}(\Omega). \tag{3.2}$$

This assumption is needed in the proof of Lemma 3.1 below. In the following we will show that (A4) is a consequence of Assumption 1.1, cf. proof of Theorem 1.3 below.

The construction of the approximate resolvent is based on the following lemma.

Lemma 3.1. *Let* $\Omega \subseteq \mathbb{R}^n$, $n \geq 2$, $1 < q < \infty$, *and* $\delta \in (0,\pi)$ *be as in Assumption* 1.1. *Moreover, assume that* (A4) *holds and that* $(\lambda + A_{0,q})^{-1}$ *exists for some* $\lambda \in \Sigma_\delta$ *with* $|\lambda| \geq R$. *Then* $(\lambda + A_q)^{-1}$ *exists and*

$$A_{0,q}|_{L^q_\sigma(\Omega)} = A_q, \qquad (\lambda + A_{0,q})^{-1}|_{L^q_\sigma(\Omega)} = (\lambda + A_q)^{-1}. \tag{3.3}$$

Proof. The first statement can be seen as follows: If $u \in \mathcal{D}(A_{0,q}) \cap L^q_\sigma(\Omega)$, then $\operatorname{div}(-\Delta u + \nabla K_N T u) = 0$ in the sense of distributions and

$$\gamma_\nu(-\Delta u + \nabla K_N T u) = -\gamma_\nu \Delta u + \partial_\nu K_N T u|_{\partial\Omega} = 0$$

in the sense of (2.4). Hence $-\Delta u = (-\Delta + \nabla K_N T)u - \nabla K_N T u$ is the Helmholtz decomposition of $-\Delta u$ by (A2), i.e., $(-\Delta + \nabla K_N T)u = P_q(-\Delta)u = A_q u$.

In order to prove the second relation let $u = (\lambda + A_{0,q})^{-1} f$ with $f \in L^q_\sigma(\Omega)$. Then multiplying $(\lambda + A_{0,q})u = f$ by ∇v, $v \in W^1_{q'}(\Omega)$, and using (2.8) and (2.4) we obtain that $g = \operatorname{div} u$ solves (3.2), which implies $\operatorname{div} u = 0$ by (A4). Therefore, $u \in L^q_\sigma(\Omega)$ by (1.4) and $(\lambda + A_q)u = (\lambda + A_{0,q})u = f$. Since by the first statement $\lambda + A_q = (\lambda + A_{q,0})|_{L^q_\sigma(\Omega)}$ is injective, we finally conclude that $(\lambda + A_q)^{-1} f = u = (\lambda + A_{q,0})^{-1} f$ for every $f \in L^q_\sigma(\Omega)$. □

Lemma 3.2. *Let* $\Omega \subset \mathbb{R}^n$, $n \geq 2$, *and* $1 < q < \infty$ *be as in Assumption* 1.1. *If* $\lambda + A_{q'}$ *is surjective for* $\lambda \notin (-\infty, 0]$, *then there is no non-trivial solution of* (3.2).

Proof. Let $f \in L^{q'}(\Omega)$ be arbitrary and let $u \in \mathcal{D}(A_{0,q'})$ such that $(\lambda + A_{0,q'})u = f$. Then multiplying f with ∇g we observe that $\operatorname{div} u \in W^1_{q'}(\Omega)$ solves

$$-\lambda(\operatorname{div} u, g) - (\nabla \operatorname{div} u, \nabla g) = (f, \nabla g), \qquad \text{for all } g \in W^1_q(\Omega).$$

Hence, if $g \in W^1_q(\Omega)$ solves (3.2), then $(f, \nabla g) = 0$ for all $f \in L^{q'}(\Omega)$ and therefore $\nabla g = 0$. Because of (3.2) and $\lambda \neq 0$, we conclude $g = 0$. □

4. Construction of the approximative resolvent

The proof of Theorem 1.3 is based on the following result.

Theorem 4.1. *Let $\mathbb{R}^n_\gamma$, $n \geq 2$, $\gamma \in C^{1,1}(\mathbb{R}^{n-1})$, be a curved half-space, $1 < q < \infty$, and let $\delta \in (0,\pi)$. Then there is a bounded operator $R_{\gamma,\lambda}\colon L^q(\mathbb{R}^n_\gamma)^n \to W^2_q(\mathbb{R}^n_\gamma)^n$ such that*

$$(\lambda - \Delta + \nabla \widetilde{K}_{\gamma,N} T) R_{\gamma,\lambda} f = f + S_{\gamma,\lambda} f \quad \text{in } \mathbb{R}^n_\gamma, \tag{4.1}$$

$$R_{\gamma,\lambda} f = 0 \quad \text{on } \partial\mathbb{R}^n_\gamma \tag{4.2}$$

for every $f \in L^q(\mathbb{R}^n_\gamma)^n$ and $\lambda \in \mathbb{C}\backslash(-\infty,0]$, where $\|S_{\gamma,\lambda}\|_{\mathcal{L}(L^q(\mathbb{R}^n_\gamma))} \leq C_{q,\delta}(1+|\lambda|)^{-\varepsilon}$ uniformly in $\lambda \in \Sigma_\delta$ for some $\varepsilon > 0$. Here $\widetilde{K}_{\gamma,N}\colon W_q^{s-\frac{1}{q}}(\partial\mathbb{R}^n_\gamma) \to W^{s+1}_q(\mathbb{R}^n_\gamma)$, $s \in (-\frac{1}{q'}, 1+\frac{1}{q})$, is a bounded operator satisfying

$$\Delta \widetilde{K}_{\gamma,N} a = R'_\gamma a \quad \text{in } \mathbb{R}^n_\gamma, \tag{4.3}$$

$$\partial_\nu \widetilde{K}_{\gamma,N} a|_{\partial\Omega} = a + S'_\gamma a \quad \text{on } \partial\mathbb{R}^n_\gamma, \tag{4.4}$$

where $R'_\gamma\colon W_q^{-\frac{1}{q}-\varepsilon}(\partial\mathbb{R}^n_\gamma) \to W^{-1}_{q,0}(\mathbb{R}^n_\gamma)$ and $S'_\gamma\colon W_q^{-\frac{1}{q}-\varepsilon}(\partial\mathbb{R}^n_\gamma) \to W_q^{-\frac{1}{q}}(\partial\mathbb{R}^n_\gamma)$ are bounded operators. Moreover, for every $R > 0$

$$(1+|\lambda|)\|R_{\gamma,\lambda}\|_{\mathcal{L}(L^q(\mathbb{R}^n_\gamma))} + \|\nabla^2 R_{\gamma,\lambda}\|_{\mathcal{L}(L^q(\mathbb{R}^n_\gamma))} \leq C_{q,\delta}, \qquad \lambda \in \Sigma_\delta, \tag{4.5}$$

$$\left\| \int_{\Gamma_R} h(-\lambda) R_{\gamma,\lambda}\, d\lambda \right\|_{\mathcal{L}(L^q(\mathbb{R}^n_\gamma))} \leq C_{q,\delta}\|h\|_\infty, \quad h \in H(\delta). \tag{4.6}$$

Proof. The theorem is proved in [5, Theorem 4.1] but only for $s = 0$ in the mapping properties of $\widetilde{K}_{\gamma,N}$, which remains to be extended.

The operator $\widetilde{K}_{\gamma,N} \equiv \widetilde{K}_1$ is defined in [3, Section 5.5] as

$$\widetilde{K}_{\gamma,N} \quad = \quad F^{*,-1} \underline{k}_1(D_x, x') F^*_0,$$

where $\underline{k}_1(D_x,x')$ is a Poisson operator of order -1 in R-form with $C^{0,1}$-coefficients in the sense of [3, Section 4] and $F^{*,-1}$, F^*_0 are as in Section 2. By duality

$$(\partial^\alpha_x \underline{k}_1(D_x,x')a, f)_\Omega = (-1)^{|\alpha|}(a, t(x',D_x)\partial^\alpha_x f)_{\partial\Omega} \tag{4.7}$$

for $a \in C_0^\infty(\mathbb{R}^{n-1})$, $f \in C^\infty_{(0)}(\overline{\mathbb{R}^n_+})$, and $\alpha \in \mathbb{N}^n_0$, where

$$t(x',D_x)f := \mathcal{F}^{-1}_{\xi'\mapsto x'}\left[\int_0^\infty \overline{\tilde{\underline{k}}_1(x',\xi',y_n)}\tilde{f}(\xi',y_n)dy_n\right],$$

$\tilde{f}(\xi',x_n) = \mathcal{F}_{x'\mapsto\xi'}[f(.,x_n)]$, and $\tilde{\underline{k}}_1(x',\xi',y_n)$ denotes the symbol-kernel of $\underline{k}_1(D_x,x')$, cf. [3, Section 4]. Using (4.7) for suitable α and [1, Theorem 4.8] for $t(x',D_x)$, the mapping properties of $\underline{k}_1(D_x,x')$ are obtained by duality and interpolation. $\square$

Now we define the approximate resolvent R_λ on a domain Ω satisfying (A1) as

$$R_\lambda f = \sum_{j=1}^{m} \psi_j R_{\gamma_j,\lambda}(\varphi_j f), \qquad f \in L^q(\Omega)^n,$$

where $R_{\gamma_j,\lambda}$, $j = 1, \ldots, m$, is the approximate resolvent on $\mathbb{R}^n_{\gamma_j}$ due to Theorem 4.1. Moreover, we define the approximate Poisson operator

$$\widetilde{K}_N a = \sum_{j=1}^{m} \psi_j \widetilde{K}_{\gamma_j,N}(\varphi_j a), \qquad a \in W_q^{-\frac{1}{q}}(\partial\Omega),$$

where $\widetilde{K}_{\gamma_j,N}$ is the operator due to Theorem 4.1 for $\mathbb{R}^n_{\gamma_j}$. Now we have

Lemma 4.2. *Let $1 < q < \infty$ and $\Omega \subseteq \mathbb{R}^n$ be as in Assumption 1.1 and let $r = q$ or $r = q'$. Moreover, let K_N be the Poisson operator of the Neumann problem as defined in (2.8). Then there is some $\varepsilon > 0$ such that*

$$\|\nabla(K_N - \widetilde{K}_N)Tu\|_r \le C_r \|u\|_{2-\varepsilon,r} \qquad \textit{for all } u \in W_r^{2-\varepsilon}(\Omega)^n \cap W^1_{r,0}(\Omega)^n.$$

Proof. For simplicity let $r = q$. Let $f \in L^{q'}(\Omega)^n$ be arbitrary and let $f = f_0 + \nabla p$, $f_0 \in L^{q'}_\sigma(\Omega)$, $p \in \dot{W}^1_{q'}(\Omega)$ be its Helmholtz decomposition. By (A3) we have $p = p_1 + p_2$, where $p_1 \in W^1_{q'}(\Omega)$ and $p_2 \in L^{q'}_{\mathrm{loc}}(\overline{\Omega})$ with $\nabla p_2 \in W^1_{q'}(\Omega)$ and $\|(p_1, \nabla p_2)\|_{1,q'} \le C_{q'} \|\nabla p\|_{1,q'}$. Then by (4.3)–(4.4)

$$\begin{aligned}(\nabla(K_N - \widetilde{K}_N)Tu, f)_\Omega &= (\nabla(K_N - \widetilde{K}_N)Tu, \nabla p)_\Omega \\ &= (\nabla(K_N - \widetilde{K}_N)Tu, \nabla p_2)_\Omega + \langle (I - \partial_\nu \widetilde{K}_N)Tu, p_1\rangle_{\partial\Omega} - (\Delta \widetilde{K}_N Tu, p_1)_\Omega.\end{aligned}$$

The term $\langle (I - \partial_\nu \widetilde{K}_N)Tu, p_1\rangle_{\partial\Omega}$ can be estimated by $\|u\|_{2-\varepsilon,q}\|f\|_{q'}$ for some $\varepsilon > 0$ in a straightforward manner using Theorem 4.1 and (3.1). Moreover,

$$\begin{aligned}|(\Delta \widetilde{K}_N Tu, p_1)_\Omega| &\le \sum_{j=1}^{m} |(\psi_j R'_{\gamma_j}(\varphi_j Tu), p_1)_\Omega| \\ &+ \sum_{j=1}^{m} |(2(\nabla\psi_j)\cdot\nabla\widetilde{K}_{\gamma_j,N}(\varphi_j Tu) + (\Delta\psi_j)\widetilde{K}_{\gamma_j,N}(\varphi_j Tu), p_1)_\Omega|,\end{aligned}$$

where

$$|(\psi_j R'_{\gamma_j}(\varphi_j Tu), p_1)_\Omega| \le C \|R'_{\gamma_j}\varphi_j Tu\|_{W^{-1}_{q,0}(\mathbb{R}^n_{\gamma_j})} \|p_1\|_{1,q'} \le C\|u\|_{2-\varepsilon,q}\|f\|_{q'}$$

for some $\varepsilon > 0$. Since $p_1 \in W^1_{q'}(\Omega) \hookrightarrow W^\varepsilon_{q'}(\Omega)$ with $0 < \varepsilon < \frac{1}{q'}$, by the mapping property $\widetilde{K}_{\gamma_j,N} \colon W_q^{-\varepsilon-\frac{1}{q}}(\partial\mathbb{R}^n_{\gamma_j}) \to W_q^{1-\varepsilon}(\mathbb{R}^n_\gamma)$, and by (2.3)

$$|(2(\nabla\psi_j)\cdot\nabla\widetilde{K}_{\gamma_j,N}(\varphi_j Tu) + (\Delta\psi_j)\widetilde{K}_{\gamma_j,N}(\varphi_j Tu), p_1)_\Omega| \le C\|u\|_{2-\varepsilon,q}\|f\|_{q'}$$

for $0 < \varepsilon < \frac{1}{q'}$. The term $(\nabla \widetilde{K}_N T u, \nabla p_2)_\Omega$ is estimated in the same way using $\nabla p_2 \in W^1_q(\Omega)$. Finally, by the definitions of K_N and T

$$\begin{aligned} |(\nabla K_N T u, \nabla p_2)_\Omega| &= |((\Delta - \nabla \operatorname{div})u, \nabla p_2)_\Omega| \\ &\leq |(\nabla u, \nabla^2 p_2)_\Omega| + |(\partial_\nu u, \nabla p_2)_{\partial\Omega}| + |(\operatorname{div} u, \Delta p_2)_\Omega| + |(\operatorname{div} u, \partial_\nu p_2)_{\partial\Omega}| \\ &\leq C\|u\|_{2-\varepsilon,q}\|\nabla p_2\|_{1,q'} \leq C\|u\|_{2-\varepsilon,q}\|f\|_{q'} \end{aligned}$$

for some $\varepsilon > 0$. – The proof for $r = q'$ is done in the same way. □

Proof of Theorem 1.3. The proof is the same as in [5, Theorem 4.4] with minor modifications. We include it for the convenience of the reader.

First of all, by (4.5) and interpolation

$$\|R_{j,\lambda}(\varphi_j f)\|_{s,q} \leq C_{q,\delta,R}(1 + |\lambda|)^{-1+\frac{s}{2}}\|f\|_q, \qquad \lambda \in \Sigma_\delta, \tag{4.8}$$

for all $s \in [0, 2]$ and $f \in L^q(\Omega)^n$, $j = 1, \ldots, m$. Moreover, by (4.1)

$$\begin{aligned} &(\lambda - \Delta + \nabla K_N T)R_\lambda f \\ &= f + \sum_{j=1}^m \psi_j S_{\gamma_j,\lambda}(\varphi_j f) - \sum_{j=1}^m (2(\nabla\psi_j) \cdot \nabla R_{j,\lambda}(\varphi_j f) + (\Delta\psi_j)R_{j,\lambda}(\varphi_j f)) \\ &\quad + (\nabla K_N T - \nabla \widetilde{K}_N T)R_\lambda f. \end{aligned}$$

Hence (4.8), Theorem 4.1, and Lemma 4.2 imply

$$(\lambda - \Delta + \nabla K_N T)R_\lambda = I + S'_\lambda,$$

where $\|S'_\lambda\|_{\mathcal{L}(L^q(\Omega))} \leq C_{q,\delta}(1+|\lambda|)^{-\varepsilon}$ uniformly in $\lambda \in \Sigma_\delta$ for some $\varepsilon > 0$. Therefore $(\lambda + A_{0,q})^{-1}$ exists for all $\lambda \in \Sigma_\delta$ with $|\lambda| \geq R$ for some $R > 0$ and

$$(\lambda + A_{0,q})^{-1} = R_\lambda + S_\lambda,$$

where $\|S_\lambda\|_{\mathcal{L}(L^q(\Omega))} \leq C_{q,\delta}(1+|\lambda|)^{-1-\varepsilon}$ uniformly in $\lambda \in \Sigma_\delta$, $|\lambda| \geq R$. Furthermore, for $B_\lambda = R_\lambda, S_\lambda$ we have

$$\left\|\int_{\Gamma_R} h(-\lambda)B_\lambda \, d\lambda\right\|_{\mathcal{L}(L^q(\Omega))} \leq C_{q,\delta}\|h\|_\infty, \qquad h \in H(\delta),$$

because of (4.6) and $\|S_\lambda\| \leq C_{q,\delta}(1 + |\lambda|)^{-1-\varepsilon}$.

The same arguments apply to $A_{0,q'}$ instead of $A_{0,q}$. Hence we can assume that also $(\lambda + A_{0,q'})^{-1}$ exists for all $\lambda \in \Sigma_\delta$ with $|\lambda| \geq R$. In particular, this implies that (A4) is valid because of Lemma 3.2. Now $(\lambda + A_{0,q})^{-1}|_{L^q_\sigma(\Omega)} = (\lambda + A_q)^{-1}$ by Lemma 3.1, which proves (1.5). The rest of the theorem follows from (1.5) and the fact that $f(\lambda) = (\lambda + A_q)^{-1}$ is uniformly bounded on compact subsets of $\rho(-A_q)$. □

5. Bounded H_∞-calculus for an aperture domain

Roughly speaking an aperture domain is a domain separated by a wall with a hole (aperture) inside. More precisely, an aperture domain $\Omega \subset \mathbb{R}^n$, $n \geq 2$, is a domain such that

$$\Omega \cup B = \mathbb{R}^n_+ \cup \mathbb{R}^n_- \cup B$$

for some ball B, where $\mathbb{R}^n_- := \{x \in \mathbb{R}^n : x_n < -d\}$ for some $d > 0$. Concerning the Stokes equations the aperture domain is of particular interest since under certain circumstances depending on the dimension n and the integral exponent q an additional condition has to be posed to get unique solutions, cf. [16]. This may be done by prescribing the *flux* $\phi(u)$ of the velocity field u through the hole, i.e.,

$$\phi(u) := \int_{M'} \nu \cdot u d\sigma,$$

where M' is an $(n-1)$-dimensional compact manifold dividing Ω into an upper and lower part Ω_+, Ω_-, resp., such that $\Omega_\pm \cup B = \mathbb{R}^n_\pm \cup B$. In the following let $\varphi_\pm, \psi_\pm \in C_b^\infty(\overline{\mathbb{R}^n_\pm})$ be cut-off functions such that $\varphi_\pm(x) = 1$ for $x \in \mathbb{R}^n_\pm$ with $|x| \geq R$ for some suitable $R > 0$, $\psi_\pm \equiv 1$ on $\operatorname{supp} \varphi_\pm$, and $\operatorname{supp} \psi_\pm \subset \overline{\mathbb{R}^n_\pm} \setminus B$.

In order to apply Theorem 1.3, it remains to verify Assumption 1.1.

Proposition 5.1. *Let $\Omega \subset \mathbb{R}^n$, $n \geq 2$, be an aperture domain with $C^{1,1}$-boundary. Then Assumption 1.1 is valid for Ω and every $1 < q \leq \frac{n}{n-1}$ and $\delta \in (0, \pi)$.*

Proof. Obviously, (A1) is true. The first part of (A2) is true for every $1 < q < \infty$ but (1.4) only holds if $1 < q \leq \frac{n}{n-1}$, cf. [16, Theorem 2.6]. The condition (A3) can be easily verified by using the cut-off functions $\varphi_\pm$ and the corresponding statement for the half-spaces $\mathbb{R}^n_\pm$. □

We will analyze the resolvent $(\lambda + A_q)^{-1}$ near the origin by comparing it with the resolvents of the Stokes operator on $\mathbb{R}^n_\pm$. Hence let $R^\pm_\lambda = (\lambda + A_{\mathbb{R}^n_\pm, q})^{-1}$, $1 < q < \infty$, denote the resolvent of the Stokes operator on $\mathbb{R}^n_\pm$. Then

$$|\lambda| \|u\|_q + \|\nabla^2 u\|_q \leq C_{q,\delta} \|f\|_q, \qquad \lambda \in \Sigma_\delta, f \subset L^q(\mathbb{R}^n_\pm), \tag{5.1}$$

cf., e.g., [15, Theorem 1.3]. Moreover,

Lemma 5.2. *Let $1 < q < n$, $\delta \in (0, \pi)$, and let $B^\pm_M = B_M(0) \cap \mathbb{R}^n_\pm$. Then for every $M > 0$*

$$\|(R^\pm_\lambda f, \nabla R^\pm_\lambda f)\|_{L^q(B^\pm_M)} \leq C_{q,\delta,M} \|f\|_q$$

uniformly in $\lambda \in \Sigma_\delta$ and $f \in L^q(\mathbb{R}^n_\pm)^n$.

Proof. By Poincaré's inequality, $\|R^\pm_\lambda f\|_{L^q(B^\pm_M)} \leq C_{q,M} \|\nabla R^\pm_\lambda f\|_{L^q(B^\pm_M)}$. Hence it remains to estimate $\nabla R^\pm_\lambda$. Then, if $\frac{1}{q^*} = \frac{1}{q} - \frac{1}{n}$,

$$\|\nabla R^\pm_\lambda f\|_{L^q(B^\pm_M)} \leq C_M \|\nabla R^\pm_\lambda f\|_{L^{q^*}(\mathbb{R}^n_\pm)} \leq C_{q,M} \|\nabla^2 R_\lambda f\|_q \leq C_{q,\delta,M} \|f\|_q$$

uniformly in $\lambda \in \Sigma_\delta$ and $f \in L^q(\mathbb{R}^n_\pm)^n$ by (5.1). □

In order to analyze (1.2) for $A = A_q$ with Γ replaced by $\Gamma'_R := \Gamma \cap B_R(0)$, $R > 0$, we consider

$$u_\lambda := (\lambda + A_q)^{-1} f - \psi^+ R^+_\lambda \varphi^+ f - \psi^- R^-_\lambda \varphi^- f, \qquad f \in L^q_\sigma(\Omega). \tag{5.2}$$

Moreover, let $p_\lambda = q_\lambda - \psi^+ p^+_\lambda - \psi^- p^-_\lambda$, where $\nabla q_\lambda = (I - P_q)\Delta(\lambda + A_q)^{-1} f$ and $\nabla p^\pm_\lambda = (I - P^\pm_q)\Delta R^\pm_\lambda \varphi^\pm f$ is chosen such that $\int_{B^\pm_M} p^\pm_\lambda = 0$ for $M > 0$ so large that $\psi^\pm \equiv 1$ on $\mathbb{R}^n_\pm \setminus B_M(0)$. Then (u_λ, p_λ) solves

$$\begin{aligned} (\lambda - \Delta)u_\lambda + \nabla p_\lambda &= (1 - \varphi^+ - \varphi^-)f + f^+_\lambda + f^-_\lambda =: \tilde{f}_\lambda && \text{in } \Omega,\\ \operatorname{div} u_\lambda &= g^+_\lambda + g^-_\lambda =: \tilde{g}_\lambda && \text{in } \Omega,\\ u_\lambda|_{\partial\Omega} &= 0 && \text{on } \partial\Omega, \end{aligned}$$

where $g^\pm_\lambda = -\nabla\psi^\pm \cdot R^\pm_\lambda \varphi^\pm f$ and

$$f^\pm_\lambda = 2(\nabla\psi^\pm) \cdot \nabla R^\pm_\lambda \varphi^\pm f + (\Delta\psi^\pm) R^\pm_\lambda \varphi^\pm f - (\nabla\psi^\pm) p^\pm_\lambda.$$

Moreover, $\phi(u_\lambda) = \phi((\lambda + A_q)^{-1} f) = 0$.

Lemma 5.3. *For every $1 < q < \infty$ there is some $0 < a < 1$ such that $|\lambda|^a \|u_\lambda\|_q \leq C\|\tilde{f}_\lambda\|_q$ uniformly in $\lambda \in \Sigma_\delta$, $|\lambda| \leq 1$.*

Proof. By [16, Theorem 2.1] for every $1 < q < n$

$$|\lambda|\|u_\lambda\|_q + \|\nabla^2 u_\lambda\|_q \leq C_{q,\delta}\left(\|\tilde{f}_\lambda\|_q + \|\nabla\tilde{g}_\lambda\|_q + (1 + |\lambda|)\|\tilde{g}_\lambda\|_{\dot{W}^{-1}_{q,0}}\right) \tag{5.3}$$

uniformly in $\lambda \in \Sigma_\delta$. Moreover,

$$\int_{B^\pm_M} g^\pm_\lambda dx = \int_{B^\pm_M} \operatorname{div}\left((1 - \psi^\pm) R^\pm_\lambda \varphi^\pm f\right) dx = 0.$$

Hence $\|g^\pm_\lambda\|_{\dot{W}^{-1}_{q,0}(\mathbb{R}^n_\pm)} \leq C_{R,q}\|g^\pm_\lambda\|_q$ by Poincaré's inequality. Furthermore, if $n < r < \infty$ and $1 < q < \frac{n}{n-1}$ is defined by $\frac{1}{q} = \frac{1}{r} + \frac{1}{n}$, then by [16, Corollary 2.4]

$$\begin{aligned} \|\nabla u_\lambda\|_r &\leq C_{r,\delta}\left(|\lambda|\|u_\lambda\|_q + \|\tilde{f}_\lambda\|_q + \|\nabla\tilde{g}_\lambda\|_q + \|\tilde{g}_\lambda\|_{\dot{W}^{-1}_{q,0}}\right)\\ &\leq C_{r,\delta}\left(\|\tilde{f}_\lambda\|_r + \|\tilde{g}_\lambda\|_{1,r}\right) \end{aligned}$$

for $\lambda \in \Sigma_\delta$, $|\lambda| \leq 1$, since $\operatorname{supp}\tilde{f}_\lambda, \operatorname{supp}\tilde{g}_\lambda \subset B_M(0)$. Moreover, since $\|\nabla u_\lambda\|_q \leq C_q\|(u_\lambda, \nabla^2 u_\lambda)\|_q$, interpolation of the latter inequality with (5.3) yields that for every $1 < q < \infty$ there is some $0 < a < 1$ such that

$$|\lambda|^a\|\nabla u_\lambda\|_q \leq C_q\left(\|\tilde{f}_\lambda\|_q + \|\tilde{g}_\lambda\|_{1,q}\right), \qquad \lambda \in \Sigma_\delta, |\lambda| \leq 1.$$

This implies that for $n < r < \infty$ and $1 < q < \frac{n}{n-1}$ defined by $\frac{1}{q} = \frac{1}{r} + \frac{1}{n}$ there is some $0 < a < 1$ such that

$$|\lambda|^a\|u_\lambda\|_r \leq C_r|\lambda|^a\|\nabla u_\lambda\|_q \leq C_{r,\delta}\left(\|\tilde{f}_\lambda\|_q + \|\tilde{g}_\lambda\|_{1,q}\right) \leq C_{r,\delta}\left(\|\tilde{f}_\lambda\|_r + \|\tilde{g}_\lambda\|_{1,r}\right)$$

for $\lambda \in \Sigma_\delta$ with $|\lambda| \leq 1$ using Sobolev's inequality, cf. [16, Lemma 3.1]. Interpolating again finishes the proof. □

Combining Lemma 5.2 and Lemma 5.3 we obtain:

Corollary 5.4. *Let* u_λ *be defined as in* (5.2)*. Then for every* $1 < q < n$ *there is some* $0 < a < 1$ *such that* $\|u_\lambda\|_q \leq C|\lambda|^{-a}\|f\|_q$ *for all* $f \in L^q_\sigma(\Omega)$ *uniformly in* $\lambda \in \Sigma_\delta$ *with* $|\lambda| \leq 1$.

Proof of Theorem 1.4. First of all, $(\lambda + A_q)^{-1}$ exists for every $\lambda \in \mathbb{C} \setminus (-\infty, 0]$ and $(A_q)' = A_{q'}$, cf. [16, Theorem 2.5]. Hence by duality and interpolation it is sufficient to prove the statement for $1 < q < \frac{n}{n-1}$. Moreover, $\mathcal{R}(A_q)$ is dense, cf. Franzke [17, Theorem 6]. Therefore A_q is sectorial and it is sufficient to prove (1.3) for $h \in H(\delta)$, cf. Denk, Hieber, and Prüss [11, Section 2.4]. Moreover, because of Theorem 1.3 it remains to prove

$$\left\| \int_{\Gamma'_R} h(-\lambda)(\lambda + A_q)^{-1} d\lambda \right\|_{\mathcal{L}(L^q_\sigma(\Omega))} \leq C_{\delta,q}\|h\|_\infty \qquad \text{for all } h \in H(\delta). \tag{5.4}$$

Since $(\lambda + A_q)^{-1}$ is bounded on compact subsets of $\mathbb{C} \setminus (-\infty, 0]$, it is sufficient to consider $R = 1$. Because of (5.2) and Corollary 5.4, we can replace A_q by $A_{q,\mathbb{R}^n_\pm}$. Moreover, using

$$\frac{1}{2\pi i}\int_{\Gamma'_1} h(-\lambda)(\lambda + A_{\mathbb{R}^n_\pm,q})^{-1} d\lambda = h(A_{\mathbb{R}^n_\pm,q}) - \frac{1}{2\pi i}\int_{\Gamma_1} h(-\lambda)(\lambda + A_{\mathbb{R}^n_\pm,q})^{-1} d\lambda$$

with $\Gamma_1 = \Gamma \setminus \overline{B_1(0)}$ the estimate (5.4) is a consequence of the bounded H_∞-calculus for the Stokes operator on $\mathbb{R}^n_\pm$, cf. Desch, Hieber, and Prüss [12], and of Theorem 1.3 for $\Omega = \mathbb{R}^n_\pm$. □

References

[1] H. Abels. *Pseudodifferential Boundary Value Problems with Non-Smooth Coefficients.* Preprint No. 2343, TU Darmstadt, 2004, to appear in Comm. Part. Diff. Eq.

[2] H. Abels. *Reduced and generalized Stokes resolvent equations in asymptotically flat layers, part I: unique solvability.* J. Math. Fluid. Mech. 7, No. 2, 201–222, 2005.

[3] H. Abels. *Reduced and generalized Stokes resolvent equations in asymptotically flat layers, part II: H_∞-calculus.* J. Math. Fluid. Mech. 7, No. 2, 223–260, 2005.

[4] H. Abels. *Bounded imaginary powers of the Stokes operator in an infinite layer.* J. Evol. Eq. 2, 439–457, 2002.

[5] H. Abels. *Bounded Imaginary Powers and H_∞-Calculus of the Stokes Operators in Two-Dimensional Exterior Domains.* Preprint No. 2338, TU Darmstadt, 2004, to appear in Math. Z.

[6] H. Abels and M. Wiegner. *Resolvent estimates for the Stokes operator on an infinite layer.* Preprint TU Darmstadt, 2004 to appear in Diff. Int. Eq.

[7] R.A. Adams. *Sobolev Spaces.* Academic Press, New York, San Francisco, London, 1975.

[8] H. Amann, M. Hieber, and G. Simonett. *Bounded H_∞-calculus for elliptic operators.* Diff. Int. Eq., Vol. 7, No. 3, 613–653, 1994.

[9] W. Borchers and H. Sohr. *On the semigrup of the Stokes operator for exterior domains in L^q-spaces.* Math. Z. 196, 415–425, 1987.

[10] S.K. Chua. *Extension theorems on weighted Sobolev space.* Indiana Univ. Math. J., 41, 1027–1076, 1992.

[11] R. Denk, M. Hieber, and J. Prüss. *$\mathcal{R}$-boundedness, Fourier multipliers and problems of elliptic and parabolic Type.* Mem. Am. Math. Soc. 788, 114 p., 2003.

[12] W. Desch, M. Hieber, and J. Prüss. *L^p-theory of the Stokes equation in a half-space.* J. Evol. Eq. 1, 115–142, 2001.

[13] G. Dore and A. Venni. *On the closedness of the sum of two closed operators.* Math. Z. 196, 189–201, 1987.

[14] R. Farwig. *Weighted L^q-Helmholtz decompositions in infinite cylinders and in infinite layers.* Adv. Diff. Eq. 8, 357–384, 2003.

[15] R. Farwig and H. Sohr. *Generalized resolvent estimates for the Stokes system in bounded and unbounded domains.* J. Math. Soc. Japan 46, No. 4, 607–643, 1994.

[16] R. Farwig and H. Sohr. *Helmholtz Decomposition and Stokes Resolvent System for Aperture Domains in L^q-Spaces.* Analysis 16, 1–26, 1996.

[17] M. Franzke. *Strong L^q-theory of the Navier-Stokes equations in aperture domains.* Ann. Univ. Ferrara, Nuova Ser., Sez. VII 46, 161–173, 2000.

[18] A. Fröhlich. *Maximal regularity for the non-stationary Stokes system in an aperture domain.* J. Evol. Eq. 2, 471–493, 2002.

[19] Y. Giga. *Analyticity of the semigroup generated by the Stokes operator in L_r spaces.* Math. Z. 178, 297–329, 1981.

[20] Y. Giga. *Domains of fractional powers of the Stokes operator in L_r Spaces.* Arch. Ration. Mech. Anal. 89, 251–265, 1985.

[21] Y. Giga and H. Sohr. *On the Stokes operator in exterior domains.* J. Fac. Sci. Univ. Tokyo Sect. IA Math. 36, 103–130, 1989.

[22] Y. Giga and H. Sohr. *Abstract L^p estimates for the Cauchy problem with applications to the Navier-Stokes equations in exterior domains.* J. Funct. Anal. 102, 72–94, 1991.

[23] G. Grubb. *Functional Calculus of Pseudodifferential Boundary Problems, 2nd Edition.* Birkhäuser, Basel - Boston - Berlin, 1996.

[24] G. Grubb and V.A. Solonnikov. *Boundary value problems for the nonstationary Navier-Stokes equations treated by pseudo-differential methods.* Math. Scand. 69, 217–290, 1991.

[25] A. McIntosh. *Operators which have an H_∞-calculus.* In "Miniconference on Operator Theory and Partial Differential Equations", B. Jefferies, A. McIntosh, W. Ricker, editors Proc. Center Math. Anal. A.N.U., 14, 210–231, 1986.

[26] T. Miyakawa. *The Helmholtz decomposition of vector fields in some unbounded domains.* Math. J. Toyama Univ. 17, 115–149, 1994.

[27] A. Noll and J. Saal. *H^∞-calculus for the Stokes operator on L_q-spaces.* Math. Z. 244, 651–688, 2003.

[28] C.G. Simader and H. Sohr. *A new approach to the Helmholtz decomposition and the Neumann problem in L_q-spaces for bounded and exterior domains.* Series on Advances in Mathematics for Applied Sciences, Vol. 11, Singapore, World Scientific, 1–35, 1992.

[29] H. Triebel. *Interpolation Theory, Function Spaces, Differential Operators.* North-Holland Publishing Company, Amsterdam, New York, Oxford, 1978.

Helmut Abels
Max Planck Institute for Mathematics in the Sciences
Inselstr. 22
D-04103 Leipzig, Germany
e-mail: abels@mis.mpg.de

Progress in Nonlinear Differential Equations
and Their Applications, Vol. 64, 17–35

Exact Estimates of Solutions to the Robin Boundary Value Problem for Elliptic Non-divergent Second-order Equations in a Neighborhood of the Boundary Conical Point

Mikhail Borsuk and Agneszka Zawadzka

Abstract. We investigate the behavior of strong solutions to the Robin boundary value problem for linear and quasilinear elliptic non-divergent second-order equations in a neighborhood of the boundary conical point. We establish precise exponent of the solution decreasing rate.

Mathematics Subject Classification (2000). 35J65, 35J70, 35B05, 35B45, 35B65.

Keywords. Elliptic equations, the Robin problem, conical points.

1. Introduction

Let $G \subset \mathbb{R}^n$, $n \geq 2$ be a bounded domain with boundary ∂G that is a smooth surface everywhere except at the origin $\mathcal{O} \in \partial G$ and near the point $\mathcal{O}$ it is a **convex** conical surface with its vertex at $\mathcal{O}$. We consider the elliptic value problems

$$\begin{cases} \mathcal{L}[u] \equiv a^{ij}(x)u_{x_ix_j} + a^i(x)u_{x_i} + a(x)u = f(x), \ x \in G, \\ \mathcal{B}[u] \equiv \frac{\partial u}{\partial \vec{n}} + \frac{1}{|x|}\gamma(x)u = g(x), \ x \in \partial G \setminus \mathcal{O}. \end{cases} \tag{L}$$

and

$$\begin{cases} a_{ij}(x, u, u_x)u_{x_ix_j} + a(x, u, u_x) = 0, \ a^{ij} = a^{ji}, \ x \in G, \\ \frac{\partial u}{\partial \vec{n}} + \frac{1}{|x|}\gamma(x)u = g(x), \ x \in \partial G \setminus \mathcal{O} \end{cases} \tag{QL}$$

(summation over repeated indices from 1 to n is understood), $\vec{n}$ denotes the unite outward normal to $\partial G \setminus \mathcal{O}$. We obtain **the best possible estimates** of the problems (L) and (QL) strong solutions near a conical boundary point. Analogous results were established in [2, 3, 4, 1] for the Dirichlet problem. Many mathematicians have considered the third boundary value problem. For the first time, V.A. Kondrat'ev [10] studied general elliptic boundary value problems in domains with conical points. He proved the solvability of such problems with infinitely differentiable coefficients in weighted and usual L_2-Sobolev spaces. Later V.G. Maz'ya and B.A. Plamenevskiy extended the results of V.A. Kondrat'ev to the L_p-Sobolev weighted spaces (see, e.g., [11, 12]). The oblique derivative problem for elliptic equations in non-smooth domains investigated M. Faierman [6], G. Lieberman [13],[15], M. Garroni, V.A. Solonnikov and M. Vivaldi [7], H. Reisman [18]. G. Lieberman [13, 15] proved the local and global maximum principles for general second-order linear and quasi-linear elliptic equations. He studied the problems of the solutions existence and regularity in Lipschitz domains for equation with the Hölder continuous coefficients. Recently G. Lieberman [15] proved the Hölder continuity of strong solutions under weak hypotheses on the (L) coefficients. M.Faierman [6] investigated the regularity in the Sobolev space $W^{2,p}(G)$ of generalized solutions of (L) in a rectangle. H.Reisman [18] investigated such problem in weighted Sobolev spaces for the domain with dihedral edges and the equation coefficients being infinitely smooth. At last, authors of [7] considered the problem for the Poisson equation on the infinite angle. A principal new feature of this article is the consideration of our estimates for equations with minimal smooth coefficients. Our examples demonstrate this fact.

Let $\mathcal{C}$ be the rotational cone $\{x_1 > r\cos\frac{\omega_0}{2}\}$. For a domain G which has a conical point at $\mathcal{O} \in \partial G$ we introduce the notations:

- $\Omega := \mathcal{C} \cap S^{n-1}$; $d\Omega$ denotes the $(n-1)$-dimensional area element of Ω;
- $G_a^b := G \cap \{(r,\omega) : 0 \le a < r < b, \omega \in \Omega\}$: a layer in $\mathbb{R}^n$;
- $\Gamma_a^b := \partial G \cap \{(r,\omega) : 0 \le a < r < b, \omega \in \partial\Omega\}$: the lateral surface of the layer G_a^b;
- $G_d := G \setminus G_0^d$; $\Gamma_d := \partial G \setminus \Gamma_0^d$;
- $\Omega_\rho := \overline{G_0^d} \cap \partial B_\varrho(0)$, $\varrho \le d$.

We use the standard function spaces:

- $C^k(\overline{G}), C_0^k(G)$ with the norm $|u|_{k,G}$,
- Lebesgue space $L_p(G), p \ge 1$ with the norm $\|u\|_{p,G}$,
- the Sobolev space $W^{k,p}(G)$.

We define the weighted Sobolev spaces: $V_{p,\alpha}^k(G)$ for integer $k \ge 0$ and real α as the closure of $C_0^\infty(\overline{G} \setminus \mathcal{O})$ with respect to the norm

$$\|u\|_{V_{p,\alpha}^k(G)} = \left(\int_G \sum_{|\beta|=0}^{k} r^{\alpha+p(|\beta|-k)} |D^\beta u|^p \, dx\right)^{\frac{1}{p}}$$

and $V_{p,\alpha}^{k-\frac{1}{p}}(\partial G)$ as the space of functions φ, given on ∂G, with the norm

$$\|\varphi\|_{V_{p,\alpha}^{k-\frac{1}{p}}(\partial G)} = \inf \|\Phi\|_{V_{p,\alpha}^k(G)},$$

where the infimum is taken over all functions Φ such that $\Phi\Big|_{\partial G} = \varphi$ in the sense of traces. We denote

$$W^k(G) \equiv W^{k,2}(G), \quad \overset{\circ}{W}{}^k_\alpha(G) \equiv V^k_{2,\alpha}(G),$$
$$\overset{\circ}{W}{}^{k-\frac{1}{2}}_\alpha(\partial G) \equiv V^{k-\frac{1}{2}}_{2,\alpha}(\partial G).$$

Definition 1. *A strong solution of problem (L) is a function* $u(x) \in W^{2,n}_{\mathrm{loc}}(G) \cap W^2(G_\varepsilon) \cap C^0(\overline{G})$ *that for each* $\varepsilon > 0$ *satisfies an equation for almost all* $x \in G_\varepsilon$ *and the boundary condition in the sense of traces on* Γ_ε.

Definition 2. *A strong solution of problem (QL) is a function* $u(x) \in C^0(\overline{G}) \cap W^1(G) \cap W^{2,q}_{\mathrm{loc}}(\overline{G} \setminus \mathcal{O})$, $q \ge n$ *that satisfies the equation for almost all* $x \in G$, *and the boundary condition in the sense of traces on* $\partial G \setminus \mathcal{O}$.

We assume that $M_0 = \max_{x \in \overline{G}} |u(x)|$ is known (see, e.g., [13, 15]).

We assume the existence $d > 0$ such that G_0^d is *the* **convex** *rotational cone* with the vertex at $\mathcal{O}$ and the aperture $\omega_0 \in (\frac{\pi}{2}, \pi)$, thus

$$\Gamma_0^d = \left\{ (r,\omega) \Big| x_1^2 = \cot^2 \frac{\omega_0}{2} \sum_{i=2}^n x_i^2; \ |\omega_1| = \frac{\omega_0}{2}, \ \omega_0 \in \left(\frac{\pi}{2}, \pi\right) \right\}. \tag{1.1}$$

Linear problem. Regarding the (L) equation we assume that the following **conditions** are satisfied:

(a) *the condition of the uniform ellipticity:*

$$\nu\xi^2 \le a^{ij}(x)\xi_i\xi_j \le \mu\xi^2, \quad \forall x \in \overline{G}, \quad \forall \xi \in \mathbb{R}^n; \quad \nu, \mu = \text{const} > 0, \ \textit{and}\ a^{ij}(0) = \delta_i^j;$$

(b) $a^{ij} \in C^0(\overline{G}), \quad a^i \in L_p(G),\ p > n, \quad a, f \in L_n(G)$; *for these the inequalities*

$$\left(\sum_{i,j=1}^n |a^{ij}(x) - a^{ij}(y)|^2 \right)^{\frac{1}{2}} \le \mathcal{A}(|x-y|); \quad |x| \left(\sum_{i=1}^n |a^i(x)|^2 \right)^{\frac{1}{2}} + |x|^2 |a(x)| \le \mathcal{A}(|x|)$$

hold for $x, y \in \overline{G}$, *where* $\mathcal{A}(r)$ *is the monotonically increasing, nonnegative function,* **continuous at zero**, $\mathcal{A}(0) = 0$,

(c) *there exist numbers* $f_1 \ge 0,\ g_1 \ge 0,\ s > 1,\ \beta \ge s-2,\ \gamma_0 > \tan\frac{\omega_0}{2}$ *such that*

$$|f(x)| \le f_1|x|^\beta, \ |g(x)| \le g_1|x|^{s-1}, \ \gamma(x) \ge \gamma_0;$$

(d) $a(x) \le 0$ *in* G, $\gamma(x) \in L^\infty(\partial G) \cap C^1(\partial G \setminus \mathcal{O})$;

Let $\Omega \subset S^{n-1}$ be a bounded domain with the smooth boundary $\partial\Omega$. Let $\vec{\nu}$ be the exterior normal to $\partial\Omega$. Let $\gamma(x) \in C^0(\partial\Omega)$, $\gamma(x) \ge \gamma_0 > 0$. We consider the eigenvalue problem for the Laplace-Beltrami operator $\triangle_\omega$ on the unit sphere:

$$\begin{cases} \triangle_\omega u + \vartheta u = 0, \ \omega \in \Omega, \\ \frac{\partial u}{\partial \vec{\nu}} + \gamma(x) u \Big|_{\partial\Omega} = 0, \end{cases} \tag{EVP}$$

which consists of the determination of all values ϑ (eigenvalues) for which (EVP) has a non-zero weak solutions (eigenfunctions).

Remark 1. The eigenvalue problem (EVP) was studied in Section VI [5] and in §2.5 [19] (see as well §2.4 [12]). We observe that $\vartheta = 0$ is **not** an eigenvalue of (EVP).

We denote for the smallest positive eigenvalue ϑ

$$\lambda = \frac{2 - n + \sqrt{(n-2)^2 + 4\vartheta}}{2}. \tag{1.2}$$

Our main results are following theorems.

Theorem 1.1. *Let u be a strong solution of the problem (L) and assumptions* (a)–(d) *are satisfied with $\mathcal{A}(r)$* **Dini-continuous** *at zero. Suppose, in addition, that*

$$g(x) \in \overset{\circ}{W}{}^{\frac{1}{2}}_{4-n}(\partial G) \quad \text{and} \quad a(x) \in \overset{\circ}{W}{}^{0}_{4-n}(G), \gamma(x) \in \overset{\circ}{W}{}^{\frac{1}{2}}_{2-n}(\partial G) \text{ if } u(0) \neq 0$$

and there exist numbers

$$\sup_{\varrho>0} \varrho^{-s}\Big(\|f\|_{\overset{\circ}{W}{}^{0}_{4-n}(G_0^\varrho)} + \|g\|_{\overset{\circ}{W}{}^{1/2}_{4-n}(\Gamma_0^\varrho)} + |u(0)|\big(\|a\|_{\overset{\circ}{W}{}^{0}_{4-n}(G_0^\varrho)} + \|\gamma\|_{\overset{\circ}{W}{}^{1/2}_{2-n}(\Gamma_0^\varrho)}\big)\Big) =: k_s, \tag{1.3}$$

$$\sup_{\varrho>0} \varrho^{1-s}\Big(\|f\|_{n,G^\varrho_{\varrho/2}} + |u(0)|\|a\|_{n,G^\varrho_{\varrho/2}}\Big) =: \varkappa_s.$$

Then there are $d \in (0,1)$ and a constant $C > 0$ depending only on ν, μ, d, s, n, λ, γ_0, $\|\gamma\|_{C^1(\partial G\setminus\mathcal{O})}$, meas G and on the quantity $\int_0^d \frac{\mathcal{A}(r)}{r}dr$ such that $\forall x \in G_0^d$

$$\begin{aligned} &|u(x) - u(0)| \\ &\le C\Big(|u|_{0,G} + k_s + \varkappa_s + \|f\|_{\overset{\circ}{W}{}^{0}_{4-n}(G)} + \|g\|_{\overset{\circ}{W}{}^{1/2}_{4-n}(\partial G)} + g_1 \\ &+ |u(0)|\big(1 + \|a\|_{\overset{\circ}{W}{}^{0}_{4-n}(G)} + \|\gamma\|_{\overset{\circ}{W}{}^{1/2}_{2-n}(\partial G)}\big)\Big) \begin{cases} |x|^\lambda, & \text{if } s > \lambda, \\ |x|^\lambda \ln^{3/2}\left(\frac{1}{|x|}\right), & \text{if } s = \lambda, \\ |x|^s, & \text{if } s < \lambda. \end{cases} \end{aligned} \tag{1.4}$$

If, in addition, there is a number

$$\sup_{\varrho>0} \varrho^{-s}\left(\|f + u(0)a\|_{V^0_{p,2p-n}(G^\varrho_{\varrho/2})} + \|g\|_{V^{1-\frac{1}{p}}_{p,2p-n}(\Gamma^\varrho_{\varrho/2})} + |u(0)|\|\gamma\|_{V^{1-\frac{1}{p}}_{p,p-n}(\Gamma^\varrho_{\varrho/2})}\right) =: \tau_s, \tag{1.5}$$

then

$$\begin{aligned} |\nabla u(x)| &\le C\Big(|u|_{0,G} + k_s + \varkappa_s + \tau_s + \|f\|_{\overset{\circ}{W}{}^{0}_{4-n}(G)} + \|g\|_{\overset{\circ}{W}{}^{1/2}_{4-n}(\partial G)} + g_1 \\ &+ |u(0)|\big(1 + \|a\|_{\overset{\circ}{W}{}^{0}_{4-n}(G)} + \|\gamma\|_{\overset{\circ}{W}{}^{1/2}_{2-n}(\partial G)}\big)\Big) \begin{cases} |x|^{\lambda-1}, & \text{if } s > \lambda, \\ |x|^{\lambda-1} \ln^{3/2}\left(\frac{1}{|x|}\right), & \text{if } s = \lambda, \\ |x|^{s-1}, & \text{if } s < \lambda. \end{cases} \end{aligned} \tag{1.6}$$

Theorem 1.2. *Let u be a strong solution of the problem (L) and assumptions of Theorem* 1.1 *are satisfied with $\mathcal{A}(r)$ that is a function continuous at zero but* **not** *Dini-continuous at zero. Then there are $d \in (0,1)$ and for each $\varepsilon > 0$ a constant $C_\varepsilon > 0$ depending only on ν, μ, d, s, n, λ, γ_0, $\|\gamma\|_{C^1(\partial G \setminus \mathcal{O})}$, ε, meas G and on $\mathcal{A}(\operatorname{diam} G)$ such that $\forall x \in G_0^d$*

$$|u(x) - u(0)| \le C_\varepsilon\Big(|u|_{0,G} + \|f\|_{\overset{\circ}{W}{}^0_{4-n}(G)} + \|g\|_{\overset{\circ}{W}{}^{1/2}_{4-n}(\partial G)} + g_1 + k_s + \varkappa_s \qquad (1.7)$$
$$+ |u(0)|\big(1 + \|a\|_{\overset{\circ}{W}{}^0_{4-n}(G)} + \|\gamma\|_{\overset{\circ}{W}{}^{1/2}_{2-n}(\partial G)}\big)\Big) \begin{cases} |x|^{\lambda-\varepsilon}, & \text{if } s > \lambda, \\ |x|^{s-\varepsilon}, & \text{if } s \le \lambda \end{cases}$$

and

$$|\nabla u(x)| \le C_\varepsilon\Big(|u|_{0,G} + \|f\|_{\overset{\circ}{W}{}^0_{4-n}(G)} + \|g\|_{\overset{\circ}{W}{}^{1/2}_{4-n}(\partial G)} + g_1 + k_s + \varkappa_s + \tau_s$$
$$+ |u(0)|\big(1 + \|a\|_{\overset{\circ}{W}{}^0_{4-n}(G)} + \|\gamma\|_{\overset{\circ}{W}{}^{1/2}_{2-n}(\partial G)}\big)\Big) \begin{cases} |x|^{\lambda-1-\varepsilon}, & \text{if } s > \lambda, \\ |x|^{s-1-\varepsilon}, & \text{if } s \le \lambda. \end{cases} \qquad (1.8)$$

Theorem 1.3. *Let u be a strong solution of the problem (L) and assumptions of Theorem* 1.1 *are satisfied with $s \ge \lambda$, $\mathcal{A}(r) \ln \frac{1}{r} \le \text{const}$, $r > 0$ and $\mathcal{A}(0) = 0$. Then there are $d \in (0,1)$ and the constants $C > 0, c > 0$ depending only on $\nu, \mu, d, n, \lambda, \gamma_0, \|\gamma\|_{C^1(\partial G \setminus \mathcal{O})}$, meas G and on $\mathcal{A}(\operatorname{diam} G)$ such that $\forall x \in G_0^d$*

$$|u(x) - u(0)| \le C\Big(|u|_{0,G} + \|f\|_{\overset{\circ}{W}{}^0_{4-n}(G)} + \|g\|_{\overset{\circ}{W}{}^{1/2}_{4-n}(\partial G)} + k_s + \varkappa_s + g_1$$
$$+ |u(0)|\big(1 + \|a\|_{\overset{\circ}{W}{}^0_{4-n}(G)} + \|\gamma\|_{\overset{\circ}{W}{}^{1/2}_{2-n}(\partial G)}\big)\Big)|x|^\lambda \ln^{c+1} \frac{1}{|x|} \qquad (1.9)$$

and

$$|\nabla u(x)| \le C\Big(|u|_{0,G} + \|f\|_{\overset{\circ}{W}{}^0_{4-n}(G)} + \|g\|_{\overset{\circ}{W}{}^{1/2}_{4-n}(\partial G)} + g_1 + k_s + \varkappa_s + \tau_s$$
$$+ |u(0)|\big(1 + \|a\|_{\overset{\circ}{W}{}^0_{4-n}(G)} + \|\gamma\|_{\overset{\circ}{W}{}^{1/2}_{2-n}(\partial G)}\big)\Big)|x|^{\lambda-1} \ln^{c+1} \frac{1}{|x|}. \qquad (1.10)$$

These Theorems are proved in the same way as in [2] for the Dirichlet problem with corresponding modifications considering the other boundary condition. There the idea of the proof is based on the deduction a new inequality of the Hardy-Friedrichs-Wirtinger type with the exact constant adapted to the Robin problem. The precise exponent of the solution decreasing rate depends on this exact constant. Therefore we restrict oneself here only to the deduction of the similar inequality (see below).

Quasilinear problem. Let $\mathfrak{M} = \{(x, u, z) | x \in \overline{G}, u \in \mathbb{R}, z \in \mathbb{R}^n\}$. Regarding the problem (QL) we assume that the following **conditions** are satisfied:

(A) $a_{ij}(x, u, z) \in W^{1,q}(\mathfrak{M})$, $q > n$; $\gamma(x) \in L^\infty(\partial G) \cap C^1(\partial G \setminus \mathcal{O})$;
the condition of **Carathéodory**: *functions $a(x, u, z)$, $\frac{\partial a(x,u,z)}{\partial u} \in CAR$, that is:*
- (i) *they are measurable on G as functions of variable x for $\forall u, z$;*
- (ii) *they are continuous with respect to u, z for almost all $x \in G$;*

(AA) $a_{ij}(0, u(0), 0) = \delta_i^j$ $(i, j = 1, \ldots, n)$ – *the Kronecker symbol;*

(B) *the condition of the uniform ellipticity:*

$$\nu\xi^2 \le a_{ij}(x, u, z)\xi_i\xi_j \le \mu\xi^2, \quad \forall \xi \in \mathbb{R}^n, (x, u, z) \in \mathfrak{M}; \; \nu, \mu = \text{const} > 0;$$

(C) $\frac{\partial a(x,u,z)}{\partial u} \le 0, \; \forall (x, u, z) \in \mathfrak{M};$

(D) *there exist numbers* $\beta > -1, \gamma_0 > \tan\frac{\omega_0}{2}$, $\gamma_1 \ge \gamma_0$, *positive constants* δ, μ_1, k_1, $g_0 \ge 0$ *and nonnegative functions* $b(x), f(x) \in L^q_{\mathrm{loc}}(\overline{G} \setminus \mathcal{O}), q \ge n$ *such that on* $\mathfrak{M}$ *the inequalities*

$$|a(x, u, z)| + \left|\frac{\partial a(x, u, z)}{\partial u}\right| \le \mu_1|z|^2 + b(x)|z| + f(x),$$
$$b(x) + f(x) \le k_1|x|^\beta, \; |g(x)| \le g_0|x|^\delta; \; \gamma_0 \le \gamma(x) \le \gamma_1$$

hold;

(E) *the coefficients of the problem* (QL) *satisfy such conditions that guarantee* $u \in C^{1+\widetilde{\varkappa}}(G')$ *and the existence of a local a priori estimate* $|u|_{1+\widetilde{\varkappa},G'} \le M_1$, $\widetilde{\varkappa} \in (0, 1)$ *for any smooth* $G' \subset\subset \overline{G} \setminus \{\mathcal{O}\}$ *(see Theorems* 13.13, 13.14 [17] *and Lemma* 2.3 [16]*).*

Our main result about the problem (QL) is the following

Theorem 1.4. *Let* $\lambda > 1$ *be define by* (1.2) *and let* $u(x)$ *be a solution of problem* (QL), $q > n$ *and the assumptions* $(A), (AA), (C) - -(E)$ *are satisfied with* $\beta > \lambda - 2 > -1$, $\delta \ge \lambda - 1 > 0$. *Suppose, in addition, that* $\gamma(x) \in \overset{\circ}{W}{}^{1/2}_{2-n}(\partial G)$, $g(x) \in \overset{\circ}{W}{}^{1/2}_{4-n}(\partial G)$ *and for each* $\varepsilon > 0$ *there is*

$$\sup_{\rho>0} \rho^{-\lambda-\varepsilon}\left(\|g\|_{\overset{\circ}{W}{}^{1/2}_{4-n}(\Gamma_0^\rho)} + |u(0)|\|\gamma\|_{\overset{\circ}{W}{}^{1/2}_{2-n}(\Gamma_0^\rho)}\right) := k_2. \tag{1.11}$$

Then there exist numbers $d > 0, \overline{C}_0, \overline{C}_1$ *not depending on* u *but only on* n, λ, ν, μ, μ_1, β, k_1, q, M_0 *and the domain* G *such that*

1) $|u(x) - u(0)| \le \overline{C_0}|x|^\lambda$; $\quad |\nabla u(x)| \le \overline{C_1}|x|^{\lambda-1}$, $x \in G_0^{d/2}$.

In addition, if $g(x) \in V^{1-1/q}_{q,\alpha}(\partial G)$ *and* $\gamma(x) \in V^{1-1/q}_{q,\alpha-q}(\partial G)$ *if* $u(0) \ne 0$ *and*

$$\|g(x)\|_{V^{1-1/q}_{q,\alpha}(\Gamma_0^\varrho)} + |u(0)| \cdot \|\gamma(x)\|_{V^{1-1/q}_{q,\alpha-q}(\Gamma_0^\varrho)} \le C\varrho^{\lambda-2+\frac{\alpha+n}{q}}, \; 0 < \rho < d/2, \tag{1.12}$$

then there exist numbers $d > 0$, $\overline{C}_2$, *not depend on* $u(x)$, *but only on* n, λ, ν, μ, μ_1, β, k_1, q, M_0 *and the domain* G, *such that:*

2) *if* $\alpha + q(\lambda - 2) + n > 0$, *then* $u(x) - u(0) \in V^2_{q,\alpha}(G)$ *and*

$$\|u(x) - u(0)\|_{V^2_{q,\alpha}(G_0^\rho)} \le \overline{C_2}\rho^{\lambda-2+\frac{n+\alpha}{q}}; \; 0 < \rho < d/2;$$

3) *if* $1 < \lambda < 2$, $q > \frac{n}{2-\lambda}$, *then* $u(x) - u(0) \in C^\lambda(\overline{G_0^{d/2}})$.

For the proof of this Theorem we use the barrier techniques and the comparison principle.

2. The Hardy-Friedrichs-Wirtinger inequality

Definition 3. *Function u is called a* **weak** *solution of the problem (EVP) provided that $u \in W^1(\Omega)$ and satisfies the integral identity*

$$\int_\Omega \left\{ \frac{1}{q_i} \frac{\partial u}{\partial \omega_i} \frac{\partial \eta}{\partial \omega_i} - \vartheta u \eta \right\} d\Omega + \int_{\partial\Omega} \gamma(x) u \eta d\sigma = 0 \tag{II}$$

for all $\eta(x) \in W^1(\Omega)$.

Let us introduce functionals on $W^1(\Omega)$:

$$F[u] = \int_\Omega |\nabla_\omega u|^2 d\Omega + \int_{\partial\Omega} \gamma(x) u^2 d\sigma, \quad G[u] = \int_\Omega u^2 d\Omega,$$

$$H[u] = \int_\Omega \left\langle |\nabla_\omega u|^2 - \vartheta u^2 \right\rangle d\Omega + \int_{\partial\Omega} \gamma(x) u^2 d\sigma$$

and corresponding to them bilinear forms

$$F(u,\eta) = \int_\Omega \frac{1}{q_i} \frac{\partial u}{\partial \omega_i} \frac{\partial \eta}{\partial \omega_i} d\Omega + \int_{\partial\Omega} \gamma(x) u \eta d\sigma, \quad G(u,\eta) = \int_\Omega u \eta d\Omega.$$

We define the set $K = \{u \in W^1(\Omega) \big| \, G[u] = 1\}$. Since $K \subset W^1(\Omega)$, $F[u]$ is bounded from below for $u \in K$. The greatest lower bound of $F[u]$ for this family we denote by $\vartheta : \inf_{u \in K} F[u] = \vartheta$. We formulate the following statement:

Theorem 2.1. (see Theorem of Subsection 4 §2.5, p. 123 [19]). *Let $\Omega \subset S^{n-1}$ be bounded domain with smooth boundary $\partial\Omega$. Let $\gamma(x) \in C^0(\partial\Omega)$, $\gamma(x) \geq \gamma_0 > 0$. There exist $\vartheta > 0$ and a function $u \in K$ such that $F(u,\eta) - \vartheta G(u,\eta) = 0$ for arbitrary $\eta \in W^1(\Omega)$. In particular, $F[u] = \vartheta$. In addition, on Ω, u has continuous derivatives of arbitrary order and satisfies the equation $\triangle_\omega u + \vartheta u = 0$, $\omega \in \Omega$ and thc boundary condition of (EVP) in the weak sense (see Remark on p. 121–122 [19]).*

Next from the variational principle we obtain the Friedrichs-Wirtinger inequality:

Theorem 2.2. *Let ϑ be the smallest positive eigenvalue of problem (EVP) (it there exists according to Theorem 2.1). Let $\Omega \subset S^{n-1}$ be bounded domain. Let $u \in W^1(\Omega)$ and $\gamma(x) \in C^0(\partial\Omega)$, $\gamma(x) \geq \gamma_0 > 0$. Then*

$$\vartheta \int_\Omega u^2(\omega) d\Omega \leq \int_\Omega |\nabla_\omega u(\omega)|^2 d\Omega + \int_{\partial\Omega} \gamma(\omega) u^2(\omega) d\sigma. \tag{2.1}$$

Proof. Consider functionals $F[u], G[u], H[u]$ on $W^1(\Omega)$ described above. We will find the element u that gives the minimum of the functional $F[u]$ in the set K. For this we investigate the minimization of the quadratic functional $H[u]$ on all functions $u(\omega)$, for which the integrals exist and which satisfy the boundary condition from (EVP). The necessary condition of existence of the functional extremum is $\delta H[u] = 0$. By the calculation of the first variation δH we have

$$\delta H[u] = -2\int\limits_{\Omega}(\triangle_\omega u + \vartheta u)\delta u d\Omega + 2\int\limits_{\partial\Omega}\frac{\partial u}{\partial\overrightarrow{\nu}}\delta u d\sigma + 2\int\limits_{\partial\Omega}\gamma(x)u\delta u d\sigma.$$

Hence we obtain the Euler equation and boundary condition that are our (EVP).

Backwards, let $u(\omega)$ be the solution of (EVP). By Theorem 2.1 $u \in C^2(\Omega)$. Therefore we can multiply both sides of equation (EVP) by u and integrate over Ω using the Gauss-Ostrogradskiy formula:

$$0 = \int\limits_{\Omega}(u\triangle_\omega u + \vartheta u^2)d\Omega = \vartheta\int\limits_{\Omega}u^2 d\Omega - \int\limits_{\Omega}|\nabla_\omega u|^2 d\Omega$$
$$+\int\limits_{\Omega}\frac{\partial}{\partial\omega_i}\left(u\frac{J}{q_i}\frac{\partial u}{\partial\omega_i}\right)d\omega = \vartheta\int\limits_{\Omega}u^2 d\Omega - \int\limits_{\Omega}|\nabla_\omega u|^2 d\Omega + \int\limits_{\partial\Omega}u\frac{\partial u}{\partial\overrightarrow{\nu}}d\sigma$$
$$= \vartheta\int\limits_{\Omega}u^2 d\Omega - \int\limits_{\Omega}|\nabla_\omega u|^2 d\Omega - \int\limits_{\partial\Omega}\gamma(x)u^2 d\sigma \underset{(\text{by } K)}{=} \vartheta - F[u] \Rightarrow \vartheta = F[u].$$

Consequently the required minimum is the least eigenvalue of (EVP). The existence a function $u \in K$ such that $F[u] \le F[v]$ for all $v \in K$ was proved in Theorem 2.1. □

Throughout what follows we work with the value λ defined by (1.2). Therefore the Friedrichs-Wirtinger inequality will be written in the following form

$$\lambda(\lambda+n-2)\int\limits_{\Omega}\psi^2 d\Omega \le \int\limits_{\Omega}|\nabla_\omega\psi|^2 d\Omega + \int\limits_{\partial\Omega}\gamma(x)\psi^2 d\sigma, \tag{2.2}$$
$$\forall\psi \in W^1(\Omega),\ \gamma(x) \in C^0(\partial\Omega),\ \gamma(x) \ge \gamma_0 > 0.$$

The Hardy-Friedrichs-Wirtinger inequality
Let $u \in C^0(\overline{G}) \cap W^1(G)$ and $\gamma(x) \in C^0(\partial G \setminus \mathcal{O})$, $\gamma(x) \ge \gamma_0 > 0$. Then

$$\int\limits_{G_0^d} r^{\alpha-4}u^2 dx \le H(\lambda,n,\alpha)\left\{\int\limits_{G_0^d} r^{\alpha-2}|\nabla u|^2 dx + \int\limits_{\Gamma_0^d} r^{\alpha-3}\gamma(x)u^2(x)ds\right\}, \tag{2.3}$$

$$H(\lambda,n,\alpha) = \frac{1}{\lambda(\lambda+n-2)+\frac{1}{4}(4-n-\alpha)^2},\quad \alpha \le 4-n \tag{2.4}$$

provided that integrals on the right are finite.

Proof. By Theorem 2.2 the inequality (2.2) holds. Multiplying it by $r^{n-5+\alpha}$ and integrating over $r \in (0, d)$ we obtain

$$\int\limits_{G_0^d} r^{\alpha-4} u^2 dx \le \frac{1}{\lambda(\lambda + n - 2)} \left\{ \int\limits_{G_0^d} r^{\alpha-2} \frac{1}{r^2} |\nabla_\omega u|^2 dx + \int\limits_{\Gamma_0^d} r^{\alpha-3} \gamma(x) u^2(x) ds \right\},$$
$$\forall \alpha \le 4 - n. \tag{2.5}$$

Hence (2.3) follows for $\alpha = 4 - n$. Now, let $\alpha < 4 - n$. We show that $u(0) = 0$. In fact, from the representation $u(0) = u(x) - \big(u(x) - u(0)\big)$ by the Cauchy inequality we have $\frac{1}{2}|u(0)|^2 \le |u(x)|^2 + |u(x) - u(0)|^2$. Putting $v(x) = u(x) - u(0)$ we obtain

$$\frac{1}{2}|u(0)|^2 \int\limits_{G_0^d} r^{\alpha-4} dx \le \int\limits_{G_0^d} r^{\alpha-4} u^2(x) dx + \int\limits_{G_0^d} r^{\alpha-4} |v|^2 dx < \infty \tag{2.6}$$

(the first integral from the right is finite by (2.5) and the second integral is also finite, in virtue of the Hardy inequality (see Theorem 330 [9])). Since $\int\limits_{G_0^d} r^{\alpha-4} dx =$ $\text{meas}\Omega \int\limits_0^d r^{\alpha+n-5} dr = \infty$, by $\alpha + n - 4 < 0$, the assumption $u(0) \ne 0$ contradicts (2.6). Thus $u(0) = 0$.

Therefore we can use the Hardy inequality and we obtain

$$\int\limits_{G_0^d} r^{\alpha-4} u^2 dx \le \frac{4}{|4 - n - \alpha|^2} \int\limits_{G_0^d} r^{\alpha-2} u_r^2 dx. \tag{2.7}$$

Adding the inequalities (2.5), (2.7) and using the formula

$$|\nabla u|^2 = \left(\frac{\partial u}{\partial r}\right)^2 + \frac{1}{r^2} |\nabla_\omega u|^2,$$

we get the desired (2.3). □

Lemma 2.3. *Let G_0^d be the conical domain and $\nabla v(\varrho, \cdot) \in L_2(\Omega)$ a.e. $\varrho \in (0, d)$. Let*

$$V(\rho) = \int\limits_{G_0^\varrho} r^{2-n} |\nabla v|^2 dx + \int\limits_{\Gamma_0^\varrho} r^{1-n} \gamma(x) v^2(x) ds < \infty, \ \varrho \in (0, d).$$

Then

$$\int\limits_\Omega \left(\varrho v \frac{\partial v}{\partial r} + \frac{n-2}{2} v^2 \right)\Bigg|_{r=\varrho} d\Omega \le \frac{\varrho}{2\lambda} V'(\varrho).$$

Proof. Writing $V(\varrho)$ in spherical coordinates

$$\begin{aligned} V(\varrho) &= \int_0^{\varrho} r^{2-n}\left(\int_\Omega |\nabla v|^2 d\Omega\right) r^{n-1}dr + \int_0^{\varrho} r^{1-n}\left(\int_{\partial\Omega} \gamma(x)|v|^2 d\sigma\right) r^{n-2}dr \\ &= \int_0^{\varrho} r\left(\int_\Omega |\nabla v|^2 d\Omega\right) dr + \int_0^{\varrho} \frac{1}{r}\left(\int_{\partial\Omega} \gamma(x)|v|^2 d\sigma\right) dr \end{aligned}$$

and differentiating with respect to ϱ we obtain

$$V'(\varrho) = \int_\Omega \left(\varrho\left(\frac{\partial v}{\partial r}\right)^2 + \frac{1}{\varrho}|\nabla_\omega v|^2\right)\Bigg|_{r=\varrho} d\Omega + \frac{1}{\varrho}\int_{\partial\Omega} \gamma(\varrho,\omega)v^2(\varrho,\omega)d\sigma.$$

Moreover, by Cauchy's inequality, we have for all $\varepsilon > 0 : \rho v\frac{\partial v}{\partial r} \le \frac{\varepsilon}{2}v^2 + \frac{1}{2\varepsilon}\rho^2\left(\frac{\partial v}{\partial r}\right)^2$. Thus choosing $\varepsilon = \lambda$ we obtain, by the Friedrichs-Wirtinger inequality (2.2),

$$\begin{aligned} &\int_\Omega \left(\varrho v\frac{\partial v}{\partial r} + \frac{n-2}{2}v^2\right)\Bigg|_{r=\varrho} d\Omega \le \frac{\varepsilon+n-2}{2}\int_\Omega v^2 d\Omega + \frac{\varrho^2}{2\varepsilon}\int_\Omega \left(\frac{\partial v}{\partial r}\right)^2 d\Omega \\ &\le \frac{\varepsilon+n-2}{2\lambda(\lambda+n-2)}\int_\Omega |\nabla_\omega v|^2 d\Omega \\ &+ \frac{\varrho^2}{2\varepsilon}\int_\Omega \left(\frac{\partial v}{\partial r}\right)^2 d\Omega + \frac{\varepsilon+n-2}{2\lambda(\lambda+n-2)}\int_{\partial\Omega} \gamma(x)v^2(x)d\sigma = \frac{\varrho}{2\lambda}V'(\varrho). \end{aligned}$$

□

Proposition 2.4. The comparison principle. *Let G_0^d be a convex rotational cone with vertex at $\mathcal{O}$ and the aperture $\omega_0 \in (\frac{\pi}{2}, \pi)$. Let $\mathcal{L}$ be uniformly elliptic in G_0^d; $a^i(x), a(x) \in L^\infty_{\rm loc}(G_0^d)$, $a(x) \le 0$ in G_0^d. Let $\gamma(x) \in C^0(\Gamma_0^d)$, $\gamma(x) \ge \gamma_0 > 0$ on Γ_0^d. Suppose that v and w are functions in $W^{2,n}_{\rm loc}(G_0^d) \cap C^0(\overline{G_0^d})$ satisfying*

$$\begin{cases} \mathcal{L}[w(x)] \le \mathcal{L}[v(x)], & x \in G_0^d; \\ \mathcal{B}[w(x)] \ge \mathcal{B}[v(x)], & x \in \Gamma_0^d; \\ \quad w(x) \ge v(x), & x \in \Omega_d \cup \mathcal{O}. \end{cases} \tag{2.8}$$

Then $v(x) \le w(x)$ in $\overline{G_0^d}$.

This proposition is the direct consequence of the Lieberman global and strong maximum principles for Lipschitz domains (Lemma 1.1 [13], Proposition 2.1 [15], Corollary 3.2 [15]). Here it should be noted that our condition $\omega_0 \in (\frac{\pi}{2}, \pi)$ guarantees the obliqueness of the operator $\mathcal{B}$ at $\forall x_0 \in \Gamma_0^d$ with the modulus of obliqueness lesser than 1.

3. The barrier function. The preliminary estimate of the solution modulus

Let G_0^d be a convex rotational cone with a solid angle $\omega_0 \in (0,\pi)$ and the lateral surface Γ_0^d, such that $G_0^d \subset \{x_1 \geq 0\}$. Let define the linear elliptic operator: $\mathcal{L}_0 \equiv a^{ij}(x)\frac{\partial^2}{\partial x_i \partial x_j}; a^{ij}(x) = a^{ji}(x),\ x \in G_0^d;\ \nu\xi^2 \leq a^{ij}(x)\xi_i\xi_j \leq \mu\xi^2, \forall x \in G_0^d, \forall \xi \in \mathbb{R}^n; \nu,\mu = \text{const} > 0$ and the boundary operator: $\mathcal{B} \equiv \frac{\partial}{\partial \vec{n}} + \frac{1}{|x|}\gamma(x),\ \gamma(x) \geq \gamma_0 > 0,\ x \in \Gamma_0^d$.

Lemma 3.1. (Existence of the barrier function). *Fix the numbers* $\gamma_0 > \tan\frac{\omega_0}{2}, \delta > 0, g_1 \geq 0, d \in (0,1)$. *There exist* $h > 0$ *depending only on* ω_0, *the number* $\varkappa_0 \in (0, \gamma_0 \cot\frac{\omega_0}{2} - 1)$, *a number* $B > 0$ *and a function* $w(x) \in C^1(\overline{G_0}) \cap C^2(G_0)$ *that depend only on* ω_0, *the ellipticity constants* ν,μ *of the operator* $\mathcal{L}_0$ *and the quantities* $\gamma_0, \delta, g_1, \varkappa_0$, *such that for any* $\varkappa \in (0; \min(\delta, \varkappa_0))$ *the following hold:*

$$\mathcal{L}_0[w(x)] \leq -\nu h^2 |x|^{\varkappa-1};\quad x \in G_0^d; \tag{3.1}$$

$$\mathcal{B}[w(x)] \geq g_1 |x|^{\delta};\quad x \in \Gamma_0^d; \tag{3.2}$$

$$0 \leq w(x) \leq C_0(\varkappa_0, B, \omega_0)|x|^{\varkappa+1};\quad x \in \overline{G_0^d}; \tag{3.3}$$

$$|\nabla w(x)| \leq C_1(\varkappa_0, B, \omega_0)|x|^{\varkappa};\quad x \in \overline{G_0^d}. \tag{3.4}$$

Proof. Let $(x,y,x') \in \mathbb{R}^n$, where $x = x_1, y = x_2, x' = (x_3,\dots,x_n)$. In $\{x_1 \geq 0\}$ we consider the cone K with the vertex in $\mathcal{O}$, such that $K \supset G_0^d$ (we recall that $G_0^d \subset \{x_1 \geq 0\}$). Let ∂K be the lateral surface of K and let $\partial K \cap yOx = \Gamma_\pm$ be $x = \pm hy$, where $h = \cot\frac{\omega_0}{2}, 0 < \omega_0 < \pi$, such that in the interior of K the inequality $x > h|y|$ holds. We shall consider the function:

$$w(x;y,x') \equiv x^{\varkappa-1}(x^2 - h^2y^2) + Bx^{\varkappa+1},\quad \text{with some } \varkappa \in (0;1),\ B > 0. \tag{3.5}$$

Let coefficients of the operator $\mathcal{L}_0$ be: $a^{2,2} = a,\ a^{1,2} = b,\ a^{1,1} = c$. Then we have:

$$\mathcal{L}_0 w = a w_{yy} + 2b w_{xy} + c w_{xx}; \tag{3.6}$$

$$\nu\eta^2 \leq a\eta_1^2 + 2b\eta_1\eta_2 + c\eta_2^2 \leq \mu\eta^2;\quad \eta^2 = \eta_1^2 + \eta_2^2;\ \forall \eta_1,\eta_2 \in \mathbb{R}. \tag{3.7}$$

Let us calculate the operator $\mathcal{L}_0$ on the function (3.5). For $t = \frac{y}{x}, |t| < \frac{1}{h}$ we obtain: $\mathcal{L}_0 w = -h^2 x^{\varkappa-1}\phi(\varkappa)$, where

$$\begin{aligned}\phi(\varkappa) &= 2a - 4bt + 4bt\varkappa - ch^{-1}(1+B)(\varkappa^2+\varkappa) + ct^2\varkappa^2 - 3ct^2\varkappa + 2ct^2\\ &= c(t^2 - h^{-2}(1+B))\varkappa^2 + (4bt - ch^{-2}(1+B) - 3ct^2)\varkappa + 2(ct^2 - 2bt + a);\end{aligned}$$

$$c(t^2 - h^{-2}(1+B)) = c\left(\frac{y^2}{x^2} - \frac{1+B}{h^2}\right) \leq -c\frac{B}{h^2} < 0.$$

Because of (3.7), we have $\phi(0) = 2(ct^2 - 2bt + a) \geq 2\nu$ and since $\phi(\varkappa)$ is the square function there exists the number $\varkappa_0 > 0$ depending only on ν, μ, h such that $\phi(\varkappa) \geq \nu$ for $\varkappa \in [0;\varkappa_0]$. Therefore we obtain (3.1).

Now, let us notice that

$$\Gamma_\pm : x = \pm hy,\ h = \cot\frac{\omega_0}{2},\ 0 < \omega_0 < \pi. \tag{3.8}$$

Then we have

$$\text{on } \Gamma_+ : \begin{cases} x &= r\cos\frac{\omega_0}{2}, \\ y &= r\sin\frac{\omega_0}{2} \end{cases} \quad \begin{cases} \angle(\vec{n},x) = \frac{\pi}{2} + \frac{\omega_0}{2}, \\ \angle(\vec{n},y) = \frac{\omega_0}{2} \end{cases}$$

$$\text{on } \Gamma_- : \begin{cases} x &= r\cos\frac{\omega_0}{2}, \\ y &= -r\sin\frac{\omega_0}{2} \end{cases} \quad \begin{cases} \angle(\vec{n},x) = \frac{\pi}{2} + \frac{\omega_0}{2}, \\ \angle(\vec{n},y) = \pi + \frac{\omega_0}{2} \end{cases}$$

$$\sin\frac{\omega_0}{2} = \frac{1}{\sqrt{1+h^2}}, \quad \cos\frac{\omega_0}{2} = \frac{h}{\sqrt{1+h^2}}.$$

Therefore we obtain:

$$\begin{aligned} w_x = (1+\varkappa)x^{\varkappa}(1+B) - (\varkappa-1)h^2y^2x^{\varkappa-2} &\Rightarrow w_x\big|_{\Gamma_\pm} = [2 + B(1+\varkappa)]x^{\varkappa}, \\ w_y = -2h^2yx^{\varkappa-1} &\Rightarrow w_y\big|_{\Gamma_\pm} = \mp 2hx^{\varkappa}. \end{aligned} \tag{3.9}$$

Because of

$$\frac{\partial w}{\partial \vec{n}}\bigg|_{\Gamma_\pm} = w_x\cos\angle(\vec{n},x)\bigg|_{\Gamma_\pm} + w_y\cos\angle(\vec{n},y)\bigg|_{\Gamma_\pm}$$

and (3.9), we get:

$$\frac{\partial w}{\partial \vec{n}}\bigg|_{\Gamma_\pm} = -r^{\varkappa}\frac{h^{\varkappa}}{(1+h^2)^{\frac{\varkappa+1}{2}}}[2(1+h^2) + B(1+\varkappa)].$$

Hence it follows:

$$\mathcal{B}[w]\Big|_{\Gamma_\pm} \geq \frac{h^{\varkappa}}{\left(1+h^2\right)^{\frac{\varkappa+1}{2}}} r^{\varkappa}\left[Bh\gamma_0 - B(1+\varkappa) - 2(1+h^2)\right].$$

Since $h > \frac{1}{\gamma_0}$ for $\varkappa \leq \varkappa_0$ we obtain:

$$\mathcal{B}[w]\Big|_{\Gamma_\pm^d} \geq \frac{h^{\varkappa_0}r^{\varkappa}}{\left(1+h^2\right)^{\frac{\varkappa_0+1}{2}}}\left[B(h\gamma_0 - 1 - \varkappa_0) - 2(1+h^2)\right] \geq g_1 r^{\delta}, \ 0 < r < d < 1$$

if we take $\varkappa \leq \delta$ and

$$B \geq \left\{\frac{g_1\left(1+h^2\right)^{\frac{\varkappa_0+1}{2}}}{h^{\varkappa_0}} + 2(1+h^2)\right\} \cdot \frac{1}{h\gamma_0 - 1 - \varkappa_0} \tag{3.10}$$

(it should be pointed out that we can choose (if it is necessary) $\varkappa_0$ so small that $\varkappa_0 < h\gamma_0 - 1$). Now we'll show (3.3). Let us rewrite the function (3.5) in spherical coordinates. Recalling that $h = \cot\frac{\omega_0}{2}$ we obtain:

$$\begin{aligned} w(x;y,x') &= (1+B)(r\cos\omega)^{1+\varkappa} - h^2r^2\sin^2\omega(r\cos\omega)^{\varkappa-1} \\ &= r^{1+\varkappa}\cos^{\varkappa-1}\omega\left(B\cos^2\omega + \frac{\chi(\omega)}{\sin^2\frac{\omega_0}{2}}\right), \quad \forall\omega \in \left[-\frac{\omega_0}{2}; \frac{\omega_0}{2}\right], \end{aligned}$$

II. The function $u(r,\omega) = r^\lambda \left(\ln \frac{1}{r}\right)^{\frac{\lambda-1}{\lambda+1}} \psi(\omega)$ with λ and $\psi(\omega)$ defined by (4.1)–(4.2) is a solution of the problem

$$\begin{cases} \sum\limits_{i,j=1}^{n} a^{ij}(x) u_{x_i x_j} = 0, \ x \in G_0, \\ \left(\frac{\partial u}{\partial n} + \frac{1}{r}\gamma_\pm u\right)\Big|_{\Gamma_\pm} = 0, \ \gamma_\pm > 0 \end{cases}$$

in the corner G_0, where

$$a^{11}(x) = 1 - \frac{2}{\lambda+1} \cdot \frac{x_2^2}{r^2 \ln 1/r}, \qquad r > 0;$$
$$a^{12}(x) = a^{21}(x) = \frac{2}{\lambda+1} \cdot \frac{x_1 x_2}{r^2 \ln 1/r}, \qquad r > 0;$$
$$a^{22}(x) = 1 - \frac{2}{\lambda+1} \cdot \frac{x_1^2}{r^2 \ln 1/r}, \qquad r > 0;$$
$$a^{ij}(0) = \delta_i^j, \ (i,j = 1,2).$$

In the domain G_0^d, $d < e^{-2}$ the equation is uniformly elliptic with ellipticity constants $\mu = 1$ and $\nu = 1 - \frac{2}{\ln(1/d)}$. Further, $\mathcal{A}(r) = \frac{2}{\lambda+1} \ln^{-1}\left(\frac{1}{r}\right)$, i.e., the function $\mathcal{A}(r)$ does not satisfy the Dini condition at zero. Moreover, $a^{ij}(x)$ are continuous at the point $\mathcal{O}$. This example shows that the condition of Theorem 1.1 about Dini-continuity of leading coefficients of (L) are essential, and it illustrates the precision of assumptions of Theorem 1.3 as well.

III. The function $u(r,\omega) = r^\lambda \ln \frac{1}{r} \psi(\omega)$ with λ and $\psi(\omega)$ defined by (4.1)–(4.2) is a solution of the problem

$$\begin{cases} \triangle u + \frac{2\lambda}{r^2 \ln \frac{1}{r}} u = 0, \ x \in G_0, \\ \left(\frac{\partial u}{\partial n} + \frac{1}{r}\gamma_\pm u\right)\Big|_{\Gamma_\perp} = 0, \ \gamma_\pm > 0 \end{cases}$$

in the corner G_0. This example shows that assumptions of Theorems 1.1 and 1.3 on lowest coefficients of (L) are precise and essential.

IV. The function $u(r,\omega) = r^\lambda \ln \frac{1}{r} \psi(\omega)$ with λ and $\psi(\omega)$ defined by (4.1)–(4.2) is a solution of the problem

$$\begin{cases} \triangle u = -2\lambda r^{\lambda-2} \psi(\omega), \ x \in G_0, \\ \left(\frac{\partial u}{\partial n} + \frac{1}{r}\gamma_\pm u\right)\Big|_{\Gamma_\pm} = 0, \ \gamma_\pm > 0 \end{cases}$$

in the corner G_0. All assumptions of Theorem 1.1 are fulfilled with $s = \lambda$. This example shows the precision of assumptions for the right-hand side of (L) in Theorem 1.1.

References

[1] M. Bochniak, M.Borsuk, *Dirichlet problem for linear elliptic equations degenerating at a conical boundary point,* Analysis. München, Germany **23**, 3 (2003), pp. 225–248.

[2] M.V. Borsuk, Best-possible estimates of solutions of the Dirichlet problem for linear elliptic nondivergence equations of second-order in a neighborhood of a conical point on the boundary, *Math. USSR Sbornik* **74** (1993), 185–201.

[3] M.V. Borsuk, *Estimates of solutions of Dirichlet problem for elliptic nondivergence second-order equations in a neighbourhood of a conical boundary point,* Differ. Uravn. **30**, 1 (1994), pp. 104–108.

[4] M.V. Borsuk, *On the solvability of the first boundary value problem for second-order elliptic equations in a domain with a conical point on the boundary,* Mat. Fiz. Anal. Geom. **4**, 4 (1997), pp. 428–452.

[5] R. Courant, D. Hilbert, "Methoden der mathematischen Physik", Bd. 1, Springer-Verlag, Berlin, 1931.

[6] M. Faierman, Regularity of solutions of an elliptic boundary value problem in a rectangle, *Comm. in PDE* **12** (1987), 285–305.

[7] M.G. Garroni, V.A. Solonnikov and Vivaldi M.A., On the oblique derivative problem in an infinite angle, *Topological methods in nonlinear analysis* **7** (1996), 299–325.

[8] D. Gilbarg and N.S. Trudinger, "Elliptic Partial Differential Equations of Second Order", Springer-Verlag, Berlin/Heidelberg/New York, 1977. Revised Third Printing, 1998.

[9] G.H. Hardy, J.E. Littlewood and G. Pólya, "Inequalities", University Press, Cambridge, 1952.

[10] V.A. Kondrat'ev, Boundary value problem for elliptic equations in domains with conical or angular points, *Trudy Moscov. Mat. Obshch.* **16** (1967), 209–292.

[11] V.A. Kozlov, V.G. Maz'ya and Rossman J., "Elliptic boundary value problem in domains with point singularities", AMS. Mathematical surveys and monographs, **52** (1997).

[12] V.A. Kozlov, V.G. Maz'ya and Rossman J., "Spectral problems associated with corner singularities of solutions to elliptic equations", AMS. Mathematical surveys and monographs, **85** (2001).

[13] G.M. Lieberman, Local estimates for subsolutions and supersolutions of oblique derivative problems for general second-order elliptic equations, *Trans. of AMS.* **304** (1987), 343–353.

[14] G.M. Lieberman, Oblique derivative problems in Lipschitz domains. Continuous boundary data, *Bull. Un. Mat. Ital. B (7)* **1** (1987), 1185–1210.

[15] G.M. Lieberman, Pointwise estimates for oblique derivative problems in nonsmooth domains, *J. Diff. Equat.* **173** (2001), no. 1, 178–211.

[16] Gary M. Lieberman, *The nonlinear oblique derivative problem for quasilinear elliptic equations.* Nonlinear Analysis. Theory, Methods and Applications, **8** (1984), p. 49–65.

[17] Gary M. Lieberman, *Second-order parabolic differential equations,* World Scientific, Singapore – New Jersey – London – Hong Kong, 1996, 439 p.

[18] H. Reisman, Second-order elliptic boundary value problem in a domain with edges, *Comm. in PDE* **6** (1995), 1023–1042.

[19] S.L. Sobolev, "Some applications of functional analysis in mathematical physics", Trans. of Math. Monographs, **90** (1991). AMS, Providence, Rhode Island.

Mikhail Borsuk and Agneszka Zawadzka
Department of Mathematics and Informatics,
University of Warmia and Mazury in Olsztyn,
PL-10-957 Olsztyn-Kortowo, Poland
e-mail: borsuk@uwm.edu.pl

Progress in Nonlinear Differential Equations
and Their Applications, Vol. 64, 37–61

Well-posedness of a Two-phase Flow with Soluble Surfactant

Dieter Bothe, Jan Prüss and Gieri Simonett

Dedicated to Herbert Amann

Abstract. The presence of surfactants, ubiquitous at most fluid/liquid interfaces, has a pronounced effect on the surface tension, hence on the stress balance at the phase boundary: local variations of the capillary forces induce transport of momentum along the interface – so-called Marangoni effects. The mathematical model governing the dynamics of such systems is studied for the case in which the surfactant is soluble in one of the adjacent bulk phases. This leads to the two-phase balances of mass and momentum, complemented by a species equation for both the interface and the relevant bulk phase. Within the model, the motions of the surfactant and of the adjacent bulk fluids are coupled by means of an interfacial momentum source term that represents Marangoni stresses. Employing L_p-maximal regularity we obtain well-posedness of this model for a certain initial configuration. The proof is based on recent L_p-theory for two-phase flows without surfactant.

Mathematics Subject Classification (2000). Primary 35R35, Secondary 35Q30, 76D45, 76T10.

Keywords. Navier-Stokes equations, surface tension, Marangoni forces, maximal regularity, surface transport theorem.

1. Introduction

Surfactants are surface active agents that are adsorbed at gas/liquid or liquid/liquid interfaces, where they lower the surface tension. The presence of surfactants in multiphase systems, either in the form of unavoidable impurities, or as specific substances like emulsifiers to affect the system's behavior, has a significant influence on the shape and the dynamics of the interfaces. Indeed, resulting variations of the capillary force induce Marangoni stresses which increase the drag of fluid particles and hinder the internal circulation [7]. For example, the terminal rise velocity of bubbles or droplets can be drastically reduced, which results in larger contact times between the fluid phases and, hence, alters mass transfer.

Surfactants are usually soluble in at least one of the adjacent bulk phases, hence there is also exchange of surfactants between the relevant bulk phases and the interface by adsorption and desorption [8]. Along the interface surfactant is transported by convection and diffusion. Furthermore, changes of the interfacial area due to compression or stretching cause corresponding changes in the surfactant concentration.

Mathematically, the analysis of two-phases flows with surfactant corresponds to the study of free boundary problems for the Navier-Stokes equations, coupled with a diffusion equation having a dynamic condition on the free boundary.

Existence results for the *one-phase Navier-Stokes equations with a free surface* were obtained in [2, 5, 31, 32, 33]. In a series of publications, Solonnikov considered the motion of an isolated viscous incompressible drop subject to capillary forces, see [27] and the references contained therein.

There is significantly less work for the *one-phase Navier-Stokes equations with Marangoni effects*, which refers to situations where the surface endures tangential stresses caused by variations of surface tension. The latter can be caused by temperature gradients, for instance. This case is considered in [1], where the flow of a viscous liquid with its free surface subjected to a nonuniform thermal flux is studied; see also [19]. Assuming the Boussinesq approximation to account for density variations, stationary solutions are obtained in weighted Hölder spaces. The instationary situation is considered in [34], where small time existence of the free surface flow of a heat conducting viscous droplet of constant density in three dimensions is proved.

Problems involving the *two-phase Navier-Stokes equations with surface tension* have been investigated in [28, 29, 10, 9]. In [28, 29] small time existence for smooth initial data as well as global existence and convergence to the equilibrium configuration for nearby initial data is proven. Small time existence in weighted Hölder spaces is obtained in [10] for incompressible fluids, while combinations of compressible and incompressible fluids are allowed in [9]. For compressible two-phase flow without surface tension but with additional energy balance see [30].

There is a large amount of literature in the Engineering Sciences on two-phase flows in the presence of surface active substances, but a rigorous mathematical analysis is still missing so far. However, interesting first numerical results in 2D are obtained in [18], while analytical calculations for bubbles in plane extensional flow of low velocity are given in [25]. In these contributions, the surfactant is considered as insoluble, a common idealization.

In the present paper we consider the two-phase Navier-Stokes equations with surfactant dependent surface tension complemented by a convection-diffusion equation inside one phase with another convection-diffusion equation on the free surface which are nonlinearly coupled. We give a derivation of the mathematical model, for which an appropriate surface transport theorem is employed. To keep the paper self-contained, a proof of the latter together with an introduction of the notion of surface divergence is included in two appendices. The main result of this paper is the existence of a unique classical solution to this free boundary problem for a certain initial configuration of the phases. The latter corresponds to situations encountered by the usual localization procedure which leads to "deformed half-spaces" for the two phases. The strategy here is to build up the proof based on recent results on two-phase flows without surfactant. This works, since, luckily, the surfactant equations are decoupled from the Navier-Stokes equations in the associated linear problem.

Our proof makes use of recent developments in the theory of maximal regularity for parabolic problems. For a general reference to this subject, we refer to the monograph [11]. Of particular importance for this work are the papers [12, 13, 14, 22, 23], where basic notations used throughout this paper can be found.

2. The mathematical model

The subsequent mathematical model describes two-phase flows with a free boundary to capture the motion of a fluid particle (droplet or bubble) in an ambient fluid phase (gas or liquid). We consider isothermal flows of two immiscible Newtonian fluids with constant densities. Based on continuum mechanics, balance of mass and momentum inside the phases lead to the Navier-Stokes equations, i.e.,

$$\nabla \cdot u = 0, \qquad \partial_t(\rho_\pm u) + \nabla \cdot (\rho_\pm u \otimes u - S) = 0$$

with the stress tensor

$$S = -\pi I + \mu_\pm \Big(\nabla u + (\nabla u)^{\mathsf{T}} \Big).$$

These balances hold in $\Omega_+(t) \cup \Omega_-(t)$, where $\Omega_\pm(t)$ are the domains occupied at time t by two fluids with density $\rho_\pm$ and viscosity $\mu_\pm$, respectively. Hence the material parameters depend on the phase. Whenever a distinction between the different phases is not necessary, the phase indices $+$ and $-$ are omitted. The phases are separated by a phase boundary $\Gamma(t)$ at which at least one of the material parameters has a jump-discontinuity. At this interface, the additional jump conditions

$$\begin{aligned} [\rho(u - u_{\text{int}})]\, n &= 0, \\ [\rho u \otimes (u - u_{\text{int}}) - S]\, n &= \sigma \kappa n + \text{grad}_\Gamma \sigma \end{aligned}$$

appear, where u_{int} is the interfacial velocity, $\text{grad}_\Gamma \sigma$ denotes the surface gradient of the surface tension σ (cf. Appendix A) and

$$\kappa = -\text{div}_\Gamma n$$

is the sum of the (local) principal curvatures of the interface. More precisely, it should read $\text{div}_{\Gamma(t)}$, but the dependence on t is dropped whenever this is reasonable. Throughout the paper, n denotes the unit normal at the interface directed into the phase $\Omega_+(t)$, say, in which case $\kappa(t,x) > 0$ at $x \in \Gamma(t)$ iff $\Omega_+(t) \cap B_r(x)$ is convex for small $r > 0$. The notation $[\phi]$ stands for the jump of a physical quantity ϕ across the interface, i.e.,

$$[\phi](t,x) = \lim_{h \to 0+} \Big(\phi(t, x + h\,n) - \phi(t, x - h\,n) \Big), \quad x \in \Gamma(t).$$

Below, it is assumed that the two-fluid system under consideration does not endure phase changes like melting or evaporation. Consequently, there are no convective fluxes across the interface, and hence the normal components of the fluid velocities are continuous at the phase boundary, and the interface itself is advected with the flow, i.e.,

$$(u_+ \cdot n)_{|\Gamma(t)} = (u_- \cdot n)_{|\Gamma(t)} = V,$$

where V denotes the normal velocity of the interface. Furthermore, we assume no-slip at the interface. Then the jump conditions simplify to

$$[u] = 0, \qquad [-S]\,n = \sigma\kappa n + \mathrm{grad}_\Gamma \sigma.$$

Throughout this paper, we consider the case of a soluble surfactant, i.e., a chemical species which is adsorbed at the interface but also lives in at least one of the bulk phases. For technical simplicity, we concentrate on the case in which this species is only present in phase Ω_+, which is also of practical relevance. The local concentration c of a dilute chemical species is governed by the balance equation

$$\partial_t c + \nabla \cdot (cu + J) = 0,$$

where we assume that the molecular flux is given by Fick's law,

$$J = -D\nabla c,$$

with constant diffusivity $D > 0$. This leads to the convection-diffusion equation

$$\partial_t c + u \cdot \nabla c - D\Delta c = 0$$

inside Ω_+. For more details about the continuum mechanical modeling of two-phase flows up to this point see [16, 26].

At the phase boundary, mass balance for a material control area, i.e., $M(t) = \Gamma(t) \cap V(t)$ with a material control volume $V(t)$, implies

$$\frac{d}{dt} \int_{M(t)} c_\Gamma \, d\sigma_t = - \int_{\partial M(t)} J_\Gamma \cdot N \, ds_t - \int_{M(t)} [J] \cdot n \, d\sigma_t \,.$$

Here c_Γ denotes the surface specific concentration of the surfactant on the interface Γ and J_Γ is the interfacial molecular flux of surfactant which will be modeled by Fick's law again, i.e.,

$$J_\Gamma = -D_\Gamma \, \mathrm{grad}_\Gamma c_\Gamma$$

with $D_\Gamma > 0$. Observe also that $[J] \cdot n$ reduces to $-D(\nabla c \cdot n)_{|\Gamma(t)}$, understood as the Ω_+-sided limit, here, since the species c is only present in Ω_+. Application of the surface divergence theorem (Theorem A in Appendix A), the surface transport theorem (Theorem B in Appendix B) and the usual localization procedure leads to the differential formulation of the surfactant balance. A completely Eulerian formulation is problematic since c_Γ is only defined on $\mathrm{gr}(\Gamma)$. An intermediate Eulerian/Lagrangian formulation reads as

$$\partial_{t,n} c_\Gamma + \mathrm{div}_\Gamma(c_\Gamma u_\Gamma) - c_\Gamma \kappa V - D_\Gamma \Delta_\Gamma c_\Gamma = D(\nabla c \cdot n)_{|\Gamma(t)},$$

where $\partial_{t,n} c_\Gamma$ stands for the derivative of c_Γ along a purely normal path, u_Γ is the tangential part of u and Δ_Γ denotes the Laplace-Beltrami operator on Γ; cf. Appendix B.

One more constitutive equation is needed to determine the surfactant distribution. Here we assume that the process of ad- and desorption of surfactant at the interface is instantaneous if viewed on the time scale of convective and diffusive transport. This leads to the additional relation

$$c_\Gamma = \gamma(c_{|\Gamma(t)}),$$

where it is reasonable to assume that $\gamma : \mathbb{R}_+ \to \mathbb{R}_+$ is C^2 with $\gamma' > 0$. In concrete constitutive equations, γ will usually satisfy $\gamma(\infty) = c_\Gamma^\infty$, since the phase boundary has a finite capacity concerning adsorption of surfactant. This relation gets more complicated in case of liquid/liquid/surfactant-systems in which the surfactant is soluble in both phases; a survey on adsorption and partitioning of surfactants is given in [24].

Finally, the surface tension σ is a function of c_Γ and it can be significantly lowered due to the adsorption of surfactant. Below, we assume that $\sigma : \mathbb{R}_+ \to \mathbb{R}_+$ is C^2 with $\sigma > 0$. In this situation, a nonhomogeneous surfactant distribution leads to nonzero surface gradients of the surface tension. This corresponds to so-called Marangoni forces which cause tangential stress at the phase boundary and can trigger currents along the interface, the so-called Marangoni convection. This back-effect of the surfactant distribution on the hydrodynamics is mediated via the surface gradient of σ within the momentum jump condition. The full model of a two-phase flow with soluble surfactant now reads as

Balance equations

$$\begin{aligned} \nabla \cdot u &= 0 \quad \text{for} \quad t > 0,\ x \in \Omega_\pm(t), \\ \partial_t(\rho_\pm u) + \nabla \cdot (\rho_\pm u \otimes u - S) &= 0 \quad \text{for} \quad t > 0,\ x \in \Omega_\pm(t), \\ \partial_t c + u \cdot \nabla c - D\Delta c &= 0 \quad \text{for} \quad t > 0,\ x \in \Omega_+(t). \end{aligned} \tag{1}$$

Free boundary conditions

$$\begin{aligned} [u] = 0, \quad V = u \cdot n \quad & \text{for} \quad t > 0,\ x \in \Gamma(t), \\ [-S]\, n = \sigma(c_\Gamma)\kappa n + \sigma'(c_\Gamma)\, \mathrm{grad}_\Gamma c_\Gamma \quad & \text{for} \quad t > 0,\ x \in \Gamma(t), \\ D(\nabla c \cdot n)_{|\Gamma(t)} \qquad\qquad & \\ = \partial_{t,n} c_\Gamma + \mathrm{div}_\Gamma(c_\Gamma u_\Gamma) - c_\Gamma \kappa V - D_\Gamma \Delta_\Gamma c_\Gamma \quad & \text{for} \quad t > 0,\ x \in \Gamma(t), \\ c_\Gamma(t,x) = \gamma(c(t,x)) \quad & \text{for} \quad t > 0,\ x \in \Gamma(t). \end{aligned} \tag{2}$$

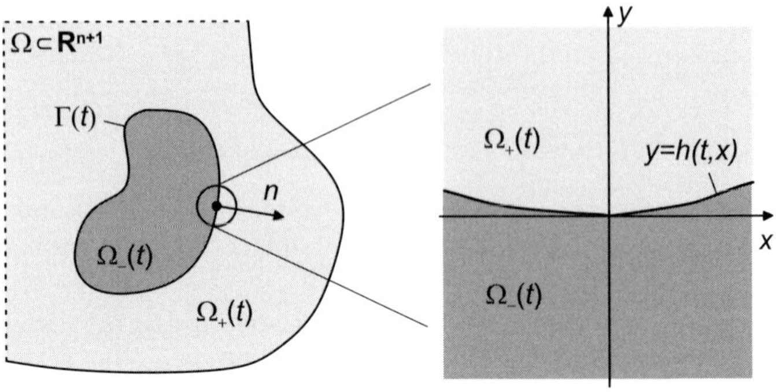

FIGURE 1. Configuration of the phases at the interface and localization.

Initial conditions

$$\begin{aligned} u(0,x) &= u_0(x), && \text{for} \quad x \in \Omega_\pm(0), \\ c(0,x) &= c_0(x) && \text{for} \quad x \in \Omega_+(0), \\ \Gamma(0)=\Gamma_0, \quad c_\Gamma(0,x) &= c_\Gamma^0(x) && \text{for} \quad x \in \Gamma_0. \end{aligned} \tag{3}$$

Recall that S is given by

$$S = -\pi I + S^0 \quad \text{with} \quad S^0 = \mu_\pm\Big(\nabla u + (\nabla u)^\mathsf{T}\Big). \tag{4}$$

3. Main result: Small time existence in a deformed half-space

The principal aim of this investigation is to obtain local existence of classical solutions for Γ_0 being a closed manifold such that the bounded component of $\mathbb{R}^{n+1} \setminus \Gamma_0$ represents a fluid particle like a drop or a bubble. This general situation can be treated by a localization procedure, which will be carried out in a forthcoming paper. Here, we are interested in the model problem which occurs by localization at points on Γ_0; cf. Figure 1. In this case, $\Omega_\pm$ are deformed half-spaces and Γ_0 is close to a plane, say $\mathbb{R}^n$, i.e., Γ_0 is a graph over $\mathbb{R}^n$ given by a function h_0 with $|\nabla h_0|_\infty$ small. The main result for this model problem reads as follows.

Theorem 1. *Let $\rho_\pm, \mu_\pm, D, D_\Gamma > 0$ and $\gamma, \sigma \in C^2(\mathbb{R}_+)$ with $\sigma > 0$. Fix $p > n+3$ and suppose the following conditions to hold.*

Regularity of initial data:

$$h_0 \in W_p^{3-2/p}(\mathbb{R}^n), \quad \Gamma_0 = \mathrm{gr}(h_0), \quad u_0 \in W_p^{2-2/p}(\mathbb{R}^{n+1} \setminus \Gamma_0; \mathbb{R}^{n+1}),$$

$$c_0 - \bar{c} \in W_p^{2-2/p}(\Omega_+^0), \quad c_\Gamma^0 - \bar{c}_\Gamma \in W_p^{2-2/p}(\Gamma_0)$$

with appropriate $\bar{c}$, $\bar{c}_\Gamma \in \mathbb{R}$ and $\Omega_+^0 = \{(x,y) \in \mathbb{R}^{n+1} : x \in \mathbb{R}^n,\, y > h_0(x)\}$.

Compatibility condition:

$$\nabla \cdot u_0 = 0, \quad [S^0_{|t=0} n - \langle n, S^0_{|t=0} n\rangle\, n] = -\sigma'(c_\Gamma^0)\,\mathrm{grad}_{\Gamma_0} c_\Gamma^0,$$

$$[u_0] = 0, \quad c_\Gamma^0 = \gamma(c_{0|\Gamma_0}), \quad \bar{c}_\Gamma = \gamma(\bar{c}).$$

Then there is $\eta > 0$ *such that the following holds. Given* h_0 *with* $|\nabla h_0| \leq \eta$ *and* $c_\Gamma^0 \geq 0$ *with*

$$\sup_{x\in\Gamma_0} c_\Gamma^0 - \inf_{x\in\Gamma_0} c_\Gamma^0 \leq \eta,$$

there exist $t_0 = t_0(u_0, h_0, c_0, c_\Gamma^0)$ *and a classical solution* (u, π, c, c_Γ) *of* (1)–(4) *in* $(0, t_0)$*. The solution is unique within the class of maximal regularity as described in Theorem* 2 *below.*

The proof of Theorem 1 consists of three main steps as given in the subsections to follow.

3.1. Reduction to a half-space

Suppose that $\Gamma(t)$ is a graph over $\mathbb{R}^n$, parameterized as

$$\Gamma(t) = \{(x, h(t,x)) : x \in \mathbb{R}^n\} \quad \text{for } t \in J = [0,T]$$

with $\Omega_+(t)$ lying "above" $\Gamma(t)$, i.e.,

$$\Omega_\pm(t) = \{(x,y) \in \mathbb{R}^n \times \mathbb{R} : y \gtrless h(t,x)\} \quad \text{for } t \in J\,.$$

Reduction from deformed into true half-spaces is achieved by means of the transformations

$$\begin{aligned} v(t,x,y) &= \begin{bmatrix} u_1(t,x,h(t,x)+y) \\ \vdots \\ u_n(t,x,h(t,x)+y) \end{bmatrix} \\ w(t,x,y) &= u_{n+1}(t,x,h(t,x)+y) \\ p(t,x,y) &= \pi(t,x,h(t,x)+y) \end{aligned}$$

where $t \in J$, $x \in \mathbb{R}^n$, $y \in \mathbb{R}$ and

$$\begin{aligned} \tilde{c}(t,x,y) &= c(t,x,h(t,x)+y) - \bar{c} \\ \tilde{c}_\Gamma(t,x) &= c_\Gamma(t,x,h(t,x)) - \gamma(\bar{c})\,, \end{aligned}$$

where $t \in J$, $x \in \mathbb{R}^n$, $y \in \mathbb{R}_+ = [0,\infty)$. Notice that we passed to concentration differences w.r. to the "values at infinity" and that now (x,y) instead of x denotes a point in $\mathbb{R}^{n+1}$. For the two-phase balance equations for mass and momentum, this conversion has already been carried out in [14]. Therefore, we concentrate on those calculations needed to transform the surfactant balance. Since the orientation of $\Gamma(t)$ is such that the normal field is exterior to $\Omega_-(t)$, the unit normal at the point $(x, h(t,x))$ is given by

$$n(t,x) = \beta(t,x)^{-1/2} \begin{bmatrix} -\nabla_x h(t,x) \\ 1 \end{bmatrix} \quad \text{with } \beta(t,x) = 1 + |\nabla_x h(t,x)|^2\,.$$

The normal velocity V of $\Gamma(\cdot)$ is

$$V(t,x) = \beta(t,x)^{-1/2}\,\partial_t h(t,x)\,.$$

Together with the kinematic condition $V = (u \cdot n)_{|\Gamma}$ this yields the transport equation

$$\partial_t h + v_{|y=0} \cdot \nabla_x h = w_{|y=0}\,, \tag{5}$$

which describes the evolution of h. Note that u is continuous across $\Gamma(t)$, hence $v_{|y=0}$ and $w_{|y=0}$ are well defined.

According to (24) in Appendix A, the curvature of $\Gamma(t)$ at $x \in \Gamma(t)$ can be calculated as

$$\kappa(t,x) = -\mathrm{div}_{\Gamma(t)} n(t,x) = \nabla_x \cdot (\beta(t,x)^{-1/2} \nabla_x h(t,x))\,, \tag{6}$$

hence

$$\kappa = \Delta_x h - G_\kappa(h)$$

with

$$G_\kappa(h) = (1-\beta^{-1/2})\Delta_x h + \beta^{-3/2} \langle \nabla_x h\,,\, \nabla_x^2 h\, \nabla_x h\rangle\,. \tag{7}$$

Differentiation of $c(t,x,y) = \tilde{c}(t,x,y-h(t,x)) + \bar{c}$ yields

$$u \cdot \nabla c = v \cdot \nabla_x \tilde{c} + (w - v \cdot \nabla_x h)\, \partial_y \tilde{c}$$

as well as

$$\Delta c = \Delta_x \tilde{c} + \beta\, \partial_y^2 \tilde{c} - 2\, \langle \nabla_x \partial_y \tilde{c}\,,\, \nabla_x h \rangle - \partial_y \tilde{c}\, \Delta_x h\,,$$

where the corresponding arguments have to be used. Therefore, the surfactant balance in $\Omega_+(t)$ transforms into the quasilinear partial differential equation

$$\partial_t \tilde{c} - D\, \Delta_x \tilde{c} - D\, \partial_y^2 \tilde{c} = F_c(v,\tilde{c},h)\,, \quad t \in J,\ x \in \mathbb{R}^n,\ y \in \mathbb{R}\,.$$

Here, the nonlinearity is given by

$$F_c(v,\tilde{c},h) = -v \cdot \nabla_x \tilde{c} + D\Big((\beta-1)\,\partial_y^2 \tilde{c} - 2\,\langle \nabla_x \partial_y \tilde{c}\,,\, \nabla_x h\rangle - \partial_y \tilde{c}\,\Delta_x h\Big)\,. \tag{8}$$

For the transformation of the surfactant balance on $\Gamma(t)$, we prefer to use the formulation

$$\frac{Dc_\Gamma}{Dt} + c_\Gamma\, \mathrm{div}_\Gamma u_\Gamma - c_\Gamma\, \kappa\, V - D_\Gamma\, \Delta_\Gamma\, c_\Gamma = D\,(\nabla c \cdot n)_{|\Gamma}\,,$$

i.e., the one which employs the full Lagrangian derivative

$$\frac{Dc_\Gamma}{Dt}\,(t,x,h(t,x)) = \frac{d}{ds}\, c_\Gamma(t+s, \xi(t+s), \eta(t+s))_{|s=0}$$

with (ξ,η) the solution of

$$\begin{aligned} \dot{\xi}_k(s) &= u_k(s,\xi(s),\eta(s))\,, & \xi_k(t) &= x_k \text{ for } k=1,\dots,n\,,\\ \dot{\eta}(s) &= u_{n+1}(s,\xi(s),\eta(s))\,, & \eta(t) &= h(t,x)\,. \end{aligned}$$

Exploiting (5) it follows that $\eta(s) = h(s, \xi(s))$, hence

$$\begin{aligned}\frac{Dc_\Gamma}{Dt}(t,x,h(t,x)) &= \frac{d}{ds}\,c_\Gamma(t+s,\xi(t+s),h(t+s,\xi(t+s)))_{|s=0}\\ &= \frac{d}{ds}\,\tilde c_\Gamma(t+s,\xi(t+s))_{|s=0}\\ &= \partial_t\tilde c_\Gamma(t,x)+v(t,x)\cdot\nabla_x\tilde c_\Gamma(t,x)\,.\end{aligned}$$

We will now compute $\mathrm{div}_\Gamma u = \mathrm{div}_\Gamma u_\Gamma - \kappa V$ in the new variables. According to (19), the surface divergence $\mathrm{div}_\Gamma u$ of the vector field u is given by

$$\begin{aligned}\mathrm{div}_\Gamma u(x,h(x)) &= \sum_{i,j=1}^n g^{ij}(x)\langle\partial_i u(x,h(x)),(e_j,\partial_j h(x))\rangle\\ &= \sum_{i,j=1}^n \Big(\delta_{ij}-\beta^{-1}\partial_i h(x)\partial_j h(x)\Big)\Big(\partial_i v_j(x)+\partial_i w(x)\partial_j h(x)\Big)\\ &= \nabla_x\cdot v(x)-\beta^{-1}\langle\nabla_x w(x)-\nabla_x v(x)\nabla_x h(x),\nabla_x h(x)\rangle.\end{aligned}$$

In the computations above we have suppressed the variable t.

The Laplace-Beltrami operator (see Appendix A) transforms as

$$\Delta_\Gamma\, c_\Gamma = \beta^{-1/2}\,\nabla_x\cdot\big(\beta^{1/2}\,\nabla_x\tilde c_\Gamma-\beta^{-1/2}\langle\nabla_x h\,,\,\nabla_x\tilde c_\Gamma\rangle\,\nabla_x h\big)\,.$$

A direct calculation shows that

$$\begin{aligned}\Delta_\Gamma\, c_\Gamma = \Delta_x\tilde c_\Gamma &-\beta^{-1}\,\langle\nabla_x h\,,\,\nabla_x^2\tilde c_\Gamma\,\nabla_x h+\Delta_x h\,\nabla_x\tilde c_\Gamma\rangle\\ &+\beta^{-2}\,\langle\nabla_x^2 h\,\nabla_x h\,,\,\nabla_x h\rangle\,\nabla_x h\cdot\nabla_x\tilde c_\Gamma\,.\end{aligned}$$

Here, $\nabla_x^2\tilde c_\Gamma$ and $\nabla_x^2 h$ denote the matrix of all (mixed) second-order derivatives of $\tilde c_\Gamma$ and h, respectively. Finally, exploitation of

$$\nabla c(t,x,y) = (\nabla_x\tilde c-\partial_y\tilde c\,\nabla_x h\,,\partial_y\tilde c)(t,x,y-h(t,x))$$

yields

$$(\nabla c\cdot n)_{|\Gamma} = (\beta^{1/2}\,\partial_y\tilde c-\beta^{-1/2}\,\nabla_x\tilde c\cdot\nabla_x h)_{|y=0}\,.$$

Altogether, this leads to the quasilinear partial differential equation

$$\partial_t\tilde c_\Gamma - D_\Gamma\,\Delta_x\tilde c_\Gamma = G_c(v,w,\tilde c_\Gamma,\tilde c,h)\,,\quad t\in J\,,x\in\mathbb{R}^n$$

with

$$\begin{aligned}G_c(v,w,\tilde c_\Gamma,\tilde c,h) = &-v\cdot\nabla_x\tilde c_\Gamma-(\tilde c_\Gamma+\bar c_\Gamma)(\nabla_x\cdot v+\beta^{-1}\,\langle\nabla_x w-\nabla_x v\,\nabla_x h\,,\,\nabla_x h\rangle)\\ &-D_\Gamma\,\beta^{-1}\,\langle\nabla_x h\,,\,\nabla_x^2\tilde c_\Gamma\,\nabla_x h+\Delta_x h\,\nabla_x\tilde c_\Gamma\rangle\\ &+D_\Gamma\,\beta^{-2}\,\langle\nabla_x^2 h\,\nabla_x h\,,\,\nabla_x h\rangle\,\nabla_x h\cdot\nabla_x\tilde c_\Gamma\\ &+D\,\beta^{1/2}\,\partial_y\tilde c_{|y=0}-D\,\beta^{-1/2}\,\nabla_x\tilde c_{|y=0}\cdot\nabla_x h\,.\end{aligned}\tag{9}$$

The nonlinear boundary condition for the transformed concentration reads as

$$\tilde{c}_\Gamma = \gamma(\bar{c} + \tilde{c}_{|y=0}) - \gamma(\bar{c}) .$$

To obtain the linearized version, this is rewritten as

$$\tilde{c}_\Gamma - \gamma_0 \, \tilde{c}_{|y=0} = G_\gamma(\tilde{c}_{|y=0})$$

with $\gamma_0 = \gamma'(\bar{c})$ and

$$G_\gamma(\tilde{c}_{|y=0}) = \gamma(\bar{c} + \tilde{c}_{|y=0}) - \gamma(\bar{c}) - \gamma'(\bar{c})\,(\tilde{c}_{|y=0}) . \tag{10}$$

To exploit the results from [14] for constant surface tension, those terms have to be specified that correspond to the changes in the momentum jump condition in case σ depends on the surfactant concentration c_Γ. For this purpose, let $\sigma_0 = \sigma(\bar{c}_\Gamma)$ and rewrite the second equation in (2) as

$$[-S]\, n - \sigma_0 \kappa n = (\sigma(c_\Gamma) - \sigma_0)\kappa n + \sigma'(c_\Gamma)\,\mathrm{grad}_\Gamma c_\Gamma .$$

The right-hand side consists of a tangential part corresponding to v and a normal part corresponding to w. After application of the above transformations, this leads to nonlinearities G_v^{surf} and G_w^{surf} which are given by

$$\begin{aligned} G_v^{\mathrm{surf}}(\tilde{c}_\Gamma, h) &= (\sigma(\tilde{c}_\Gamma) - \sigma_0)(G_\kappa(h) - \Delta_x h)\nabla_x h \\ &\quad + \sigma'(\tilde{c}_\Gamma)\nabla_x \tilde{c}_\Gamma - \beta^{-1}\sigma'(\tilde{c}_\Gamma)(\nabla_x \tilde{c}_\Gamma \cdot \nabla_x h)\nabla_x h, \\ G_w^{\mathrm{surf}}(\tilde{c}_\Gamma, h) &= (\sigma(\tilde{c}_\Gamma) - \sigma_0)(\Delta_x h - G_\kappa(h)) + \beta^{-1}\sigma'(\tilde{c}_\Gamma)\nabla_x \tilde{c}_\Gamma \cdot \nabla_x h. \end{aligned}$$

Before writing down the full system of transformed equations, we will introduce some more notation. We set

$$\begin{aligned} &\mathbb{R}^n \times \dot{\mathbb{R}} := \{(x, y) \in \mathbb{R}^n \times \mathbb{R} : x \in \mathbb{R}^n,\ y \neq 0\}, \\ &\mathbb{H}_+^{n+1} := \mathbb{R}^n \times (0, \infty), \\ &\mathbb{H}_-^{n+1} := \mathbb{R}^n \times (-\infty, 0). \end{aligned}$$

Suppose that $f : \mathbb{R}^n \times \dot{\mathbb{R}} \to \mathbb{R}^k$ is some given function. Then

$$f_+ := f|_{\mathbb{H}_+^{n+1}}, \qquad f_- := f|_{\mathbb{H}_-^{n+1}}$$

denote the restrictions of f to $\mathbb{H}_+^{n+1}$ and $\mathbb{H}_-^{n+1}$, respectively. It is clear that f completely determines the functions f_+ and f_-, and vice versa. Moreover, the mapping

$$f \mapsto (f_+, f_-), \quad W_p^s(\mathbb{R}^n \times \dot{\mathbb{R}}; \mathbb{R}^k) \to W_p^s(\mathbb{H}_+^{n+1}; \mathbb{R}^k) \times W_p^s(\mathbb{H}_-^{n+1}; \mathbb{R}^k)$$

defines an isometric isomorphism, where W_p^s are the usual Sobolev-Slobodeskii spaces.

Next suppose that we are given a pair of positive numbers, say $\mu = (\mu_+, \mu_-)$, and a function $f : \mathbb{R}^n \times \dot{\mathbb{R}} \to \mathbb{R}^k$. Then

$$\mu f(x, y) := \begin{cases} \mu_+ \, f_+(x, y) & \text{if } y > 0, \\ \mu_- \, f_-(x, y) & \text{if } y < 0. \end{cases}$$

If the function f has one-sided limits at $y = 0$ these are denoted by $f_+(x,0)$ and $f_-(x,0)$, respectively. Then we set

$$[\mu f](x) := \mu_+ f_+(x,0) - \mu_- f_-(x,0), \qquad x \in \mathbb{R}^n.$$

If $[f] = 0$ we write

$$f(x) := f(x,0) := f_+(x,0) = f_-(x,0).$$

Note that in this case we have $[\mu f] = [\mu]f$ for any pair $\mu = (\mu_+, \mu_-) \in \mathbb{R}^2$.

Combining the transformed equations derived above with those from [14], we end up with the following coupled system of nonlinear equations.

Differential Equations

$$\begin{aligned}
\rho\,\partial_t v - \mu\,\Delta v + \nabla_x p &= F_v(v,w,p,h) && \text{in} \quad J \times \mathbb{R}^n \times \dot{\mathbb{R}} \\
\rho\,\partial_t w - \mu\,\Delta w + \partial_y p &= F_w(v,w,h) && \text{in} \quad J \times \mathbb{R}^n \times \dot{\mathbb{R}} \\
\nabla_x \cdot v + \partial_y w &= F_p(v,h) && \text{in} \quad J \times \mathbb{R}^n \times \dot{\mathbb{R}} \\
\partial_t \tilde{c} - D\,\Delta \tilde{c} &= F_c(v,\tilde{c},h) && \text{in} \quad J \times \mathbb{H}^{n+1}_+ \\
\partial_t h - w &= G_h(v,h) && \text{on} \quad J \times \mathbb{R}^n \\
\partial_t \tilde{c}_\Gamma - D_\Gamma\,\Delta_x \tilde{c}_\Gamma &= G_c(v,w,\tilde{c}_\Gamma,\tilde{c},h) && \text{on} \quad J \times \mathbb{R}^n \\
-[\mu\,\partial_y v] - [\mu]\,\nabla_x w &= G_v(v,w,[p],h) + G_v^{\mathrm{surf}}(\tilde{c}_\Gamma,h) && \text{on} \quad J \times \mathbb{R}^n \\
-2\,[\mu\,\partial_y w] + [p] - \sigma_0\,\Delta h &= G_w(v,w,h) + G_v^{\mathrm{surf}}(\tilde{c}_\Gamma,h) && \text{on} \quad J \times \mathbb{R}^n.
\end{aligned} \tag{11}$$

Boundary conditions

$$[v] = 0\,, \quad [w] = 0\,, \quad \tilde{c}_\Gamma - \gamma_0\,\tilde{c}_{|y=0} = G_\gamma(\tilde{c}_{|y=0}) \quad \text{on} \quad J \times \mathbb{R}^n\,. \tag{12}$$

Here G_κ, F_c, G_c and G_γ are given by (7), (8), (9) and (10), respectively. The other nonlinearities have been computed in [14] and are given by

$$\begin{aligned}
F_v(v,w,p,h) &= \mu\big(-2(\nabla_x h\cdot\nabla_x)\partial_y v + |\nabla_x h|^2\partial_y^2 v - \Delta_x h\partial_y v\big) + \nabla_x h\partial_y p \\
&\quad + \rho\big(-(v\cdot\nabla_x)v + (v\cdot\nabla_x h)\partial_y v - w\partial_y v + (w - v\cdot\nabla_x h)\partial_y v\big), \\
F_w(v,w,p,h) &= \mu\big(-2(\nabla_x h\cdot\nabla_x)\partial_y w + |\nabla_x h|^2\partial_y^2 w - \Delta_x h\partial_y w\big) \\
&\quad + \rho\big(-(v\cdot\nabla_x)w + (v\cdot\nabla_x h)\partial_y w - w\partial_y w + (w - v\cdot\nabla_x h)\partial_y w\big), \\
F_p(v,h) &= \nabla_x h\cdot\partial_y v, \\
G_h(v,h) &= -\nabla_x h\cdot v, \\
G_v(v,w,[p],h) &= \sigma_0(G_\kappa(h) - \Delta_x h)\nabla_x h - [\mu\partial_y w]\nabla_x h + [p]\nabla_x h \\
&\quad - [\mu]\big(\nabla_x v + (\nabla_x v)^{\mathsf{T}}\big)\nabla_x h + |\nabla_x h|^2[\mu\partial_y v] + \nabla_x h\cdot[\mu\partial_y v]\nabla_x h, \\
G_w(v,w,h) &= -\sigma_0 G_\kappa(h) - [\mu]\nabla_x h\cdot\nabla_x w + |\nabla_x h|^2[\mu\partial_y w] - \nabla_x h\cdot[\mu\partial_y v].
\end{aligned}$$

3.2. The associated linear problem

According to Section 3.1, the linear system associated with (11), (12) has the following form.

Differential Equations

$$
\begin{aligned}
\rho\,\partial_t v - \mu\,\Delta v + \nabla_x p &= f_v && \text{in} \quad J\times\mathbb{R}^n\times\dot{\mathbb{R}}\\
\rho\,\partial_t w - \mu\,\Delta w + \partial_y p &= f_w && \text{in} \quad J\times\mathbb{R}^n\times\dot{\mathbb{R}}\\
\nabla_x\cdot v + \partial_y w &= f_p && \text{in} \quad J\times\mathbb{R}^n\times\dot{\mathbb{R}}\\
\partial_t\tilde{c} - D\,\Delta\tilde{c} &= f_c && \text{in} \quad J\times\mathbb{H}^{n+1}_+\\
\partial_t h - w &= g_h && \text{on} \quad J\times\mathbb{R}^n\\
\partial_t\tilde{c}_\Gamma - D_\Gamma\,\Delta_x\tilde{c}_\Gamma &= g_c && \text{on} \quad J\times\mathbb{R}^n\\
-[\mu\,\partial_y v] - [\mu]\,\nabla_x w &= g_v && \text{on} \quad J\times\mathbb{R}^n\\
-2\,[\mu\,\partial_y w] + [p] - \sigma_0\,\Delta h &= g_w && \text{on} \quad J\times\mathbb{R}^n.
\end{aligned}
\tag{13}
$$

Boundary Conditions

$$
[v]=0\,,\quad [w]=0\,,\quad \tilde{c}_\Gamma-\gamma_0\,\tilde{c}=g_\gamma \quad \text{on} \quad J\times\mathbb{R}^n\,. \tag{14}
$$

Initial Conditions

$$
\begin{aligned}
v_{|t=0}=v_0\,,\quad w_{|t=0} &= w_0 && \text{on} \quad \mathbb{R}^n\times\dot{\mathbb{R}}\,,\\
\tilde{c}_{|t=0} &= \tilde{c}_0 && \text{on} \quad \mathbb{H}^{n+1}_+\,,\\
h_{|t=0}=h_0\,,\quad \tilde{c}_{\Gamma\,|t=0} &= \tilde{c}^0_\Gamma && \text{on} \quad \mathbb{R}^n\,.
\end{aligned}
\tag{15}
$$

Maximal regularity in an appropriate L_p-setting for this problem is given by the following result. Before stating this result we introduce some function spaces. Let $\dot{H}^1_p(\mathbb{H}^{n+1}_\pm)$ denote the closure of $C_c^\infty(\overline{\mathbb{H}^{n+1}_\pm})$ w.r. to the L_p-norm of the gradient, factored over the constants, and let ${}_0\dot{H}^1_p(\mathbb{H}^{n+1}_\pm)$ be the closure of $C_c^\infty(\mathbb{H}^{n+1}_\pm)$ w.r. to the same norm. Moreover, let $\dot{H}^{-1}_p(\mathbb{H}^{n+1}_\pm)$ be the dual of ${}_0\dot{H}^1_{p'}(\mathbb{H}^{n+1}_\pm)$.

Theorem 2. *Let $\rho_\pm,\mu_\pm,D,D_\Gamma,\gamma_0,\sigma_0>0$. Let $1<p<\infty$ with $p\neq 3/2,3$ and $J=[0,T]$. Then problem (13)–(15) admits a unique solution $(v,w,p,h,\tilde{c},\tilde{c}_\Gamma)$ such that*

$$
\begin{aligned}
&(v,w)\in H^1_p\big(J;L_p(\mathbb{R}^n\times\dot{\mathbb{R}};\mathbb{R}^{n+1})\big)\cap L_p\big(J;H^2_p(\mathbb{R}^n\times\dot{\mathbb{R}};\mathbb{R}^{n+1})\big)\,,\\
&p_\pm\in L_p\big(J;\dot{H}^1_p(\mathbb{H}^{n+1}_\pm;\mathbb{R})\big)\,,\\
&[p]\in W^{1/2-1/2p}_p\big(J;L_p(\mathbb{R}^n)\big)\cap L_p\big(J;W^{1-1/p}_p(\mathbb{R}^n)\big)\,,\\
&h\in W^{2-1/2p}_p\big(J;L_p(\mathbb{R}^n)\big)\cap H^1_p\big(J;W^{2-1/p}_p(\mathbb{R}^n)\big)\cap L_p\big(J;W^{3-1/p}_p(\mathbb{R}^n)\big)\,,\\
&\tilde{c}\in H^1_p\big(J;L_p(\mathbb{H}^{n+1}_+)\big)\cap L_p\big(J;H^2_p(\mathbb{H}^{n+1}_+)\big)\,,\\
&\tilde{c}_\Gamma\in H^1_p\big(J;L_p(\mathbb{R}^n)\big)\cap L_p\big(J;H^2_p(\mathbb{R}^n)\big)
\end{aligned}
$$

if and only if the data of the problem satisfy

a) $(f_v, f_w) \in L_p\big(J; L_p(\mathbb{R}^n \times \dot{\mathbb{R}}; \mathbb{R}^{n+1})\big)$,

b) $(v_0, w_0) \in W_p^{2-2/p}(\mathbb{R}^n \times \dot{\mathbb{R}}; \mathbb{R}^{n+1})$, $h_0 \in W_p^{3-2/p}(\mathbb{R}^n)$,

$\tilde{c}_0 \in W_p^{2-2/p}(\mathbb{H}_+^{n+1})$, $\tilde{c}_\Gamma^0 \in W_p^{2-2/p}(\mathbb{R}^n)$,

c) $f_p^\pm \in H_p^1\big(J; \dot{H}_p^{-1}(\mathbb{H}_\pm^{n+1})\big) \cap L_p\big(J; H_p^1(\mathbb{H}_\pm^{n+1})\big)$,

d) $f_c \in L_p\big(J; L_p(\mathbb{H}_+^{n+1})\big)$,

e) $(g_v, g_w) \in W_p^{1/2-1/2p}\big(J; L_p(\mathbb{R}^n; \mathbb{R}^{n+1})\big) \cap L_p\big(J; W_p^{1-1/p}(\mathbb{R}^n; \mathbb{R}^{n+1})\big)$,

f) $g_h \in W_p^{1-1/2p}\big(J; L_p(\mathbb{R}^n)\big) \cap L_p\big(J; W_p^{2-1/p}(\mathbb{R}^n)\big)$,

g) $g_c \in L_p\big(J; L_p(\mathbb{R}^n)\big)$,

h) $g_\gamma \in W_p^{1-1/2p}\big(J; L_p(\mathbb{R}^n)\big) \cap L_p\big(J; W_p^{2-1/p}(\mathbb{R}^n)\big)$,

and the compatibility conditions

i) $[v_0] = 0,\ [w_0] = 0$ *if* $p > 3/2$,

j) $\tilde{c}_{0\,|y=0} - \tilde{c}_\Gamma^0 = g_{\gamma\,|t=0}$ *if* $p > 3/2$,

k) $\nabla_x \cdot v_0^\pm + \partial_y w_0^\pm = f_{p\,|t=0}^\pm$ *in* $\dot{H}_p^{-1}(\mathbb{H}_\pm^{n+1})$,

l) $-[\mu\, \partial_y v_0] - [\mu]\, \nabla_x w_0 = g_{v\,|t=0}$ *if* $p > 3$.

Proof. Concerning the proof, notice first that the "only if"-part follows by taking traces; see section 5 in [12]. For the "if"-part observe that problem (13)–(15) consists of two decoupled subsystems. The first one involves only (v, w, p, h) and the corresponding result is Theorem 3.3 in [14]. The second one is built by the surfactant equations, i.e., those for $\tilde{c}$ and $\tilde{c}_\Gamma$. Here the desired maximal regularity follows from that of the diffusion equation on $\mathbb{R}^n$ and on a half-space with Dirichlet boundary condition; cf. [11] and [22]. □

3.3. Solution of the quasilinear problem

We do not intend to give the detailed proof here, but rather present the main ideas of the proof. In fact, the arguments are very similar to those employed in [12] for the Stefan problem with surface tension, and in [14] for the free boundary value problem of the Navier-Stokes equations.

Introduce function spaces $X_T \times Y_T$ for the right-hand side of the differential equations in (13) and Z_T for the corresponding solutions as follows. The elements of X_T are functions

$$f = (f_v, f_w, f_p, f_c), \quad \text{defined on } [0, T] \times \mathbb{R}^{n+1} \times \dot{\mathbb{R}},$$

with regularity given by a), c), d) of Theorem 2. The elements of Y_T are functions

$$g = (g_v, g_w, g_h, g_c, g_\gamma), \quad \text{defined on } [0, T] \times \mathbb{R}^n,$$

with regularity defined in e), f), g), h) of Theorem 2. The solution space Z_T consists of functions $z = (v, w, p, h, \tilde{c}, \tilde{c}_\Gamma)$ with regularity as described in Theorem 2. Z_T additionally also contains the first of the boundary conditions in (14), i.e., $[v] = 0$, $[w] = 0$. Equipped with their natural norms these spaces are Banach spaces. The corresponding spaces of functions with zero trace at $t = 0$ (if they exist) will be denoted by a lower left subscript 0.

Next it is convenient to introduce an auxiliary function $z^* \in Z_T$ which resolves the initial conditions and the compatibilities j), k) and l) of Theorem 2. By the assumptions of Theorem 1 such a function z^* exists, it can be constructed via Theorem 2 applied to special right-hand sides. This way we can now work in the spaces ${}_0Z_T$ and ${}_0X_T \times {}_0Y_T$, which allow for embeddings with embedding constants independent of T. The choice of the exponent $p > n+3$ yields

$$(v_\pm, w_\pm) \in C([0,T]; C_0^1(\mathbb{H}_\pm^{n+1})), \qquad \tilde{c} \in C([0,T]; C_0^1(\mathbb{H}_+^{n+1})),$$
$$\tilde{c}_\Gamma \in C([0,T]; C_0^1(\mathbb{R}^{n+1})), \qquad h \in C([0,T]; C_0^2(\mathbb{R}^n)) \cap C^1([0,T]; C_0^1(\mathbb{R}^n)),$$

and the spaces for the g's are Banach algebras, except for that of g_c.

Introducing the linear operator $L : {}_0Z_T \to {}_0X_T \times {}_0Y_T$, defined by the left-hand side of (13) and the last term in (14), Theorem 2 shows that L is linear, bounded, and invertible. The norm of L^{-1} is bounded by some constant M, which does not depend on $T \in (0, T_0]$, where $T_0 < \infty$ is fixed. We now define $H : Z_T \to X_T \times Y_T$ by the right-hand side of (11) and by the last boundary condition in (12).

Then the nonlinear problem can be rewritten as

$$Lz = H(z + z^*) - Lz^* := H_0(z), \quad z \in {}_0Z_T.$$

Note that H_0 maps ${}_0Z_T$ into ${}_0X_T \times {}_0Y_T$ by construction. Inverting L, this problem becomes the fixed point problem

$$z = L^{-1}H_0(z), \quad z \in {}_0Z_T.$$

Fix a closed ball $\bar{B}_r(0) \subset {}_0Z_T$. The aim is to show that $L^{-1}H_0$ maps $\bar{B}_r(0)$ into itself and is a strict contraction on $\bar{B}_r(0)$. Thus we have to show that

$$|H_0(z) - H_0(\bar{z})|_{{}_0X_T \times {}_0Y_T} \le \frac{1}{2M}|z - \bar{z}|_{{}_0Z_Z}, \quad |H_0(z)|_{{}_0X_T \times {}_0Y_T} \le \frac{r}{M}, \quad z, \bar{z} \in \bar{B}_r(0).$$

Among the finitely many terms appearing in the definition of the F's and G's we encounter three types:

- terms containing the factor $\nabla_x h$;
- terms containing the factor $\sigma(\tilde{c}_\Gamma) - \sigma_0$;
- lower order terms.

Next observe that terms containing derivatives of highest order are linear in these derivatives and are of the first two types. They can be made small by choice of $\eta > 0$. On the other hand, lower order terms can be made small by the choice of $T > 0$ and $r > 0$. These estimates are of the same character as in the two papers cited above.

In conclusion, we can manage this way to show that $L^{-1}H_0 : \bar{B}_r(0) \to \bar{B}_r(0)$ is a strict contraction, and hence admits a unique fixed point, thanks to the contraction mapping principle. This yields a unique solution on a possibly small time interval in Z_T, the optimal regularity class of type L_p.

Further regularity of the solutions can be obtained by the parameter trick; cf. [23] and [13]. Under the assumptions of Theorem 1 we obtain in this way classical solutions for $t \in (0, T]$, and even real analyticity in case the given functions γ and σ have this property. We refrain from giving more details here.

Appendix A. Surface divergence

Let M be an m-dimensional C^k-submanifold of $\mathbb{R}^n$. Then, locally around any point $p \in M$, the submanifold is given by a parametrization of class C^k. Hence there is an open parameter region $B \subset \mathbb{R}^m$ and a C^k-function $g : B \to \mathbb{R}^n$ such that $g(B) = M \cap U$ for some open neighborhood $U \subset \mathbb{R}^n$ of p. Furthermore, the set

$$\{\partial_1 g(y), \ldots, \partial_m g(y)\}$$

is a linearly independent subset of $\mathbb{R}^n$, and g is a homeomorphism from B onto $M \cap U$. The tangent space T_pM at $p \in M$ is given by

$$T_pM = \{\tau \in \mathbb{R}^n : \exists\, C^1\text{-curve } \gamma : (-\epsilon, \epsilon) \to M \text{ s.t. } \gamma(0) = p\,,\ \gamma'(0) = \tau\}\,.$$

Then,

$$T_pM = \operatorname{span}\{\partial_1 g(y), \ldots, \partial_m g(y)\} \quad \text{for } p \in M, y = g^{-1}(p)\,.$$

Let f be a vector field on M, meaning that $f : M \to \mathbb{R}^n$ is everywhere tangential to M, i.e., $f(p) \in T_pM$ for all $p \in M$. Such an f is a C^1-vector field on M if $f \circ g$ is continuously differentiable for every C^1-parametrization g of M. It then follows that $f \circ \gamma$ is continuously differentiable for every C^1-curve $\gamma : (-\epsilon, \epsilon) \to M$. In this case, given $p \in M$ and $\tau \in T_pM$ with $\tau \neq 0$, the directional derivative of f at p in direction of τ is

$$\frac{\partial f}{\partial \tau}(p) = \frac{d}{dt}(f \circ \gamma)(0)$$

for a C^1-curve $\gamma : (-\epsilon, \epsilon) \to M$ such that $\gamma(0) = p$, $\gamma'(0) = \tau$. Correspondingly, a scalar field $\phi : M \to \mathbb{R}$ is C^1 if $\phi \circ g$ is C^1 for every C^1-parametrization g. In this case, the *surface gradient* $\operatorname{grad}_M \phi$ is given by

$$\operatorname{grad}_M \phi(p) = \sum_{i=1}^{m} \frac{\partial \phi}{\partial \tau_i}\, \tau_i\,,$$

where $\{\tau_1, \ldots, \tau_m\}$ is an orthonormal basis of T_pM.

Given a C^1-vector field f on M, a coordinate-free definition of the *surface divergence* $\operatorname{div}_M f$ of f is

$$\operatorname{div}_M f(p) = \lim_{r \to 0+} \frac{1}{|M_r(p)|} \int_{\partial M_r(p)} \langle f, N \rangle\, d\sigma_{m-1}\,, \tag{16}$$

where

$$M_r(p) = M \cap V_r(p)$$

with a family of open sets $V_r(p) \subset \mathbb{R}^n$ such that

$$p \in V_r(p)\,,\ \operatorname{diam}(V_r(p)) = \mathcal{O}(r) \quad \text{as } r \to 0+$$

and $\partial V_r(p)$ is piecewise of class C^2; notice that $M_r(p)$ is an m-dimensional submanifold with boundary $\partial M_r(p)$ and the latter is itself an $(m-1)$-dimensional submanifold of M without boundary. In (16), $\langle \cdot\,, \cdot \rangle$ is the Euclidean (inner) product on $\mathbb{R}^n$ and N denotes the exterior unit normal field to $\partial M_r(p)$, i.e., $N(q) \in T_qM$, $N(q) \perp T_q\partial M_r(p)$ at $q \in \partial M_r(p)$ and $N(q)$ points outward w.r. to $M_r(p)$. Further, $|M_r(p)|$ denotes the surface area of $M_r(p)$. If M has parametrization $g : B \subset \mathbb{R}^m \to \mathbb{R}^n$ near $p \in M$, then

$$|M_r(p)| = \int_{B_r} \sqrt{\det G(y)}\, dy$$

with $B_r = g^{-1}(M_r(p))$ and the Gramian

$$G(y) = \big(\langle \partial_i g(y), \partial_j g(y)\rangle\big)_{1\le i,j\le m}\,.$$

In the following, $d\sigma_m = \sqrt{\det G(y)}\, dy$ denotes the surface measure on M and $d\sigma_{m-1}$ is the corresponding surface measure on the respective $(m-1)$-dimensional boundary-submanifold.

A more common formulation of $\operatorname{div}_M f$ is obtained as follows. Given $p \in M$ and a basis $\{\tau_1, \dots, \tau_m\}$ of T_pM, let $\{\tau_1^*, \dots, \tau_m^*\} \subset T_pM$ denote the co-basis such that

$$\langle \tau_i, \tau_j^* \rangle = \delta_{ij}\,,$$

i.e., $\{\tau_i\}, \{\tau_j^*\} \subset T_pM$ form a biorthogonal system. Extend $\{\tau_1, \dots, \tau_m\}$ by means of $\tau_{m+1}, \dots, \tau_n \in (T_pM)^\perp$ to a basis $\{\tau_1, \dots, \tau_n\}$ of $\mathbb{R}^n$. We now exploit the fact that M is locally a graph over T_pM: Let $\varphi : M \cap B_\epsilon(p) \to T_pM$ be given by

$$\varphi(q) = \sum_{i=1}^m y_i\,\tau_i \quad \text{for } q \in M \text{ with } \; q = p + \sum_{i=1}^m y_i\,\tau_i + \sum_{k=m+1}^n t_k\,\tau_k\,.$$

For sufficiently small $\epsilon > 0$, the map $\varphi : M \cap B_\epsilon(p) \to \varphi(M \cap B_\epsilon(p))$ is a diffeomorphism. Now, let $B_r = (-r, r)^m \subset \mathbb{R}^m$ with $r > 0$ and define the parametrization $g : B_r \to \mathbb{R}^n$ by

$$g(y) = \varphi^{-1}\Big(\sum_{i=1}^m y_i\,\tau_i\Big)\,.$$

Then

$$\partial_i g(y) = \tau_i + o(|y|) \quad \text{as } y \to 0\,,$$

since φ approximates the identity near the point p. Let

$$V_r(p) = \Big\{p + \sum_{i=1}^n t_i\,\tau_i \;:\; t_i \in (-r, r)\Big\}$$

and

$$M_r(p) = M \cap V_r(p) = g(B_r) \quad \text{for small } r > 0\,.$$

Then

$$|M_r(p)| = \int_{B_r} \sqrt{\det\big(\langle \partial_i g(y)\,,\, \partial_j g(y)\rangle\big)}\, dy = (2r)^m \sqrt{\det G} + o(r^m) \quad \text{as } r \to 0+$$

with $G = \big(\langle \tau_i, \tau_j \rangle\big)_{1\le i,j\le m}$. The boundary of $M_r(p)$ is

$$\partial M_r(p) = \bigcup_{j=1}^{m} W_r^j(p) \quad \text{with} \quad W_r^j(p) = W_r^{j,+}(p) \cup W_r^{j,-}(p)\,,$$

where

$$W_r^{j,\pm}(p) = \big\{ g(y) \,:\, y_i \in [-r,r] \text{ for } i \neq j\,,\; y_j = \pm r \big\}\,.$$

On $\partial M_r(p)$ the outer unit normal field N satisfies

$$N(q) = \pm \frac{\tau_j^*}{|\tau_j^*|} + o(r) \quad \text{for } q \in W_r^{j,\pm}(p)\,.$$

Hence

$$\begin{aligned}
\int_{W_r^j(p)} \langle f\,,\, N\rangle\, d\sigma_{m-1} &= \int_{W_r^{j,+}(p)} \langle f\,,\, N\rangle\, d\sigma_{m-1} + \int_{W_r^{j,-}(p)} \langle f\,,\, N\rangle\, d\sigma_{m-1} \\
&= \int_{[-r,r]^{m-1}} \left(\left\langle (f\circ g)(y)\Big|_{y_j=-r}^{y_j=r}, \frac{\tau_j^*}{|\tau_j^*|} \right\rangle + o(r) \right) \sqrt{\det G^j(y)}\, dy^j \\
&= \int_{[-r,r]^{m-1}} \left(\int_{-r}^{r} \left\langle \partial_j (f\circ g)(0) + o(r), \frac{\tau_j^*}{|\tau_j^*|} \right\rangle dy_j + o(r) \right) \sqrt{\det G^j(y)}\, dy^j \\
&= (2r)^m \left\langle \frac{\partial f}{\partial \tau_j}(p), \frac{\tau_j^*}{|\tau_j^*|} \right\rangle \sqrt{\det G^j} + o(r^m) \quad (\text{as } r \to 0+)\,,
\end{aligned}$$

where $dy^j = dy_1 \dots dy_{j-1} dy_{j+1} \dots dy_m$ and the matrix G^j is obtained from G by elimination of the jth row and column. Due to the basic geometric fact that "volume = base area × height", it follows that

$$\sqrt{\det G} = \sqrt{\det G^j} \left\langle \tau_j\,, \frac{\tau_j^*}{|\tau_j^*|} \right\rangle = \frac{1}{|\tau_j^*|} \sqrt{\det G^j} \quad \text{for } j = 1, \dots, m\,.$$

Hence,

$$\operatorname{div}_M f(p) = \sum_{j=1}^{m} \left\langle \frac{\partial f}{\partial \tau_j}(p)\,,\, \tau_j^* \right\rangle, \tag{17}$$

where $\{\tau_i\}, \{\tau_j^*\} \subset T_pM$ is *any* biorthogonal system. In case $\{\tau_1, \dots, \tau_m\}$ is an orthonormal basis of T_pM, (17) resembles the well-known definition of the divergence in $\mathbb{R}^m$. For concrete computations it is more useful to write $\operatorname{div}_M f(p)$ in the local coordinates that correspond to a given parametrization, i.e., to use

$$\tau_j = \partial_j g(y) \quad \text{with } y = g^{-1}(p)\,.$$

By definition of G, it follows that $\{\tau_j^*\}$ is then given by

$$\big(\tau_1^* \big| \dots \big| \tau_m^*\big) = \big(\tau_1 \big| \dots \big| \tau_m\big) G^{-1}\,,$$

i.e.,

$$\tau_j^* = \sum_{i=1}^{m} g^{ij}\,\tau_i \quad \text{in case } (g^{ij}) := G^{-1}\,. \tag{18}$$

Inserting this into (17) yields

$$\operatorname{div}_M f(p) = \sum_{i,j=1}^{m} g^{ij}\,\langle \partial_j (f\circ g)(y)\,,\, \partial_i g(y)\rangle \quad \text{for } y = g^{-1}(p)\,. \tag{19}$$

Using the fact that

$$\operatorname{div}_M\Big(\frac{1}{\sqrt{\det G}}\,\tau_k\Big) = 0 \quad \text{for } k = 1,\ldots,m\,, \tag{20}$$

a significant simplification of (19) can be obtained. Indeed,

$$\operatorname{div}_M f = \frac{1}{\sqrt{\det G}}\sum_{j=1}^{m}\partial_j\big((v^j\circ g)\,\sqrt{\det G}\big)\,, \tag{21}$$

where the v^j denote the components of f w.r. to the basis $\{\tau_j\}$, i.e.,

$$f(p) = \sum_{j=1}^{m} v^j(p)\,\tau_j(p) \quad \text{on } M\,.$$

Let us briefly justify equation (20). By (17) we have

$$\begin{aligned}
\operatorname{div}_M\Big(\frac{1}{\sqrt{\det G}}\,\tau_k\Big) &= \sum_{j=1}^{m}\Big\langle \partial_j\Big(\frac{1}{\sqrt{\det G}}\,\tau_k\Big), \tau_j^*\Big\rangle \\
&= \frac{1}{\sqrt{\det G}}\sum_{j=1}^{m}\Big(-\frac{\partial_j \det G}{2\det G}\,\langle\tau_k,\tau_j^*\rangle + \langle\partial_j\tau_k,\tau_j^*\rangle\Big) \\
&= \frac{1}{\sqrt{\det G}}\Big(-\frac{\partial_k \det G}{2\det G} + \sum_{j=1}^{m}\langle\partial_j\tau_k,\tau_j^*\rangle\Big).
\end{aligned}$$

The matrix $G = (\langle\tau_i,\tau_j\rangle)$ satisfies the equation $\partial_k G = QG$, with $Q = (\partial_k G)G^{-1}$. According to Liouville's Theorem, the determinant $\det G$ then satisfies

$$\partial_k \det G = (\operatorname{tr} Q)\,\det G.$$

The matrix Q is given by $Q = (\partial_k\langle\tau_i,\tau_j\rangle)G^{-1}$. For the trace of Q one obtains

$$\begin{aligned}
\operatorname{tr} Q &= \sum_{i,j=1}^{m}\big(\langle\partial_k\tau_i,\tau_j\rangle + \langle\tau_i,\partial_k\tau_j\rangle\big)g^{ij} = \sum_{i,j=1}^{m}\big(\langle\partial_k\tau_i,g^{ij}\tau_j\rangle + \langle g^{ij}\tau_i,\partial_k\tau_j\rangle\big) \\
&= \sum_{i=1}^{m}\langle\partial_k\tau_i,\tau_i^*\rangle + \sum_{j=1}^{m}\langle\tau_j^*,\partial_k\tau_j\rangle \;=\; 2\sum_{j=1}^{m}\langle\tau_j^*,\partial_k\tau_j\rangle.
\end{aligned}$$

Consequently, $\partial_k \det G = 2\sum_j\langle\tau_j^*,\partial_k\tau_j\rangle\,\det G$ and (20) follows.

The *divergence theorem* is inherited from the definition of the surface divergence. It reads as (see Theorem 2.1 in Chapter XII of [20]).

Theorem A (Divergence theorem). *Let M be a compact m-dimensional C^2-submanifold of $\mathbb{R}^n$ with continuous normal field, and let N be the exterior unit normal field to its boundary ∂M. Then, for every C^1-vector field f on M,*

$$\int_M \operatorname{div}_M f \, d\sigma_m = \int_{\partial M} \langle f, N \rangle \, d\sigma_{m-1} \,. \tag{22}$$

By means of (17) or (19) it is possible to define $\operatorname{div}_M f$ also for nontangential functions $f : M \to \mathbb{R}^n$. One important example is the surface divergence of the unit normal field of an $(n-1)$-dimensional submanifold (i.e., a hypersurface). In fact, the sum of the principal curvatures of M at p is given by (see, for instance, p. 223 in [4])

$$\kappa = -\operatorname{div}_M n_M \,. \tag{23}$$

Indeed, equation (19) then becomes

$$-\operatorname{div}_M n_M = -\sum_{i,j=1}^{m} g^{ij} \left\langle \partial_j (n_M \circ g) \,,\, \partial_i g \right\rangle = \sum_{i,j=1}^{m} g^{ij} h_{ji},$$

where

$$h_{ji} := -\langle \partial_j (n_M \circ g) \,,\, \partial_i g \rangle = \langle n_M \circ g, \partial_j \partial_i g \rangle$$

is the second fundamental form of M (with respect to the parametrization g). The expression $\sum_{i,j=1}^{m} g^{ij} h_{ji}$ is the trace of the second fundamental form, and it gives the sum of the principal curvatures of M. In case M is the graph of a function $h : B \to \mathbb{R}$ for some open set $B \subset \mathbb{R}^m$, that is, $M := \{(x, h(x)) \;:\; x \in B\}$, one easily shows that

$$g^{ij} = \left(\delta_{ij} - \frac{\partial_i h \, \partial_j h}{1 + |\nabla h|^2} \right)$$

and

$$\kappa = \frac{1}{\sqrt{1+|\nabla h|^2}} \sum_{i,j=1}^{m} \left(\delta_{ij} - \frac{\partial_i h \, \partial_j h}{1+|\nabla h|^2} \right) \partial_i \partial_j h - \operatorname{div}_{\mathbb{R}^m} \left(\frac{1}{\sqrt{1+|\nabla h|^2}} \nabla h \right). \tag{24}$$

Let us note in passing that for $n = 3$, Gauss called $H = \frac{1}{2}\kappa$ the mean curvature of M. Notice also that the divergence theorem does not apply to $f = n_M$. For a not necessarily tangent field $f : M \to \mathbb{R}^n$ the decomposition

$$f = f_M + \langle f, n_M \rangle \, n_M \,,$$

where f_M is the tangential part of f, leads to the formula

$$\operatorname{div}_M f = \operatorname{div}_M f_M - \kappa \, \langle f, n_M \rangle \,; \tag{25}$$

observe that

$$\operatorname{div}_M \big(\langle f, n_M \rangle \, n_M \big) = \langle \operatorname{grad}_M \langle f, n_M \rangle \,,\, n_M \rangle + \langle f, n_M \rangle \operatorname{div}_M n_M$$

and $\mathrm{grad}_M\phi(p) \in T_pM$ for any scalar field ϕ. Finally, recall that in local coordinates given by a parametrization g, the Laplace-Beltrami operator Δ_M, defined by $\Delta_M\phi = \mathrm{div}_M\,\mathrm{grad}_M\phi$, is given as

$$\Delta_M\phi(p) = \frac{1}{\sqrt{\det G(y)}}\sum_{j,k=1}^{m}\partial_j\Big(\sqrt{\det G(y)}\,g^{jk}\,\partial_k(\phi\circ g)(y)\Big)\,,\quad y = g^{-1}(p)\,,$$

for $y \in B$.

For more information concerning vector analysis on (sub-)manifolds see, e.g., [3, 4, 20].

Appendix B. Surface transport theorem

With the notation from A, we are in a position to prove the following result.

Theorem B (Surface transport theorem). *Let $I \subset \mathbb{R}$ open, $\Omega \subset \mathbb{R}^n$ open and the velocity field $v : J \times \Omega \to \mathbb{R}^n$ be C^1. Given $t_0 \in I$, $x_0 \in \Omega$ let $\phi(t; t_0, x_0)$ denote the solution of the ODE-system*

$$\dot{\phi}(t) = v(t, \phi(t))\,,\quad \phi(t_0) = x_0\,.$$

Given a compact C^2-hypersurface $M_0 \subset \Omega$, let

$$M_t = \{\phi(t; t_0, x_0) : x_0 \in M_0\}$$

denote the advected (material) surface; due to the assumptions on v, Ω and M_0 the surface M_t exists for all $t \in I_\delta = (t_0 - \delta, t_0 + \delta)$ for some $\delta > 0$. Let

$$c : \mathrm{gr}(M_t) = \{(t, x) : t \in I_\delta, x \in M_t\} \to \mathbb{R}$$

be continuously differentiable. Then,

$$\begin{aligned}\frac{d}{dt}\int_{M_t} c\,d\sigma_t &= \int_{M_t}\Big(\frac{Dc}{Dt} + c\,\mathrm{div}_M\,v\Big)\,d\sigma_t\\ &= \int_{M_t}\Big(\frac{Dc}{Dt} + c\,\mathrm{div}_M\,v_M - c\,\kappa\,V\Big)\,d\sigma_t\end{aligned} \tag{26}$$

for all $t \in I_\delta$. Here, $d\sigma_t$ denotes the surface measure on M_t, $\kappa = \kappa(t, x)$ the curvature of $\Gamma(t)$ at $x \in \Gamma(t)$ according to (23), v_M the tangential part of v, $V = \langle v\,, n_M\rangle$ the normal component of v, and

$$\frac{Dc}{Dt}(t, x) = \frac{d}{ds}\,c(t + s, \phi(t + s; t, x))_{|s=0}$$

is the Lagrangian derivative of c.

Proof. Let $m = n - 1$, $g : B \to \mathbb{R}^n$, with $B \subset \mathbb{R}^m$ open and bounded, be a parametrization of M_0 and $\Phi^t(y) = \phi(t; t_0, y)$ be the flow map associated with v. Then,

$$g^t : B \to \mathbb{R}^n \quad \text{with } g^t := \Phi^t \circ g$$

is a parametrization of M_t. Hence,

$$\int_{M_t} c(t,x)\,d\sigma_t = \int_B c(t,\Phi^t(y))\,\sqrt{\det G(t,y)}\,dy\,,$$

where $G(t,y)$ is the Gramian

$$G(t,y) = \big(\langle \partial_i g^t(y)\,,\,\partial_j g^t(y)\rangle\big)_{1\le i,j\le m}\,.$$

Therefore,

$$\begin{aligned}\frac{d}{dt}&\int_{M_t} c(t,x)\,d\sigma_t\\ &= \int_B \Big(\frac{d}{dt}\,c(t,\Phi^t(y))\Big)\,\sqrt{\det G(t,y)}\,dy + \int_B \frac{c(t,\Phi^t(y))}{\sqrt{\det G(t,y)}}\,\frac{1}{2}\,\partial_t \det G(t,y)\,dy\end{aligned} \tag{27}$$

and, by the definition of $\frac{Dc}{Dt}$, the first term on the right is

$$\int_B \Big(\frac{d}{dt}\,c(t,\Phi^t(y))\Big)\,\sqrt{\det G(t,y)}\,dy = \int_{M_t} \frac{Dc}{Dt}\,(t,x)\,d\sigma_t\,.$$

The derivative of $\partial_t \det G(t,y)$ can be computed by using Liouville's Theorem. However, we prefer here to give a self-contained derivation. For this purpose, notice first that

$$\begin{aligned}&\partial_t \det G(t,y)\\ &\quad= \sum_{j=1}^m \det \begin{pmatrix} \langle \partial_1 g^t\,,\,\partial_1 g^t\rangle & \cdots & \partial_t\langle \partial_1 g^t\,,\,\partial_j g^t\rangle & \cdots & \langle \partial_1 g^t\,,\,\partial_m g^t\rangle \\ \vdots & & \vdots & & \vdots \\ \langle \partial_m g^t\,,\,\partial_1 g^t\rangle & \cdots & \partial_t\langle \partial_m g^t\,,\,\partial_j g^t\rangle & \cdots & \langle \partial_m g^t\,,\,\partial_m g^t\rangle \end{pmatrix}.\end{aligned}$$

Using $\partial_t\partial_j g^t(y) = \nabla_x v(t,g^t(y))\,\partial_j g^t(y)$ we obtain

$$\frac{1}{2}\,\partial_t\langle \partial_i g^t(y)\,,\,\partial_j g^t(y)\rangle = \langle D\,\partial_i g^t(y)\,,\,\partial_j g^t(y)\rangle$$

with

$$D = \frac{1}{2}\,(\nabla_x v + (\nabla_x v)^T)\,,\quad (\nabla_x v)^T = \big(\nabla_x v_1\big|\ \ldots\ \big|\nabla_x v_n\big)\,.$$

Consequently, if $\tau_i = \partial_i g^t$, then

$$\frac{1}{2}\,\partial_t \det G(t,y) = \sum_{j=1}^m \det \begin{pmatrix} \langle \tau_1\,,\,\tau_1\rangle & \cdots & \langle D\tau_1\,,\,\tau_j\rangle & \cdots & \langle \tau_1\,,\,\tau_m\rangle \\ \vdots & & \vdots & & \vdots \\ \langle \tau_m\,,\,\tau_1\rangle & \cdots & \langle D\tau_m\,,\,\tau_j\rangle & \cdots & \langle \tau_m\,,\,\tau_m\rangle \end{pmatrix}.$$

Define $Q \in \mathbb{R}^{m\times m}$ by means of

$$QG = Q\big(\langle \tau_i\,,\,\tau_j\rangle\big) = \big(\langle D\tau_i\,,\,\tau_j\rangle\big)\,. \tag{28}$$

Let $a_{ij} = \langle \tau_i\,,\, \tau_j\rangle$ and $\alpha^i = (\alpha_{i1}, \ldots, \alpha_{im})^T$. Then

$$\frac{1}{2}\,\partial_t \det G(t,y) = \sum_{j=1}^{m} \det\left(\alpha^1\big|\ldots\big|Q\alpha^j\big|\ldots\big|\alpha^m\right).$$

Due to the properties of the determinant it is now easy to obtain

$$\frac{1}{2}\,\partial_t \det G(t,y) = \operatorname{tr} Q\, \det\left(\alpha^1\big|\ldots\big|\alpha^m\right) = \operatorname{tr} Q\, \det G(t,y)\,.$$

By (28), the matrix Q satisfies $Q = \big(\langle D\tau_i\,,\,\tau_j\rangle\big)\, G(t,y)^{-1}$, hence

$$Q_{ij} = \sum_{k=1}^{m} \langle D\tau_i\,,\, \tau_k\rangle\, g^{kj} \quad \text{if } (g^{ij}) := G^{-1}\,.$$

This yields

$$Q_{ij} = \left\langle D\tau_i\,, \sum_{k=1}^{m} g^{kj}\,\tau_k \right\rangle = \langle D\tau_i\,,\, \tau_j^*\rangle$$

with τ_j^* according to (18), i.e., $\{\tau_1^*, \ldots, \tau_m^*\} \subset \operatorname{span}\{\tau_1, \ldots, \tau_m\}$ is a co-basis. Therefore

$$\operatorname{tr} Q = \sum_{i=1}^{m} \frac{1}{2}\left(\langle(\nabla_x v)\,\tau_i\,,\,\tau_i^*\rangle + \langle(\nabla_x v)\,\tau_i^*\,,\,\tau_i\rangle\right),$$

hence

$$\frac{1}{2}\,\partial_t \det G(t,y) = \operatorname{div}_{M_t} v(t, g^t(y))\, \det G(t,y)$$

by (17). Insertion into (27) finally yields

$$\begin{aligned}\frac{d}{dt}\int_{M_t} c(t,x)\, d\sigma_t &= \int_B \left(\frac{d}{dt}\, c(t, \Phi^t(y))\right) \sqrt{\det G(t,y)}\, dy \\ &\quad + \int_B c(t,\Phi^t(y))\, \operatorname{div}_{M_t} v(t, g^t(y))\, \sqrt{\det G(t,y)}\, dy \\ &= \int_{M_t} \left(\frac{Dc}{Dt}\,(t,x) + c(t,x)\, \operatorname{div}_{M_t} v(t,x)\right) d\sigma_t\,.\end{aligned}$$ □

The literature contains several variants of Theorem B (cf. [35]), sometimes with uncommon notations. One such variant of (26) is

$$\frac{d}{dt}\int_{M_t} c\, d\sigma_t = \int_{M_t} \left(\left[\frac{\partial c}{\partial t}\right]_n + \operatorname{div}_M(c\, v_M) - c\,\kappa\, V\right) d\sigma_t\,, \tag{29}$$

where

$$\left[\frac{\partial c}{\partial t}\right]_n (t,x) = \frac{d}{ds}\, c(t+s, \phi_n(t+s; t, x))_{|s=0}$$

with ϕ_n the purely normal interface motion. In the sections above, we used the notation $\partial_{t,n} c$ instead of $\left[\frac{\partial c}{\partial t}\right]_n$ to avoid confusion with the interfacial jump of a quantity. The function ϕ_n is the solution of

$$\dot{\phi}_n(t) = v_n(t, \phi_n(t))\,, \quad \phi_n(t_0) = x_0\,, \tag{30}$$

where

$$v_n(t,x) = \langle v(t,x)\,,\, n_M(x)\rangle\, n_M(x) \quad \text{on } \mathrm{gr}(M_t)\,.$$

The existence of a solution ϕ_n of (30) is not obvious, since $v_n(t,x)$ is only defined on $\mathrm{gr}(M_t)$. In such a situation, the tangency condition

$$\lim_{h\to 0+} \frac{1}{h}\, \mathrm{dist}(x_0 + hv_n(t_0,x_0), M_{t_0+h}) = 0 \quad \text{for all } t_0 \text{ and } x_0 \in M_{t_0}$$

is necessary and sufficient for local existence of solutions, given that M_t enjoys some mild regularity which holds if $\mathrm{gr}(M_t)$ is a C^1-submanifold of $\mathbb{R}^{n+1}$ and v_n is locally Lipschitz continuous; see, e.g., [6]. Now, given $x_0 \in M_{t_0}$, the C^1-regularity of M_{t_0} and the fact that $v - v_n$ with $v = v(t_0,x_0)$ and $v_n = v_n(t_0,x_0)$ is tangent to M_{t_0} at x_0 yields

$$x_0 - h(v - v_n) + e_h \in M_{t_0} \quad \text{for small } h > 0$$

with certain $e_h \in \mathbb{R}^n$ such that $|e_h| = o(h)$ as $h \to 0+$. This implies

$$\begin{aligned}
&\mathrm{dist}(x_0 + hv_n, M_{t_0+h})\\
&\quad \le |x_0 + hv_n - \phi(t_0+h; t_0, x_0 - h(v-v_n) + e_h)|\\
&\quad = |x_0 + hv_n - (x_0 - h(v-v_n) + hv(t_0, x_0 - h(v-v_n)))| + o(h)\\
&\quad = h|v(t_0,x_0) - v(t_0, x_0 - h(v-v_n))| + o(h) = o(h) \quad \text{as } h \to 0+
\end{aligned}$$

if v is continuous. Hence the required tangency condition holds.

Application of Theorem A yields the formulation

$$\frac{d}{dt}\int_{M_t} c\, d\sigma_t = \int_{M_t} \left(\left[\frac{\partial c}{\partial t}\right]_n - c\,\kappa\, V \right) d\sigma_t + \int_{\partial M_t} c\,\langle v\,,\, N\rangle\, ds_t\,,$$

where ds_t denotes the surface measure on ∂M_t. This form of the surface transport theorem nicely displays the different contributions to the change in surfactant mass due to transport of species and geometrical changes. A more formal derivation of Theorem B in the language of Differential Geometry can be found in [4], p. 228ff, while a more heuristic version for $n = 3$ is given in [35]. For purely tangential surface motion Theorem B reduces to Theorem 2.11 in Chapter XII of [3], while for purely normal surface motion a generalization to $M_t \cap V$ with fixed $V \subset \mathbb{R}^n$ instead of M_t is given in [15]; for extensions cf. Remark 3 in [15] and [17, 21].

References

[1] F. Abergel, C. Dupaix: Existence of smooth, stationary interfaces for Marangoni-type flow. Nonl. Anal., Theory Meth. Appl. **27**(11), 1329–1350 (1996).

[2] G. Allain: Small-time existence for the Navier-Stokes equations with a free surface. Appl. Math. Optim. **16**, 37–50 (1987).

[3] H. Amann, J. Escher: *Analysis III.* Birkhäuser, 2001.

[4] R. Aris: *Vectors, tensors, and the basic equations of fluid mechanics.* Dover Publications, 1989.

[5] J.T. Beale: Large-time regularity of viscous surface waves. Arch. Rational Mech. Anal. **84**, 307–352 (1984).

[6] D. Bothe: Multivalued differential equations on graphs. Nonl. Anal., Theory Meth. Appl. **18**(3), 245–252 (1992).

[7] R. Clift, J.R. Grace, M.E. Weber: *Bubbles, Drops, and Particles.* Academic Press, New York, 1978.

[8] R. Defay, I. Prigogine, A. Bellemans: *Surface tension and adsorption.* Wiley and Sons, New York, 1966.

[9] I.V. Denisova: Evolution of compressible and incompressible fluids separated by a closed interface. Interfaces Free Bound. **2**(3), 283–312 (2000).

[10] I.V. Denisova, V.A. Solonnikov: Classical solvability of the problem on the motion of two viscous incompressible fluids. St. Petersburg Math. J. **7**(5), 755–786 (1996); translation from Algebra Anal. **7**(5), 101–142 (1995).

[11] R. Denk, M. Hieber, J. Prüss: *$\mathcal{R}$-boundedness, Fourier multipliers and problems of elliptic and parabolic type.* Mem. Amer. Math. Soc. **166**(788), 2003.

[12] J. Escher, J. Prüss, G. Simonett: Analytic solutions for a Stefan problem with Gibbs-Thomson correction. J. reine angew. Math. **563**, 1–52 (2003).

[13] J. Escher, J. Prüss, G. Simonett: A new approach to the regularity of solutions for parabolic equations. In: *Evolution Equations.* Lecture Notes in Pure and Appl. Math. **234**, 167–190. Dekker, New York, 2003.

[14] J. Escher, J. Prüss, G. Simonett: Analytic solutions of the free boundary value problem for the two-phase Navier-Stokes system. Preprint.

[15] M.E. Gurtin, A. Struthers, W.O. Williams: A transport theorem for moving interfaces. Quarterly of Applied Mathematics **47**(4), 773–777 (1989).

[16] M. Ishii: *Thermo-Fluid Dynamic Theory of Two-Phase Flow.* Eyrolles, Paris, 1975.

[17] J.P. Jaric: On a transport theorem for moving interface. Int. J. Engng. Sci. **30**(10), 1535–1542 (1992).

[18] A.J. James, J. Lowengrub: A surfactant-conserving volume-of-fluid method for interfacial flows with insoluble surfactant. J. Comp. Phys. **201**(2), 685–722 (2004).

[19] M.V. Lagunova: On the solvability of a three-dimensional problem of thermocapillary convection. J. Sov. Math. **64**(6), 1233–1240 (1993); translation from Probl. Math. Anal. **11**, 18–27 (1990).

[20] S. Lang: *Differential and Riemannian Manifolds.* Springer, New York, 1995.

[21] H. Petryk, Z. Mroz: Time derivates of integrals and functionals defined on varying volume and surface domains. Arch. Mech. **38**(5-6), 697–724 (1986).

[22] J. Prüss: Maximal regularity for abstract parabolic problems with inhomogeneous boundary data in L_p-spaces. Math. Bohem. **127**(2), 311–327 (2002).

[23] J. Prüss: Maximal regularity for evolution equations in L_p-spaces. Conf. Semin. Mat. Univ. Bari (2002) **285**, 1–39 (2003).

[24] F. Ravera, M. Ferrari, L. Liggieri: Adsorption and partitioning of surfactants in liquid-liquid systems. Adv. Coll. Int. Sci. **88**, 129–177 (2000).

[25] M. Siegel: Influence of surfactant on rounded and pointed bubbles in two-dimensional Stokes flow. SIAM J. Appl. Math. **59**(6), 1998–2027 (1999).

[26] J.C. Slattery: *Advanced Transport Phenomena*. Cambridge University Press, Cambridge 1999.

[27] V.A. Solonnikov: On the quasistationary approximation in the problem of motion of a capillary drop. pp. 643–671 in *Topics in Nonlinear Analysis. The Hermann Amann Anniversary Volume*, (J. Escher, G. Simonett, eds.). Birkhäuser, Basel, 1999.

[28] N. Tanaka: Global existence of two-phase non-homogeneous viscous incompressible fluid. Commun. Partial Differ. Equations **18**(1-2), 41–81 (1993).

[29] N. Tanaka: Two-phase free boundary problem for viscous incompressible thermo-capillary convection. Jap. J. Math., New Ser. 21(1), 1–42 (1995).

[30] A. Tani: Two-phase free boundary problem for compressible viscous fluid motion. J. Math. Kyoto Univ. **24**, 243–267 (1984).

[31] A. Tani: Small-time existence for the three-dimensional Navier-Stokes equations for an incompressible fluid with a free surface. Arch. Rat. Mech. Anal. **133**, 299–331 (1996).

[32] A. Tani, N. Tanaka: Large-time existence of surface waves in incompressible viscous fluids with or without surface tension. Arch. Rat. Mech. Anal. **130**, 303–314 (1995).

[33] Y. Teramato: On the Navier-Stokes flow down an inclined plane. J. Math. Kyoto Univ. **32**, 593–619 (1992).

[34] A. Wagner: Nonstationary Marangoni convection. Appl. Math. **26**(2), 195–220 (1999).

[35] H. Wong, D. Rumschitzki, C. Maldarelli: On the surfactant mass balance at a deforming fluid interface. Phys. Fluids **8**(11), 3203–3204 (1996).

Dieter Bothe
Institut für Mathematik & Department Chemie,
Technische Chemie und Chemische Verfahrenstechnik
Universität Paderborn
Warburger Str. 100
D-33098 Paderborn, Germany
e-mail: `bothe@tc.uni-paderborn.de`

Jan Prüss
Fachbereich Mathematik und Informatik
Martin-Luther-Universität Halle-Wittenberg
D-60120 Halle, Germany
e-mail: `pruess@mathematik.uni-halle.de`

Gieri Simonett
Department of Mathematics
Vanderbilt University
Nashville, TN 37240, USA
e-mail: `simonett@math.vanderbilt.edu`

Progress in Nonlinear Differential Equations
and Their Applications, Vol. 64, 63–70

Resolvent Differences for General Obstacles

J. Brasche and M. Demuth*

Abstract. In case of potential perturbations the second resolvent equation transforms the resolvent difference into a product of operators. For obstacle perturbations this behavior is maintained due to Dynkin's formula. In the present article we study generalized obstacle perturbations, e.g., perturbations by measures with infinite weight. It turns out that the resolvent difference equals a product of two operators, one factor is the free resolvent, the second factor contains all the interaction. This result is applicable to differential operators of arbitrary order and to a wide class of perturbations.

1. Introduction

Resolvent differences play an essential role in spectral theory of selfadjoint operators. For instance if the difference is compact the essential spectra are stable (Weyl's Theorem), or if it is trace class the absolutely continuous spectra are the same (Birman-Kuroda Theorem).

For studying resolvent differences the simplest example is the Schrödinger operator $L = L_0 + M_V$ in $L^2(\mathbb{R}^d)$, where L_0 is the selfadjoint realization of the Laplacian and M_V is the multiplication operator with a real-valued potential function $V(.)$, such that $L_0 + M_V + 1$ is a selfadjoint, positive operator. Let $\mathrm{dom}(L_0) = \mathrm{dom}(L_0 + M_V)$, set $H_0 = L_0 + 1$ and $H = L_0 + M_V + 1$. Then the second resolvent equation says

$$H_0^{-1} - H^{-1} = P_V\ H_0^{-1}\ , \tag{1}$$

with the denotation

$$P_V = (L_0 + M_V + 1)^{-1} M_V \tag{2}$$

defined on $\mathrm{dom}(H_0)$. That means the resolvent difference is equal to a product of the free resolvent with an operator P_V which contains all the information on the perturbation.

*Supported by the Volkswagenstiftung.

This procedure becomes more difficult, if L_0 is perturbed by imposing Dirichlet boundary conditions. One possibility comes from stochastic analysis. For simplicity, let L_0 be again the Laplacian in $L^2(\mathbb{R}^d)$. It generates the semigroup $\{e^{-tL_0}, t \geq 0\}$. The operators are given by

$$(e^{-tL_0} f)(x) = E_x\{f\big(X(t)\big)\}, \ f \in L^2(\mathbb{R}^d),$$

where $X(.)$ are the trajectories of the Wiener process and $E_x\{.\}$ is the associated expectation. For investigating the perturbation we introduce an obstacle region, called Γ. Γ is assumed to be a closed set in $\mathbb{R}^d$ with positive volume and smooth boundary. Define

$$\tau_\Gamma = \inf\{s, \ s > 0, \ X(s) \in \Gamma\} ,$$

the first hitting time of Γ. Set $\Sigma = \mathbb{R}^d \setminus \Gamma$. Then the family of operators

$$\big(T_\Sigma(t) f\big)(x) = E_x\{f\big(X(t)\big), \ \tau_\Gamma \geq t\}$$

restricted to $L^2(\Sigma)$ forms a strongly continuous semigroup. We denote its generator by L_∞. In this simple situation L_∞ is known to be the Friedrichs extension of $L_0|_{L^2(\Sigma)}$, or L_∞ is given by L_0 with Dirichlet boundary conditions on the boundary of Γ.

Now we set $H_0 = L_0 + 1, \ H_\infty = L_\infty + 1$. Extending the resolvent H_∞^{-1} to $L^2(\mathbb{R}^d)$ the resolvent difference is given by

$$H_0^{-1} - (H_\infty^{-1} \oplus 0) = P_\Gamma \ H_0^{-1} \ . \tag{3}$$

Formula (3) is known as Dynkins's formula (see, e.g., [FOT] or [DvC]). The operator P_Γ contains again all the information about the interaction. Here it is the perturbation by an obstacle on Γ. This can be modelled by a potential barrier with infinite height, because if $H_\beta = H_0 + \beta\chi_\Gamma$, then

$$\operatorname*{s-lim}_{\beta\to\infty} H_\beta^{-1} \ = \ H_\infty^{-1} \ .$$

P_Γ is given by

$$\begin{aligned} \operatorname{dom}(P_\Gamma) \ &= \ \operatorname{dom}(H_0) \ , \\ (P_\Gamma f)(x) \ &= \ E_x\Big\{e^{-\tau_\Gamma} f\big(X(\tau_\Gamma)\big)\Big\} \ . \end{aligned}$$

It is a projection operator in a new Hilbert space $\mathfrak{H}_1$ consisting of $f \in \operatorname{dom}(H_0^{1/2})$ having the scalar product

$$\langle f, g\rangle_1 \ = \ \langle H_0^{1/2} f, \ H_0^{1/2} g\rangle \ .$$

P_Γ projects onto the orthogonal complement of the subspace $\mathfrak{F}_\Sigma$ of $\mathfrak{H}_1$, given by

$$\mathfrak{F}_\Sigma \ = \ \{f, \ f \in \operatorname{dom}(H_0^{1/2}), \ \ \tilde{f}(x) = 0 \ \text{ quasi everywhere on } \Gamma\} \ .$$

Here $\tilde{f}$ is the quasi-continuous version of $f \in \operatorname{dom}(H_0^{1/2})$.

The stochastic representation and Dynkin's formula (3) is restricted to L_0 which are generators of Markov processes, i.e., to second-order differential or pseudo-differential operators. And it is restricted to obstacles which are in some sense regular enough.

2. Resolvent differences for general free operators

In a previous article ([BD], Theorem 2) an essentially more general situation was studied. There L_0 was only assumed to be a selfadjoint nonnegative operator in a separable Hilbert space $\mathfrak{H}$ with the scalar product $\langle .,.\rangle$. Setting $H_0 = L_0 + 1$ we define a further Hilbert space $\mathfrak{H}_1$ with vectors $f \in \mathrm{dom}(L_0^{1/2})$ and with the scalar product

$$\begin{aligned} \langle f, g\rangle_1 &= \langle L_0^{1/2} f,\ L_0^{1/2} g\rangle \ + \langle f, g\rangle \\ &= \langle H_0^{1/2} f,\ H_0^{1/2} g\rangle \ . \end{aligned}$$

Let J be a bounded linear operator with dense range, mapping $\mathfrak{H}_1$ into another Hilbert space $\mathfrak{H}_2$. ($\mathfrak{H}_2$ contains the perturbation. It is arbitrary in the present framework. If $\mathfrak{H} = L^2(\mathbb{R}^d, |.|)$, then $\mathfrak{H}_2$ can be $L^2(\mathbb{R}^d, \mu)$ with some measure μ. In this case J describes perturbations by measure valued potentials.)

Now we define in $\mathfrak{H}$ a selfadjoint operator H_β, $H_\beta \geq 1$, by

$$\begin{aligned} \mathrm{dom}(H_\beta^{1/2}) &= \mathfrak{H}_1 \subset \mathfrak{H} \\ \|H_\beta^{1/2} f\|^2 &= \|H_0^{1/2} f\|^2 + \beta\, \|Jf\|^2_{\mathfrak{H}_2} \ . \end{aligned}$$

Then there is a unique nonnegative, invertible selfadjoint operator H_∞ given by

$$\begin{aligned} \mathrm{dom}(H_\infty^{1/2}) &= \ker(J)\ , \\ \|H_\infty^{1/2} f\| &= \|H_0^{1/2} f\| \ . \end{aligned}$$

H_∞ models a generalized obstacle perturbation, because

$$\begin{aligned} (H_\beta^{-1})g \ &\underset{\beta\to\infty}{\to}\ H_\infty^{-1} g\ , \quad g \in \overline{\ker(J)} \\ H_\beta^{-1} h \ &\underset{\beta\to\infty}{\to}\ 0 \qquad , \quad h \in \overline{\ker(J)}^{\perp}\ , \end{aligned}$$

where $\overline{\ker(J)}$ is the closure of $\ker(J)$ in $\mathfrak{H}$.

It turns out that the resolvent difference is given by

$$H_0^{-1} - (H_\infty^{-1} \oplus 0) = P_J\ H_0^{-1}\ . \tag{4}$$

P_J is an orthogonal projection operator in $\mathfrak{H}_1$ with $P_J \mathfrak{H}_1 = [\ker(J)]^{\perp}$, where $\perp$ means orthogonal w.r.t. $\langle .,.\rangle_1$.

Equation (4) can be considered as a generalization of Dynkin's formula or as a generalized second resolvent equation. Again P_J contains all the information due to the interaction. For instance if $\mathfrak{H}_2 = L^2(\mathbb{R}, \mu)$ where μ is some singular measure the selfadjoint operator L_0 is perturbed by infinitely high obstacles on the support of this measure μ.

3. Resolvent differences for unbounded interaction operators

The aim of this article is to extend the last result to densely defined closed operators J mapping $\mathfrak{H}_1$ into $\mathfrak{H}_2$. We will prove the following result.

Theorem 1. Let L_0 be a nonnegative selfadjoint operator in the separable Hilbert space $\mathfrak{H}$ endowed with the scalar product $\langle .\,,.\rangle$.

Let $\mathfrak{H}_1$ be the Hilbert space $\{\mathrm{dom}(L_0^{1/2}),\ \langle .\,,.\rangle_1\ \}$ with

$$\langle f,g\rangle_1 = \langle L_0^{1/2}f,\ L_0^{1/2}g\rangle + \langle f,g\rangle\ .$$

Let J be a densely defined closed operator mapping $\{\mathfrak{H}_1,\ \langle .\,,.\rangle_1\ \}$ into a further separable Hilbert space $\{\mathfrak{H}_2,\ \langle .\,,.\rangle_2\ \}$. Assume $\mathrm{ran}(J)$ to be dense in $\mathfrak{H}_2$.

Define $H_\beta,\ \beta > 0$, by

$$\begin{aligned} \mathrm{dom}(H_\beta^{1/2}) &= \mathrm{dom}(J) \\ \|H_\beta^{1/2}f\|^2 &= \|(L_0+1)^{1/2}f\|^2 + \beta\ \|Jf\|_2^2\ . \end{aligned}$$

That means H_β is the perturbation of L_0 given by J with the coupling constant β. Define a nonnegative selfadjoint operator H_∞ by

$$\begin{aligned} \mathrm{dom}(H_\infty^{1/2}) &= \ker J\ , \\ \|H_\infty^{1/2}f\|^2 &= \|(L_0+1)^{1/2}f\|^2\ . \end{aligned}$$

(In other words: H_∞ is defined in $\ker(J) \subset \mathrm{dom}(L_0^{1/2})$ via strong resolvent convergence of H_β as β tends to infinity.)

Setting $H_0 = L_0 + 1$ the second resolvent equation reads

$$H_0^{-1} - (H_\infty^{-1} \oplus 0) = P_J\ H_0^{-1}\ .$$

Decomposing $\mathfrak{H}_1 = \ker(J) \oplus_{\langle .\,,.\rangle_1} \big(\ker(J)\big)^\perp$ then P_J is the orthogonal projector onto $\big(\ker(J)\big)^\perp$.

Before we are able to prove this theorem, we formulate a lemma which will imply finally the result.

Lemma 2. Assume $L_0,\ J,\ H_\beta,\ H_\infty$ and $\mathfrak{H},\ \mathfrak{H}_1,\ \mathfrak{H}_2$ are as given in Theorem 1. Then there is a compact operator $J_c : \mathfrak{H}_1 \to \mathfrak{H}_2$ such that

$$\begin{aligned} \mathrm{dom}(J_c) &= \mathfrak{H}_1\ , \\ \ker(J_c) &= \ker(J)\ , \end{aligned}$$

and $\mathrm{ran}(J_c)$ is dense in $\mathfrak{H}_2$.

Proof of Lemma 2. J^* is a closed operator from the separable Hilbert space $\{\mathfrak{H}_2,\ \langle .\,,\,.\rangle_2\}$ to the Hilbert space $\{\mathfrak{H}_1,\ \langle .\,,\,.\rangle_1\}$. Hence $\{\mathrm{dom}(J^*),\ \|.\|_{J^*}\}$ is separable where $\|g\|_{J^*}^2 = \|g\|^2 + \|J^*g\|^2$.

Let $\{f_n\}_{n\in\mathbb{N}}$ be a dense set in $\{\mathrm{dom}(J^*),\ \|.\|_{J^*}\}$. From that we can select a set $\{g_n\}_{n\in\mathbb{N}}$, such that g_n are linearly independent and

$$\mathrm{span}\ \{g_n,\ n\in\mathbb{N}\} = \mathrm{span}\ \{f_n,\ n\in\mathbb{N}\}\ .$$

Using the Gram-Schmidt orthogonalisation we get an orthonormal system $\{e_n\}_{n\in\mathbb{N}}$ in $\{\mathfrak{H}_2,\ \langle .\,,\,.\rangle_2\}$ with

$$\operatorname{span}\{e_n,\ n\in\mathbb{N}\} = \operatorname{span}\{g_n,\ n\in\mathbb{N}\}\ ,$$

and $\operatorname{span}\{e_n,\ n\in\mathbb{N}\}$ is dense in $\{\operatorname{dom}(J^*),\ ||.||_{J^*}\}$.

J was assumed to be closed, so $\operatorname{dom}(J^*)$ is dense in $\mathfrak{H}_2$, i.e., $\{e_n,\ n\in\mathbb{N}\}$ is an orthonormal basis in $\mathfrak{H}_2$. With this basis we are able to define the compact operator J_c.

Set

$$\lambda_k = 2^{-k}\frac{1}{1+||J^*e_k||_1}\ , k\in\mathbb{N}\ .$$

Define an operator J_0 by $\operatorname{dom}(J_0)=\operatorname{dom}(J)$

$$J_0 f = \sum_{k\in\mathbb{N}} \lambda_k\ \langle e_k,\ Jf\rangle_2\ e_k\ .$$

By definition J_0 is densely defined and bounded, mapping $\mathfrak{H}_1$ into $\mathfrak{H}_2$. Its bounded extension is denoted by J_c, i.e.,

$$\bar{J_0} = J_c\ ,\ \operatorname{dom}(J_c) = \mathfrak{H}_1\ .$$

J_c is a Hilbert-Schmidt operator. To show that take an orthonormal basis $\{f_j,\ j\in\mathbb{N}\}$ in $\operatorname{dom}(J)\subset\mathfrak{H}_1$. Then

$$\begin{aligned}
&\sum_{j\in\mathbb{N}} ||J_c f_j||_2^2\\
&= \sum_{j\in\mathbb{N}} ||\sum_{k\in\mathbb{N}} \lambda_k\ \langle e_k,\ Jf_j\rangle_2\ e_k||_2^2\\
&= \sum_{j\in\mathbb{N}}\sum_{k\in\mathbb{N}} \lambda_k^2\ |\langle J^*e_k,\ f_j\rangle_1\ |^2\\
&= \sum_{k\in\mathbb{N}} \lambda_k^2\ ||J^*e_k||_1^2\ < \infty\ .
\end{aligned}$$

Next we show $\ker(J)=\ker(J_c)$. Obviously, $\ker(J)\subseteqq\ker(J_c)$, because of the definition of J_0 and J_c, $Jf=0$ implies $J_0f=J_cf=0$. On the other hand, J is densely defined and closed. Hence $\ker(J)=[\operatorname{ran}(J^*)]^\perp$. Take an $f\in\ker(J_c)$. Then there is a sequence $\{f_n\}$, $f_n\in\operatorname{dom}(J_0)$ such that $f_n\to f$ and $J_0f_n\to J_cf=0$. Let $\{e_k,\ k\in\mathbb{N}\}$ be the orthonormal basis in $\mathfrak{H}_2$ introduced above. Then

$$\begin{aligned}
\langle J_cf,\ e_k\rangle_2\ &= 0\\
&= \lim_{n\to\infty}\ \langle J_0f_n,\ e_k\rangle_2\\
&= \lim_{n\to\infty}\ \Big(\sum_{m\in\mathbb{N}} \lambda_m\langle e_m, Jf_n\rangle_2\ \langle e_m,\ e_k\rangle_2\Big)\\
&= \lim_{n\to\infty}\ \lambda_k\langle e_k, Jf_n\rangle_2\\
&= \lambda_k\langle J^*e_k, f\rangle_1\ .
\end{aligned}$$

Therefore $f\in\ker(J_c)$ is orthogonal to any $J^*e_k,\ k\in\mathbb{N}$, implying $f\perp\operatorname{ran}(J^*)$ or $f\in\ker(J)$.

It remains to prove the density of $\operatorname{ran}(J_c)$ in $\mathfrak{H}_2$. The range of J was assumed to be dense. Fix a $k_0 \in \mathbb{N}$ and an $\varepsilon > 0$. Choose $f \in \operatorname{dom}(J)$ in such a way that

$$||Jf - \frac{e_{k_0}}{\lambda_{k_0}}||_2^2 < \varepsilon .$$

Then $||J_c f - e_{k_0}||_2^2 < \varepsilon$, because

$$\begin{aligned}
&||J_c f - e_{k_0}||_2^2 \\
&= \Big\| \sum_{k\in\mathbb{N}} \lambda_k \, \langle e_k, Jf\rangle_2 \, e_k - e_{k_0} \Big\|_2^2 \\
&= \sum_{\substack{k\in\mathbb{N}\\ k\neq k_0}} \lambda_k^2 \, |\langle e_k, Jf\rangle_2|^2 \; + \; \lambda_{k_0}^2 \, \Big| \langle e_{k_0}, Jf\rangle_2 - \frac{1}{\lambda_{k_0}} \Big|^2 \\
&\leqq \sum_{\substack{k\in\mathbb{N}\\ k\neq k_0}} |\langle e_k, Jf\rangle_2|^2 \; + \; \Big| \langle e_{k_0}, Jf\rangle_2 - \frac{1}{\lambda_{k_0}} \Big|^2 \\
&= \Big\| \sum_{k\in\mathbb{N}} \langle e_k, Jf\rangle_2 \, e_k - \frac{e_{k_0}}{\lambda_{k_0}} \Big\|_2^2 \\
&= \Big\| Jf - \frac{e_{k_0}}{\lambda_{k_0}} \Big\|_2^2 \; < \; \varepsilon .
\end{aligned}$$

Thus $e_{k_0} \in \overline{\operatorname{ran}(J_c)}$. Because clospan $\{e_k,\ k \in \mathbb{N}\} = \mathfrak{H}_2$, $\overline{\operatorname{ran}(J_c)} = \mathfrak{H}_2$.

This proves Lemma 2. □

Proof of Theorem 1. In [BD] Theorem 1 the result was shown for a compact operator J_c given in such a way that

$$J_c \, H_0^{-1/2} g = \sum_{k\in\mathbb{N}} \mu_k \, \langle v_k, g\rangle \, w_k$$

with an orthonormal basis $\{v_k, k \in \mathbb{N}\}$ in $\mathfrak{H}$ and with an orthonormal basis $\{w_k, k \in \mathbb{N}\}$ in $\mathfrak{H}_2$, and with $\mu_k \to 0$ as $k \to \infty$. The result was

$$\begin{aligned}
&\big[H_0^{-1} - (H_\infty^{-1} \oplus 0)\big] g \\
&= \sum_{k\in\mathbb{N}} \langle H_0^{-1/2} v_k, g\rangle \; H_0^{-1/2} v_k \\
&= \sum_{k\in\mathbb{N}} \langle H_0^{-1/2} v_k, H_0^{-1} g\rangle_1 \; H_0^{-1/2} v_k \, .
\end{aligned}$$

Here $H_0^{-1/2} v_k$, $k \in \mathbb{N}$, is an orthonormal system in $\big(\ker(J_c)\big)^\perp \subset \mathfrak{H}_1$. Defining the projection operator to $\big(\ker(J_c)\big)^\perp \subset \mathfrak{H}_1$ by

$$P_{J_c} \cdot \; = \; \sum_{k\in\mathbb{N}} \langle H_0^{-1/2} v_k, \; \cdot \; \rangle_1 \; H_0^{-1/2} v_k$$

we get

$$\big[H_0^{-1} - (H_\infty^{-1} \oplus 0)\big] g = P_{J_c} H_0^{-1} g \, . \tag{5}$$

However, this representation is not unique. In the Lemma 2 above we have shown that for any closed J there is always a compact J_c such that $\ker(J_c) = \ker(J)$. Thus $P_{J_c} = P_J$ and the definition of H_∞ is unchanged as long as $\ker(J_c) = \ker(J)$. $\square$

4. Applications

The general second resolvent formula (5) can be used to study the spectrum of H_∞ in this abstract framework. Possible applications can be explained in analogy to stochastic spectral analysis. Assume L_0 to be a differential operator of second order in $L^2(\mathbb{R}^d) = \mathfrak{H}$, generating a Hunt process (for details see [FOT]). Let L_∞ be the generator of the strongly continuous semigroup

$$\left(e^{-tL_\infty} f\right)(x) = E_x\left\{f\left(X(t)\right),\ \tau_\Gamma \geqq t\right\} .$$

τ_Γ is the first hitting time of Γ similar to the description in Section 1 for Wiener trajectories. Then Dynkin's formula (3) can be used in the following way:

$$\left[H_0^{-1} - (H_\infty^{-1} \oplus 0)\right] \mathbb{1} = P_{J_c} H_0^{-1} \mathbb{1} .$$

If the process is conservative, i.e., if $\left(e^{-tL_0}\mathbb{1}\right)(x) = \mathbb{1}(x)$, then $(H_0^{-1}\mathbb{1})(x) = \mathbb{1}(x)$. On the other hand

$$\begin{aligned}
&\left[H_0^{-1}\mathbb{1} - (H_\infty^{-1} \oplus 0)\mathbb{1}\right](x) \\
&= \int_0^\infty d\lambda\, e^{-\lambda}\left[E_x\{\mathbb{1}\} - E_x\{\tau_\Gamma \geq \lambda\}\right] \\
&= E_x\left\{\int_{\tau_\Gamma}^\infty e^{-\lambda} d\lambda\right\} \\
&= E_x\left\{e^{-\tau_\Gamma}\right\} ,
\end{aligned}$$

where $E_x\{.\}$ is the expectation of the process having in mind. $E_x\left\{e^{-\tau_\Gamma}\right\} = v_\Gamma(x)$ is known to be the one-equilibrium potential of Γ. This is related to the capacity of Γ. For a conservative process we have

$$\operatorname{cap}(\Gamma) = \int_{\mathbb{R}^d} v_\Gamma(x) dx .$$

Sets of finite capacity play an essential role in spectral theory of obstacle perturbations (see [DvC]). If the capacity is finite the essential and the absolutely continuous spectra of L_0 and L_∞ are the same. Certain integral conditions for v_Γ guarantee the absence of singularly continuous spectra for H_∞.

In the general case of Equation (5) the situation is similar. Consider $\mathfrak{H} = L^2(\mathbb{R}^d)$ and assume that H_0^{-1} maps $L^\infty(\mathbb{R}^d)$ to $L^\infty(\mathbb{R}^d)$. Then $(P_J\mathbb{1})(x)$ can be considered as a generalized equilibrium potential and $\int_{\mathbb{R}^d} (P_J\mathbb{1})(x)dx$ as a general notion of a capacity. If H_∞^{-1} and H_0^{-1} are integral operators it is known that the comparison function

$$\begin{aligned} D(x) &= \int |H_0^{-1}(x,y) - (H_\infty^{-1} \oplus 0)(x,y)|dy \\ &= (P_J \mathbb{1})(x) \end{aligned}$$

plays the essential role for the spectral behavior of H_∞ (see, e.g., [D]).

More detailed applications in $\mathfrak{H} = L^2(\mathbb{R})$ and for $\mathfrak{H}_2 = L^2(\mathbb{R}, \mu)$ with different measures μ are given in [BD]. Also the large coupling behavior as $\beta \to \infty$ is studied there in an optimal way.

References

[BD] J. Brasche, M. Demuth, Dynkin's formula and large coupling convergence, J. Funct. Analysis 219 (2005), 34–69.

[D] M. Demuth, Integral conditions for the asymptotic completeness of two-space scattering systems, Helv. Phys. Acta 103 (1993), 333–339.

[DvC] M. Demuth, J. van Casteren, Stochastic Spectral Theory for Selfadjoint Feller Operators, Birkhäuser Verlag, Basel, 2000.

[FOT] M. Fukushima, Y. Oshima, M. Takeda, Dirichlet Forms and Symmetric, Markov Processes, de Gruyter, Berlin, 1994.

J. Brasche and M. Demuth
Institute of Mathematics
TU Clausthal
Erzstr. 1
D-38678 Clausthal-Zellerfeld
Germany

Progress in Nonlinear Differential Equations
and Their Applications, Vol. 64, 71–97

Special Finite Time Extinction in Nonlinear Evolution Systems: Dynamic Boundary Conditions and Coulomb Friction Type Problems

J.I. Díaz

Dedicated to Herbert Amann on the occasion of his 65th birthday

1. Introduction

The finite extinction time phenomenon (the solution reaches an equilibrium after a finite time) is peculiar to certain nonlinear problems which solutions exhibit an asymptotic behavior entirely different from the typical behavior of solutions of linear problems.

After recalling, in Section 2, some general results and methods on this property, we shall consider several examples of nonlinear systems where finite extinction time is not an universal property of all the solutions of the problem: a feature very different from the case of scalar dissipative equations.

For instance, sometimes the vector solution has some components which stabilize in finite time, and others for which this phenomenon does not occur. This is the case for the linear heat equation with a suitable nonlinear dynamical boundary condition. In Section 3 we present an unpublished result by H. Amann and the author (Madrid, October 1988), the vector solution has two components, one is the solution of the pde in the interior, the other one its trace on the boundary which vanishes after a finite time.

In other nonlinear systems, finite extinction time is peculiar to a finite set of orbits. Such behavior arises, for instance, in the case of oscillation under strong friction which is close to Coulomb dry friction for small values of the velocity. This will be presented in Section 4, where, in particular, we shall recall the results by H. Amann and the author on this type of problems.

2. A survey on finite extinction time properties

In order to fix ideas, let $\Omega \subset \mathbf{R}^N$, $N \geq 1$, be a general open set, let $Q_\infty = \Omega \times (0, +\infty)$, $\Sigma_\infty = \partial\Omega \times (0, +\infty)$, and consider an evolution boundary value problem formulated as

$$\begin{cases} u_t + Au &= f(x,t) \quad \text{in } Q_\infty, \\ Bu &= g(x,t) \quad \text{on } \Sigma_\infty, \\ u(x,0) &= u_0(x) \quad \text{on } \Omega. \end{cases} \tag{1}$$

Here, Au denotes a nonlinear operator and Bu denotes a boundary operator (we assume, for simplicity, that A and B are *autonomous operators*).

In the study of the stabilization of solutions, as $t \to +\infty$, it is usually assumed that $f(x,t) \to f_\infty(x)$ and $g(x,t) \to g_\infty(x)$ as $t \to +\infty$, in some functional spaces and the main task is to prove that $u(x,t) \to u_\infty(x)$, as $t \to +\infty$, in some topology of a suitable functional space, with $u_\infty(x)$ solution of

$$\begin{cases} Au_\infty &= f_\infty(x) \quad \text{in } \Omega, \\ Bu_\infty &= g_\infty(x) \quad \text{on } \partial\Omega. \end{cases} \tag{2}$$

Here we are interested in a stronger property. We start by assuming that

$$\begin{array}{ll} f(x,t) = f_\infty(x) & \text{for } t \geq T_f, \\ g(x,t) = g_\infty(x) & \text{for } t \geq T_g. \end{array} \tag{3}$$

We arrive to the following natural concept

Definition 2.1. *Let u be a solution of problem* (1). *We will say that $u(x,t)$ possesses the property of "stabilization in a finite time" to the stationary solution $u_\infty(x)$ if there exists $t^* < \infty$ such that $u(x,t) \equiv u_\infty(x)$, on Ω, for any $t \geq t^*$.*

In many cases $f_\infty(x) \equiv 0$, $g_\infty(x) \equiv 0$, and then $u_\infty(x) \equiv 0$. Then, the above property is known as the *finite time extinction property.*

Most of the material collected in this Section is devoted to the case of the extinction property. For a different survey on this property and other applications, in particular to some problems in Fluid Mechanics, see Chapters 2 and 4 of the monograph [5].

2.1. Some abstract results for multivalued operators: finite extinction time and a dichotomy for hyperbolic equations

Perhaps, the pioneer abstract result on finite extinction time of solutions is due to H. Brezis [41]. He proved that if $X = H$ is a Hilbert space, and $A : D(A) \to \mathcal{P}(H)$ is a maximal monotone operator multivalued at 0 (with $0 \in \operatorname{int} D(A)$) then the solution of the abstract Cauchy problem

$$\begin{cases} \frac{du}{dt}(t) + Au(t) \ni f(t) \quad \text{in } X, \\ u(0) = u_0, \end{cases}$$

possesses the property of finite extinction in finite time once we assume $f(t)$ such that

$$B(f(t), \epsilon) \subset A0, \text{ for a.e. } t \geq t_f, \text{ for some } \epsilon > 0 \text{ and } t_f \geq 0. \tag{4}$$

It turns out that if, for instance, $H = L^2(\Omega)$ this property is difficult to be checked for most multivalued operators. This was the motivation of the work [50] in which the property of finite extinction time was proved for Banach spaces X and $A : D(A) \to \mathcal{P}(X)$ a multivalued m-accretive operator. Several applications for the special case of $X = L^\infty(\Omega)$, to some parabolic problems of the type

$$u_t - \Delta u + \beta(u) \ni f,$$

with β a maximal monotone graph of $\mathbb{R}^2$ (including second-order parabolic obstacle problems) were given in that paper. The application to multivalued nonlinear diffusion equations (of the porous media type)

$$u_t - \Delta\beta(u) \ni f,$$

was carried out in [54].

The finite extinction property can be proved also (via this abstract result) for other nonlinear multivalued parabolic problems of the type

$$\begin{cases} u_t - \nu\Delta u - g\mathrm{div}\left(\frac{\nabla u}{|\nabla u|}\right) = f\,(t,x) & \text{in } Q_\infty, \\ u = 0 & \text{on } \Sigma_\infty, \\ u(0,x) = u_0(x) & \text{on } \Omega, \end{cases}$$

for $\nu \geq 0$, $g > 0$ and $f\,(t,x) \neq 0$. Such problems arise in several applied contexts (non-Newtonian fluids of Bingham type, image processing, microgranular structures: see references in [2]). The above multivalued operator is also related to some old works in Differential Geometry ([81]).

A problem which looks quite similar to the previous ones but for which the above abstract results does not apply is the multivalued (second-order in time) hyperbolic dry friction type problem as, for instance,

$$(\mathrm{DSP}) \quad \begin{cases} u_{tt} - u_{xx} + \beta(u_t) \ni 0 & \text{in } (0,1) \times (0,+\infty), \\ u(t,0) = u(t,1) = 0 & t \geq 0, \\ u(0,.) = u_0(.) & \text{in } (0,1), \\ u_t(0,.) = v_0(.) & \text{in } (0,1), \end{cases}$$

where now β denotes the maximal monotone graph of $\mathbb{R}^2$ given by

$$\beta(u) = \{1\} \text{ if } u > 0,\ \beta(0) = [-1,1] \text{ and } \beta(u) = \{-1\} \text{ if } u < 0. \tag{5}$$

According well-known results (see, e.g., Haraux [70] and also [39]) $u(t,x) \to \zeta(x)$ in $H_0^1(0,1)$ as $t \to +\infty$, with ζ verifying $-1 \leq \zeta_{xx} \leq 1$. Since the beginnings of the seventies, H. Brezis proposed a conjecture: the equilibrium position is reached after a finite time (*stabilization in finite time*). Some partial results in this direction were obtained by H. Cabannes [42], [43] for some special initial data u_0 and v_0. The case of arbitrary initial data, u_0 and v_0, seems to be still an open problem.

Due to the difficulty of the above problem (and also suggested by the numerical approach of solutions) some easier formulations were considered in the literature, as, for instance, the spatially discretized vibrating string via finite differences:

$$\text{(DDSP)}\quad \begin{cases} \dfrac{d^2u_i}{dt^2} - \dfrac{u_{i+1} - 2u_i + u_{i-1}}{h^2} + \beta\left(\dfrac{du_i}{dt}\right) \ni 0 \quad i = 1, \dots, n, \\ u_0 = u_{n+1} = 0, \\ u_i(0) = a_i \quad \text{and} \quad \dfrac{du_i}{dt}(0) = b_i. \end{cases}$$

This system also arises in the study of the dynamics vibration of N-particles of equal mass m. Indeed, we denote the located positions, along the interval $(0, 1)$ of the x axis, by $x_i(t)$ and we assume that each particle is connected to its neighbors by two harmonic springs of strength k. We also assume that the particles are subject to a resultant friction force which is the composition of a Coulomb (or solid) friction and other type of frictions such as, for instance, the one due to the viscosity of an surrounding fluid. Then, the equations of motion for this mechanical system are

$$(P_N)\begin{cases} m\ddot{x}_i(t) + k(-x_{i-1}(t) + 2x_i(t) - x_{i+1}(t)) + \mu_\beta\beta(\dot{x}_i(t)) + \mu_g g(\dot{x}_i(t)) \ni 0 \\ x_i(0) = u_{0,i}, \ \dot{x}_i(0) = v_{0,i} \end{cases}$$

$i = 1, \dots, N$, where we are assuming that

$$x_0(t) = 0, \ x_{N+1}(t) = 1 \quad \text{for any } t \in (0, +\infty),$$

μ_β, μ_g are positive constants, the term $\mu_\beta\beta(\dot{x}_i(t))$ represents the Coulomb friction, g is a Lipschitz continuous function such that $g(0) = 0$ and the initial data $(u_{0,i})$, $(v_{0,i})$ are given in $\mathbb{R}^N$.

Notice that if we write, for simplicity, $k = \frac{1}{h^2}$ (with $h = 1/(N+1)$) and $m = 1$, then problem (P_N) coincides with the spatial discretization, by finite differences, of the damped string equation

$$(P_\infty)\begin{cases} u_{tt} - u_{xx} + \mu_\beta\beta(u_t) + \mu_g g(u_t) \ni 0 & \text{in } (0,1) \times (0, +\infty), \\ u(0,t) = u(1,t) = 0, & t \in (0, +\infty), \\ u(x,0) = u_0(x), u_t(x,0) = v_0(x) & x \in (0,1). \end{cases} \tag{6}$$

In fact, it was by passing to the limit, $N \to \infty$ in (P_N) (in absence of any friction) how the wave equation was obtained in 1746 by Jean Le Rond D'Alembert.

In order to give several criteria to have the stabilization in a finite time for (P_N) it is useful to start with the study of the special case of a single oscillator, $N = 1$, without viscous friction

$$m\ddot{x} + 2kx + \mu_\beta\beta(\dot{x}) \ni 0. \tag{7}$$

The study of this elementary equation can be found in many textbooks (see, for instance, [73]). It is easy to see then that the motion stops definitively after a finite time: i.e., there exists $T_e < +\infty$ and $x_\infty \in [-\frac{\mu_\beta}{2k}, \frac{\mu_\beta}{2k}]$ such that $x(t) \equiv x_\infty$ for any $t \geq T_e$.

Concerning the case of N-particles we can mention the work by Bamberger and Cabannes [29] in which they prove the stabilization in a finite time in absence of viscous friction ($\mu_g = 0$). We point out that this type of friction arises very often in the applications and that its consideration was already proposed by Lord Rayleigh (see, e.g., [80]). Concrete expressions for g can be found also in [73].

System (P_N) can be written, in short, as a vectorial problem

$$(\mathbf{P}_N)\begin{cases} m\ddot{\mathbf{x}}(t) + k\mathbf{A}\mathbf{x}(t) + \mu_\beta\mathbf{B}(\dot{\mathbf{x}}(t)) + \mu_\beta\mathbf{G}(\dot{\mathbf{x}}(t)) \ni \mathbf{0}, \\ \mathbf{x}(0) = \mathbf{x}_0,\ \dot{\mathbf{x}}(0) = \mathbf{v}_0 \end{cases}$$

where $\mathbf{x}(t) := (x_1(t), x_2(t), \ldots, x_N(t))^T$, $\mathbf{A}$ is the symmetric positive definite matrix of $\mathbb{R}^{N\times N}$ given by

$$\mathbf{A} = \begin{pmatrix} 2 & -1 & 0 & \ldots & 0 \\ -1 & 2 & -1 & 0 & \ldots \\ 0 & -1 & 2 & -1 & 0 \\ \ldots & 0 & -1 & 2 & -1 \\ 0 & \ldots & 0 & -1 & 2 \end{pmatrix},$$

$\mathbf{B} : \mathbb{R}^N \to \mathcal{P}(\mathbb{R}^N)$ and $\mathbf{G} : \mathbb{R}^N \to \mathbb{R}^N$ denote the (multivalued) maximal monotone operator and the Lipschitz continuous function given by

$$\mathbf{B}(y_1, \ldots, y_N) = (\beta(y_1), \ldots, \beta(y_N))^T \text{ and } \mathbf{G}(y_1, \ldots, y_N) = (g(y_1), \ldots, g(y_N))^T$$

(here $\mathbf{h}^T$ means the transposed vector of $\mathbf{h}$). As in the case of the damped wave equation, it is not difficult to prove ([63]) that for any $(\mathbf{x}_0, v_0) \in \mathbb{R}^{2N}$, problem (P_N) admits a unique weak solution $\mathbf{x} \in C^1([0, +\infty) : \mathbb{R}^N)$ and that there exists a unique equilibrium state $\mathbf{x}_\infty \in \mathbb{R}^N$ (i.e., satisfying that $A\mathbf{x}_\infty \in ([-\frac{\mu_\beta}{2k}, \frac{\mu_\beta}{2k}]^N)^T)$ such that $\| \dot{\mathbf{x}}(t) \| + \| \mathbf{x}(t) - \mathbf{x}_\infty \| \to 0$ as $t \to +\infty$.

Sharper results on the asymptotic behavior were obtained in ([63]). It was proved there that the presence of a viscous friction (with a suitable behavior of g near 0) may originate a qualitative distinction among the orbits in the sense that the state of the system may reach an equilibrium state in a finite time or merely in an asymptotic way (as $t \to +\infty$), according the initial data $\mathbf{x}(0) = \mathbf{x}_0$ and $\dot{\mathbf{x}}(0) = \mathbf{v}_0$. This dichotomy seems to be new in the literature and contrasts with the phenomena of *finite extinction time* for first-order (in time) ordinary and parabolic nonlinear equations. More precisely, the following was proved in ([63]): i) if $g(r)r \le 0$ in some neighborhood of 0, then all solutions of (P_N) stabilize in a finite time, ii) if $g(r) = \lambda r$ with $\lambda \ge 2\sqrt{\lambda_1 mk}/(\mu_\beta\mu_g)$, where λ_1 denotes the first eigenvalue of $\mathbf{A}$ then there exist solutions of (P_N) which do not stabilize in any finite time, and iii) if $N = 1$, $A = 1 \in \mathbb{R}$ and $g'(0) < 2\sqrt{mk}/(\mu_\beta\mu_g)$ any solution stabilize in finite time but if $g'(0) \ge 2\sqrt{mk}/(\mu_\beta\mu_g)$ there exist solutions which do not stabilize in any finite time.

We point out that the positive results on stabilization in a finite time remain true under the presence of some *impulsive forces* $\mathbf{f}(t)$ leading to the system

$$m\ddot{\mathbf{x}}(t) + k\mathbf{A}\mathbf{x}(t) + \mu_\beta\mathbf{B}(\dot{\mathbf{x}}(t)) + \mu_\beta\mathbf{G}(\dot{\mathbf{x}}(t)) \ni \mathbf{f}(t)$$

assuming that their amplitude is small enough: more precisely

$$\mathbf{f}(t)^T \in [-\frac{\mu_\beta}{2k} + \epsilon, \frac{\mu_\beta}{2k} - \epsilon]^N \text{ for a.e. } t \geq T_f, \text{ for some } T_f, \ \epsilon > 0.$$

Some sharper results are the main goal of some work in progress with G. Hetzer [59]. This behavior contrasts with the case in which the amplitude of $\mathbf{f}(t)$ becomes large and $g'(v) < 0$ for any $v \neq 0$. It was proved in [45] that, then, the dynamics generates a wide range of events leading to the chaos.

As mentioned before, the simultaneous possibility of the occurrence of stabilization in a finite or infinite time does not hold for solutions of scalar first-order in time equations of the form

$$u_t - d\Delta u + \beta(u) \ni 0 \tag{8}$$

for $\beta(u)$ multivalued at $u = 0$ and $d \geq 0$ (see, for instance, [50] and its references). We assume given some Dirichlet and initial conditions. Moreover, if we add an extra term of the form $g(u)$, such that, $g(u)u \geq 0$ for any $u \in \mathbb{R}$, then the solutions of

$$U_t - d\Delta U + \beta(U) + g(U) \ni 0 \tag{9}$$

satisfy that $\|u(t,.)\|_{L^p(\Omega)} \geq \|U(t,.)\|_{L^p(\Omega)}$ and, so, the extinction in a finite time of $u(t,.)$ implies the same property for $U(t,.)$. The opposed comparison holds when $g(u)u \leq 0$. This explain the important different behaviors among the solutions of problems of first- and second-order in time. Notice that if we assume $k = 0$ in $(\mathbf{P}_1)$ then we get that $U(t) = \dot{x}(t)$ satisfies an equation similar to (9) with $d = 0$. Notice, also, that if m is very small then problem $(\mathbf{P}_1)$ becomes a *quasi-static problem* (in the terminology of [69]) and then the solutions are closed to the solutions of the first-order in time problem

$$(QSP_1) \begin{cases} 2kx + \mu_\beta \beta(\dot{x}) + \mu_g g(\dot{x}(t)) \ni 0, \\ x(0) = x_0. \end{cases}$$

In that case, $g(u)u \geq 0$ implies a comparison opposite to the above mentioned one with respect the solutions with $g = 0$. Nevertheless, now the multivalued character of β at $u = 0$ does not imply the stabilization in a finite time for the solutions of (QSP_1).

The mentioned dichotomy may arise for other dynamical systems, as, for instance, the damped wave equation

$$u_{tt} - u_{xx} + \beta(u_t) + \lambda u_t \ni 0 \quad \text{in } (0,1) \times (0, +\infty), \tag{10}$$

with Dirichlet boundary conditions $u(0,t) = u(1,t) = 0$ for $t \in (0, +\infty)$. If we assume $\lambda \geq 2\pi$ then we can find some solutions of (10) which does not stabilize in any finite time although there are solutions which stabilize in a finite time. Indeed: we can construct solutions of the first type in the form $u(x,t) = a(t)sin(\pi x) + \frac{1}{2}x(x-1)$, with $a(t)$ such that

$$\ddot{a} + \pi^2 a + \lambda \dot{a} = 0, \tag{11}$$

and $\dot{a}(t) > 0$ for all time (which is possible since $\lambda \geq 2\pi$). By the contrary, if we choose $b(t)$ solution of (11) such that $\dot{b}(t) > 0$ for all $t \in [0,1)$, $\dot{b}(1) = 0$,

$b(1) = 1/\pi^2$ and take $a(t) = b(t)$ if $t \leq 1$ and $a(t) = 1/\pi^2$ for $t \geq 1$ then we get a solution which attains the stationary state $u_\infty \equiv 1$ after $t = 1$.

It is possible to obtain some abstract results for the localization in a finite time of solutions of some second-order in time Cauchy problem of the type

$$\begin{cases} \frac{d^2u}{dt^2}(t) + \partial\Psi(u(t)) + \partial\Phi(\frac{du}{dt}(t)) \ni 0 \quad \text{in } H, \\ u(0) = u_0, \ \frac{du}{dt}(0) = v_0, \end{cases}$$

in the framework of the subdifferential operators of convex functions Ψ and Φ on an Hilbert space H when we assume

$$\Phi(u) \geq \alpha \|u\| \text{ for some } \alpha > 0,$$

$$0 \in \operatorname{int} \partial\Phi(0) \text{ and } \partial\Psi \text{ is single-valued.}$$

As before, we assume proved, as a previous step, that there exists $u_\infty \in H$ such that

$$\| \frac{du}{dt}(t) \| + \| u(t) - u_\infty \| \to 0 \quad \text{as } t \to +\infty. \tag{12}$$

Then, if

$$-\partial\Psi(u_\infty) \in \operatorname{int} \partial\Phi(0) \tag{13}$$

there exists $t^* \geq 0$ such that $u(t) = u_\infty$ for any $t \geq t^*$.

This abstract theorem ([64]) can be applied, once we assume (13), not only to the finite-dimensional system of the vibration of N damped particles (already given in [1] and [44]) but also to the damped string equation (6) and to some viscoelastic Bingham materials leading to the multivalued equation

$$u_{tt} - \Delta u - \Delta\beta(u_t) \ni 0$$

(this time with the choice $H = H^{-1}(\Omega)$ as in [54]).

The last remark of this section concerns the case of periodic solutions of scalar second-order in time equations as (7). Several results on solutions presenting a *dead core* can be found in [47], [37] and the monograph [75] where many other references can be found. On the other hand, when the friction is replaced by $\beta(\dot{x} - \widehat{v})$ containing a given excitation $\widehat{v}$ it is possible to prove the existence of a periodic solution (a limit cycle) which is *attaint in a finite* time by other orbits. This behavior can be also present for damped wave equations of the form (10) [64].

2.2. Finite extinction time for single-valued operators via semidiscretization

One of the possible motivations of a different set of results on the extinction in finite time property for single-valued operators comes from the study of one of the archetype of quasilinear partial differential operators: the p-Laplacian

$$\Delta_p u = \operatorname{div}(|\nabla u|^{p-2} \nabla u), \quad 1 < p < \infty.$$

During some time, this operator was considered as an academic illustration of nonlinear diffusion operators but without any special relevant role in applied frameworks. Perhaps this was the reason (and because the often use of it in the J.L. Lions' school literature: see, e.g., [78]) why the operator was sometimes called as the "French nonlinear Laplacian". As we shall mention later, the situation changed soon and many important applications were modeled in terms of such quasilinear operator.

If, to fix ideas, we consider the problem

$$(P^p)\begin{cases} \dfrac{\partial u}{\partial t} = \Delta_p u & \text{in } Q_\infty, \\ u(t,x) = 0 & \text{on } \Sigma_\infty, \\ u(0,x) = u_0(x) & \text{on } \Omega\,. \end{cases}$$

Very different qualitative properties were shown for its solutions according the degenerate or singular nature of the operator. So, for instance, if $p > 2$ there is *finite speed of propagation* (i.e., if $supp(u_0) \subset B(0,r) \subset\subset \Omega$, then the solution of problem (P^p) satisfies that $supp(u(t))$ is a compact set for any $t > 0$ ([57], [58]). In contrast to that, if $1 < p \leq 2$ and $u_0 \geq 0$, $u_0 \neq 0$, then $u(t) > 0$ or $u(t) = 0$ in Ω for all $t > 0$ ([58]). The finite time extinction of the solutions of (P^p) when $\frac{2N}{N+2} \leq p < 2$, $N \geq 2$ was proved in [28], and, for $1 < p < \frac{2N}{N+1}$, in [72] (see also [5]).

Problems of this type are connected to some problems in fluid mechanics as, for instance, the discharge of a turbulent and perfect gas in a pipeline. Indeed, if we assume the pipeline occupying the interval $(0, L)$ and with a section of diameter D very small in comparison with L, the *hydraulic approximation* leads to a system of equations for the density ρ, velocity u, pressure p and temperature T

$$\begin{cases} \dfrac{\partial \rho}{\partial t} + \dfrac{\partial(\rho u)}{\partial x} = 0, \\ \rho\dfrac{\partial u}{\partial t} + \rho u\dfrac{\partial u}{\partial x} = -\dfrac{\partial p}{\partial x} - \dfrac{1}{2}\rho\,|u|\,u, \\ (\dfrac{\partial u}{\partial t} + u\dfrac{\partial}{\partial x})(\dfrac{\gamma}{\gamma-1}T + \dfrac{1}{2}u^2) - \dfrac{\partial p}{\partial t} = -\dfrac{1}{2}\,|u|\,(\dfrac{\gamma}{\gamma-1}T - \dfrac{1}{2}u^2), \\ \dfrac{p}{\rho} = T. \end{cases} \tag{14}$$

Replacing the third equation (the enthalpy equation) by

$$(\frac{\partial u}{\partial t} + u\frac{\partial}{\partial x})Ln(p/\rho^\gamma) = -\frac{|u|}{2}\frac{\gamma}{T}(T - 1 - \frac{\gamma-1}{2\gamma}u^2) \tag{15}$$

and assuming the initial and boundary conditions corresponding to an static initially full pipeline with one closed boundary point and other in which the discharge make take place at the pressure p_0 $(p_0 > p_a)$ and temperature T_0 (for any time) it was shown in ([60]) that when $t >> 1/f$ the second and fourth equation can be simplified, by neglecting lower-order terms and using some suitable variable scales), to

$$0 = -\frac{\partial p}{\partial x} - \frac{1}{2}\rho\,|u|\,u \quad \text{and} \quad \frac{p}{\rho} = T = 1.$$

Then, from the first equation we deduce that p satisfies that

$$\begin{cases} \dfrac{\partial p}{\partial t} - \dfrac{\partial}{\partial x}\left(\left|\dfrac{\partial p^2}{\partial x}\right|^{1/2} \operatorname{sign}\left(\dfrac{\partial p^2}{\partial x}\right)\right) = 0 & t > 0,\ x \in (0,1), \\ \dfrac{\partial p}{\partial x}(0,t) = 0,\ p(1,t) = p_a & t > 0, \\ p(x,0) = 1 & x \in (0,1). \end{cases}$$

Notice that since $u \geq 0$, making $p^2 - p_a^2 := w$ we arrive to the doubly nonlinear parabolic problem

$$\begin{cases} \dfrac{\partial \psi(w)}{\partial t} - \Delta_q w = 0 & t > 0,\ x \in (0,1), \\ \dfrac{\partial w}{\partial x}(0,t) = 0,\ w(1,t) = 0 & t > 0, \\ w(x,0) = w_0 & x \in (0,1), \end{cases} \tag{16}$$

with $\psi(w) = (w + p_a^m)^{1/m}$ which is a nondecreasing function of w. The correct exponents here are $m = 2$ and $q = 3/2$ nevertheless other interesting cases are $m = 7/4$ and $q = 11/7$ (case of very polished pipes) and $m = 1$ and $q = 3/2$ (laminar regime). The existence and uniqueness of solutions of a larger class of problems of this type was the main motivation of the paper [68]. The *finite extinction property* (here $w(x,t) \equiv p_a$ for any $t \geq t_0$ and $x \in (0,1)$) was proved in [60].

A family of more general problems, including as special case problem (16), corresponds to the following formulation, also arising in the study of the nonlinear heat equation with absorption

$$(\mathcal{P}) \begin{cases} \frac{\partial}{\partial t}\left(u\,|u|^{\gamma-1}\right) - \Delta_p u + |u|^{\sigma-1} u = f + \operatorname{div} \mathbf{g} & \text{in } Q_\infty, \\ u = 0 & \text{on } \Sigma_\infty, \\ u(x,0) = u_0(x) & \text{in } \Omega, \end{cases}$$

where $\Omega \subset \mathbb{R}^N$ is a bounded open set, $\gamma > 0$, $\sigma > 0$, $1 \leq p < \infty$, $\lambda \geq 0$ (if $p = 1$, ∇u represents the *total variation*). In order to study the extinction in finite time of solutions we assume that there exists $T_0 \geq 0$ such that $f(t,.) = 0$, $\mathbf{g}(t,.) = \mathbf{0}$ in Ω, if $t > T_0$.

Under suitable conditions, the "solution" (notion to be made precise in each case) satisfy an integral energy inequality leading to the extinction in a finite time of the function

Theorem 1. *Let $u \in L^1_{\mathrm{loc}}(T_0, +\infty : W_0^{1,p}(\Omega))$ for some $p > 1$ (or $u \in L^1_{\mathrm{loc}}(T_0, +\infty : BV_0(\Omega))$, if $p = 1$) such that $\exists\, \gamma, k, c > 0, \lambda \geq 0, \sigma > k - 1$ for which*

$$|u|^{\gamma+k}, |u|^{\sigma+k}, |\nabla u|^p\, |u|^{k-1} \in L^1_{\mathrm{loc}}(T_0, +\infty : L^1(\Omega))$$

and

$$y(t) + c\int_s^t \int_\Omega |\nabla u|^p\, |u|^{k-1} + \lambda \int_{t_f}^t \int_\Omega |u|^{\sigma+k} \tag{17}$$

$$\leq y(s) \text{ a.e. } s,t \in (T_0, +\infty), \tag{18}$$

where $y(t) = \int_\Omega |u|^{\gamma+k}\,dx$. *Assume that*

$$1 \le p < \gamma + 1 \quad \textit{and} \quad \lambda = 0 \tag{19}$$

or

$$1 \le p, \quad \sigma < \gamma \quad \textit{and} \quad 0 < \lambda, \tag{20}$$

and let

$$k = \begin{cases} 1 & \textit{if } N \le p \textit{ or } (\gamma+1) \le \dfrac{Np}{N-p}, \\ \dfrac{N-p}{p}\left(1+\gamma-\dfrac{p(N-1)}{N-p}\right) > 1 & \textit{if } 1 < p < N \textit{ and } \gamma+1 > \dfrac{Np}{(N-p)}. \end{cases}$$

Then $u \in C^{0,\alpha}_{\mathrm{loc}}([T_0,+\infty) : L^{\gamma+k}(\Omega))$ *for some* $\alpha \in (0,1)$ *and there exists a* $T_e \in (T_0,+\infty)$ *such that* $u(t,.) \equiv 0$ *in* Ω $\forall t \ge T_e$.

The proof uses an integral version of the Torricelli-Bernoulli energy inequality found in [5] (Proposition 1.1. or Theorem 2.1, Chapter 2) for $p > 1$ and [2] for $p = 1$. More precisely $y(t) + C\int_s^t y(\tau)^\mu dt \le y(s)$, for a.e. $s,t \in (T_0,+\infty)$ for some $\mu \in (0,1)$. Some relevant choices of the parameters γ, p, σ which provide the fulfillment of the above conditions are: $p = 2$, $\gamma = 1$ and $\sigma < 1$; $\sigma = 1$, $p = 2$ and $\gamma > 1$; $\sigma = 1$, $\gamma = 1$ and $p < 2$.

Several notions of solutions are possible (for simplicity, we assume now $p > 1$). The "variational theory" search for solutions in the "energy space" $u \in L^{p'}(0,T; W_0^{1,p}(\Omega))$, and use that (if $p \ge \frac{2N}{N+2}$) $W_0^{1,p}(\Omega) \subset L^2(\Omega) \subset W^{-1,p'}(\Omega)$. At least for $k = 1, u \in L^p(0,T; W_0^{1,p}(\Omega)), \forall T > 0$ implies that $|\nabla u|^p |u|^{k-1} \in L^1_{\mathrm{loc}}(0,+\infty : L^1(\Omega))$. A first problem arises with the zero-order term $|u|^{\sigma-1} u$ since $u \in L^{p`}(0,T; W_0^{1,p}(\Omega)) \Rightarrow |u|^{\sigma+k} \in L^1_{\mathrm{loc}}(0,+\infty : L^1(\Omega))$. Then, if the equation takes place in $D'(\Omega)$ the natural regularity for u_t is $|u_t|^{\gamma-1} u_t \in L^{p'}_{\mathrm{loc}}(0,+\infty : W^{-1,p'}(\Omega)) + L^1_{\mathrm{loc}}(0,+\infty : L^1(\Omega))$. In that case the test functions must be taken in $L^p(0,T; W_0^{1,p}(\Omega)) \cap L^\infty(0,T : L^\infty(\Omega))$. The existence of solutions in the above framework is due to many authors (see references in [5]) assumed $|u_0|^{(\gamma-1)} u_0 \in L^2(\Omega)$ and $f, \mathbf{g} \in L^{p'}_{\mathrm{loc}}(0,+\infty : L^{p'}(\Omega))$ and the, so-called, *weak solution* satisfies that $u \in C([0,+\infty) : L^2(\Omega))$. The regularity $|u|^{\gamma+k}, |u|^{\sigma+k}, |\nabla u|^p |u|^{k-1} \in L^1_{\mathrm{loc}}(T_0,+\infty : L^1(\Omega))$ can be obtained by asking some extra regularity to the data (see, e.g., [36]). A nontrivial fact is the justification of the time integration by parts formula

$$\left\langle (|u|^{\gamma-1} u)_t, |u|^{k-1} u \right\rangle = \frac{\gamma}{\gamma+k}\int_0^T \left[\frac{d}{dt}\int_\Omega |u(t,.)|^{\gamma+k}\,dx\right] dt.$$

It could be easily justified for the case of *strong solutions* (i.e., $\frac{\partial}{\partial t}\left(u\,|u|^{\gamma-1}\right) \in L^1(Q)$) but it is know that this class of solutions are quite exceptional. More in general (but for $\gamma = k = 1, \lambda = 0$) this was proved in a pioneering paper by J.L. Lions [77]. See other references, in particular for $\gamma \ne 1$ and $k \ne 1$, in [46].

In some cases the *extinction energy* $y(t) = \int_\Omega |u|^{\gamma+k}\,dx$ may be not well defined for solutions $u(t) \in W_0^{1,p}(\Omega) \subset L^2(\Omega)$. For instance, this is the case if

$\gamma = 1, \lambda = 0$ and $1 \le p < 2N/(N+2)$. Due to this difficulty, following to [34], it is useful to justify the energy inequality (for $k = 1$) by working in the space $W = W_0^{1,p}(\Omega) \cap L^{\gamma+1}(\Omega)$ if $\gamma < p-1$ or $1 < \sigma \le p$ or Ω bounded, otherwise W is defined as the closure of $C_0^\infty(\Omega)$ in the Banach space $\{u \in L^{\gamma+1}(\Omega), \nabla u \in L^p(\Omega)\}, \|u\| = \|u\|_{\gamma+1} + \|\nabla u\|_p$. The existence of *an energy solution* (i.e., $u \in C([0,+\infty) : L^{\gamma+1}(\Omega)) \cap L^p(0,T : W) \cap L^{\sigma+1}((0,T)\times\Omega))$ for any finite T, satisfying the equation in D' and with $u(0,.) = u_0(.)$) was proved by assuming that $u_0 \in L^{\gamma+1}(\Omega)$, and $f + \operatorname{div} \mathbf{g} \in L^{p'}(0,T : W') + L^{(\sigma+1)'}((0,T)\times\Omega)$.

But the notion of solution can be found out of the energy space W. Among the several types of solutions in this framework we could mention, specially, the so called *mild solutions* motivated by the numerical analysis and the *abstract semigroup theory*: given $\epsilon > 0$ and a time discretization $t_0 = 0 < t_1 < \cdots < t_n \le T, t_i - t_{i-1} < \epsilon, T - t_n < \epsilon$, and given $f_i \in L^\infty(\Omega)$, $w_0 \in L^\infty(\Omega)$ we consider the implicit time-discretization,

$$(DP)\begin{cases} \frac{b(w_i)-b(w_{i-1})}{t_i - t_{i-1}} - \Delta_p w_i + \lambda |w_i|^{\sigma-1} w_i = f_i & \text{in } \Omega, \\ w_i = 0 & \text{on } \partial\Omega, \end{cases}$$

where $b(u) = |u|^{\gamma-1} u$. Notice that $w_i \in W_0^{1,p}(\Omega) \cap L^\infty(\Omega)$. Now, let $u_0 \in L^\gamma(\Omega)$, $f \in L^1(0,T : L^1(\Omega)), \mathbf{g} = \mathbf{0}$.

Definition. A *mild solution of (P)* is a function u such that $b(u) \in C([0,+\infty) : L^1(\Omega))$, $u(0,.) = u_0(.)$, and, for any $\epsilon > 0$ there exists $(t_0, t_1, \ldots, t_n, f_0, f_1, \ldots, f_n, w_0, w_1, \ldots, w_n)$ satisfying (DP) with

$$\|b(u_0) - b(w_0)\|_1 \le \epsilon,\ \sum_i \int_{t_{i-1}}^{t_i} \|f(t) - f_i\|_1 \, dt \le \epsilon \text{ and } \|b(u(t)) - b(w_0)\|_1 \le \epsilon$$

for any $t \in (t_{i-1}, t_i]$, $i = 1, \ldots, n$.

The existence of a mild solution was due to [33]. Moreover, it was proved there that if, in addition, $u_0 \in L^{\gamma+1}(\Omega), f \in L^{p'}(0,T : W') + L^{(\sigma+1)'}((0,T)\times\Omega)$ then the mild solution is also an energy solution.

Now we can study the finite extinction time for the step function $w_\epsilon(t) := w_i$ if $t \in (t_{i-1}, t_i]$, $i - 1, \ldots, n$.

Definition. *We say that $w_\epsilon(t)$ extincts in a finite time if there exists $T_{\epsilon,e} = t_j$, for some $j \le n$ such that $\|w_\epsilon(t)\|_\infty > 0$ for $t \in [0, T_{\epsilon,e})$ and $\|w_\epsilon(t)\|_\infty = 0$ for $t \in [T_{\epsilon,e}, T]$.*

Since $w_\epsilon(t)$ satisfies the integral energy inequality (17) we get to the following result (due to [32] for $p = 2, \lambda = 0$ and [72] for $p > 1$ and $\lambda = 0$)

Corollary 1. *Assume that there exists $T_0 = t_m$, $m \le n$ such that $f_\epsilon(t,.) = 0$, in Ω, if $t > T_0$ ($f_\epsilon(t,.)$ defined in a similar way to $w_\epsilon(t)$). Then, under the assumption of Theorem 1 on $\gamma, p, k,$ and σ, function $w_\epsilon(t)$ extincts in a finite time $T_{\epsilon,e}$. Moreover, if u is a mild solution and assume that there exists $T_0 \ge 0$ such that $f(t,.) = 0, \mathbf{g}(t,.) = \mathbf{0}$ in Ω, if $t > T_0$, and that $u(T_0) \in L^{\gamma+k}(\Omega)$) then $u(t)$ extincts in a finite time T_e(only dependent on $\|u(T_0)\|_{\gamma+k}$).*

Notice that due to the regularizing effects (see [33]), it is possible to have a finite energy at time T_0 ($u(T_0) \in L^{\gamma+k}(\Omega)$)) even if $u_0 \in L^{\gamma}(\Omega)$.

An unpleasant fact of mild solutions is the lake of an easy characterization in terms of test functions and the lake of information on their spatial regularity. A different notion of solutions corresponds to the so called *renormalized solutions* (introduced, for second-order equations, in [36]). Since the general integral energy holds, the finite extinction time phenomenon can be obtained also for such solutions assumed, again, $u(T_0) \in L^{\gamma+k}(\Omega)$).

The assumption $u(T_0) \in L^{\gamma+k}(\Omega)$ is, in some sense, necessary. A counterexample can be done in other case: take $\gamma = 1, \lambda = 0$, $p = 1$, assume that $0 \in \Omega$ and $u(0,.) = \delta_0$ (the Dirac delta at the origin). Then, it is possible to show ([3]) that there is not any regularizing effect and $u(t,.) = C(t)\delta_0$ with $C(t) > 0$.

The extinction time also exists for other time-discretizations (now of semi-implicit type). We write (assuming now $w \geq 0$)

$$(w^{\gamma})_t = \frac{\gamma}{\gamma+1-p}(w^{p-1})(w^{\gamma+1-p})_t \approx \frac{\gamma}{\gamma+1-p}(w_i^{p-1})\frac{(w_i^{\gamma+1-p} - w_{i-1}^{\gamma+1-p})}{t_i - t_{i-1}}.$$

Given $\epsilon > 0$, a time discretization $t_0 = 0 < t_1 < \cdots < t_n \leq T, t_i - t_{i-1} < \epsilon, T - t_n < \epsilon$, and given $f_i \in L^{\infty}(\Omega)$, $w_0 \in L^{\infty}(\Omega)$ we consider the semi-implicit time-discretization,

$$\frac{\gamma}{\gamma+1-p}(w_i^{p-1})\frac{(w_i^{\gamma+1-p} - w_{i-1}^{\gamma+1-p})}{t_i - t_{i-1}} - \Delta_p w_i + \lambda |w_i|^{\sigma-1} w_i = f_i \text{ in } \Omega.$$

When f_i and w_0 are nonnegative, the existence and uniqueness of a nonnegative w_i is consequence of the results of [67]. The convergence of the scheme was given in [76] for $p = 2, \lambda = 0$ and in [79] for $p \neq 2$ and $\lambda = 0$. We have

Corollary 2. *Assume that there exists $T_0 = t_m$, $m \leq n$ such that $f_\epsilon(t,.) = 0$, in Ω, if $t > T_0$ ($f_\epsilon(t,.)$ defined as done for $w_\epsilon(t)$). Then, under the assumption of the Theorem 1 on γ, p, k, and σ function $w_\epsilon(t)$ extincts in a finite time $T_{\epsilon,e}$.*

The existence of a finite extinction time can be also proved for another type of semi-implicit time-discretization (see [35])

$$\frac{b(w_i) - b(w_{i-1})}{t_i - t_{i-1}} - \operatorname{div}(|\nabla w_{i-1}|^{p-2}\nabla w_i) + \lambda |w_i|^{\sigma-1} w_i = f_i \text{ in } \Omega.$$

A collection of results on finite extinction time for other type of schemes in which the discretization is not in time but in space can be found in [53].

We end this section by pointing out that, some times, the finite extinction time property can be proved by many other methods that the mentioned above. For instance, it is quite usual to get it via comparison principle (see references, e.g., in [5] and the contribution by Ph. Souplet to this volume). The comparison of symmetrical rearrangements allows, sometimes, to improve pointwise criteria (see, e.g., [65]). Spectral arguments can be also used to this end ([31]). Finally, we mention that the finite extinction time property is clearly connected with the question of the controllability to zero property in control theory (see, for instance, [66]).

3. Extinction by components

In this section we present an unpublished result by H. Amann and the author [26], lost for sometime among the files of both authors but found by one of them recently (joking, almost as the case of the 1519 Leonardo's manuscript about the anatomy lost since 1630 discovered, by chance, on 1966 in the archives of the Spanish National Library at Madrid). The main goal of this study was to exhibit some nonlinear system for which only one of the two components satisfies the finite extinction time property.

One of such systems can be formulated as follows

$$\begin{cases} u_t - \Delta u = f(t,x) & \text{in } Q_\infty, \\ u_t + u_\nu + \beta(u) \ni g(t,x) & \text{on } \Sigma_\infty, \\ u(0) = u_0^\Omega & \text{in } \Omega, \\ u(0) = u_0^\Gamma & \text{in } \Gamma. \end{cases}$$

Here Ω denotes a *convex* bounded domain in $\mathbb{R}^N$ and $\beta : D(\beta) \to \mathcal{P}(\mathbb{R})$ is the Signorini maximal monotone graph

$$\beta(r) = \begin{cases} \phi & \text{if } r < 0, \\ (-\infty, 0] & \text{if } r = 0, \\ 0 & \text{if } r > 0, \end{cases}$$

$u_0^\Omega \in L^\infty(\Omega)$, $u_0^\Gamma \in L^\infty(\Gamma)$ and u_ν is the conormal derivative of u at points of Γ.

The existence of solutions is well known in the literature, specially, after the deep contributions by H. Amann and collaborators (a large list of references can be found, for instance, in [30]). The dynamics of the problem becomes quite peculiar when β is not Lipschitz continuous near the origin.

Theorem 2. *Assume that $f(t,x) \le -\varepsilon^2$ near $\partial\Omega$ for $t \ge t_f, g(t,x) \le -\varepsilon^2$* on $\partial\Omega \times$ *$(t_g, +\infty)$. Then the trace $u(t,\cdot)$ on Γ vanishes after a finite time t_0.*

The proof uses the explicit construction of a local barrier function of the form $\overline{u}\,(t,x) = \Phi(t-t_0) + C\,|x - x_0|^2$ for any $x_0 \in \partial\Omega$ and for some suitable Φ, t_0 and C (see some related coincidence results in [55] and [56]).

Notice that we can reformulate the problem in terms of a vectorial abstract problem: take the Hilbert spaces V and H, with $V = \{(u,v) \in H^1(\Omega) \times H^{1/2}(\Gamma);\, u_{|\Gamma} = v\}$ which is a real separable Hilbert space isomorphic to $H^1(\Omega)$ where the latter is endowed with the equivalent norm

$$\|u\|_{H^1(\Omega)} = \left(\|\nabla u\|^2_{L^2(\Omega)} + \|u_{|\Gamma}\|^2_{L^2(\Gamma)} \right)^{1/2},$$

$H = L^2(\Omega) \times L^2(\Gamma)$ which endowed with the usual inner product

$$\langle (u,v), (\tilde{u}, \tilde{v}) \rangle = \langle u, \tilde{u} \rangle_{L^2(\Omega)} + \langle v, \tilde{v} \rangle_{L^2(\Gamma)},$$

is a real Hilbert space. Let us define $A : V \to V^*$ by

$$(A(u,v), (\varphi, \psi)) = \int_\Omega \nabla u \cdot \nabla \varphi \, dx,$$

where $(\cdot,\cdot)$ is the usual pairing between V and V^*. We define the restriction $A_H : D(A_H) \subset H \to H$ of A to H by $D(A_H) = \{(u,v) \in V;\ A(u,v) \in H\}$ and $A_H(u,v) = A(u,v)$, for each $(u,v) \in D(A_H)$. It is easy to see that

$$D(A_H) = \{(u,v) \in L^2(\Omega) \times L^2(\Gamma);\ \Delta u \in L^2(\Omega),\ u_\nu \in L^2(\Gamma),\ u_{|\Gamma} = v\}$$

and $A_H(u,v) = (-\Delta u, u_\nu)$. Then the solution is associated to the vector $U(t,.) = (u(t,\cdot), u(t,\cdot)_{|\Gamma}) \in L^2(\Omega) \times L^2(\Gamma)$ can be formulated as

$$\begin{gathered} \tfrac{d}{dt}U(t,.) + AU(t,.) + BU(t,.) \ni F(t,.) \\ U(0,.) = U_0 \end{gathered}$$

with $BU(.) = (0, \beta(u))$ and the above theorem shows that the component $u(t,\cdot)_{|\Gamma}$ vanishes in a finite time although, by the strong maximum principle for the linear heat equation, $u(t,\cdot)$ cannot vanish in any subset of positive measure of Ω for any time $t > 0$.

Another nonlinear system for which only one of the two components satisfies the finite extinction time property is the hybrid problem

$$\begin{cases} v_{tt}(t,x) - v_{xx}(t,x) = 0, & \text{in } Q_\infty, \\ v(t,0) = 0,\ v(t,1) = z(t), & t > 0, \\ v(0,\cdot) = 0, v_t(0,\cdot) = 0, & \text{in } \Omega, \\ \ddot{z}(t) + \alpha v_x(t,1) + z(t) + \beta(\dot{z}) \ni 0, & \\ z(0) = 0,\ \dot{z}(0) = 1, & \end{cases} \tag{21}$$

where $\Omega = (0,1)$ and β is given by (5). This system corresponds, for instance, to the modelling of a point block which lies on a rough surface and is connected to a fixed support by a spring and one end of a flexible string is attached to the block, the other end is fixed above the surface so that the string is tense. For small horizontal displacements $v(t,x)$ the dynamics of the string are governed by the wave equation and $z(t)$ denotes the displacement of the block which is assume to be in contact, without any lubrication with a surface, so dry friction occurs (see [38]).

4. Finite extinction time for a finite set of orbits

A very special dynamics arise in the case of the damped oscillator

$$m\ddot{x} + \mu\,|\dot{x}|^{\alpha-1}\,\dot{x} + kx = 0, \tag{22}$$

when $\alpha \in (0,1)$. Here μ and $k > 0$ are fixed parameters. In fact we can simplify the above formulation to

$$\ddot{x} + |\dot{x}|^{\alpha-1}\,\dot{x} + x = 0, \tag{23}$$

by dividing by k and by introducing the rescaling $\widetilde{x}(\widetilde{t}) = \beta^{1/(\alpha-1)}x(\lambda\widetilde{t})$ where $\lambda = \sqrt{m}/\sqrt{k}$ and $\beta = \mu/(k^{(2-\alpha)/2}m^{\alpha/2})$. Notice that the x-rescaling fails for the linear case $\alpha = 1$ since there is not any defined scale for x and the equation is merely reduced to $\ddot{x} + \beta\dot{x} + x = 0$ with $\beta = \mu/(\sqrt{km})$, a parameter which characterizes the dynamics. Notice also that the limit case $\alpha \to 0$ corresponds to the Coulomb friction equation (7).

We recall that, even if the nonlinear term $|\dot{x}|^{\alpha-1}\dot{x}$ is not a Lipschitz continuous function of $\dot{x}$, the existence and uniqueness of solutions of the associate Cauchy problem

$$(P_\alpha)\begin{cases} \ddot{x}+|\dot{x}|^{\alpha-1}\dot{x}+x=0 & t>0, \\ x(0)=x_0,\ \dot{x}(0)=v_0 \end{cases}$$

is well known in the literature: see, e.g., Brezis [39]. The asymptotic behavior, for $t \to \infty$, of solutions of the Coulomb and linear problems (P_0) and (P_1) (limit cases when $\alpha \to 0$ or $\alpha \to 1$) is well known. In the second case the decay is exponential. In the first one, as mentioned in Section 2, given x_0 and v_0 there exist a finite time $T = T(x_0, v_0)$ and $\zeta \in [-1, 1]$ such that $x(t) \equiv \zeta$ for any $t \geq T(x_0, v_0)$. When $\alpha \in (0, 1)$ it is well known that the solutions of (P_α) verify that $(x(t), \dot{x}(t)) \to (0, 0)$ as $t \to \infty$ (see, e.g., Haraux [70]).

In a series of papers ([61], [62] and [25]) it was shown that the generic asymptotic behavior above described for the limit case (P_0) is only exceptional for the sublinear case $\alpha \in (0, 1)$ since the generic orbits $(x(t), \dot{x}(t))$ decay to $(0, 0)$ in a infinite time and only two one-parameter families of them decay to $(0, 0)$ in a finite time: in other words, when $\alpha \to 0$ the exceptional behavior becomes generic.

In order to present the main arguments of those papers, we can start with some formal results via asymptotic arguments. We can rewrite the equation (23) as the planar system

$$\begin{cases} \dot{x}=y \\ \dot{y}=-x-|y|^{\alpha-1}y \end{cases} \tag{24}$$

which, by eliminating the time variable, for $y \neq 0$, leads to the differential equation of the orbits in the phase plane

$$y_x = \frac{-x-|y|^{\alpha-1}y}{y} \tag{25}$$

and that allows us to carry out a phase plane description of the dynamics.

The plane phase is antisymmetric since if $y = \sigma(x)$ is a solution of (25) then the function $y = -\sigma(-x)$ is also solution. So, it is enough to describe a semiplane (for instance $x \geq 0$). By multiplying by x and y, respectively, we get that $(x^2+y^2)_t = 2|y|^{\alpha+1}$ and $(1/x, 1/y)$ satisfy a system which has the point $(0, 0)$ as a spiral unstable point. For values of $x^2+y^2 >> 1$ the orbits of the system are given, in first approximation, by $x^2 + y^2 = C$ because $|y|^{\alpha-1}y$ is small compared with x. The effect of this term is to decrease slowly C with time giving the trajectory a spiral character. Notice that for $\alpha = 1$ the character of the trajectories close to the origin depends on the parameter β: for $\beta > \beta_c := 2$ the origin is a stable mode and for $\beta < \beta_c$ is a stable spiral (underdamped oscillations).

As we shall see, there are two modes of approach to the origin and so that the origin $(0, 0)$ is a node for the system (24). The lines of zero slope are given by

$$-x = |y|^{\alpha-1}y. \tag{26}$$

So the convergence to $(0,0)$ is only possible through the regions $\{(x,y) : x > 0,\ y < -x^{1/\alpha}\} \cup \{(x,y) : x < 0,\ y > (-x)^{1/\alpha}\}$. Let us see (formally) that the "ordinary" mode corresponds to orbits that are very close to the ones corresponding to small effects of the inertia. Due to the symmetry it is enough to describe this behavior for the orbits approaching the origin with values of $x > 0$ and $y < 0$. Let $-y = \widetilde{y} > 0$. Equation (25) takes the form

$$\widetilde{y}\widetilde{y}_x = -x + \widetilde{y}^{\alpha}. \tag{27}$$

The line of zero slope is $\widetilde{y} = x^{1/\alpha}$ and we search for orbits obeying, for $0 < x << 1$, to the expression $\widetilde{y} = x^{1/\alpha} + z(x)$ for some function $z(x)$. If we anticipate the condition $0 < z(x) << x^{1/\alpha}$, equation (25) takes the "linearized form" $\frac{1}{\alpha}x^{(\frac{1}{\alpha}-1)}z + x^{\frac{1}{\alpha}}z_x - \alpha x^{(1-\frac{1}{\alpha})}z = 0$. Thus the first term can be neglected, compared with the last one, and then the solution can be written as $z(x) \sim C\exp\{-[\alpha^2/2(1-\alpha)]x^{-\frac{2(1-\alpha)}{\alpha}}\}$ with C an arbitrary constant (which explain the name of "ordinary" orbits). This type of orbits are given, close to the origin, by the approximate equation (26), which for the orbits that reach the origin from below implies that $\widetilde{y} \sim x^{1/\alpha} \sim -\frac{dx}{dt}$ and so, integrating the simplified equation

$$\frac{dx}{dt} = -x^{1/\alpha} \tag{28}$$

we get that

$$x(t) \sim [\frac{\alpha}{(1-\alpha)(t+t_1)}]^{\alpha/(1-\alpha)} \tag{29}$$

and so that it takes an infinite time to reach the origin.

Some different orbits approaching the origin can be found by searching among solutions with large values of $|y|$ compared with $|x|^{1/\alpha}$. Thus, close to the origin, the orbits with negative y are "very close" to the solutions of the equation found by replacing (27) by the simplified the equation

$$\widetilde{y}\widetilde{y}_x = \widetilde{y}^{\alpha} \tag{30}$$

corresponding to a balance of inertia and damping. The solution ending at the origin ($\widetilde{y}(0) = 0$) is given by

$$\widetilde{y}(x) = -\{(2-\alpha)x\}^{1/(2-\alpha)}. \tag{31}$$

Notice that it involves no arbitrary constant. So this curve is unique (a symmetric curve arises for $y > 0$ and $x < 0$) which justifies the term of "extraordinary" orbit. The time evolution of this orbit is given, for $x << 1$, by integrating the equation

$$-\frac{dx}{dt} = [(2-\alpha)x]^{1/(2-\alpha)} \tag{32}$$

and so

$$x(t) = \frac{1}{(2-\alpha)}[\frac{(2-\alpha)(1-\alpha)}{2\alpha}(t_0-t)_+]^{(2-\alpha)/(1-\alpha)},$$

where, in general, $h(t)_+ = \max\{0, h(t)\}$. This indicates that the motion (of this approximated solution) ends at a finite time, t_0, determined by the initial conditions which, by (32) must satisfy that $v_0 \sim \pm[(2-\alpha)\,|x_0|]^{1/(2-\alpha)}$. We point out that the two exceptional orbits emanating from the origin spiral around the origin when $x^2 + y^2$ grows toward infinity and so each of them is a separatrix curve in the phase plane. Notice that due to the autonomous nature of the equation, if $x(t)$ is the solution of the Cauchy problem (P_α) of initial data (x_0, v_0) then for any parameter $\tau \geq 0$ the function $\widetilde{x}(t) := x(t+\tau)$ coincides with the solution of (P_α) of initial data $(x(\tau), \dot{x}(\tau))$. In this way, the above extraordinary orbits give rise to two curves of initial data for which the corresponding solutions of (P_α) vanish after a finite time.

In order to make more rigorous the arguments on the extraordinary orbits we can use a fixed point argument to show that there exists two curves Γ_+ and Γ_- of initial data (x_0, v_0) for which the solutions $x(t)$ of the corresponding Cauchy problem (P_α) vanish after a finite time.(see [61]). Moreover, we can obtain some additional results on these two curves ([60]):

(i) Near the origin the curves Γ_+ and Γ_- can be represented by two functions, $y = \psi_+(x)$ and $y = \psi_-(x)$, solutions of the equation (25), where $\psi_+ : [0, \varepsilon] \to (-\infty, 0]$ and $\psi_- : [-\varepsilon, 0] \to [0, +\infty)$, for some $\varepsilon > 0$.

(ii) Functions ψ_+ and ψ_-, satisfy that $\psi_\pm(0) = 0$

$$-\infty < \int_0^\varepsilon \frac{ds}{\psi_+(s)} \text{ and } \int_{-\varepsilon}^0 \frac{ds}{\psi_-(s)} < +\infty. \tag{33}$$

In particular, $\psi_+'(x) \downarrow -\infty$ when $x \downarrow 0$ and $\psi_-'(x) \uparrow +\infty$ when $x \uparrow 0$.

(iii) We have

$$\begin{aligned} -Cx^{\frac{1}{2-\alpha}} &\leq \psi_+(x) \leq -x^{\frac{1}{\alpha}} \text{ for } x \in [0, \varepsilon] \text{ and} \\ (-x)^{\frac{1}{\alpha}} &\leq \psi_-(x) \leq C(-x)^{\frac{1}{2-\alpha}} \text{ for } x \in [-\varepsilon, 0], \end{aligned}$$

for some $C > 0$.

(iv) There exists a $x_s \in (0, \varepsilon]$ such that $\psi_+(x_s) = -(x_s)^{\frac{1}{\alpha}}$ and $(-x_s)^{\frac{1}{\alpha}} = \psi_-(x_s)$. Moreover the regions $D_+ := \{(x, y) \colon x \in [0, x_s] \text{ and } \psi_+(x) \leq y \leq -x^{\frac{1}{\alpha}}\}$, $D_- := \{(x, y) \colon x \in [-x_s, 0] \text{ and } (-x)^{\frac{1}{\alpha}} \leq y \leq \psi_-(x)\}$ are time invariants for equation (25).

Curiously, the rigorous proof of the other decay to zero (now in an infinite time) is harder than for the exceptional case. Although an invariant subset M (such that each orbit meeting M needs infinite time to reach the rest) was given in [60], a more careful study of this case was carried out in [25] for an equation slightly more general

$$\ddot{x} + \varphi(\dot{x}) + x = 0, \tag{34}$$

where the friction function φ is assumed now to be continuous on $\mathbb{R}$ and locally Lipschitz continuous on $\mathbb{R} \setminus \{0\}$, odd, positive for positive arguments and such that

$\varphi(\eta) = a\eta^\alpha$ for $0 < \eta \leq \varepsilon$, for some $a, \varepsilon > 0$ and $\alpha \in (0,1)$. In addition we assume that

$$\lim_{\eta \to \infty} \varphi(\eta)/\eta < \frac{1}{2}. \tag{35}$$

It is clear that the above case $\varphi(\eta) = |\eta|^{\alpha-1}\eta$ is included as well as some nonmonotone functions such as

$$\varphi(\eta) = \begin{cases} a\eta^\alpha, & 0 \leq \eta \leq \varepsilon, \\ 1 + 1/\eta, & \varepsilon < \eta < \infty, \\ -\varphi(-\eta), & -\infty < \eta < 0, \end{cases}$$

where $a = \varepsilon^{-1-\alpha} + \varepsilon^{-\alpha}$. This nonlinearity appears in the literature since the pioneering work of Lord Rayleigh (see [73], [80]). The continuous approximations of the Coulomb equation also leads often to such type of formulations.

As in the previous case, is easy to see that for any $R > 0$, the disc

$$\bar{\mathbb{B}}_R := \{ (x,y) \in \mathbb{R}^2 \ ; \ x^2 + y^2 \leq R^2 \}$$

is positively invariant and the origin is globally asymptotically stable. In [25] we show, by elementary techniques from the theory of ordinary differential equations, that problem (34) possesses also at least two orbits converging to the rest state in finite time. Furthermore, we give a rather detailed analysis of the flow in the neighborhood of the critical point, showing that most orbits reach it in infinite time only. In fact, on the basis of our results we conjecture that there are precisely two orbits converging to zero in finite time.

By the oddness of φ, the phase portrait of

$$\begin{cases} \dot{x} = y \\ \dot{y} = -x - \varphi(y) \end{cases} \tag{36}$$

is, again, invariant under the reflection $(x,y) \mapsto (-x,-y)$. Thus it suffices to study the semiflow on the closure of the half-plane $\mathbb{H} := \{ (x,y) \in \mathbb{R}^2 \ ; \ y > 0 \}$, induced by restriction from the flow generated by 36. Consequently, in the rest of this section all assertions on invariant regions etc. pertain to this semiflow. We refer to [6] for the elementary facts about semiflows and invariant regions.

From (36) we see that every orbit meets the real axis off zero vertically. For $0 < s \leq 1$ we denote by Γ_s the curve

$$\Gamma_s := \{ (x,y) \in \mathbb{H} \ ; \ -x = s\varphi(y) \},$$

oriented so that y is decreasing. Then every orbit in $\mathbb{H}$ meets $\Gamma := \Gamma_1$ horizontally, and the direction field has a positive x-component and a negative y-component on the right of Γ. Thus no orbit approaches the origin through the first quadrant. Furthermore, given any $\eta > 0$,

$$\{ (x,y) \in \mathbb{H} \ ; \ -x \leq \varphi(y), \ y \leq \eta \}$$

is positively invariant.

We denote by $C(x,y)$ the solution curve of (36) passing through $(x,y) \in \mathbb{H}$ and being oriented in positive time direction. It is not difficult to prove $C(x,y)$ meets the positive y-axis whenever $(x,y) \in \Gamma_s$ for some $s \in (0,1]$.

In order to built a sharper invariant region, for small values of $s > 0$, we define the functions

$$k_1(s) := -s\varphi(k_2(s)) \text{ and } k_2(s) := \lambda(s)^\beta$$

with

$$\lambda(s) := \alpha a^2 s(1-s) \text{ and } \beta := \frac{1}{2}(1-\alpha).$$

Then we define the oriented curve $K := K_\varepsilon$ in $\mathbb{H}^-$ by means of the parametrization

$$(0, s_\varepsilon] \to \mathbb{H}^-, \quad s \mapsto k(s) := \big(k_1(s), k_2(s)\big),$$

for some suitable $s_\varepsilon > 0$. Then, by construction

$$k(s) \in \Gamma_s, \qquad 0 < s \leq s_\varepsilon,$$

and the positive unit tangent vectors to $C(x,y)$ and Γ_s at $(x,y) = k(s)$ coincide for $0 < s \leq s_\varepsilon$. In the following picture we depict $\mathbb{H}^-$ in the neighborhood of the origin together with K, Γ, and Γ_s for three values of s, as well as seven orbits for problem (36) with $\alpha = 0.4$ and $\varepsilon = 1.2$.

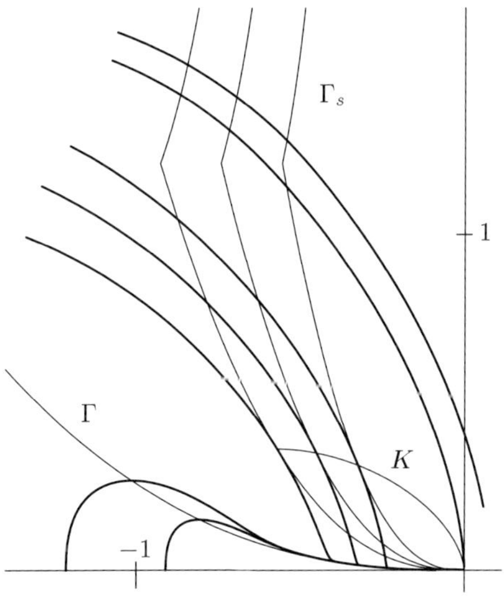

Then we prove in [4] that

a) the region M in $\mathbb{H}^-$, bounded by $\big\{ \big(k_1(s_\varepsilon), \eta\big) \; ; \; 0 < \eta \leq k_2(s_\varepsilon) \big\}$ and K, is positively invariant,
b) there exists a positive semiorbit in $\mathbb{H}^-$ reaching the origin in finite time and
c) every orbit meeting M needs infinite time to reach the origin.

The main theorem of [4] gives a rather precise picture of the flow on $\mathbb{R}^2$ generated by (36)

Theorem 3. *The origin is the only critical point of the phase flow of* (36), *and it is globally asymptotically stable. More precisely, given any* $R > 0$, *the disc* $\bar{\mathbb{B}}_R$ *is positively invariant and every orbit converges to the rest state. The phase portrait is symmetric with respect to the origin. There exist real numbers satisfying* $0 < \overline{x} \leq \widehat{x}$ *with the following:*

(i) *Every orbit meets* $[-\widehat{x}, \widehat{x}\,] \times \{0\}$.

(ii) *If an orbit meets* $[-\widehat{x}, 0) \times \{0\}$, *resp.* $(0, \widehat{x}\,] \times \{0\}$, *then it stays in* $\mathbb{H}^-$, *resp.*

$$(-\mathbb{H})^+ := \{\, (x,y) \in \mathbb{R}^2 \ ; \ x > 0, \ y < 0\,\}.$$

If an orbit meets $(-\infty, -\widehat{x}\,) \times \{0\}$, *resp.* $(\widehat{x}, \infty) \times \{0\}$, *then it leaves* $\mathbb{H}^-$, *resp.* $(-\mathbb{H})^+$, *after finite time.*

(iii) *The orbits through* $(-\widehat{x}, 0)$ *and* $(0, \widehat{x}\,)$ *reach the rest point in finite time.*

(iv) *Every orbit meeting* $\big[(-\overline{x}, 0) \cup (0, \overline{x})\big] \times \{0\}$ *needs infinite time to reach the critical point.*

In the next picture a portion of the phase plane of problem (36) is depicted where φ is given as in (35) with $\alpha = 0.4$ and $\varepsilon = 1.2$.

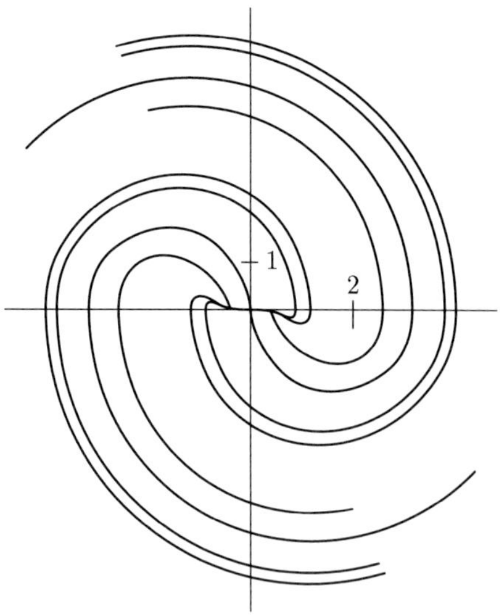

In the mentioned paper it was conjectured that there are precisely two orbits reaching the origin in finite time. This was proved in the paper Vázquez [82] where other related equations were also considered.

5. Final conclusions

The methods and results of most of the results presented here have many common points with the exposition made by H. Amann ([24]) in occasion of his nomination as Academico Extranjero de la Real Academia de Ciencias de España. At the Introduction of his lecture he says:

"In the following, I shall try to give an idea of some of my research interests of the last twenty years. During that period I was predominantly concerned with a functional-analytical approach to parabolic evolution equations. In my opinion, functional analysis, combined with so-called "hard analysis" and many other mathematical subjects, is particularly well suited for providing an abstract, powerful, and sufficient general framework for the study of nonlinear partial differential equations."

Of course, it is well known that there are intimate connections between functional analysis and partial differential equations. In fact, large parts of linear functional analysis have been developed in order provide the abstract tools for an efficient and unified study of linear partial differential equations. The point I want to make is that functional analysis is also very useful for the investigation of nonlinear different equations.

The central exposition was concerning some nonlinear systems for which he explained some of his many works (he mentioned, in particular, [7], [8], [10], [11], [11], [13], [14], [15], [9], [12], [16], [17], [19], [20], [21], [22], [23] among them. At the end of his paper he included some final remarks: a splendid sample of the beauty and depth reached by the Mathematical Analysis and the Applied Mathematics of the last third of the last century and beginnings of the present one thanks to his immense work.

"In order to summarize and to give a somewhat broader view I discuss now some of the interrelationships of the theory of parabolic evolution equations with other fields of analysis, as indicated in the following diagram.

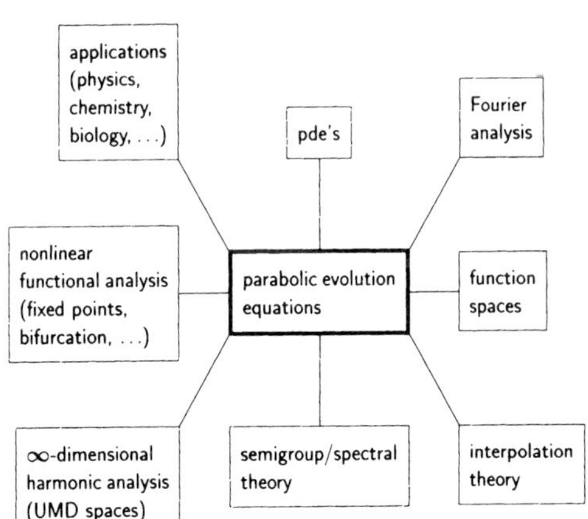

I have put parabolic evolution equations in the middle, since they are in the center of my present interest, and have grouped around them several other subjects. I did not put arrows on the connecting lines since in many cases the interaction is bilateral.

Let us start at the left upper corner. It is well known – and I have taken reaction-diffusion systems as an example – that many concrete models for the understanding of phenomena in science lead to parabolic evolution equations and, vice versa, results on parabolic evolution equations have immediate interpretations and consequences for those models.

Partial differential equations are, of course, intimately connected with parabolic evolution equations. However, neither forms a subfield of the other. For example, parabolic evolution equations encompass also other systems like integro-differential equations or infinite systems of reaction-diffusion equations involving even uncountably many unknowns, as they occur in statistical physics (see [15], [23]).

The connection between parabolic evolution equations and Fourier analysis lies on a more technical level and can be described adequately by more detailed explanations only.

As pointed out earlier, the choice of the correct state space is fundamental when studying partial differential equations, parabolic evolution equations in particular. The well-developed theory of function spaces provides us with a wide variety of possibilities. Spaces more refined than integer order Sobolev spaces like Besov and Bessel potential spaces have become increasingly important during the last years. This is true, in particular, in the study of the Navier-Stokes equations (cf. [16], [19], [20], and the references therein).

Interpolation theory provides us, on a more abstract level, with the right tool for measuring very precisely regularity properties which are the key to a successful approach to nonlinear equations.

As I have explained, semigroup theory is precisely what is needed – on the abstract level – to derive the most general local existence theory for parabolic evolution equations. Spectral theory comes in when one starts to study stability questions and the long-time behavior.

I could not into detail on the relation between parabolic evolution equations and infinite-dimensional harmonic analysis. Among other things, it has to do with "maximal regularity" questions and has, in particular during the last few years, stimulated much research in Banach space theory. I only want to mention recent results of Weis [83], Kalton [74], and others on Fourier multiplier theorems with operator-valued symbols in vector-valued Lp-spaces. Those results are tied to the theory of UMD spaces, for example. (Also see [48], [49] and, for results valid for arbitrary Banach spaces, [16], [17].)

Finally, methods from nonlinear functional analysis, fixed point theorems, bifurcation theory, etc., play an important role in the difficult and fascinating investigation of qualitative properties of the semiflows generated by parabolic evolution equations.

I hope that this enumeration of subjects, which is far from being complete, shows that the field of parabolic evolution equations is a fascinating one, invoking a lot of deep and beautiful mathematics."

Acknowledgments

Research partially supported by project REN2003-0223-C03 of the DGISGPI (Spain) and RTN HPRN-CT-2002-00274 of the EC.

References

[1] S. Adly, H. Attouch and A. Cabot, Finite time stabilization of nonlinear oscillators subject to dry friction, to appear in *Progresses in Nonsmooth Mechanics and Analysis* (edited by P. Alart, O. Maisonneuve and R.T. Rockafellar), Advances in Mathematics and Mechanics, Kluwer.

[2] F. Andreu, V. Caselles, J.I. Díaz and J.M. Mazón, Some Qualitative properties for the Total Variation, *Journal of Functional Analysis*, **188**, 516–547, 2002.

[3] F. Andreu, V. Caselles, J. M. Mazón and S. Moll, The Minimizing Total Variation Flow with Measure Initial Conditions, To appear in *ARMA*.

[4] S.N. Antonsev and J.I. Diaz, *New results on space and time localization of solutions of nonlinear elliptic or parabolic equations via energy methods*, Soviet Math. Dokl. **303** 524–528 (1988).

[5] S.N. Antontsev, J.I. Díaz and S.I. Shmarev, *Energy Methods for Free Boundary Problems: Applications to Nonlinear PDEs and Fluid Mechanics*, Progress in Nonlinear Differential Equations and Their Applications, 48, Birkhäuser, Boston, 2002.

[6] H. Amann, *Ordinary Differential Equations,* Walter de Gruyter, Berlin, 1990.

[7] H. Amann, Dynamic theory of quasilinear parabolic equations - II. Reaction-diffusion systems. *Diff. Int. Equ.*, **3**, 1990, 13–75.

[8] H. Amann, Hopf bifurcation in quasilinear reaction-diffusion systems. In S. Busenberg, M. Martelli, editors, *Delay Differential Equations and Dynamical Systems*, Proceedings, Claremont 1990, pages New York, Lecture Notes in Math. #1475. Springer Verlag. 1991, 53–63.

[9] H. Amann, Nonhomogeneous linear and quasilinear elliptic and parabolic boundary valué problems. In H.J. Schmeisser, H. Triebel, editors, *Function Spaces, Differential Operators and Nonlinear Analysis*, Teubner, Stuttgart, Leipzig. 1993, 9–126.

[10] H. Amann, Stability of the rest state of a viscous incompressible fluid. *Arch. Rat. Mech. Anal*, **126**, 1994, 231–242.

[11] H. Amann, Heat-conducting incompressible viscous fluids. In A. Sequeira, editor, *Navier-Stokes Equations and Related Non-Linear Problems*, Plenum Press, New York, 1995, 231–243.

[12] H. Amann, Linear and Quasilinear Parabolic Problems, Volume I: Abstract Linear Theory. Birkhäuser, Basel. 1996.

[13] H. Amann, Stability and bifurcation in viscous incompressible fluids. *Zapiski Nauchn. Seminar. POMI*, **233**, 1996, 9–29.

[14] H. Amann, Operator-valued Fourier multipliers, vector-valued Besov spaces, and applications. *Math. Nachr.*, **186**, 1997, 5–56.

[15] H. Amann, Coagulation-fragmentation processes. *Arch. Rat. Mech. Anal*, **151**, 2000, 339–366.

[16] H. Amann, On the strong solvability of the Navier-Stokes equations. *J. Math. Fluid Mech.*, **2**, 2000, 16–98.

[17] H. Amann, Elliptic operators with infinite-dimensional state spaces. *Evol. Equ.*, **1**, 2001, 143–188

[18] H. Amann, Linear parabolic problems involving measures.*Rev. R. Acad. Cien. Serie A Matem.* RACSAM, **95**, 2001, 85–119 (http://racefyn.insde.es/Publicaciones/racsam/artículos/racsam%2095_1/Amann.pdf)

[19] H. Amann, Nonhomogeneous Navier-Stokes equations in spaces of low regularity. *Quaderni di matemática*, **12** (2002), 13–31.

[20] H. Amann, Nonhomogeneous Navier-Stokes equations with integrable low-regularity data. In *Nonlinear Problems in Mathematical Physics and Related Topics II. In Honor of Professor O.A. Ladyzhenskaya* (M.S. Birman, St. Hildebrandt, V.A. Solonnikov, N.N. Uraltseva, eds.), Kluwer/Plenum, in print 2002.

[21] H. Amann and P. Quittner, Elliptic boundary valué problems involving measures: existence, regularity, and multiplicity. *Advances in Diff. Equ.*, **3**, 1998, 753–813.

[22] H. Amann and P. Quittner, Semilinear parabolic equations involving measures and low-regularity data, *Trans. AMS*, **356** (2004), 1045–1119.

[23] H. Amann, and F. Weber, On a quasilinear coagulation-fragmentation model with diffusion. *Adv. Math. Sel. Appl.*, **11**, 2001, 227–263.

[24] H. Amann, Abstract methods in differential equations, *Rev. R. Acad. Cien. Serie A Matem.* RACSAM, **97** (1), 2003, 89–105, (http://racefyn.insde.es/Publicaciones/racsam/artículos/racsam%2097_1/2002-20-Amann.pdf)

[25] H. Amann and J.I. Diaz, A note on the dynamics of an oscillator in the presence of strong friction, *Nonlinear Anal.* **55**, 209–216 (2003)

[26] H. Amann and J.I. Díaz, Unpublished. Madrid, October 1988.

[27] L. Alvarez and J.I. Díaz, The waiting time property for parabolic problems trough the nondiffusion of the support for the stationary discretization problems. *Rev. R. Acad. Cien. Serie A Matem.* RACSAM, **97** (1), 2003, 83–88. (http://racefyn.insde.es/Publicaciones/racsam/artículos/racsam%2097_1/2002-7-Alvarez-Diaz.pdf)

[28] A. Bamberger, Etude d'une équation doublement non linéaire,*J. Funct. Anal.* **24** 148–155 (1977).

[29] A. Bamberger and H. Cabannes, *Mouvements d'une corde vibrante soumise à un frottement solide*, C. R. Acad. Sc. Paris, **292**, (1981), 699–702.

[30] I. Bejenaru, J.I. Díaz and I.I. Vrabie, An abstract approximate controllability result and applications to elliptic and parabolic systems with dynamical boundary conditions, *Electr. J. Diff. Eqns.* 2001 (No. 50), 2001, 1–19.

[31] Y. Belaud and J.I. Díaz: Finite extinction time for solutions of some parabolic equations with a singular term. Lecture at *Nonlinear Partial Differential Equations describing Front Propagation and other Singular Phenomena*, Leiden, 8–12 November, 2004, and paper in preparation.

[32] Ph. Bénilan and M. Crandall, The continuous dependence on φ of solutions of $u_t - \Delta\varphi(u) = 0$, *Indiana Univ. Math. J.*, **30**, 1981, 161–177.

[33] Ph. Benilan and P. Wittbold, On mild and weak solutions of elliptic-parabolic problems, *Adv. Diff. Eq.*, **1**, 1996, 1053–1073.

[34] F. Bernis, Finite speed of propagation and asymptotic rates for some nonlinear higher-order parabolic equations with absorption, *Proc. Roy. Soc. Edinburgh Sect. A*, **104**, 1986, 1–19.

[35] R. Bermejo, J.I. Díaz and L. Tello, Mathematical and Numerical Analysis of a Nonlinear Diffusive Climatological Energy Balance Model, to appear.

[36] L. Boccardo, D. Giachetti, J.I. Díaz and F. Murat, Existence of a solution for a weaker form of a nonlinear elliptic equation, in *Recent Advances in Nonlinear Elliptic and Parabolic Problems*, Pitman Res. Notes Math., Ser. 208, Longman Sc. Tech., Harlow, (1989), 229–246.

[37] D. Bothe, Periodic solutions of non-smooth friction oscillations, *Z. angew. Math. Phys.* **50**, 1990, 779–808.

[38] D. Bothe and J.I. Díaz. Article in progress.

[39] H. Brézis, Problèmes unilatéraux, *J. Math. Pures Appl.* **51**, 1–168 (1972).

[40] H. Brezis, *Opérateurs maximaux monotones et semigroupes de contractions dans les espaces de Hilbert*, North-Holland, Amsterdam 1972.

[41] H. Brezis, Monotone operators, nonlinear semigroups and applications. *Proceedings of the International Congress of Mathematicians* (Vancouver, B. C., 1974), Vol. **2**, Canad. Math. Congress, Montreal, Que., 1975, 249–255.

[42] H. Cabannes, Mouvement d'une corde vibrante soumise à un frottement solide, *C. R. Acad. Sci. Paris Sér.* A-B 287 (1978), 671–673.

[43] H. Cabannes, *Study of motions of a vibrating string subject to a solid friction, Math. Meth. in the Appl. Sci.*, **3**, 1981, 287–300.

[44] A. Cabot, Stabilization of oscillators subject to dry friction: finite time convergence versus exponential decay results, to appear.

[45] J.M. Carlson and J.S. Langer, Properties of Earthquakes Generated by Fault Dynamics, *Physical Review Letters*, **62**, 1989, 2632–2635.

[46] J. Carrillo and P. Wittbold, Uniqueness of renormalized solutions of degenerate elliptic-parabolic problems, *J. Diff. Eq.*, **156**, 1999, 93–121.

[47] K. Deimling, G. Hetzer and W.X. Shen, Almost periodicity enforced by Coulomb friction, *Adv. Differential Equations*, **1**, 1996, 265–281.

[48] R. Denk, G. Dore, M. Hieber, J. Prüss and A. Venni, New thoughts on old results of R.T. Seeley, preprint. 2002.

[49] R. Denk, G. Dore, M. Hieber and J. Prüss, $\mathcal{R}$-boundedness, Fourier multipliers and problems of elliptic and parabolic type, preprint. 2001.

[50] J.I.Díaz, Anulación de soluciones para operadores acretivos en espacios de Banach. Aplicaciones a ciertos problemas parabólicos no lineales. *Rev. Real. Acad. Ciencias Exactas, Físicas y Naturales de Madrid*, Tomo **LXXIV**, 865–880, 1980.

[51] J.I. Díaz, *Nonlinear Partial Differential Equations and Free Boundaries.* Research Notes in Mathematics n° 106, Pitman, London 1985.

[52] J.I. Díaz, Qualitative study of the total variation flow and a related formulation for Bingham fluids. Lecture at the Université Pierre et Marie Curie, June, 13, 2001.

[53] J.I. Díaz, Energy methods for free boundary problems: new results and some remarks on numerical algorithms. To appear in *Esaim Proc.*

[54] G. Díaz and J.I. Díaz Finite extinction time for a class of nonlinear parabolic equations. *Comm. Part. Diff. Equat.* **4**, No. 11, 1213–1231, 1979.

[55] J.I. Díaz and R. Jimenez, Behaviour on the boundary solutions of parabolic equations with nonlinear boundary conditions: the parabolic Signorini problem. Lecture at the VIII CEDYA, Univ. de Santander, 1986. Appeared in *Análisis no lineal*, R.F. Jiménez and H. Mennickent Eds., Concepción (Chile), 69–80, 1986.

[56] J.I. Díaz and R. Jiménez, Boundary behaviour of solutions of the Signorini problem: The elliptic case, *Bolletino dell'Unione Matemática Italiana*, **7**, No. 2 B, 1988, 127–139.

[57] J.I. Díaz and M.A. Herrero, Propertiés de support compact pour certaines équations elliptiques et paraboliques non linéaires, *C. R. Acad. Sci. Paris* Sér. *I Math.* **286**, 815–817, 1978.

[58] J.I. Díaz and M.A. Herrero, Estimates on the support of the solutions of some nonlinear elliptic and parabolic problems, *Proc. Roy. Soc. Edinburgh*, **89A**, 249–258, 1981.

[59] J.I. Díaz and G. Hetzer, Article in preparation.

[60] J.I. Díaz and A. Liñán, Movimiento de descarga de gases en conductos largos: modelización y estudio de una ecuación doblemente no lineal. In the book *Reunión Matemática en Honor de A. Dou* (J.I. Díaz and J.M. Vegas eds.) Universidad Complutense de Madrid, 1989.

[61] J.I. Díaz and A. Liñán, On the asymptotic behavior of solutions of a damped oscillator under a sublinear friction term: from the exceptional to the generic behaviors. In.the *Proceedings of the Congress on nonlinear Problems* (Fez, May 2000), Lecture Notes in Pure and Applied Mathematics (A. Benkirane and A. Touzani. eds.), Marcel Dekker, New York, 2001, 163–170 .

[62] J.I. Díaz and A. Liñán, On the asymptotic behaviour of solutions of a damped oscillator under a sublinear friction term, *Rev. R. Acad. Cien.Serie A Matem.* (RACSAM), **95**, 2001, 155–160. (http://racefyn.insde.es/Publicaciones/racsam/comunicaciones/vol95_1/DiazLinan.pdf)

[63] J.I. Díaz and V. Millot, Coulomb friction and oscillation: stabilization in finite time for a system of damped oscillators. *CD-Rom Actas XVIII CEDYA / VIII CMA*, Servicio de Publicaciones de la Univ. de Tarragona 2003.

[64] J.I. Díaz and V. Millot, Article in preparation.

[65] J.I.Díaz and J. Mossino, Isoperimetric inequalities in the parabolic obstacle problems. *Journal de Mathématiques Pures et Appliqueés*, Vol. 71, 233–266. 1992.

[66] J.I. Díaz, A.M. Ramos, Some results about the approximate controllability property for quasilinear diffusion equations, *Comptes Rendus Acad. Sci. Paris*, **324**, Série I, 1997, 1243–1248.

[67] J.I. Díaz and J.E. Saa, Existence et unicité de solutions positives pour certaines équations elliptiques quasilinéaires, *Comptes Rendus Acad. Sc. Paris,* t. **305**, Série I, 1987, 521–524.

[68] J.I. Díaz and F. de Thelin, On a nonlinear parabolic problems arising in some models related to turbulence flows, *SIAM J. Math. Anal.* **25**, 1994, 1085–1111.

[69] G. Duvaut and J. L. Lions, *Les Inéquations en Mécanique et en Physique*, Dunod, Paris, 1972.

[70] A. Haraux , Comportement à l'infini pour certains systèmes dissipatifs non linéaires, *Proc. Roy. Soc. Edinburgh, Sect. A* **84A**, 1979, 213–234.

[71] A. Haraux, *Systèmes dynamiques et applications*, Masson, Paris, 1991.

[72] M. Herrero and J.L. Vázquez, Asymptotic behaviour of the solutions of a strongly nonlinear parabolic problem, *Ann. Fac. Sci. Toulouse* **3**, 1981, 113–127.

[73] D.W. Jordan and P. Smith, *Nonlinear Ordinary Differential Equations*, (Second Edition), Clarendon Press, Oxford 1979.

[74] N. Kalton, Applications of Banach space theory to sectorial operators. In *Recent progress in functional analysis* (Valencia, 2000), Amsterdam. North-Holland Math. Stud. 189, North-Holland, 2001, 61–74.

[75] M. Kunze, *Non-Smooth Dynamical Systems,* Lecture Notes in Mathematics 1744, Springer, Berlin, 2000.

[76] M.N. Le Roux, Semi-discretization in Time of a Fast Diffusion Equation, *Journal of Mathematical Analysis and Applications*, **137**, 1989, 354–370.

[77] J.L. Lions, Quelques remarques sur les équations differentielles opérationnelles du 1er ordre, *Rend. Sem. Math. Padova*, **33**, 1963, 213–225.

[78] J.L. Lions, *Quelques Méthodes de Résolution des Problèmes aux Limites non Linéaires*, Dunod, Gauthier-Villars, Paris 1969.

[79] E. Melendez, *Estudio numérico de una ecuación parabólica doblemente no lineal con datos mixtos de contorno: Propiedad de extinción en tiempo finito de su solución.* Tésis de Licenciatura, Univ. Complutense de Madrid, 1990.

[80] J.W. Rayleigh, B. Strutt, *The theory of sound*, Dover Publications, New York, 2d ed., 1945.

[81] B. Riemann: Über die Fläche vom kleinsten Inhalt bei gegebener Begrenzung, *Abh. Königl. Ges. d. Wiss. Göttingen, Mathem. Cl.* **13**, 3–52, 1867.

[82] J.L. Vázquez, The nonlinearly damped oscillator, *ESAIM Control Optim. Calc. Var.* **9** 231–246 (2003).

[83] L. Weis, Operator-valued Fourier multiplier theorems and maximal L_p-regularity. *Math. Ann.*, **319**, 2001, 735–758.

J.I. Díaz
Departamento de Matemática Aplicada
Facultad de Matemáticas
Universidad Complutense de Madrid
E-28040 Madrid, Spain
e-mail: ji_diaz@mat.ucm.es

Progress in Nonlinear Differential Equations
and Their Applications, Vol. 64, 99–111

Spectral Projections, Riesz Transforms and H^∞-calculus for Bisectorial Operators

Markus Duelli and Lutz Weis

1. Introduction

For a bisectorial operator A with its spectrum in a double cone around the imaginary axis, the spectral projections P_1, P_2 corresponding to the upper and lower cone in general cannot be defined by the Dunford calculus and, indeed, they may be unbounded. E.g., for $A = \frac{d}{dx}$ on $L_p(\mathbb{R})$ the projection P_1 is only bounded for $1 < p < \infty$ and projects $L_p(\mathbb{R})$ onto the Hardy space H_p. However, the boundedness of these spectral projections is closely connected to some important inequalities in analysis. We study these questions in the framework of the H^∞-functional calculus, which allows us to give more direct proofs for some known results, improve on them and also to discover new connections.

E.g., it is a result of Dore and Venni that, for $0 \in \rho(A)$, the boundedness of these spectral projections is equivalent to the "symmetry" of the operators A and $-A$ in the sense that for $w \in \mathbb{C} \setminus \mathbb{Z}$ we have

$$\|A^w x\| \approx \|(-A)^w x\| \qquad \text{for } x \in \mathcal{D}(A^w) = \mathcal{D}((-A)^w).$$

In Section 3 we give a different proof using the H^∞-functional calculus that also works for $0 \notin \rho(A)$.

More importantly, the boundedness of the spectral projection is equivalent to the estimate (see Section 4)

$$\|Ax\| \approx \left\|(-A^2)^{1/2} x\right\| \qquad \text{for } x \in \mathcal{D}(A) = \mathcal{D}((-A^2)^{1/2}),$$

i.e., to the boundedness and invertibility of $A(-A^2)^{1/2}$ which equals the Hilbert transform if $A = \frac{d}{dx}$ on $L_p(\mathbb{R})$. In the case of Hilbert space and operators A with a bounded H^∞-calculus this was first shown in [2].

If $A_1, \dots, A_n$ is a commuting family of bisectorial operators, the classical example $A_i = \frac{d}{dx_i}$, $i = 1, \dots, n$, on $L_p(\mathbb{R}^n)$ motivates us to call the operators $A_i(-A)^{-1/2}$, where $A = \sum_{i=1}^n A_i^2$, Riesz transforms for $A_1, \dots, A_n$. If $A_1, \dots, A_n$

have a joint H^∞-functional calculus one can not only show that these Riesz transforms are bounded but also obtains inequalities of the form

$$\mathbb{E}\left\|\sum_{i=1}^n r_i A_i x\right\| \approx \left\|(-A)^{1/2}x\right\|, \qquad \mathbb{E}\left\|\sum_{i,j=1}^n r_{ij} A_i A_j x\right\| \approx \|Ax\|$$

where $(r_i), (r_{ij})$ are Bernoulli random variables, and which extend the classical inequalities

$$\left\|\left(\sum_{i=1}^n \left|\frac{\partial}{\partial x_i} f\right|^2\right)^{1/2}\right\|_{L_p} \approx \left\|(-\Delta)^{1/2} f\right\|_{L_p}, \quad \left\|\left(\sum_{i,j=1}^n \left|\frac{\partial^2}{\partial x_i \partial x_j} f\right|^2\right)^{1/2}\right\|_{L_p} \approx \|\Delta f\|_{L_p}$$

for $f \in L_p(\mathbb{R}^n)$ with $1 < p < \infty$. See Section 4.

For bisectorial invertible operators that generate a bounded group, it is already shown in [12] that the boundedness of the spectral projection is equivalent to the boundedness of the Hilbert transform of the group. Instead of using the *analytic generator* of the group as in [12] we give in Section 5 an alternative more direct approach to this result based on the H^∞-functional calculus. In this way we can avoid the assumption $0 \in \rho(A)$.

The H^∞-functional calculus provides a convenient and unifying framework for all results in this paper. We recall its construction shortly in Section 2.

2. H^∞-calculus for bisectorial operators

For $\omega \in (0, \pi)$ we define the sector S_ω as the set $\{re^{i\phi} : r > 0, \phi \in (-\omega, \omega)\}$. For $\omega \in (0, \pi/2)$ we define the bisector $S^\omega = \{re^{i\phi} : r > 0, |\phi \pm \pi/2| < \omega\}$. Given two sectors (bisectors) Σ_1, Σ_2 we write $\Sigma_1 < \Sigma_2$ if there are $0 < \omega_1 < \omega_2$ such that $\Sigma_k = S_{\omega_k}$ ($\Sigma_k = S^{\omega_k}$), $k = 1, 2$.

In the following X will always denote a complex Banach space. An operator $A : \mathcal{D}(A) \to X$ having dense domain and range is called *sectorial*, if there exists $\omega \in (0, \pi)$ such that $\sigma(A) \subset \overline{S_\omega}$ and $\sup\{\|zR(z, A)\| : z \notin \overline{S_{\omega'}}\} < \infty$ for all $\omega' \in (\omega, \pi)$. We will write $A \in \mathrm{Sect}(S_\omega)$ in this case. The infimum of all such ω is called the spectral angle of A and denoted by $\omega(A)$. We will say that $A \in \mathrm{Sect}(e^{i\eta} S_\omega)$ if $e^{-i\eta} A \in \mathrm{Sect}(S_\omega)$.

We will call A *bisectorial*, if there exists $\omega \in (0, \pi/2)$ such that both A and $-A$ are sectorial of angle $\pi/2 + \omega$. The infimum of all such ω is the spectral angle of the bisectorial operator A and is denoted by $\omega(A)$.

In the following we will sketch the construction of a *holomorphic functional calculus* for sectorial (bisectorial) operators. Assume that A is sectorial (bisectorial) of angle ω. Let $\Sigma = S_\omega$ ($\Sigma = S^\omega$) and $\Sigma < \Sigma'$. The contour $\Gamma = \partial\tilde{\Sigma}$, where $\Sigma < \tilde{\Sigma} < \Sigma'$, is orientated in the positive sense. In a first step we define $f(A)$ for holomorphic functions satisfying an estimate

$$|f(z)| \le C\frac{|z|^s}{(1+z)^{2s}}, \qquad z \in \Sigma' \tag{1}$$

for some $s > 0$. We denote this class by $H_0^\infty(\Sigma')$. This estimate implies that the integral

$$f(A) = \frac{1}{2\pi i}\int_\Gamma f(z)R(z,A)dz \tag{2}$$

exists as a Bochner integral in $\mathcal{L}(X)$. In a second step we consider functions in $H^\infty(\Sigma')$ (or, more generally, polynomially bounded functions). Consider the *regularizing* function $\phi(z) = z(1+z)^{-2}$. Since $\phi \in H_0^\infty(\Sigma')$ we can define $\phi(A)$ by (1) and check that it is injective with $\phi(A)^{-1} = 2I + A + A^{-1}$, and that $\mathcal{D}(\phi(A)^{-1}) = \mathcal{R}(\phi(A)) = \mathcal{D}(A) \cap \mathcal{R}(A)$. To define $f(A)$ for $f \in H^\infty(\Sigma')$ we write $f(z) = \phi(z)^{-1}(\phi(z)f(z))$. The function ϕf is in $H_0^\infty(\Sigma')$; hence we may define

$$\mathcal{D}(f(A)) = \{x \in X : (\phi f)(A)x \in \mathcal{D}(\phi(A)^{-1})\}$$
$$f(A) = \phi(A)^{-1}(\phi f)(A).$$

One can show that this definition is consistent in case $f \in H_0^\infty(\Sigma')$ and that the usual properties of a holomorphic functional calculus hold. In particular we have the *algebraic* properties for $f, g \in H^\infty(\Sigma')$

- $\mathcal{D}(f(A)) \cap \mathcal{D}(g(A)) \subset \mathcal{D}((f+g)(A))$ and $(f+g)(A)x = f(A)x + g(A)x$, for $x \in \mathcal{D}(f(A)) \cap \mathcal{D}(g(A))$
- $\mathcal{D}(f(A)g(A)) = \mathcal{D}((fg)(A)) \cap \mathcal{D}(g(A))$ and $(fg)(A)x = f(A)[g(A)x]$ for $x \in \mathcal{D}(f(A)g(A))$

and the important 'convergence' property

- If $f_n \in H^\infty(\Sigma')$ with $\sup_n \|f_n\|_\infty < \infty$ and $f_n(z) \to f(z)$ for $z \in \Sigma'$, then for all $x \in \mathcal{D}(A) \cap \mathcal{R}(A)$

$$\lim_{n\to\infty} f_n(A)x = f(A)x.$$

We say that A has a *bounded $H^\infty(\Sigma')$-calculus*, if the operator $f(A)$ is bounded for all $f \in H^\infty(\Sigma')$. In this case the map $f \in H_0^\infty(\Sigma') \mapsto f(A) \in \mathcal{L}(X)$ extends to a bounded algebra homomorphism on $H^\infty(\Sigma')$. The infimum of all angles ω' such that A has a bounded $H^\infty(\Sigma')$-calculus ($\Sigma' = S_{\omega'}$ or $\Sigma' = S^{\omega'}$) is denoted by $\omega_{H^\infty}(A)$. For details of this construction for sectorial operators, see, e.g., [7], [11, Chapter II], for bisectorial operators [5] and [6].

We call the operator A R-sectorial, if there is a $\omega \le \tilde\omega < \pi/2$ such that the set $\{zR(z,A) : z \in \mathbb{C} \setminus \overline{S_{\tilde\omega}}\}$ is R-bounded. If the set $\{zAR(z,A)^2 : z \in \mathbb{C} \setminus \overline{S_{\tilde\omega}}\}$ is R-bounded, we call A almost R-sectorial. In general, the notion of almost R-sectoriality is strictly weaker than R-sectoriality.

Following [9], we say that A is R-bisectorial (almost R-bisectorial) of angle ω, if $\pm A$ are R-sectorial (almost R-sectorial) of angle $\pi/2 + \omega$. The infimum of all such angles is denoted by $\omega_r(A)$. We will write $A \in \mathrm{RSect}(\Sigma)$ ($A \in \mathrm{aRSect}(\Sigma)$)

in this case. It is remarkable, that, if A has a bounded H^∞-calculus, we have $\omega_{H^\infty}(A) = \omega_r(A)$ (cf. [9]).

Let $\mathcal{A} \subset \mathcal{L}(X)$ be an algebra of operators that commute with the resolvents of A. We denote by $RH^\infty(\Sigma, \mathcal{A})$ the set of all $\mathcal{A}$-valued holomorphic functions on Σ that have R-bounded range. For $F \in RH^\infty(\Sigma, \mathcal{A})$ we may define $F(A)$ by (2). If A has a bounded scalar $H^\infty(\Sigma)$-calculus, then the map $F \mapsto F(A)$ extends to an *operator-valued functional calculus* mapping $RH^\infty(\Sigma', \mathcal{A})$ into $\mathcal{L}(X)$ for all $\Sigma' > \Sigma$ with analogous properties, see [5], [10] and [11] for details.

If the Banach space X has Pisier's property (α), as an application of the operator-valued functional calculus one may construct the *joint functional calculus* of resolvent commuting operators each having a bounded H^∞-calculus. If the Banach space has property (α) the joint functional calculus is bounded from $H^\infty(\Sigma_1 \times \cdots \times \Sigma_n)$ into $\mathcal{L}(X)$, see [11, Thm. 12.12] and [6] for details.

3. Spectral projections and fractional powers

In this section let A always be a bisectorial operator of angle ω. Let $\Sigma = S^\omega$ and $\Sigma' = S^{\omega'}$ for some $\omega < \omega' < \pi/2$. We assume that A is densely defined and has dense range. Recall that this implies that A is injective. We will denote by Σ_1 and Σ_1' the component of Σ and Σ' lying in the upper half-plane; similarly, Σ_2 and Σ_2' will denote the component in the lower half-plane. We say that $X = X_1 \oplus X_2$ is a *spectral decomposition* of X if X_i is invariant under A and if the restriction A_i of A to X_i is in $\mathrm{Sect}(\Sigma_i)$. In particular, $\sigma(A_i) = \sigma(A) \cap \overline{\Sigma_i}$. We will call A_i the *reduced operators.*

The following proposition gives uniqueness of such a decomposition and links it to the functional calculus. Clearly, if the operator A would not be injective, there could be no hope to obtain uniqueness, as we could choose to incorporate the kernel of A into either of the two subspaces.

3.1. Proposition.

1. *Let $X = X_1 \oplus X_2$ be a spectral decomposition of X and denote by P the canonical projection onto X_1, that is $P : x = x_1 \oplus x_2 \mapsto x_1$. Then $P = p(A)$, where $p \in H^\infty(\Sigma')$ is the indicator function of the set Σ_1'.*
2. *The spectral decomposition exists if and only if the operator $P = p(A)$ is bounded.*

Proof. 1. Since $R(z, A_i) = R(z, A)|_{X_i}$ for all $z \in \rho(A) = \rho(A_1) \cap \rho(A_2)$, it is easy to see that

$$f(A) = f(A_1) \oplus f(A_2) \tag{3}$$

with $\mathcal{D}(f(A)) = \mathcal{D}(f(A_1)) \oplus \mathcal{D}(f(A_2))$ for all $f \in H^\infty(\Sigma')$.

The fact that $A_i \in \mathrm{Sect}(\Sigma_i)$ together with Cauchy's integral theorem implies that the $H^\infty(\Sigma_i')$-functional calculus for A_i extends its $H^\infty(\Sigma')$-functional calculus. As p is constant on both Σ_1' and Σ_2' we obtain that $p(A) = p(A_1) \oplus p(A_2) = I_{X_1} \oplus 0 = P$.

2. Assume that $p(A)$ is bounded. It follows readily from elementary properties of the functional calculus that $P := p(A)$ is then a bounded projection commuting with A. Thus $X_1 = \mathcal{R}(P)$ and $X_2 = \ker P$ are A-invariant subspaces of X. We denote by A_i the restriction of A to X_i. It remains to show that $A_i \in \mathrm{Sect}(\Sigma_i)$.

First, we verify that the spectrum of A_i is situated where it should be: $\sigma(A_i) = \sigma(A) \cap \overline{\Sigma_i}$. It suffices to consider the case $i = 1$ ($i = 2$ follows by symmetry, as $p(A)$ is bounded if and only if $(1-p)(A)$ is bounded) and to show that $0 \neq \mu \in \overline{\Sigma_2}$ is in the resolvent set of A_1. Fix $\lambda \in \mathbb{C} \setminus \overline{\Sigma'}$. Then the identity

$$(\mu - z)^{-1} = \lambda\mu^{-1}(\lambda - z)^{-1} + (\lambda - \mu)\mu^{-1}z(\lambda - z)^{-1}(\mu - z)^{-1}$$

plugged into the functional calculus gives

$$(r_\mu p)(A) = (c_1 r_\lambda p + c_2 hp)(A) = c_1 R(\lambda, A)P + c_2(hp)(A), \tag{4}$$

where $h(z) = z(\lambda - z)^{-1}(\mu - z)^{-1}$, $r_c(z) = (c-z)^{-1}$, $c_1 = \lambda\mu^{-1}$ and $c_2 = (\lambda-\mu)\mu^{-1}$. As $hp \in H_0^\infty(\Sigma')$, the operator $(hp)(A)$ is bounded and hence also $(r_\mu p)(A)$. The identity $(\mu - A)(r_\mu p)(A) = p(A)$ and the fact that A and $(r_\mu p)(A)$ commute imply that $\mu \in \rho(A_1)$ and $R(\mu, A_1) = (r_\mu p)(A)|_{X_1}$.

It remains to establish the bound on the resolvent. By (4) we have

$$\|\mu R(\mu, A_1)\| \leq \|P\| \, \|\mu(r_\mu p)(A)\| \leq \|P\| \left(\|\lambda R(\lambda, A)\| + |\lambda - \mu| \, \|(hp)(A)\|\right).$$

We will fix the argument of λ and choose λ such that $|\lambda| = |\mu|$. Then the first term is bounded by our sectoriality assumption and as $|\lambda - \mu| \leq 2|\mu|$. Thus, we only have to find an upper bound for $|\mu| \, \|(hp)(A)\|$, $0 \neq \mu \in \Sigma_2'$. Denote by Γ_+ the part of the contour lying in the upper half-plane. A direct estimate gives

$$\begin{aligned} |\mu| \, \|hp(A)\| &\leq C \int_{\Gamma_+} \frac{|\mu|}{|\lambda - z| \, |\mu - z|} \, |dz| \\ &= C \int_{\Gamma_+} \frac{1}{\left|\frac{\lambda}{|\lambda|} - \frac{z}{|\mu|}\right| \left|\frac{\mu}{|\mu|} - \frac{z}{|\mu|}\right|} \frac{|dz|}{|\mu|} = C \int_{\Gamma_+} \frac{|dz|}{|e^{i\rho} - z| \, |e^{i\nu} - z|} \end{aligned}$$

where $\rho = \arg\lambda$ and $\nu = \arg\mu$, and this integral is uniformly bounded for $\mu \in \Sigma_2'$.

The resolvent estimate for A_2 follows by symmetry, as $p(A)$ is bounded if and only if $(1-p)(A)$ is bounded. □

If A has a bounded $H^\infty(\Sigma')$-functional calculus it is easy to see that the spectral projection is bounded and induces a spectral decomposition.

Conversely, the boundedness of the spectral projection connects the spectral properties of A with those of the reduced operators A_1, A_2.

3.2. Corollary. *Assume that the spectral projections of A are bounded and $\Sigma \subset \tilde{\Sigma}$. Then A has a bounded $H^\infty(\tilde{\Sigma})$-calculus [or $A \in \mathrm{RSect}(\tilde{\Sigma})$, $A \in \mathrm{aRSect}(\tilde{\Sigma})$] if and only if A_i has a bounded $H^\infty(\tilde{\Sigma}_i)$-calculus [or $A_i \in \mathrm{RSect}(\tilde{\Sigma}_i)$, $A_i \in \mathrm{aRSect}(\tilde{\Sigma}_i)$] for $i = 1, 2$.*

Proof. For the H^∞-calculus use again the first part of the proof of Proposition 3.1. For the (almost) R-sectoriality use in addition Example 2.16 in [11]. □

However, the spectral projection is not always bounded, not even in Hilbert space. In [6] is constructed a counterexample using a conditional Schauder basis. See also 4.1 below.

Now we will link the notions of fractional powers and the spectral projection.

3.3. Theorem. *The following statements are equivalent:*

1. *$P = p(A)$ is bounded.*
2. *For one (all) $w \in \mathbb{C} \setminus \mathbb{Z}$ we have $\mathcal{D}(A^w) = \mathcal{D}((-A)^w)$ and the equivalence of norms:*
$$\|A^w x\| \sim \|(-A)^w x\|, \qquad x \in \mathcal{D}(A^w).$$
3. *For one (all) $w \in \mathbb{C} \setminus \mathbb{Z}$ the operator $A^w(-A)^{-w}$ admits a bounded extension to X.*
4. *The operator $\log A - \log(-A)$ admits a bounded extension to all of X.*

If A is invertible, we may add the following equivalent statements to the list:

5. *There exists $w \in \mathbb{C} \setminus \mathbb{Z}$ with $\mathcal{R}e(w) > 0$ such that $\mathcal{D}(A^w) \supset \mathcal{D}((-A)^w)$.*
6. *For all $w \in \mathbb{C}$ with $\mathcal{R}e(w) > 0$ we have $\mathcal{D}(A^w) = \mathcal{D}((-A)^w)$.*

3.4. Remark. For invertible operators this was first shown in [5].

Proof. Let $w \in \mathbb{C}$. The function $h = h_w : z \mapsto z^w(-z)^{-w}$ is holomorphic on $\mathbb{C} \setminus \mathbb{R}$. As the sum of the exponents is zero, we have, for $z \notin \mathbb{R}$,
$$h(z) = e^{iw(\arg(z) - \arg(-z))},$$
where $\arg : \mathbb{C} \setminus (-\infty, 0] \to (-\pi, \pi]$ denotes the argument function associated to the principal branch of the logarithm used to define the complex powers z^w. Thus, h is constant on the upper and lower half-planes. It is easy to see that, if $w \notin \mathbb{Z}$, the function h takes two different values c_1 and c_2.
$$p = (c_1 - c_2)^{-1}(h - c_2).$$
Applying the H^∞-functional calculus to this identity gives immediately
$$P = p(A) = (c_1 - c_2)^{-1}(h - c_2)(A) = (c_1 - c_2)^{-1}h(A) - c_2(c_1 - c_2)^{-1}, \tag{5}$$
that is, the spectral projection is bounded if and only if $h(A)$ is bounded.

The equivalence of 1. and 2. is immediate from the identity (5) and the fact that the densely defined operator $A^w(-A)^{-w}$ is a restriction of $h(A)$ (the operator is densely defined, as its domain contains the range of $\phi(A)$). The equivalence of 2. and 3. is clear. The other statements follow analogously. We give some details concerning the last two statements. The implication 7. $\Rightarrow$ 1. is true, as the relation $\mathcal{R}((-A)^{-w}) = \mathcal{D}((-A)^w)$ and the fact that the invertibility of A implies the boundedness of $(-A)^{-w}$ tell us $\mathcal{D}(h(A)) \supset \mathcal{D}(A^w(-A)^{-w}) = X$. Finally, we show 1. $\Rightarrow$ 8.: Let P be bounded and denote by $A_k \in \mathrm{Sect}(\Sigma_k)$ the reduced operators, then $\mathcal{D}((\pm A)^w) = \mathcal{D}((\pm A_1)^w) \oplus \mathcal{D}((\pm A_2)^w)$ (compare (3)). Now it suffices to observe that, as h is constant on Σ_k', we have $h(A_k) = c_k \neq 0$, hence $A_k^{-w} = c_k(-A_k)^{-w}$. In particular, the ranges agree, which proves the claim. $\square$

4. Boundedness of Riesz transforms

Our results are motivated by the following well-known example.

4.1. Example. Let $p \in [1, \infty)$ and define $A = \frac{d}{dt}$ on $L_p(\mathbb{R}, X)$ with domain $\mathcal{D}(A) = W^{1,p}(\mathbb{R}, X)$. As A is the generator of the bounded group of translations on $L_p(\mathbb{R}, X)$, it follows that $A \in \mathrm{Sect}(\Sigma)$ for any bisector Σ. In order to decide whether the spectral projection corresponding to the upper complex half-plane is bounded we compute the operator $A^{1/2}(-A)^{-1/2}$ using Fourier multipliers.

Since A is the Fourier multiplier associated to the symbol $m(x) = ix$, the operators $A^{1/2}$ and $(-A)^{-1/2}$ have the symbols

$$(ix)^{1/2} = e^{1/2(\log|x| + i \arg(ix))} \text{ and } (-ix)^{-1/2} = e^{-1/2(\log|x| + i \arg(-ix))}.$$

Hence $A^{1/2}(-A)^{-1/2}$ has the symbol

$$(ix)^{1/2}(-ix)^{-1/2} = e^{i/2(\arg(ix) - \arg(-ix))} = i \operatorname{sign}(x).$$

Thus, up to a constant $A^{1/2}(-A)^{-1/2}$ is the Hilbert transform. On the other hand, the spectral projection P_1 corresponding to the upper half-plane, has symbol $\chi_{\mathbb{R}_+}$.

As a consequence of Theorem 3.3 we obtain the equivalence of the following statements.

1. The spectral projection P_1 (the Poisson projection) is bounded;
2. $\left\|\left(\frac{d}{dx}\right)^{1/2} f\right\| \approx \left\|\left(-\frac{d}{dx}\right)^{1/2} f\right\| \qquad \forall f \in \mathcal{D}\left(\left(\frac{d}{dx}\right)^{1/2}\right) = \mathcal{D}\left(\left(-\frac{d}{dx}\right)^{1/2}\right)$.
3. the operator $\left(-\frac{d^2}{dx^2}\right)^{1/2}\left(\frac{d}{dx}\right)^{-1}$, the Hilbert transform, is bounded;
4. $\left\|\left(-\frac{d^2}{dx^2}\right)^{1/2} f\right\| \approx \left\|\frac{d}{dx} f\right\| \qquad \forall f \in W^{1,p}(\mathbb{R}, X)$.

It is well known that the Hilbert transform is bounded, if and only if $p \in (1, \infty)$ and if X is a UMD space.

With the same arguments as in the proof of Theorem 3.3 we can generalize the equivalence of 1. $\Leftrightarrow$ 4. Note that for a bisectorial operator $A \in \mathrm{Sect}(S^\omega)$, the operator $-A^2$ is sectorial in $\mathrm{Sect}(S_{2\omega})$.

4.2. Theorem. *Let A be a bisectorial operator on X. Then the spectral projection P corresponding to the upper half-plane is bounded if and only if $A(-A^2)^{-1/2}$ extends to a bounded operator, i.e.,*

$$\left\|(-A)^2 x\right\| \approx \|Ax\| \qquad \textit{for } x \in \mathcal{D}(A) = \mathcal{D}((-A^2)^{1/2}).$$

For the Hilbert space case it was noted in [2] that the boundedness of the H^∞-calculus implies the above estimate.

Let $A_1, \ldots, A_n$ now be a family of commuting bisectorial operators on X. Motivated by the example $A_i = \frac{d}{dx_i}$ on $L_p(\mathbb{R}^n)$ and the classical Riesz transforms $\frac{d}{dx_i}(-\Delta)^{-1/2}$ we call the operators $A_i(-A)^{-1/2}$ with $A = \sum_{i=1}^n A_i^2$ *Riesz transforms* of the family $A_1, \ldots, A_n$. By means of the H^∞-calculus we obtain the boundedness of such Riesz-transforms and related inequalities.

4.3. Theorem. *Let $A_1, \dots, A_n$, be commuting bisectorial operators with a bounded joint $H^\infty(\Sigma^n)$-calculus, $\Sigma = S^\omega$, $\omega \in (0, \pi/4)$. Then we have the inequalities*

$$\mathbb{E}\left\|\sum_{i=1}^n r_i A_i x\right\| \approx \left\|(-A)^{1/2}x\right\|, \qquad \mathbb{E}\left\|\sum_{i,j=1}^n r_{ij} A_i A_j x\right\| \approx \|Ax\|$$

where $(r_i), (r_{ij})$ are iid Bernoulli random variables.

Proof. Consider the holomorphic functions $h(z) = (\sum_{i=1}^n \epsilon_i z_i)(-\sum_{i=1}^n z_i^2)^{-1/2}$ and $g(z) = (\sum_{i,j=1}^n \epsilon_{ij} z_i z_j)(\sum_{i=1}^n z_i^2)^{-1}$. It is easy to see that $g, h \in H^\infty(\Sigma^n)$. Moreover, the functions are bounded uniformly with respect the choice of signs $\epsilon_i, \epsilon_{ij} \in \{-1, 1\}$. Hence, by the boundedness of the joint functional calculus, we obtain the inequalities

$$\left\|\sum_{i=1}^n \epsilon_i A_i x\right\| \leq C\,\|h\|_\infty \left\|(-A)^{1/2}x\right\|, \qquad \left\|\sum_{i,j=1}^n \epsilon_{ij} A_i A_j x\right\| \leq C\,\|g\|_\infty \|Ax\|$$

where C denotes the norm of the joint calculus. Since this estimate is uniform in the choice of signs, we obtain one of the two desired inequalities by integration.

We will obtain the reverse inequalities combining randomization and a duality argument. We will give the details for the first equivalence, the argument for the second is similar. We denote by $X^\#$ the *moon-dual* of X, that is the set $\overline{\mathcal{D}(A^*)} \cap \overline{\mathcal{R}(A^*)}$ where the closure is taken in the norm of X^*. The moon-dual operator $A^\#$ of A is the part of A^* in $X^\#$; if A is sectorial or has a bounded H^∞-calculus then so does $A^\#$. If X is reflexive we have $A^\# = A^*$. We note that $X^\#$ is *norming* for X (see [11, Section 15A] for details). Let $x \in \mathcal{D}((-A)^{1/2})$, $y \in \mathcal{D}((-A^\#)^{1/2})$, then

$$\left|\langle(-\sum A_i^2)^{1/2}x, ((-\sum A_i^2)^{1/2})^\# y\rangle\right| = \left|\langle\sum A_i^2 x, y\rangle\right| = \left|\sum\langle A_i x, A_i^\# y\rangle\right|$$
$$\leq \left(\mathbb{E}\left\|\sum r_i A_i x\right\|^2\right)^{1/2}\left(\mathbb{E}\left\|\sum r_i A_i^\# y\right\|^2\right)^{1/2} \leq c\mathbb{E}\left\|\sum r_i A_i x\right\| \left\|(-A^\#)^{1/2}y\right\|;$$

in the last step we used the Khinchine-Kahane inequality and made use the inequality proved above (substituting $A^\#$ for A). Therefore, since the set $\mathcal{R}((-A^\#)^{1/2})$ is norming, we find

$$\left\|(-A)^{1/2}x\right\| = \sup_{\|z\|=1}\left|\langle(-A)^{1/2}x, z\rangle\right| \leq c\mathbb{E}\left\|\sum r_i A_i x\right\|$$

which completes the proof. □

4.4. Example. Let $X = L_p(\mathbb{R}^n, \frac{dx_1}{x_1} \cdots \frac{dx_n}{x_n})$. For $i = 1, \dots, n$, define on X the semigroup $T_i(t)f(x) = f(x_1, \dots, x_{i-1}, e^t x_i, x_{i+1}, \dots, x_n)$. As the measure of the L_p space X is invariant under scaling, we find that T_i are contraction semigroups. Moreover, it is easy to see that they are positive and commute. Their generators, given by $A_i = x_i \frac{d}{dx_i}$, have thus a bounded bisectorial functional calculus (apply the result of Hieber and Prüß [8]). Since the L_p space X has property (α), we find that $(A_1, \dots, A_n)$ have a bounded joint bisectorial functional calculus.

The operator A_i^2 is just $x_i \frac{d}{dx_i}(x_i \frac{d}{dx_i}) = x_i \frac{d}{dx_i} + x_i^2 \frac{d^2}{dx_i^2}$. Therefore, the preceding theorem gives the equivalence

$$\left\| \left(\sum_{i,j=1}^{n} \left| x_i x_j \frac{d}{dx_i} \frac{d}{dx_j} f \right|^2 \right)^{1/2} \right\| \approx \left\| \sum_{i=1}^{n} x_i^2 \frac{d^2}{dx_i^2} f + x_i \frac{d}{dx_i} f \right\| .$$

Making use of [11, Cor. 10.15] we may extend this result to the vector-valued setting, allowing for functions taking values in a UMD space.

5. Spectral projection and Hilbert transform of a bounded C_0-group

Let p denote the indicator function of the upper complex half-plane. The example $A = \frac{d}{dt}$ shows that the spectral projection $P = p(A)$ is bounded if and only if the Hilbert transform is a bounded operator on L_p. In general, let $(U(t))_{t\in\mathbb{R}}$ be a bounded C_0-group on the Banach space X with generator A. For $0 < \epsilon < T < \infty$ consider the operators

$$H^U_{\epsilon,T} x := \frac{i}{\pi} \int_{\epsilon<|s|<T} \frac{U(s)x}{s} \, ds;$$

if the strong limit of these operators (as $\epsilon \to 0$ and $T \to \infty$) exists, it defines a bounded operator H^U which we will call the (generalized) Hilbert transform associated to U. We recover the classical Hilbert transform if U is the group of translations and $X = L_p(\mathbb{R})$.

In this paragraph we will assume that the set $\mathcal{D}(A) \cap \mathcal{R}(A)$ is dense in X, i.e., that A is bisectorial. Integrating by parts, we obtain the following result.

5.1. Lemma. *For all $x \in \mathcal{D}(A) \cap \mathcal{R}(A)$ and all $d \geq 0$ the principal value integral*

$$H^U_d x := \frac{i}{\pi} PV\text{-}\int_{\mathbb{R}} e^{-d|s|} \frac{U(s)x}{s} \, ds := \frac{i}{\pi} \lim_{r\to 0, R\to\infty} \int_{r\leq|s|\leq R} e^{-d|s|} \frac{U(s)x}{s} \, ds$$

exists and $H^U_d x \to H^U x := H^U_0 x$ for $d \to 0+$.

We will say that H^U is bounded, if the densely defined operator H^U has a bounded extension to all of X. This is the case if, for example, the principal value integral exists for all $x \in X$.

Next we will relate the Hilbert transform with the resolvent of A by means of the Fourier transform. Observe that for $x \in \mathcal{D}(A)$ we have the inequality

$$\|(\pm d + is) R(\pm d + is, A)x\| = \|x + R(\pm d + is, A)Ax\| \leq \|x\| + Cd^{-1} \|Ax\| .$$

By iteration and using the resolvent equation this implies that, for $x \in \mathcal{D}(A^2)$, the function $s \mapsto \|R(d+is,A)x - R(-d+is,A)x\|$ is majorized by $s \mapsto Cd^{-1}|d+is|^{-2}$, with $C = C(x) \geq 0$, which is integrable on $\mathbb{R}$.

5.2. Lemma. *Let $d > 0$ and $x \in \mathcal{D}(A^2)$, then*

$$H_d^U x = \frac{1}{2\pi}\int_{\mathbb{R}} \operatorname{sign} s\,(R(-d+is,A)x - R(d+is,A)x)\,ds. \tag{6}$$

Proof. We have $\int_{r\leq|s|\leq R} e^{-d|s|}\frac{U(s)x}{s}\,ds = \int_{-\infty}^{\infty} h_{r,R}(s)e^{-d|s|}U(s)x\,ds$, where $h_{r,R}$ denotes the function $s \mapsto \frac{1}{s}1_{\{r\leq|u|\leq R\}}(s) \in L_p(\mathbb{R})$ for all $p \in [1,\infty]$. As the resolvent is given by the Laplace transform of its semigroup, we obtain

$$\mathcal{F}(e^{-d|\cdot|}U(\cdot))(t) = \int_{\mathbb{R}} e^{-ist}e^{-d|s|}(s)U(s)\,ds = R(d+it,A) - R(-d+it,A). \tag{7}$$

Recall that, given two functions $f \in L_1(\mathbb{R})$, $g \in L_1(\mathbb{R},Y)$ for some Banach space Y, we have as a consequence of Fubini's theorem the relation $\int_{\mathbb{R}} \check{f}(t)g(t)\,dt = \int_{\mathbb{R}} f(t)\check{g}(t)\,dt$, where $\check{f}$ denotes the inverse Fourier transform of f. By the inversion formula, applying this relation to the identity (7) we find

$$\int_{r\leq|s|\leq R} e^{-d|s|}\frac{U(s)x}{s}\,ds = \int_{-\infty}^{\infty} \check{h}_{r,R}(s)((R(d+is,A)x - R(-d+is,A)x))\,ds.$$

We compute the inverse Fourier transform of $h_{r,R}$: for $s \in \mathbb{R}\setminus\{0\}$ we have

$$\check{h}_{r,R}(s) = (2\pi)^{-1}\int_r^R t^{-1}(e^{ist} - e^{-ist})\,dt = i\pi^{-1}\int_{r/s}^{R/s}\frac{\sin u}{u}\,du;$$

it is well known from calculus that the integral $\int_0^\infty u^{-1}\sin u\,du$ converges and has as value $\pi/2$. Therefore, the functions $\check{h}_{r,R}$ are uniformly bounded on $\mathbb{R}$ and converge pointwise (as $r \to 0$ and $R \to \infty$) to the limit $i2^{-1}\operatorname{sign}(\cdot)$. Hence, Lebesgue's theorem implies

$$H_d^U x = \frac{1}{2\pi}\int_{\mathbb{R}} \operatorname{sign} s\,(R(-d+is,A)x - R(d+is,A)x)\,ds,$$

which proves the claim. □

Using this representation, we will compare the Hilbert transform H^U and the spectral projection P. Denote by Σ the bisector S^ω with $\omega \in (\pi/4,\pi/2)$.

Denote by $p \in H^\infty(\mathbb{C}\setminus\mathbb{R})$ the indicator function of the upper complex half-plane and define the signum-function s by $s(z) = p(z) - p(-z) = 2p(z) - 1$; then $p = (1+s)/2$. Thus, the projection $P = p(A)$ is bounded if and only if the operator $s(A)$ is bounded, where both operators are defined by means of the $H^\infty(\Sigma)$-functional calculus. We choose the following approximations of the identity $\varphi_n(z) = e^{is(z)z/n}$; the functions φ_n are in $H^\infty(\Sigma)$ with norms uniformly bounded by 1 and converge pointwise to the function which is identically 1. The operator $s(A)$ is approximated by the bounded operators $S_n = (s\varphi_n)(A)$: for $x \in \mathcal{D}(A)\cap\mathcal{R}(A) \subset \mathcal{D}(P) = \mathcal{D}(s(A))$ we have

$$s(A)x = \lim_{n\to\infty} S_n x \tag{8}$$

by the Convergence Lemma. The vector $S_n x$ is given by the Cauchy integral

$$S_n x = \frac{1}{2\pi i}\int_\Gamma s(z)\varphi_n(z)R(z,A)x\,dz.$$

For $d > 0$ we define also the contour $\tilde\Gamma_d$ as $\{w \in \mathbb{C} : |\mathcal{R}e(w)| = d\}$ with positive orientation, and we consider the operators

$$Q_{d,n}x = \frac{1}{2\pi i}\int_{\tilde\Gamma_d} s(z)\varphi_n(z)R(z,A)x\,dz.$$

Let $x \in \mathcal{D}(A^2)\cap\mathcal{R}(A)$. Since $x \in \mathcal{R}(A)$, say $x = Ay$, we have $R(z,A)x = -y + zR(z,A)y$. Consequently, the function $R(\cdot,A)x$ is bounded on the complement of $\overline{S^\theta}$ for all $\theta > 0$. Hence estimating the difference of the two integrals implies the existence of a constant $C'_x > 0$, independent of n, such that

$$\|Q_{d,n}x - S_n x\| \le C'_x d. \tag{9}$$

As n tends to infinity, $S_n x$ converges to $s(A)x = (2P - I)x$. Hence, it remains to study the behavior of the expression $Q_{d,n}x$:

$$\begin{aligned} Q_{d,n}x = (2\pi)^{-1}\int_{-\infty}^{\infty} & s(is)\varphi_n(d+is)R(d+is,A)x \\ & -s(-is)\varphi_n(-d-is)R(-d-is,A)x\,ds. \end{aligned}$$

As $s(-is) = \operatorname{sign}(-s) = -s(is)$ and as $\varphi_n(-z) = \varphi_n(z)$ we obtain

$$\begin{aligned} Q_{d,n}x &= (2\pi)^{-1}\int_{-\infty}^{\infty} s(is)\varphi_n(d+is)\big[R(d+is,A)x - R(-d+is,A)x\big]\,ds \qquad (10)\\ &+ (2\pi)^{-1}\int_{-\infty}^{\infty} s(is)\varphi_n(d+is)\big[R(-d+is,A)x + R(-d-is,A)x\big]\,ds. \end{aligned}$$

We consider the first integral in (10): as $x \in \mathcal{D}(A^2)$, the function in the parentheses is integrable; hence we may apply Lebesgue's theorem, which tells us that the first term converges to $-H_d^U x$ as n tends to infinity. In order to find an estimate for the second integral in (10) note that by substituting s by $-s$ we have

$$\begin{aligned} &(2\pi)^{-1}\int_{-\infty}^{\infty} s(is)\varphi_n(d+is)\big[R(-d+is,A)x + R(-d-is,A)x\big]\,ds \\ &\quad = (2\pi)^{-1}\int_{-\infty}^{\infty} s(is)\big[\varphi_n(d+is) - \varphi_n(-d+is)\big]R(-d+is,A)x\,ds. \end{aligned}$$

Rewriting the term in the squared brackets

$$\varphi_n(d+is) - \varphi_n(-d+is) = \varphi_n(-d+is)\big(e^{2is(is)d/n} - 1\big)$$

the second integral in (10) can be rewritten as

$$-(e^{-2id/n}-1)(2\pi)^{-1}\int_{-\infty}^{0}\varphi_n(-d+is)R(-d+is,A)x\,ds \\ +(e^{2id/n}-1)(2\pi)^{-1}\int_{0}^{\infty}\varphi_n(-d+is)R(-d+is,A)x\,ds.$$

By Cauchy's theorem we may deform the path of integration $i(-\infty,0)-d$ into the path $\gamma=(-1-i)(0,\infty)-d$ without changing the value of the integral, and similarly we may deform the path $i(0,\infty)-d$ into $(-1+i)(0,\infty)-d$. Thus, we obtain by the sectoriality of A and as $x\in\mathcal{R}(A)$ a constant C''_x independent of d such that

$$\left\|\int_{-\infty}^{0}\varphi_n(-d+is)R(-d+is,A)x\,ds\right\| = \left\|\int_{\gamma}\varphi_n(z)R(z,A)x\,dz\right\| \\ \leq C''_x\int_{\gamma}|\varphi_n(z)|\,|dz| = C''_x n. \tag{11}$$

The same estimate also holds for the second term. Hence, the second integral in (10) is estimated by

$$\left(\left|(e^{2id/n}-1)\right|+\left|(e^{-2id/n}-1)\right|\right)(2\pi)^{-1}C''_x n = 2\sin(d/n)\pi^{-1}C''_x n, \tag{12}$$

which implies

$$\limsup_{n\to\infty}\left\|Q_{d,n}x+H_d^U x\right\| \leq 2\pi^{-1}C''_x d. \tag{13}$$

Putting all this together, we find for $x\in\mathcal{D}(A^2)\cap\mathcal{R}(A)$ and for all $d>0$ and $n\in\mathbb{N}$

$$\left\|s(A)x+H^U x\right\| \leq \|s(A)x-S_nx\|+\|S_nx-Q_{d,n}x\| \\ +\left\|Q_{d,n}x+H_d^U x\right\|+\left\|H_d^U x-H^U x\right\|;$$

taking the limes superior with respect to n, we obtain by (8), (9) and (13)

$$\left\|s(A)x+H^U x\right\| \leq C'_x d+2C''_x\pi^{-1}d+\left\|H_d^U x-H^U x\right\|,$$

from which we conclude applying Lemma 5.1, by letting $d\to 0$, that

$$\left\|s(A)x+H^U x\right\| = 0.$$

Hence, the operators H^U and $I-2P$ agree on the dense set $\mathcal{D}(A^2)\cap\mathcal{R}(A)$. We summarize these considerations in the following theorem.

5.3. Theorem. *Let A be the generator of the bounded C_0-group U. Assume that A has dense range. Then the following assertions are equivalent:*

1. *The spectral projection $P=p(A)$ is bounded;*
2. *The Hilbert transform H^U has a bounded extension to X; more precisely $H^U=I-2P$.*

It was shown in [12] that the spectral projection is bounded if the principal value integral defining the Hilbert transform converges strongly. The converse was established for invertible operators.

References

[1] W. Arendt and S. Bu, *Sums of Bisectorial Operators and applications*, Integral Equations and Operator Theory, submitted 2004.

[2] P. Auscher, A. McIntosh and A. Nahmod, *Holomorphic functional calculi of operators, quadratic estimates and interpolation*, Indiana Univ. Math. J. **46** (1997), no. 2, 375–403.

[3] G. Da Prato and P. Grisvard, *Sommes d'opérateurs linéaires et équations différentielles opérationnelles*, J. Math. Pures Appl. (9) **54** (1975), no. 3, 305–387.

[4] G. Dore and A. Venni, *Separation of two (possibly unbounded) components of the spectrum of a linear operator*, Integral Equations and Operator Theory, Vol. **12** (1989), 470–485.

[5] G. Dore and A. Venni, *H^∞ functional calculus for sectorial and bisectorial operators*, Preprint 2001.

[6] M. Duelli, *Spectral calculus for bisectorial operators and applications to linear and non-linear evolution equations*, doctoral dissertation, Universität Ulm, in preparation.

[7] M. Haase, *The functional calculus for sectorial operators and similarity methods*, doctoral dissertation, Universität Ulm, 2003.

[8] M. Hieber and J. Prüss, *Functional calculi for linear operators in vector-valued L^p-spaces via the transference principle*, Adv. Differential Equations **3** (1998), no. 6, 847–872.

[9] N.J. Kalton, P.C. Kunstmann and L. Weis, *Perturbation and Interpolation Theorems for the H^∞-calculus with Applications to Differential Operators*, submitted.

[10] N.J. Kalton and L. Weis, *The H^∞-calculus and sums of closed operators*, Math. Ann. **321** (2001), no. 2, 319–345.

[11] P.C. Kunstmann and L. Weis, *Maximal L_p-regularity for Parabolic equations, Fourier multiplier theorems and H^∞-functional calculus*, in: Functional Analytic Methods for Evolution Equations, M. Ianelli, R. Nagel and S. Piazzera, eds., Lecture Notes for Mathematics, Springer, 2004.

[12] S. Monniaux, *Générateur analytique et régularite maximale*, Thèse, Université de Franche-Compté, Besançon, 1995.

[13] J. Prüss, *Evolutionary Integral Equations and Applications*, Birkhäuser Verlag, Monographs in Mathematics, 87, 1993.

Markus Duelli and Lutz Weis
Mathematisches Institut I
Universität Karlsruhe
Englerstraße 2
D-76128 Karlsruhe
e-mail: markus.duelli@math.uni-karlsruhe.de
e-mail: lutz.weis@math.uni-karlsruhe.de

Progress in Nonlinear Differential Equations
and Their Applications, Vol. 64, 113–136

Very Weak Solutions of Stationary and Instationary Navier-Stokes Equations with Nonhomogeneous Data

Reinhard Farwig, Giovanni P. Galdi and Hermann Sohr

Dedicated to Professor Dr. H. Amann on the Occasion of his 65th Birthday

Abstract. We investigate several aspects of *very weak solutions* u to stationary and nonstationary Navier-Stokes equations in a bounded domain $\Omega \subseteq \mathbb{R}^3$. This notion was introduced by Amann [3], [4] for the nonstationary case with nonhomogeneous boundary data $u_{|_{\partial\Omega}} = g$ leading to a new and very large solution class. Here we are mainly interested to investigate the "largest possible" class for the more general problem with arbitrary divergence $k = \operatorname{div} u$, boundary data $g = u_{|_{\partial\Omega}}$ and an external force f, as weak as possible. In principle, we will follow Amann's approach.

Mathematics Subject Classification (2000). Primary 35Q30, 76D05; Secondary 35J25, 35J65, 35K60.

Keywords. Stokes and Navier-Stokes equations, very weak solutions, nonhomogeneous data.

1. Introduction

Throughout this paper $\Omega \subset \mathbb{R}^3$ is a bounded domain with boundary $\partial\Omega$ of class $C^{2,1}$ and $N = N(x) \in \mathbb{R}^3$ denotes the unit outer normal at $x = (x_1, x_2, x_3) \in \partial\Omega$. In $\Omega \times [0, T)$, where $0 < T \leq \infty$, we consider the system of Navier-Stokes equations in the very general form

$$u_t - \Delta u + u \cdot \nabla u + \nabla p = f, \quad \operatorname{div} u = k, \quad u_{|_{\partial\Omega}} = g, \quad u_{|_{t=0}} = u_0 \tag{1.1}$$

with nonhomogeneous data $f = \operatorname{div} F$ and k, g, u_0 satisfying

$$\begin{aligned} F &= (F_{ij})_{i,j=1,2,3} \in L^s\big(0, T; L^r(\Omega)\big), & k \in L^s\big(0, T, L^r(\Omega)\big), \\ g &= (g_1, g_2, g_3) \in L^s\big(0, T; W^{-\frac{1}{q},q}(\partial\Omega)\big), & u_0 \in \mathcal{J}^{q,s}(\Omega), \end{aligned} \tag{1.2}$$

where $3 < q < \infty$, $2 < s < \infty$, $1 < r < q$ such that $\frac{1}{3} + \frac{1}{q} = \frac{1}{r}$, $\frac{2}{s} + \frac{3}{q} = 1$. For simplicity, we assume that the coefficient of viscosity equals 1. See (2.19) concerning the space of initial values $\mathcal{J}^{q,s}(\Omega)$ as a space of functionals. Further we suppose the compatibility condition

$$\int_\Omega k(t)dx = \int_{\partial\Omega} N \cdot g(t)dS \quad \text{for a.a. } t \in [0,T), \tag{1.3}$$

where $\int_{\partial\Omega} N \cdot g(t)dS := \langle N \cdot g, 1\rangle_{\partial\Omega}$ must be understood as the functional $N \cdot g \in W^{-\frac{1}{q},q}(\partial\Omega)$ applied to the test function $1 \in W^{1-1/q',q'}(\partial\Omega)$. The largest possible class in the context of very weak solutions seems to be Serrin's uniqueness class $L^s\big(0,T;L^q(\Omega)\big)$ for u defined by $\frac{2}{s} + \frac{3}{q} = 1$. Indeed, we cannot expect, up to now, to obtain the desired uniqueness and regularity properties for any larger class.

To obtain the relation which defines a very weak solution u of (1.1), we follow Amann [3], [4] in principle and apply formally to (1.1) the test function $w \in C_0^1\big([0,T); C^2_{0,\sigma}(\overline{\Omega})\big)$, where $C^2_{0,\sigma}(\overline{\Omega}) = \{v \in C^2(\overline{\Omega})^3 : \operatorname{div} v = 0 \text{ in } \Omega, v_{|_{\partial\Omega}} = 0\}$. Then an integration by parts yields the relation

$$\begin{aligned}\int_0^T \big(-\langle u, w_t\rangle_\Omega - \langle u, \Delta w\rangle_\Omega + \langle g, N\cdot\nabla w\rangle_{\partial\Omega} - \langle uu, \nabla w\rangle_\Omega - \langle ku, w\rangle_\Omega\big)dt \quad (1.4)\\ = \langle u_0, w(0)\rangle_\Omega - \int_0^T \langle F, \nabla w\rangle_\Omega dt, \quad w \in C_0^1\big([0,T); C^2_{0,\sigma}(\overline{\Omega})\big).\end{aligned}$$

Here $\langle\cdot,\cdot\rangle_\Omega$ is the usual $L^q - L^{q'}$-pairing in Ω and $\langle g(t), N \cdot \nabla w(t)\rangle_{\partial\Omega}$ denotes the value of the boundary distribution $g(t) \in W^{-\frac{1}{q},q}(\partial\Omega)$ at the normal derivative $w_N(t) = N \cdot \nabla w(t)_{|_{\partial\Omega}}$; furthermore, $\langle u_0, w(0)\rangle_\Omega$ means the value of the functional $u_0 \in \mathcal{J}^{q,s}(\Omega)$ at $w(0) = w_{|_{t=0}} \in C^2_{0,\sigma}(\overline{\Omega})$, see (2.19), and $uu = u \otimes u = (u_iu_j)_{i,j=1,2,3}$ for $u = (u_1, u_2, u_3)$. We also use the relation $u \cdot \nabla u = (u\cdot\nabla)u = \operatorname{div}(uu) - ku$. Note that $\operatorname{div} F = \big(\sum_{i=1}^3 (\partial/\partial x_i)F_{ij}\big)_{j=1,2,3}$.

Using in (1.4) in particular the test function $w \in C_0^1\big([0,T); C^2_{0,\sigma}(\Omega)\big)$ with $C^2_{0,\sigma}(\Omega) = \{v \in C_0^2(\Omega)^3;\ \operatorname{div} v = 0\}$, we obtain, together with some appropriate distribution p, the validity of the first equation in (1.1) in the sense of distributions. The second equation $\operatorname{div} u = k$ in (1.1) must be supposed additionally to (1.4). To explain the boundary condition $u_{|_{\partial\Omega}} = g$ in (1.1) we observe that the normal derivative $w_N(t)$ of $w(t)$ at $\partial\Omega$ has the form

$$w_N(t) = N \cdot \nabla w(t) = \big(\operatorname{curl} w(t)\big) \times N, \tag{1.5}$$

and therefore, the relation (1.4) contains only the tangential component $N \times g$ of u at $\partial\Omega$. Indeed, we will show that the tangential component of u is well defined by (1.4) as a distribution on $\partial\Omega$, and we will derive an explicit formula, see (4.5). The (well-defined) condition $N \cdot u(t)_{|_{\partial\Omega}} = N \cdot g(t)$ for the normal component of u at $\partial\Omega$ must be supposed additionally. This leads to a precise formulation of the boundary condition $u_{|_{\partial\Omega}} = g$, see Remarks 3(2) in Section 4 below. Thus we suppose, additionally to (1.4), the conditions

$$\operatorname{div} u(t) = k(t), \quad N \cdot u(t)_{|_{\partial\Omega}} = N \cdot g(t) \quad \text{for a.a. } t \in [0,T). \tag{1.6}$$

This leads to the following

Definition 1. Assume that the data F, k, g and u_0 satisfy (1.2) and (1.3). Then $u \in L^s(0,T; L^q(\Omega))$ is called a *very weak solution* of the Navier-Stokes system (1.1) if the conditions (1.4) and (1.6) are satisfied.

Note that a very weak solution u need not have any differentiability property in space and time. In particular, u need not satisfy any energy inequality with finite energy $\|u\|^2_{L^\infty(0,T;L^2(\Omega))} + \|\nabla u\|^2_{L^2(0,T;L^2(\Omega))} < \infty$ like weak solutions in the sense of Hopf. This justifies the notation "very weak solution". On the other hand, a very weak solution is unique – a fact which is not known in general for weak solutions in the sense of Hopf.

The notion of very weak solutions is not new for homogeneous data $k = 0$ and $g = 0$, see [3], [4], [15], and the literature therein. However, Amann's notion of very weak solutions in [3], [4] for $k = 0$ and boundary values $g \neq 0$ introduces a completely new aspect. It leads to new solution classes of very low smoothness in space such that the boundary condition $u_{|_{\partial\Omega}} = g$ is not defined by usual trace theorems but more generally by the conditions (1.3), (1.4). This will have interesting applications.

Following Amann's approach our aim is to extend the solution classes for (1.1) to $\operatorname{div} u = k \neq 0$ and to the "weakest" possible case $u_{|_{\partial\Omega}} = g \in W^{-\frac{1}{q},q}(\partial\Omega)$ for a.a. $t \in [0,T]$, leading to the solution class in Definition 1 without any smoothness in space and only satisfying Serrin's condition, see Theorem 1. We will develop such a theory also for the linear nonstationary Stokes system, see Theorem 4. Further we improve the results on very weak solutions of stationary Stokes and Navier-Stokes systems developed in [15] to the more general case that $k \in L^r(\Omega)$, $\frac{1}{3} + \frac{1}{q} = \frac{1}{r}$, see Theorems 2 and 3.

Concerning the initial condition $u_{|_{t=0}} = u_0$ in (1.1) we note, see Theorem 1, that $A_q^{-1} P_q u$ is well defined as a continuous function on $[0,T)$ with values in $L^q_\sigma(\Omega)$; here, A_q means the Stokes operator and P_q the Helmholtz projection. We obtain the well-defined condition

$$A_q^{-1} P_q u_{|_{t=0}} = A_q^{-1} P_q u_0, \tag{1.7}$$

which can be understood as the precise meaning of $u_{|_{t=0}} = u_0$, see (2.19) for $A_q^{-1} P_q u_0$ and the proof of Theorem 1.

Thus each condition in the system (1.1) has a well-defined direct meaning for a very weak solution u. The first two equations hold in the sense of distributions on $\Omega \times (0,T)$, $u_{|_{\partial\Omega}} = g$ holds in the sense of distributions in $\partial\Omega$, and $u_{|_{t=0}} = u_0$ holds in the sense of (1.7). Moreover, if the data are sufficiently smooth, u coincides with the usual strong solution.

Our main theorem on the system (1.1) reads as follows.

Theorem 1. *Suppose the data F, k, g and u_0 satisfy (1.2) and (1.3) with $3 < q < \infty$, $2 < s < \infty$, $1 < r < q$, $\frac{1}{3} + \frac{1}{q} = \frac{1}{r}$, $\frac{2}{s} + \frac{3}{q} = 1$. Then there exists some $T' = T'(F, k, g, u_0) > 0$, $0 < T' \le T$, and a uniquely determined very weak solution $u \in L^s(0, T'; L^q(\Omega))$ of the system (1.1) satisfying $A_q^{-1} P_q u_t \in L^{s/2}(0, T''; L^q_\sigma(\Omega))$ for all $0 < T'' < T'$, and $A_q^{-1} P_q u \in C([0, T'); L^q_\sigma(\Omega))$. The existence interval $[0, T')$ is determined by the condition (4.23), depending on the data, and includes the case $T' = T = \infty$ if the data are sufficiently small.*

Up to now we cannot prove that there exists an open maximal existence interval as in [3], [4] for the case $k = 0$. The reasons are the very weak assumptions on g and k in (1.2).

In the linearized case $u \cdot \nabla u \equiv 0$ we have to omit the term $\langle u \cdot \nabla u, w \rangle_\Omega = -\langle uu, \nabla w \rangle_\Omega - \langle ku, w \rangle_\Omega$ in (1.4), and we may omit the restriction $\frac{2}{s} + \frac{3}{q} = 1$ in (1.2), which is caused by the nonlinear term. Then we can show the existence and uniqueness of a very weak solution $u \in L^s(0, T; L^q(\Omega))$ of the linearized system (1.1) satisfying the estimate

$$\begin{aligned} &\|A_q^{-1} P_q u_t\|_{L^s(0,T;L^q(\Omega))} + \|u\|_{L^s(0,T;L^q(\Omega))} \\ &\le C\big(\|u_0\|_{\mathcal{J}^{q,s}(\Omega)} + \|F\|_{L^s(0,T;L^r(\Omega))} + \|k\|_{L^s(0,T;L^r(\Omega))} + \|g\|_{L^s(0,T;W^{-\frac{1}{q},q}(\partial\Omega))}\big) \end{aligned} \tag{1.8}$$

with $C = C(\Omega, q, s) > 0$, see Theorem 4 in Section 4 below.

In the stationary case we consider the system

$$-\Delta u + u \cdot \nabla u + \nabla p = f, \quad \operatorname{div} u = k, \quad u_{|_{\partial\Omega}} = g \tag{1.9}$$

with data $f = \operatorname{div} F$ and k, g satisfying

$$F = (F_{ij}) \in L^r(\Omega),\ k \in L^r(\Omega),\ g \in W^{-\frac{1}{q},q}(\partial\Omega),\ \int_\Omega k\, dx = \langle N \cdot g, 1 \rangle_{\partial\Omega}\,, \tag{1.10}$$

with $3 < q < \infty$, $1 < r < q$, $\frac{1}{3} + \frac{1}{q} = \frac{1}{r}$.

An obvious modification of the nonstationary case yields the following

Definition 2. Assume that the data F, k and g satisfy (1.10). Then $u \in L^q(\Omega)$ is called a *very weak solution* of the stationary Navier-Stokes system (1.9) if the relation

$$-\langle u, \Delta w \rangle_\Omega + \langle g, N \cdot \nabla w \rangle_{\partial\Omega} - \langle uu, \nabla w \rangle_\Omega - \langle ku, w \rangle_\Omega = -\langle F, \nabla w \rangle_\Omega, \tag{1.11}$$

holds for all $w \in C^2_{0,\sigma}(\overline{\Omega})$, and the conditions

$$\operatorname{div} u = k, \quad N \cdot u_{|_{\partial\Omega}} = N \cdot g \tag{1.12}$$

are satisfied.

In this case we obtain the following result.

Theorem 2. *Suppose the data F, k and g satisfy (1.10) with $3 < q < \infty$, $1 < r < q$, $\frac{1}{3} + \frac{1}{q} = \frac{1}{r}$. There exists a constant $K = K(\Omega, q) > 0$ such that if*

$$\|F\|_{L^r(\Omega)} + \|k\|_{L^r(\Omega)} + \|g\|_{W^{-\frac{1}{q},q}(\partial\Omega)} \leq K, \tag{1.13}$$

then we obtain a uniquely determined very weak solution $u \in L^q(\Omega)$ to the stationary Navier-Stokes system (1.9). This solution satisfies the estimate

$$\|u\|_{L^q(\Omega)} \leq C\big(\|F\|_{L^r(\Omega)} + \|k\|_{L^r(\Omega)} + \|g\|_{W^{-\frac{1}{q},q}(\partial\Omega)}\big) \tag{1.14}$$

with $C = C(\Omega, q) > 0$.

Similarly as in (1.5) we obtain for $x \in \partial\Omega$ the identity

$$w_N = N \cdot \nabla w = (\operatorname{curl} w) \times N, \quad w \in C^2_{0,\sigma}(\overline{\Omega}). \tag{1.15}$$

Setting $w \in C^2_{0,\sigma}(\Omega)$ in (1.11) we obtain that $-\Delta u + u \cdot \nabla u + \nabla p = f$ holds in the sense of distributions with some distribution p. In the stationary case we can also prove that each very weak solution u has a well-defined trace $u_{|\partial\Omega}$ even in the space $W^{-\frac{1}{q},q}(\partial\Omega)$, and there is an explicit representation formula for $u_{|\partial\Omega}$, see (3.6).

Note that Theorem 2 improves the result in [15] where $\|k\|_{L^r(\Omega)}$ is replaced by the stronger norm $\|k\|_{L^q(\Omega)}$.

In the linearized case $u \cdot \nabla u \equiv 0$, we omit $\langle uu, \nabla w\rangle_\Omega$, $\langle ku, w\rangle_\Omega$ in relation (1.11), and the existence result together with estimate (1.14) holds without any smallness condition for every $1 < q < \infty$, $1 < r < q$, $\frac{1}{3} + \frac{1}{q} \geq \frac{1}{r}$, see Theorem 3 in Section 3 below.

The improvement concerning $\|k\|_{L^r(\Omega)}$ leads to a certain scaling invariance in the following sense. Let $\lambda > 0$, consider some ball $B_a(x_0) \subseteq \mathbb{R}^3$ with radius $a > 0$ and center $x_0 \in \mathbb{R}^3$, and let F, k, g be data as in (1.10) with $\Omega = B_a(x_0)$, $3 < q < \infty$, $1 < r < q$, $\frac{1}{3} + \frac{1}{q} = \frac{1}{r}$. Then it is easy to show that $u \in L^q(B_a(x_0))$ is a very weak solution to the system (1.9) on Ω with data F, k, g iff $\tilde{u} \in L^q(\tilde{\Omega})$, $\tilde{\Omega} = B_{a/\lambda}(x_0)$, is a very weak solution of the system

$$-\Delta\tilde{u} + \tilde{u} \cdot \nabla\tilde{u} + \nabla\tilde{p} = \operatorname{div}\tilde{F}, \quad \operatorname{div}\tilde{u} = \tilde{k}, \quad \tilde{u}_{|\partial\tilde{\Omega}} = \tilde{g} \tag{1.16}$$

on $\tilde{\Omega}$, where $\tilde{u}, \tilde{p}, \tilde{F}, \tilde{k}, \tilde{g}$ are defined by

$$\begin{aligned} \tilde{u}(x) &= \lambda u\big(\lambda(x - x_0) + x_0\big), & \tilde{p}(x) &= \lambda^2 p\big(\lambda(x - x_0) + x_0\big), \\ \tilde{F}(x) &= \lambda^2 F\big(\lambda(x - x_0) + x_0\big), & \tilde{k}(x) &= \lambda^2 k\big(\lambda(x - x_0) + x_0\big), \\ \tilde{g}(x) &= \lambda g\big(\lambda(x - x_0) + x_0\big). \end{aligned}$$

Then we conclude that if $K_{a,q} = K\big(B_a(x_0), q\big) > 0$ is the constant in (1.13) for $B_a(x_0)$, then $\lambda^{1-\frac{3}{q}} K_{a,q}$ can be chosen as the corresponding constant for $B_{a/\lambda}(x_0)$. We also conclude that (1.14) holds for u, F, k, g in $\Omega = B_a(x_0)$ with

$C = C(B_a(x_0), q)$ iff (1.14) holds for $\tilde{u}, \tilde{F}, \tilde{k}, \tilde{g}$ in $\tilde{\Omega} = B_{a/\lambda}(x_0)$ with the same constant C. These properties have several applications in the local regularity theory. Similar results hold in the nonstationary theory.

The proofs of these theorems are organized as follows. First we consider the linearized stationary and nonstationary system and prove explicit representation formulas. Applying these formulas to the nonlinear system leads to a fixed point problem which can be solved by Banach's fixed point theorem for sufficiently small data. In Section 2 we prepare several preliminaries.

2. Notations and preliminaries

Classical Function Spaces. Let $1 < q < \infty$ and $q' = \frac{q}{q-1}$ such that $\frac{1}{q} + \frac{1}{q'} = 1$. We need the usual spaces $L^q(\Omega)$ and $W^{\alpha,q}(\Omega)$, $W_0^{\alpha,q}(\Omega)$, $0 \le \alpha \le 2$, with norms $\|\cdot\|_{L^q(\Omega)} = \|\cdot\|_{q,\Omega}$ and $\|\cdot\|_{W^{\alpha,q}(\Omega)} = \|\cdot\|_{\alpha;q,\Omega}$, resp. The space $W^{-\alpha,q}(\Omega) = W_0^{\alpha,q'}(\Omega)'$ is the dual space of $W_0^{\alpha,q'}(\Omega)$ with the natural duality pairing $\langle\cdot,\cdot\rangle_\Omega$ and the norm $\|\cdot\|_{W^{-\alpha,q}(\Omega)} = \|\cdot\|_{-\alpha;q,\Omega}$. Thus, e.g., $\langle f, h\rangle_\Omega$ means the value of the functional $f \in W^{-\alpha,q}(\Omega)$ at $h \in W_0^{\alpha,q'}(\Omega)$. Similarly, for functions on the boundary, $L^q(\partial\Omega)$ and $W^{\alpha,q}(\partial\Omega), W^{-\alpha,q'}(\partial\Omega)$, with norms $\|\cdot\|_{L^q(\partial\Omega)} = \|\cdot\|_{q,\partial\Omega}$ and $\|\cdot\|_{W^{\alpha,q}(\partial\Omega)} = \|\cdot\|_{\alpha;q,\partial\Omega}$, $\|\cdot\|_{W^{-\alpha,q}(\partial\Omega)} = \|\cdot\|_{-\alpha;q,\partial\Omega}$, resp., and the duality pairing $\langle\cdot,\cdot\rangle_{\partial\Omega}$ are the corresponding notions for $\partial\Omega$. In particular, the pairing between $L^q(\partial\Omega)$ and its dual space $L^{q'}(\partial\Omega) = L^q(\partial\Omega)'$ is given by

$$\langle f, g\rangle_{\partial\Omega} = \int_{\partial\Omega} f \cdot g \, dS$$

where $\int_{\partial\Omega} \dots dS$ means the surface integral on $\partial\Omega$, see [25] and [27], p. 33, p. 40. For more details on these spaces cf. [1], [12], [13], [27], [29]. In general, we use the same symbol for scalar, vector, and tensor-valued spaces.

By $C_0^\nu(\Omega)$, $C^\nu(\Omega)$, $C^\nu(\overline{\Omega})$, $\nu = 0, 1, \dots$ and $\nu = \infty$, we denote the usual spaces of smooth functions. We set $C_0^\nu(\overline{\Omega}) = \{v \in C^\nu(\overline{\Omega});\, v|_{\partial\Omega} = 0\}$. The space of distributions $C_0^\infty(\Omega)'$ is the dual space of the test space $C_0^\infty(\Omega)$ with the usual topology, the duality pairing is again denoted by $\langle\cdot,\cdot\rangle_\Omega$. Similarly, the space $C^1(\partial\Omega)'$ of boundary distributions is the dual space of the test space $C^1(\partial\Omega)$ with the duality pairing $\langle\cdot,\cdot\rangle_{\partial\Omega}$. This test space has the form $C^1(\partial\Omega)$ since $\partial\Omega$ has no boundary.

Spaces of solenoidal vector-valued functions are denoted by appending "σ". Thus we have $C_{0,\sigma}^\nu(\Omega) = \{v \in C_0^\nu(\Omega);\, \operatorname{div} v = 0\}$ and $C_{0,\sigma}^\nu(\overline{\Omega}) = \{v \in C_0^\nu(\overline{\Omega});\, \operatorname{div} v = 0\}$. The corresponding functional space for solenoidal test functions $C_{0,\sigma}^\infty(\Omega)$ is the dual space $C_{0,\sigma}^\infty(\Omega)'$, again with pairing $\langle\cdot,\cdot\rangle_\Omega$. By a theorem of de Rham, [28], Chapter I, Proposition 1.1, a distribution $d = (d_1, d_2, d_3) \in C_0^\infty(\Omega)'$ with $\langle d, v\rangle_\Omega = 0$ for all $v \in C_{0,\sigma}^\infty(\Omega)$ has the form $d = \nabla h$ with some scalar distribution h.

Let $L^q_\sigma(\Omega)$ be the closure of $C^\infty_{0,\sigma}(\Omega)$ in the norm $\|\cdot\|_{q,\Omega}$. Then $L^{q'}_\sigma(\Omega) = L^q_\sigma(\Omega)'$ is the dual space of $L^q_\sigma(\Omega)$ with pairing $\langle\cdot,\cdot\rangle_\Omega$.

Traces and Extensions. Let $\alpha = 1, 2$. Then the *trace map* $f \mapsto f|_{\partial\Omega}$ is a well-defined bounded operator from $W^{\alpha,q}(\Omega)$ onto $W^{\alpha-\frac{1}{q},q}(\partial\Omega)$. Conversely, there exists a linear and bounded *extension operator* E: $h \mapsto E_h$ from $W^{\alpha-\frac{1}{q},q}(\partial\Omega)$ into $W^{\alpha,q}(\Omega)$ satisfying $E_h|_{\partial\Omega} = h$. Thus it holds $\|E_h\|_{\alpha;q,\Omega} \le C\|h\|_{\alpha-\frac{1}{q};q,\partial\Omega}$ with $C = C(\Omega,\alpha,q) > 0$.

Let $1 < r \le q$, $\frac{1}{3}+\frac{1}{q} \ge \frac{1}{r}$, and let $f = (f_1, f_2, f_3) \in L^q(\Omega)$ with $\operatorname{div} f \in L^r(\Omega)$. Then we use Green's identity

$$\langle \operatorname{div} f, E_h\rangle_\Omega = \langle N\cdot f, h\rangle_{\partial\Omega} - \langle f, \nabla E_h\rangle_\Omega$$

for $h \in W^{1-\frac{1}{q'},q'}(\partial\Omega) = W^{\frac{1}{q},q'}(\partial\Omega)$ and with the extension operator $E : W^{\frac{1}{q},q'}(\partial\Omega) \to W^{1,q'}(\Omega)$. This leads, using the embedding property $\|E_h\|_{r',\Omega} \le C(\|E_h\|_{q',\Omega} + \|\nabla E_h\|_{q',\Omega}), C = C(\Omega,q,r) > 0$, to the estimate

$$|\langle N\cdot f, h\rangle_{\partial\Omega}| \le C(\|f\|_{q,\Omega} + \|\operatorname{div} f\|_{r,\Omega})\|h\|_{\frac{1}{q};q',\partial\Omega} \tag{2.1}$$

with $C = C(\Omega,q,r) > 0$. Hence the trace $N\cdot f|_{\partial\Omega} \in W^{-\frac{1}{q},q}(\partial\Omega)$ of the normal component of f at $\partial\Omega$ is well defined and it holds the estimate

$$\|N\cdot f\|_{-\frac{1}{q};q,\partial\Omega} \le C(\|f\|_{q,\Omega} + \|\operatorname{div} f\|_{r,\Omega}). \tag{2.2}$$

Using the corresponding identity

$$\langle \operatorname{curl} f, E_h\rangle_\Omega = \langle N\times f, E_h\rangle_{\partial\Omega} + \langle f, \operatorname{curl} E_h\rangle_\Omega$$

now for $h = (h_1,h_2,h_3) \in W^{1-\frac{1}{q'},q'}(\partial\Omega)$, we obtain the following trace property: If $f = (f_1,f_2,f_3) \in L^q(\Omega)$, $\operatorname{curl} f \in L^r(\Omega)$, then the trace $N\times f|_{\partial\Omega} \in W^{-\frac{1}{q},q}(\partial\Omega)$ of the tangential component of f at $\partial\Omega$ is well defined and it holds the estimate

$$\|N\times f\|_{-\frac{1}{q};q,\partial\Omega} \le C(\|f\|_{q,\Omega} + \|\operatorname{curl} f\|_{r,\Omega}) \tag{2.3}$$

with $C = C(\Omega,q,r) > 0$. The identity $f = (N\cdot f)N + (N\times f)\times N$ at $\partial\Omega$ shows that it is justified to call $N\times f|_{\partial\Omega}$ the tangential component of f at $\partial\Omega$.

Let $f \in L^q(\Omega)$ with $\int_\Omega f\,dx = 0$. Then there exists some $b = b^f \in W^{1,q}_0(\Omega)$ with $\operatorname{div} b^f = f$ such that $f \mapsto b^f$ is a linear mapping satisfying

$$\|b^f\|_{1;q,\Omega} \le C\|f\|_{q,\Omega}, \quad C = C(\Omega,q) > 0. \tag{2.4}$$

If moreover $f \in W^{1,q}_0(\Omega)$, then $b^f \in W^{2,q}_0(\Omega)$ and

$$\|b^f\|_{2;q,\Omega} \le C\|\nabla f\|_{q,\Omega}, \quad C = C(\Omega,q) > 0, \tag{2.5}$$

see [6], [12], Theorem III.3.2 and [27], p. 68.

Using properties of the weak Neumann problem in $L^q(\Omega)$, see [24], we find for each $h \in W^{-\frac{1}{q},q}(\partial\Omega)$ some $E^h = (E_1^h, E_2^h, E_3^h) \in L^q(\Omega)$ with $\operatorname{div} E^h \in L^r(\Omega)$, $N \cdot E^h|_{\partial\Omega} = h$, such that $h \mapsto E^h$ is a linear map satisfying

$$\|E^h\|_{q,\Omega} + \|\operatorname{div} E^h\|_{r,\Omega} \le C\|h\|_{-\frac{1}{q};q,\partial\Omega} \tag{2.6}$$

with $C = C(\Omega, q, r) > 0$.

Let $h = (h_1, h_2, h_3) \in W^{1-\frac{1}{q},q}(\partial\Omega)$. Then we find an extension $w^h \in W^{2,q}(\Omega) \cap W_0^{1,q}(\Omega)$ such that $N \cdot \nabla w^h|_{\partial\Omega} = h$ depending linearly on h; moreover,

$$\|w^h\|_{2;q,\Omega} \le C\|h\|_{1-1/q;q,\partial\Omega} \tag{2.7}$$

with $C = C(\Omega, q) > 0$, see [23], Theorem 5.8, p. 104, or [29], 5.4.4, p. 385.

If additionally $N \cdot h|_{\partial\Omega} = 0$, then we can show that $\operatorname{div} w^h|_{\partial\Omega} = 0$ and $N \cdot \nabla w^h|_{\partial\Omega} = -N \times \operatorname{curl} w^h|_{\partial\Omega} = h$, see [15]. This yields $\int_\Omega \operatorname{div} w^h dx = 0$, $\operatorname{div} w^h \in W_0^{1,q}(\Omega)$, and we find $b = b(w^h) \in W_0^{2,q}(\Omega)$ satisfying $\operatorname{div} b = \operatorname{div} w^h$ and (2.4), (2.5). Setting $\hat{w}^h = w^h - b(w^h)$ we see that $\hat{w}^h \in W^{2,q}(\Omega)$ satisfies

$$\hat{w}^h|_{\partial\Omega} = 0, \quad N \cdot \nabla \hat{w}^h|_{\partial\Omega} = -N \times \operatorname{curl} \hat{w}^h|_{\partial\Omega} = h, \quad \operatorname{div} \hat{w}^h = 0. \tag{2.8}$$

Moreover, the mapping $h \mapsto \hat{w}^h$ is linear and

$$\|\hat{w}^h\|_{2;q,\Omega} \le C\|h\|_{1-1/q;q,\partial\Omega}, \quad C = C(\Omega, q) > 0, \tag{2.9}$$

see [15], (2.14).

The Stokes Operator. Let $f = (f_1, f_2, f_3) \in L^q(\Omega)$. Then, the weak Neumann problem $\Delta H = \operatorname{div} f$, $N \cdot (\nabla H - f)|_{\partial\Omega} = 0$, has a unique solution $\nabla H \in L^q(\Omega)$ satisfying the estimate

$$\|\nabla H\|_{q,\Omega} \le C\|f\|_{q,\Omega} \tag{2.10}$$

with $C = C(\Omega, q) > 0$, see [24]. Setting $P_q f = f - \nabla H$ we define the *Helmholtz projection operator* P_q as a bounded operator from $L^q(\Omega)$ onto $L^q_\sigma(\Omega)$, satisfying $P_q^2 = P_q$ and

$$\langle P_q f, g\rangle_\Omega = \langle f, P_{q'} g\rangle_\Omega, \quad \text{for all } f \in L^q(\Omega), g \in L^{q'}(\Omega).$$

Hence $P_q' = P_{q'}$ holds for the dual operator P_q' of P_q.

The *Stokes operator* A_q, with domain

$$D(A_q) = L^q_\sigma(\Omega) \cap W_0^{1,q}(\Omega) \cap W^{2,q}(\Omega) \subseteq L^q_\sigma(\Omega)$$

and range $R(A_q) = L^q_\sigma(\Omega)$, is defined by $A_q u = -P_q \Delta u$, $u \in D(A_q)$. The fractional power $A_q^\alpha : D(A_q^\alpha) \to L^q_\sigma(\Omega) = R(A_q^\alpha)$ with $D(A_q) \subseteq D(A_q^\alpha) \subseteq L^q_\sigma(\Omega)$, $0 \le \alpha \le 1$, is well defined, bijective and its inverse $(A_q^\alpha)^{-1} = A_q^{-\alpha}$ is a bounded operator from $L^q_\sigma(\Omega)$ into $L^q_\sigma(\Omega)$ with range $R(A_q^{-\alpha}) = D(A_q^\alpha)$; furthermore, the operator $(A_q^\alpha)' = A_{q'}^\alpha$ is the dual operator of A_q^α. The norms $\|u\|_{2;q,\Omega}$ and $\|A_q u\|_{q,\Omega}$ are equivalent for $u \in D(A_q)$; analogously, the norms $\|u\|_{1;q,\Omega}$ and $\|A_q^{1/2} u\|_{q,\Omega}$ are equivalent for $u \in D(A_q^{\frac{1}{2}}) = L^q_\sigma(\Omega) \cap W_0^{1,q}(\Omega)$. Note that the space $D(A_q^\alpha)$ endowed with the graph norm $\|A_q^\alpha u\|_{q,\Omega}$, $u \in D(A_q^\alpha)$, is a Banach space. Furthermore, we mention the important embedding property

$$\|u\|_{q,\Omega} \le C\|A_\gamma^\alpha u\|_{\gamma,\Omega}, \quad u \in D(A_\gamma^\alpha), \quad 1 < \gamma \le q, \quad 2\alpha + \frac{3}{q} = \frac{3}{\gamma}, \tag{2.11}$$

with $C = C(\Omega, q) > 0$. See [2], [10], [14], [18], [19], [22], [26], [27], [29] concerning proofs and further properties of the Stokes operator. Finally we observe that $A_q u = A_\rho u$ holds if $u \in D(A_q) \cap D(A_\rho)$, $1 < q < \infty$, $1 < \rho < \infty$.

It is well known that $-A_q$ generates a bounded analytic semigroup $\{e^{-tA_q} : t \ge 0\}$, see [2], [17], [20], [26], [27], and that

$$\|A_q^\alpha e^{-tA_q} v\|_{q,\Omega} \le C e^{-\delta t} t^{-\alpha} \|v\|_{q,\Omega}, \quad v \in L^q_\sigma(\Omega),\ t > 0, \tag{2.12}$$

with constants $C = C(\Omega, q) > 0$, $\delta = \delta(\Omega, q) > 0$.

Let $0 < \alpha \le 1$, $1 < q < \infty$, and let $d = (d_1, d_2, d_3) \in C_0^\infty(\Omega)'$ be a distribution. Assume that $\langle d, v\rangle_\Omega$ is well defined for all $v \in D(A_{q'}^\alpha)$ and is continuous in the norm $\|A_{q'}^\alpha v\|_{q',\Omega}$, i.e., there exists a constant $C > 0$ such that $|\langle d, v\rangle_\Omega| \le C\|A_{q'}^\alpha v\|_{q',\Omega}$. In other words, the functional $\langle d, v\rangle_\Omega$, $v \in D(A_{q'}^\alpha)$, is a well-defined element of the dual space $D(A_{q'}^\alpha)'$ of $D(A_{q'}^\alpha)$. Writing formally $\langle d, v\rangle_\Omega = \langle d, P_{q'} v\rangle_\Omega = \langle P_q d, v\rangle_\Omega$, we call $P_q d = \langle P_q d, \cdot\rangle_\Omega$ the restriction of the functional d to test functions $v \in D(A_{q'}^\alpha)$, giving P_q a generalized meaning; in short, we write $P_q d \in D(A_{q'}^\alpha)'$.

Let $d \in C_0^\infty(\Omega)'$ with $P_q d \in D(A_{q'}^\alpha)'$. Since $R(A_{q'}^\alpha) = L^{q'}_\sigma(\Omega)$, there exists a uniquely determined element $d^* \in L^q_\sigma(\Omega)$ satisfying the relation $\langle d, v\rangle_\Omega = \langle d^*, A_{q'}^\alpha v\rangle_\Omega$. We set $d^* = A_q^{-\alpha} P_q d$, giving the operator $A_q^{-\alpha}$ a generalized meaning. Thus $A_q^{-\alpha} P_q d \in L^q_\sigma(\Omega)$ is well defined by the relation

$$\langle d, v\rangle_\Omega = \langle P_q d, v\rangle_\Omega = \langle P_q d, A_{q'}^{-\alpha} A_{q'}^\alpha v\rangle_\Omega = \langle A_q^{-\alpha} P_q d, A_{q'}^\alpha v\rangle_\Omega\,, \tag{2.13}$$

$v \in D(A_{q'}^\alpha)$, similarly as in the theory of distributions. We conclude that the operation $A_q^{-\alpha} P_q d \in L^q_\sigma(\Omega)$ is well defined by $\big(2.13\big)$ if $d \in C_0^\infty(\Omega)'$ and $P_q d \in D(A_{q'}^\alpha)'$.

To obtain examples, we consider a functional f in the (vector-valued) Bessel potential space $H_q^{-2\alpha}(\Omega)$ for $0 < \alpha \le \frac{1}{2}$, see [1], [2], [29]. Then $P_q f \in D(A_{q'}^\alpha)'$, since $C_{0,\sigma}^\infty(\Omega)$ is dense in the Banach space $D(A_{q'}^\alpha)$ because of $\alpha \le \frac{1}{2}$. Therefore, $A_q^{-\alpha} P_q f \in L^q_\sigma(\Omega)$ is well defined in this case.

Let $1 < r \le q$, $\frac{1}{3} + \frac{1}{q} \ge \frac{1}{r}$, and let $u = (u_1, u_2, u_3) \in L^q(\Omega)$. Assume that the distribution $d = \Delta u$ is continuous in the norm $\|A_{r'}^{\frac{1}{2}} \cdot \|_{r',\Omega}$. Then the element $A_r^{-1/2} P_r \Delta u \in L^r_\sigma(\Omega)$ is well defined by the relation

$$\langle A_r^{-\frac{1}{2}} P_r \Delta u,\, A_{r'}^{\frac{1}{2}} v\rangle_\Omega = \langle u, \Delta v\rangle_\Omega, \quad v \in C_{0,\sigma}^\infty(\Omega), \tag{2.14}$$

according to $\big(2.13\big)$; see also [15], Section 2. Here we use that $C_{0,\sigma}^\infty(\Omega)$ is dense in the Banach space $D(A_{r'}^{\frac{1}{2}})$.

Let $F = (F_{ij}) \in L^r(\Omega)$ where $1 < r \le q$, $\frac{1}{3} + \frac{1}{q} \ge \frac{1}{r}$, and set $d = \operatorname{div} F$. Then using (2.11) and the estimate

$$\|\nabla v\|_{r',\Omega} \le C_1 \|A_{r'}^{\frac{1}{2}} v\|_{r',\Omega} \le C_2 \|A_{q'} v\|_{q',\Omega}\,, \quad v \in D(A_{q'})\,,$$

$C_i = C_i(\Omega, q, r) > 0$, $i = 1, 2$, we see that the distribution $v \mapsto -\langle F, \nabla v\rangle_\Omega$ is contained in $D(A_{q'})'$. Therefore, the element $\hat{F} = -A_q^{-1} P_q \operatorname{div} F \in L^q_\sigma(\Omega)$ is well defined by the relation

$$\langle \hat{F}, A_{q'} v\rangle_\Omega = -\langle A_q^{-1} P_q \operatorname{div} F, A_{q'} v\rangle_\Omega = -\langle \operatorname{div} F, v\rangle_\Omega \tag{2.15}$$

and it holds

$$\|\hat{F}\|_{q,\Omega} = \|A_q^{-1} P_q \operatorname{div} F\|_{q,\Omega} \le C \|F\|_{r,\Omega} \tag{2.16}$$

with $C = C(\Omega, q, r) > 0$; see [27], III, Lemma 2.6.1 concerning similar operations.

Spaces $L^q(0,T;X)$. Let $1 < q, s < \infty$. Then we introduce the usual Bochner space $L^s(0,T;X)$ with norm $\|\cdot\|_{L^s(0,T;X)} = \big(\int_0^T \|\cdot\|_X^s\, dt\big)^{1/s}$ where X is any Banach space with norm $\|\cdot\|_X$.

In the case $X = W^{\alpha,q}(\Omega)$, $-1 \le \alpha \le 1$, we set $\|\cdot\|_{L^s(0,T;W^{\alpha,q}(\Omega))} = \|\cdot\|_{\alpha;q,s,\Omega}$, and for $X = W^{\alpha,q}(\partial\Omega)$ let $\|\cdot\|_{L^s(0,T;W^{\alpha,q}(\partial\Omega))} = \|\cdot\|_{\alpha;q,s,\partial\Omega}$. Finally, if $X = L^q(\Omega)$ or $X = L^q(\partial\Omega)$, we set $\|\cdot\|_{L^s(0,T;L^q(\Omega))} = \|\cdot\|_{q,s,\Omega}$, and $\|\cdot\|_{L^s(0,T;L^q(\partial\Omega))} = \|\cdot\|_{q,s,\partial\Omega}$, resp. As duality pairing we define

$$\langle f, g\rangle_{\Omega,T} = \int_0^T \langle f, g\rangle_\Omega\, dt \quad \text{with} \quad \langle f, g\rangle_\Omega = \int_\Omega f \cdot g\, dx \tag{2.17}$$

for $f = (f_1, f_2, f_3) \in L^s(0,T;L^q(\Omega))$, $g = (g_1, g_2, g_3) \in L^{s'}(0,T;L^{q'}(\Omega))$. Similarly, we use the notation

$$\langle f, g\rangle_{\partial\Omega,T} = \int_0^T \langle f, g\rangle_{\partial\Omega}\, dt \quad \text{with} \quad \langle f, g\rangle_{\partial\Omega} = \int_{\partial\Omega} f \cdot g\, dS.$$

We also need the spaces $C^\nu\big([0,T);X\big)$, $\nu = 0, 1, 2, \ldots$, of X-valued functions $v(t)$, such that $v, (d/dt)v, \ldots, (d/dt)^\nu v$ are continuous on $[0,T)$.

We set $C^0\big([0,T);X\big) = C\big([0,T);X\big)$. The space $C_0^1\big([0,T);X\big)$ is the subspace of $C^1\big([0,T);X\big)$ consisting of functions v with compact support contained in $[0,T)$, whereas $C_0^1\big((0,T);X\big)$ is the subspace of $C^1\big([0,T);X\big)$-functions v with compact support contained in $(0,T)$.

Let $f \in L^s\big(0,T;L^q_\sigma(\Omega)\big)$. Then there exists a unique function

$$v \in L^s\big(0,T;D(A_q)\big) \text{ with } v_t \in L^s\big(0,T;L^q_\sigma(\Omega)\big) \text{ and } v \in C\big([0,T);L^q_\sigma(\Omega)\big),$$

satisfying the evolution system

$$v_t + A_q v = f, \quad 0 \le t < T, \quad v(0) = 0.$$

To be more precise,

$$v(t) = \int_0^t e^{-(t-\tau)A_q} f(\tau) d\tau, \quad 0 \le t < T,$$

and it holds the 'maximal regularity' estimate

$$\|v_t\|_{q,s,\Omega} + \|A_q v\|_{q,s,\Omega} \le C\|f\|_{q,s,\Omega}, \quad C = C(\Omega, q, s) > 0, \tag{2.18}$$

see [20], [26].

The space of initial values. Let $1 < q, s < \infty$. The space of initial values $\mathcal{J}^{q,s}(\Omega)$ consists of distributions u_0 satisfying $A_q^{-1} P_q u_0 \in L^q_\sigma(\Omega)$, see (2.13), and an additional integrability condition in time. To be more precise, let

$$\mathcal{J}^{q,s}(\Omega) = \{u_0 \in C_0^\infty(\Omega)' : \; A_q^{-1} P_q u_0 \in L^q_\sigma(\Omega), \; \int_0^\infty \|A_q e^{-tA_q} A_q^{-1} P_q u_0\|_q^s \, dt < \infty\} \tag{2.19}$$

and

$$\|u_0\|_{\mathcal{J}^{q,s}(\Omega)} = \|A_q^{-1} P_q u_0\|_{q,\Omega} + \Big(\int_0^\infty \|A_q e^{-tA_q} A_q^{-1} P_q u_0\|_q^s \, dt\Big)^{\frac{1}{s}}.$$

Obviously, $\|\cdot\|_{\mathcal{J}^{q,s}(\Omega)}$ defines a seminorm in $\mathcal{J}^{q,s}(\Omega)$ which becomes a norm if we identify two elements $u_0, v_0 \in \mathcal{J}^{q,s}(\Omega)$ satisfying $\|A_q^{-1} P_q(u_0 - v_0)\|_{\mathcal{J}^{q,s}(\Omega)} = 0$, i.e., $u_0 - v_0$ is a gradient. Of course, since $w(0) \in C^2_{0,\sigma}(\overline{\Omega})$ in (1.4) is solenoidal, initial values can be prescribed only modulo gradients.

As an example, let $u_0 \in C_0^\infty(\Omega)'$ and assume that $A_q^{-\frac{1}{s}+\varepsilon} P_q u_0 \in L^q_\sigma(\Omega)$ where $0 < \varepsilon < \frac{1}{s}$. Then by (2.12) $u_0 \in \mathcal{J}^{q,s}(\Omega)$. Using similar calculations as in [4] we can show that $u_0 \in B^{-2/s}_{q,s}(\Omega)$ is sufficient for $u_0 \in \mathcal{J}^{q,s}(\Omega)$ in the case $s \ge 2$; here $B^{-2/s}_{q,s}(\Omega)$ denotes a Besov space, see [2], [3], [4], [29].

Consider $u \in L^s\big(0,T; L^q(\Omega)\big)$ such that its time derivative $(A_q^{-1} P_q u)_t = A_q^{-1} P_q u_t \in L^s\big(0,T; L^q_\sigma(\Omega)\big)$ in the sense of distributions. Then, after redefining on a null set of $[0,T)$, we get $A_q^{-1} P_q u \in C\big([0,T); L^q_\sigma(\Omega)\big)$. Thus $A_q^{-1} P_q u(t) \in L^q_\sigma(\Omega)$ is well defined for each $t \in [0,T)$, and therefore $A_q^{-1} P_q u(0) = A_q^{-1} P_q u_0$ in (1.7) is well defined.

Let $v_0 \in L^q_\sigma(\Omega)$ such that $\int_0^\infty \|A_q e^{-tA_q} v_0\|_q^s \, dt < \infty$, let $f \in L^s\big(0,T; L^q_\sigma(\Omega)\big)$ and consider the general system $v_t + A_q v = f, v(0) = v_0$. Then we apply (2.18) to $\hat{v}(t) = v(t) - e^{-tA_q} v_0$, obtain the estimate

$$\|v_t\|_{q,s,\Omega} + \|A_q v\|_{q,s,\Omega} \le C\left(\Big(\int_0^T \|A_q e^{-tA_q} v_0\|_q^s \, dt\Big)^{\frac{1}{s}} + \|f\|_{q,s,\Omega}\right) \tag{2.20}$$

with $C = C(\Omega, q, s) > 0$, and the representation formula

$$v(t) = e^{-tA_q} v_0 + \int_0^t e^{-(t-\tau)A_q} f(\tau) \, d\tau, \quad 0 \le t < T. \tag{2.21}$$

3. Stationary very weak solutions

First we consider the linearized stationary system

$$-\Delta u + \nabla p = f, \quad \operatorname{div} u = k, \quad u\big|_{\partial\Omega} = g \tag{3.1}$$

with data $f = \operatorname{div} F$ and k, g satisfying

$$F \in L^r(\Omega),\ k \in L^r(\Omega),\ g \in W^{-\frac{1}{q},q}(\partial\Omega),\ \int_\Omega k\,dx = \langle N \cdot g\rangle_{\partial\Omega}\,, \tag{3.2}$$

where $1 < q < \infty$, $1 < r \leq q$, $\frac{1}{3} + \frac{1}{q} \geq \frac{1}{r}$. Here we follow [15] in principle, but in [15] the stronger condition $k \in L^q(\Omega)$ is supposed.

Modifying Definition 2 in an obvious way for the linearized case, $u \in L^q(\Omega)$ is called a *very weak solution* of the system (3.1) with data (3.2) if

$$-\langle u, \Delta w\rangle_\Omega + \langle g, N \cdot \nabla w\rangle_{\partial\Omega} = -\langle F, \nabla w\rangle_\Omega, \quad w \in C^2_{0,\sigma}(\overline{\Omega}), \tag{3.3}$$

and additionally the conditions (1.12) are satisfied, i.e., $\operatorname{div} u = k$ and $N \cdot u = N \cdot g$ on $\partial\Omega$. Our main result on (3.1), improving [15], Theorem 3, reads as follows.

Theorem 3. *Suppose the data F, k, g satisfy (3.2) with $1 < q < \infty$, $1 < r \leq q$, $\frac{1}{3} + \frac{1}{q} \geq \frac{1}{r}$. Then there exists a unique very weak solution $u \in L^q(\Omega)$ of the system (3.1) satisfying the estimate*

$$\|u\|_{q,\Omega} \leq C(\|F\|_{r,\Omega} + \|k\|_{r,\Omega} + \|g\|_{-1/q;q,\partial\Omega}) \tag{3.4}$$

with $C = C(\Omega, q, r) > 0$.

Remarks 1. **(1)** Setting $w \in C^\infty_{0,\sigma}(\Omega)$ in (3.3) and using de Rham's argument we find some pressure $p \in W^{-1,q}(\Omega)$ such that $-\Delta u + \nabla p = f$ holds in the sense of distributions, and that $\|p\|_{-1;q,\Omega}$ satisfies the same estimate as $\|u\|_{q,\Omega}$ in (3.4). Moreover we conclude from (3.3) that

$$A_r^{-\frac{1}{2}} P_r \Delta u = -A_r^{-\frac{1}{2}} P_r \operatorname{div} F, \tag{3.5}$$

where $A_r^{-\frac{1}{2}} P_r \operatorname{div} F$ is defined by the relation $\langle A_r^{-\frac{1}{2}} P_r \operatorname{div} F, A_{r'}^{\frac{1}{2}} v\rangle_\Omega = -\langle F, \nabla v\rangle_\Omega$, $v \in C^\infty_{0,\sigma}(\Omega)$, see (2.14) and (2.15).

(2) Assume for a moment that u is sufficiently smooth. Then, inserting $w = \hat{w}^h$ from (2.8), (2.9) in the expression $\langle u, \Delta w\rangle_\Omega$, and using integration by parts, we obtain an explicit trace formula for $u_{|_{\partial\Omega}}$ as a functional in $W^{-\frac{1}{q},q}(\partial\Omega)$. To be more precise, the map $h \mapsto \langle u_{|_{\partial\Omega}}, h\rangle_{\partial\Omega}$ for $h \in W^{1-1/q',q'}(\partial\Omega) = W^{\frac{1}{q},q'}(\partial\Omega)$, $N \cdot h_{|_{\partial\Omega}} = 0$ is defined by

$$\langle u_{|_{\partial\Omega}}, h\rangle_{\partial\Omega} = \langle u, \Delta\hat{w}^h\rangle_\Omega - \langle A_r^{-\frac{1}{2}} P_r \Delta u, A_{r'}^{\frac{1}{2}} \hat{w}^h\rangle_\Omega, \quad h \in W^{-\frac{1}{q},q'}(\partial\Omega). \tag{3.6}$$

Using (2.9) with q replaced by q' we then obtain the estimate

$$|\langle u_{|_{\partial\Omega}}, h\rangle_{\partial\Omega}| \leq C(\|u\|_{q,\Omega} + \|A_r^{-\frac{1}{2}} P_r \Delta u\|_{r,\Omega})\|h\|_{1/q;q';\partial\Omega} \tag{3.7}$$

with $C = C(\Omega, q, r) > 0$. Formula (3.6) is well defined for each very weak solution $u \in L^q(\Omega)$ and yields an explicit formula for the tangential component of $u_{|_{\partial\Omega}}$. The normal component $N \cdot u_{|_{\partial\Omega}}$ is well defined by (1.12), (2.1). This shows that the trace $u_{|_{\partial\Omega}} \in W^{-\frac{1}{q},q}(\partial\Omega)$ is well defined for a very weak solution u, and (2.1), (3.7) yield the estimate

$$\|u_{|_{\partial\Omega}}\|_{-\frac{1}{q};q,\partial\Omega} \leq C(\|u\|_{q,\Omega} + \|A_r^{-\frac{1}{2}} P_r \Delta u\|_{r,\Omega} + \|\operatorname{div} u\|_{r,\Omega}) \tag{3.8}$$

with $C = C(\Omega, q, r) > 0$.

(3) We conclude that a very weak solution $u \in L^q(\Omega)$ of (3.1) satisfies the conditions (1.12), (3.5) and the condition $u_{|_{\partial\Omega}} = g$ as elements of $W^{-\frac{1}{q},q}(\partial\Omega)$. Conversely, if $u \in L^q(\Omega)$ satisfies (1.12), (3.5), and the (well-defined) trace $u_{|_{\partial\Omega}}$ is equal to g, then u is a very weak solution of (3.1).

Proof of Theorem 3. Following in principle [15], we first assume that $u \in L^q(\Omega)$ is a given very weak solution u of (3.1), and prepare some estimates.

Using the trace map $W^{1,q'}(\Omega) \to W^{1-1/q',q'}(\partial\Omega)$ and the embedding estimate (2.11) we obtain that

$$|\langle g, N \cdot \nabla w\rangle_{\partial\Omega}| \leq C\|g\|_{-\frac{1}{q};q,\partial\Omega} \|v\|_{q',\Omega}, \quad w \in C^2_{0,\sigma}(\overline{\Omega}),\ v = A_{q'}w \tag{3.9}$$

with $C = C(\Omega, q) > 0$. Therefore, the functional $v \mapsto \langle g, N \cdot \nabla A_{q'}^{-1} v\rangle_{\partial\Omega}$ is continuous in $\|v\|_{q',\Omega}$, and it holds $\langle g, N \cdot \nabla A_{q'}^{-1} v\rangle_{\partial\Omega} = \langle G, v\rangle_\Omega$ with some unique $G \in L^q_\sigma(\Omega)$ satisfying $\|G\|_{q,\Omega} \leq C\|g\|_{-\frac{1}{q};q,\partial\Omega}$.

Similarly, $v \mapsto \langle F, \nabla A_{q'}^{-1} v\rangle_\Omega$, $v = A_{q'}w$, is continuous in $\|v\|_{q',\Omega}$, and with $\hat{F} = -A_q^{-1} P_q \operatorname{div} F$, see (2.15), we get that $\langle F, \nabla A_{q'}^{-1} v\rangle_\Omega = \langle \hat{F}, v\rangle_\Omega$ and $\|\hat{F}\|_{q,\Omega} \leq C\|F\|_{r,\Omega}$ with $C = C(\Omega, q, r) > 0$ by (2.16).

Using E^h for $h = N \cdot g$, cf. (2.6), the compatibility condition in (3.2) yields $\int_\Omega (\operatorname{div} E^h - k) dx = 0$. Hence there exists $b \in W_0^{1,r}(\Omega)$ satisfying $\operatorname{div} b = \operatorname{div} E^h - k$ and, due to (2.6), (2.4),

$$\|b\|_{q,\Omega} \leq C_1 \|\nabla b\|_{r,\Omega} \leq C_2(\|\operatorname{div} E^h\|_{r,\Omega} + \|k\|_{r,\Omega}) \tag{3.10}$$

where $C_i = C_i(\Omega, q, r) > 0$, $i = 1, 2$. Then we use the solution $\nabla H \in L^q(\Omega)$ of the weak Neumann problem

$$\Delta H = \operatorname{div}(E^h - b) = k = \operatorname{div} u, \quad N \cdot \nabla H_{|_{\partial\Omega}} = N \cdot (E^h - b)_{|_{\partial\Omega}},$$

and applying (2.10), (2.6), (3.10) leads to the estimate

$$\|\nabla H\|_{q,\Omega} \leq C_1 \|E^h - b\|_{q,\Omega} \leq C_2(\|g\|_{-\frac{1}{q};q,\partial\Omega} + \|k\|_{r,\Omega}) \tag{3.11}$$

with $C_i = C_i(\Omega, q, r) > 0$. Further we get from (2.10) that $\|\nabla H\|_{q,\Omega} \leq C\|u\|_{q,\Omega}$ with $C = C(\Omega, q) > 0$. Obviously, ∇H only depends on the data k, g.

Using (2.2) and (2.3) with $\operatorname{curl} \nabla H = 0$ we conclude that $\nabla H_{|_{\partial\Omega}} \in W^{-\frac{1}{q},q}(\partial\Omega)$ is well defined, and that

$$\|\nabla H\|_{-\frac{1}{q};q,\partial\Omega} \leq C(\|g\|_{-\frac{1}{q};q,\partial\Omega} + \|k\|_{r,\Omega}) \tag{3.12}$$

with $C = C(\Omega, q, r) > 0$.

Set $\hat{u} = P_q u = u - \nabla H \in L^q_\sigma(\Omega)$ and $\hat{g} = g - \nabla H_{|\partial\Omega} \in W^{-\frac{1}{q},q}(\partial\Omega)$. Then

$$|\langle \hat{g}, N \cdot \nabla w\rangle_{\partial\Omega}| \le C(\|g\|_{-\frac{1}{q};q,\partial\Omega} + \|k\|_{r,\Omega})\|v\|_{q',\Omega}, \quad w \in C^2_{0,\sigma}(\overline{\Omega}),\ v = A_{q'}w, \tag{3.13}$$

cf. (3.9). As above, we construct $\hat{G} \in L^q_\sigma(\Omega)$ satisfying $\langle \hat{g}, N \cdot \nabla A_{q'}^{-1} v\rangle_{\partial\Omega} = \langle \hat{G}, v\rangle_\Omega$, and the estimate

$$\|\hat{G}\|_{q,\Omega} \le C(\|g\|_{-\frac{1}{q};q,\partial\Omega} + \|k\|_{r,\Omega}) \tag{3.14}$$

holds with $C = C(\Omega, q, r) > 0$.

In the next step we use the relation $\langle \nabla H, \Delta w\rangle_\Omega = \langle \nabla H, N \cdot \nabla w\rangle_{\partial\Omega}, w$ as in (3.13), which follows using an approximation of H by smooth functions and an integration by parts. Then by (3.3) a calculation leads to $\langle \hat{g}, N \cdot \nabla w\rangle_{\partial\Omega} + \langle F, \nabla w\rangle_\Omega = -\langle \hat{u}, v\rangle_\Omega$, and inserting $\hat{G}, \hat{F}$ yields $\langle \hat{G}, v\rangle_\Omega + \langle \hat{F}, v\rangle_\Omega = -\langle \hat{u}, v\rangle_\Omega$. From the regularity properties of the Stokes operator $A_{q'}$, see [26], we know that the set of all $v = A_{q'}w$ with $w \in C^2_{0,\sigma}(\overline{\Omega})$ is dense in $L^{q'}_\sigma(\Omega)$. Thus we may use the last relation for all $v \in L^{q'}_\sigma(\Omega)$, and we get the representation formula

$$u = \nabla H - \hat{G} - \hat{F} \tag{3.15}$$

for the given very weak solution u.

Since the right-hand side of (3.15) depends only on the data F, k, g, we can use (3.15) to construct $u \in L^q(\Omega)$. Then the same arguments as above show that u satisfies (3.3) and (1.12). Thus u defined by (3.15) is a very weak solution of (3.1). Since each given very weak solution of (3.1) has the form (3.15), we obtain the uniqueness assertion. The estimate (3.4) follows from (3.12), (3.14), and (2.16). This proves Theorem 3. □

Remarks 2. **(1)** Suppose that the data F, k, g in (3.1), (3.2) satisfy the stronger conditions $F \in L^q(\Omega)$, $k \in L^q(\Omega)$, $g \in W^{1-1/q,q}(\partial\Omega)$, $1 < q < \infty$. Then the very weak solution u in Theorem 3 satisfies $u \in W^{1,q}(\Omega)$ and estimate

$$\|u\|_{1;q,\Omega} \le C(\|F\|_{q,\Omega} + \|k\|_{q,\Omega} + \|g\|_{1-1/q;q,\partial\Omega}) \tag{3.16}$$

with $C = C(\Omega, q) > 0$. The existence of such a solution u of (3.1) is well known, see [10], [12], [14]. Since u is obviously also a very weak solution which is unique, we conclude this regularity property; see also [15], Lemma 4.

(2) In the same way we conclude that if the data in (3.1), (3.2) satisfy the conditions $f = \operatorname{div} F \in L^q(\Omega)$, $k \in W^{1,q}(\Omega)$, $g \in W^{2-1/q,q}(\partial\Omega)$, then this solution satisfies

$$\|u\|_{2;q,\Omega} \le C(\|f\|_{q,\Omega} + \|k\|_{1;q,\Omega} + \|g\|_{2-1/q;q,\partial\Omega}) \tag{3.17}$$

with $C = C(\Omega, q) > 0$. Thus (3.16) and (3.17) are regularity properties of the very weak solution u if the data are sufficiently smooth.

Proof of Theorem 2. Following [15] we first consider a given very weak solution $u \in L^q(\Omega)$ of the system (1.9). Using similar arguments as in the previous proof we obtain, since $u\cdot\nabla u = \operatorname{div}(uu) - ku$ and $\frac{1}{3}+\frac{1}{q}=\frac{1}{r}$, that

$$|\langle u\cdot\nabla u, w\rangle_\Omega| \le C(\|u\|^2_{q,\Omega} + \|k\|_{r,\Omega}\,\|u\|_{q,\Omega})\|\nabla w\|_{r',\Omega}, \quad w\in C^2_{0,\sigma}(\overline{\Omega}). \tag{3.18}$$

Hence we find $W(u)\in L^r(\Omega)$ satisfying

$$\langle u\cdot\nabla u, w\rangle_\Omega = \langle \operatorname{div}(uu) - ku, w\rangle_\Omega = \langle \operatorname{div} W(u), w\rangle_\Omega = -\langle W(u), \nabla w\rangle_\Omega, \tag{3.19}$$

and

$$\|W(u)\|_{r,\Omega} \le C(\|u\|^2_{q,\Omega} + \|k\|_{r,\Omega}\,\|u\|_{q,\Omega}) \tag{3.20}$$

with $C = C(\Omega,q,r) > 0$, similarly as in (2.15), (2.16). We see that u is a very weak solution of the linear system

$$-\Delta u + \nabla p = \operatorname{div}\big(F - W(u)\big), \quad \operatorname{div} u = k, \quad u_{|\partial\Omega} = g. \tag{3.21}$$

Then (3.15) leads to the formula

$$u = \nabla H - \hat{G} - \hat{F} - \hat{W}(u) \tag{3.22}$$

where $\nabla H, \hat{G}, \hat{F}, \hat{W}(u)$ are determined by

$$\begin{aligned}
\Delta H &= k, \quad N\cdot(\nabla H - g)_{|\partial\Omega} = 0,\\
\langle \hat{G}, v\rangle_\Omega &= \langle g - \nabla H_{|\partial\Omega}, N\cdot\nabla A^{-1}_{q'} v\rangle_{\partial\Omega},\\
\langle \hat{F}, v\rangle_\Omega &= \langle F, \nabla A^{-1}_{q'} v\rangle_\Omega,\\
\langle \hat{W}(u), v\rangle_\Omega &= -\langle uu, \nabla A^{-1}_{q'} v\rangle_\Omega - \langle ku, A^{-1}_{q'} v\rangle_\Omega \quad \text{for all } w\in C^2_{0,\sigma}(\overline{\Omega}),\ v = A_{q'}w.
\end{aligned}$$

Setting $\hat{u} = u - \nabla H$, $\mathcal{F}(\hat{u}) = -\hat{F} - \hat{G} - \hat{W}(\hat{u} + \nabla H)$, we obtain from (3.22) the equation $\hat{u} = \mathcal{F}(\hat{u})$, which can be solved by Banach's fixed point theorem. This leads to the desired solution $u = \hat{u} + \nabla H$.

For this purpose we use similar estimates as in the previous proof, and obtain

$$\begin{aligned}
\|\mathcal{F}(\hat{u})\|_{q,\Omega} \le\ & C(\|\hat{u}\|_{q,\Omega} + \|g\|_{-\frac{1}{q};q,\partial\Omega} + \|k\|_{r,\Omega})^2 \\
&+ C\|k\|_{r,\Omega}\big(\|\hat{u}\|_{q,\Omega} + \|g\|_{-\frac{1}{q};q,\partial\Omega} + \|k\|_{r,\Omega}\big) \\
&+ C\big(\|F\|_{r,\Omega} + \|g\|_{-\frac{1}{q};q,\partial\Omega} + \|k\|_{r,\Omega}\big)
\end{aligned} \tag{3.23}$$

with $C = C(\Omega,q) > 0$. Setting $a = C$, $\beta = \|g\|_{-\frac{1}{q};q,\partial\Omega} + \|k\|_{r,\Omega}$, $\gamma = C(\|F\|_{r,\Omega} + \beta) + \beta$, $\delta = C\|k\|_{r,\Omega}$, we obtain the estimate

$$\|\mathcal{F}(\hat{u})\|_{q,\Omega} + \beta \le a(\|\hat{u}\|_{q,\Omega} + \beta)^2 + \delta(\|\hat{u}\|_{q,\Omega} + \beta) + \gamma. \tag{3.24}$$

Then we consider the closed ball $\mathcal{B} = \{\hat{u}\in L^q_\sigma(\Omega);\ \|\hat{u}\|_{q,\Omega} + \beta \le y_1\}$ where $y_1 > 0$ means the smallest root of the equation $y = ay^2 + \delta y + \gamma$. Supposing the smallness condition $4a\gamma + 2\delta < 1$ we get $y_1 > \beta$ and $\|\mathcal{F}(\hat{u}) - \mathcal{F}(\hat{v})\|_{q,\Omega} \le \hat{a}\|\hat{u} - \hat{v}\|_{q,\Omega}$ with some $0 < \hat{a} < 1$. Now Banach's fixed point theorem yields a unique solution $\hat{u}\in\mathcal{B}$ with $\hat{u} = \mathcal{F}(\hat{u})$, see [27], V.4.2, for details. Then $u = \hat{u} + \nabla H$ solves (3.22) and is a very weak solution of (1.9). The smallness condition $4a\gamma + 2\delta < 1$ can be written in the form (1.13).

To prove uniqueness we follow [15] and assume that there exists another very weak solution $v \in L^q(\Omega)$ of the system (1.9) with the same data F, k, g as for u. Setting $U = u - v$ we can show that the equation

$$\langle U, \Delta w + v \cdot \nabla w + u \cdot (\nabla w)^T + kw \rangle_\Omega = 0 \tag{3.25}$$

is satisfied for all $w \in C^2_{0,\sigma}(\overline{\Omega})$, where "$T$" denotes the transpose. Note that v need not satisfy any smallness condition, but that $\|u\|_q$ is small. Then, by standard arguments, we solve for each $G \in C_0^\infty(\Omega)$ the modified Stokes system

$$-\Delta w - v \cdot \nabla w - u \cdot (\nabla w)^T - kw + \nabla \pi = G, \quad \operatorname{div} w = 0, \; w_{|_{\partial\Omega}} = 0$$

to get a solution $(w, \nabla\pi) \in \big(W^{2,3/2}(\Omega) \cap W_0^{1,3/2}(\Omega)\big) \times L^{3/2}(\Omega)$. Since $C^2_{0,\sigma}(\overline{\Omega})$ is dense in $\big(W^{2,3/2}(\Omega) \cap W_0^{1,3/2}(\Omega)\big) \cap L^{3/2}_\sigma(\Omega)$, we may insert this w in (3.25) which finally shows that $U = 0$, $u = v$. This proves Theorem 2. The estimate (1.14) is an easy consequence. □

4. Nonstationary very weak solutions

Consider the linearized nonstationary system

$$u_t - \Delta u + \nabla p = f, \quad \operatorname{div} u = k, \quad u_{|_{\partial\Omega}} = g, \quad u_{|_{t=0}} = u_0 \tag{4.1}$$

with data $f = \operatorname{div} F$ and k, g, u_0 satisfying

$$\begin{aligned} &F \in L^s\big(0,T; L^r(\Omega)\big), \quad k \in L^s\big(0,T; L^r(\Omega)\big), \quad g \in L^s\big(0,T; W^{-\frac{1}{q},q}(\partial\Omega)\big), \\ &u_0 \in \mathcal{J}^{q,s}(\Omega), \quad \int_\Omega k\,dx = \langle N \cdot g, 1\rangle_{\partial\Omega} \quad \text{for a.a. } t \in [0,T), \end{aligned} \tag{4.2}$$

where $1 < r \le q < \infty$, $\frac{1}{3} + \frac{1}{q} \ge \frac{1}{r}$, $1 < s < \infty$.

Modifying Definition 1 in an obvious way for the linearized case, a vector field $u \in L^s\big(0,T; L^q(\Omega)\big)$ is called a *very weak solution* of the system (4.1) with data (4.2) if

$$-\langle u, w_t\rangle_{\Omega,T} - \langle u, \Delta w\rangle_{\Omega,T} + \langle g, N \cdot \nabla w\rangle_{\partial\Omega,T} = \langle u_0, w(0)\rangle_\Omega - \langle F, \nabla w\rangle_{\Omega,T}, \tag{4.3}$$

$w \in C_0^1\big([0,T); C^2_{0,\sigma}(\overline{\Omega})\big)$, and additionally the conditions (1.6) are satisfied, i.e., $\operatorname{div} u = k$ and $N \cdot u_{|_{\partial\Omega}} = N \cdot g$ a.e. in $(0,T)$.

Our main result on this system reads as follows.

Theorem 4. *Suppose that the data F, k, g and u_0 satisfy (4.2) with $1 < r \le q < \infty$, $\frac{1}{3} + \frac{1}{q} \ge \frac{1}{r}$, $1 < s < \infty$. Then there exists a unique very weak solution $u \in L^s\big(0,T; L^q(\Omega)\big)$ of the system (4.1), satisfying*

$$A_q^{-1} P_q u_t \in L^s\big(0,T; L^q_\sigma(\Omega)\big), A_q^{-1} P_q u \in C\big([0,T); L^q_\sigma(\Omega)\big), A_q^{-1} P_q u_{|_{t=0}} = A_q^{-1} P_q u_0$$

and

$$\|A_q^{-1}P_q u_t\|_{q,s,\Omega} + \|u\|_{q,s,\Omega} \le C\big(\|u_0\|_{\mathcal{J}^{q,s}(\Omega)} + \|F\|_{r,s,\Omega} + \|k\|_{r,s,\Omega} + \|g\|_{-\frac{1}{q};q,s,\partial\Omega}\big) \tag{4.4}$$

with $C = C(\Omega, q, r, s) > 0$.

Remarks 3. (1) Setting in particular $w \in C_0^\infty\big((0,T); C_{0,\sigma}^\infty(\Omega)\big)$ in $\big(4.3\big)$, we obtain, see [27], p. 248, p. 202, [28], the existence of a distribution p such that $u_t - \Delta u + \nabla p = f$ holds in $\Omega \times (0,T)$ in the sense of distributions.

(2) Let $h = (h_1, h_2, h_3) \in C_0^1\big((0,T); W^{1-1/q',q'}(\partial\Omega)\big)$ with $N \cdot h_{|_{\partial\Omega}} = 0$. Then $h(t) \mapsto \hat{w}^{h(t)}$, see $\big(2.8\big)$, $\big(2.9\big)$, is a linear mapping satisfying $(\hat{w}^h)_t = \hat{w}^{h_t}$. We may insert $w = \hat{w}^h$ in $\big(4.3\big)$ and obtain the formula

$$\langle g, h\rangle_{\partial\Omega,T} = \langle u, \hat{w}^{h_t}\rangle_{\Omega,T} + \langle u, \Delta\hat{w}^h\rangle_{\Omega,T} - \langle F, \nabla\hat{w}^h\rangle_{\Omega,T}. \tag{4.5}$$

Since the normal component $N \cdot h$ of the test function h is zero, this formula yields a well-defined expression for the tangential component $N \times g$ of the boundary values. It is easy to see using integration by parts that $N \times g$ coincides with the usual trace $N \times u_{|_{\partial\Omega}}$ if u is sufficiently smooth. Therefore, we may call the right-hand side of $\big(4.5\big)$ the trace $N \times u_{|_{\partial\Omega}}$ of the tangential component of u at $\partial\Omega$ in the sense of distributions. Since the normal component $N \cdot u_{|_{\partial\Omega}}$ of u at $\partial\Omega$ is well defined by $\big(1.6\big)$, we get an explicit trace formula for $u_{|_{\partial\Omega}}$ in the sense of distributions at $\partial\Omega$ which coincides with the usual trace of u at $\partial\Omega$ if u is smooth. This yields a precise meaning of the general boundary condition $u_{|_{\partial\Omega}} = g$ in the sense of boundary distributions.

(3) Since $w(0)$ in $\big(4.3\big)$ is solenoidal we expect that the initial condition $u_{|_{t=0}} = u_0$ only makes sense "modulo gradients". Therefore, the condition $A_q^{-1}P_q u_{|_{t=0}} = A_q^{-1}P_q u_0$, see $\big(1.7\big)$, seems to be the adequate precise formulation of the initial condition $u_{|_{t=0}} = u_0$. If u is sufficiently smooth, we need additional (necessary) compatibility conditions in order to reach that $u(0) = u_0$, see $\big(4.14\big)$.

Proof of Theorem 4. Let $E(t) = E^{k(t),g(t)} \in L^q(\Omega)$ be the very weak solution of the stationary system

$$-\Delta E(t) + \nabla p(t) = 0, \quad \operatorname{div} E(t) = k(t),\ E(t)_{|_{\partial\Omega}} = g(t) \text{ for a.a. } t \in [0,T]. \tag{4.6}$$

Then from $\big(3.3\big)$ we obtain the relation $\langle g, N\cdot\nabla w\rangle_{\partial\Omega,T} = \langle E, \Delta w\rangle_{\Omega,T}$ for every $w \in C_0^1\big([0,T); C_{0,\sigma}^2(\overline{\Omega})\big)$. Therefore, given a very weak solution u, $\big(4.3\big)$ can be written in the form

$$-\langle A_q^{-1}P_q u, v_t\rangle_{\Omega,T} - \langle u - E, \Delta A_{q'}^{-1} v\rangle_{\Omega,T} = \langle A_q^{-1}P_q u_0, v(0)\rangle_\Omega - \langle F, \nabla A_{q'}^{-1} v\rangle_{\Omega,T} \tag{4.7}$$

where $w \in C_0^1\big([0,T); C_{0,\sigma}^2(\overline{\Omega})\big)$, $v = A_{q'}w$. Since $A_q^{-1}P_q u_0 \in L_\sigma^q(\Omega)$, see $\big(2.19\big)$, and since $\operatorname{div}(u - E) = 0$, $N \cdot (u - E)_{|_{\partial\Omega}} = 0$ yielding $u - E = P_q(u - E)$ we obtain that

$$\begin{aligned}\langle u-E,\Delta A_{q'}^{-1}v\rangle_{\Omega,T} &= \langle P_q(u-E),\Delta A_{q'}^{-1}v\rangle_{\Omega,T}\\ &= \langle u-E,P_{q'}\Delta A_{q'}^{-1}v\rangle_{\Omega,T} = -\langle u-E,v\rangle_{\Omega,T}.\end{aligned}$$

Further we use (2.15) for a.a. $t\in[0,T)$ and get a unique $\hat F = -A_q^{-1}P_q \operatorname{div} F \in L^s\big(0,T;L^q_\sigma(\Omega)\big)$ satisfying the relation

$$\langle F,\nabla A_{q'}^{-1}v\rangle_{\Omega,T} = \langle \hat F, v\rangle_{\Omega,T} \quad \text{for all } v = A_{q'}w,\ w\in C_0^1\big([0,T);C^2_{0,\sigma}(\overline{\Omega})\big). \tag{4.8}$$

This leads to the relation

$$\begin{aligned}-\langle A_q^{-1}P_q u, v_t\rangle_{\Omega,T} &= \langle A_q^{-1}P_q u_0, v(0)\rangle_\Omega - \langle P_q u, v\rangle_{\Omega,T}\\ &\quad + \langle P_q E, v\rangle_{\Omega,T} + \langle A_q^{-1}P_q \operatorname{div} F, v\rangle_{\Omega,T}.\end{aligned} \tag{4.9}$$

Then a standard argument shows, see [28], III, 1.1 or [27], IV, 1.3, that $A_q^{-1}P_q u_t \in L^s\big(0,T;L^q_\sigma(\Omega)\big)$ is well defined, that

$$A_q^{-1}P_q u \in C\big([0,T);L^q_\sigma(\Omega)\big), \quad A_q^{-1}P_q u(0) = A_q^{-1}P_q u_0,$$

and that the evolution system

$$(A_q^{-1}P_q u)_t + A_q(A_q^{-1}P_q u) = A_q^{-1}P_q \operatorname{div} F + P_q E^{k,g}, \quad (A_q^{-1}P_q u)(0) = A_q^{-1}P_q u_0 \tag{4.10}$$

is satisfied. From (2.21) we now obtain the representation formula

$$\begin{aligned}\hat u(t) &\equiv P_q u(t)\\ &= A_q e^{-tA_q}A_q^{-1}P_q u_0 + \int_0^t A_q e^{-(t-\tau)A_q}\big(A_q^{-1}P_q \operatorname{div} F + P_q E^{k,g}\big)d\tau\end{aligned} \tag{4.11}$$

for the very weak solution u. As in (3.22) we get $\hat u(t) = P_q u(t) = u(t) - \nabla H(t)$ where $\nabla H(t)$ is determined by $\Delta H(t) = k(t)$, $N\cdot\big(\nabla H(t) - g(t)\big)_{|\partial\Omega} = 0$ for a.a. $t\in[0,T)$. Since $\nabla H(t)$ only depends on $g(t), k(t)$, see (3.11), we obtain by (4.11) a formula for $u = \hat u + \nabla H$ which determines u uniquely by the data F, k, g and u_0.

Now use (4.11) to construct a very weak solution u. Using the same calculation as above we obtain the existence assertion of u, the uniqueness of which follows from the representation (4.11). The estimate (4.4) is based on (2.20) and the estimates of $E, \hat F$, see (3.4) and (2.16).

Using (2.20) we conclude that the term $\|u_0\|_{\mathcal{J}^{q,s}(\Omega)}$ in (4.4) can be replaced by the weaker norm

$$\|u_0\|_{\mathcal{J}^{q,s}_T(\Omega)} \equiv \Big(\int_0^T \|A_q e^{-tA_q}A_q^{-1}P_q u_0\|^s_{q,\Omega}\,dt\Big)^{\frac{1}{s}}. \tag{4.12}$$

Now the proof of Theorem 4 is complete. □

Next we consider some regularity properties. Suppose the data $f = \operatorname{div} F$ and k, g, u_0 of the system (4.1) satisfy the stronger conditions

$$\begin{gathered} F \in L^s\big(0,T;W^{1,q}(\Omega)\big),\ k \in L^s\big(0,T;W^{1,q}(\Omega)\big),\ k_t \in L^s\big(0,T;L^r(\Omega)\big), \\ g \in L^s\big(0,T;W^{2-1/q,q}(\partial\Omega)\big),\ g_t \in L^s\big(0,T;W^{-\frac{1}{q},q}(\partial\Omega)\big),\ u_0 \in W^{2,q}(\Omega) \end{gathered} \tag{4.13}$$

with $1 < r \le q < \infty$, $\frac{1}{3} + \frac{1}{q} \ge \frac{1}{r}$, $1 < s < \infty$, and the compatibility conditions

$$\int_\Omega k\,dx = \langle N \cdot g, 1\rangle_{\partial\Omega} \text{ for } t \in [0,T), \quad u_0\big|_{\partial\Omega} = g(0), \quad \operatorname{div} u_0 = k(0). \tag{4.14}$$

Observe that $g(0)$ and $k(0)$ are well defined because of the assumptions on k_t, g_t. For simplicity the assumption on u_0 is not optimally chosen.

Then we will show the existence of a unique solution $u \in L^s\big(0,T;W^{2,q}(\Omega)\big)$ with $u_t \in L^s\big(0,T;L^q(\Omega)\big)$ of the system (4.1), together with a pressure function p such that $\nabla p \in L^s\big(0,T;L^q(\Omega)\big)$, satisfying the estimate

$$\begin{aligned} &\|u_t\|_{q,s,\Omega} + \|u\|_{2;q,s,\Omega} + \|\nabla p\|_{q,s,\Omega} \\ &\quad \le C\Big(\|u_0\|_{2;q,\Omega} + \|f\|_{q,s,\Omega} + \|k\|_{1;q,s,\Omega} + \|k_t\|_{r,s,\Omega} \\ &\qquad + \|g\|_{2-1/q;q,s,\partial\Omega} + \|g_t\|_{-1/q;q,s,\partial\Omega}\Big) \end{aligned} \tag{4.15}$$

with $C = C(\Omega, q, s) > 0$. The equations $u_t - \Delta u + \nabla p = f$, $\operatorname{div} u = k$ and $u\big|_{\partial\Omega} = g$ in (4.1) are satisfied in the strong sense for a.a. $t \in [0,T)$; hence $u \in C\big([0,T);L^q(\Omega)\big)$ and $u(0) = u_0$ is well defined. The initial value $u_0 \in W^{2,q}(\Omega)$ can be treated as a functional from $\mathcal{J}^{q,s}(\Omega)$, see (2.19). Thus we see that (4.13), (4.14) are stronger than the conditions (4.2).

Therefore, Theorem 4 yields a unique very weak solution $u \in L^s\big(0,T;L^q(\Omega)\big)$ to (4.1) which coincides with each more regular solution by the uniqueness property.

To show the existence of a regular solution u satisfying (4.15) we first suppose that such a solution is given. Let $E = E^{k,g}$ be chosen as in (4.6). Then $E(0) = E^{k(0),g(0)}$ satisfies the system

$$-\Delta E(0) + \nabla p(0) = 0, \quad \operatorname{div} E(0) = k(0), \quad E(0)\big|_{\partial\Omega} = g(0)$$

and (3.17) shows that $E(0) \in W^{2,q}(\Omega)$. Using (4.14) we see that $u_0 - E(0)\big|_{\partial\Omega} = 0$, $\operatorname{div}\big(u_0 - E(0)\big) = 0$ which leads to $u_0 - E(0) \in D(A_q)$. Further, using $E_t = (E^{k,g})_t = E^{k_t,g_t}$, the assumptions on k, k_t, g, g_t, estimate (3.4) with u, F, k, g replaced by $E_t, 0, k_t, g_t$, and the estimate (3.17) with u, f, k, g replaced by $E, 0, k, g$, we obtain the estimate

$$\begin{aligned} &\|E_t\|_{q,s,\Omega} + \|E\|_{2;q,s,\Omega} \\ &\quad \le C\big(\|k\|_{1;q,s,\Omega} + \|k_t\|_{r,s,\Omega} + \|g\|_{2-1/q;q,s,\Omega} + \|g_t\|_{-\frac{1}{q};q,s,\partial\Omega}\big) \end{aligned} \tag{4.16}$$

with $C = C(\Omega, q, s) > 0$.

Setting $\tilde{u}(t) = u(t) - E(t)$ we obtain the evolution system

$$\tilde{u}_t + A_q\tilde{u} = P_q f - P_q E_t,\ \operatorname{div}\tilde{u} = 0,\ \tilde{u}_{|_{\partial\Omega}} = 0,\ \tilde{u}_{|_{t=0}} = u_0 - E(0).$$

Then (2.21) yields the representation formula

$$u(t) = E(t) + e^{-tA_q}(u_0 - E(0)) + \int_0^t e^{-(t-\tau)A_q}(P_q f - P_q E_\tau)d\tau, \quad 0 \le t < T, \tag{4.17}$$

for the given regular solution u.

In the next step we use (4.17) to construct the desired solution u, and we use (2.20), (4.16). Further we apply (3.17) to $E(0)$, and use that $k(0) = \operatorname{div} u_0$, $g(0) = u_{0|_{\partial\Omega}}$. This yields the regularity properties of u, the estimate (4.15) for u, and its uniqueness. The pressure term ∇p, constructed by de Rham's argument, can be written in the form $\nabla p = f - u_t + \Delta u$ proving (4.15) for p.

Proof of Theorem 1. First let u be a given solution of (1.1) for some $0 < T' \le T$ with the properties of this theorem. Further we consider the solution $E = E^{F,k,g,u_0}$ of the corresponding linearized system

$$E_t - \Delta E + \nabla\hat{p} = \operatorname{div} F, \quad \operatorname{div} E = k, \quad E_{|_{\partial\Omega}} = g, \quad E_{|_{t=0}} = u_0$$

according to Theorem 4. Setting $\tilde{u} = u - E$, the calculation as in (3.19) shows that $\tilde{u}$ is a very weak solution of the linear system

$$\tilde{u}_t - \Delta\tilde{u} + \nabla\tilde{p} = -\operatorname{div} W(u), \quad \operatorname{div}\tilde{u} = 0, \quad \tilde{u}_{|_{\partial\Omega}} = 0, \quad \tilde{u}_{|_{t=0}} = 0, \tag{4.18}$$

where $W(u)$ is defined as in (3.19); in particular, $\operatorname{div} W(u) = \operatorname{div}(uu) - ku$. Using (3.20) we conclude that $W(u) \in L^{s/2}(0,T''; L^r(\Omega))$ for $0 < T'' < T'$. If $T' < \infty$ we set $T'' = T'$. Thus we may use the representation formula (4.11) with $k = 0$, $g = 0$, $u_0 = 0$, $F = -W(u) = -W(\tilde{u} + E)$ and with s replaced by $s/2$. Hence

$$\tilde{u}(t) = (\mathcal{F}(\tilde{u})(t) := -\int_0^t A_q e^{-(t-\tau)A_q} A_q^{-1} P_q \operatorname{div} W(u)\, d\tau, \quad 0 \le t < T'. \tag{4.19}$$

To solve (4.19) by Banach's fixed point theorem we have to estimate $\|\mathcal{F}(\tilde{u})\|_{q,s,\Omega}$ where $\|\cdot\|_{q,s,\Omega} = \|\cdot\|_{q,s,\Omega,T'} = \left(\int_0^{T'} \|\cdot\|_{q,\Omega}^s\, dt\right)^{\frac{1}{s}}$. Let $\alpha' = \frac{3}{2q}$ yielding $\frac{1}{2} - \alpha' + \frac{1}{s} = \frac{1}{s/2}$ and $2\alpha' + \frac{3}{q} = \frac{3}{q/2}$. Using (2.12) and (2.11) with $\alpha = \alpha'$, $\gamma = \frac{q}{2}$, we get that

$$\|\mathcal{F}(\tilde{u})(t)\|_{q,\Omega} \le C \int_0^t \frac{1}{(t-\tau)^{1/2+\alpha'}} \|A_{q/2}^{-1/2} P_{q/2} \operatorname{div} W(u)\|_{q/2}\, d\tau\,.$$

Looking at the integrand, we apply the estimate

$$\|A_{q/2}^{-1/2} P_{q/2}(ku)\|_{q/2} \le C\|P_\gamma(ku)\|_\gamma \le C\|k\|_r\|u\|_q$$

which is based on (2.11) with $\alpha = \frac{1}{2}$ and $\frac{1}{\gamma} = \frac{1}{3} + \frac{2}{q} = \frac{1}{r} + \frac{1}{q}$ and on Hölder's inequality. Furthermore, $\|A_{q/2}^{-1/2} P_{q/2} \operatorname{div}(uu)\|_{q/2} \leq C\|uu\|_{q/2} \leq C\|u\|_q^2$ since

$$|\langle A_{q/2}^{-1/2} P_{q/2} \operatorname{div}(uu), \varphi\rangle_\Omega| = | - \langle uu, \nabla A_{(q/2)'}^{-1/2} P_{(q/2)'} \varphi\rangle_\Omega| \leq C\|uu\|_{q/2}\|\varphi\|_{(q/2)'}$$

for all $\varphi \in L^{(q/2)'}(\Omega)$. Summarizing we conclude that

$$\|\mathcal{F}(\tilde{u})(t)\|_{q,\Omega} \leq C \int_0^t \frac{1}{(t-\tau)^{1/2+\alpha'}} \left(\|u\|_q^2 + \|k\|_r \|u\|_q\right) d\tau\,.$$

Then the Hardy-Littlewood inequality, see [27], p. 103, [29],

$$\left(\int_0^T |\int_0^t (t-\tau)^{\alpha-1} h(\tau)\, d\tau|^s \, dt\right)^{\frac{1}{s}} \leq C \left(\int_0^T |h(t)|^{\tilde{s}} \, dt\right)^{\frac{1}{\tilde{s}}}$$

with $\alpha = \frac{1}{2} - \alpha'$ and $\tilde{s} = \frac{s}{2}$ yields the estimate

$$\|\mathcal{F}(\tilde{u})\|_{q,s,\Omega} \leq C\Big(\big(\|\tilde{u}\|_{q,s,\Omega} + \|E\|_{q,s,\Omega}\big)^2 + \|k\|_{r,s,\Omega}\big(\|\tilde{u}\|_{q,s,\Omega} + \|E\|_{q,s,\Omega}\big)\Big) \quad (4.20)$$

with $C = C(\Omega, q, s) > 0$.

Setting $a = C$, $\beta = \|E\|_{q,s,\Omega}$, $\delta = C\|k\|_{r,s,\Omega}$ and $\gamma = \beta$, (4.20) is equivalent to the estimate

$$\|\mathcal{F}(\tilde{u})\|_{q,s,\Omega} + \beta \leq a(\|\tilde{u}\|_{q,s,\Omega} + \beta)^2 + \delta(\|\tilde{u}\|_{q,s,\Omega} + \beta) + \gamma; \quad (4.21)$$

cf. (3.24) in the proof of Theorem 2 for the stationary case. Thus, in the same way as in that proof, we obtain a solution $\tilde{u} \in L^s\big(0, T'; L^q(\Omega)\big)$ of the fixed point equation $\tilde{u} = \mathcal{F}(\tilde{u})$ if the condition $4a\gamma + 2\delta < 1$, i.e.,

$$4C\Big(\int_0^{T'} \|E\|_{q,\Omega}^s dt\Big)^{\frac{1}{s}} + 2C\Big(\int_0^{T'} \|k\|_{r,\Omega}^s dt\Big)^{\frac{1}{s}} < 1\,. \quad (4.22)$$

is satisfied. Using (4.4) and (4.12) we may use also the (weaker) smallness condition

$$\begin{aligned}\Big(\int_0^{T'} \|A_q e^{-tA_q} A_q^{-1} P_q u_0\|_{q,\Omega}^s dt\Big)^{\frac{1}{s}} + \Big(\int_0^{T'} \|F\|_{r,\Omega}^s dt\Big)^{\frac{1}{s}} \quad (4.23)\\ + \Big(\int_0^{T'} \|k\|_{r,\Omega}^s dt\Big)^{\frac{1}{s}} + \Big(\int_0^{T'} \|g\|_{-\frac{1}{q};q,\partial\Omega}^s dt\Big)^{\frac{1}{s}} < c\end{aligned}$$

with $c = c(\Omega, q, s) > 0$. This condition is always satisfied if $T' > 0$ is sufficiently small; note that the case $T' = \infty$ is possible.

Writing (4.19) in the form

$$\begin{aligned} A_q^{-1}\tilde{u}(t) &= A_q^{-1} P_q u(t) - A_q^{-1} P_q E(t) \quad (4.24)\\ &= -\int_0^t e^{-(t-\tau)A_q} A_q^{-1} P_q \operatorname{div} W(u) d\tau, \quad 0 \leq t < T', \end{aligned}$$

we conclude using (2.20), (2.16), (3.20), together with Hölder's inequality, that

$$\begin{aligned}\|(A_q^{-1}\tilde{u})_t\|_{q,s/2,\Omega} &\le C_1\|A_q^{-1}P_q \operatorname{div} W(u)\|_{q,s/2,\Omega} \le C_2\|W(u)\|_{r,s/2,\Omega}\\ &\le C_3\big(\|u\|^2_{q,s,\Omega} + \|k\|_{r,s,\Omega}\,\|u\|_{q,s,\Omega}\big) < \infty\end{aligned}$$

with C_i, $i = 1, 2, 3$, depending on Ω, q, s.

Furthermore, we obtain from (4.4) that

$$A_q^{-1}P_qE_t \in L^s\big(0, T''; L^q_\sigma(\Omega)\big) \subseteq L^{s/2}\big(0, T''; L^q_\sigma(\Omega)\big), \quad 0 < T'' < T'.$$

This proves that $A_q^{-1}P_qu_t \in L^{s/2}\big(0, T''; L^q_\sigma(\Omega)\big)$ for all T'' with $0 < T'' < T'$, and all $0 < T'' < \infty$ if $T' = \infty$.

A calculation shows that u defined by $u = \tilde{u} + E$ is a very weak solution of (1.1). To prove the uniqueness of u we assume that $v \in L^s\big(0, T'; L^q(\Omega)\big)$ is another very weak solution of (1.1). Setting $U = u - v$ we obtain in the same way as in (4.18) that U is a very weak solution of the system

$$\begin{aligned}&U_t - \Delta U + \nabla P = -\operatorname{div}(Uu) - \operatorname{div}(vU) + kU,\\ &\operatorname{div} U = 0, \quad U_{|_{\partial\Omega}} = 0, \quad U_{|_{t=0}} = 0.\end{aligned} \tag{4.25}$$

The same method as used for (4.18) and (4.19) then leads to the estimate

$$\|U\|_{q,s,\Omega} \le C(\|u\|_{q,s,\Omega} + \|v\|_{q,s,\Omega} + \|k\|_{r,s,\Omega})\,\|U\|_{q,s,\Omega} \tag{4.26}$$

with $C = C(\Omega, q, s) > 0$. Since $\|\cdot\|_{q,s,\Omega} = \|\cdot\|_{q,s,\Omega,T'}$, we observe that C does not depend on T'. Thus we can choose $T'' \in (0, T')$ such that

$$\|u\|_{q,s,\Omega,T''} + \|v\|_{q,s,\Omega,T''} + \|k\|_{r,s,\Omega,T''} \le \frac{1}{2C}\,.$$

This leads to $\|U\|_{q,s,\Omega,T''} \le 0$, hence $U = 0$ and $u = v$ on the interval $[0, T'']$. If $T'' < T$ we can continue this procedure and get $u = v$ on $[0, T')$ in finitely many steps. The proof of Theorem 1 is complete. □

References

[1] Adams, R.A.: Sobolev Spaces, Academic Press, New York 1975

[2] Amann, H.: Linear and Quasilinear Parabolic Equations, Birkhäuser Verlag, Basel 1995

[3] Amann, H.: Navier-Stokes Equations with Nonhomogeneous Dirichlet Data, Journal Nonlinear Math. Phys., 10, Suppl. 1 (2003), 1–11

[4] Amann, H.: Nonhomogeneous Navier-Stokes Equations with Integrable Low-Regularity Data, Int. Math. Ser., Kluwer Academic/Plenum Publishing, New York (2002), 1–26

[5] Berselli, L.C., and Galdi, G.P.: On the space-time regularity of $C(0, T; L^n)$-very weak solutions to the Navier-Stokes equations. Nonlinear Anal., 58 (2004), 703–717

[6] Bogowski, M.E.: Solution of the First Boundary Value Problem for the Equation of Continuity of an Incompressible Medium, Soviet Math. Dokl., 20 (1979), 1094–1098

[7] Cannone, M.: Viscous flows in Besov Spaces, Advances in Mathematical Fluid Mechanics, Springer-Verlag, Berlin (2002), 1–34

[8] Fabes, E.B., Jones, B.F., and Rivière, N.M.: The Initial Value Problem for the Navier-Stokes Equations with Data in L^p, Arch. Rational Mech. Anal., 45 (1972), 222–240

[9] Farwig, R., and Sohr, H.: The stationary and nonstationary Stokes system in exterior domains with non-zero divergence and non-zero boundary values, Math. Methods Appl. Sci., 17 (1994), 269–291

[10] Farwig, R., and Sohr, H.: Generalized Resolvent Estimates for the Stokes System in Bounded and Unbounded Domains, J. Math. Soc. Japan, 46 (1994), 607–643

[11] Fujiwara, D., and Marimoto, H.: An L_r-theorem of the Helmholtz Decomposition of Vector Fields, J. Fac. Sci. Univ. Tokyo (1A), 24 (1977), 685–700

[12] Galdi, G.P.: An Introduction to the Mathematical Theory of the Navier-Stokes Equations; Linearized Steady Problems, Springer Tracts in Natural Philosophy, Vol. 38, Springer-Verlag, New York 1998

[13] Galdi, G.P.: An Introduction to the Mathematical Theory of the Navier-Stokes Equations; Nonlinear Steady Problems, Springer Tracts in Natural Philosophy, New York 1998

[14] Galdi, G.P., Simader, C.G., and Sohr, H.: On the Stokes Problem in Lip-schitz Domains, Ann. Mat. Pura Appl., 167 (1994), 147–163

[15] Galdi, G.P., Simader, C.G., and Sohr, H.: A Class of Solutions to Stationary Stokes and Navier-Stokes Equations with Boundary Data in $W^{-\frac{1}{q},q}(\partial\Omega)$, Math. Ann., 331 (2005), 41–74

[16] Galdi, G.P., and Simader, C.G.: Existence, uniqueness and L^q-estimates for the Stokes problem in exterior domains, Arch. Rational Mech. Anal., 112 (1990), 147–163

[17] Giga, Y.: Analyticity of the Semigroup Generated by the Stokes Operator in L_r-spaces, Math. Z., 178 (1981), 287–329

[18] Giga, Y.: Domains of Fractional Powers of the Stokes Operator in L_r-spaces, Arch. Rational Mech. Anal., 89 (1985), 251–265

[19] Giga, Y., and Sohr, H.: On the Stokes Operator in Exterior Domains, J. Fac. Sci. Univ. Tokyo, Sec. IA, 36 (1989), 103–130

[20] Giga, Y., and Sohr, H.: Abstract L^q-estimates for the Cauchy problem with applications to the Navier-Stokes equations in exterior domains, J. Funct. Anal., 102 (1991), 72–94

[21] Kato, T.: Strong L^p-solutions to the Navier-Stokes Equations in $\mathbb{R}^m$, with Applications to Weak Solutions, Math. Z., 187 (1984), 471–480

[22] Kozono, H., and Yamazaki, M.: Local and Global Solvability of the Navier-Stokes Exterior Problem with Cauchy Data in the Space $L^{n,\infty}$, Houston J. Math., 21 (1995), 755–799

[23] Nečas, J.: Les Méthodes Directes en Théorie des Équations Elliptiques, Academia, Prag 1967

[24] Simader, C.G., and Sohr, H.: A new approach to the Helmholtz decomposition and the Neumann problem in L^q-spaces for bounded and exterior domains, Adv. Math. Appl. Sci., 11, World Scientific (1992), 1–35

[25] Simader, C.G., and Sohr, H.: The Dirichlet Problem for the Laplacian in Bounded and Unbounded Domains, Pitman, Longman Scientific, Vol. 360, 1997

[26] Solonnikov, V.A.: Estimates for Solutions of Nonstationary Navier-Stokes Equations, J. Soviet Math., 8 (1977), 467–528

[27] Sohr, H.: The Navier-Stokes Equations. An Elementary Functional Analytic Approach, Birkhäuser Advanced Texts, Birkhäuser Verlag, Basel 2001

[28] Temam, R.: Navier-Stokes Equations, North-Holland, Amsterdam, New York, Tokyo 1977

[29] Triebel, H.: Interpolation theory, function spaces, differential operators, North-Holland, Amsterdam 1978

Reinhard Farwig
Fachbereich Mathematik
Darmstadt University of Technology
D-64283 Darmstadt, Germany
e-mail: farwig@mathematik.tu-darmstadt.de

Giovanni P. Galdi
Department of Mechanical Engineering
University of Pittsburgh
15261 Pittsburgh, USA
e-mail: galdi@engrng.pitt.edu

Hermann Sohr
Fakultät für Elektrotechnik
Informatik und Mathematik
University of Paderborn
D-33098 Paderborn, Germany
e-mail: hsohr@math.uni-paderborn.de

Progress in Nonlinear Differential Equations
and Their Applications, Vol. 64, 137–150

Grow-up on the Boundary for a Semilinear Parabolic Problem

Marek Fila, Juan J.L. Velázquez and Michael Winkler

Dedicated to Herbert Amann on the occasion of his 65th birthday

Abstract. We study the asymptotic behavior of positive solutions of a semilinear parabolic equation with a nonlinear boundary condition. This problem admits a unique stationary solution which is not bounded and attracts all positive solutions. We find their growth rate at the singular point on the boundary.

Mathematics Subject Classification (2000). 35K60, 35B40.

Keywords. Semilinear parabolic equation, nonlinear boundary condition, asymptotic behavior.

1. Introduction

Consider the problem

$$\begin{aligned} u_t &= \Delta u - au^p, && x \in \Omega,\ t > 0,\\ \frac{\partial u}{\partial \nu} &= u^q, && x \in \partial\Omega,\ t > 0,\\ u(x,0) &= u_0(x) \geq 0, && x \in \Omega, \end{aligned}$$

where $a > 0$, $p, q > 1$ and Ω is a smoothly bounded domain in $\mathbb{R}^N$. The study of this problem was initiated in [5] (and independently in [17]) and continued in [2], [3], [4], [6], [7], [8], [10], [11], [13], [16], [18], [19], [20], [21], [22], for example. Earlier results on more general problems can be found in [1] and [12], for instance. Concerning blow-up and global existence, the results of [5], [21] and [2] can be summarized as follows. If

$$p < 2q - 1 \qquad \text{or} \qquad a < q \text{ and } p = 2q - 1$$

then there are solutions which blow up in finite time, while for

$$p > 2q - 1 \qquad \text{or} \qquad a > q \text{ and } p = 2q - 1$$

all solutions are global and bounded. It was also shown in [5] that for

$$a = q, \quad p = 2q - 1, \quad N = 1 \text{ and } \Omega = (-1, 1)$$

all nontrivial solutions are global and converge pointwise (as $t \to \infty$) to the unique solution w of the problem

$$\begin{aligned} w_{xx} - qw^{2q-1} &= 0, \qquad |x| < 1, \\ w(\pm 1) &= \infty. \end{aligned}$$

In this paper we study the growth rate of these global unbounded solutions as $t \to \infty$. We consider the problem

$$\begin{aligned} u_t &= u_{xx} - pu^{2p-1}, & 0 < x < 1, t > 0, \\ -u_x &= u^p, & x = 0, \\ u_x &= 0, & x = 1, \\ u|_{t=0} &= u_0 \geq 0, & 0 \leq x \leq 1, \end{aligned} \tag{1.1}$$

where $p > 1$. Our main result is:

Theorem 1.1.

i) *There is $C_1 > 0$ (depending on p and $\|u_0\|_{L^\infty((0,1))}$ only) such that*

$$\|u(\cdot, t)\|_{L^\infty((0,1))} \leq C_1(t+1)^{\frac{1}{2}}, \qquad t > 0.$$

ii) *For any $\nu > 0$ there is $C_2 > 0$ (depending on p, u_0 and ν) such that*

$$\|u(\cdot, t)\|_{L^\infty((0,1))} \geq C_2(t+1)^{\frac{1}{2}-\nu}, \qquad t > 0.$$

This implies in particular that the large-time behavior is not selfsimilar.

Non-selfsimilar behavior of global unbounded solutions was discussed in [9] for the problems

$$\begin{aligned} &u_t = \Delta u + u^p, && x \in B_1 = \{x \in \mathbb{R}^N \mid |x| < 1\},\ t > 0, \\ &u = \left[\frac{2}{p-1}\left(N - 2 - \frac{2}{p-1}\right)\right]^{\frac{1}{p-1}}, && x \in \partial B_1,\ t > 0, \\ &p \geq 1 + \frac{4}{N - 4 - 2\sqrt{N-1}}, && N > 10, \end{aligned}$$

and

$$\begin{aligned} &u_t = \Delta u + e^u, && x \in B_R = \{x \in \mathbb{R}^N \mid |x| < R\},\ t > 0, \\ &u = 0, && x \in \partial B_R,\ t > 0, \\ &R = \sqrt{2(N-2)}, && N > 10. \end{aligned}$$

These problems possess stationary solutions which are singular at the origin and stable from below. The rate at which the solutions starting below the singular stationary solution tend to infinity at the origin was found in [9].

For the Cauchy problem for the equation

$$\begin{aligned} u_t &= \Delta u + u^p, && x \in \mathbb{R}^N,\ t > 0, \\ & p \geq 1 + \frac{4}{N-4-2\sqrt{N-1}}, && N > 10, \end{aligned}$$

this was done in [14] but there the growth rate depends on the spatial decay rate of initial data.

The growth rate of global unbounded solutions of the problem

$$\begin{aligned} u_t &= \Delta u + u^p, && x \in B_1,\ t > 0, \\ u &= 0, && x \in \partial B_1,\ t > 0, \\ & p = \frac{N+2}{N-2}, && N \geq 3, \end{aligned}$$

was found in [15]. In this case the global unbounded solutions do not converge to a singular stationary solution but, again, their behavior is not of selfsimilar nature.

In Section 2 we find the growth rate of solutions of (1.1) formally by matching an asymptotic development of the solution (obtained by a suitable rescaling and linearization) with the asymptotics of the singular steady state w. This matching procedure is different from those in [9], [14] and [15].

Section 3 contains a rigorous proof of Theorem 1.1. The proof is based on comparison with suitable super- and subsolutions. Their form is suggested by the formal matched asymptotics from Section 2.

2. Matched asymptotics

We begin with an analysis of the solution of the steady state problem

$$\begin{aligned} w_{xx} - pw^{2p-1} &= 0, && 0 < x < 1, \\ w &= \infty, && x = 0, \\ w_x &= 0, && x = 1. \end{aligned} \tag{2.2}$$

By Lemma 4.1 in [5], it is unique and we can write it as

$$x = \int_{w(x)}^{\infty} \frac{du}{\sqrt{u^{2p} - m^{2p}}}, \qquad m = w(1). \tag{2.3}$$

Since

$$1 = \int_{m}^{\infty} \frac{du}{\sqrt{u^{2p} - m^{2p}}},$$

an obvious change of variables yields

$$m^{p-1} = \int_1^\infty \frac{dz}{\sqrt{z^{2p}-1}}.$$

Taylor's expansion gives

$$\frac{1}{\sqrt{u^{2p}-m^{2p}}} = \frac{1}{u^p}\left(1+\frac{1}{2}\frac{m^{2p}}{u^{2p}}+\cdots\right) \qquad \text{as } u \to \infty.$$

From (2.3) it then follows that

$$x = \frac{1}{p-1}\frac{1}{w^{p-1}(x)} + \frac{m^{2p}}{2(3p-1)}\frac{1}{w^{3p-1}(x)} + \cdots \qquad \text{as } x \to 0; \tag{2.4}$$

therefore,

$$w(x) \approx \alpha_p x^{-\frac{1}{p-1}} \qquad \text{as } x \to 0, \qquad \alpha_p = (p-1)^{-\frac{1}{p-1}}.$$

Now, (2.4) implies

$$\frac{1}{p-1}\frac{1}{w^{p-1}} = x - \frac{m^{2p}}{2(3p-1)}(p-1)^{\frac{3p-1}{p-1}} x^{\frac{3p-1}{p-1}} + \cdots \qquad \text{as } x \to 0.$$

Hence,

$$w(x) = \alpha_p x^{-\frac{1}{p-1}} + K_p m^{2p} x^{\frac{2p-1}{p-1}} + \cdots \qquad \text{as } x \to 0, \tag{2.5}$$

where

$$K_p = \frac{(p-1)^{2p-1}}{2(3p-1)}.$$

Now we make a local analysis in a region near $x = 0$. We introduce the following change of variables:

$$u(x,t) = (\varepsilon(t))^{-\frac{1}{p-1}} v(y,t), \qquad y = \frac{x}{\varepsilon(t)},$$

where $\varepsilon(t)$ is defined by

$$u(0,t) = \alpha_p (\varepsilon(t))^{-\frac{1}{p-1}}.$$

Then v satisfies

$$\begin{aligned} -\frac{1}{p-1}\varepsilon\dot{\varepsilon}v - \varepsilon\dot{\varepsilon}y v_y + \varepsilon^2 v_t &= v_{yy} - p v^{2p-1}, && 0 < y < \frac{1}{\varepsilon(t)},\ t > 0, \\ v &= \alpha_p, && y = 0,\ t > 0, \\ -v_y &= v^p, && y = 0,\ t > 0, \\ v_y &= 0, && y = \frac{1}{\varepsilon(t)},\ t > 0. \end{aligned}$$

Since $\varepsilon(t) \to 0$ as $t \to \infty$ and we expect that yv_y and v_t are bounded, we may assume that $v(y,t) \to \tilde{v}(y)$ as $t \to \infty$, where $\tilde{v}$ satisfies

$$\begin{aligned} \tilde{v}_{yy} - p\tilde{v}^{2p-1} &= 0, & y > 0, \\ -\tilde{v}_y &= \tilde{v}^p, & y = 0, \\ \tilde{v} &= \alpha_p, & y = 0. \end{aligned}$$

Note that $\tilde{v}$ is given explicitly by

$$\tilde{v}(y) = \alpha_p (y+1)^{-\frac{1}{p-1}}, \qquad y > 0.$$

Writing

$$v(y,t) = \tilde{v}(y) + z(y,t) \tag{2.6}$$

and linearizing formally, we obtain that z satisfies the equation

$$-\varepsilon\dot{\varepsilon}\Big(\frac{1}{p-1}\tilde{v} + y\tilde{v}_y\Big) + \varepsilon^2 z_t = z_{yy} - p(2p-1)\tilde{v}^{2(p-1)} z,$$

that suggests the asymptotics

$$z(y,t) \approx \varepsilon\dot{\varepsilon} G(y), \tag{2.7}$$

where G solves the problem

$$\begin{aligned} -\Big(\frac{1}{p-1}\tilde{v} + y\tilde{v}_y\Big) &= G_{yy} - p(2p-1)\tilde{v}^{2(p-1)} G, \quad & y > 0, \\ G &= 0, & y = 0, \\ -G_y &= p\tilde{v}^{p-1} G, & y = 0, \end{aligned}$$

which can be written as

$$G_{yy} - \frac{p(2p-1)}{(p-1)^2}\frac{G}{(y+1)^2} = -[(p-1)(y+1)]^{-\frac{p}{p-1}}, \qquad y > 0, \tag{2.8}$$

with initial conditions

$$G(0) = G_y(0) = 0. \tag{2.9}$$

The general solution of (2.8) is

$$G(y) = -A_p (y+1)^{\frac{2p-1}{p-1}} - B_p (y+1)^{-\frac{p}{p-1}} + \frac{1}{2(p+1)}\alpha_p (y+1)^{\frac{p-2}{p-1}}.$$

The constants A_p and B_p for which (2.9) holds are

$$A_p = \frac{1}{(p+1)(3p-1)}(p-1)^{\frac{p-2}{p-1}}, \quad B_p = \frac{1}{2(3p-1)}\alpha_p.$$

It follows then that

$$G(y) \approx -A_p y^{\frac{2p-1}{p-1}} \qquad \text{as } y \to \infty. \tag{2.10}$$

From (2.10) we now obtain (as $y \to \infty$)

$$\begin{aligned} u(x,t) &= \varepsilon^{-\frac{1}{p-1}} v(y,t) \\ &\approx \varepsilon^{-\frac{1}{p-1}} (\tilde{v}(y) + \varepsilon\dot{\varepsilon} G(y)) \\ &\approx \varepsilon^{-\frac{1}{p-1}} \Big(\alpha_p y^{-\frac{1}{p-1}} - \varepsilon\dot{\varepsilon} A_p y^{\frac{2p-1}{p-1}} \Big) \\ &= \alpha_p x^{-\frac{1}{p-1}} - A_p \dot{\varepsilon} \varepsilon^{-\frac{p+1}{p-1}} x^{\frac{2p-1}{p-1}}. \end{aligned}$$

Since we want this asymptotics to match with (2.5), we require

$$\dot{\varepsilon} = -\frac{K_p m^{2p}}{A_p} \varepsilon^{\frac{p+1}{p-1}}.$$

Therefore,

$$\varepsilon(t) \approx \lambda_p t^{-\frac{p-1}{2}}, \qquad \lambda_p = \Big(\frac{2K_p m^{2p}}{(p-1)A_p} \Big)^{-\frac{p-1}{2}}.$$

3. The rigorous proof of the blow-up rate

In this section we prove Theorem 1.1. The main idea is to construct sub- and supersolutions of the form (2.6), (2.7) that is suggested by matched asymptotics. To obtain a subsolution we modify the function G slightly and because of this modification we are not able to reach the power $1/2$ in Theorem 1.1 ii).

i) As u_x satisfies a linear parabolic equation with smooth coefficients and $u_x \le 0$ in $\{0,1\} \times (0,\infty)$, the maximum principle shows that $u_x \le 0$ if u_0 is nonincreasing and smooth. Replacing u_0 by $\|u_0\|_{L^\infty((0,1))}$ if necessary, we thus may assume $u_x \le 0$ in $(0,1) \times (0,\infty)$.
With a differentiable function $\varepsilon : [0,\infty) \to (0,\infty)$ to be specified below, we define

$$Q := \Big\{ (y,t) \in (0,\infty)^2 \mid 0 < y < \frac{1}{\varepsilon(t)} \Big\} \tag{3.11}$$

and the function $v : Q \to \mathbb{R}$ by

$$u(x,t) =: \varepsilon^{-\frac{1}{p-1}}(t) v\Big(\frac{x}{\varepsilon(t)}, t \Big). \tag{3.12}$$

Then

$$\begin{aligned} Pv := v_t - \frac{1}{\varepsilon^2}(v_{yy} - pv^{2p-1}) - \frac{1}{p-1}\frac{\dot{\varepsilon}}{\varepsilon} v - \frac{\dot{\varepsilon}}{\varepsilon} y v_y &= 0, \qquad && (y,t) \in Q, \\ -v_y &= v^p, && y = 0, \\ v_y &= 0, && y = \frac{1}{\varepsilon(t)}, \\ v(y,0) = \varepsilon^{-\frac{1}{p-1}}(0) u_0(\varepsilon(0)y), & && 0 < y < \frac{1}{\varepsilon(0)}. \end{aligned}$$

We first claim that

$$v(y,t) \le w(y,t) := \tilde{v}(y) + \frac{1}{2}\varepsilon(t)\dot{\varepsilon}(t)G(y) \qquad \text{in } Q$$

if we choose

$$\varepsilon(t) := \lambda(t+1)^{-\frac{p-1}{2}}$$

with $\lambda > 0$ small enough (see below).
Indeed, we have

$$\begin{aligned} Pw &= \frac{1}{2}(\dot{\varepsilon}^2 + \varepsilon\ddot{\varepsilon})G - \frac{1}{\varepsilon^2}\Big[\tilde{v}_{yy} + \frac{1}{2}\varepsilon\dot{\varepsilon}G_{yy} - p(\tilde{v} + \frac{1}{2}\varepsilon\dot{\varepsilon}G)^{2p-1}\Big] \\ &\quad -\frac{\dot{\varepsilon}}{\varepsilon}\Big(\frac{1}{p-1}\tilde{v} + y\tilde{v}_y\Big) + \dot{\varepsilon}^2\Big(-\frac{1}{2(p-1)}G - \frac{1}{2}yG_y\Big), \end{aligned} \tag{3.13}$$

where

$$\frac{1}{p-1}\tilde{v} + y\tilde{v}_y = (p-1)^{-\frac{p}{p-1}}(y+1)^{-\frac{p}{p-1}}.$$

Assume first that $p \ge \frac{3}{2}$. Since $\dot{\varepsilon}G \ge 0$, we have

$$p\Big(\tilde{v} + \frac{1}{2}\varepsilon\dot{\varepsilon}G\Big)^{2p-1} \ge p\tilde{v}^{2p-1} + \frac{p(2p-1)}{2}\varepsilon\dot{\varepsilon}\tilde{v}^{2p-2}G + \frac{p(2p-1)(p-1)}{4}\varepsilon^2\dot{\varepsilon}^2\tilde{v}^{2p-3}G^2 \tag{3.14}$$

and hence, since $\tilde{v}_{yy} = p\tilde{v}^{2p-1}$ and due to the equation defining G,

$$\begin{aligned} Pw &\ge -\frac{1}{2}\frac{\dot{\varepsilon}}{\varepsilon}\Big[G_{yy} - \frac{p(2p-1)}{(p-1)^2}(y+1)^{-2}G + 2(p-1)^{-\frac{p}{p-1}}(y+1)^{-\frac{p}{p-1}}\Big] \\ &\quad +\frac{1}{2}\dot{\varepsilon}^2\Big[\Big(\frac{p-2}{p-1} + \frac{\varepsilon\ddot{\varepsilon}}{\dot{\varepsilon}^2}\Big)G - yG_y + \frac{1}{2}p(2p-1)(p-1)^{-\frac{p-2}{p-1}}(y+1)^{-\frac{2p-3}{p-1}}G^2\Big] \\ &= \frac{1}{2}\dot{\varepsilon}^2\Big[\Big(\frac{p-2}{p-1} + \frac{\varepsilon\ddot{\varepsilon}}{\dot{\varepsilon}^2}\Big)G - yG_y + \frac{1}{2}p(2p-1)(p-1)^{-\frac{p-2}{p-1}}(y+1)^{-\frac{2p-3}{p-1}}G^2 \\ &\quad -\frac{1}{\varepsilon\dot{\varepsilon}}(p-1)^{-\frac{p}{p-1}}(y+1)^{-\frac{p}{p-1}}\Big] \\ &= \frac{1}{2}\dot{\varepsilon}^2\Big[c_1G - yG_y + c_2(y+1)^{-\frac{2p-3}{p-1}}G^2 + c_3\lambda^{-2}(t+1)^p(y+1)^{-\frac{p}{p-1}}\Big] \\ &\ge \frac{1}{2}\dot{\varepsilon}^2\Big[c_1G + c_2(y+1)^{-\frac{2p-3}{p-1}}G^2 + c_3\lambda^{-2}(y+1)^{-\frac{p}{p-1}}\Big] \end{aligned} \tag{3.15}$$

with $c_1 = \frac{2p-1}{p-1}$, $c_2 = \frac{1}{2}p(2p-1)(p-1)^{-\frac{p-2}{p-1}} > 0$ and $c_3 = 2(p-1)^{-\frac{2p-1}{p-1}} > 0$. As

$$|G(y)| \le c(y+1)^{\frac{2p-1}{p-1}}$$

and

$$(y+1)^{-\frac{2p-3}{p-1}}G^2(y) \ge c(y+1)^{\frac{2p+1}{p-1}}$$

for all $y \ge 0$, we can find $y_0 > 0$ such that $c_1G(y) + c_2(y+1)^{-\frac{2p-3}{p-1}}G^2(y) \ge 0$ for $y \ge y_0$ and hence

$$Pw \ge 0 \quad \text{whenever } y \ge y_0. \tag{3.16}$$

On the other hand, if

$$\lambda^2 \le \frac{c_3}{c_1} \cdot \Big| \min_{y\in[0,y_0]} G(y) \Big|^{-1} (y_0+1)^{-\frac{p}{p-1}}$$

then $c_1 G(y) + c_3\lambda^{-2}(y+1)^{-\frac{p}{p-1}} \ge 0$ for $y \in (0, y_0)$, implying $Pw \ge 0$ for $y \in (0, y_0)$. We thus have proved

$$Pw \ge 0 \quad \text{in } Q. \tag{3.17}$$

In the case $1 < p < \frac{3}{2}$ we proceed similarly, except that in (3.13) we use the estimate

$$(1+\xi)^\gamma \ge 1 + \gamma\xi + c_4 \min\{\xi^\gamma, \xi^2\}$$

valid for $\gamma := 2p-1 \in (1,2)$, $c_4 = \min\{2^{\gamma-3}\gamma(\gamma-1), 2^\gamma - \gamma - 1\} > 0$ and all $\xi \ge 0$. Accordingly, in the set

$$Q_1 := \Big\{ (y,t) \in Q \mid \frac{1}{2}\varepsilon(t)\dot\varepsilon(t)\tilde v^{-1}(y)G(y) \le 1 \Big\},$$

we have an estimate similar to (3.14) – except for a smaller but still positive constant factor in the last term – and we conclude as before that if λ is small enough then

$$Pw \ge 0 \quad \text{in } Q_1. \tag{3.18}$$

If $(y,t) \in Q \setminus Q_1$, however, (3.14) is replaced by

$$p\Big(\tilde v + \frac{1}{2}\varepsilon\dot\varepsilon G\Big)^{2p-1} \ge p\tilde v^{2p-1} + \frac{p(p-1)}{2}\varepsilon\dot\varepsilon\tilde v^{2p-2}G + pc_4|\varepsilon\dot\varepsilon|^{2p-1}|G|^{2p-1},$$

so that instead of (3.15) we find

$$\begin{aligned} Pw \quad \ge \quad & -\frac{1}{2}\frac{\dot\varepsilon}{\varepsilon}\Big[G_{yy} - \frac{p(2p-1)}{(p-1)^2}(y+1)^{-2}G + 2(p-1)^{-\frac{p}{p-1}}(y+1)^{-\frac{p}{p-1}}\Big] \\ & + \frac{1}{2}\dot\varepsilon^2\Big[\Big(\frac{p-2}{p-1} + \frac{\varepsilon\ddot\varepsilon}{\dot\varepsilon^2}\Big)G - yG_y + 2pc_4|\varepsilon\dot\varepsilon|^{2p-3}|G|^{2p-1}\Big]. \end{aligned}$$

Using that $|\varepsilon\dot\varepsilon|^{2p-3} = (\frac{2}{p-1})^{3-2p}\lambda^{-2(3-2p)}(t+1)^{p(3-2p)} \ge (\frac{2}{p-1})^{3-2p}$ for $\lambda \le 1$ and $t \ge 0$, we arrive at

$$Pw \ge \frac{1}{2}\dot\varepsilon^2\Big[c_1 G + \tilde c_2|G|^{2p-1} + c_3\lambda^{-2}(y+1)^{-\frac{p}{p-1}}\Big]$$

with c_1, c_3 as above and $\tilde c_2 = 2p(\frac{2}{p-1})^{3-2p}c_4 > 0$. As $2p-1>1$, we again have $c_1 G + \tilde c_2|G|^{2p-1} \ge 0$ for $y \ge \tilde y_0$ and some $\tilde y_0 > 0$, while $c_1 G + c_3\lambda^{-2}(y+1)^{-\frac{p}{p-1}} \ge 0$ for $y \in (0, \tilde y_0)$ if λ is small enough. Hence,

$$Pw \ge 0 \quad \text{in } Q \setminus Q_1, \tag{3.19}$$

so that in fact (3.17) continues to hold for $p \in (1, \frac{3}{2})$.

To check the corresponding inequalities on the boundary, we first note that

$$-w_y = -\tilde v_y = \tilde v^p = w^p \quad \text{for } y = 0, \tag{3.20}$$

while at $y = \frac{1}{\varepsilon(t)}$,

$$w_y = \tilde{v}_y + \frac{1}{2}\varepsilon\dot{\varepsilon}G_y.$$

Due to the asymptotics of G, we can find y_1 large and $c_5 > 0$ such that $G_y(y) \leq -c_5(y+1)^{\frac{p}{p-1}}$ for all $y > y_1$. Thus, if $\lambda < \frac{1}{y_1}$ then $y = \frac{1}{\varepsilon(t)} > y_1$ and

$$\begin{aligned} w_y &\geq (p-1)^{-\frac{p}{p-1}}(y+1)^{-\frac{p}{p-1}}\left[-1 - c_6\varepsilon\dot{\varepsilon}(y+1)^{\frac{2p}{p-1}}\right] \\ &\geq (p-1)^{-\frac{p}{p-1}}(y+1)^{-\frac{p}{p-1}}\left[-1 - c_6\varepsilon^{-\frac{p+1}{p-1}}\dot{\varepsilon}\right] \\ &\geq (p-1)^{-\frac{p}{p-1}}(y+1)^{-\frac{p}{p-1}}\left[-1 + \frac{p-1}{2}c_6\lambda^{-\frac{2}{p-1}}\right] \\ &\geq 0, \end{aligned} \tag{3.21}$$

provided that λ is small enough.
Finally, at $t = 0$ we have

$$\begin{aligned} v(y,0) &= \lambda^{\frac{1}{p-1}}u_0(\lambda y) \\ &\leq \lambda^{\frac{1}{p-1}}\|u_0\|_{L^\infty((0,1))}, \end{aligned}$$

and, since by interpolation $G(y) \leq -\frac{A_p}{2}(y+1)^{\frac{2p-1}{p-1}} + c_7(y+1)^{-\frac{1}{p-1}}$,

$$\begin{aligned} w(y,0) &= (p-1)^{-\frac{1}{p-1}}(y+1)^{-\frac{1}{p-1}} - \frac{p-1}{4}\lambda^2 G(y) \\ &\geq \left[(p-1)^{-\frac{1}{p-1}} - \frac{p-1}{4}c_7\lambda^2\right](y+1)^{-\frac{1}{p-1}} + \frac{(p-1)A_p}{4}\lambda^2(y+1)^{\frac{2p-1}{p-1}}. \end{aligned}$$

Hence, for λ small enough we have

$$v \leq w \quad \text{at } t = 0 \tag{3.22}$$

if

$$\begin{aligned} \|u_0\|_{L^\infty((0,1))} &\leq c_8\lambda^{-\frac{1}{p-1}}(y+1)^{-\frac{1}{p-1}} + c_9\lambda^{\frac{2p-3}{p-1}}(y+1)^{\frac{2p-1}{p-1}} \\ &=: \varphi(y). \end{aligned}$$

Minimizing with respect to $y \in [0,\infty)$, we obtain that

$$\varphi(y) \geq c_{10}\lambda^{-\frac{1}{p(p-1)}}$$

with $c_{10} = \left[(2p-1)^{\frac{1}{2p}} + (2p-1)^{-\frac{2p-1}{2p}}\right]c_8^{\frac{2p-1}{2p}}c_9^{\frac{1}{2p}}$ and thus (3.22) is satisfied if λ lies below some positive constant depending on p and $\|u_0\|_{L^\infty((0,1))}$ only.
Altogether, we gain from (3.17), (3.20)–(3.22) and the comparison principle that

$$u(x,t) \leq \varepsilon^{-\frac{1}{p-1}}(t)w\left(\frac{x}{\varepsilon(t)}, t\right) \quad \text{in } (0,1)\times(0,\infty),$$

whence in accordance with our assumption on the monotonicity of u with respect to x,

$$\|u(t)\|_{L^\infty((0,1))} = u(0,t) \le \varepsilon^{-\frac{1}{p-1}}(t) = \lambda^{-\frac{1}{p-1}}(t+1)^{\frac{1}{2}},$$

which completes the proof of part i).

ii) We first note that for small positive δ, the solutions G_δ of

$$G_{\delta yy} - (1+\delta)\frac{p(2p-1)}{(p-1)^2}(y+1)^{-2}G_\delta + (p-1)^{-\frac{p}{p-1}}(y+1)^{-\frac{p}{p-1}} = 0 \quad \text{in } (0,\infty),$$
$$G_\delta(0) = G_{\delta y}(0) = 0,$$

have the form

$$G_\delta(y) = -A_{p\delta}(y+1)^{\frac{2p-1}{p-1}+\eta_\delta} - B_{p\delta}(y+1)^{-\frac{p}{p-1}-\eta_\delta} + C_{p\delta}(y+1)^{\frac{p-2}{p-1}},$$

where $\eta_\delta = \sqrt{\frac{1}{4} + (1+\delta)\frac{p(2p-1)}{(p-1)^2}} - \sqrt{\frac{1}{4} + \frac{p(2p-1)}{(p-1)^2}} \searrow 0$ and $A_{p\delta}, B_{p\delta}$ and $C_{p\delta}$ converge to A_p, B_p and C_p, respectively, as $\delta \searrow 0$. Therefore, setting $\beta := \frac{p-1}{2} - (p-1)\nu < \frac{p-1}{2}$, we can fix $\delta > 0$ small such that $\frac{1}{2} \le \frac{A_{p\delta}}{A_p}, \frac{B_{p\delta}}{B_p}, \frac{C_{p\delta}}{C_p} \le 2$ and $\eta_\delta < \frac{1}{\beta} - \frac{2}{p-1}$. Using

$$\varepsilon(t) := \lambda(t+\lambda^\kappa)^{-\beta}, \quad t \ge 0,$$

with λ sufficiently large and $\kappa \in (\frac{2}{2\beta+1}, \frac{1}{\beta})$ arbitrary but fixed, we shall this time compare v in Q (see (3.11) and (3.12)) with

$$w(y,t) := \tilde{v}(y) + 2\varepsilon(t)\dot{\varepsilon}(t)G_\delta(y).$$

Then

$$\begin{aligned} Pw &= 2(\dot{\varepsilon}^2 + \varepsilon\ddot{\varepsilon})G_\delta - \frac{1}{\varepsilon^2}\Big[\tilde{v}_{yy} + 2\varepsilon\dot{\varepsilon}G_{\delta yy} - p(\tilde{v} + 2\varepsilon\dot{\varepsilon}G_\delta)^{2p-1}\Big] \\ &\quad -\frac{\dot{\varepsilon}}{\varepsilon}\Big(\frac{1}{p-1}\tilde{v} + y\tilde{v}_y\Big) + \dot{\varepsilon}^2\Big(-\frac{2}{p-1}G_\delta - 2yG_{\delta y}\Big). \end{aligned} \tag{3.23}$$

Since for $\gamma > 1$ we have $(1+\xi)^\gamma \le 1 + (1+\delta)\gamma\xi$ for $0 \le \xi \le \xi_\delta := (1+\delta)^{\frac{1}{\gamma-1}} - 1$, the estimate

$$p(\tilde{v} + 2\varepsilon\dot{\varepsilon}G_\delta)^{2p-1} \le p\tilde{v}^{2p-1} + 2(1+\delta)\cdot p(2p-1)\varepsilon\dot{\varepsilon}\tilde{v}^{2p-2}G_\delta \tag{3.24}$$

holds whenever $|2\varepsilon\dot{\varepsilon}\tilde{v}^{-1}G_\delta| \le \xi_\delta$. But

$$|2\varepsilon\dot{\varepsilon}\tilde{v}^{-1}G_\delta| \le c_{11}\lambda^{-\frac{1}{\beta}}\varepsilon^{\frac{2\beta+1}{\beta}}(y+1)^{\frac{2p}{p-1}+\eta_\delta}, \tag{3.25}$$

so that if $(y,t) \in Q$ fulfils $\varepsilon(t) \le 1$ then, since $y+1 \le \frac{1}{\varepsilon(t)} + 1 \le \frac{2}{\varepsilon(t)}$ in this case,

$$\begin{aligned} |2\varepsilon\dot{\varepsilon}\tilde{v}^{-1}G_\delta| &\le c_{12}\lambda^{-\frac{1}{\beta}}\cdot\varepsilon^{\frac{2\beta+1}{\beta}-\frac{2p}{p-1}-\eta_\delta} \\ &\le c_{12}\lambda^{-\frac{1}{\beta}} \\ &\le \xi_\delta \end{aligned} \tag{3.26}$$

for λ sufficiently large, where we have used that $\frac{2\beta+1}{\beta} - \frac{2p}{p-1} - \eta_\delta > 0$ due to our restriction on δ (and thus of η_δ). If, conversely, $(y,t) \in Q$ with $\varepsilon(t) > 1$ then $y+1 \leq 2$ and therefore

$$\begin{aligned} |2\varepsilon\dot{\varepsilon}\tilde{v}^{-1}G_\delta| &\leq c_{13}\lambda^{-\frac{1}{\beta}} \cdot \varepsilon^{\frac{2\beta+1}{\beta}} \\ &= c_{13}\lambda^2(t+\lambda^\kappa)^{-2\beta-1} \\ &\leq c_{13}\lambda^{2-(2\beta+1)\kappa} \\ &\leq \xi_\delta \end{aligned} \tag{3.27}$$

for large λ, since $\kappa > \frac{2}{2\beta+1}$.

Having thus proved (3.24) in all of Q, we use it in (3.23) to infer that

$$\begin{aligned} Pw &\leq -2\frac{\dot{\varepsilon}}{\varepsilon}\Big[G_{\delta yy} - (1+\delta)\frac{p(2p-1)}{(p-1)^2}(y+1)^{-2}G_\delta + \frac{1}{2}(p-1)^{-\frac{p}{p-1}}(y+1)^{-\frac{p}{p-1}}\Big] \\ &\quad +2\dot{\varepsilon}^2\Big[\Big(\frac{p-2}{p-1} + \frac{\varepsilon\ddot{\varepsilon}}{\dot{\varepsilon}^2}\Big)G_\delta - yG_{\delta y}\Big] \\ &= 2\dot{\varepsilon}^2\Big[\Big(\frac{p-2}{p-1} + \frac{\varepsilon\ddot{\varepsilon}}{\dot{\varepsilon}^2}\Big)G_\delta - yG_{\delta y} + \frac{1}{2\varepsilon\dot{\varepsilon}}(p-1)^{-\frac{p}{p-1}}(y+1)^{-\frac{p}{p-1}}\Big]. \end{aligned}$$

Since

$$\begin{aligned} \frac{p-2}{p-1} + \frac{\varepsilon\ddot{\varepsilon}}{\dot{\varepsilon}^2} - \frac{yG_{\delta y}}{G_\delta} &\to \frac{p-2}{p-1} + \frac{\beta+1}{\beta} - \frac{2p-1}{p-1} - \eta_\delta \\ &= \frac{1}{\beta} - \frac{2}{p-1} - \eta_\delta > 0 \qquad \text{as } y \to \infty, \end{aligned}$$

we can find $y_2 > 0$ such that

$$Pw \leq 0 \quad \text{whenever } y \geq y_2. \tag{3.28}$$

But $-\varepsilon\dot{\varepsilon} = \beta\lambda^2(t+\lambda^\kappa)^{-2\beta-1} \leq \beta\lambda^{2-(2\beta+1)\kappa}$ for $t \geq 0$, so that, since $\kappa > \frac{2}{2\beta+1}$,

$$\begin{aligned} Pw &\leq 2\dot{\varepsilon}^2\Big[\max_{y\in[0,y_2]}(-yG_{\delta y}(y)) - \frac{1}{2\beta}(p-1)^{-\frac{p}{p-1}}\lambda^{(2\beta+1)\kappa-2}\Big] \\ &\leq 0 \qquad \text{for } y < y_2, \end{aligned} \tag{3.29}$$

provided that λ is large enough.

Evidently,

$$-w_y = w^p \quad \text{for } y = 0, \tag{3.30}$$

while if $y = \frac{1}{\varepsilon(t)}$ then we use the estimate $G_{\delta y}(y) \geq -c_{14}(y+1)^{\frac{p}{p-1}+\eta_\delta}$, valid for any $y > 0$, in deriving

$$\begin{aligned} w_y &= \tilde{v}_y + 2\varepsilon\dot{\varepsilon}G_{\delta y} \\ &\leq (p-1)^{-\frac{p}{p-1}}(y+1)^{-\frac{p}{p-1}}\Big[-1 + c_{15}\lambda^{-\frac{1}{\beta}}\varepsilon^{\frac{2\beta+1}{\beta}}(y+1)^{\frac{2p}{p-1}+\eta_\delta}\Big]. \end{aligned}$$

The positive term in the large brackets is a multiple of the right-hand side in (3.25), so that the same ideas which lead to (3.26) and (3.27) also apply here to yield for large λ

$$w_y \leq 0 \quad \text{if } y = \frac{1}{\varepsilon(t)}. \tag{3.31}$$

To achieve the 'suitable' inequality also at $t = 0$, we observe that if $\lambda \geq 1$ and $(y, 0) \in Q$ then $y + 1 \leq \frac{1}{\varepsilon(0)} + 1 = \lambda^{\beta\kappa - 1} + 1 \leq 2$ due to $\kappa < \frac{1}{\beta}$, so that

$$\begin{aligned} w(y,0) &= \tilde{v}(y) + 2\varepsilon(0)\dot{\varepsilon}(0)G_\delta(y) \\ &\leq (p-1)^{-\frac{1}{p-1}}(y+1)^{-\frac{1}{p-1}} + c_{16}\lambda^{2-(2\beta+1)\kappa}(y+1)^{\frac{2p-1}{p-1}+\eta\delta} \\ &\leq (p-1)^{-\frac{1}{p-1}} + c_{16} \cdot 2^{\frac{2p-1}{p-1}+\eta\delta}\lambda^{2-(2\beta+1)\kappa} \\ &\leq c_{17}, \end{aligned}$$

where we have once more used $\kappa > \frac{2}{2\beta+1}$. On the other hand, the strong maximum principle guarantees that without loss of generality we may assume $m := \min\limits_{x\in[0,1]} u_0(x) > 0$ (otherwise we take $t+1$ as initial time) and hence obtain

$$\begin{aligned} v(y,0) &= \lambda^{\frac{1-\beta\kappa}{p-1}} u_0(\lambda^{1-\beta\kappa}y) \\ &\geq \lambda^{\frac{1-\beta\kappa}{p-1}} m \\ &\geq c_{17} \end{aligned}$$

for any $(y, 0) \in Q$ and λ sufficiently large, because $\kappa < \frac{1}{\beta}$. Thus, $v \geq w$ at $t = 0$ and accordingly, due to (3.28)–(3.31) and the comparison principle, $v \geq w$ in Q. In particular,

$$\|u(t)\|_{L^\infty((0,1))} \geq \varepsilon^{-\frac{1}{p-1}}(t)w(0) = \lambda^{-\frac{1}{p-1}}(t+\lambda^\kappa)^{\frac{\beta}{p-1}},$$

which yields the assertion, since $\frac{\beta}{p-1} = \frac{1}{2} - \nu$. □

Acknowledgement

Part of this work was done while the third author was a postdoc at the Comenius University supported by the European Community's Human Potential Programme under contract HPRN-CT-2002-00274, Fronts - Singularities. The first author was partially supported by VEGA Grant 1/0259/03. The authors thank Victor A. Galaktionov for a stimulating discussion.

References

[1] H. Amann, *Parabolic evolution equations and nonlinear boundary conditions,* J. Differ. Equations **72** (1988), 201–269.

[2] F. Andreu, J.M. Mazón, J. Toledo and J.D. Rossi, *Porous medium equation with absorption and a nonlinear boundary condition,* Nonlinear Anal. TMA **49** (2002), 541–563.

[3] J.M. Arrieta and A. Rodríguez-Bernal, *Localisation near the boundary of blow-up for reaction-diffusion equations with nonlinear boundary conditions,* Comm. Part. Diff. Equations **29** (2004), 1127–1148.

[4] T.K. Boni, *Sur l'explosion et le comportement asymptotique de la solution d'une équation parabolique semi-linéaire du second ordre,* C. R. Acad. Sci. Paris, Série I **326** (1998), 317–322.

[5] M. Chipot, M. Fila and P. Quittner, *Stationary solutions, blow-up and convergence to stationary solutions for semilinear parabolic equations with nonlinear boundary conditions,* Acta Math. Comenianae **60** (1991), 35–103.

[6] M. Chipot and P. Quittner, *Equilibria, connecting orbits and a priori bounds for semilinear parabolic equations with nonlinear boundary conditions,* J. Dynamics Differ. Equations **16** (2004), 91–138.

[7] M. Chipot and M. Ramaswamy, *Semilinear elliptic problems with nonlinear boundary conditions,* Differ. Equ. Dyn. Syst **6** (1998), 51–75.

[8] M. Chipot and F. Voirol, *On a class of nonlinear elliptic problems with Neumann boundary conditions growing like a power,* Z. Anal. Anwend. **14** (1995), 853–868.

[9] J.W. Dold, V.A. Galaktionov, A.A. Lacey and J.L. Vázquez, *Rate of approach to a singular steady state in quasilinear reaction-diffusion equations,* Ann. Scuola Norm. Sup. Pisa Cl. Sci **26** (1998), 663–687.

[10] Y.V. Egorov and V.A. Kondratiev, *Two theorems on blow-up solutions for semilinear parabolic equations of second order,* C. R. Acad. Sci. Paris, Série I **327** (1998), 47–52.

[11] Y.V. Egorov and V.A. Kondratiev, *On blowing-up solutions to second-order parabolic equations,* Doklady Math. **59** (1999), 90–92.

[12] J. Escher, *Global existence and nonexistence for semilinear parabolic systems with nonlinear boundary conditions,* Math. Ann. **284** (1989), 285–305.

[13] M. Fila and B. Kawohl, *Large time behavior of solutions to a quasilinear parabolic equation with a nonlinear boundary condition,* Adv. Math. Sci. Appl. **11** (2001), 113–126.

[14] M. Fila, M. Winkler and E. Yanagida, *Grow-up rate of solutions for a supercritical semilinear diffusion equation,* J. Differ. Equations **205** (2004), 365–389.

[15] V.A. Galaktionov and J.R. King, *Composite structure of global unbounded solutions of nonlinear heat equations with critical Sobolev exponents,* J. Differ. Equations **189** (2003), 199–233.

[16] A.A. Lacey, J.R. Ockendon, J. Sabina and D. Salazar, *Perturbation analysis of a semilinear parabolic problem with nonlinear boundary conditions,* Rocky Mt. J. of Math. **26** (1996), 195–212.

[17] J. López Gómez, V. Márquez and N. Wolanski, *Dynamic behavior of positive solutions to reaction-diffusion problems with nonlinear absorption through the boundary,* Rev. Unión Mat. Argent. **38** (1993), 196–209.

[18] K. Pflüger, *On indefinite nonlinear Neumann problems,* Partial differential and integral equations (H.G.W. Begehr et al, eds.), Kluwer Academic Publishers: Dordrecht (1999), 335–346.

[19] P. Quittner, *On global existence and stationary solutions for two classes of semilinear parabolic problems,* Comment. Math. Univ. Carolinae **34** (1993), 105–124.

[20] A. Rodríguez-Bernal, *Attractors for parabolic equations with nonlinear boundary conditions, critical exponents, and singular initial data,* J. Differ. Equations **181** (2002), 165–196.

[21] A. Rodríguez-Bernal and A. Tajdine, *Nonlinear balance for reaction-diffusion equations under nonlinear boundary conditions: dissipativity and blow-up*, J. Differ. Equations **169** (2001), 332–372.

[22] J.D. Rossi, *The blow-up rate for a semilinear parabolic equation with a nonlinear boundary condition,* Acta Math. Univ. Comenianae **67** (1998), 343–350.

Marek Fila
Institute of Applied Mathematics
Comenius University, Mlynská dolina
84248 Bratislava, Slovakia

Juan J.L. Velázquez
Departamento de Matemática Aplicada
Facultad de Ciencias Matemáticas
Universidad Complutense
E-28040 Madrid, Spain

Michael Winkler
Department of Mathematics I
RWTH Aachen
D-52056 Aachen, Germany

Progress in Nonlinear Differential Equations
and Their Applications, Vol. 64, 151–188

Existence and Uniqueness Results for Reaction-diffusion Processes of Electrically Charged Species

H. Gajewski and I.V. Skrypnik

Dedicated to Prof. Herbert Amann on the occasion of his 65th birthday

Abstract. We study initial-boundary value problems for elliptic-parabolic systems of nonlinear partial differential equations describing drift-diffusion- reaction processes of electrically charged species in N-dimensional bounded Lipschitzian domains. We include Fermi-Dirac statistics and admit nonsmooth material coefficients. We prove existence and uniqueness of bounded global solutions.

Mathematics Subject Classification (2000). 35B45, 35K15, 35K20, 35K65.

Keywords. Nonlinear elliptic-parabolic systems, nonlocal drift, global bounded solutions, uniqueness, nonstandard assumptions, degenerate type.

1. Introduction

We study a mathematical model describing drift-diffusion-reaction processes of electrically charged species. Such processes play an important role in many branches of modern technology (see [4], [13], [14], [17]). The classical drift-diffusion model of charged carrier transport in semiconductors was established by van Roosbroeck [16]. It consists of a Poisson equation for the electrostatic potential v_0 and continuity equations for the densities u_1, u_2 of electrons and holes. The classical van Roosbroeck model describes processes in homogeneous semiconductor materials (e.g., silicon). Modern devices are often heterostructures where complex reactions take place. By this reason we admit nonsmooth data and $n\,(\geq 2)$ species with densities u_i and specific charges q_i.

The mathematical model is formulated below in Section 2. It consists of a Poisson equation (2.1) and n continuity equations (2.2). The equation (2.1) expresses the Gauss law, the system (2.2) means local carrier conservation. The system (2.1), (2.2) is completed by current relations (2.3), which suppose the antigradients of the electrochemical potentials ζ_i from (2.5) to be driving forces for carrier transport. We consider the system (2.1), (2.2) in a bounded Lipschitzian domain $\Omega \subset \mathbb{R}^N$, $N \geq 3$, completed by boundary and initial conditions (2.7)–(2.9).

The initial-boundary value problem (2.1), (2.2), (2.7)–(2.9) was formulated and studied in [8]. In that paper the free energy was recovered to be Lyapunov functional of the system and was used for proving a priori estimates, existence and uniqueness results. However, for space dimension $N \geq 3$ a gap remained between existence and uniqueness results in particular for the physically most relevant case that the state relation (2.4) is according to Fermi-Dirac statistics. Actually for $N \geq 3$ the paper [8] rests on following restricting hypotheses:

The existence result holds for dielectric permittivity ϵ from (2.1) and conductivity functions d_i from (2.3) such that

$$\epsilon = \text{constant}\ , \quad d_i(t,x,z,\xi) = \sigma_i(z)\,\xi\ , \tag{1.1}$$

that means, $J_i = -\sigma_i(v_i)\,\nabla\zeta_i$. Moreover, except for the special case of Boltzmann statistics (i.e., e_i in (2.4) is specified as exponential function), the uniqueness result in [8] supposes the electrostatic potential v_0 to satisfy the regularity condition

$$\nabla v_0 \in L^\infty\big((0,T); L^p(\Omega)\big) \quad \text{for some} \quad p > N\ . \tag{1.2}$$

(As to the validity of (1.2) in some nonsmooth situations comp. [5].)

The present paper mainly aims to fill that gap by proving global existence and uniqueness results without the restricting hypotheses (1.1), (1.2). To this end we apply to problem (2.1), (2.2), (2.7)–(2.9) an approach developed for model situations in our papers [9], [10], [11], [12]. The key role play sophisticated test functions in integral identities for proving a priori estimates and the uniqueness result.

The paper is organized as follows. Formulations of all hypotheses and main results are contained in Section 2. Integral estimates for the chemical potentials v_i and the electrostatical potential v_0 are proved in Section 3. In Section 4 we study the boundedness of the potentials v_i and v_0. A proof of the existence result is sketched in Section 5, where also a detailed proof of our main result, the uniqueness theorem, is given. Finally, Section 6 is devoted to the special case of functions d_i being linear with respect to ζ. This case is studied without growth conditions for the reaction terms.

2. Formulation of model and main results

The drift-diffusion model describing n species with densities u_i and specific charges q_i was formulated in [6], [7], [8] and reads as follows

$$-\nabla\cdot(\epsilon\nabla v_0) = f + \sum_{i=1}^n q_i\, u_i \quad \text{in} \quad Q_T = \Omega\times(0,T)\ , \tag{2.1}$$

$$\frac{\partial u_i}{\partial t} + \nabla\cdot J_i + R_i = 0 \quad \text{in} \quad Q_T\ , \quad i = 1,\dots,n\ , \tag{2.2}$$

where T is a finite time and Ω is a bounded Lipschitzian domain in $\mathbb{R}^N$. We suppose later on that $N \geq 3$. In (2.1) v_0 is the electrostatical potential, ϵ is the dielectric permittivity, f describes external sources (impurities). The currents J_i are given in the form

$$J_i = -d_i(\cdot, v_i, \nabla\zeta_i), \quad i = 1, \dots, n\,, \tag{2.3}$$

where v_i are chemical potentials related to the densities u_i by the state equations

$$u_i = u_i^*\, e_i(v_i), \quad i = 1, \dots, n\,, \tag{2.4}$$

with given, strictly positive state densities $u_i^* \in L^\infty(\Omega)$. The electrochemical potentials ζ_i from (2.3) are defined by

$$\zeta_i = q_i\, v_0 + v_i, \quad i = 1, \dots, n. \tag{2.5}$$

Remark 2.1. *The state equations 2.4 are chosen for simplicity. The results of the paper remain true for state equations like*

$$u_i = u_i^*\, e_i(v_i + q_i g_i), \quad i = 1, \dots, n\,,$$

with given band edges $g_i \in H^{1,q}(\Omega)$, $q > N$. This can be seen by replacing the argument $v_i = \zeta_i - q_i\, v_0$ of e_i and its derivatives by $v_i = \zeta_i - q_i\, \bar{v}_0$, where $\bar{v}_0 = v_0 - g_i$. However, an extension of our results to piece-wise constant g_i's in view of heterostructures is not trivial. That makes a difference to the peculiar situation of Boltzmann statistics, i.e., $e_i = \exp$, where even the case $g_i \in L^\infty(\Omega)$ can be simply included by setting $u_i^* := u_i^* \exp(q_i g_i) \in L^\infty(\Omega)$.

The reaction terms R_i in (2.2) have the form:

$$R_i(\cdot, v, \zeta) = \sum_{(\alpha,\beta)\in\mathcal{R}} \big[r_{\alpha\beta}(\cdot, v, \alpha\cdot\zeta) - r_{\alpha\beta}(\cdot, v, \beta\cdot\zeta)\big](\alpha_i - \beta_i)\,, \tag{2.6}$$

where $\alpha = (\alpha_1, \dots, \alpha_n)$, $\beta = (\beta_1, \dots, \beta_n) \in \mathcal{R} \subset \mathbb{R}^n$ are vectors of stoichiometric coefficients and the finite set $\mathcal{R}$ denotes the reactions actually taking place in the volume Ω occupied by the species.

Remark 2.2. *There are modified drift-diffusion models of charged species. So in the papers* [3] *and* [4] *Poisson's equation* (2.1) *is replaced by the neutrality condition*

$$f + \sum_{i=1}^{n} q_i\, u_i = 0 \quad on \quad Q_T = \Omega \times (0, T)\,.$$

We complete system (2.1), (2.2) by boundary and initial conditions :

$$\nu \cdot J_i + R_i^\Gamma = 0 \quad \text{on} \quad \Gamma_T = (0, T) \times \partial\Omega, \tag{2.7}$$

$$\nu \cdot (\epsilon\nabla v_0) + \kappa v_0 = f^\Gamma \quad \text{on} \quad \Gamma_T, \tag{2.8}$$

$$u_i(0, \cdot) = h_i \quad \text{in} \quad \Omega\,, \quad i = 1, \dots, n, \tag{2.9}$$

where $\nu(x')$ is the outer unit normal at $x' \in \partial\Omega$, R_i^Γ represents reactions taking place on the boundary $\partial\Omega$ of Ω. We assume that

$$R_i^\Gamma = \sum_{(\alpha,\beta)\in\mathcal{R}^\Gamma} \big(r_{\alpha\beta}^\Gamma(\cdot, v, \alpha\cdot\zeta) - r_{\alpha\beta}^\Gamma(\cdot, v, \beta\cdot\zeta)\big)(\alpha_i - \beta_i)\,, \tag{2.10}$$

where $\mathcal{R}^\Gamma$ is a finite set of vector pairs of stoichiometric coefficients and the functions $r^\Gamma_{\alpha\beta}$ model surface reaction rates.

Remark 2.3. *As a special feature the boundary condition* (2.8) *with* (2.10) *allows thermal equilibria, i.e., steady states with vanishing driving forces* $\nabla\zeta_i$. *However, the results of the paper remain true for other kinds of boundary conditions, for example*

$$\nu \cdot J_i + \kappa_i\,(\zeta_i - f_i^\Gamma) = 0 \quad on \quad \Gamma_T = (0,T)\times\partial\Omega,$$

with $\kappa_i,\ f_i^\Gamma \in L^\infty(\partial\Omega),\ \kappa_i \geq 0$.

The system (2.1), (2.2), (2.7)–(2.9) will be solved for the unknown vector $v = (v_0, v_1, \ldots, v_n)$ taking into account the relations (2.3)–(2.5) between v and $J = (J_1,\ldots,J_n)$, $u = (u_1,\ldots,u_n)$, $\zeta = (\zeta_1,\ldots,\zeta_n)$, respectively.

We assume the data of problem (2.1), (2.2), (2.7)–(2.9) to satisfy following hypotheses:

i) $d_{ij}(t,x,z,\xi)$, $i = 1,\ldots,n$, $j = 1,\ldots,N$, $r_{\alpha\beta}(t,x,v,y)$, $r^\Gamma_{\gamma\delta}(t,x',v,y)$, $(\alpha,\beta)\in\mathcal{R}$, $(\gamma,\delta)\in\mathcal{R}^\Gamma$, are measurable functions of $(t,x)\in Q_T$, $(t,x')\in\Gamma_T$ with respect to Lebesgue and surface measures respectively for every $z,y\in\mathbb{R}^1$, $\xi\in\mathbb{R}^N$, $v\in\mathbb{R}^{n+1}$ and continuous functions with respect to y,z,ξ,v for almost every $(t,x)\in Q_T$, $(t,x')\in\Gamma_T$, $d_i(t,x,z,0)=0$ for $i=1,\ldots,n$; ϵ and u_i^* are measurable functions on Ω; $\kappa\in L^\infty(\partial\Omega)$; $\kappa\geq 0$, $\kappa\neq 0$; $q_i\in\mathbb{R}^1$;

ii) $e_i \in (\mathbb{R}^1\to\mathbb{R}^1)$ is continuously differentiable such that $e_i'(z)>0$, $z\in\mathbb{R}^1$; $\lim\limits_{z\to-\infty} e_i(z) = 0$, $\lim\limits_{z\to+\infty} e_i(z) = +\infty$, $\int_{-\infty}^0 e_i(z)\,dz < \infty$, $i=1,\ldots,n$;

iii) there exist positive constants ν_1,ν_2 such that for arbitrary $\xi,\xi',\xi''\in\mathbb{R}^N$, $(t,x)\in Q_T, z\in\mathbb{R}^1$

$$\sum_{j=1}^N \left[d_{ij}(t,x,z,\xi') - d_{ij}(t,x,z,\xi'')\right](\xi_j' - \xi_j'') \geq \nu_1\, e_i'(z)|\xi'-\xi''|^2,$$
$$\left|d_{ij}(t,x,z,\xi)\right| \leq \nu_2(1+|\xi|)e_i'(z);\ e_i'(z)\leq \nu_2\, e_i(z) \quad \text{for} \quad z<0,$$
$$\nu_1\leq\epsilon(x)\leq\nu_2,\ \nu_1\leq u_i^*(x)\leq\nu_2, \quad i=1,\ldots,n; \quad j=1,\ldots,N;$$

iv) the functions $r_{\alpha\beta}(t,x,v,y)$, $r^\Gamma_{\gamma\delta}(t,x',v,y)$, $(\alpha,\beta)\in\mathcal{R}$, $(\gamma,\delta)\in\mathcal{R}^\Gamma$, are increasing in $y\in\mathbb{R}^1$ for $(t,x)\in Q_T$, $(t,x')\in\Gamma_T$, $v\in\mathbb{R}^{n+1}$ and there exist convex functions $M:\mathbb{R}^1\to\mathbb{R}^1_+ = \{z\in\mathbb{R}^1 : z>0\}$, $M^\Gamma:\mathbb{R}^1\to\mathbb{R}^1_+$ such that

$$[r_{\alpha\beta}(t,x,v,\alpha\cdot\zeta) - r_{\alpha\beta}(t,x,v,\beta\cdot\zeta)](\alpha-\beta)\cdot\zeta \leq M(|v|),$$
$$[r^\Gamma_{\gamma\delta}(t,x',v,\gamma\cdot\zeta) - r^\Gamma_{\gamma\delta}(t,x',v,\delta\cdot\zeta)](\gamma-\delta)\cdot\zeta \leq M^\Gamma(|v|),$$
$$\zeta_i = v_i + q_i v_0.$$

Finally, we assume the data f, f^Γ, h_i to satisfy:

$$
\begin{aligned}
&f \in C\big([0,T]; L^{p_1}(\Omega)\big), \quad \frac{\partial f}{\partial t} \in L^2\big(0,T;[W^{1,2}(\Omega)]^*\big), \qquad p_1 > \frac{N}{2},\\
&f^\Gamma \in C\big([0,T]; L^{p_2}(\partial\Omega)\big), \quad \frac{\partial f^\Gamma}{\partial t} \in L^2\big(0,T;[W^{\frac{1}{2},2}(\partial\Omega)]^*\big), \quad p_2 > N-1\,,\\
&\log(h_i) \in L^\infty(\Omega), \quad i = 1,\dots,n.
\end{aligned} \tag{2.11}
$$

Definition 2.1. *A vector $v = (v_0,\dots,v_n)$ is called solution of problem* (2.1), (2.2), (2.7)–(2.9), *if for $i = 1,\dots,n$:*

i) $v_0 \in C\big([0,T]; W^{1,2}(\Omega)\big), \quad v_i \in L^2\big(0,T; W^{1,2}(\Omega)\big)$,
$u_i = u_i^* e_i(v_i) \in C\left([0,T]; L^2(\Omega)\right), \quad \frac{\partial}{\partial t} u_i \in L^2\left(0,T; \left[W^{1,2}(\Omega)\right]^*\right)$,
where the time derivative is to be understood in the sense of distributions,

$$
\begin{aligned}
&\iint\limits_{Q_T} \Big\{ e_i'(v_i)\Big[\Big|\frac{\partial v_i}{\partial x}\Big|^2 + \Big|\frac{\partial v_0}{\partial x}\Big|^2\Big] + M(|v|)\Big\}\,dx\,dt < \infty\,,\\
&\iint\limits_{\Gamma_T} M^\Gamma(|v|)\,dx\,dt < \infty\,;
\end{aligned} \tag{2.12}
$$

ii) *for arbitrary test functions $\varphi \in C^\infty(\overline{Q}_T)$, $\psi \in C^\infty(\overline{\Omega})$, almost every $\tau \in (0,T)$ and $i = 1,\dots,n$ the following integral identities hold:*

$$
\begin{aligned}
&\int_0^\tau \Big\{ \langle \frac{\partial u_i}{\partial t}, \varphi\rangle + \int_\Omega \Big[\sum_{j=1}^N d_{ij}\Big(t,x,v_i,\frac{\partial(v_i + q_i v_0)}{\partial x}\Big)\frac{\partial \varphi}{\partial x_j}\\
&+ R_i(t,x,v,\zeta)\,\varphi\Big]\,dx + \int_{\partial\Omega} R_i^\Gamma(t,x,v,\zeta)\,\varphi\,ds\Big\}\,dt = 0\,,
\end{aligned} \tag{2.13}
$$

$$
\begin{aligned}
&\int_\Omega \Big\{\epsilon(x)\sum_{j=1}^N \frac{\partial v_0}{\partial x_j}\frac{\partial \psi}{\partial x_j} - \Big[\sum_{i=1}^n q_i\,u_i + f(t,x)\Big]\psi\Big\}\,dx\\
&\qquad + \int_{\partial\Omega} \big(\kappa(x)v_0 - f^\Gamma\big)\psi\,ds = 0\,,
\end{aligned} \tag{2.14}
$$

where $\zeta(t,x) = \big(\zeta_1(t,x),\dots,\zeta_n(t,x)\big)$, $\zeta_i(t,x) = v_i(t,x) + q_i\,v_0(t,x)$;

iii) *for test functions $\varphi \in C^\infty\left(\overline{Q}_T\right)$ with $\varphi(\tau,x) = 0$, $x \in \Omega$, the integral identity*

$$
\int_0^\tau \langle \frac{\partial u_i}{\partial t}, \varphi\rangle dt + \int_0^\tau\int_\Omega [u_i - h_i]\frac{\partial \varphi}{\partial t}\,dx\,dt = 0 \tag{2.15}
$$

holds for $\tau \in (0,T)$, $\quad i = 1,\dots,n$.

Besides of (2.2), (2.7) we shall consider for $\delta \in [0,1]$ the regularized equations

$$
\frac{\partial u_i}{\partial t} + \nabla\cdot J_i^{(\delta)} + R_i = 0 \quad \text{in} \quad Q_T\,, \tag{2.16}
$$

$$
\nu\cdot J_i^{(\delta)} + R_i^\Gamma = 0\,, \quad \text{on} \quad \Gamma_T\,, \tag{2.17}
$$

$$J_i^{(\delta)} = -d_i\Big(\cdot, v_i^{(\delta)}, \nabla\zeta_i\Big)\,, \quad v_i^{(\delta)} = \max\Big\{v_i, -\frac{1}{\delta}\Big\},\ \ J_i^{(0)} = J_i\,, \quad i = 1,\dots,n. \tag{2.18}$$

Solutions of problem (2.1), (2.16), (2.17), (2.8), (2.9) are defined as in Definition 2.1.

In what follows we understand as known parameters the numbers ν_1, ν_2, n, N, T, vectors in $\mathcal{R}$, $\mathcal{R}^\Gamma$, norms of the data f, f^Γ, h_i in respective spaces and numbers that depend only on Ω, M, M^Γ and κ. Moreover, we denote by c_k, $k = 1,\dots,$ constants depending only on known parameters.

Theorem 2.1. *Let the conditions i)–iv),* (2.11) *be satisfied. Then there exists a constant K_1 depending only on known parameters and independent of $\delta \in [0,1]$ such that each solution v of problem* (2.1), (2.16), (2.17), (2.8), (2.9) *satisfies*

$$\operatorname{ess\,sup}_{t\in(0,T)}\Big\{\int_\Omega\Big[\Lambda_i\big(v_i(t,x)\big) + \Big|\frac{\partial v_0(t,x)}{\partial x}\Big|^2\Big]\,dx \ + \int_{\partial\Omega}\kappa(x)v_0^2(t,x)\,ds\Big\}$$
$$+\iint\limits_{Q_T} e_i'(v_i)\Big|\frac{\partial(v_i+q_iv_0)}{\partial x}\Big|^2\,dx\,dt \le K_1\,, \tag{2.19}$$

where

$$\Lambda_i(v) = \int_0^v s\,e_i'(s)\,ds, \quad i = 1,\dots,n. \tag{2.20}$$

For establishing further integral estimates we need growth conditions for the functions e_i', R_i, R_i^Γ, $i = 1,\dots,n$:

$$\nu_3(v^{\gamma_i}+1) \le e_i'(v) \le \nu_4(v^{\gamma_i}+1), \quad v>0, \quad 0\le\gamma_i<\frac{4}{N-2}\,, \tag{2.21}$$

$$R_i(t,x,v,\zeta) \ge -\nu_4\Big(\sum_{j=1}^n [v_j]_+^{p_3} + |v_0|^{p_3}\Big) - \alpha_1(t,x), \quad \text{for} \quad v_i>0\,, \tag{2.22}$$

$$R_i^\Gamma(t,x',v,\zeta) \ge -\nu_4\Big(\sum_{j=1}^n [v_j]_+^{p_4} + |v_0|^{p_4}\Big) - \alpha_2(t,x'), \quad \text{for} \quad v_i>0\,, \tag{2.23}$$

where ν_3, ν_4 are positive constants and

$$p_3 < \gamma_*+1+\frac{2}{N}(\gamma_*+2), \quad p_4 < \gamma_*+1+\frac{1}{N}(\gamma_*+2), \quad \gamma_* = \min(\gamma_1,\dots,\gamma_n),$$
$$\alpha_1 \in L^{r_1}(Q_T), \quad r_1 > \frac{N+2}{2}, \quad \alpha_2 \in L^{r_2}(\Gamma_T), \quad r_2 > N+1.$$

Remark 2.4. *The growth condition* (2.21) *is satisfied by functions e_i according to Fermi-Dirac statistics, i.e., by Fermi Integrals:*

$$e_i(v) = \mathcal{F}_{\gamma_i}(v) = \frac{1}{\Gamma(\gamma_i+1)}\int_0^\infty \frac{s^{\gamma_i}\,ds}{1+\exp(s-v)}\,.$$

Note that the exponential function (Boltzmann statistics) violates (2.21). *Standard reaction terms like Shockley-Read and Auger recombination/generation* [8] *satisfy* (2.22)–(2.23).

We understand numbers from conditions (2.21)–(2.23) and norms of the functions α_1, α_2 as known parameters too.

Theorem 2.2. *Let the assumptions of Theorem* 2.1 *and the conditions* (2.21)–(2.23) *be satisfied. Then there exists a constant* K_2 *depending only on known parameters and independent of* $\delta \in [0,1]$ *such that each solution* v *of problem* (2.1), (2.16), (2.17), (2.8), (2.9) *satisfies*

$$\iint_{Q_T} e_i'(v_i)\left\{\left|\frac{\partial v_i}{\partial x}\right|^2 + \left|\frac{\partial v_0}{\partial x}\right|^2\right\} dx\, dt \le K_2, \quad i = 1, \dots, n. \tag{2.24}$$

Theorem 2.3. *Let the assumptions of Theorem* 2.2 *be satisfied. Then there exist constants* K_3 *and* $\eta \in (0,1)$ *depending only on known parameters and independent of* δ *such that for arbitrary* $t \in [0,T]$, $x, y \in \Omega$

$$\|v_0\|_{L^\infty(Q_T)} \le K_3, \quad |v_0(t,x) - v_0(t,y)| \le K_3\,|x-y|^\eta\,. \tag{2.25}$$

In view of controlling $v_i(t,x)$ from below we suppose additionally to (2.22) and (2.23) that for $v_i < 0$, $i = 1, \dots, n$:

$$R_i(t,x,v,\zeta) \le \nu_4\, e_i(v_i)\big[F\big(v_0, e(v)\big) + \alpha_1(t,x)\big]\,, \tag{2.26}$$

$$R_i^\Gamma(t,x',v,\zeta) \le \nu_4\, e_i(v_i)\big[F\big(v_0, e(v)\big) + \alpha_2(t,x')\big] \tag{2.27}$$

with ν_4, $\alpha_1(t,x)$, $\alpha_2(t,x')$ as in (2.21), (2.22), $e(v) = (e_1(v_1), \dots, e_n(v_n))$ and some continuous function $F : \mathbb{R}^{n+1} \to \mathbb{R}^1$.

Theorem 2.4. *Let the assumptions of Theorem* 2.2 *and the conditions* (2.26), (2.27) *be satisfied. Then there exists a constant* K_4 *depending only on known parameters and independent of* $\delta \in [0,1]$ *such that for each solution* $v = (v_0, v_1, \dots, v_n)$ *of problem* (2.1), (2.16), (2.17), (2.8), (2.9)

$$\operatorname{ess\,sup}\big\{|v_i(t,x)| : (t,x) \in Q_T\big\} \le K_4\,, \quad i = 0, \dots, n\,. \tag{2.28}$$

Theorem 2.5. *Let the conditions i)–iv),* (2.11), (2.21)–(2.23), (2.26), (2.27) *be satisfied. Then the initial-boundary value problem* (2.1), (2.2), (2.7)–(2.9) *has at least one solution in the sense of the Definition* 2.1.

Theorem 2.6. *Let the conditions of Theorem* 2.5 *be satisfied. Assume additionally that for* $i = 1, \dots, n$, $j = 1, \dots, N$*:*

(i) *the functions* $d_{ij}(t,x,z,\xi)$ *have the special structure*

$$d_{ij}(t,x,z,\xi) = e_i'(z)\gamma_{ij}(t,x,\xi) \tag{2.29}$$

where $e_i' \circ e_i^{-1} : (0,\infty) \to (0,\infty)$ *is piece-wise differentiable and concave;*

(ii) *the functions* e_i'', $\gamma_{ij}(t,x,\xi)$, $r_{\alpha\beta}(t,x,v,y)$, $r^\Gamma_{\gamma\delta}(t,x',v,y)$ *are locally Lipschitzian with respect to* ξ, v, y.

Then the initial-boundary value problem (2.1), (2.2), (2.7)–(2.9) *has a unique solution in the sense of the Definition* 2.1.

Remark 2.5. *The Fermi integrals from Remark* 2.4 *satisfy the respective assumptions of Theorem* 2.6 [1].

Corollary 2.1. *Let the conditions of Theorem* 2.6 *be satisfied and assume additionally that the functions* f, f^Γ, d_{ij}, $r_{\alpha\beta}$, $r^\Gamma_{\alpha\beta}$ *are Lipschitzian with respect to* t. *Then the solution* v *of problem* (2.1), (2.2), (2.7)–(2.9) *is regular in the sense that*

$$t \to t\,\frac{\partial v_i}{\partial t} \in L^\infty(0,T;L^2(\Omega)) \cap L^2(0,T;W^{1,2}(\Omega)),\ i = 1,\dots,n.$$

Remark 2.6. *Corollary* 2.1 *and Theorem* 2.4 *imply that*

$$t \to t\,\frac{\partial u_i}{\partial t} \in L^\infty(0,T;L^2(\Omega)).$$

Consequently, (2.2) *can be understood not only in the sense of distributions, but even as an equation in* $L^2(0,T;L^2(\Omega))$.

We conclude this section considering the special case that the currents J_i are linear with respect to the gradients of the electrochemical potentials ζ_i. This case is interesting in so far as we do not need the growth restrictions (2.22), (2.23) for the reaction terms.

Theorem 2.7. *Let the conditions i)–iv),* (2.11), (2.21), (2.26), (2.27) *be satisfied. Suppose that the reference densities from* (2.4) *and the exponents* γ_i *from* (2.21) *satisfy*

$$u_i^* = u^*\,, \quad \gamma_i = \gamma\,, \quad i = 1,\dots,n \tag{2.30}$$

and that for $(\alpha,\beta) \in \mathcal{R}$, $(\gamma,\delta) \in \mathcal{R}^\Gamma$

$$\alpha \cdot q = \beta \cdot q, \quad \gamma \cdot q = \delta \cdot q\,. \tag{2.31}$$

Moreover, assume the functions d_{ij} *to have the structure*

$$d_{ij}(t,x,z,\xi) = \sum_{k=1}^{N} e_i'(z) a_{kj}(t,x)\xi_k\,, \quad i = 1,\dots,n, \quad j = 1,\dots,N. \tag{2.32}$$

Then all assertions of the Theorems 2.2–2.6 *are valid.*

Remark 2.7. *We assumed the coincidence of the* γ_i*'s for simplicity. It is possible to replace it by some restriction on* $\max\{|\gamma_i - \gamma_j|,\ 1 \le i,\ j \le n\}$. *Analogously to* [11] *it is possible to prove Theorem* 2.7 *for* γ_i *satisfying only* $0 \le \gamma_i < \frac{2}{N-2}$, $i = 1,\dots,n$.

We shall prove the Theorems 2.1, 2.2 in Section 3, the Theorems 2.3, 2.4 in Section 4 and the Theorems 2.5, 2.6 in Section 5. Finally we shall make some comments with respect to the proof of Theorem 2.7 in Section 6.

3. Proof of integral estimates

The proof of a priori estimates in this section rests on testing the integral identities (2.13), (2.14) by suitable functions. For that purpose the following remark is useful:

Remark 3.1. *Let $F : \mathbb{R}^{n+1} \to \mathbb{R}^1$ be an arbitrary piece-wise differentiable function with bounded gradient and let $v(t,x)$ be a solution of problem* (2.1), (2.16), (2.17), (2.8), (2.9). *Then the equality* (2.13) *holds for $\varphi(t,x) = F(v(t,x))$. Moreover,* (2.14) *holds for arbitrary functions $\psi \in W^{1,2}(\Omega)$. That follows from (2.12) after approximating $v(t,x)$ by smooth functions.*

Proof of Theorem 2.1. Let v be a solution of problem (2.1), (2.16), (2.17), (2.8), (2.9). Denote by $g_0(x)$ a solution of the problem

$$-\nabla \cdot (\epsilon \nabla g_0) = f(0,x) + \sum_{i=1}^{n} q_i h_i(x) \quad \text{in} \quad \Omega\,, \tag{3.1}$$

$$\nu \cdot (\epsilon \nabla g_0) + \kappa g_0 = f^{\Gamma}(0,x') \quad \text{on} \quad \partial\Omega\,. \tag{3.2}$$

We extend $v_i(t,x)$ for $t<0$, $x \in \Omega$ by setting $v_i(t,x) = g_i(x)$, where $g_i(x) = e_i^{-1}\left(\frac{h_i(x)}{u_i^*(x)}\right)$, $i = 1,\dots,n$. In an analogous way we extend $f(t,x)$ and $f^{\Gamma}(t,x')$. Testing the integral identity (2.14) with $\psi(x) = v_0(t+s,x) - v_0(t,x)$ and integrating on t, we obtain for $\tau \in (0,T)$, $s \in (0, T-\tau)$,

$$\begin{aligned}
&\int_{-s}^{\tau}\int_{\Omega}\Big\{\epsilon(x)\sum_{j=1}^{N}\frac{\partial}{\partial x_j}\Big[v_0(t+s,x)+v_0(t,x)\Big]\frac{\partial}{\partial x_j}\big[v_0(t+s,x)-v_0(t,x)\big]\\
&-\Big[\sum_{i=1}^{n} q_i\big(u_i(t+s,x)+u_i(t,x)\big)+f(t+s,x)+f(t,x)\Big]\\
&\times\big[v_0(t+s,x)-v_0(t,x)\big]\Big\}\,dx\,dt+\int_{-s}^{\tau}\int_{\partial\Omega}\Big\{\kappa(x)\big[v_0(t+s,x)+v_0(t,x)\big]\\
&-f^{\Gamma}(t+s,x)-f^{\Gamma}(t,x)\Big\}\big[v_0(t+s,x)-v_0(t,x)\big]\,ds\,dt=0\,.
\end{aligned} \tag{3.3}$$

Arguing as in the proof of Theorem 2.3 in [11], we infer from (3.3)

$$\begin{aligned}
&\int_{\Omega}\epsilon(x)\left|\frac{\partial v_0(\tau,x)}{\partial x}\right|^2 dx+\int_{\partial\Omega}\kappa(x)v_0^2(\tau,x)\,ds-\sum_{i=1}^{n}q_i\int_0^{\tau}\langle\frac{\partial u_i}{\partial t},v_0\rangle\,dt\\
&\le c_1\Big\{1+\int_0^{\tau}\int_{\Omega}\epsilon(x)\left|\frac{\partial v_0(\tau,x)}{\partial x}\right|^2 dx\,dt+\int_0^{\tau}\int_{\partial\Omega}\kappa(x)v_0^2(t,x)\,ds\,dt\Big\}.
\end{aligned} \tag{3.4}$$

Note that $W^{1,2}(\Omega)$ can be normed equivalently by

$$\left\{\int_{\Omega}\left|\frac{\partial u(x)}{\partial x}\right|^2 dx+\int_{\partial\Omega}\kappa(x)u^2(x)\,ds\right\}^{\frac{1}{2}}.$$

Remark 3.1 allows us to test the regularized version of 2.13 with $\varphi = v_i + q_i v_0$:

$$\int_0^\tau \Big\{\langle\frac{\partial u_i}{\partial t},\varphi\rangle + \int_\Omega \Big[\sum_{j=1}^N d_{ij}\Big(t,x,v_i^\delta,\frac{\partial(v_i+q_iv_0)}{\partial x}\Big)\frac{\partial\varphi}{\partial x_j} + R_i(t,x,v,\zeta)\varphi\Big]\,dx\Big\}\,dt + \int_0^\tau\int_{\partial\Omega} R_i^\Gamma(t,x,v,\zeta)\varphi\,ds\,dt = 0. \tag{3.5}$$

So, using (3.4), we get

$$\begin{aligned}&\int_\Omega \epsilon(x)\left|\frac{\partial v_0(\tau,x)}{\partial x}\right|^2 dx + \int_{\partial\Omega}\kappa(x)v_0^2(\tau,x)\,ds + \sum_{i=1}^n\int_0^\tau\Big\{\langle\frac{\partial u_i}{\partial t},v_i\rangle\\ &+\int_\Omega\Big[\sum_{j=1}^N d_{ij}\Big(t,x,v_i^{(\delta)},\frac{\partial(v_i+q_iv_0)}{\partial x}\Big)\frac{\partial}{\partial x_j}(v_i+q_iv_0)\\ &+R_i(t,x,v,\zeta)(v_i+q_iv_0)\Big]\,dx\Big\}\,dt\\ &+\int_0^\tau\int_{\partial\Omega}R_i^\Gamma(t,x,v,\zeta)(v_i+q_iv_0)\,ds\,dt\\ &\le c_2\Big\{1+\int_0^\tau\int_\Omega\epsilon(x)\left|\frac{\partial v_0(t,x)}{\partial x}\right|^2 dx\,dt + \int_0^\tau\int_{\partial\Omega}\kappa(x)v_0^2(t,x)\,ds\,dt\Big\}.\end{aligned} \tag{3.6}$$

We transform the integral with $\frac{\partial u_i}{\partial t}$ by means of Lemma 1 and Lemma 3 from [10] and obtain

$$\int_0^\tau <\frac{\partial u_i}{\partial t},v_i>\,dt = \int_\Omega u_i^*(x)\big[\Lambda_i\big(v_i(\tau,x)\big)-\Lambda_i\big(g_i(x)\big)\big]\,dx. \tag{3.7}$$

Estimating terms with R_i, R_i^Γ by means of condition iv), we get

$$\begin{aligned}&\sum_{i=1}^n R_i(t,x,v,\zeta)(v_i+q_iv_0)\\ &=\sum_{(\alpha,\beta)\in R}\big[r_{\alpha,\beta}(t,x,v,\alpha\cdot\zeta)-r_{\alpha\beta}(t,x,v,\beta\cdot\zeta)\big]\cdot(\alpha-\beta)\cdot\zeta\ge 0,\\ &\sum_{i=1}^n R_i^\Gamma(t,x',v,\zeta)(v_i+q_iv_0)\\ &=\sum_{(\alpha,\beta)\in R^\Gamma}\big[r_{\alpha\beta}^\Gamma(t,x',v,\alpha\cdot\zeta)-r_{\alpha\beta}^\Gamma(t,x',v,\beta\cdot\zeta)\big](\alpha-\beta)\cdot\zeta\ge 0.\end{aligned} \tag{3.8}$$

By condition iii) we obtain from (3.6)–(3.8)

$$\begin{aligned}
&\int_\Omega \epsilon(x)\left|\frac{\partial v_0(\tau,x)}{\partial x}\right|^2 dx + \int_{\partial\Omega}\kappa(x)v_0^2(\tau,x)\,dx \\
&+\sum_{i=1}^n\int_0^\tau\int_\Omega e_i'(v_i)\left|\frac{\partial(v_i+q_iv_0)}{\partial x}\right|^2 dx\,dt \qquad (3.9)\\
&\le c_2\Big\{1+\int_0^\tau\int_\Omega \epsilon(x)\left|\frac{\partial v_0(t,x)}{\partial x}\right|^2 dx\,dt + \int_0^\tau\int_{\partial\Omega}\kappa(x)v_0^2(t,x)\,ds\,dt\Big\}.
\end{aligned}$$

The last inequality and Gronwall's lemma imply (2.19) and the proof of Theorem 2.1 is complete. □

Lemma 3.1. *Let the conditions of Theorem* 2.1 *be satisfied. Suppose that*

$$\operatorname{ess\,sup}_{t\in(0,T)}\int_\Omega u_i^r(t,x)\,dx \le L_1 \quad \textit{for} \quad i=1,\dots,n\,, \tag{3.10}$$

with numbers $r\in\left(\frac{2N}{N+2},\frac{N}{2}\right)$ *and* L_1 *depending only on known parameters. Then*

$$\begin{aligned}
\operatorname{ess\,sup}_{t\in(0,T)}\Big\{&\int_\Omega\left(\left|v_0(t,x)\right|^{\frac{pN}{N-2}}+\left|v_0(t,x)\right|^{p-2}\left|\frac{\partial v_0(t,x)}{\partial x}\right|^2\right)dx \\
&+\int_{\partial\Omega}\left|v_0(t,x)\right|^{\frac{p(N-1)}{N-2}}ds\Big\}\le L_2\,,
\end{aligned}\tag{3.11}$$

where the constant L_2 *depends only on known parameters and* p *is defined by*

$$p\cdot\frac{N}{N-2}=(p-1)\frac{r}{r-1}\,. \tag{3.12}$$

Proof. For arbitrary functions w we define

$$w_k(t,x)=\min\{w(t,x),k\},\quad k\in\mathbb{R}^1,\ (t,x)\in Q_T\,. \tag{3.13}$$

Testing the integral identity (2.14) with $\psi(t,x)=|v_0(t,x)|_k^{p-1}\operatorname{sign}v_0(t,x)$, $k>0$, using the conditions iii), (2.11), (3.10) and Hölder's inequality, we obtain

$$\begin{aligned}
&\int_\Omega |v_0|_k^{p-2}\left|\frac{\partial|v_0|_k}{\partial x}\right|^2 dx + \int_{\partial\Omega}\kappa(x)|v_0|_k^p\,ds \\
&\quad\le c_4\left(\int_\Omega |v_0|_k^{(p-1)\frac{r}{r-1}}dx\right)^{\frac{r-1}{r}} + c_4\left(\int_{\partial\Omega}|v_0|_k^{(p-1)\frac{r(N-1)}{N(r-1)}}ds\right)^{\frac{N(r-1)}{r(N-1)}}.
\end{aligned}\tag{3.14}$$

Hence Sobolev's embedding theorem yields

$$\begin{aligned}
&\left(\int_\Omega |v_0|_k^{p\frac{N}{N-2}}dx\right)^{\frac{N-2}{N}} + \left(\int_{\partial\Omega}|v_0|_k^{p\frac{N-1}{N-2}}ds\right)^{\frac{N-2}{N-1}} \\
&\quad\le c_5\left(\int_\Omega |v_0|_k^{(p-1)\frac{r}{r-1}}dx\right)^{\frac{r-1}{r}} + c_5\left(\int_{\partial\Omega}|v_0|_k^{(p-1)\frac{r(N-1)}{N(r-1)}}ds\right)^{\frac{N(r-1)}{r(N-1)}}.
\end{aligned}\tag{3.15}$$

In view of the restriction on r and (3.12) we infer (3.11) from (3.14) and (3.15) letting $k \to \infty$. The proof of Lemma 3.1 is completed. □

In what follows we suppose the conditions (2.21)–(2.23) to be satisfied. We fix a $\Delta \in (0,1)$ such that

$$\begin{gathered}\Delta \leq 1 + \gamma_* + \frac{2}{N}(\gamma_* + 2) - p_3, \\ \Delta \leq 1 + \gamma_* + \frac{1}{N}(\gamma_* + 2) - p_4, \\ \gamma^* = \max\{\gamma_1, \dots, \gamma_n\} \leq \frac{4}{N-2} - \frac{\Delta N}{N-2}\end{gathered} \tag{3.16}$$

and define

$$r(m) = \Delta m\,, \quad m = 0, 1, 2 \dots . \tag{3.17}$$

Lemma 3.2. *Let the conditions of Theorem 2.2 be satisfied. Suppose that for some nonnegative integer m*

$$\begin{gathered}\iint\limits_{Q_T} |v_0(t,x)|^{r(m)} \left|\frac{\partial v_0}{\partial x}\right|^2 dx\, dt + \iint\limits_{\Gamma_T} \kappa(x)|v_0(t,x)|^{2+r(m)} ds\, dt \leq L_3\,, \\ \iint\limits_{Q_T} [v_i(t,x)]_+^{r(m)} \left|\frac{\partial v_i}{\partial x}\right|^2 dx\, dt \leq L_3, \quad i = 1, \dots, n\,,\end{gathered} \tag{3.18}$$

with $[v_i(t,x)]_+ = \max\{v_i(t,x), 0\}$ and a constant L_3 depending only on known parameters and m. Then there exists a constant L_4 depending only on known parameters and m such that

$$\iint\limits_{Q_T} |v_0(t,x)|^{r(m+1)} \left|\frac{\partial v_0}{\partial x}\right|^2 dx\, dt + \iint\limits_{\Gamma_T} \kappa(x)|v_0(t,x)|^{r(m+1)+2} ds\, dt \leq L_4\,. \tag{3.19}$$

Proof. Remark that by condition (2.21)

$$\begin{gathered}e_i(v) \leq c_6\, v^{\gamma_i+1}, \quad \Lambda_i(v) \geq c_6\, v^{\gamma_i+2}, \\ v \geq 1, \quad i = 1, \dots, n,\end{gathered} \tag{3.20}$$

where the function Λ_i is defined by (2.20). From (2.19), (3.20) we have

$$\operatorname{ess\,sup}_{t\in(0,T)} \int_\Omega [v_i(t,x)]_+^{\gamma_i+2} dx \leq c_7, \quad i = 1, \dots, n. \tag{3.21}$$

Testing the integral identity (2.14) with $\psi(t,x) = \left|v_0(t,x)\right|_k^{r(m+1)+1}$ sign $v_0(t,x)$ and using condition iii) and (2.11) we have

$$\begin{aligned}
&\iint\limits_{Q_T} |v_0|_k^{r(m+1)} \left|\frac{\partial |v_0|_k}{\partial x}\right|^2 dx\,dt + \iint\limits_{\Gamma_T} \kappa(x)|v_0|_k^{r(m+1)+2}\,ds\,dt \\
&\le c_8\Bigg\{\sum_{i=1}^{n}\iint\limits_{Q_T} u_i|v_0|_k^{r(m+1)+1}\,dx\,dt \\
&\quad + \int_0^T\left\{\int_\Omega \left|v_0(t,x)\right|_k^{[r(m+1)+1]p_1'}\,dx\right\}^{\frac{1}{p_1'}} dt \\
&\quad + \int_0^T\left\{\int_{\partial\Omega} \left|v_0(t,x)\right|_k^{[r(m+1)+1]p_2'}\,ds\right\}^{\frac{1}{p_2'}} dt\Bigg\}
\end{aligned} \tag{3.22}$$

with $p_i' = \frac{p_i}{p_i-1}$, $i=1,2$. The embedding theorem and (3.18) imply

$$\begin{aligned}
&\int_0^T\left\{\int_\Omega |v_0|^{[r(m)+2]\frac{N}{N-2}}\,dx\right\}^{\frac{N-2}{N}} dt \\
&\qquad + \int_0^T\left\{\int_{\partial\Omega} |v_0|^{[r(m)+2]\frac{N-1}{N-2}}\,ds\right\}^{\frac{N-2}{N-1}} dt \le c_9.
\end{aligned} \tag{3.23}$$

Hence we can estimate the second and the third integral on the right-hand side of (3.22) by a constant depending only on known parameters.

In order to estimate the first integral on the right-hand side of (3.22) we derive firstly an auxiliary estimate for $v_i(t,x)$. By Hölder's inequality, the embedding theorem, (3.18) and (3.21) we obtain with an arbitrary number $q \in \left(0, \frac{N}{N-2}\right)$:

$$\begin{aligned}
&\int_0^T\left\{\int_\Omega \left[v_i(t,x)\right]_+^{(\gamma_i+2)[1-q\frac{N-2}{N}]+[r(m)+2]q}\,dx\right\}^{\frac{1}{q}} dt \\
&\le \int_0^T\left\{\int_\Omega \left[v_i(t,x)\right]_+^{\gamma_i+2}\,dx\right\}^{\frac{1}{q}-\frac{N-2}{N}} \\
&\qquad \cdot\left\{\int_\Omega \left[v_i(t,x)\right]_+^{\frac{[r(m)+2]N}{N-2}}\,dx\right\}^{\frac{N-2}{N}} dt \\
&\le \int_0^T\left\{\int_\Omega \left[v_i(t,x)\right]_+^{\gamma_i+2}\,dx\right\}^{\frac{1}{q}-\frac{N-2}{N}} \\
&\qquad \cdot\int_\Omega \left[v_i(t,x)\right]_+^{r(m)}\left|\frac{\partial v_i}{\partial x}\right|^2 dx\ dt \le c_{10}\,.
\end{aligned} \tag{3.24}$$

Let us choose the number q_* such that

$$\left[r(m+1)+1\right]q_*' = \left[r(m)+2\right]\frac{N}{N-2}, \quad q_*' = \frac{q_*}{q_*-1}. \tag{3.25}$$

Since $\Delta \in (0,1)$ and $r(m) = m\Delta$, we have $q'_* > \frac{N}{N-2}$. Using Hölder's inequality, (2.4) and (2.21), we get

$$\begin{aligned}
&\iint\limits_{Q_T} u_i |v_0|_k^{r(m+1)+1}\, dx\, dt \\
&\qquad \le c_{11} \int_0^T \left\{ \int_\Omega |v_0|_k^{[r(m)+2]\frac{N}{N-2}}\, dx \right\}^{\frac{1}{q'_*}} \left\{ 1 + \int_\Omega [v_i]_+^{(\gamma_i+1)q_*}\, dx \right\}^{\frac{1}{q_*}} dt \\
&\qquad \le c_{11} \left\{ \int_0^T \left\{ \int_\Omega |v_0|_k^{[r(m)+2]\frac{N}{N-2}}\, dx \right\}^{\frac{N-2}{N}} dt \right\}^{\frac{N}{q'_*(N-2)}} \\
&\qquad\quad \times \left\{ \int_0^T \left\{ 1 + \int_\Omega [v_i]_+^{(\gamma_i+1)q_*}\, dx \right\}^{\frac{1}{q_* - \frac{N(q_*-1)}{N-2}}} dt \right\}^{1-\frac{N}{q'_*(N-2)}} .
\end{aligned} \tag{3.26}$$

Let $q = q_* - \frac{N}{N-2}(q_* - 1) = \frac{N}{N-2} - \frac{2}{N-2}\, q_* \in \left(0, \frac{N}{N-2}\right)$ with q_* defined by (3.25). Since

$$\big[r(m)+2\big] q = q_* \big[r(m)+2\big] \left(1 - \frac{N}{N-2} \cdot \frac{1}{q'_*}\right) = q_* \big(r(m) - r(m+1) + 1\big) = q_*(1-\Delta) ,$$

we have by (3.16)

$$\begin{aligned}
&(\gamma_i+2)\left[1 - q\frac{N-2}{N}\right] + \big[r(m)+2\big] q - (\gamma_i+1) q_* \\
&= \left[\frac{2}{N}(\gamma_i+2) - \gamma_i - \Delta\right] q_* = \frac{q_*}{N}\big[4 - N\Delta - \gamma_i(N-2)\big] \ge 0 .
\end{aligned} \tag{3.27}$$

The inequalities (3.18), (3.21), (3.24), (3.26), (3.27) imply

$$\iint\limits_{Q_T} u_i |v_0|_k^{r(m+1)+1}\, dx\, dt \le c_{12} . \tag{3.28}$$

So we obtain the desired estimate (3.19) from (3.22), (3.23) and (3.28). This ends the proof of Lemma 3.2. □

Lemma 3.3. *Suppose that the assumptions of Theorem 2.2 and the inequalities (3.18) are satisfied for a nonnegative integer m such that*

$$\gamma_{i_0} \ge r(m+1) , \quad i_0 \in \{1, \dots, n\} . \tag{3.29}$$

Then there exists a constant L_5 depending only on known parameters such that

$$\iint\limits_{Q_T} \big[v_{i_0}(t,x)\big]_+^{r(m+1)} \left| \frac{\partial v_{i_0}(t,x)}{\partial x} \right|^2 dx\, dt \le L_5 . \tag{3.30}$$

Proof. For arbitrary functions $w_1(t,x)$, $w_2(t,x)$ defined on Q_T we define the set

$$\big\{ w_1 \le w_2 \big\} = \big\{ (t,x) \in Q_T : w_1(t,x) \le w_2(t,x) \big\} .$$

By (2.19) and (3.19) we have

$$\iint\limits_{\{[v_{i_0}]_+\le 2|v_0|\}} \left[v_{i_0}(t,x)\right]_+^{r(m+1)} \left|\frac{\partial v_{i_0}(t,x)}{\partial x}\right|^2 dx\,dt \le c_{13}. \tag{3.31}$$

To complete the proof we need an analogous estimate with respect to $\{[v_{i_0}]_+ > 2|v_0|\}$. Testing the identity (2.14) with

$$\psi = |v_0|_k \Big\{ \big[[v_{i_0} - |v_0|_k]_+\big]_k + |v_0|_k \Big\}^{r(m)+\varepsilon} \operatorname{sign} v_0\,, \quad \varepsilon \in (0,\Delta],\ k > 1,$$

and using condition iii) and (2.11), we obtain

$$I_1 + I_2 \le c_{14}\big[I_3(1) + I_4 + I_5\big] \tag{3.32}$$

where

$$I_1 = \iint\limits_{\{|v_0|<k\}} \Big\{ \big[[v_{i_0} - |v_0|_k]_+\big]_k + |v_0|_k \Big\}^{r(m)+\varepsilon} \left|\frac{\partial v_0}{\partial x}\right|^2 dx\,dt,$$

$$I_2 = \iint\limits_{\Gamma_T} \kappa(x)|v_0|_k^2 \Big\{ \big[[v_{i_0} - |v_0|_k]_+\big]_k + |v_0|_k \Big\}^{r(m)+\varepsilon} ds\,dt,$$

$$I_3(l) = \iint\limits_{\{|v_0|_k<v_{i_0}\}} |v_0|_k^l \Big\{ \big[[v_{i_0} - |v_0|_k]_+\big]_k + |v_0|_k \Big\}^{r(m)+\varepsilon-l} \left|\frac{\partial v_{i_0}}{\partial x}\right| \left|\frac{\partial v_0}{\partial x}\right| dx\,dt,$$

$$I_4 = \iint\limits_{Q_T} \left[\sum_{i=1}^n u_i + \big|f(t,x)\big|\right] |v_0|_k \Big\{ \big[[v_{i_0} - |v_0|_k]_+\big]_k + |v_0|_k \Big\}^{r(m)+\varepsilon} dx\,dt,$$

$$I_5 = \iint\limits_{\Gamma_T} |f^\Gamma(t,x)||v_0|_k \Big\{ \big[[v_{i_0} - |v_0|_k]_+\big]_k + |v_0|_k \Big\}^{r(m)+\varepsilon} ds\,dt.$$

Up to the end of Lemma 3.3 we choose $\varepsilon = \Delta$.
We estimate $I_3(l)$ for natural numbers $l < r(m+1)$ by Young's inequality

$$I_3(l) \le \varepsilon_1 I_1 + \frac{c_{15}}{\varepsilon_1} \iint\limits_{\{|v_0|_k<v_{i_0}\}} |v_0|_k^{2l} \Big\{ \big[[v_{i_0} - |v_0|_k]_+\big]_k + |v_0|_k \Big\}^{r(m+1)-2l} \left|\frac{\partial v_{i_0}}{\partial x}\right|^2 dx\,dt$$
$$+\, c_{15} \iint\limits_{\{k\le|v_0|<v_{i_0}\}} \left\{ |v_{i_0}|^{r(m+1)} \left|\frac{\partial(v_{i_0} + q_{i_0}v_0)}{\partial x}\right|^2 + |v_0|^{r(m+1)} \left|\frac{\partial v_0}{\partial x}\right|^2 \right\} dx\,dt, \tag{3.33}$$

where ε_1 is an arbitrary positive number. Using the simple inequality

$$\left|\frac{\partial v_{i_0}}{\partial x}\right|^2 \le c_{16}\left(\left|\frac{\partial(v_{i_0} + q_{i_0}v_0)}{\partial x}\right|^2 + \left|\frac{\partial v_{i_0}}{\partial x}\right| \left|\frac{\partial v_0}{\partial x}\right| \right),$$

we have from (3.33), (2.19), (3.19) and (3.29)

$$
\begin{aligned}
I_3(l) \le \varepsilon_1 I_1 + \frac{c_{17}}{\varepsilon_1}\Big\{1 + I_3(2l) + \iint\limits_{Q_T} |v_0|^{r(m+1)} \left|\frac{\partial v_0}{\partial x}\right|^2 \, dx\, dt \\
+ \iint\limits_{Q_T} [v_{i_0}]_+^{\gamma_i} \left|\frac{\partial (v_{i_0} + q_{i_0} v_0)}{\partial x}\right|^2 \, dx\, dt\Big\} \le \varepsilon_1 I_1 + \frac{c_{18}}{\varepsilon_1}\big(1 + I_3(2l)\big).
\end{aligned}
\tag{3.34}
$$

The inequalities (2.19), (3.19) imply also $I_3(l) \le c_{19}$ for $l > r(m+1)$. Therefore, iterating (3.34), we get

$$
I_3(1) < \frac{1}{2c_{14}} I_1 + c_{19}. \tag{3.35}
$$

Next we estimate the term I_4 by Hölder's inequality and condition (2.11):

$$
\begin{aligned}
I_4 \le & c_{20}\Big\{\sum_{i=1}^{n} \iint\limits_{Q_T} |v_i|^{\gamma_i+2+r(m+1)} \, dx\, dt + \int_0^T \Big\{\int_\Omega |v_i|^{[r(m+1)+1]p_1'} \, dx\Big\}^{\frac{1}{p_1'}} dt \\
& + \sum_{i=1}^{n} \iint\limits_{Q_T} |v_0|_k^{r(m+1)+1} u_i \, dx\, dt + \int_0^T \Big\{\int_\Omega |v_0|_k^{[r(m+1)+1]p_1'} \, dx\Big\}^{\frac{1}{p_1'}} dt + 1\Big\}.
\end{aligned}
\tag{3.36}
$$

Now all integrals in (3.36) can be estimated from above by a constant depending only on known parameters. Indeed, since by (3.16)

$$
\begin{aligned}
&\frac{2}{N}(\gamma_i + 2) + r(m) + 2 - \big[\gamma_i + 2 + r(m+1)\big] = \\
&\qquad = \frac{2}{N}(\gamma_i + 2) - \gamma_i - \Delta = \frac{1}{N}\big[4 - (N-2)\gamma_i - N\Delta\big] \ge 0,
\end{aligned}
$$

an estimate of the first integral in (3.36) follows from (3.24) with $q = 1$. In analogous way the second integral in (3.36) can be estimated by means of (3.24) with $q = p_1'$. Estimates for the third and the fourth integral in (3.36) follow from (3.28) and (3.23), respectively. So we have shown that

$$
I_4 \le c_{21}. \tag{3.37}
$$

Further, condition (2.11), (3.18), (3.23), (3.24) and the embedding theorem yield

$$
\begin{aligned}
I_5 &\le c_{22}\Big\{\int_0^T \Big[\int_{\partial\Omega} \big[|v_{i_0}| + |v_0|_k\big]^{[r(m)+2]\frac{N-1}{N-2}} \, ds\Big]^{\frac{N-2}{N-1}} dt + 1\Big\} \\
&\le c_{23}\Big\{\iint\limits_{Q_T} \Big[|v_{i_0}|^{r(m)}\Big(\left|\frac{\partial v_{i_0}}{\partial x}\right|^2 + |v_{i_0}|^2\Big) \\
&\quad + |v_0|^{r(m)}\Big(\left|\frac{\partial v_0}{\partial x}\right|^2 + |v_0|^2\Big)\Big] \, dx\, dt + 1\Big\} \le c_{24}.
\end{aligned}
\tag{3.38}
$$

Now (3.32), (3.35), (3.37), (3.38) and (3.19) imply

$$\iint\limits_{\{v_{i_0}>|v_0|\}} [v_{i_0}]^{r(m+1)} \left|\frac{\partial v_0}{\partial x}\right|^2 dx\, dt \le c_{25}. \tag{3.39}$$

Finally, the desired inequality (3.30) follows from (2.19), (3.31), (3.39) and the proof of Lemma 3.3 is completed. □

Lemma 3.4. *Let the assumptions of Theorem 2.2 be satisfied and suppose that the inequalities* (3.18) *hold for some nonnegative integer* m. *Moreover, let* m *and* i_0 *be such that*

$$\gamma_{i_0} < r(m+1) \tag{3.40}$$

and suppose that

$$\operatorname{ess\,sup}_{t\in(0,T)} \int_\Omega [v_i(t,x)]_+^{r(m)+2} dx + \iint\limits_{Q_T} [v_i(t,x)]_+^{r(m)} \left|\frac{\partial v_0}{\partial x}\right|^2 dx\, dt \le L_6\,, \tag{3.41}$$

with a constant L_6 *depending only on* m *and known parameters. Then there exists a constant* L_7 *depending only on the same parameters such that*

$$\begin{aligned} &\operatorname{ess\,sup}_{t\in(0,T)} \int_\Omega [v_{i_0}(t,x)]_+^{r(m+1)+2} dx \\ &+ \iint\limits_{Q_T} [v_{i_0}(t,x)]_+^{r(m+1)} \left\{ \left|\frac{\partial v_0}{\partial x}\right|^2 + \left|\frac{\partial v_{i_0}}{\partial x}\right|^2 \right\} dx\, dt \le L_7\,. \end{aligned} \tag{3.42}$$

Proof. We start by proving that

$$\iint\limits_{Q_T} [v_i(t,x)]_+^{r(m)+\varepsilon} \left|\frac{\partial v_0}{\partial x}\right|^2 dx\, dt \le c_{26} \tag{3.43}$$

for $i = i_0$ and $\varepsilon = \varepsilon_1 = \frac{\Delta}{r(m)+2}$. We want to apply (3.32) with this ε. Since in the proof of Lemma 3.3 I_4, I_5 have been estimated without using assumption (3.29), we can suppose (3.38) to be hold. Further, (3.19) holds also true. To estimate $I_3(l)$ with the chosen ε we apply Young's inequality, (3.18), (3.41) and (3.19):

$$\begin{aligned} I_3(1) \le c_{27} \iint\limits_{Q_T} &\left\{ [v_{i_0}]_+^{r(m)} \left(\left|\frac{\partial v_{i_0}}{\partial x}\right|^2 + \left|\frac{\partial v_0}{\partial x}\right|^2 \right) \right. \\ &\left. + |v_0|_k^{r(m)+1} \Big\{ \big[[v_{i_0} - |v_0|_k]_+\big]_k + |v_0|_k \Big\}^{[r(m)+1]\varepsilon-1} \left|\frac{\partial v_0}{\partial x}\right|^2 \right\} dx\, dt \le c_{28}. \end{aligned} \tag{3.44}$$

Now the inequalities (3.31), (3.32), (3.37), (3.38) and (3.44) imply (3.43) for $i = i_0$. Note that this estimate follows in the same way for $i = 1, \dots, n$.

The key for continuing our previous discussions is following estimate

$$\operatorname{ess\,sup}_{t\in(0,T)}\int_\Omega \left[v_i(t,x)\right]_+^{r(m)+\varepsilon+2}dx+\iint\limits_{Q_T}\left[v_i(t,x)\right]^{r(m)+\varepsilon}\left|\frac{\partial v_i}{\partial x}\right|^2 dx\,dt\le c_{29}. \tag{3.45}$$

Indeed, it can be seen from the proof of the Lemma 3.2, that (3.19) and (3.45) imply (3.19) with $r(m+1)+\varepsilon$ instead of $r(m+1)$. This ensures that (3.43) remains true for $\varepsilon=\varepsilon_2=\frac{2\Delta}{r(m)+2}$ and even for further steps.

The estimate (3.45) follows immediately from (3.43) and (2.19) provided $r(m)+\varepsilon\le\gamma_i$. So it remains to prove (3.45) for the case that $r(m)+\varepsilon>\gamma_i$. To this end we test integral identity (3.5) with

$$\varphi=\Big[\left[e_i(v_i)-e_i(m_0)\right]_+\Big]_{k(i)}\Big\{a(\rho)+\Big[\left[e_i(v_i)-e_i(m_0)\right]_+\Big]_{k(i)}^2\Big\}^\rho,\quad \rho>-\frac12,$$

where

$$\begin{aligned} m_0&=\operatorname{ess\,sup}\left\{\left|e_i^{-1}\left(\frac{h_i(x)}{u_i^*(x)}\right)\right|;\quad x\in\Omega,\quad i=1,\dots,n\right\},\\ z_+&=\max(z,0),\ [s]_{k(i)}=\min(s,k(i)),\\ k(i)&=e_i(k)-e_i(m_0)\text{ for }k>m_0;\quad a(\rho)=1\text{ for }\rho\le1,\ a(\rho)=0\text{ for }\rho>1\,. \end{aligned}$$

Then, using Lemma 2 from [10], we can evaluate the first term:

$$\int_0^\tau <\frac{\partial u_i}{\partial t},\varphi>dt=\int_\Omega u^*(x)\Lambda_{k,i}^{(\rho)}\Big(e_i(v_i(\tau,x))-e_i(m_0)\Big)\,dx\,, \tag{3.46}$$

where

$$\Lambda_{k,i}^{(\rho)}(z)=\left[\int_0^z[s]_{k(i)}\Big\{a(\rho)+[s]_{k(i)}^2\Big\}^\rho ds\right]_+\ge\frac{1}{2(\rho+1)}[z_+]_{k(i)}^{2\rho+2}. \tag{3.47}$$

We write the space derivative of φ in the form

$$\frac{\partial\varphi}{\partial x_j}=\Phi_{k,i}^{(\rho)}(v_i)\frac{\partial v_i}{\partial x_j}\chi(m_0<v_i<k)\,, \tag{3.48}$$

where $\chi(m_0<v_i<k)$ is the characteristic function of the set $\{m_0<v_i<k\}$ and the function $\Phi_{k,i}^{(\rho)}(v_i)$ satisfies for $\rho>-\frac12$ the estimate

$$\begin{aligned}\rho_*e_i'(v_i)\Big\{a(\rho)+\left[e_i(v_i)-e_i(m_0)\right]^2\Big\}^\rho&\le\Phi_{k,i}^{(\rho)}(v_i)\\ &\le c_{31}(1+\rho)\,e_i'(v_i)\Big\{a(\rho)+\left[e_i(v_i)-e_i(m_0)\right]^2\Big\}^\rho\end{aligned} \tag{3.49}$$

for $m_0<v_i<k$ with $\rho_*=\min(1;1+2\rho)$.

Using (3.46)–(3.49) and the conditions iii), (2.20)–(2.22), we obtain from (3.5) with the chosen test function φ

$$\begin{aligned}
&\int_\Omega \Big[\big[v_i(\tau,x)\big]_+\Big]_k^{2(\gamma_i+1)(\rho+1)}\,dx \\
&\quad + \int_0^\tau\int_\Omega \Big[\big[v_i(t,x)\big]_+\Big]_k^{2\gamma_i+2(\gamma_i+1)\rho}\left|\frac{\partial v_i}{\partial x}\right|^2 \chi(m_0<v_i<k)\,dx\,dt \\
&\le c_{32}\Big(\frac{\rho+1}{\rho_*}\Big)^2\Big\{\int_0^\tau\int_\Omega \Big[\big[v_i(t,x)\big]_+\Big]_k^{2\gamma_i+2(\gamma_i+1)\rho}\left|\frac{\partial v_0}{\partial x}\right|^2 \\
&\qquad \times(1+\rho)^2\chi(m_0<v_i<k)\,dx\,dt \\
&\quad + \sum_{j=1}^n\int_0^\tau\int_\Omega\Big(\big[v_j(t,x)\big]_+^{(\gamma_i+1)(1+2\rho)+p_3}+\big|v_0(t,x)\big|^{(\gamma_i+1)(1+2\rho)+p_3}\Big)\,dx\,dt \\
&\quad + \Big\{\int_0^\tau\int_\Omega\Big[\big[v_i(t,x)\big]_+\Big]_k^{(\gamma_i+1)(1+2\rho)r_1'}\,dx\,dt\Big\}^{\frac{1}{r_1'}} \\
&\quad + \int_0^\tau\int_{\partial\Omega}\sum_{j=0}^n\big[v_j(t,x)\big]_+^{(\gamma_i+1)(1+2\rho)+p_4}\,ds\,dt \\
&\quad + \Big\{\int_0^\tau\int_{\partial\Omega}\Big[\big[v_i(t,x)\big]_+\Big]_k^{(\gamma_i+1)(1+2\rho)r_2'}\,ds\,dt\Big\}^{\frac{1}{r_2'}}+1\Big\}.
\end{aligned} \tag{3.50}$$

To continue the proof of the inequality (3.45) we choose ρ such that

$$2\gamma_i+2(\gamma_i+1)\rho = r(m)+\varepsilon \tag{3.51}$$

and estimate the right-hand side of (3.50) integral by integral. An estimation of the first one follows from (3.43). Note that by (3.51) and (3.16)

$$\begin{aligned}
(\gamma_i+1)(1+2\rho)+p_3 &= \Big[r(m)+2+\frac{2}{N}(\gamma_*+2)\Big] \\
&\quad + \Big[p_3-1-\gamma_i-\frac{2}{N}(\gamma_*+2)\Big]+\varepsilon \\
&< r(m)+2+\frac{2}{N}(\gamma_*+2).
\end{aligned}$$

Hence estimates for the v_j terms, $j=1,\dots,n$, of the second integral on the right-hand side of (3.50) follow from (3.24) with $q=1$. Taking into account (3.11), the v_0 term can be estimated by the same arguments.

In order to estimate the third integral we use the next inequality that follows analogously to the inequality (3.24):

$$
\begin{aligned}
&\iint\limits_{Q_T} \big[v_i(t,x)\big]^{(r(m)+2)(1+\frac{2}{N})}\,dx\,dt \\
&\quad\le \int_0^T \Big\{\int_\Omega \big[v_i(t,x)\big]_+^{r(m)+2}\,dx\Big\}^{\frac{2}{N}} \Big\{\int_\Omega \big[v_i(t,x)\big]_+^{(r(m)+2)\frac{N}{N-2}}dx\Big\}^{\frac{N-2}{N}}\,dt \\
&\quad\le c_{33}\big[r(m)+2\big]^2 \operatorname{ess\,sup}_{t\in(0,T)} \Big\{\int_\Omega \big[v_i(t,x)\big]_+^{r(m)+2}\,dx\Big\}^{\frac{2}{N}} \\
&\qquad\times \iint\limits_{Q_T} \Big\{\big[v_i(t,x)\big]_+^{r(m)}\Big|\frac{\partial v_i}{\partial x}\Big|^2 + \big[v_i(t,x)\big]_+^{r(m)+2}\Big\}\,dx\,dt.
\end{aligned}
\tag{3.52}
$$

It is simple to check that $(\gamma_i+1)(1+2\rho)r_1' < \big(r(m)+2\big)(1+\frac{2}{N})$, such that the third integral can be estimated by means of (3.52), (3.41), (3.18).
To estimate the last integrals in (3.50) we note firstly following auxiliary inequality that follows analogously to the inequality (3.52):

$$
\begin{aligned}
&\iint\limits_{\Gamma_T} \big[v_i(t,x)\big]_+^{(r(m)+2)(1+\frac{1}{N})}\,ds\,dt \le c_{34}\big[r(m)+2\big]^2 \\
&\qquad\times \int_0^T \Big\{\int_\Omega \big[v_i(t,x)\big]_+^{r(m)+\frac{2}{N+1}}\Big(\Big|\frac{\partial v_i}{\partial x}\Big| + [v_i]_+\Big)^{\frac{2N}{N+1}}dx\Big\}^{\frac{N+1}{N}}\,dt \\
&\quad\le c_{35}\big[r(m)+2\big]^2 \operatorname{ess\,sup}_{t\in(0,T)} \Big\{\int_\Omega \big[v_i(t,x)\big]_+^{r(m)+2}dx\Big\}^{\frac{1}{N}} \\
&\qquad\times \iint\limits_{Q_T} \big[v_i(t,x)\big]_+^{r(m)}\Big(\Big|\frac{\partial v_i}{\partial x}\Big|^2 + \big[v_i(t,x)\big]_+^2\Big)\,dx\,dt.
\end{aligned}
\tag{3.53}
$$

Then analogous arguments as used for proving (3.24) and (3.53) lead to

$$
\iint\limits_{\Gamma_T} \big[v_i(t,x)\big]_+^{r(m)+2+\frac{1}{N}(\gamma_i+2)}\,ds\,dt \le c_{36}. \tag{3.54}
$$

Since by (3.51) and (3.16)

$$
(\gamma_i+1)(1+2\rho)+p_4 < r(m)+2+\frac{1}{N}(\gamma_*+2)\,,
$$

(3.54) implies an estimation for the fourth integral on the right-hand side of (3.50). Finally, (3.53) implies an estimate for the last integral in (3.50). With (3.50) the key estimate (3.45) is fully proved. This ends the proof of Lemma 3.4. □

Proof of Theorem 2.2. Remark that for $m=0$ the conditions (3.18), (3.41) follow from Lemma 3.1 and Theorem 2.1. Starting from $m=0$, we can iterate the application of the Lemmas 3.2–3.4. After $M+1$ steps we arrive at the inequalities (3.19) and (3.42) with $m=M$. Taking M so large that $\Delta\cdot(M+1)\ge\gamma^*$, we get Theorem 2.2. □

4. Proof of L^∞-estimates

Proof of Theorem 2.3. We apply Lemma 3.4 with $m = M$ and M such that $r(M+1)+2 > \frac{N}{2}(\gamma^*+1)$, $\gamma^* = \max(\gamma_1,\dots,\gamma_n)$. Then Theorem 2.3 follows immediately from (3.42), conditions i), iii), (2.11) and well-known results on the regularity of solutions of linear elliptic equations (see, for example [15]) to Poisson's equation (2.1). □

In what follows we assume the conditions of Theorem 2.4 to be satisfied. We shall estimate for v_i, $i = 1,\dots,n$, separately on the sets $\{v_i > 0\}$ (Lemma 4.1) and $\{v_i < 0\}$ (Lemma 4.2).

Lemma 4.1. *Let the condition of Theorem 2.4 be satisfied. Then there exists a constant L_8 depending only on known parameters such that for $i = 1,\dots,n$,*

$$\operatorname{ess\,sup}\big\{v_i(t,x) : (t,x) \in Q_T\big\} \le L_8 . \tag{4.1}$$

Proof. Using Lemma 3.4 and (3.50) we get for $r \ge r_* = 2 + 4\max(\gamma_1,\dots,\gamma_n)$

$$\begin{aligned} &\int_\Omega \big[v_i(\tau,x)\big]_+^{r+2}\,dx + \int_0^\tau\!\!\int_\Omega \big[v_i(t,x)\big]_+^{r}\left|\frac{\partial v_i}{\partial x}\right|^2 dx\,dt \\ &\le c_{37}\, r^2\Big\{1 + \int_0^\tau\!\!\int_\Omega \big[v_i(t,x)\big]_+^{r}\left|\frac{\partial v_0}{\partial x}\right|^2 dx\,dt \\ &+ \left[\int_0^\tau\!\!\int_\Omega \big[v_i(t,x)\big]_+^{(r+1)r_1'}\,dx\,dt\right]^{\frac{1}{r_1'}} + \left[\int_0^\tau\!\!\int_{\partial\Omega} \big[v_i(t,x)\big]_+^{(r+1)r_2'}\,ds\,dt\right]^{\frac{1}{r_2'}}\Big\}. \end{aligned} \tag{4.2}$$

Remark only that Lemma 3.4 gives us the estimate of $[v_j]_+^{p_3}$ in $L^{r_1}(Q_T)$, $j = 1,\dots,n$.
We start estimating the first integral on the right-hand side of (4.2). Let

$$\{\varphi_j \in C^\infty(\mathbb{R}^N),\ j - 1,\dots,J,\}$$

be a partition of unity such that

$$\begin{gathered} \sum_{j=1}^{J}\varphi_j^2(x) = 1, \quad \left|\frac{\partial\varphi_j}{\partial x}\right| \le \frac{c_0}{R} \quad \text{for} \quad x \in \Omega, \quad \operatorname{supp}\varphi_j \subset B(x_j,R), \\ JR^N \le c_0\big[d(\Omega)\big]^N, \quad R < 1, \quad \sum_{j=1}^{J}\chi\big(B(x_j,R)\big) \le c_0 , \end{gathered} \tag{4.3}$$

where $B(x_j,R)$ is the ball of radius R with center $x_j \in \Omega$, c_0 is a constant depending only on N, $d(\Omega)$ is the diameter of Ω, $\chi(B(x_j,R))$ is the characteristic function of $B(x_j,R)$. The radius R will be fixed later.

We test the integral identity (2.14) with

$$\Psi(t,x) = \big[v_i(t,x)\big]_+^r \cdot \big[v_0(t,x) - v_{0,l}(t)\big]\varphi_l^2(x),\ v_{0,l}(t) = v_0(t,x_l). \tag{4.4}$$

Integration with respect to t and summing up on l yield

$$\int_0^\tau \int_\Omega \big[v_i(t,x)\big]_+^r \Big|\frac{\partial v_0}{\partial x}\Big|^2 dx\,dt \le c_{38}\, r\Big\{ I_1(r) + I_2(r) + \int_0^\tau \Big\{\int_\Omega \big[v_i(t,x)\big]_+^{rp_1'} dx\Big\}^{\frac{1}{p_1'}} dt + \int_0^\tau \Big\{\int_{\partial\Omega} \big[v_i(t,x)\big]_+^{rp_2'} ds\Big\}^{\frac{1}{p_2'}} dt\Big\}, \tag{4.5}$$

where

$$I_1(r) = \sum_{l=1}^J \int_0^\tau \int_\Omega \big[v_i(t,x)\big]_+^{r-1} \big|v_0(t,x) - v_{0,l}(t)\big| \varphi_l^2(x) \Big|\frac{\partial v_i}{\partial x}\Big| \Big|\frac{\partial v_0}{\partial x}\Big|\, dx\,dt\,,$$

$$I_2(r) = \frac{1}{R}\int_0^\tau \int_\Omega \big[v_i(t,x)\big]_+^r \Big|\frac{\partial v_0}{\partial x}\Big|\, dx\,dt.$$

Since by (2.25)

$$\big|v_0(t,x) - v_{0,l}(t)\big| \le K_3\, R^\eta \quad \text{for} \quad x \in B(x_l, R)\,, \tag{4.6}$$

we obtain

$$c_{38}\, r I_1(r) \le \frac{1}{2}\int_0^\tau \int_\Omega \big[v_i(t,x)\big]_+^r \Big|\frac{\partial v_0}{\partial x}\Big|^2 dx\,dt + c_{39}\Big[r^2 R^{2\eta} \int_0^\tau \int_\Omega \big[v_i(t,x)\big]_+^r \Big|\frac{\partial v_i}{\partial x}\Big|^2 dx\,dt + 1\Big]. \tag{4.7}$$

We fix R such that $4\, c_{39}\, c_{37}\, r^4\, R^{2\eta} = 1$. Estimating $I_2(r)$ by Cauchy's inequality and using (4.5), (4.7), we deduce from (4.2)

$$\begin{aligned}
&\int_\Omega \big[v_i(\tau,x)\big]_+^{r+2} dx + \int_0^\tau \int_\Omega \big[v_i(t,x)\big]_+^r \Big|\frac{\partial v_i}{\partial x}\Big|^2 dx\,dt \\
&\le c_{40}\, r^{2+\frac{4}{\eta}} \Big\{1 + \Big[\int_0^\tau \int_\Omega \big[v_i(t,x)\big]_+^{(r+1)r_1'} dx\,dt\Big]^{\frac{1}{r_1'}} \\
&+ \int_0^\tau \Big\{\int_\Omega \big[v_i(t,x)\big]_+^{rp_1'} dx\Big\}^{\frac{1}{p_1'}} dt + \Big[\int_0^\tau \int_{\partial\Omega} \big[v_i(t,x)\big]^{(r+1)r_2'} ds\,dt\Big]^{\frac{1}{r_2'}} \\
&+ \int_0^\tau \Big\{\int_{\partial\Omega} \big[v_i(t,x)\big]_+^{rp_2'} ds\Big\}^{\frac{1}{p_2'}} dt\Big\}.
\end{aligned}$$

Hence Sobolev's embedding theorem and standard Moser iteration lead to (4.1) and the proof of Lemma 4.1 is completed. □

For $\varepsilon > 0$ and arbitrary functions g defined on Q_T we use the notations

$$g^{(\varepsilon)}(t,x) = \max\big\{g(t,x), \varepsilon\big\},\ g_-(t,x) = \min\big\{g(t,x), 0\big\}\,. \tag{4.8}$$

Lemma 4.2. *Let the conditions of the Theorem 2.4 be satisfied. Then there exists a constant L_9 depending only on known parameters such that*

$$\operatorname{ess\,inf}\left\{v_i(t,x):(t,x)\in Q_T\right\}\geq -L_9,\quad i=1,\dots,n. \tag{4.9}$$

Proof. Denote

$$\begin{aligned}
m_0 &= \operatorname{ess\,sup}\left\{\left|e_i^{-1}\left(\frac{h_i(x)}{u_i^*(x)}\right)\right|:x\in\Omega,\quad i=1,\dots,n\right\},\\
\widetilde{e}_i(v) &= \frac{e_i(v)}{e_i(-m_0)}\,,\\
\Psi^{(r)}(z) &= -\big(\frac{1}{z^2}\,|ln\,z|^r+\frac{r}{z^2}\,|ln\,z|^{r-1}\big)e_i^2(-m_0)\,,z>0\,.
\end{aligned}$$

We test the integral identity (3.5) with

$$\varphi=\frac{1}{e_i^{(\varepsilon)}(v_i)}\big|ln_-\,\widetilde{e}_i^{(\varepsilon)}(v_i)\big|^r,\quad 0<\varepsilon<1,\quad r\geq 1\,,$$

to get

$$\begin{aligned}
&\int_0^\tau\langle\frac{\partial u_i}{\partial t},\varphi\rangle dt+\sum_{j=1}^n\iint_{Q_\tau}d_{ij}\left(t,x,v_i^{(\delta)},\frac{\partial(v_i+q_iv_0)}{\partial x}\right)\Psi^{(r)}\big(\widetilde{e}_i(v_i)\big)\\
&\qquad\times e_i'(v_i)\frac{\partial v_i}{\partial x_j}\chi\big(\widetilde{e}_i^{-1}(\varepsilon)<v_i<-m_0\big)\,dx\,dt\\
&+\iint_{Q_\tau}R_i(t,x,v,\zeta)\frac{1}{e_i^{(\varepsilon)}(v_i)}\big|ln_-\,\widetilde{e}_i^{(\varepsilon)}(v_i)\big|^r\,dx\,dt\\
&+\iint_{\Gamma_\tau}R_i^\Gamma(t,x,v,\zeta)\frac{1}{e_i^{(\varepsilon)}(v_i)}\,\big|ln_-\,\widetilde{e}_i^{(\varepsilon)}(v_i)\big|^r\,ds\,dt=0\,.
\end{aligned} \tag{4.10}$$

Evaluating the first integral in (4.10) analogously to equality (40) in [10] yields

$$\begin{aligned}
\int_0^\tau\langle\frac{\partial u_i}{\partial t},\,\varphi\rangle\,dt &= \frac{1}{r+1}\int_{\{\widetilde{e}_i(v_i)>\varepsilon\}}\big|ln_-\,\widetilde{e}_i^{(\varepsilon)}\big(v_i(\tau,x)\big)\big|^{r+1}u_i^*(x)\,dx\\
&\quad-\int_{\{\widetilde{e}_i(v_i)\leq\varepsilon\}}|ln\,\varepsilon|^r\big|ln_-\,\widetilde{e}_i^{(\varepsilon)}\big(v_i(\tau,x)\big)\big|u_i^*(x)\,dx\\
&\leq-\frac{1}{r+1}\int_\Omega\big|ln_-\,\widetilde{e}_i^{(\varepsilon)}\big(v_i(\tau,x)\big)\big|^{r+1}u_i^*(x)\,dx\,.
\end{aligned} \tag{4.11}$$

We estimate the second integral in (4.10) by using the condition iii) to obtain

$$\sum_{j=1}^{n} \iint_{Q_\tau} d_{ij}\Big(t,x, v_i^{(\delta)}, \frac{\partial(v_i + q_i v_0)}{\partial x}\Big) e_i'(v_i) \Psi^{(r)}\big(\widetilde{e}_i(v_i)\big) \frac{\partial v_i}{\partial x_j}$$

$$\times \chi\big(\widetilde{e}_i^{-1}(\varepsilon) < v_i < -m_0\big)\, dx\, dt$$

$$\le -c_{41}\, r \iint_{Q_\tau} \big|ln_-\, \widetilde{e}_i^{(\varepsilon)}(v_i)\big|^{r-1} \Big|\frac{\partial}{\partial x} ln_-\, \widetilde{e}_i^{(\varepsilon)}(v_i)\Big|^2 dx\, dt \tag{4.12}$$

$$+ c_{42}\, r \iint_{Q_\tau} \Big(1 + \big|ln_-\, \widetilde{e}_i^{(\varepsilon)}(v_i)\big|^r\Big) \Big|\frac{\partial v_0}{\partial x}\Big|^2 dx\, dt\,.$$

Estimating the two last integrals in (4.10) by using (2.26), (2.27) and Lemma 4.1, we get

$$\iint_{Q_\tau} R_i(t,x,v,\zeta) \frac{1}{e_i^{(\varepsilon)}(v_i)} \big|ln_-\, \widetilde{e}_i^{(\varepsilon)}(v_i)\big|^r dx\, dt$$

$$+ \iint_{\Gamma_\tau} R_i^{\Gamma}(t,x,v,\zeta) \frac{1}{e_i^{(\varepsilon)}(v_i)} \big|ln_-\, \widetilde{e}_i^{(\varepsilon)}(v_i)\big|^r ds\, dt$$

$$\le c_{43} \Big\{ \iint_{Q_\tau} \big[1 + \alpha_1(t,x)\big] \big|ln_-\, \widetilde{e}_i^{(\varepsilon)}(v_i)\big|^r dx\, dt \tag{4.13}$$

$$+ \iint_{\Gamma_\tau} \big[1 + \alpha_2(t,x)\big] \big|ln_-\, \widetilde{e}_i^{(\varepsilon)}(v_i)\big|^r ds\, dt \Big\}.$$

By (4.10), (4.11) and (4.13) we find for $w^{(\varepsilon)}(t,x) = \big|ln_-\, \widetilde{e}_i^{(\varepsilon)}\big((v_i)(t,x)\big)\big|$

$$\int_\Omega \big[w^{(\varepsilon)}(\tau,x)\big]^{r+1} dx + \iint_{Q_\tau} \big[w^{(\varepsilon)}(t,x)\big]^{r-1} \Big|\frac{\partial w^\varepsilon}{\partial x}\Big|^2 dx\, dt$$

$$\le c_{44}\, r^2 \Big\{ \iint_{Q_\tau} \Big(1 + \big[w^{(\varepsilon)}(t,x)\big]^r\Big) \Big(\Big|\frac{\partial v_0}{\partial x}\Big|^2 + \alpha_1(t,x) + 1\Big) dx\, dt \tag{4.14}$$

$$+ \iint_{\Gamma_\tau} \big[w^{(\varepsilon)}(t,x)\big]^r \big(\alpha_2(t,x) + 1\big)\, ds\, dt \Big\}\,.$$

To estimate that term in (4.14) with the derivative of v_0, we test the integral identity (2.14) with

$$\Psi(t,x) = \big[w^{(\varepsilon)}(t,x)\big]^2 \big[v_0(t,x) - v_{0,l}(t)\big] \varphi_l^2(x)\,, \tag{4.15}$$

where $v_{0,l}(t), \varphi_l(x)$ are the functions from (4.4). By integration on t and taking the sum on l we get

$$\iint_{Q_\tau} \left[w^{(\varepsilon)}(t,x)\right]^r \left|\frac{\partial v_0}{\partial x}\right|^2 dx\,dt \le c_{45}\, r \left\{ I(r) + \frac{1}{R^2} \iint_{Q_\tau} \left[w^\varepsilon(t,x)\right]^r dx\,dt \right.$$
$$\left. + \int_0^\tau \left\{ \int_\Omega \left[w^{(\varepsilon)}(t,x)\right]^{rp_1'} dx \right\}^{\frac{1}{p_1'}} dt + \int_0^\tau \left\{ \int_{\partial\Omega} \left[w^{(\varepsilon)}(t,x)\right]^{rp_2'} ds \right\}^{\frac{1}{p_2'}} dt \right\}, \tag{4.16}$$

where

$$I(r) = \sum_{l=1}^{J} \iint_{Q_\tau} \left[w^{(\varepsilon)}(t,x)\right]^{r-1} \left|v_0(t,x) - v_{0,l}(t)\right| \varphi_l^2(x) \left|\frac{\partial w^{(\varepsilon)}}{\partial x}\right| \left|\frac{\partial v_0}{\partial x}\right| dx\, dt.$$

Using (4.7), we can estimate the last integral

$$c_{45}\, r\, I(r) \le \frac{1}{2} \iint_{Q_\tau} \left[w^{(\varepsilon)}(t,x)\right]^r \left|\frac{\partial v_0}{\partial x}\right|^2 dx\, dt$$
$$+ c_{46} \left\{ 1 + r^2 R^{2\eta} \iint_{Q_\tau} \left|w^{(\varepsilon)}(t,x)\right|^r \left|\frac{\partial w^{(\varepsilon)}}{\partial x}\right|^2 dx\, dt \right\}. \tag{4.17}$$

Fixing the number R such that $4\, c_{44}\, c_{46}\, r^4\, R^{2\eta} = 1$, we obtain from (4.14), (4.16) and (4.17)

$$\int_\Omega \left[w^{(\varepsilon)}(\tau,x)\right]^{r+1} dx + \iint_{Q_\tau} \left[w^{(\varepsilon)}(t,x)\right]^{r-1} \left|\frac{\partial w^{(\varepsilon)}}{\partial x}\right|^2 dx\, dt$$
$$\le c_{47}\, r^{2+\frac{4}{\eta}} \left\{ 1 + \left[\iint_{Q_\tau} \left[w^{(\varepsilon)}(t,x)\right]^{r\cdot r_1'} dx\, dt \right]^{\frac{1}{r_1'}} \right.$$
$$+ \left[\iint_{\Gamma_\tau} \left[w^{(\varepsilon)}(t,x)\right]^{r\, r_2'} ds\, dt \right]^{\frac{1}{r_2'}} + \int_0^\tau \left[\int_\Omega \left[w^{(\varepsilon)}(t,x)\right]^{r\, p_1'} dx \right]^{\frac{1}{p_1'}} dt \tag{4.18}$$
$$\left. + \int_0^\tau \left[\int_{\partial\Omega} \left[w^{(\varepsilon)}(t,x)\right]^{r\, p_2'} ds \right]^{\frac{1}{p_2'}} dt \right\}.$$

Remark also that (4.18) implies

$$\int_\Omega \left[w^{(\varepsilon)}(\tau,x)\right]^2 dx + \iint_{Q_\tau} \left|\frac{\partial w^{(\varepsilon)}(t,x)}{\partial x}\right|^2 dx\, dt \le c_{48} \tag{4.19}$$

with a constant c_{48} depending only on known parameters. To verify (4.19) we have to estimate the integrals on the right-hand side of (4.18) with $r = 1$ and then to apply Gronwall's Lemma. As an example we consider the third integral. Define $\overline{p}$

by $p_1' = \frac{N}{N-\overline{p}}$. Then $\overline{p} < 2$ and we can assume $\overline{p} > 1$. Using Sobolev's embedding theorem we have with $\overline{p}^* = \frac{N\overline{p}}{N-\overline{p}}$

$$\begin{aligned} c_{47} \int_0^\tau \Big[\int_\Omega \big[w^{(\varepsilon)}(t,x)\big]^{p_1'}\, dx \Big]^{\frac{1}{p_1'}} dt \\ \leq c_{49} \Big\{ 1 + \int_0^\tau \Big[\int_\Omega \big[w^{(\varepsilon)}(t,x)\big]^{\overline{p}^*}\, dx \Big]^{\frac{\overline{p}}{\overline{p}^*}} dt \Big\} \\ \leq c_{50} \Big\{ 1 + \iint\limits_{Q_\tau} \Big(\Big| \frac{\partial w^{(\varepsilon)}}{\partial x} \Big|^{\overline{p}} |w^{(\varepsilon)}|^{\overline{p}} \Big)\, dx\, dt \Big\} \\ \leq \frac{1}{8} \iint\limits_{Q_\tau} \Big| \frac{\partial w^{(\varepsilon)}}{\partial x} \Big|^2 dx\, dt + c_{51} \Big\{ 1 + \iint\limits_{Q_\tau} \big| w^{(\varepsilon)}(t,x) \big|^2\, dx\, dt \Big\}. \end{aligned} \tag{4.20}$$

Now (4.18), (4.19) and standard Moser iterations give

$$w^{(\varepsilon)}(t,x) \leq c_{52}\,, \tag{4.21}$$

with a constant c_{52} depending only on known parameters and independent of ε. The estimate (4.21) implies that the measure of the set $\{\widetilde{e}_i(v_i(t,x)) < \varepsilon\}$ is equal to zero if $|ln\,\varepsilon| > c_{52}$, i.e.,

$$v_i(t,x) > \widetilde{e}_i^{\,-1}\big(e^{-c_{52}-1}\big)$$

and the proof of Lemma 4.2 is complete. □

Proof of Theorem 2.4. Theorem 2.4 follows immediately from the inequalities (4.1) and (4.9). □

5. Proof of existence and uniqueness

Proof of Theorem 2.5. We modify the functions $e_i(z)$, $d_{ij}(t,x,z,\xi)$, $r(t,x,v,y)$, $r^\Gamma_{\gamma\delta}(t,x',v,y)$ in following way:

$$\begin{aligned} \widetilde{e}_i(z) &= \int_{-\infty}^{z} e_i'(\min[s,K_4])\, ds\,, \\ \widetilde{d}_{ij}(t,x,z,\xi) &= d_{ij}\big(t,x,\min[z,K_4],\xi\big), \\ \widetilde{r}_{\alpha\beta}(t,x,v,y) &= r_{\alpha\beta}\big(t,x,\min[v,K_3+K_4],\min[y,K_{\alpha\beta}]\big), \\ \widetilde{r}^\Gamma_{\gamma\delta}(t,x',v,y) &= r^\Gamma_{\gamma\delta}\big(t,x',\min[v,K_3+K_4],\min[y,K_{\gamma\delta}]\big), \end{aligned} \tag{5.1}$$

where K_3, K_4 are the constants from Theorem 2.3, 2.4 and

$$\min[v,K_3+K_4] = \min_{j=0,\dots,n}[v_j,K_3+K_4], \quad K_{\alpha\beta} = \sum_{i=1}^{n}\big(|\alpha_i|+|\beta_i|\big)(K_3+K_4).$$

Now we consider the system

$$-\nabla\cdot(\epsilon\nabla v_0) = f + \sum_{i=1} q_i\widetilde{u}_i \quad \text{in} \quad Q_T, \tag{5.2}$$

$$\frac{\partial\widetilde{u}_i}{\partial t} + \nabla\cdot\widetilde{J}_i^{(\delta)} + \widetilde{R}_i = 0, \quad i=1,\dots,n \quad \text{in} \quad Q_T\,, \tag{5.3}$$

where $\widetilde{u}_i, \widetilde{J}_i^{\delta}, \widetilde{R}_i$ are defined by (2.4), (2.18), (2.6) with $\widetilde{e}_i, \widetilde{d}_i, \widetilde{r}_{\alpha\beta}$ instead of $e_i, d_i, r_{\alpha\beta}$. We assume further that $\delta = \frac{1}{K_4}$.
In an analogous way we modify the boundary condition (2.17):

$$\nu\cdot\widetilde{J}_i^{(\delta)} + \widetilde{R}_i^{\Gamma} = 0, \quad i=1,\dots,n \quad \text{on} \quad \Gamma_T. \tag{5.4}$$

The solvability of the nondegenerate problem (5.2)–(5.4), (2.8), (2.9) can be simply shown by using backward time discretization (see, for example [2]). By Theorems 2.3, 2.4 each solution $v=(v_0,v_1,\dots,v_n)$ of that nondegenerate problem, satisfies the a priori estimates (2.25), (2.28). But, because of (5.1), v is automatically solution of the original problem (2.1), (2.2), (2.7)–(2.9). So theorem 2.5 is proved. □

We want now to prepare the proof of the uniqueness result. Let us to this aim suppose contradictionarily the existence of two solutions

$$v^{(1)} = \left(v_0^{(1)}, v_1^{(1)},\dots,v_n^{(1)}\right),\ v^{(2)} = \left(v_0^{(2)}, v_1^{(2)},\dots,v_n^{(2)}\right)$$

of problem (2.1), (2.2), (2.7)–(2.9). Remark that both solutions necessarily fulfill the a priori estimates (2.25), (2.28). We shall show that $v^{(1)} = v^{(2)}$.
We start by proving auxiliary Lemmas.

Lemma 5.1. *Let the assumptions of Theorem* 2.6 *be satisfied. Then there exists a constant* L_{10} *depending only on known parameters such that for arbitrary* $\tau\in(0,T]$

$$\begin{aligned}
&\sum_{i=1}^{n}\left\{\int_\Omega\left|v_i^{(1)}(\tau,x)-v_i^{(2)}(\tau,x)\right|^2dx + \iint\limits_{Q_\tau}\left|\frac{\partial(v_i^{(1)}-v_i^{(2)})}{\partial x}\right|^2 dx\,dt\right\}\\
&\qquad\le L_{10}\iint\limits_{Q_\tau}\left[\left|\frac{\partial(v_0^{(1)}-v_0^{(2)})}{\partial x}\right|^2 + \sum_{l=0}^{n}\left|v_l^{(1)}-v_l^{(2)}\right|^2\right.\\
&\qquad\left.+\sum_{i=1}^{n}\left|v_i^{(1)}-v_i^{(2)}\right|^2\left(1+\left|\frac{\partial(v_i^{(1)}+q_iv_0^{(1)})}{\partial x}\right|\right)\left|\frac{\partial v_0^{(1)}}{\partial x}\right|\right]dx\,dt\,.
\end{aligned}\tag{5.5}$$

Proof. We test the integral identity (2.13) for the solution $v^{(k)}$, $k=1,2$, with $\varphi^{(k)}$, where

$$\varphi^{(1)} = \frac{1}{e_i'(v_i^{(1)})}\left[e_i(v_i^{(1)})-e_i(v_i^{(2)})\right], \quad \varphi^{(2)} = v_i^{(2)}-v_i^{(1)}.$$

Taking the sum of the obtained equalities, we get

$$\sum_{i=1}^{n}\sum_{k=1}^{2}\Big\{\int_0^\tau <\frac{\partial u_i^{(k)}}{\partial t},\varphi^{(k)}> dt + \iint\limits_{Q_\tau}\Big[\sum_{j=1}^{N} e_i'(v_i^{(k)})\gamma_{ij}\Big(t,x,\frac{\partial \zeta_i^{(k)}}{\partial x}\Big)\frac{\partial \varphi^{(k)}}{\partial x_j} + R_i\big(t,x,v^{(k)},\zeta^{(k)}\big)\varphi^{(k)}\Big]\,dx\,dt + \iint\limits_{\Gamma_\tau} R_i^\Gamma\big(t,x,v^{(k)},\zeta^{(k)}\big)\varphi^{(k)}\,ds\,dt\Big\} = 0. \tag{5.6}$$

We evaluate the first integral in (5.6) analogously to Lemma 2 from [10] and obtain

$$\begin{aligned}\sum_{k=1}^{2}\int_0^\tau <\frac{\partial u_i^{(k)}}{\partial t},\varphi^{(k)}> dt &= \int_\Omega u_i^*(x)\int_{v_i^{(2)}(\tau,x)}^{v_i^{(1)}(\tau,x)}\big[v_i^{(1)}(\tau,x)-z\big]e_i'(z)\,dz\,dx\\ &\ge c_{53}\int_\Omega \big|v_i^{(1)}(\tau,x)-v_i^{(2)}(\tau,x)\big|^2\,dx.\end{aligned} \tag{5.7}$$

The second one can be estimated by the assumptions (i), (ii) of Theorem 2.6:

$$e_i'(z_1)-\frac{e_i''(z_1)}{e_i'(z_1)}\big[e_i(z_1)-e_i(z_2)\big] \ge e_i'(z_1)-\int_{z_2}^{z_1}\frac{e_i''(s)}{e_i'(s)}e_i'(s)\,ds = e_i'(z_2)\,, \tag{5.8}$$

$$e_i'(z_1)-e_i'(z_2)-\frac{e_i''(z_1)}{e_i'(z_1)}\big[e_i(z_1)-e_i(z_2)\big] \le c_{54}|z_1-z_2|^2. \tag{5.9}$$

The last inequalities, conditions (2.29), iii) and the local Lipschitz continuity of γ_{ij} imply

$$\begin{aligned}&\sum_{k=1}^{2}\sum_{j=1}^{N} e_i'(v_i^{(k)})\gamma_{ij}\Big(t,x,\frac{\partial\zeta_i^{(k)}}{\partial x}\Big)\frac{\partial\varphi^{(k)}}{\partial x_j}\\ &\ge e_i'(v_i^{(2)})\sum_{j=1}^{N}\Big[\gamma_{ij}\Big(t,x,\frac{\partial\zeta_i^{(1)}}{\partial x}\Big)-\gamma_{ij}\Big(t,x,\frac{\partial\zeta_i^{(2)}}{\partial x}\Big)\Big]\frac{\partial\big(\zeta_i^{(1)}-\zeta_i^{(2)}\big)}{\partial x_j}\\ &\quad - c_{55}\big|v_i^{(1)}-v_i^{(2)}\big|^2\Big(1+\Big|\frac{\partial\zeta_i^{(1)}}{\partial x}\Big|\Big)\Big|\frac{\partial v_0^{(1)}}{\partial x}\Big|\\ &\quad - c_{55}\Big|\frac{\partial\big(\zeta_i^{(1)}-\zeta_i^{(2)}\big)}{\partial x}\Big|\Big|\frac{\partial\big(v_0^{(1)}-v_0^{(2)}\big)}{\partial x}\Big|\\ &\ge c_{56}\Big|\frac{\partial\big(v_i^{(1)}-v_i^{(2)}\big)}{\partial x}\Big|^2\\ &\quad - c_{57}\Big\{\big|v_i^{(1)}-v_i^{(2)}\big|^2\Big(1+\Big|\frac{\partial\zeta_i^{(1)}}{\partial x}\Big|\Big)\Big|\frac{\partial v_0^{(1)}}{\partial x}\Big| + \Big|\frac{\partial\big(v_0^{(1)}-v_0^{(2)}\big)}{\partial x}\Big|^2\Big\}.\end{aligned} \tag{5.10}$$

By the local Lipschitz continuity of R_i, R_i^Γ we get

$$\left|\sum_{k=1}^{2} R_i\big(t,x,v^{(k)},\zeta^{(k)}\big)\varphi^{(k)}\right| \le c_{58}\sum_{l=0}^{n}\big|v_l^{(1)}-v_l^{(2)}\big|^2, \tag{5.11}$$

$$\left|\sum_{k=1}^{2} R_i^\Gamma\big(t,x,v^{(k)},\zeta^{(k)}\big)\varphi^{(k)}\right| \le c_{58}\sum_{l=0}^{n}\big|v_l^{(1)}-v_l^{(2)}\big|^2. \tag{5.12}$$

Using the interpolation inequality

$$\iint_{\Gamma_\tau} v^2(t,x)\,ds\,dt \le \iint_{Q_\tau}\left\{\varepsilon\left|\frac{\partial v}{\partial x}\right|^2 + c_\varepsilon|v|^2\right\}dx\,dt$$

for functions $v_l^{(1)}-v_l^{(2)}$, $l=0,1,\dots,n$, and suitable $\varepsilon>0$, we obtain (5.5) from (5.6), (5.7), (5.10), (5.12) and the proof of Lemma 5.1 is completed. □

Lemma 5.2. *Let the conditions of Theorem* 2.6 *be satisfied. Then a constant* L_{11} *depending only on known parameters exists such that*

$$\int_\Omega\left(\left|\frac{\partial(v_0^{(1)}-v_0^{(2)})}{\partial x}\right|^2 + \big|v_0^{(1)}-v_0^{(2)}\big|^2\right)dx \le L_{11}\sum_{i=1}^{n}\int_\Omega\big|v_i^{(1)}-v_i^{(2)}\big|^2\,dx\,. \tag{5.13}$$

Proof. We test the integral identity (2.14) associated with the solutions $v^{(k)}$, $k=1,2$, with $\psi^{(1)}=v_0^{(1)}-v_0^{(2)}$, $\psi^{(2)}=v_0^{(2)}-v_0^{(1)}$. The sum of the obtained equalities reads:

$$\int_\Omega \epsilon(x)\left|\frac{\partial(v_0^{(1)}-v_0^{(2)})}{\partial x}\right|^2 dx + \int_{\partial\Omega}\kappa(x)\big|v_0^{(1)}-v_0^{(2)}\big|^2\,ds = \sum_{k=1}^{2}\sum_{i=1}^{n} q_i\int_\Omega u_i^{(k)}\psi^{(k)}\,dx. \tag{5.14}$$

Now (5.13) follows from (5.14), Cauchy and embedding inequalities. □

Lemma 5.3. *Let the conditions of Theorem* 2.6 *be satisfied. Then a constant* L_{12} *depending only on known parameters exists such that for arbitrary* $\tau\in(0,T]$

$$\begin{aligned}
&\sum_{i=1}^{n}\left\{\int_\Omega\big|v_i^{(1)}(\tau,x)-v_i^{(2)}(\tau,x)\big|^2 dx + \iint_{Q_\tau}\big|v_i^{(1)}-v_i^{(2)}\big|^2\left[\left|\frac{\partial v_i^{(1)}}{\partial x}\right|^2\right.\right.\\
&\qquad\left.\left.+\left|\frac{\partial v_i^{(2)}}{\partial x}\right|^2\right]dx\,dt\right\} \le L_{12}\iint_{Q_\tau}\left\{\sum_{i=1}^{n}\left(\left|\frac{\partial(v_i^{(1)}-v_i^{(2)})}{\partial x}\right|^2\right.\right.\\
&\qquad\left.+\left[1+\left|\frac{\partial v_0^{(1)}}{\partial x}\right|+\left|\frac{\partial v_0^{(2)}}{\partial x}\right|\right]^2\big|v_i^{(1)}-v_i^{(2)}\big|^2\right)\\
&\qquad\left.+\left|\frac{\partial}{\partial x}\big(v_0^{(1)}-v_0^{(2)}\big)\right|^2+\big|v_0^{(1)}-v_0^{(2)}\big|^2\right\}dx\,dt\,.
\end{aligned} \tag{5.15}$$

Proof. We test the integral identity (2.13) for the solution $v^{(k)}$, $k = 1, 2$, with

$$\widetilde{\varphi}^{(1)} = \frac{\exp\left(A\, e_i(v_i^{(1)})\right) - \exp\left(A e_i(v_i^{(2)})\right)}{e_i'(v_i^{(1)})},$$
$$\widetilde{\varphi}^{(2)} = A\big[v_i^{(2)} - v_i^{(1)}\big] \exp\left(A\, e_i(v_i^{(2)})\right),$$

where A is a positive number, depending only on known parameters, such that

$$A\big[e_i'(s)\big]^2 + 2\, e_i''(s) \geq 1 \quad \text{for} \quad |s| \leq K_4, \quad i = 1, \dots, n\,, \tag{5.16}$$

with K_4 from (2.28). Taking the sum of the obtained equalities, we get

$$\sum_{i=1}^{n}\sum_{k=1}^{2}\Big\{\int_0^\tau < \frac{\partial u_i^{(k)}}{\partial t}, \widetilde{\varphi}^{(k)} > dt + \iint\limits_{Q_\tau}\Big[\sum_{j=1}^{N} e_i'(v_i^{(k)})\gamma_{ij}\Big(t, x, \frac{\partial \zeta_i^{(k)}}{\partial x}\Big)\frac{\partial \widetilde{\varphi}^{(k)}}{\partial x_j}$$
$$+ R_i\big(t, x, v^{(k)}, \zeta^{(k)}\big)\widetilde{\varphi}^{(k)}\Big]\, dx\, dt + \iint\limits_{\Gamma_\tau} R_i^\Gamma\big(t, x, v^{(k)}, \zeta^{(k)}\big)\widetilde{\varphi}^{(k)}\, ds\, dt\Big\} = 0. \tag{5.17}$$

We transform the first integral in (5.17) analogously to the inequality (5.7) to obtain

$$\sum_{i=1}^{2}\int_0^\tau < \frac{\partial u_i^{(k)}}{\partial t}, \widetilde{\varphi}^{(k)} > dt = A\int_\Omega u_i^*(x)\int_{v_i^{(2)}(\tau,x)}^{v_i^{(1)}(\tau,x)} \big[v_i^{(1)}(\tau, x) - z\big]$$
$$\times\, e_i'(z) \exp A e_i(z)\, dz\, dt \geq c_{59}\int_\Omega \big[v_i^{(1)}(\tau, x) - v_i^{(2)}(\tau, x)\big]^2 dx. \tag{5.18}$$

To estimate the second term in (5.17) we use the inequality

$$-\frac{e_i''(z_1)}{e_i'(z_1)}\big[\exp[A e_i(z_1)] - \exp[A e_i(z_2)]\big] \geq -A\int_{z_2}^{z_1} e_i''(z)\exp[A e_i(z)]\, dz$$
$$= A\big[e_i'(z_2)\exp[A e_i(z_2)] - e_i'(z_1)\exp[A e_i(z_1)]\big] + A^2\int_{z_2}^{z_1}\big[e_i'(z)\big]^2\exp[A e_i(z)]\, dz\,, \tag{5.19}$$

that follows for $z_1, z_2 \in \mathbb{R}^1$ from condition 1) of Theorem 2.6. So we obtain

$$\sum_{j=1}^{N}\iint\limits_{Q_\tau} e_i'(v_i^{(1)})\gamma_{ij}\Big(t, x, \frac{\partial \zeta_i^{(1)}}{\partial x}\Big)\frac{\partial \widetilde{\varphi}^{(1)}}{\partial x_j}\, dx\, dt \geq I^{(1)} + I^{(2)} + I^{(3)}\,, \tag{5.20}$$

where

$$I^{(1)} = A^2 \sum_{j=1}^{N} \iint_{Q_\tau} \gamma_{ij}\Big(t, x, \frac{\partial \zeta_i^{(1)}}{\partial x}\Big) \frac{\partial \zeta_i^{(1)}}{\partial x_j} \int_{v_i^{(2)}}^{v_i^{(1)}} \big[e_i'(z)\big]^2 \exp[Ae_i(z)]\, dz\, dx\, dt,$$

$$I^{(2)} = A \sum_{j=1}^{N} \iint_{Q_\tau} \gamma_{ij}\Big(t, x, \frac{\partial \zeta_i^{(1)}}{\partial x}\Big) \frac{\partial \big(v_i^{(1)} - v_i^{(2)}\big)}{\partial x_j} e_i'\big(v_i^{(2)}\big) \exp[Ae_i\big(v_i^{(2)}\big]\big)\, dx\, dt,$$

$$I^{(3)} = A\, q_i \sum_{j=1}^{N} \iint_{Q_\tau} \gamma_{ij}\Big(t, x, \frac{\partial \zeta_i^{(1)}}{\partial x}\Big) \frac{\partial v_0^{(1)}}{\partial x_j} \Big\{ \frac{e_i''\big(v_i^{(1)}\big)}{e_i'\big(v_i^{(1)}\big)} \int_{v_i^{(2)}}^{v_i^{(1)}} e_i'(z) \exp[A\, e_i(z)]\, dz$$

$$- e_i'\big(v_i^{(1)}\big) \exp[Ae_i\big(v_i^{(1)}\big)] + e_i'\big(v_i^{(2)}\big) \exp[Ae_i\big(v_i^{(2)}\big)]\Big\}\, dx\ dt.$$

We rewrite the second term in (5.17) with $k = 2$ as follows

$$\sum_{j=1}^{N} \iint_{Q_\tau} e_i'\big(v_i^{(2)}\big) \gamma_{ij}\Big(t, x, \frac{\partial \zeta_i^{(2)}}{\partial x}\Big) \frac{\partial \widetilde{\varphi}^{(2)}}{\partial x_j}\, dx\, dt = I^{(4)} + I^{(5)} + I^{(6)} + I^{(7)}\,, \qquad (5.21)$$

$$I^{(4)} = -A^2 \sum_{j=1}^{N} \iint_{Q_\tau} \gamma_{ij}\Big(t, x, \frac{\partial \zeta_i^{(1)}}{\partial x}\Big) \frac{\partial \zeta_i^{(1)}}{\partial x_j} \big[e_i'\big(v_i^{(2)}\big)\big]^2 \exp[Ae_i\big(v_i^{(2)}\big)]\big(v_i^{(1)} - v_i^{(2)}\big)\, dx\, dt,$$

$$I^{(5)} = q_i A^2 \sum_{j=1}^{N} \iint_{Q_\tau} \gamma_{ij}\Big(t, x, \frac{\partial \zeta_i^{(1)}}{\partial x}\Big) \frac{\partial v_0^{(1)}}{\partial x_j} \big[e_i'\big(v_i^{(2)}\big)\big]^2 \exp[Ae_i\big(v_i^{(2)}\big)]\big(v_1^{(1)} - v_i^{(2)}\big)\, dx\, dt,$$

$$I^{(6)} = A^2 \sum_{j=1}^{N} \iint_{Q_\tau} \Big[\gamma_{ij}\Big(t, x, \frac{\partial \zeta_i^{(1)}}{\partial x}\Big) \frac{\partial v_i^{(1)}}{\partial x_j} - \gamma_{ij}\Big(t, x, \frac{\partial \zeta_i^{(2)}}{\partial x}\Big) \frac{\partial v_i^{(2)}}{\partial x_j}\Big]$$

$$\times \big[e_i'\big(v_i^{(2)}\big)\big]^2 \exp[Ae_i\big(v_i^{(2)}\big)]\big(v_i^{(1)} - v_i^{(2)}\big)\, dx\, dt,$$

$$I^{(7)} = -A \sum_{j=1}^{N} \iint_{Q_\tau} \gamma_{ij}\Big(t, x, \frac{\partial \zeta_i^{(2)}}{\partial x}\Big) e_i'\big(v_i^{(2)}\big) \exp[Ae_i\big(v_i^{(2)}\big)] \frac{\partial \big(v_i^{(1)} - v_i^{(2)}\big)}{\partial x_j}\, dx\, dt.$$

We want to estimate sums of terms from (5.20) and (5.21). Note that by (5.16)

$$\begin{aligned}
&\int_{z_2}^{z_1} \big[e_i'(z)\big]^2 \exp[Ae_i(z)]\, dz \;-\; \big[e_i'(z_2)\big]^2 \exp[Ae_i(z_2)](z_1 - z_2) \\
&= \int_{z_2}^{z_1} \int_{z_2}^{z} \Big(2\, e_i''(\theta) + A\big[e_i'(\theta)\big]^2\Big) e_i'(\theta) \exp[Ae_i(\theta)]\ d\theta\ dz \\
&\geq c_{60} |z_1 - z_2|^2 \qquad \text{for} \qquad |z_1|, |z_2| \leq K_4
\end{aligned}$$

and hence

$$I^{(1)} + I^{(4)} \geq c_{61} \iint_{Q_\tau} \big|v_i^{(1)} - v_i^{(2)}\big|^2 \Big|\frac{\partial \zeta_i^{(1)}}{\partial x}\Big|^2\, dx\ dt\ . \qquad (5.22)$$

The next estimate follows from conditions iii) and (2.29)

$$I^{(2)} + I^{(7)} \geq c_{62} \iint\limits_{Q_\tau} \left| \frac{\partial\big(\zeta_i^{(1)} - \zeta_i^{(2)}\big)}{\partial x} \right|^2 dx\, dt \;- c_{63} \iint\limits_{Q_\tau} \left| \frac{\partial\big(v_0^{(1)} - v_0^{(2)}\big)}{\partial x} \right|^2 dx\, dt\,. \tag{5.23}$$

The local Lipschitz continuity of the function e_i'' implies

$$\Big| A\left[e_i'(z_2)\right]^2 \exp[Ae_i(z_2)](z_1 - z_2) + \frac{e_i''(z_1)}{e_i'(z_1)} \int_{z_2}^{z_1} e_i'(z) \exp[Ae_i(z)]\, dz$$
$$- e_i'(z_1) \exp[Ae_i(z_1)] + e_i'(z_2) \exp[Ae_i(z_2)] \Big| \leq c_{64} |z_1 - z_2|^2$$

for arbitrary numbers $z_1, z_2 \in [-K_4, K_4]$ and consequently

$$\big|I^{(3)} + I^{(5)}\big| \leq c_{65} \iint\limits_{Q_\tau} \left(1 + \left|\frac{\partial \zeta_i^{(1)}}{\partial x}\right|\right) \left|\frac{\partial v_0^{(1)}}{\partial x}\right| \cdot \big|v_i^{(1)} - v_i^{(2)}\big|^2\, dx\, dt. \tag{5.24}$$

Further, the local Lipschitz condition for γ_{ij} yields:

$$\begin{aligned} \big|I^{(6)}\big| \leq & c_{66} \iint\limits_{Q_\tau} \Bigg[\left| \frac{\partial\big(v_i^{(1)} - v_i^{(2)}\big)}{\partial x} \right| \left(\left|\frac{\partial v_i^{(1)}}{\partial x}\right| + \left|\frac{\partial \zeta_i^{(2)}}{\partial x}\right| \right) \\ & + \left|\frac{\partial v_i^{(1)}}{\partial x}\right| \cdot \left| \frac{\partial\big(v_0^{(1)} - v_0^{(2)}\big)}{\partial x} \right| \Bigg] \big|v_i^{(1)} - v_i^{(2)}\big|\, dx\, dt. \end{aligned} \tag{5.25}$$

Finally, we obtain from (5.17), (5.18), (5.22)–(5.25) with view of (5.11), (5.12)

$$\begin{aligned} & \sum_{i=1}^n \left\{ \int_\Omega \left[v_i^{(1)}(\tau, x) - v_i^{(2)}(\tau, x)\right]^2 dx + \iint\limits_{Q_\tau} \big|v_i^{(1)} - v_i^{(2)}\big|^2 \left|\frac{\partial v_i^{(1)}}{\partial x}\right|^2 dx\, dt \right\} \\ & \qquad \leq c_{67} \iint\limits_{Q_\tau} \Bigg\{ \sum_{i=1}^n \Bigg(\left| \frac{\partial\big(v_i^{(1)} - v_i^{(2)}\big)}{\partial x} \right|^2 + \left[1 + \left|\frac{\partial v_0^{(1)}}{\partial x}\right|^2\right] \big|v_i^{(1)} - v_i^{(2)}\big|^2 \\ & \qquad\quad + \left| \frac{\partial\big(v_i^{(1)} - v_i^{(2)}\big)}{\partial x} \right| \left|\frac{\partial \zeta_i^{(2)}}{\partial x}\right| \big|(v_i^{(1)} - v_i^{(2)}\big| \Bigg) \\ & \qquad\quad + \left| \frac{\partial\big(v_0^{(1)} - v_0^{(2)}\big)}{\partial x} \right|^2 + \big|v_0^{(1)} - v_0^{(2)}\big|^2 \Bigg\}\, dx\, dt. \end{aligned} \tag{5.26}$$

Changing the places of $v_i^{(1)}$ and $v_i^{(2)}$ in (5.26) and applying Cauchy's inequality, we arrive at the desired estimate (5.15) and the proof of Lemma 5.3 is complete. □

Lemma 5.4. *Under the conditions of Theorem 2.6 a constant L_{13} depending only on known parameters exists such that for all $\tau \in (0,T],\ R \in (0,1]$,*

$$
\begin{aligned}
&\iint\limits_{Q_\tau} |v_i^{(1)} - v_i^{(2)}|^2 \left[\left|\frac{\partial v_0^{(1)}}{\partial x}\right| + \left|\frac{\partial v_0^{(2)}}{\partial x}\right|\right]^2 dx\,dt \\
&\le L_{13}\Bigg\{R^{2\eta} \iint\limits_{Q_\tau} \left[\tfrac{1}{R^2}|v_i^{(1)} - v_i^{(2)}|^2 + \left|\frac{\partial\big(v_i^{(1)} - v_i^{(2)}\big)}{\partial x}\right|^2\right] dx\,dt \\
&\qquad + \int_0^\tau \Bigg[\left\{\int_\Omega |v_i^{(1)} - v_i^{(2)}|^{2p_1'}\,dx\right\}^{\frac{1}{p_1'}} \\
&\qquad + \left\{\int_{\partial\Omega} |v_i^{(1)} - v_i^{(2)}|^{2p_2'}\,ds\right\}^{\frac{1}{p_2'}}\Bigg] dt\Bigg\}, \qquad (5.27)
\end{aligned}
$$

where η is the Hölder exponent from Theorem 2.3.

Proof. Let $\{\varphi_j(x)\},\ j = 1,\dots,J$, be a partition of unity satisfying (4.3) with a number R to be chosen later on. We test the integral identity (2.14) associated with the solution $v^{(k)},\ k = 1,2$, with

$$
\Psi^{(k)}(t,x) = \sum_{l=1}^{J} \big[v_0^{(k)}(t,x) - v_{0,l}^{(k)}(t)\big]\varphi_l^2(x)\big|v_i^{(1)}(t,x) - v_i^{(2)}(t,x)\big|^2,
$$

$$
v_{0,l}^{(k)}(t) = v_0^{(k)}(x_l).
$$

We obtain after integration with respect to t and using the Hölder inequality

$$
\begin{aligned}
&\iint\limits_{Q_\tau} \epsilon(x)|v_i^{(1)} - v_i^{(2)}|^2\left|\frac{\partial v_0^{(k)}}{\partial x}\right|^2 dx\,dt \\
&\le I^{(8)}(k) + I^{(9)}(k) \\
&\quad + c_{68}\int_0^\tau \Bigg[\left\{\int_\Omega |v_i^{(1)} - v_i^{(2)}|^{2p_1'}\,dx\right\}^{\frac{1}{p_1'}} \qquad (5.28)\\
&\quad + \left\{\int_{\partial\Omega} |v_i^{(1)} - v_i^{(2)}|^{2p_2'}\,ds\right\}^{\frac{1}{p_2'}}\Bigg] dt\,,
\end{aligned}
$$

$$
I^{(8)}(k) = -2\sum_{l=1}^{J}\sum_{j=1}^{N}\iint\limits_{Q_\tau} \epsilon(x)\frac{\partial v_0^{(k)}}{\partial x_j}\frac{\partial\varphi_l}{\partial x_j}\varphi_l\cdot\big[v_0^{(k)} - v_{0,l}^{(k)}\big]\big|v_i^{(1)} - v_i^{(2)}\big|^2\,dx\,dt,
$$

$$
I^{(9)}(k) = -2\sum_{l=1}^{J}\sum_{j=1}^{N}\iint\limits_{Q_\tau} \epsilon(x)\frac{\partial v_0^{(k)}}{\partial x_j}\cdot\frac{\partial\big(v_i^{(1)} - v_i^{(2)}\big)}{\partial x_j}\varphi_l^2\big[v_0^{(k)} - v_{0,l}^{(k)}\big]\big[v_i^{(1)} - v_i^{(2)}\big]\,dx\,dt.
$$

Estimating $I^{(8)}(k), I^{(9)}(k)$ by Cauchy's inequality and using (4.6), we obtain (5.27) immediately from (5.28) and the proof of Lemma 5.4 is complete. $\square$

Proof of Theorem 2.6. From (5.5) we get by applying Cauchy's inequality to the second term and using (5.13), (5.15), (5.27) with suitable R

$$\sum_{i=1}^{n}\Big\{\int_{\Omega}\big|v_i^{(1)}(\tau,x)-v_i^{(2)}(\tau,x)\big|^2\,dx+\iint\limits_{Q_\tau}\Big|\frac{\partial(v_i^{(1)}-v_i^{(2)})}{\partial x}\Big|^2\,dx\,dt\Big\}$$
$$\le c_{69}\sum_{i=1}^{n}\Big\{\iint\limits_{Q_\tau}|v_i^{(1)}-v_i^{(2)}|^2\,dx\,dt+\int_0^{\tau}\Big\{\int_{\Omega}|v_i^{(1)}-v_i^{(2)}|^{2p_1'}\,dx\Big\}^{\frac{1}{p_1'}}dt \tag{5.29}$$
$$+\int_0^{\tau}\Big\{\int_{\partial\Omega}|v_i^{(1)}-v_i^{(2)}|^{2p_2'}\,ds\Big\}^{\frac{1}{p_2'}}dt\Big\}.$$

Here the second integral on the right-hand side can by estimated by the interpolation inequality

$$\int_0^{\tau}\Big\{\int_{\Omega}|v_i^{(1)}-v_i^{(2)}|^{2p_1'}dx\Big\}^{\frac{1}{p_1'}}dt$$
$$\le\varepsilon\Big\{\operatorname{ess\,sup}_{0<\theta<\tau}\int_{\Omega}\big|v_i^{(1)}(\theta,x)-v_i^{(2)}(\theta,x)\big|^2\,dx \tag{5.30}$$
$$+\iint\limits_{Q_\tau}\Big|\frac{\partial\big(v_i^{(1)}-v_i^{(2)}\big)}{\partial x}\Big|^2\,dx\,dt\Big\}+c(\varepsilon)\iint\limits_{Q_\tau}|v_i^{(1)}-v_i^{(2)}|^2\,dx\,dt\,,$$

where $p_1'=\frac{p_1}{p_1-1}<\frac{N}{N-2}$ and $\varepsilon>0$. Analogously we get for the last integral in (5.29)

$$\sum_{i=1}^{n}\int_{\Omega}\big|v_i^{(1)}(\tau,x)-v_i^{(2)}(\tau,x)\big|^2\,dx\le c_{70}\sum_{i=1}^{n}\iint\limits_{Q_\tau}|v_i^{(1)}-v_i^{(2)}|^2\,dx\,dt. \tag{5.31}$$

Hence Gronwall's lemma yields $v_i^{(1)}=v_i^{(2)}$ for $i=1,\dots,n$. Finally, $v_0^{(1)}=v_0^{(2)}$ follows from (5.13) and Theorem 2.6 is proved. □

Proof of Corollary 2.1. Let v be the solution of (2.1), (2.2), (2.7)–(2.9). Set

$$v_i^{(1)}=v_i(t,x),\quad v_i^{(2)}(t,x)=v_i(t+\delta,x),\quad i=0,\dots,n\,,\ \delta\in(0,T-t).$$

We test integral identity (3.5) with the functions

$$\varphi^{(1)}=t^2\frac{1}{e_i'(v_i^{(1)})}\big[e_i(v_i^{(1)})-e_i(v_i^{(2)})\big],\quad \varphi^{(2)}=t^2(v_i^{(2)}-v_i^{(1)}).$$

Then, arguing essentially as in the proof of (5.29) and (5.31), we obtain

$$\tau^2 \sum_{i=1}^{n} \Big\{ \int_\Omega \big|v_i^{(1)}(\tau,x) - v_i^{(2)}(\tau,x)\big|^2 \,dx + \iint\limits_{Q_\tau} t^2 \Big|\frac{\partial(v_i^{(1)} - v_i^{(2)})}{\partial x}\Big|^2 \,dx\,dt \Big\}$$

$$\leq c_{71} \Big\{ \sum_{i=1}^{n} \iint\limits_{Q_\tau} t^2 \big|v_i^{(1)} - v_i^{(2)}\big|^2 \,dx\,dt + \delta^2 \Big\}.$$

Now dividing by δ^2, applying Gronwall's lemma and taking the limit $\delta \to 0$, the corollary follows. □

6. Proof of Theorem 2.7

We start from the proof of (2.24) making some changes in the proof of Theorem 2.2. In the proof of Lemmas 3.1, 3.2, 3.3 we did not use the conditions (2.22), (2.23). Consequently, the results of these Lemmas remain valid.

We replace the test function φ in the proof of Lemma 3.4 by

$$\varphi = \Big[a(\rho) + \big[\Lambda(v) - M_0\big]_+\Big]_k^\rho v_i, \quad \Lambda(v) = \sum_{i=1}^{n} \Lambda_i(v_i), \quad i = 1,\ldots,n\,, \tag{6.1}$$

where $\rho > 0$, $k > 1$, the function $\Lambda_i(z)$ is defined by (2.20), $M_0 = \sum_{i=1}^n \Lambda_i(m_0)$, m_0 and $a(\rho)$ are the same numbers as in the proofs of the Lemmas 4.2 and 3.4, respectively.
Using the equalities $\alpha \cdot q = \beta \cdot q$, $\gamma \cdot q = \delta \cdot q$ for $(\alpha,\beta) \in \mathcal{R}$, $(\gamma,\delta) \in \mathcal{R}^\Gamma$ and that the functions $r_{\alpha\beta}, r^\Gamma_{\gamma\delta}$ are increasing increasing in the last argument, we find

$$\sum_{i=1}^{n} R_i(\cdot,v,\zeta)v_i = \sum_{i=1}^{n} \sum_{(\alpha,\beta)\in R} \big[r_{\alpha\beta}(\cdot,v,\alpha\cdot\zeta) - r_{\alpha\beta}(\cdot,v,\beta\cdot\zeta)\big](\alpha_i - \beta_i)v_i$$

$$= \sum_{(\alpha,\beta)\in R} \big[r_{\alpha\beta}(\cdot,v,\alpha\cdot\zeta) - r_{\alpha\beta}(\cdot,v,\beta\cdot\zeta)\big](\alpha-\beta)\cdot\zeta \geq 0\,, \tag{6.2}$$

$$\sum_{i=1}^{n} R_i^\Gamma(\cdot,v,\zeta)v_i \geq 0\,, \quad \zeta = (\zeta_1,\ldots,\zeta_n), \quad \zeta_i = q_i v_0 + v_i\,.$$

Using the test function from (6.1) and the inequality (6.2), we obtain from (2.13)

$$\sum_{i=1}^{n} \Big\{ \int_0^\tau < \frac{\partial u_i}{\partial t}, \Big[a(\rho) + \big[\Lambda(v) - M_0\big]_+\Big]_k^\rho v_i > dt + \iint\limits_{Q_\tau} \sum_{j,k=1}^{N} e_i'(v_i) a_{jk}(t,x)$$

$$\times \frac{\partial}{\partial x_k}(v_i + q_i v_0) \frac{\partial}{\partial x_j}\Big(\Big[a(\rho) + \big[\Lambda(v) - M_0\big]_+\Big]_k^\rho v_i \Big) \,dx\,dt \Big\} \leq 0. \tag{6.3}$$

We evaluate the first integral in (6.3) following Lemma 2 in [10] to get

$$\begin{aligned}\sum_{i=1}^{n}\int_0^{\tau} &< \frac{\partial u_i}{\partial t}, \Big[a(\rho)+\big[\Lambda(v)-M_0\big]_+\Big]_k^{\rho} v_i > \, dt \\ &= \int_\Omega u^* \int_0^{\Lambda(v(\tau,x))} \Big[a(\rho)+\big[z-M_0\big]_+\Big]_k^{\rho} dz\, dx \\ &\geq \frac{1}{\rho+1}\Big\{\int_\Omega u^*\Big[a(\rho)+\big[\Lambda(v(\tau,x))-M_0\big]_+\Big]_k^{\rho+1} dx \; - c_{71}a(\rho)\Big\}.\end{aligned} \tag{6.4}$$

To estimate the second integral in (6.3) we note that:

$$\begin{aligned}&\sum_{i=1}^{n}\sum_{j,k=1}^{N} e_i'(v_i)a_{jk}\frac{\partial v_i}{\partial x_k}\frac{\partial}{\partial x_j}\Big(\Big[a(\rho)+\big[\Lambda(v)-M_0\big]_+\Big]_k^{\rho} v_i\Big) \\ &\qquad = \Big[a(\rho)+\big[\Lambda(v)-M_0\big]_+\Big]_k^{\rho}\sum_{i=1}^{n}\sum_{j,k=1}^{N} e_i'(v_i)a_{jk}\frac{\partial v_i}{\partial x_k}\frac{\partial v_i}{\partial x_j} \\ &\qquad + \rho\Big[a(\rho)+\big[\Lambda(v)-M_0\big]_+\Big]_k^{\rho-1}\sum_{j,k=1}^{N} a_{jk}\Big(\sum_{i=1}^{n} e_i'(v_i)v_i\frac{\partial v_i}{\partial x_k}\Big) \\ &\qquad \times\Big(\sum_{l=1}^{n} e_l'(v_l)v_l\frac{\partial v_l}{\partial x_j}\Big)\chi\big(0<\Lambda(v)-M_0<k-a(\rho)\big) \\ &\qquad \geq c_{72}\Big[a(\rho)+\big[\Lambda(v)-M_0\big]_+\Big]_k^{\rho}\sum_{i=1}^{n} e_i'(v_i)\Big|\frac{\partial v_i}{\partial x}\Big|^2.\end{aligned} \tag{6.5}$$

Now (6.3)–(6.5) and (2.21) imply

$$\begin{aligned}&\int_\Omega u^*\Big[a(\rho)+\big[\Lambda(v(\tau,x))-M_0\big]_+\Big]_k^{\rho+1} dx \\ &\qquad + \iint_{Q_T}\Big[a(\rho)+\big[\Lambda(v)-M_0\big]_+\Big]_k^{\rho}\sum_{i=1}^{n} e_i'(v_i)\Big|\frac{\partial v_i}{\partial x}\Big|^2 dx\, dt \\ &\qquad \leq c_{73}(\rho+1)^2\Big\{\iint_{Q_\tau}\Big\{\Big[a(\rho)+\big[\Lambda(v)-M_0\big]_+\Big]_k^{\rho}+M_0^{\rho}\Big\} \\ &\qquad \times\sum_{i=1}^{n} e_i'(v_i)\Big|\frac{\partial v_0}{\partial x}\Big|^2 dx\, dt \; + a(\rho)\Big\}.\end{aligned} \tag{6.6}$$

With view of the proof of Lemma 3.4 we want to reestablish (3.45) as a consequence of (3.43) (and (3.41)). Note that by assumption $\gamma_i = \gamma$ and let us assume that $r(m)+\varepsilon > \gamma$. Choosing ρ in (6.6) such that $(2+\gamma)\rho+\gamma = r(m)+\varepsilon$ we can estimate the right-hand side of (6.6) by (3.43) for $i = 1,\dots,n$. Hence (6.6) implies (3.45). Repeating all another discussions from the proofs of Lemma 3.4 and Theorem 2.2,

we obtain the inequality (2.24). The proof of (2.25) in the considered case coincides with that one in the proof of Theorem 2.3.
In order to prove (2.28) we need only to check (4.1). We have from (6.6) with $\rho \geq 1$

$$\int_\Omega u^*\Big[\Lambda\big(v(\tau,x)\big)-M_0\Big]_+^{\rho+1}dx+\iint\limits_{Q_\tau}\Big[\Lambda(v)-M_0\Big]_+^{\rho}\sum_{i=1}^n e_i'(v_i)\Big|\frac{\partial v_i}{\partial x}\Big|^2 dx\,dt$$
$$\leq c_{73}(\rho+1)^2\Big\{\iint\limits_{Q_\tau}\Big[\Lambda(v)-M_0\Big]_+^{\rho}\sum_{i=1}^n e_i'(v_i)\Big|\frac{\partial v_0}{\partial x}\Big|^2 dx\,dt\;+M_0^\rho\Big\}. \tag{6.7}$$

To estimate the last integral we test the identity (2.14) with

$$\psi(t,x)=\Big[\Lambda\big(v(t,x)\big)-M_0\Big]_+^{\rho}\sum_{i=1}^n\big(1+[v_i]_+\big)^\gamma\Big[v_0(t,x)-v_{0,l}(t)\Big]\varphi_l^2(x)\,,$$

where the notations from (4.4) are used. Integration on t and summing up on l give

$$\iint\limits_{Q_\tau}\Big[\Lambda(v)-M_0\Big]_+^{\rho}\sum_{i=1}^n e_i'(v_i)\Big|\frac{\partial v_0}{\partial x}\Big|^2 dx\,dt\leq c_{74}\,\rho\sum_{i=1}^n\Big\{\sum_{l=1}^J\iint\limits_{Q_\tau}e_i'(v_i)$$
$$\times\Big[\Lambda(v)-M_0\Big]_+^{\rho}\Big(|v_0-v_{0,l}|\varphi_l^2\Big|\frac{\partial v_i}{\partial x}\Big|\Big|\frac{\partial v_0}{\partial x}\Big|+\frac{1}{R^2}\Big)\,dx\,dt$$
$$+\int_0^\tau\Big\{\int_\Omega\Big(\Big[\Lambda(v)-M_0\Big]_+^{\rho}e_i'(v_i)\Big)^{p_1'}dx\Big\}^{\frac{1}{p_1'}}dt$$
$$+\int_0^\tau\Big\{\int_{\partial\Omega}\Big(\Big[\Lambda(v)-M_0\Big]_+^{\rho}e_i'(v_i)\Big)^{p_2'}ds\Big\}^{\frac{1}{p_2'}}dt\;+M_0^\rho\Big\}. \tag{6.8}$$

Using (4.6), we obtain from (6.7), (6.8)

$$\int_\Omega\Big[\Lambda\big(v(\tau,x)\big)-M_0\Big]_+^{\rho+1}dx+\iint\limits_{Q_\tau}\Big[\Lambda(v)-M_0\Big]_+^{\rho}\sum_{i=1}^n e_i'(v_i)\Big|\frac{\partial v_i}{\partial x}\Big|^2 dx\,dt$$
$$\leq c_{75}\,\rho^{2+\frac{4}{\eta}}\Big\{M_0^\rho+\int_0^\tau\Big\{\int_\Omega\Big[\Lambda(v)-M_0\Big]_+^{(\rho+1)p_1'}dx\Big\}^{\frac{1}{p_1'}}dt \tag{6.9}$$
$$+\int_0^\tau\Big\{\int_{\partial\Omega}\Big[\Lambda(v)-M_0\Big]_+^{(\rho+1)p_2'}ds\Big\}^{\frac{1}{p_2'}}dt\Big\}.$$

The last inequality implies (4.1) by standard Moser iteration.
The proofs of the existence and uniqueness (Theorems 2.5, 2.6) remain valid under the assumptions of Theorem 2.7. This ends the proof of Theorem 2.7 □

References

[1] G. Albinus, *Convex analysis of the energy model of semiconductor devices*, WIAS Preprint **285**, Berlin (1996).

[2] H.W. Alt, S. Luckhaus, *Quasilinear elliptic-parabolic differential equations*, Math. Z., **183** (1983), 311–341.

[3] H. Amann, *Nonhomogeneous linear and quasilinear elliptic and parabolic boundary value problems*, in: Function spaces, Differential Operators and Nonlinear Analysis (H.J. Scmeisser and H. Triebel, eds.), Teubner-Texte zur Mathematik, vol. 133, Teubner, Stuttgart - Leipzig, 1993, pp. 9–116.

[4] H. Amann, M. Renardy, *Reaction-diffusion problems in electrolysis*, Nonlinear Differ. Equ. Appl., **1** (1994), 91–117.

[5] C. Ebmeyer, J. Frehse, *Mixed boundary value problems for nonlinear elliptic equations in multidimensional non-smooth domains*, Math. Nachr. **203** (1999) 47–74 .

[6] H. Gajewski, *On the uniqueness of solutions to the drift-diffusion model of semiconductor devices*, Math. Models Methods Appl. Sci., **4** (1994), 121–133.

[7] H. Gajewski, K. Gröger, *Semiconductor equations for variable mobilities based on Boltzmann statistics or Fermi-Dirac statistics*, Math. Nach., **140** (1989), 7–36.

[8] H. Gajewski, K. Gröger, *Reaction-diffusion processes of electrically charged species*, Math. Nachr., **177** (1996), 109–130.

[9] H. Gajewski, I.V. Skrypnik, *To the uniqueness problem for nonlinear elliptic equations*, Nonlinear Anal., Theory Methods Appl., **52** (2003), 291–304.

[10] H. Gajewski, I.V. Skrypnik, *On the uniqueness of solutions for nonlinear parabolic equations*, Discrete Contin. Dyn. Syst., **10** (2003), 315–336.

[11] H. Gajewski, I.V. Skrypnik, *On the uniqueness of solutions for nonlinear elliptic-parabolic equations*, J. Evol. Equ., **3** (2003), 247–281.

[12] H. Gajewski, I.V. Skrypnik, *On the unique solvability of nonlocal drift-diffusion-type problems*, Nonlinear Anal. **56** (2004), 803–830.

[13] A. Glitzky, R. Hünlich, *Electro-reaction-diffusion systems including cluster reactions of higher order*, Math. Nachr. **216** (2000), 95–118.

[14] J.W. Jerome, *Analysis of Charge Transport*, Springer-Verlag, Berlin, (1996) xii+167 pp.

[15] O.A. Ladyzhenskaya, N.N. Uraltseva, *Linear and quasilinear elliptic equations*, Nauka, Moscow (1973) (Russian), 576 pp.

[16] W. van Roosbroeck, *Theory of flow of electrons and holes in germanium and other semiconductors*, Bell System Tech. J., **29** (1950), 560–607.

[17] S. Selberherr, *Analysis and simulation of semiconductor devices*, Springer-Verlag, Wien, New York (1984).

H. Gajewski
Weierstraß-Institute for Applied Analysis and Stochastics, Mohrenstr. 39
D-10117 Berlin, Germany

I.V. Skrypnik
Institute for Applied Mathematics and Mechanics, Rosa Luxemburg Str. 74
340114, Donetsk, Ukraine

Progress in Nonlinear Differential Equations
and Their Applications, Vol. 64, 189–210

An Inverse Problem for a Phase-field Model in Sobolev Spaces

Fabrizio Colombo and Davide Guidetti

Dedicated to Herbert Amann, in occasion of his retirement

Abstract. We prove an existence and uniqueness result for an inverse problem arising from a phase-field model with two memory kernels. More precisely, we identify the convolution memory kernels and the diffusion coefficient besides the temperature and the phase-field parameter. We prove our results in the framework of Sobolev spaces. Our fundamental tools are an optimal regularity result in the L^p spaces and fixed point arguments.

Mathematics Subject Classification (2000). primary 35R30, 45K05; secondary 45N05, 80A20.

Keywords. Phase-field system with memory, heat equation, Cahn-Hilliard type equation, inverse problem, optimal regularity in L^p.

1. Introduction

Phase-field models are described by systems of equations ruling the evolution of the temperature θ and the phase-field χ, that may stand for the local proportion of one of the two phases, solid and liquid, in a three-dimensional body Ω.

If we consider phase-field models with memory, the thermal memory is taken into account modifying the Fourier's heat conduction law by introducing a memory convolution kernel (see (1.4)), denoted in the sequel by b. Unfortunately, b cannot be measured directly, to determine b additional measurements on the temperature are required, so that we have to face the problem of identifying θ, χ and b simultaneously.

The model we have introduced and studied in [15] is the same we consider in this paper (it has two unknown convolution kernels), but the inverse problem differs, since the heat source is unknown in [15], but the diffusion coefficient is given, while here we have the opposite situation. More precisely, in [15], we have proved an existence and uniqueness result for the problem dealing with the determination of the state variables (temperature and phase parameter), the convolution kernels and the time dependence of the heat source. The setting in the space $C(\overline{\Omega})$, used in [15], has the advantage that it allows to make *additional boundary measurements* on the temperature, to determine the convolutions kernels, while the Sobolev setting adopted here has the good property of requiring less compatibility conditions on the data.

We point out that only in the recent literature we can find contributions related to inverse problems for phase-fields models. We refer the reader to [14] for more details.

Many important direct problems in the phase-field theory have been proposed and solved. For sake of completeness we have to mention some of the contributions in this research area, given by several authors. Among them, without claim of completeness, we mention [2],[3],[6], [7], [8], [9], [10],[11], [12], [13] and, for phase-field models with memory, the papers [16], [17] and the literature there in.

We recall here in the following the model, introduced in [17], that we are going to study.

1.1. The model

Let Ω be an open bounded set in $\mathbf{R}^3$ with sufficiently regular boundary $\partial\Omega$ occupied by an isotropic, rigid and homogeneous heat conductor. We consider only small variations of the absolute temperature and its gradient. The materials which exhibit phase transitions, due to the temperature variations, are described by two state variables: the absolute temperature Θ and the phase-field χ at each point $x \in \Omega$ and $t \in [0,T]$ for $T > 0$, where χ takes approximately value -1 in the liquid and $+1$ in the solid. In our model we prefer to work with the *temperature variational field* θ defined by:

$$\theta = \frac{\Theta - \Theta_c}{\Theta_c}, \tag{1.1}$$

where Θ_c is the reference temperature at which the transition occurs. The energy balance equation is

$$\partial_t e + \nabla \cdot \mathbf{J} = \mathcal{F}, \tag{1.2}$$

where e is the internal energy, $\mathbf{J}$ is the heat flux and $\mathcal{F}$ is the external heat supply. Taking into account a linearized version of the Coleman-Gurtin theory, we assume (as in (Coleman-Gurtin [11] and in Giorgi, Grasselli and Pata [16]) the constitutive equations:

$$e(t,x) = e_c + c_v\Theta_c\theta(t,x) + \int_0^t a(s)\theta(t-s)\,ds + \Theta_c\tilde{\lambda}(\chi(t,x)), \tag{1.3}$$

$$\mathbf{J} = -k\nabla\theta(t,x) - \int_0^t b(s)\nabla\theta(t-s)\,ds, \tag{1.4}$$

where a and b account for the memory effects, e_c, c_v and k are the internal energy at equilibrium, the specific heat and the conductivity, respectively. Moreover, $\tilde{\lambda}$ is a suitable regular given function. By (1.2), (1.3) and (1.4) we get

$$\begin{aligned}&\partial_t\Big(e_c + c_v\Theta_c\theta(t,x) + \textstyle\int_0^t a(s)\theta(t-s)\,ds + \Theta_c\tilde{\lambda}(\chi(t,x))\Big)\\ &-\nabla\cdot\Big(k\nabla\theta(t,x) + \textstyle\int_0^t b(s)\nabla\theta(t-s)\,ds\Big) = \mathcal{F}(t,x), \quad t\in[0,T],\ x\in\Omega.\end{aligned} \tag{1.5}$$

Following Novick-Cohen [23], we couple the evolution equation (1.5) with the Cahn-Hilliard type equation, which rules the phase evolution (see Brokate and Sprekels [4], Caginalp [8] and Novick-Cohen [22] for its justification):

$$\varepsilon\partial_t\chi(t,x) = \Delta[-\Delta\chi(t,x) + \varphi(\chi(t,x)) - \tilde{\lambda}'(\chi)\theta(t,x)], \quad t\in[0,T],\ x\in\Omega, \tag{1.6}$$

where φ is a smooth given function and $\varepsilon > 0$ is a parameter. We will assume that $\tilde{\lambda}$ is a linear function of its argument. Associated to the evolution equations, we will consider also the initial and the Neumann boundary conditions, to be introduced just below. As we have already observed, the kernels a and b cannot be measured directly and the physical observable that can be easily measured is the temperature, so a and b have to be indirectly determined by additional measurements on θ, made on suitable parts of the material.

We consider additional measurements on the temperature which can be represented as

$$\Phi_j(\theta)(t) := \int_\Omega \varphi_j(x)\theta(t,x)\,dx = g_j(t), \quad \forall t\in[0,T], \quad j=1,2, \tag{1.7}$$

where, for $j\in\{1,2\}$, $\varphi_j : \Omega\to\mathbf{R}$ is a given function depending on the type thermometer used for the additional measurements on θ and $g_j : [0,T]\to\mathbf{R}$ represents the result of the corresponding additional measurement on θ. We have given two conditions because we have to identify the two unknown kernels a and b.

1.2. The identification problem

Before we can precisely state the problem, we need some notations:

- $\mathbf{R}^+$ denotes positive real numbers.

- If Ω is an open subset of $\mathbf{R}^n$, lying on one side of its boundary $\partial\Omega$, which is a submanifold of $\mathbf{R}^n$ of class at least C^1, we denote by $\nu(x)$ the unit vector normal to $\partial\Omega$ in x, pointing outside Ω, and by D_ν the normal derivative.

- If X and Y are Banach spaces and $k\in\mathbf{N}$, we shall indicate with $\mathcal{L}^k(X,Y)$ the Banach space of k-linear and continuous mappings from X^k to Y. In case $k=1$, we shall simply write $\mathcal{L}(X,Y)$.

- If $s\in\mathbf{N}_0$, $p\in[1,+\infty]$, $W^{s,p}(\Omega)$ is the usual Sobolev space.
  - If $s\in\mathbf{Z}$, $s\geq 2$, $W^{s,p}_B(\Omega) := \{f\in W^{s,p}(\Omega) : D_\nu f\equiv 0\}$.
  - If $s\in\mathbf{Z}$, $s\geq 4$, $W^{s,p}_{BB}(\Omega) := \{f\in W^{s,p}(\Omega) : D_\nu f\equiv D_\nu\Delta f\equiv 0\}$.
  - $B^s_{p,q}(\Omega)$ $(s>0,\ 1\leq p,q\leq+\infty)$ is the Besov space.
  - $(.,.)_{\theta,p}$ is the real interpolation functor $(0<\theta<1,\ 1\leq p\leq+\infty)$.

- If $a \in L^1(0,T)$ and $f:(0,T)\to X$, where X is a Banach space, we set
$$(a*f)(t) := \int_0^t a(t-s)f(s)ds,$$
whenever the integral has a meaning.
- If $p\in[1,+\infty)$, $T\in\mathbf{R}^+$, $m\in\mathbf{N}_0$, X is a Banach space, $f\in W^{m,p}(0,T;X)$, we set
$$\|f\|_{W^{m,p}(0,T;X)} := \sum_{j=0}^{m-1}\|f^{(j)}(0)\|_X + \|f^{(m)}\|_{L^p(0,T;X)}.$$
- If $T\in\mathbf{R}^+$, $p\in[1,+\infty]$, we set
$$X(T,p) \quad := \quad W^{1,p}(0,T;L^p(\Omega))\cap L^p(0,T;W^{2,p}(\Omega)).$$
If $u\in X(T,p)$, we define
$$\|u\|_{X(T,p)} \quad := \quad \|u\|_{W^{1,p}(0,T;L^p(\Omega))} + \|u\|_{L^p(0,T;W^{2,p}(\Omega))}.$$
- If $T\in\mathbf{R}^+$, $p\in[1,+\infty]$, we set
$$Y(T,p) \quad := \quad W^{1,p}(0,T;L^p(\Omega))\cap L^p(0,T;W^{4,p}(\Omega)),$$
If $v\in Y(T,p)$,we define
$$\|v\|_{Y(T,p)} \quad := \quad \|v\|_{W^{1,p}(0,T;L^p(\Omega))} + \|v\|_{L^p(0,T;W^{4,p}(\Omega))}.$$

• $C(\alpha,\beta,\dots)$ will indicate a positive constant depending on α, β,.... To distinguish different constants depending on α, β, etc. in a sequence of estimates, we shall write $C_1(\alpha,\beta,\dots)$, $C_2(\alpha,\beta,\dots)$, etc.

With the above notations we can give the definition of the inverse problem we are investigating in the sequel.

Definition 1.1. *Let $p\in(1,+\infty)$. Determine θ, χ, a, b, and k with*
$$\theta\in W^{2,p}(0,\tau;L^p(\Omega))\cap W^{1,p}(0,\tau;W^{2,p}(\Omega)), \tag{1.8}$$
$$\chi\in W^{2,p}(0,\tau;L^p(\Omega))\cap W^{1,p}(0,\tau;W^{4,p}(\Omega)), \tag{1.9}$$
$$a\in W^{1,p}(0,\tau), \tag{1.10}$$
$$b\in L^p(0,\tau), \tag{1.11}$$
$$k\in\mathbf{R}, \tag{1.12}$$
satisfying the system
$$\begin{cases} D_t(\theta+\lambda\chi+a*\theta)(t,x) \\ -\Delta[k\theta(t,x)+b*\theta](t,x) = f(t,x), \quad (t,x)\in[0,\tau]\times\Omega, \\ \varepsilon D_t\chi(t,x) = \Delta[-\Delta\chi+\varphi(\chi)-\lambda\theta](t,x), \quad (t,x)\in[0,\tau]\times\Omega, \\ D_\nu\theta(t,x') = D_\nu\chi(t,x') = D_\nu\Delta\chi(t,x') = 0, \quad (t,x')\in[0,\tau]\times\partial\Omega, \\ \theta(0,x)=\theta_0(x), \quad x\in\Omega, \\ \chi(0,x)=\chi_0(x), \quad x\in\Omega, \\ \Phi_j[\theta(t)] = g_j(t), \quad j\in\{1,2\}, \quad t\in[0,\tau], \end{cases} \tag{1.13}$$
under suitable regularity and compatibility conditions on the data.

1.3. The main result

The main result of this paper states that, under suitable regularity and compatibility conditions on the data (see the list (C1)–(C14) is Section 2) the inverse problem in Definition 1.1 has a unique solution in the Sobolev setting (see Theorem 4.1). The proof of Theorem 4.1 is nontrivial an it is spread over three sections. More precisely:

- Section 2 contains an equivalent reformulation (see Theorem 2.1), in suitable Sobolev spaces, of the inverse problem in Definition 1.1.
- In Section 3 we have some optimal regularity results in the L^p spaces for the linear version of the problem (see Theorem 3.2), together with are technical facts that we shall use afterwards.
- In Section 4 thanks to the results of the previous sections, by fixed point arguments, we state and prove our main result.

2. Equivalent reformulation of the inverse problem

The main goal, when dealing with parabolic integro-differential inverse problems, is to find for them equivalent reformulations which lead to a set of regularity and compatibility conditions on the data that make the inverse problem well posed. The equivalent reformulation is also an essential tool to prove, by optimal regularity results and fixed point arguments, the existence and the uniqueness of a solution to the inverse problem. Let us introduce the set of conditions that allow us to make this reformulation:

(C1) Ω is an open bounded subset of $\mathbf{R}^n$, lying on one side of its boundary $\partial\Omega$, which is a submanifold of $\mathbf{R}^n$ of class C^4;

(C2) $\lambda \in \mathbf{R}$, $\varepsilon \in \mathbf{R}^+$;

(C3) $p \in (1, +\infty)$, $n \in \mathbf{N}$, $n < 4p$;

(C4) $\varphi \in C^\infty(\mathbf{R})$;

(C5) $\chi_0 \in W^{4,p}_{BB}(\Omega)$;

(C6) $\theta_0 \in W^{2,p}_B(\Omega)$;

(C7) for some $T \in \mathbf{R}^+$, $f \in W^{1,p}(0, T; L^p(\Omega))$;

(C8) for $j \in \{1, 2\}$, $u \in L^p(\Omega)$, $\Phi_j[u] = \int_\Omega \varphi_j(x)u(x)dx$, with $\varphi_j \in L^{p'}(\Omega)$;

(C9) for $j \in \{1, 2\}$, $g_j \in W^{2,p}(0, T)$;

(C10) $v_0 := \varepsilon^{-1}\Delta[-\Delta\chi_0 + \varphi(\chi_0) - \lambda\theta_0] \in (L^p(\Omega), W^{4,p}_{BB}(\Omega))_{1-1/p,p}$;

(C11) $\chi^{-1} := \Phi_2[\theta_0]\Phi_1[\Delta\theta_0] - \Phi_1[\theta_0]\Phi_2[\Delta\theta_0] \neq 0$;

(C12) $k_0 := \chi[\Phi_1[\theta_0]\{\Phi_2[f(0) - \lambda v_0] - g_2'(0)\} - \Phi_2[\theta_0]\{\Phi_1[f(0) - \lambda v_0] - g_1'(0)\}] \in \mathbf{R}^+$;

(C13) $\Phi_j(\theta_0) = g_j(0)$, $j \in \{1, 2\}$;

(C14) $u_0 := f(0) - \lambda v_0 - a_0\theta_0 + k_0\Delta\theta_0 \in (L^p(\Omega), W^{2,p}_B(\Omega))_{1-1/p,p}$, with a_0 defined in (2.5);

We can state and prove the following equivalence theorem.

Theorem 2.1. *Let the assumptions* (C1)–(C14) *hold.*

Let θ, χ, a, b, k *verify the conditions* (1.8)–(1.12) *and solve the system* (1.13). *We set* $u := D_t\theta$, $v := D_t\chi$, $h := D_t a$. *Then* $k = k_0$ *(defined in* (C12)*),* u, v *and* h *satisfy the following conditions* (2.8)–(2.10) *and solve the system* (2.14)–(2.15).

On the other hand,if u, v, h *satisfy the conditions* (2.8)–(2.10) *and solve the system* (2.14)–(2.15), *if we set* $\theta := \theta_0 + 1 * u$, $\chi := \chi_0 + 1 * v$, $a := a_0 + 1 * h$, $k = k_0$, *then* θ, χ, a, b, k *verify the conditions* (1.8)–(1.12) *and solve the system* (1.13).

Proof. Assume that a solution (θ, χ, a, b, k) exists, satisfying the conditions (1.8)–(1.12). From the second equation in (1.13), we obtain immediately

$$D_t\chi(0) = v_0. \tag{2.1}$$

We observe that

$$D_t(a * \theta) = a(0)\theta + D_t a * \theta, \tag{2.2}$$

$D_t a * \theta \in C([0, \tau]; L^p(\Omega))$ and $b * \Delta\theta \in C([0, \tau]; L^p(\Omega))$. So, for $t = 0$, the first equation in (1.13) becomes

$$D_t\theta(0) + \lambda v_0 + a(0)\theta_0 - k\Delta\theta_0 = f(0). \tag{2.3}$$

Applying Φ_j (for $j \in \{1, 2\}$) to the first equation in (1.13) and using the last condition (always in (1.13)), we obtain

$$a(0)\Phi_j[\theta_0] - k\Phi_j[\Delta\theta_0] = \Phi_j[f(0) - \lambda v_0] - g_j'(0), j \in \{1, 2\}. \tag{2.4}$$

From assumption (C11), it follows

$$a(0) = a_0 := \chi[\{\Phi_2[f(0) - \lambda v_0] - g_2'(0)\}\Phi_1[\theta_0] - \{\Phi_1[f(0) - \lambda v_0] - g_1'(0)\}\Phi_2[\theta_0]]. \tag{2.5}$$

$$k = k_0, \tag{2.6}$$

$$D_t\theta(0) = u_0. \tag{2.7}$$

Now we set

$$u := D_t\theta, \tag{2.8}$$

$$v := D_t\chi, \tag{2.9}$$

$$h := D_t a. \tag{2.10}$$

We have that

$$u \in W^{1,p}(0, \tau; L^p(\Omega)) \cap L^p(0, \tau; W^{2,p}(\Omega)), \tag{2.11}$$

$$v \in W^{1,p}(0, \tau; L^p(\Omega)) \cap L^p(0, \tau; W^{4,p}(\Omega)), \tag{2.12}$$

$$h \in L^p(0, \tau). \tag{2.13}$$

Differentiating the two first equations in (1.13), we see that (u, v, h, b) solves the system:

$$\begin{cases} D_t u(t,x) + \lambda D_t v(t,x) = k_0 \Delta u(t,x) - a_0 u(t,x) + D_t f(t,x) \\ -h(t)\theta_0(x) + b(t)\Delta\theta_0(x) - (h * u)(t,x) + (b * \Delta u)(t,x), \\ \qquad\qquad (t,x) \in [0,\tau] \times \Omega, \\ \varepsilon D_t v(t,x) = -\Delta^2 v(t,x) + \Delta[\varphi'(\chi_0 + 1 * v)v](t,x) - \lambda \Delta u(t,x), \\ \qquad\qquad (t,x) \in [0,\tau] \times \Omega, \\ D_\nu u(t,x') = D_\nu v(t,x') = D_\nu \Delta v(t,x') = 0, \quad (t,x') \in [0,\tau] \times \partial\Omega, \\ u(0,x) = u_0(x), \quad x \in \Omega, \\ v(0,x) = v_0(x), \quad x \in \Omega. \end{cases} \tag{2.14}$$

Applying now Φ_j to the first equation in (2.14), using the fact that $\Phi[u(t)] = g_j'(t)$, $\Phi[D_t u(t)] = g_j''(t)$ and replacing $D_t v$ with $\epsilon^{-1}\{-\Delta^2 v + \Delta[\varphi'(\chi_0 + 1 * v)v] - \lambda\Delta u\}$, we can couple (2.14) with the two further conditions

$$\begin{aligned} h(t)\Phi_j(\theta_0) - b(t)\Phi_j(\Delta\theta_0) = {} & \Phi_j[D_t f(t)] - g_j''(t) - a_0 g_j'(t) - (h * \Phi_j(u))(t) \\ & + k_0 \Phi_j[\Delta u(t)] + (b * \Phi_j(\Delta u))(t) + \epsilon^{-1}\lambda \Phi_j\{\Delta^2 v(t) \\ & - \Delta[\varphi'(\chi_0 + 1 * v)(t)v(t) + \lambda u(t)]\}, \end{aligned} \tag{2.15}$$

with $t \in [0,\tau]$, $j \in \{1,2\}$.

Assume now that u, v, h satisfy the regularity conditions (2.11), (2.12), (2.13). We set $\theta := \theta_0 + 1 * u$, $\chi := \chi_0 + 1 * v$, $a := a_0 + 1 * h$. We remark that the first equation in (2.14) can be written in the form

$$D_t[D_t\theta + \lambda D_t\chi + a_0\theta + D_t a * \theta - k_0\Delta\theta - b * \Delta\theta] = D_t f.$$

From (C14), it follows

$$D_t\theta + \lambda D_t\chi + a_0\theta + D_t a * \theta - k_0\Delta\theta - b * \Delta\theta = f,$$

so that the first equation in (1.13) is satisfied, if $k = k_0$. The second equation in (2.14) can be written in the form

$$D_t\{\varepsilon D_t\chi - \Delta[-\Delta\chi + \varphi(\chi) - \lambda\theta]\} = 0.$$

From (C10), it follows

$$\varepsilon D_t\chi - \Delta[-\Delta\chi + \varphi(\chi) - \lambda\theta] = 0.$$

The boundary and the initial conditions in (1.13) are clearly satisfied by θ and χ. It remains only to verify that $\Phi_j[\theta(t)] = g_j(t)$, $\forall t \in [0,\tau]$, $j \in \{1,2\}$. To this aim, we first observe that, by using the second equation in (2.14), we can write (2.15) in the form

$$\begin{aligned} h(t)\Phi_j(\theta_0) - b(t)\Phi_j(\Delta\theta_0) = {} & \Phi_j[D_t f(t)] - g_j''(t) - a_0 g_j'(t) - (h * \Phi_j(u))(t) \\ & + k_0\Phi_j[\Delta u(t)] + (b * \Phi_j(\Delta u))(t) - \lambda\Phi_j(D_t v(t)). \end{aligned} \tag{2.16}$$

Applying Φ_j to the first equation in (2.14) and comparing the result with (2.16), we obtain

$$D_t^2\Phi_j[\theta] + a_0 D_t\Phi_j[\theta] = g_j'' + a_0 g_j'. \tag{2.17}$$

Moreover,

$$D_t\{\Phi_j[\theta]\}(0) = \Phi_j[u_0] = \Phi_j[f(0) - \lambda v_0] - a_0\Phi_j[\theta_0] + k_0\Phi_j[\Delta\theta_0] = g_j'(0). \tag{2.18}$$

So (2.17), (C13), (2.18) imply $\Phi_j[\theta(t)] = g_j(t)$, $\forall t \in [0,\tau]$ and for $j \in \{1,2\}$. $\square$

3. Some preparatory results

We start by studying the following system

$$\begin{cases} D_t u(t,x) + \lambda D_t v(t,x) = k_0\Delta u(t,x) - a_0 u(t,x) + f(t,x), & (t,x) \in [0,T]\times\Omega, \\ \varepsilon D_t v(t,x) = -\Delta^2 v(t,x) - \lambda\Delta u(t,x) + g(t,x), & (t,x) \in [0,T]\times\Omega, \\ D_\nu u(t,x') = D_\nu v(t,x') = D_\nu\Delta\chi(t,x') = 0, & (t,x') \in [0,T]\times\partial\Omega, \\ u(0,x) = u_0(x), & x \in \Omega, \\ v(0,x) = v_0(x), & x \in \Omega. \end{cases} \tag{3.1}$$

under the following assumptions:

(A1) assumption (C1) holds;

(A2) $\lambda \in \mathbf{R}$, $k_0 \in \mathbf{R}^+$, $\varepsilon \in \mathbf{R}^+$, $a_0 \in \mathbf{R}$.

We shall use the following result, which is a consequence of Theorem 8.2 in [18]:

Theorem 3.1. *(I) Consider the problem*

$$\begin{cases} D_t u(t,x) = k_0\Delta u(t,x) - a_0 u(t,x) + f(t,x), & (t,x) \in [0,T]\times\Omega, \\ D_\nu u(t,x') = 0, & (t,x') \in [0,T]\times\partial\Omega, \\ u(0,x) = u_0(x), & x \in \Omega. \end{cases} \tag{3.2}$$

with the assumption (A1), $k_0 \in \mathbf{R}^+$, $a_0 \in \mathbf{R}$.

Then, if $p \in (1,+\infty)$, $f \in L^p(0,T;L^p(\Omega))$ and $u_0 \in (L^p(\Omega), W_B^{2,p}(\Omega))_{1-1/p,p}$, (3.2) has a unique solution $u \in X(T,p)$. Moreover, $\forall\, T_0 \in \mathbf{R}^+$, there exists $C(T_0) \in \mathbf{R}^+$, such that, if $0 < T \le T_0$,

$$\|u\|_{X(T,p)} \le C(T_0)(\|f\|_{L^p(0,T;L^p(\Omega))} + \|u_0\|_{(L^p(\Omega),W_B^{2,p}(\Omega))_{1-1/p,p}}). \tag{3.3}$$

(II) Consider the problem

$$\begin{cases} \varepsilon D_t v(t,x) = -\Delta^2 v(t,x) + \lambda^2\Delta v(t,x) + g(t,x), & (t,x) \in [0,T]\times\Omega, \\ D_\nu v(t,x') = 0, D_\nu\Delta v(t,x') = 0 & (t,x') \in [0,T]\times\partial\Omega, \\ v(0,x) = v_0(x), & x \in \Omega. \end{cases} \tag{3.4}$$

with the assumption (A1) and $\varepsilon \in \mathbf{R}^+$. Then, if $p \in (1,+\infty)$,$g \in L^p(0,T;L^p(\Omega))$ and $v_0 \in (L^p(\Omega), W_{BB}^{4,p}(\Omega))_{1-1/p,p}$, (3.4) has a unique solution $v \in Y(T,p)$. Moreover, $\forall\, T_0 \in \mathbf{R}^+$, there exists $C(T_0) \in \mathbf{R}^+$, such that, if $0 < T \le T_0$,

$$\|v\|_{Y(T,p)} \le C(T_0)(\|f\|_{L^p(0,T;L^p(\Omega))} + \|v_0\|_{(L^p(\Omega),W_{BB}^{4,p}(\Omega))_{1-1/p,p}}). \tag{3.5}$$

Proof. We show only (I). The proof of (II) is similar.
By Proposition 1.2.10 in [21], there exists $z \in X(T,p)$ such that $z(0) = u_0$. Subtracting z to u, one is reduced to the case $u_0 = 0$. The existence and uniqueness of a solution in $X(T,p)$ in case $u_0 = 0$ is a particular case of Theorem 8.2 in [18]. The estimate of the solution can be obtained employing the following argument: let $0 < T \leq T_0$. Extend f to an element $\tilde{f} \in L^p(0,T_0;L^p(\Omega))$ setting $\tilde{f}(t,.) \equiv 0$ if $t \in]T,T_0]$. Then the solution $\tilde{u}$ of (3.2), replacing T with T_0 and f with $\tilde{f}$, is an extension of u. So

$$\begin{aligned}
\|u\|_{X(T,p)} &\leq \|\tilde{u}\|_{X(T_0,p)} \\
&\leq C(T_0)(\|\tilde{f}\|_{L^p(0,T_0;L^p(\Omega))} + \|u_0\|_{(L^p(\Omega),W_B^{2,p}(\Omega))_{1-1/p,p}}) \qquad (3.6)\\
&= C(T_0)(\|f\|_{L^p(0,T;L^p(\Omega))} + \|u_0\|_{(L^p(\Omega),W_B^{2,p}(\Omega))_{1-1/p,p}}).
\end{aligned}$$

□

Remark 3.1. From Theorem 3.5 in [19], we have that, for $p \in (1,+\infty)$,

$$\begin{aligned}
&(L^p(\Omega),W_B^{2,p}(\Omega))_{1-1/p,p} \\
&= \begin{cases} B_{p,p}^{2(1-1/p)}(\Omega) & \text{if} \quad 1 < p < 3, \\ \{f \in B_{p,p}^{2(1-1/p)}(\Omega) : D_\nu f \equiv 0\} & \text{if} \quad 3 < p < +\infty. \end{cases} \qquad (3.7)
\end{aligned}$$

Moreover,

$$\begin{aligned}
&(L^p(\Omega),W_{BB}^{4,p}(\Omega))_{1-1/p,p} \\
&= \begin{cases} B_{p,p}^{4(1-1/p)}(\Omega) & \text{if} \quad 1 < p < 5/3, \\ \{f \in B_{p,p}^{4(1-1/p)}(\Omega) : D_\nu f \equiv 0\} & \text{if} \quad 5/3 < p < 5, \\ \{f \in B_{p,p}^{4(1-1/p)}(\Omega) : D_\nu f \equiv D_\nu \Delta f \equiv 0\} & \text{if} \quad 5 < p < +\infty. \end{cases} \qquad (3.8)
\end{aligned}$$

Corollary 3.1. *Consider the solution v of the problem* (3.4) *in case $v_0 = 0$. Then, if $0 < T \leq T_0$ and $s \in \mathbf{Z}$, $0 \leq s \leq 4$,*

$$\|v\|_{L^p(0,T;W^{s,p}(\Omega))} \leq C(T_0,s)T^{1-\frac{s}{4}}\|g\|_{L^p(0,T;L^p(\Omega))}.$$

Proof. If $s = 0$, we have, owing to Young's inequality,

$$\begin{aligned}
\|v\|_{L^p(0,T;L^p(\Omega))} &= \|1 * D_t v\|_{L^p(0,T;L^p(\Omega))} \\
&\leq \|1\|_{L^1(0,T)}\|D_t v\|_{L^p(0,T;L^p(\Omega))} \\
&\leq C(T_0)T\|g\|_{L^p(0,T;L^p(\Omega))}.
\end{aligned}$$

In general, there exists $C(s) > 0$ such that, $\forall v \in W^{4,p}(\Omega))$,

$$\|v\|_{W^{s,p}(\Omega)} \leq C(s)\|v\|_{L^p(\Omega)}^{1-\frac{s}{4}}\|v\|_{W^{4,p}(\Omega)}^{\frac{s}{4}}.$$

So, employing Hölder's inequality,

$$\begin{aligned}
\|v\|_{L^p(0,T;W^{s,p}(\Omega))} &= (\int_0^T \|v(t)\|_{W^{s,p}(\Omega)}^p dt)^{1/p} \\
&\leq C(s)(\int_0^T \|v(t)\|_{L^p(\Omega)}^{(1-s/4)p}\|v(t)\|_{W^{4,p}(\Omega)}^{sp/4} dt)^{1/p} \\
&\leq C(s)\|v\|_{L^p(0,T;L^p(\Omega))}^{1-s/4}\|v\|_{L^p(0,T;W^{4,p}(\Omega))}^{s/4} \\
&\leq C(T_0,s)T^{1-s/4}\|g\|_{L^p(0,T;L^p(\Omega))}.
\end{aligned}$$

Theorem 3.2. *Let $p \in (1, +\infty)$, $T \in \mathbf{R}^+$. Then the following conditions are necessary and sufficient in order that (3.1) have a solution $(u, v) \in X(T, p) \times Y(T, p)$:*

(I) $f, g \in L^p(0, T; L^p(\Omega))$;
(II) $u_0 \in (L^p(\Omega), W_B^{2,p}(\Omega))_{1-1/p,p}$;
(III) $v_0 \in (L^p(\Omega), W_{BB}^{4,p}(\Omega))_{1-1/p,p}$.

If (I)–(III) are satisfied, the solution is unique. Moreover, $\forall T_0 \in \mathbf{R}^+$ there exists $C(T_0) \in \mathbf{R}^+$, such that, if $0 < T \le T_0$,

$$\begin{aligned} \|u\|_{X(T,p)} + \|v\|_{Y(T,p)} \le C(T_0)[&\|f\|_{L^p(0,T;L^p(\Omega))} \\ &+ \|g\|_{L^p(0,T;L^p(\Omega))} \\ &+ \|u_0\|_{(L^p(\Omega),W_B^{2,p}(\Omega))_{1-1/p,p}} \\ &+ \|v_0\|_{(L^p(\Omega),W_{BB}^{4,p}(\Omega))_{1-1/p,p}}]. \end{aligned} \tag{3.9}$$

Proof. The condition (I) is obviously necessary. From the trace method to define real interpolation spaces (see [21], Section 1.2.2), we know that

$$\{u(0) : u \in X(T, p))\} \quad \text{and} \quad \{v(0) : v \in Y(T, p)\}$$

coincide with the interpolation spaces

$$(L^p(\Omega), W_B^{2,p}(\Omega))_{1-1/p,p} \quad \text{and} \quad (L^p(\Omega), W_{BB}^{4,p}(\Omega))_{1-1/p,p}$$

respectively. So, even conditions (II) and (III) are necessary.

To show that (I)–(III) are also sufficient, we set

$$z := u + \lambda v \tag{3.10}$$

and consider the new problem

$$\begin{cases} D_t z(t,x) = k_0 \Delta z(t,x) - a_0 z(t,x) \\ +(\lambda a_0 - k_0 \lambda \Delta) v(t,x) + f(t,x), \quad (t,x) \in [0,T] \times \Omega, \\ \varepsilon D_t v(t,x) = -\Delta^2 v(t,x) + \lambda^2 \Delta v(t,x) - \lambda \Delta z(t,x) + g(t,x), \\ \qquad\qquad (t,x) \in [0,T] \times \Omega, \\ D_\nu z(t,x') = D_\nu v(t,x') = D_\nu \Delta v(t,x') = 0, \quad (t,x') \in [0,T] \times \partial\Omega, \\ z(0,x) = z_0(x), \quad x \in \Omega, \\ v(0,x) = v_0(x), \quad x \in \Omega. \end{cases} \tag{3.11}$$

with

$$z_0 := u_0 + \lambda v_0. \tag{3.12}$$

We observe that $z_0 \in (L^p(\Omega), W_B^{2,p}(\Omega))_{1-1/p,p}$ and that to show that (3.1) has a unique solution (u, v) in $X(T, p) \times Y(T, p)$ is equivalent to show that (3.11) has a unique solution with the same regularity. To this aim, we indicate with $\mathcal{T}_T(u_0, f)$ the solution in $X(T, p)$ of (3.2), with $\mathcal{S}_T(v_0, g)$ the solution in $Y(T, p)$ of (3.4). We start by looking for a solution in some interval $[0, \tau]$, with $0 < \tau \le T$.

Let $X := \{z \in W^{1,p}(0,\tau;L^p(\Omega)) \cap L^p(0,T;W^{2,p}_B(\Omega)) : z(0) = z_0\}$, which is a closed subspace of $X(\tau,p)$. It is easily seen that, if a solution (z,v) exists in some interval $[0,\tau]$, then

$$v = \mathcal{S}_\tau(v_0, g - \lambda\Delta z), \tag{3.13}$$

$$\begin{aligned} z &= \mathcal{T}_\tau[u_0, f + (\lambda a_0 - k_0\lambda\Delta)\mathcal{S}_\tau(v_0, g - \lambda\Delta z)] \\ &= \mathcal{T}_\tau[u_0, f + (\lambda a_0 - k_0\lambda\Delta)\mathcal{S}_\tau(v_0, g)] - \lambda^2\mathcal{T}_\tau[0, (a_0 - k_0\Delta)\mathcal{S}_\tau(0, \Delta z)] \\ &:= \mathcal{U}_\tau(z). \end{aligned} \tag{3.14}$$

The operator $\mathcal{U}_\tau$ maps X into itself. If z_1 and z_2 are elements of X, we have, employing Theorem 3.1 and Corollary 3.1,

$$\begin{aligned} \|\mathcal{U}_\tau(z_1) - \mathcal{U}_\tau(z_2)\|_{X(\tau,p)} &= \lambda^2\|\mathcal{T}_\tau(0, (a_0 - k_0\Delta)\mathcal{S}_\tau(0, \Delta(z_1 - z_2))\|_{X(\tau,p)} \\ &\le C_1(T)\|(a_0 - k_0\Delta)\mathcal{S}_\tau(0, \Delta(z_1 - z_2))\|_{L^p(0,\tau;L^p(\Omega))} \\ &\le C_2(T)\|\mathcal{S}_\tau(0, \Delta(z_1 - z_2))\|_{L^p(0,\tau;W^{2,p}(\Omega))} \\ &\le C_3(T)\tau^{1/2}\|\Delta(z_1 - z_2))\|_{L^p(0,\tau;L^p(\Omega))} \\ &\le C_4(T)\tau^{1/2}\|z_1 - z_2\|_{L^p(0,\tau;W^{2,p}(\Omega))}. \end{aligned}$$

So, if $C_4(T)\tau^{1/2} < 1$, the operator $\mathcal{U}_\tau$ has a unique fixed point z in X. We observe that the length τ of the interval does not depend on the data of our problem. From (3.13), we can determine also v. Starting from the new data $z(\tau)$ and $v(\tau)$ (again belonging to $(L^p(\Omega), W^{2,p}_B(\Omega))_{1-1/p,p}$ and $(L^p(\Omega), W^{4,p}_{BB}(\Omega))_{1-1/p,p}$ respectively, we can extend the solution to the interval $[0, 2\tau]$. Proceeding in this way, we construct a solution in $[0,T]$.

The proof of the estimate (3.9) can be obtained with the same argument used to show (3.3) and (3.5). □

Lemma 3.1. *Let X be a Banach space, $p \in (1,+\infty)$, $\tau \in \mathbf{R}^+$, $h \in L^p(0,\tau)$, $f \in L^p(0,\tau;X)$. Then $h * f \in L^p(0,\tau;X)$ and*

$$\|h * f\|_{L^p(0,\tau;X)} \le \tau^{1-1/p}\|h\|_{L^p(0,\tau)}\|f\|_{L^p(0,\tau;X)}.$$

Proof. Using Young's inequality and Hölder's inequality, we have

$$\begin{aligned} \|h * f\|_{L^p(0,\tau;X)} &\le \|h\|_{L^1(0,\tau)}\|f\|_{L^p(0,\tau;X)} \\ &\le \tau^{1-1/p}\|h\|_{L^p(0,\tau)}\|f\|_{L^p(0,\tau;X)}. \end{aligned}$$ □

Lemma 3.2. *Under the conditions* (C1) *and* (C3), *$W^{4,p}(\Omega)$ is continuously embedded in $C(\overline{\Omega})$ and is a space of pointwise multipliers for $W^{s,p}(\Omega)$, $\forall s \in \mathbf{Z}$, $0 \le s \le 4$.*

Proof. Analogous to the proof given in [20], Lemma 4.1 in the case $p = 2$. □

Lemma 3.3. *Under the assumptions* (C1) *and* (C3), *if $S \in C^\infty(\mathbf{R})$, the map $v \to S \circ v$ is of class C^∞ from $W^{4,p}(\Omega)$ into itself.*

Moreover, $\forall k \in \mathbf{N}_0$, $(S\circ .)^{(k)}$ is bounded with values in $\mathcal{L}^k(W^{4,p}(\Omega), W^{4,p}(\Omega))$ in every bounded subset of $W^{4,p}(\Omega)$.

Proof. Analogous to the proof given in [20], Lemma 4.2 in the case $p = 2$. □

Lemma 3.4. *Assume that* (C1) *and* (C3) *are satisfied,* $S \in C^\infty(\mathbf{R})$, $\chi_0 \in W^{4,p}(\Omega)$. *Let* $R \in \mathbf{R}^+$, $0 < \tau \leq T$, V_1 *and* V_2 *be elements of* $Y(\tau, p)$,

$$\max_{j \in \{1,2\}} \|V_j\|_{L^p(0,\tau;W^{4,p}(\Omega))} \leq R.$$

Then

$$\begin{aligned} &\|S(\chi_0 + 1 * V_1)V_1 - S(\chi_0 + 1 * V_2)V_2\|_{L^p(0,\tau;W^{2,p}(\Omega))} \\ &\quad \leq C(R,T)\tau^{(1-1/p)\wedge(1/(2p))}\|V_1 - V_2\|_{Y(\tau,p)}. \end{aligned}$$

Proof. We have

$$\begin{aligned} &\|S(\chi_0 + 1 * V_1)V_1 - S(\chi_0 + 1 * V_2)V_2\|_{L^p(0,\tau;W^{2,p}(\Omega))} \\ &\quad \leq \|[S(\chi_0 + 1 * V_1) - S(\chi_0 + 1 * V_2)]V_1\|_{L^p(0,\tau;W^{2,p}(\Omega))} \\ &\qquad + \|S(\chi_0 + 1 * V_2)(V_1 - V_2)\|_{L^p(0,\tau;W^{2,p}(\Omega))} := I_1 + I_2. \end{aligned}$$

By Lemma 3.2,

$$\begin{aligned} I_1 &\leq C\|S(\chi_0 + 1 * V_1) - S(\chi_0 + 1 * V_2)\|_{L^\infty(0,\tau;W^{4,p}(\Omega))}\|V_1\|_{L^p(0,\tau;W^{2,p}(\Omega))} \\ &\leq C_1(R)\|S(\chi_0 + 1 * V_1) - S(\chi_0 + 1 * V_2)\|_{L^\infty(0,\tau;W^{4,p}(\Omega))} \end{aligned}$$

For $j \in \{1, 2\}$, by virtue of Hölder's inequality,

$$\begin{aligned} \|\chi_0 + 1 * V_j\|_{L^\infty(0,\tau;W^{4,p}(\Omega))} &\leq \|\chi_0\|_{W^{4,p}(\Omega)} + \tau^{1-1/p}\|V_j\|_{L^p(0,\tau;W^{4,p}(\Omega))} \\ &\leq \|\chi_0\|_{W^{4,p}(\Omega)} + T^{1-1/p}R. \end{aligned}$$

So, by Lemma 3.3 and Hölder's inequality,

$$\begin{aligned} I_1 &\leq C_2(R)\|1 * (V_1 - V_2)\|_{L^\infty(0,\tau;W^{4,p}(\Omega))} \\ &\leq C_2(R)\tau^{1-1/p}\|V_1 - V_2\|_{L^p(0,\tau;W^{4,p}(\Omega))}. \end{aligned} \tag{3.15}$$

With analogous arguments, we obtain

$$I_2 \leq C_3(R)\|V_1 - V_2\|_{L^p(0,\tau;W^{2,p}(\Omega))}.$$

As $W^{2,p}(\Omega) \in J_{\frac{1}{2}}(L^p(\Omega), W^{4,p}(\Omega))$, by Hölder's inequality,

$$\begin{aligned} &\left(\int_0^\tau \|V_1 - V_2)(t)\|^p_{W^{2,p}(\Omega)} dt\right)^{\frac{1}{p}} \\ &\quad \leq C\left(\int_0^\tau \|(V_1 - V_2)(t)\|^{\frac{p}{2}}_{L^p(\Omega)}\|(V_1 - V_2)(t)\|^{\frac{p}{2}}_{W^{4,p}(\Omega)} dt\right)^{\frac{1}{p}} \\ &\quad \leq C\|V_1 - V_2\|^{\frac{1}{2}}_{L^p(0,\tau;L^p(\Omega))}\|V_1 - V_2\|^{\frac{1}{2}}_{L^p(0,\tau;W^{4,p}(\Omega))}. \end{aligned}$$

We have, employing Young's inequality,

$$\begin{aligned} &\|V_1 - V_2\|_{L^p(0,\tau;L^p(\Omega))} \\ &\quad = \|V_1(0) - V_2(0) + 1 * D_t(V_1 - V_2)\|_{L^p(0,\tau;L^p(\Omega))} \\ &\quad \leq \tau^{1/p}\|V_1(0) - V_2(0)\|_{L^p(\Omega)} + \tau\|D_t(V_1 - V_2)\|_{L^p(0,\tau;L^p(\Omega))} \\ &\quad \leq (\tau^{1/p} \vee \tau)\|V_1 - V_2\|_{W^{1,p}(0,\tau;L^p(\Omega))}. \end{aligned}$$

So we obtain, if $0 < \tau \le T$,

$$\begin{aligned} I_2 &\le C_4(R)(\tau^{1/(2p)} \vee \tau^{1/2})\|V_1 - V_2\|^{1/2}_{L^p(0,\tau;W^{4,p}(\Omega))}\|V_1 - V_2\|^{1/2}_{W^{1,p}(0,\tau;L^p(\Omega))} \\ &\le C_4(R,T)\tau^{1/(2p)}\|V_1 - V_2\|_{Y(\tau,p)}. \end{aligned} \tag{3.16}$$

From (3.15) and (3.16), we get the result. □

Lemma 3.5. *Let $p \in (1, +\infty)$, Ω satisfying* (C1), *$\psi \in L^{p'}(\Omega)$, $\tau \in \mathbf{R}^+$. We define in $L^p(0,\tau; L^p(\Omega))$ the operator*

$$\Phi[f](t) := \int_\Omega \psi(x) f(t,x)dx.$$

Then, if $u \in X(\tau,p)$, $\Phi[\Delta u] \in L^p(0,\tau)$ and

$$\|\Phi[\Delta u]\|_{L^p(0,\tau)} \le \omega(\tau)\|u\|_{X(\tau,p)}, \tag{3.17}$$

with $\omega(\tau) > 0$, independent of u, and $\lim_{\tau\to 0}\omega(\tau) = 0$; if $v \in Y(\tau,p)$, $\Phi[\Delta^2 v] \in L^p(0,\tau)$ and

$$\|\Phi[\Delta^2 v]\|_{L^p(0,\tau)} \le \omega(\tau)\|v\|_{Y(\tau,p)}, \tag{3.18}$$

with $\omega(\tau) > 0$, independent of u, and $\lim_{\tau\to 0}\omega(\tau) = 0$.

Proof. We prove only the first statement; the second can be shown similarly.

It is easily seen that

$$\|\Phi[\Delta u]\|_{L^p(0,\tau)} \le C\|\psi\|_{L^{p'}(\Omega)}\|u\|_{L^p(0,\tau;W^{2,p}(\Omega))} \le C\|\psi\|_{L^{p'}(\Omega)}\|u\|_{X(\tau,p)}, \tag{3.19}$$

with $C > 0$ independent of ψ, τ, u. If $\psi \in \mathcal{D}(\Omega)$, we have

$$\Phi[\Delta u](t) = \int_\Omega \Delta\psi(x)u(t,x)dx,$$

so that

$$\begin{aligned} \|\Phi[\Delta u]\|_{L^p(0,\tau)} &\le \|\Delta\psi\|_{L^{p'}(\Omega)}\|u\|_{L^p(0,\tau;L^p(\Omega))} \\ &= \|\Delta\psi\|_{L^{p'}(\Omega)}\|u(0) + 1 * D_t u\|_{L^p(0,\tau;L^p(\Omega))} \\ &\le \|\Delta\psi\|_{L^{p'}(\Omega)}(\tau^{1/p}\|u(0)\|_{L^p(\Omega)} + \tau\|D_t u\|_{L^p(0,\tau;L^p(\Omega))}) \\ &\le \|\Delta\psi\|_{L^{p'}(\Omega)}(\tau^{1/p} \vee \tau)\|u\|_{W^{1,p}(0,\tau;L^p(\Omega))} \\ &\le \|\Delta\psi\|_{L^{p'}(\Omega)}(\tau^{1/p} \vee \tau)\|u\|_{X(\tau,p)}. \end{aligned} \tag{3.20}$$

The conclusion is a simple consequence of (3.19), (3.20) and the density of $\mathcal{D}(\Omega)$ in $L^{p'}(\Omega)$. □

4. The main result

Theorem 4.1. *Assume that the assumptions* (C1)–(C14) *are satisfied. Then there exists $\tau \in (0,T]$ such that the problem* (1.13) *has a unique solution (θ, χ, a, b, k) satisfying the conditions* (1.8)–(1.12).

Proof. We start by showing the existence of a solution with the required properties. We already know from Lemma 2.1 that, if a solution exists, necessarily $k = k_0$ and $a(0) = a_0$. Moreover, if u, v and h are the functions defined in (2.8)–(2.10), they satisfy the system (2.14)–(2.15) and vice versa. So we focalize our attention on the system (2.14)–(2.15) and look for a solution (u, v, h, b) satisfying the regularity conditions (2.11)–(2.13) and (1.11).

We start by fixing an element

$$\tilde{U} \in W^{1,p}(0,T; L^p(\Omega)) \cap L^p(0,T; W^{2,p}_B(\Omega))$$

such that $\tilde{U}(0) = u_0$ and an element

$$\tilde{V} \in W^{1,p}(0,T; L^p(\Omega)) \cap L^p(0,T; W^{4,p}_{BB}(\Omega))$$

such that $\tilde{V}(0) = v_0$.

Next, we indicate with (H, B) the solution in $L^p(0,T) \times L^p(0,T)$ of the system

$$\begin{aligned} H(t)\Phi_j(\theta_0) - B(t)\Phi_j(\Delta\theta_0) = {} & \Phi_j[D_t f(t)] - g_j''(t) - a_0 g_j'(t) \\ & + k_0\Phi_j[\Delta\tilde{U}(t)] + \epsilon^{-1}\lambda\Phi_j\{\Delta^2\tilde{V}(t) \\ & - \lambda\Delta\tilde{U}(t)]\}, t \in [0,T], j \in \{1,2\}. \end{aligned} \tag{4.21}$$

The existence and uniqueness of (H, B) follows immediately from (C11). Now we consider the problem

$$\begin{cases} D_t U(t,x) + \lambda D_t V(t,x) = k_0\Delta U(t,x) - a_0 U(t,x) + D_t f(t,x) \\ - H(t)\theta_0(x) + B(t)\Delta\theta_0(x), \quad (t,x) \in [0,\tau] \times \Omega, \\ \varepsilon D_t V(t,x) = -\Delta^2 V(t,x) - \lambda\Delta U(t,x), \quad (t,x) \in [0,T] \times \Omega, \\ D_\nu U(t,x') = D_\nu V(t,x') = D_\nu \Delta V(t,x') = 0, \quad (t,x') \in [0,T] \times \partial\Omega, \\ U(0,x) = u_0(x), \quad x \in \Omega, \\ V(0,x) = v_0(x), \quad x \in \Omega. \end{cases} \tag{4.22}$$

By Theorem 3.2, the problem (4.22) has a unique solution (U, V) in $X(T,p) \times Y(T,p)$. We set

$$C_1 := \|U\|_{X(T,p)} + \|V\|_{Y(T,p)} + \|H\|_{L^p(0,T)} + \|B\|_{L^p(0,T)}. \tag{4.23}$$

Let $\rho \in \mathbf{R}^+$. We put

$$\begin{aligned} Z(\tau,\rho) := \{ & (u, v, h, b) \in X(\tau,p) \times Y(\tau,p) \times L^p(0,\tau) \times L^p(0,\tau) \\ & : u(0) = u_0,\ v(0) = v_0,\ D_\nu u \equiv D_\nu v \equiv D_\nu \Delta v \equiv 0, \\ & \|u - U\|_{X(\tau,p)} + \|v - V\|_{Y(\tau,p)} \\ & + \|h - H\|_{L^p(0,\tau)} + \|b - B\|_{L^p(0,\tau)} \le \rho\}. \end{aligned} \tag{4.24}$$

$Z(\tau,\rho)$ is a closed subset of $X(\tau,\rho) \times Y(\tau,p) \times L^p(0,\tau) \times L^p(0,\tau)$.

If $(U_1, V_1, H_1, B_1) \in Z(\tau, \rho)$, we consider the problem

$$\begin{cases} D_t u(t,x) + \lambda D_t v(t,x) = k_0 \Delta u(t,x) - a_0 u(t,x) + D_t f(t,x) \\ -h(t)\theta_0(x) + b(t)\Delta\theta_0(x) - (h * U_1)(t,x) + (b * \Delta U_1)(t,x), \\ \qquad\qquad (t,x) \in [0,\tau] \times \Omega, \\ \varepsilon D_t v(t,x) = -\Delta^2 v(t,x) + \Delta[\varphi'(\chi_0 + 1 * V_1)V_1](t,x) - \lambda \Delta u(t,x), \\ \qquad\qquad (t,x) \in [0,\tau] \times \Omega, \\ D_\nu u(t,x') = D_\nu v(t,x') = D_\nu \Delta v(t,x') = 0, \quad (t,x') \in [0,\tau] \times \partial\Omega, \\ u(0,x) = u_0(x), \quad x \in \Omega, \\ v(0,x) = v_0(x), \quad x \in \Omega, \\ h(t)\Phi_j(\theta_0) - b(t)\Phi_j(\Delta\theta_0) = \Phi_j[D_t f(t)] - g_j''(t) - a_0 g_j'(t) \\ -(H_1 * \Phi_j(U_1))(t) + k_0 \Phi_j[\Delta U_1(t)] + (B_1 * \Phi_j(\Delta U_1))(t) \\ +\epsilon^{-1}\lambda \Phi_j\{\Delta^2 V_1(t) - \Delta[\varphi'(\chi_0 + 1 * V_1)(t)V_1(t) + \lambda U_1(t)]\}, \\ \qquad\qquad t \in (0,\tau), j \in \{1,2\}. \end{cases} \tag{4.25}$$

It is easily seen that the system (4.25) has a unique solution $S(U_1, V_1, H_1, B_1)$ belonging to $X(\tau,p) \times Y(\tau,p) \times L^p(0,\tau) \times L^p(0,\tau)$. In fact, one can determine first h and b using (C11). Next, one can find u and v, using Theorem 3.2 and the knowledge of h and b. Clearly, we have to look for a fixed point of the operator S. Let now $(U_1, V_1, H_1, B_1) \in Z(\tau, \rho)$ for some $\tau \in (0,T]$, $\rho \in \mathbf{R}^+$. We put $(u, v, h, b) := S(U_1, V_1, H_1, B_1)$.

We estimate

$$\|u - U\|_{X(\tau,p)} + \|v - V\|_{Y(\tau,p)} + \|h - H\|_{L^p(0,\tau)} + \|b - B\|_{L^p(0,\tau)}.$$

To this aim, we observe that $(u - U, v - V, h - H, b - B)$ solves the system

$$\begin{cases} D_t(u-U)(t,x) + \lambda D_t(v-V)(t,x) = k_0 \Delta(u-U)(t,x) \\ -a_0(u-U)(t,x) - (h(t) - H(t))\theta_0(x) + (b(t) - B(t))\Delta\theta_0(x) \\ -(h * U_1)(t,x) + (b * \Delta U_1)(t,x), \quad (t,x) \in [0,\tau] \times \Omega, \\ \varepsilon D_t(v-V)(t,x) = -\Delta^2(v-V)(t,x) + \Delta[\varphi'(\chi_0 + 1 * V_1)V_1](t,x) \\ -\lambda\Delta(u-U)(t,x), \quad (t,x) \in [0,\tau] \times \Omega, \\ D_\nu(u-U)(t,x') = D_\nu(v-V)(t,x') = D_\nu\Delta(v-V)(t,x') = 0, \\ \qquad\qquad (t,x') \in [0,\tau] \times \partial\Omega, \\ (u-U)(0,x) = 0, \quad x \in \Omega, \\ (v-V)(0,x) = 0, \quad x \in \Omega, \\ (h(t) - H(t))\Phi_j(\theta_0) - (b(t) - B(t))\Phi_j(\Delta\theta_0) = -(H_1 * \Phi_j(U_1))(t) + \\ k_0\Phi_j[\Delta(U_1 - \tilde{U})(t)] + (B_1 * \Phi_j(\Delta U_1))(t) + \epsilon^{-1}\lambda\Phi_j\{\Delta^2(V_1 - \tilde{V})(t) \\ -\Delta[\varphi'(\chi_0 + 1 * V_1)(t)V_1(t) + \lambda(U_1 - \tilde{U})(t)]\}, \quad t \in (0,\tau), \quad j \in \{1,2\}. \end{cases} \tag{4.26}$$

We have

$$\begin{aligned}
&\|h - H\|_{L^p(0,\tau)} + \|b - B\|_{L^p(0,\tau)} \\
&\quad \le C_2 \sum_{j=1}^{2} \{\|H_1 * \Phi_j(U_1)\|_{L^p(0,\tau)} \\
&\qquad + \|B_1 * \Phi_j(\Delta U_1))\|_{L^p(0,\tau)} \\
&\qquad + \|\Phi_j\{k_0\Delta(U_1 - \tilde{U}) + \epsilon^{-1}\lambda\Delta^2(V_1 - \tilde{V}) - \epsilon^{-1}\lambda^2\Delta(U_1 - \tilde{U})\}\|_{L^p(0,\tau)} \\
&\qquad + \|\Phi_j\{\Delta[\varphi'(\chi_0 + 1 * V_1)V_1]\}\|_{L^p(0,\tau)}\}.
\end{aligned}$$

By Lemma 3.1, it follows

$$\begin{aligned}
&\|H_1 * \Phi_j(U_1)\|_{L^p(0,\tau)} + \|B_1 * \Phi_j(\Delta U_1))\|_{L^p(0,\tau)} \\
&\le \tau^{1-1/p}(\|H_1\|_{L^p(0,\tau)}\|\Phi_j(U_1)\|_{L^p(0,\tau)} + \|B_1\|_{L^p(0,\tau)}\|\Phi_j(\Delta U_1)\|_{L^p(0,\tau)}) \\
&\le C_3(C_1 + \rho)^2\tau^{1-1/p}.
\end{aligned}$$

By Lemma 3.5, we get

$$\begin{aligned}
&\|\Phi_j\{k_0\Delta(U_1 - \tilde{U}) + \epsilon^{-1}\lambda\Delta^2(V_1 - \tilde{V}) - \epsilon^{-1}\lambda^2\Delta(U_1 - \tilde{U})\}\|_{L^p(0,\tau)} \\
&\le \omega_1(\tau)(\|U_1 - \tilde{U}\|_{X(\tau,p)} + \|V_1 - \tilde{V})\|_{Y(\tau,p)}).
\end{aligned}$$

The letter ω will be used to indicate that the constant tends to 0 as τ goes to 0. Finally, from Lemma 3.4, we have (setting $V_2 = 0$)

$$\begin{aligned}
\|\Phi_j\{\Delta[\varphi'(\chi_0 + 1 * V_1)V_1]\}\|_{L^p(0,\tau)}\} &\le C_4\|\varphi'(\chi_0 + 1 * V_1)V_1\|_{L^p(0,\tau;W^{2,p}(\Omega))} \\
&\le C_1(T,\rho)\tau^{(1-1/p)\wedge 1/(2p)}.
\end{aligned}$$

Summarizing, we have obtained an estimate of the form

$$\|h - H\|_{L^p(0,\tau)} + \|b - B\|_{L^p(0,\tau)} \le C_2(T,\rho)\omega_2(\tau). \tag{4.27}$$

So, from Theorem 3.2,

$$\begin{aligned}
\|u - U\|_{X(\tau,p)} + \|v - V\|_{Y(\tau,p)} \le C_5\{&\|h - H\|_{L^p(0,\tau)} + \|b - B\|_{L^p(0,\tau)} \\
&+ \|h * U_1\|_{L^p(0,\tau;L^p(\Omega))} + \|b * \Delta U_1\|_{L^p(0,\tau;L^p(\Omega))} \\
&+ \|\varphi'(\chi_0 + 1 * V_1)V_1\|_{L^p(0,\tau;W^{2,p}(\Omega))}\}.
\end{aligned}$$

Then we consider

$$\begin{aligned}
\|h * U_1\|_{L^p(0,\tau;L^p(\Omega))} &\le \tau^{1-1/p}\|h\|_{L^p(0,\tau)}\|U_1\|_{L^p(0,\tau;L^p(\Omega))} \\
&\le \tau^{1-1/p}[\|H\|_{L^p(0,T)} + C_2(T,\rho)\omega_2(\tau)][\|U\|_{X(T,p)} + \rho].
\end{aligned}$$

and analogously

$$\begin{aligned}
\|b * \Delta U_1\|_{L^p(0,\tau;L^p(\Omega))} &\le C_6\tau^{1-1/p}\|b\|_{L^p(0,\tau)}\|U_1\|_{L^p(0,\tau;W^{2,p}(\Omega))} \\
&\le C_6\tau^{1-1/p}[\|B\|_{L^p(0,T)} + C_2(T,\rho)\omega_2(\tau)][\|U\|_{X(T,p)} + \rho],
\end{aligned}$$

so, employing also Lemma 3.4, we obtain

$$\|u - U\|_{X(\tau,p)} + \|v - V\|_{Y(\tau,p)} \le C_3(T,\rho)\omega_3(\tau). \tag{4.28}$$

In conclusion, from (4.27) and (4.28), we have

$$\begin{aligned}&\|u-U\|_{X(\tau,p)}+\|v-V\|_{Y(\tau,p)}+\|h-H\|_{L^p(0,\tau)}+\|b-B\|_{L^p(0,\tau)}\\&\quad\le C_4(T,\rho)\omega_4(\tau).\end{aligned}\tag{4.29}$$

(4.29) implies that, if $\tau\in(0,T]$ and

$$C_4(T,\rho)\omega_4(\tau)\le\rho,\tag{4.30}$$

then S maps $Z(\tau,\rho)$ into itself.

Now we consider two elements (U_1,V_1,H_1,B_1) and (U_2,V_2,H_2,B_2) of $Z(\tau,\rho)$. Writing (u_j,v_j,h_j,b_j) instead of $S(U_j,V_j,H_j,B_j)$, we estimate

$$\|u_1-u_2\|_{X(\tau,p)}+\|v_1-v_2\|_{Y(\tau,p)}+\|h_1-h_2\|_{L^p(0,\tau)}+\|b_1-b_2\|_{L^p(0,\tau)}.$$

The difference $(u_1-u_2,v_1-v_2,h_1-h_2,b_1-b_2)$ satisfies the system

$$\begin{cases}D_t(u_1-u_2)(t,x)+\lambda D_t(v_1-v_2)(t,x)=k_0\Delta(u_1-u_2)(t,x)\\-a_0(u_1-u_2)(t,x)-[h_1(t)-h_2(t)]\theta_0(x)+[b_1(t)-b_2(t)]\Delta\theta_0(x)\\-[(h_1*U_1)-(h_2*U_2)](t,x)+[(b_1*\Delta U_1)-(b_2*\Delta U_2)](t,x),\\\qquad\qquad\qquad\qquad\qquad\qquad\qquad\qquad(t,x)\in[0,\tau]\times\Omega,\\\varepsilon D_t(v_1-v_2)(t,x)=-\Delta^2(v_1-v_2)(t,x)+\Delta[\varphi'(\chi_0+1*V_1)V_1\\-\varphi'(\chi_0+1*V_2)V_2](t,x)-\lambda\Delta(u_1-u_2)(t,x),\quad(t,x)\in[0,\tau]\times\Omega,\\D_\nu(u_1-u_2)(t,x')=D_\nu(v_1-v_2)(t,x')=0,\quad(t,x')\in[0,\tau]\times\partial\Omega,\\D_\nu\Delta(v_1-v_2)(t,x')=0,\quad(t,x')\in[0,\tau]\times\partial\Omega,\\(u_1-u_2)(0,x)=0,\quad x\in\Omega,\\(v_1-v_2)(0,x)=0,\quad x\in\Omega,\\(h_1(t)-h_2(t))\Phi_j(\theta_0)-(b_1(t)-b_2(t))\Phi_j(\Delta\theta_0)\\=-[(H_1*\Phi_j(U_1))-(H_2*\Phi_j(U_2))](t)+k_0\Phi_j[\Delta(U_1-U_2)(t)]\\+[(B_1*\Phi_j(\Delta U_1))-(B_2*\Phi_j(\Delta U_2))](t)+\epsilon^{-1}\lambda\Phi_j\{\Delta^2(V_1-V_2)(t)\\-\Delta[\varphi'(\chi_0+1*V_1)(t)V_1(t)-\varphi'(\chi_0+1*V_2)(t)V_2(t)\\+\lambda(U_1-U_2)(t)]\},\quad t\in(0,\tau),\quad j\in\{1,2\}.\end{cases}\tag{4.31}$$

So we have

$$\begin{aligned}&\|h_1-h_2\|_{L^p(0,\tau)}+\|b_1-b_2\|_{L^p(0,\tau)}\\&\quad\le C_7\sum_{j=1}^{2}\{\|H_1*\Phi_j(U_1)-H_2*\Phi_j(U_2)\|_{L^p(0,\tau)}\\&\qquad+\|B_1*\Phi_j(\Delta U_1)-B_2*\Phi_j(\Delta U_2)\|_{L^p(0,\tau)}\\&\qquad+\|\Phi_j\{k_0\Delta(U_1-U_2)+\epsilon^{-1}\lambda\Delta^2(V_1-V_2)-\epsilon^{-1}\lambda^2\Delta(U_1-U_2)\}\|_{L^p(0,\tau)}\\&\qquad+\|\Phi_j\{\Delta[\varphi'(\chi_0+1*V_1)V_1-\varphi'(\chi_0+1*V_2)V_2]\}\|_{L^p(0,\tau)}\}.\end{aligned}$$

$$
\begin{aligned}
&\|H_1 * \Phi_j(U_1) - H_2 * \Phi_j(U_2)\|_{L^p(0,\tau)} \\
&\quad \le \|(H_1 - H_2) * \Phi_j(U_1)\|_{L^p(0,\tau)} + \|H_2 * \Phi_j(U_1 - U_2)\|_{L^p(0,\tau)} \\
&\quad \le C_8 \tau^{1-1/p}[\|H_1 - H_2\|_{L^p(0,\tau)}\|U_1\|_{L^p(0,\tau;L^p(\Omega)} \\
&\qquad + \|H_2\|_{L^p(0,\tau)}\|U_1 - U_2\|_{L^p(0,\tau;L^p(\Omega))}] \\
&\quad \le C_8 \tau^{1-1/p}[\|H_1 - H_2\|_{L^p(0,\tau)}(\|U\|_{X(T,p)} + \rho) \\
&\qquad + (\|H\|_{L^p(0,T)} + \rho)\|U_1 - U_2\|_{X(\tau,p)}].
\end{aligned}
$$

Analogously we get

$$
\begin{aligned}
&\|B_1 * \Phi_j(\Delta U_1) - B_2 * \Phi_j(\Delta U_2)\|_{L^p(0,\tau)} \\
&\quad \le C_9 \tau^{1-1/p}[\|B_1 - B_2\|_{L^p(0,\tau)}(\|U\|_{X(T,p)} + \rho) \\
&\qquad + (\|B\|_{L^p(0,T)} + \rho)\|U_1 - U_2\|_{X(\tau,p)}].
\end{aligned}
$$

By Lemma 3.5, we have

$$
\begin{aligned}
&\|\Phi_j\{k_0\Delta(U_1 - U_2) + \epsilon^{-1}\lambda\Delta^2(V_1 - V_2) - \epsilon^{-1}\lambda^2\Delta(U_1 - U_2)\}\|_{L^p(0,\tau)} \\
&\quad \le \omega_5(\tau)(\|U_1 - U_2\|_{X(\tau,p)} + \|V_1 - V_2)\|_{Y(\tau,p)}).
\end{aligned}
$$

Finally, from Lemma 3.4, we have

$$
\begin{aligned}
&\|\Phi_j\{\Delta[\varphi'(\chi_0 + 1 * V_1)V_1 - \varphi'(\chi_0 + 1 * V_2)V_2]\}\|_{L^p(0,\tau)} \\
&\quad \le C_4\|\varphi'(\chi_0 + 1 * V_1)V_1 - \varphi'(\chi_0 + 1 * V_2)V_2\|_{L^p(0,\tau;W^{2,p}(\Omega))} \\
&\quad \le C_1(T,\rho)\tau^{(1-1/p)\wedge 1/(2p)}\|V_1 - V_2\|_{Y(\tau,\rho)}.
\end{aligned}
$$

Summarizing, we have obtained an estimate of the form

$$
\begin{aligned}
&\|h_1 - h_2\|_{L^p(0,\tau)} + \|b_1 - b_2\|_{L^p(0,\tau)} \\
&\quad \le C_5(T,\rho)\omega_6(\tau)[\|U_1 - U_2\|_{X(\tau,\rho)} \\
&\qquad + \|V_1 - V_2\|_{Y(\tau,\rho)} + \|H_1 - H_2\|_{L^p(0,\tau)} + \|B_1 - B_2\|_{L^p(0,\tau)}].
\end{aligned} \tag{4.32}
$$

So, from Theorem 3.2,

$$
\begin{aligned}
&\|u_1 - u_2\|_{X(\tau,p)} + \|v_1 - v_2\|_{Y(\tau,p)} \\
&\quad \le C_5\{\|h_1 - h_2\|_{L^p(0,\tau)} + \|b_1 - b_2\|_{L^p(0,\tau)} \\
&\qquad + \|h_1 * U_1 - h_2 * U_2\|_{L^p(0,\tau;L^p(\Omega))} \\
&\qquad + \|b_1 * \Delta U_1 - b_2 * \Delta U_2\|_{L^p(0,\tau;L^p(\Omega))} \\
&\qquad + \|\varphi'(\chi_0 + 1 * V_1)V_1 - \varphi'(\chi_0 + 1 * V_2)V_2\|_{L^p(0,\tau;W^{2,p}(\Omega))}\} \\
&\quad \le C_6(T,\rho)\omega_4(\tau)[\|U_1 - U_2\|_{X(\tau,\rho)} + \|V_1 - V_2\|_{Y(\tau,\rho)} \\
&\qquad + \|H_1 - H_2\|_{L^p(0,\tau)} + \|B_1 - B_2\|_{L^p(0,\tau)}].
\end{aligned} \tag{4.33}
$$

In conclusion, from (4.32) and (4.33), we have

$$
\begin{aligned}
&\|u_1 - u_2\|_{X(\tau,p)} + \|v_1 - v_1\|_{Y(\tau,p)} + \|h_1 - h_2\|_{L^p(0,\tau)} + \|b_1 - b_2\|_{L^p(0,\tau)} \\
&\quad \le C_7(T,\rho)\omega_7(\tau)[\|U_1 - U_2\|_{X(\tau,p)} + \|V_1 - V_2\|_{Y(\tau,p)} \\
&\qquad + \|h_1 - h_2\|_{L^p(0,\tau)} + \|b_1 - b_2\|_{L^p(0,\tau)}].
\end{aligned} \tag{4.34}
$$

If $\tau \in (0, T]$ is such that (4.30) holds together with

$$C_7(T,\rho)\omega_7(\tau) < 1,$$

then S maps $Z(\tau,\rho)$ into itself and is a contraction in it. It follows that the system (4.25) has a unique solution (u, v, h, b) in $Z(\tau, \rho)$.

Now we show that, if $(\theta_1, \chi_1, a_1, b_1, k_1)$ and $(\theta_2, \chi_2, a_2, b_2, k_2)$ are solutions of (1.13) both belonging to $X(\tau,p) \times Y(\tau,p) \times W^{1,p}(0,\tau) \times L^p(0,\tau) \times \mathbf{R}$, for some $\tau \in (0, T]$, then they coincide.

We have already seen that, necessarily, $k_1 = k_2 = k_0$. Setting, for $i \in \{1, 2\}$,

$$u_i := D_t\theta_i, \tag{4.35}$$

$$v_i := D_t\chi_i, \tag{4.36}$$

$$h_i := D_t a_i, \tag{4.37}$$

then (u_1, v_1, h_1, b_1) and (u_2, v_2, h_2, b_2) both satisfy the system (2.14)–(2.15). As, for a fixed $\rho \in \mathbf{R}^+$, the system (2.14)–(2.15) has a unique solution in $Z(\sigma, \rho)$ if σ is sufficiently small, necessarily (u_1, v_1, h_1, b_1) and (u_2, v_2, h_2, b_2) coincide almost everywhere in some right neighborhood of 0. So we set

$$\begin{aligned}\tau_1 := \inf\{t \in [0,\tau] :& \|u_1 - u_2\|_{X(t,p)} + \|v_1 - v_2\|_{Y(t,p)} \\ &+ \|h_1 - h_2\|_{L^p(0,t)} + \|b_1 - b_2\|_{L^p(0,t)} > 0\}.\end{aligned} \tag{4.38}$$

Of course, we assume that the set in the second term of (4.38) is not empty, so that $\tau_1 \in (0, \tau)$. We observe that

$$\|u_1 - u_2\|_{X(\tau_1,p)} + \|v_1 - v_2\|_{Y(\tau_1,p)} + \|h_1 - h_2\|_{L^p(0,\tau_1)} + \|b_1 - b_2\|_{L^p(0,\tau_1)} = 0, \tag{4.39}$$

which implies also

$$u_1(\tau_1) = u_2(\tau_1), \qquad v_1(\tau_1) = v_2(\tau_1). \tag{4.40}$$

For $t \in [0, \tau - \tau_1]$, we set

$$U(t) := u_1(\tau_1 + t) - u_2(\tau_1 + t), \tag{4.41}$$

$$V(t) := v_1(\tau_1 + t) - v_2(\tau_1 + t), \tag{4.42}$$

$$H(t) := h_1(\tau_1 + t) - h_2(\tau_1 + t), \tag{4.43}$$

$$B(t) := b_1(\tau_1 + t) - b_2(\tau_1 + t). \tag{4.44}$$

If $t \in [0, \tau - \tau_1]$ we have

$$\begin{aligned}&(h_1 * u_1)(\tau_1 + t) - (h_2 * u_2)(\tau_1 + t) \\ &= [(h_1 - h_2) * u_1](\tau_1 + t) + [h_2 * (u_1 - u_2)](\tau_1 + t) \\ &= \int_0^t [h_1(\tau_1 + t - s) - h_2(\tau_1 + t - s)]u_1(s)ds \\ &\qquad + \int_0^t h_2(t - s)[u_1(\tau_1 + s) - u_2(\tau_1 + s)]ds \\ &= (H * u_1)(t) + (h_2 * U)(t).\end{aligned} \tag{4.45}$$

Moreover, for $i \in \{1,2\}$, $t \in [0, \tau - \tau_1]$,

$$\begin{aligned} \chi_0 + (1 * v_i)(\tau_1 + t) &= \chi_0 + \int_0^{\tau_1} v_i(s)ds + \int_0^t v_i(s + \tau_1)ds \\ &= \chi_1 + (1 * v_i(. + \tau_1)), \end{aligned} \tag{4.46}$$

with

$$\chi_1 := \chi_0 + \int_0^{\tau_1} v_1(s)ds = \chi_0 + \int_0^{\tau_1} v_2(s)ds. \tag{4.47}$$

We observe that $\chi_1 \in W^{4,p}_{BB}(\Omega)$. Consequently, we have

$$\begin{cases} D_tU(t,x) + \lambda D_tV(t,x) = k_0\Delta U(t,x) - a_0U(t,x) \\ -H(t)\theta_0(x) + B(t)\Delta\theta_0(x) - (H * u_1)(t,x) \\ -(h_2 * U)(t,x) + (B * \Delta u_1)(t,x) + (b_2 * \Delta U)(t,x), \\ \qquad\qquad\qquad\qquad (t,x) \in [0, \tau - \tau_1] \times \Omega, \\ \varepsilon D_tV(t,x) = -\Delta^2 V(t,x) + \Delta[\varphi'(\chi_1 + 1 * v_1(. + \tau_1)) \\ \qquad\qquad\qquad\qquad (t,x) \in [0, \tau - \tau_1] \times \Omega, \\ v_1(. + \tau_1) - \varphi'(\chi_1 + 1 * v_2(. + \tau_1))v_2(. + \tau_1)](t,x) - \lambda\Delta U(t,x), \\ D_\nu U(t,x') = D_\nu V(t,x') = D_\nu \Delta V(t,x') = 0, \\ \qquad\qquad\qquad\qquad (t,x') \in [0, \tau - \tau_1] \times \partial\Omega, \\ U(0,x) = 0, \quad x \in \Omega, \\ V(0,x) = 0, \quad x \in \Omega \end{cases} \tag{4.48}$$

$$\begin{aligned} H(t)\Phi_j(\theta_0) - B(t)\Phi_j(\Delta\theta_0) &= -(H * \Phi_j(u_1))(t) - (h_2 * \Phi_j(U))(t) \\ &\quad + k_0\Phi_j[\Delta U(t)] + (B * \Phi_j(\Delta u_1))(t) \\ &\quad + (b_2 * \Phi_j(\Delta U))(t) + \epsilon^{-1}\lambda\Phi_j\{\Delta^2 V(t) \\ &\quad - \Delta[\varphi'(\chi_1 + 1 * v_1(. + \tau_1))(t)v_1(t + \tau_1) \\ &\quad - \varphi'(\chi_1 + 1 * v_2(. + \tau_1))(t)v_2(t + \tau_1) + \lambda U(t)]\}, \end{aligned} \tag{4.49}$$

with $t \in [0, \tau - \tau_1]$, $j \in \{1,2\}$. If we set

$$\begin{aligned} R := \|u_1\|_{X(\tau,p)} &+ \|v_1(. + \tau_1)\|_{Y(\tau - \tau_1, p)} + \|v_2(. + \tau_1)\|_{Y(\tau - \tau_1, p)} \\ &+ \|h_2\|_{L^p(0,\tau)} + \|b_2\|_{L^p(0,\tau)}, \end{aligned} \tag{4.50}$$

then, from (4.49), employing Lemmata 3.1, 3.4, 3.5, we get, for $\sigma \in (0, \tau - \tau_1]$,

$$\begin{aligned} &\|H\|_{L^p(0,\sigma)} + \|B\|_{L^p(0,\sigma)} \\ &\quad \le C_5(R)\omega_7(\sigma)[\|H\|_{L^p(0,\sigma)} + \|B\|_{L^p(0,\sigma)} + \|U\|_{X(\sigma,p)} + \|V\|_{Y(\sigma,p)}]. \end{aligned} \tag{4.51}$$

The estimate (4.51) implies that, for some $\sigma_1 \in (0, \tau - \tau_1]$, if $0 < \sigma \le \sigma_1$,

$$\|H\|_{L^p(0,\sigma)} + \|B\|_{L^p(0,\sigma)} \le C_6(R)\omega_7(\sigma)[\|U\|_{X(\sigma,p)} + \|V\|_{Y(\sigma,p)}] \tag{4.52}$$

From (4.48), using Theorem 3.2 and again Lemmata 3.1 and 3.4, we obtain

$$\begin{aligned} \|U\|_{X(\sigma,p)} + \|V\|_{Y(\sigma,p)} \le C_7(R)[&\|H\|_{L^p(0,\sigma)} + \|B\|_{L^p(0,\sigma)} + \omega_8(\sigma)(\|U\|_{X(\sigma,p)} \\ &+ \|V\|_{Y(\sigma,p)} + \|H\|_{L^p(0,\sigma)} + \|B\|_{L^p(0,\sigma)})]. \end{aligned} \tag{4.53}$$

(4.52) and (4.53) together imply

$$\|U\|_{X(\sigma,p)} + \|V\|_{Y(\sigma,p)} \leq C_8(R)\omega_9(\sigma)[\|U\|_{X(\sigma,p)} + \|V\|_{Y(\sigma,p)}]. \tag{4.54}$$

From (4.54), we have that U and V vanish in some right neighborhood of 0. From (4.52) it follows that even H and B vanish in some right neighborhood of 0. But this is in contradiction with the definition of τ_1.

With this the proof is complete. □

References

[1] R. Adams, *Sobolev spaces*, Pure and Applied Mathematics, vol. 65, Plenum Press (1975).

[2] G. Bonfanti, F. Luterotti, *Global solution to a phase-field model with memory and quadratic nonlinearity*, Adv. Math. Sci. Appl., **9** (1999), 523–538.

[3] G. Bonfanti, F. Luterotti, *Regularity and convergence results for a phase-field model with memory*, Math. Meth. Appl. Sci., **21** (1998), 1085–1105.

[4] M. Brokate, J. Sprekel, *Hysteresis and Phase Transition*, Springer, New York, 1996.

[5] D. Brandon, W.J. Hrusa, *Construction of a class of integral models for heat flow in materials with memory*, J. Integral Equations and Appl., **1** (1998), 175–201.

[6] G. Caginalp, *An analysis of a phase field model of a free boundary*, Arch. Rational Mech. Anal., **92** (1986), 205–245.

[7] G. Caginalp, *Stefan and Hele-Shaw type models as asymptotic limits of the phase field equations*, Phys. Rev. A, **39** (1989), 5887–5896.

[8] G. Caginalp, *The dynamics of a conserved phase-field system: Stefan-like, Hele-Shaw, and Cahn-Hilliard models as asymptotic limits*, IMA J. Appl. Math., **44** (1990), 77–94.

[9] G. Caginalp, X. Chen, *Convergence of the phase field model to its sharp interface limits*, European Journal of Applied Mathematics, **9** (1998), 417–445.

[10] G. Caginalp, X. Chen, *Phase field equations in the singular limit of sharp interface problems*, in: M. Gurtin and G.B. McFadden (eds.), *On the Evolution of Phase Boundaries*, IMA Volume of Mathematics and Its Applications, **43** (1992), 1-28, Springer-Verlag.

[11] B.D. Coleman, M.E. Gurtin, *Equipresence and constitutive equations for rigid heat conductors*, Z. Angew. Math. Phys., **18** (1967), 199–208.

[12] P. Colli, G. Gilardi, M. Grasselli, *Global smooth solution to the standard phase field models with memory*, Adv. Differential Equations, **2** (1997), 453–486.

[13] P. Colli, G. Gilardi, M. Grasselli, *Well-posedness of the weak formulation for the phase field models with memory*, Adv. Differential Equations, **2** (1997), 487–508.

[14] F. Colombo, *Direct and inverse problems for a phase-field model with memory*, J. Math. Anal. Appl., **260** (2001), 517–545.

[15] F. Colombo, D. Guidetti, V. Vespri, *An inverse parabolic problem for a phase-field model for boundary measurements of the temperature*, preprint.

[16] C. Giorgi, M. Grasselli, V. Pata, *Uniform attractors for a phase-field model with memory and quadratic nonlinearity*, Indiana Univ. Math. J., **48** (1999), 1395–1445.

[17] M. Grasselli, V. Pata, F.M. Vegni, *Longterm dynamics of a conserved phase-field system with memory*, Asymp. Analysis, **33** (2003), 261–320.

[18] R. Denk, M. Hieber, J. Prüss, *R-boundedness, Fourier multipliers and problems of elliptic and parabolic type*, preprint: Technische Universität Darmstadt, n. 2156 (2001).

[19] D. Guidetti, *On interpolation with boundary conditions*, Math. Zeit., **207** (1991), 439–460.

[20] D. Guidetti, A. Lorenzi, *A mixed identification problem related to a phase-field model with memory*, preprint.

[21] A. Lunardi, *Analytic semigroups and optimal regularity in parabolic problems*, Progress in Nonlinear Differential Equations and Their Applications, Birkhäuser (1995).

[22] A. Novick-Cohen, *The Cahn-Hilliard equation: mathematical and modeling perspectives*, Adv. Math. Sci. Appl., **8** (1998), 965–985.

[23] A. Novick-Cohen, *Conserved phase–field equations with memory*, in Curvature Flows and related Topics, 1994, GAKUTO Internat. Ser. Math. Sci. Appl., Vol 5, Gakkōtosho, Tokyo, 1995, pp. 179–197.

[24] A. Pazy, *Semigroups of linear operators and applications to partial differential equations*, Springer Verlag, New York, Berlin, Heidelberg, Tokyo, 1983.

[25] E. Sinestrari, *On the Cauchy problem in spaces of continuous functions,* J. Math. Anal. Appl., **107** (1985), 16–66.

[26] H.B. Stewart, *Generation of analytic semigroups by strongly elliptic operators*, Trans. Amer. Math. Soc., **199** (1974), 141–162.

[27] H. Triebel, *Theory of functions spaces*, Monogra. Math., Birkhäuser (1983).

[28] H. Triebel: *Interpolation theory, function spaces, differential operators,* North Holland, Amsterdam, New York, Oxford, 1978.

Fabrizio Colombo
Dipartimento di Matematica
Politecnico di Milano
Via Bonardi 9
I-20133 Milano, Italy

Davide Guidetti
Dipartimento di Matematica
Università di Bologna
Piazza di Porta S. Donato 5
I-40126 Bologna, Italy

Progress in Nonlinear Differential Equations
and Their Applications, Vol. 64, 211–225

Numerical Analysis of Microstructures: The Influence of Incompatibility

O. Anza Hafsa and M. Chipot

Abstract. We are concerned with the appearance of microstructures in some problems of Calculus of Variations experiencing "wells" or minimum of energy. Using piecewise linear finite elements, we give energy estimates and analyze their dependence on the incompatibility of the wells.

1. Introduction

We denote by $\mathbb{M}^{n\times N}$ the space of $n \times N$ matrices with real coefficients. Let K be a compact subset of matrices in $\mathbb{M}^{n\times N}$ and let φ be a nonnegative continuous function vanishing only on K, i.e., such that

$$\varphi(M) > 0 \ \forall M \notin K \ , \ \varphi(M) = 0 \ \forall M \in K. \tag{1}$$

It is well known that the quasiconvex envelope of φ is given by

$$Q\varphi(W) = \frac{1}{|\Omega|} \inf_{u\in W_0^{1,\infty}(\Omega;\mathbb{R}^N)} \int_\Omega \varphi(W + \nabla u(x))dx, \tag{2}$$

where $|\Omega|$ is the Lebesgue measure of Ω, ∇u denotes the gradient of u, i.e., the Jacobian matrix of the mapping

$$x \mapsto u(x)$$

from $\Omega \subset \mathbb{R}^n$ into $\mathbb{R}^N$, $W_0^{1,\infty}(\Omega;\mathbb{R}^N)$ is the space of $\mathbb{R}^N$-valued functions such that $\nabla u \in (L^\infty(\Omega))^{nN}$ and which vanish on the boundary $\Gamma = \partial\Omega$ of Ω. We refer the reader to [4] for these notions. The compatibility is described by the following definition.

Definition 1.1. *Two matrices $A, B \in \mathbb{M}^{n\times N}$ are said to be rank-one compatible if and only if*

$$\mathrm{Rk}(A - B) \leq 1.$$

Let us introduce the class of problems that we would like to address.

1.1. The continuous problem

We are interested in the following minimization problem

$$\inf_{u\in W_0^{1,\infty}(\Omega;\mathbb{R}^N)} \int_\Omega \varphi(\nabla u(x))dx. \tag{3}$$

Then we have:

Proposition 1.1. *Suppose that φ is a function satisfying* (1). *Then, if $0 \in K^{\ell c}$ where $K^{\ell c}$ denotes the laminate convex hull of K, the infimum* (3) *is 0.*

Proof. It follows from (2) that the infimum (3) is equal to 0 if and only if

$$Q\varphi(0) = 0. \tag{4}$$

From (1), (2), we have

$$Q\varphi(W) = 0 \ \forall W \in K.$$

The set $K^{\ell c}$ is the set of matrices obtained by making convex combinations of rank one compatible matrices starting with matrices in K (see [5]). Since $Q\varphi$ is rank-one convex, (see [4]), for any matrices W_1, W_2 rank-one compatible it holds that

$$0 \leq Q\varphi(\alpha W_1 + (1-\alpha)W_2) \leq \alpha Q\varphi(W_1) + (1-\alpha)Q\varphi(W_2).$$

If in particular $Q\varphi(W_1) = Q\varphi(W_2) = 0$, then it follows that

$$Q\varphi(\alpha W_1 + (1-\alpha)W_2) = 0.$$

The proposition follows then by iteration. □

1.2. The discrete problem

We will consider below an approximation to (3). Ω will be a polygonal domain of $\mathbb{R}^N$, $\{\tau_h\}_h$ a family of regular triangulations. The mesh size h is defined by

$$h = \max_{K\in\tau_h} h_K.$$

The approximation problem is

$$\inf_{v\in V_0^h} \int_\Omega \varphi(\nabla v(x))dx, \tag{5}$$

where

$$V_0^h = \{v \in V^h : v = 0 \text{ on } \partial\Omega\}$$

and

$$V^h = \{v = (v_1, \ldots, v_N) | v_i : \Omega \to \mathbb{R} \text{ is continuous, } v_i|_K \text{ is affine for } i = 1, 2 \text{ and } K \in \tau_h\}.$$

We would like to obtain, in simple cases of the proposition 1.1 different estimates of (5) in terms of the mesh size. We will restrict ourselves in this note to the case of 2×2 matrices and show somehow how the incompatibility between the wells slows down, for a given mesh size, the speed of convergence of the minimizing sequences of (3).

2. An example with three levels of laminations

A generic matrix $M \in \mathbb{M}^{2\times 2}$ is denoted by

$$M = \begin{pmatrix} x_{11} & x_{12} \\ x_{21} & x_{22} \end{pmatrix}.$$

In the x_{11}, x_{22} plane, we consider the following "wells" (element of K):

$$W_1 = \begin{pmatrix} 1/2 & 0 \\ 0 & 2 \end{pmatrix} \quad W_2 = \begin{pmatrix} 1 & 0 \\ 0 & -1 \end{pmatrix} \quad W_3 = \begin{pmatrix} -1 & 0 \\ 0 & -1 \end{pmatrix} \quad W_4 = \begin{pmatrix} -1 & 0 \\ 0 & 1 \end{pmatrix}.$$

From the above definition, it is easy to see that W_1 is not rank-one compatible with any other W_i, but (W_i, W_{i+1}) are rank-one compatible for $i = 2, 3$. Let $\varphi : \mathbb{M}^{2\times 2} \to [0; +\infty[$ be a nonnegative continuous function which satisfies (1) with

$$K = \{W_1, W_2, W_3, W_4\}. \tag{6}$$

Let us consider two intermediate matrices

$$W_1' = \begin{pmatrix} 1/2 & 0 \\ 0 & 1 \end{pmatrix} \quad W_2' = \begin{pmatrix} 1/2 & 0 \\ 0 & -1 \end{pmatrix},$$

which are represented below.

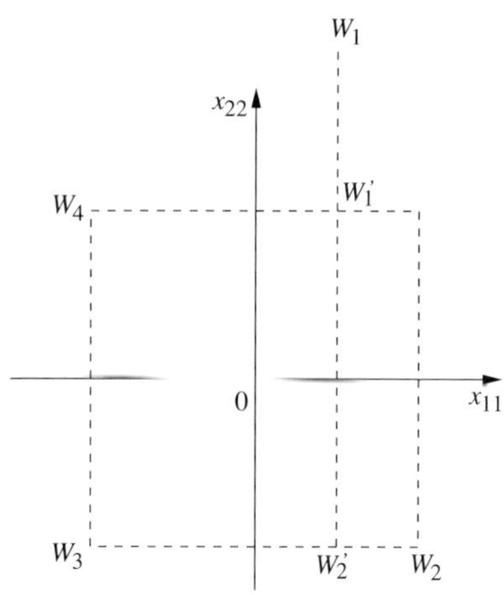

FIGURE 1

It is easy to see that $0 \in K^{\ell c}$ and thus in this case the infimum (3) is 0. In addition we have:

Proposition 2.1. *Under the above assumptions the infimum of (3) is not achieved.*

Proof. If u is a minimizer it holds that

$$\nabla u(\,\cdot\,) \in K \text{ a.e. in } \Omega.$$

If $u = (u_1, u_2)$ this implies that $\partial_{x_2} u_1 = \partial_{x_1} u_2 = 0$ and since u vanishes at the boundary $u = 0$. This is impossible since $0 \notin K$ and the proof is complete. □

In view of Proposition 2.1 we have to direct our attention on the minimizing sequences of (3). Regarding the discrete problem we have:

Theorem 2.1. *Under the above assumptions, there exists a constant $C > 0$ which does not depend on $h \ll 1$ such that*

$$\inf_{v \in V_0^h} \int_\Omega \varphi(\nabla v(x))dx \le Ch^{\frac{1}{4}}.$$

Proof. Let $\alpha \in]0;1[$. Let $v_h : [0; 2h^\alpha]^2 \to \mathbb{R}^2$ be a vector-valued function defined by $v_h = v = (v_1, v_2)$ where

$$v_1(x_1, x_2) = \begin{cases} x_1/2 & \text{if } x_1 \in [0; \frac{4}{3}h^\alpha], \\ 2h^\alpha - x_1 & \text{if } x_1 \in [\frac{4}{3}h^\alpha; 2h^\alpha], \end{cases} \tag{7}$$

$$v_2(x_1, x_2) = \begin{cases} x_2 & \text{if } x_2 \in [0; h^\alpha], \\ 2h^\alpha - x_2 & \text{if } x_2 \in [h^\alpha; 2h^\alpha]. \end{cases} \tag{8}$$

We have

$$\nabla v(x) \in \Big\{ W_1', W_2', W_3, W_4 \Big\} \text{ a.e. in } \Omega,$$

the repartition of the gradients being depicted on the Figure 2.

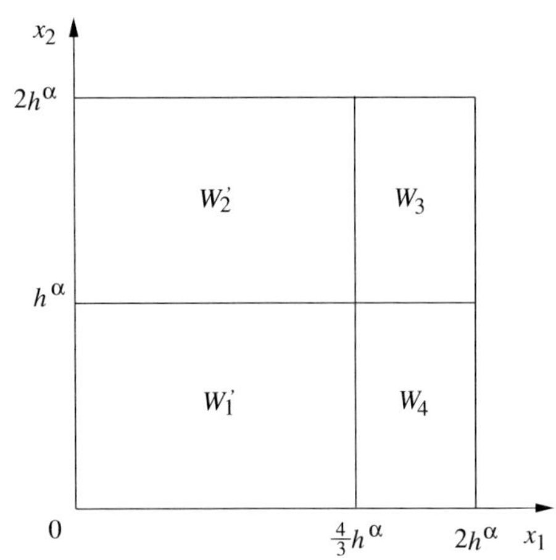

FIGURE 2

We remark that

$$W_2' = \frac{3}{4}W_2 + \frac{1}{4}W_3 \quad \text{and} \quad W_1' = \frac{2}{3}W_1 + \frac{1}{3}W_2'. \tag{9}$$

Our goal is to find a function w defined by

$$w(x) = \begin{cases} (w_1(x), w_2(x)) & \text{for } x \in [0; \frac{4}{3}h^\alpha] \times [0; h^\alpha], \\ (u_1(x), v_2(x)) & \text{for } x \in [0; \frac{4}{3}h^\alpha] \times [h^\alpha; 2h^\alpha], \\ v(x) & \text{otherwise,} \end{cases}$$

satisfying

$$\nabla w(x) \in \Big\{ W_1, W_2, W_3, W_4 \Big\} \text{ except on a small set of } [0; 2h^\alpha]^2.$$

Set

$$Q_1 = [0; \frac{4}{3}h^\alpha] \times [0; h^\alpha] \quad \text{and} \quad Q_2 = [0; \frac{4}{3}h^\alpha] \times [h^\alpha; 2h^\alpha].$$

Let $k, l \in \mathbb{N}^*$ to be chosen later on. We proceed in three steps.
Step 1. We modify the function v on Q_2. From (9) we have that it holds

$$W_2' = \frac{3}{4} W_2 + \frac{1}{4} W_3.$$

Then a lamination with W_2, W_3 could produce W_2'. Let χ be the periodic function of period 1 and defined on $]0; 1[$ by

$$\chi_\beta(t) = \begin{cases} 1 & \text{if } t \in]0; \beta], \\ 0 & \text{if } t \in]\beta; 1[, \end{cases}$$

with $\beta \in]0; 1[$. Let $\omega_1 : Q_2 \to \mathbb{R}$ be defined by

$$\omega_1(x) = \omega_1(x_1) = \int_0^{x_1} \Big(2\chi_{\frac{3}{4}}(\frac{3kt}{4h^\alpha}) - 1 \Big) dt.$$

It is easy to check that

$$\nabla \omega_1(x) = (-1, 0) \text{ or } (1, 0).$$

Let $\partial_2 Q_2 \subset \partial Q_2$ be defined by

$$\partial_2 Q_2 = \{ (x_1, x_2) \in Q_2 : x_2 = h^\alpha \text{ or } x_2 = 2h^\alpha \}.$$

One sets

$$u_1(x) = \omega_1(x) \wedge (v_1(x) + \text{dist}(x, \partial_2 Q_2)) \quad \text{for } x \in Q_2 \text{ and}$$
$$w(x) = (u_1(x), v_2(x)) \quad \text{for } x \in Q_2,$$

then $\nabla w \in \{W_2, W_3\}$ a.e. in Q_2 except on a small neighborhood of $\partial_2 Q_2$, when

$$\text{dist}(x, \partial_2 Q_2) < \frac{1}{2} \frac{h^\alpha}{k}.$$

Step 2. We modify the function v on Q_1. From (9) we have that

$$W_1' = \frac{2}{3} W_1 + \frac{1}{3} W_2'.$$

Similarly to the first step, it is easy to find a function $w_2 : Q_1 \to \mathbb{R}$ such that the function $w = (v_1, w_2)$ satisfies

$$\nabla w(x) \in \{ W_1, W_2' \} \text{ a.e. on } Q_1$$

except a small neighborhood of

$$\partial_1 Q_1 = \left\{(x_1, x_2) \in Q_1 : x_1 = 0 \text{ or } x_1 = \frac{4}{3}h^\alpha\right\}, \text{ i.e., when } \operatorname{dist}(x, \partial_1 Q_1) < \frac{2}{3}\frac{h^\alpha}{k}.$$

We can define $w_2 : Q_1 \to \mathbb{R}$ as

$$w_2(x) = \omega_2(x) \wedge (v_2(x) + \operatorname{dist}(x, \partial_1 Q_1)),$$

where

$$\omega_2(x) = \omega_2(x_2) = \int_0^{x_2} \left(3\chi_{\frac{2}{3}}(\frac{kt}{h^\alpha}) - 1\right) dt.$$

Step 3. Now, it remains to modify w on the strips

$$Q_1' = \bigcup_{p=0}^{k-1} \Sigma_p \text{ where } \Sigma_p = \left[0; \frac{4}{3}h^\alpha\right] \times \left(\left[\frac{2}{3}\frac{h^\alpha}{k}; \frac{h^\alpha}{k}\right] + \left\{p\frac{h^\alpha}{k}\right\}\right).$$

We choose the function ω_1 of the first step but with k replaced by l, i.e.,

$$\omega_1(x) = \omega_1(x_1) = \int_0^{x_1} \left(2\chi_{\frac{3}{4}}(\frac{3lt}{4h^\alpha}) - 1\right) dt.$$

We set on each strips Σ_p

$$u_1^p = \omega_1 \wedge (v_1 + \operatorname{dist}(., \partial_2 \Sigma_p)).$$

with

$$\partial_2 \Sigma_p = \left\{(x_1, x_2) \in Q_1' : x_2 = \frac{2}{3}\frac{h^\alpha}{k} + p\frac{h^\alpha}{k} \text{ or } \frac{h^\alpha}{k} + p\frac{h^\alpha}{k}\right\}.$$

Then we set

$$w(x) = \begin{cases} (u_1^p(x), w_2(x)) & \text{if } x \in \Sigma_p, \\ (v_1(x), w_2(x)) & \text{if } x \in Q_1 \setminus Q_1'. \end{cases}$$

We have $\nabla w(x) \in \{W_3, W_2\}$ a.e. on Q_1' except on a small neighborhood of $\cup\{\partial_2 \Sigma_p : 0 \le p \le k-1\}$ when

$$\operatorname{dist}(x, \partial_2 \Sigma_p) < \frac{1}{2}\frac{h^\alpha}{l}$$

for each $p = 0, \ldots, l-1$. We suppose that the following inequality holds

$$\frac{1}{2}\frac{h^\alpha}{l} < \frac{1}{6}\frac{h^\alpha}{k},$$

i.e., we choose $l > 3k$. Thus, we have

$$\nabla w(x) \in \{W_3, W_2, W_1\} \text{ a.e. on } Q_1$$

except a small neighborhood of $\partial_2 Q_1$ and $\cup\{\partial_2 \Sigma_p : 0 \le p \le k-1\}$.

Now, suppose that w is extended by periodicity of period $2h^\alpha$ in both directions x_1, x_2 (note that w is periodic on $[0, 2h^\alpha]^2$). In order to match the 0 boundary conditions one sets

$$\hat{w} = (w_1 \wedge \operatorname{dist}(., \Gamma), w_2 \wedge \operatorname{dist}(., \Gamma))$$

where Γ is the boundary of Ω and finally one defines

$$\tilde{w} = \tilde{w}_h \text{ the interpolant of } \hat{w} \text{ on } \tau_h.$$

Since $\nabla\tilde{w}$ is bounded – due to the regularity of the mesh – independently of h, we have

$$\int_\Omega \varphi(\nabla\tilde{w})dx \le C|\{x \in \Omega : \nabla\tilde{w}(x) \notin K\}| \tag{10}$$

where $|\ |$ is the Lebesgue measure of sets. Let us evaluate in terms of h the right-hand side of (10). The gradient of $\tilde{w}_h$ does not possibly belong to K on different subsets.

- On a tubular neighborhood S_1 of Γ with thickness of the order of h^α (see [1] p. 336-337).
- On a neighborhood S_2 of thickness $2h$ of the lines

  $$x_1 = 2ph^\alpha, \left(\frac{4}{3} + 2p\right) h^\alpha \text{ and } x_2 = ph^\alpha \text{ with } p \in \mathbb{Z}.$$

  (This is due to interpolation.)
- On a neighborhood S_3 of thickness

  $$\frac{1}{2}\frac{h^\alpha}{k} \text{ of } \partial_2 Q_2,\ \frac{2}{3}\frac{h^\alpha}{k} \text{ of } \partial_1 Q_1, \text{ and } \frac{1}{2}\frac{h^\alpha}{l} \text{ of } \partial_2\Sigma_p \text{ for } p = 0, \dots, k-1.$$

- On a neighborhood S_4 of thickness $2h$ of the lines
  - $x_1 = \left(1 + \dfrac{4p}{3}\right)\dfrac{h^\alpha}{k}, (1+p)\dfrac{4}{3}\dfrac{h^\alpha}{k}$ with $p \in \{0, \dots, k-1\}$ in Q_2;
  - $x_2 = \dfrac{2+3p}{3}\dfrac{h^\alpha}{k}, p\dfrac{h^\alpha}{k}$ with $p \in \{0, \dots, k-1\}$ in $Q_1 \setminus Q_1'$;
  - $x_1 = \left(1 + \dfrac{4p}{3}\right)\dfrac{h^\alpha}{l}, (1+p)\dfrac{4}{3}\dfrac{h^\alpha}{l}$ with $p \in \{0, \dots, l-1\}$ in Q_1'.

We have the following estimates:

i) $|S_1| \le Ch^\alpha, \quad |S_2| \le Ch^{1-\alpha}$;

ii) $|S_3| \le \dfrac{C}{h^{2\alpha}}\left(\dfrac{2}{3}\dfrac{h^\alpha}{k}h^\alpha + \dfrac{1}{2}\dfrac{h^\alpha}{k}\dfrac{4}{3}h^\alpha + k\dfrac{1}{2}\dfrac{h^\alpha}{l}\dfrac{4}{3}h^\alpha\right) \le C\left(\dfrac{1}{k} + \dfrac{k}{l}\right)$;

iii) $|S_4| \le \dfrac{C}{h^{2\alpha}}\left(k2hh^\alpha + k2h\dfrac{4}{3}h^\alpha + k(l2h\dfrac{1}{3}\dfrac{h^\alpha}{k})\right) \le C(k+l)h^{1-\alpha}$.

($C/h^{2\alpha}$ is the number of squares of size $2h^\alpha \times 2h^\alpha$ covering Ω.)

We deduce that

$$\int_\Omega \varphi(\nabla\tilde{w})dx \le C\Big\{h^\alpha + h^{1-\alpha} + (k+l)h^{1-\alpha} + \frac{1}{k} + \frac{k}{l}\Big\}.$$

The function $l \mapsto \frac{k}{l} + lh^{1-\alpha}$ reaches its minimum for $l = k^{\frac{1}{2}}h^{-\frac{1-\alpha}{2}}$. We obtain

$$\int_\Omega \varphi(\nabla\tilde{w})dx \le C\Big\{h^\alpha + h^{1-\alpha} + (kh^{1-\alpha})^{\frac{1}{2}} + kh^{1-\alpha} + \frac{1}{k}\Big\}.$$

Since $h \ll 1$ then $(kh^{1-\alpha})^{\frac{1}{2}} > kh^{1-\alpha}$ (see below) we get that

$$\int_\Omega \varphi(\nabla \tilde{w})dx \leq C\Big\{h^\alpha + h^{1-\alpha} + (kh^{1-\alpha})^{\frac{1}{2}} + \frac{1}{k}\Big\}.$$

The function $k \mapsto \frac{1}{k} + k^{\frac{1}{2}} h^{\frac{1-\alpha}{2}}$ is minimum for $k = 2^{-\frac{2}{3}} h^{-\frac{1-\alpha}{3}}$. It follows:

$$\int_\Omega \varphi(\nabla \tilde{w})dx \leq C\Big\{h^\alpha + h^{1-\alpha} + h^{\frac{1-\alpha}{3}}\Big\}.$$

It follows that the best possible α is

$$\alpha = \frac{1}{4},$$

which completes the proof. □

Remark 2.1. *The minimizing sequence that we constructed uses each of the wells with a fixed probability. It generates – as every bounded minimizing sequence in this case – a unique Young measure given by*

$$\nu_x = \frac{2}{9}\delta_{W_1} + \frac{1}{3}\delta_{W_2} + \frac{5}{18}\delta_{W_3} + \frac{1}{6}\delta_{W_4}, \tag{11}$$

where δ_{W_i} denotes the Dirac mass at W_i.

3. An example with four levels of laminations

In this section we consider the following four wells (see the figure below):

$$W_1 = \begin{pmatrix} 1/2 & 0 \\ 0 & 2 \end{pmatrix} \quad W_2 = \begin{pmatrix} 1 & 0 \\ 0 & -1 \end{pmatrix} \quad W_3 = \begin{pmatrix} -1 & 0 \\ 0 & -1 \end{pmatrix} \quad W_4 = \begin{pmatrix} -2 & 0 \\ 0 & 1/2 \end{pmatrix}.$$

We have:

Proposition 3.1. *The infimum* (3) *is equal to* 0*, moreover it is not achieved (i.e., the minimisation problem* (3) *does not have solutions).*

Proof. The proof is similar to the proof of Propositions 1.1 and 2.1. To see that $0 \in K^{\ell c}$ one has to consider the intermediate matrices

$$W_1' = \begin{pmatrix} 1/2 & 0 \\ 0 & 1/2 \end{pmatrix} \quad W_2' = \begin{pmatrix} 1/2 & 0 \\ 0 & -1 \end{pmatrix} \quad W_4' = \begin{pmatrix} -1 & 0 \\ 0 & 1/2 \end{pmatrix},$$

which are represented in Figure 3. □

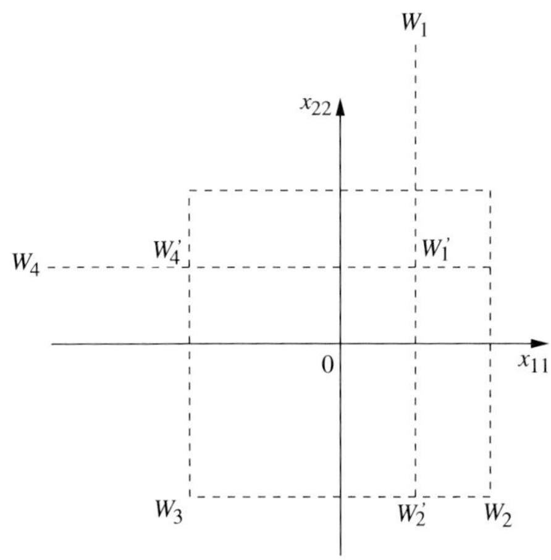

FIGURE 3

Regarding the discrete problem we have:

Theorem 3.1. *Under the above assumptions, there exists a constant $C > 0$ which does not depend on $h \ll 1$ such that*

$$\inf_{v \in V_0^h} \int_\Omega \varphi(\nabla v(x))dx \le Ch^{\frac{1}{5}}.$$

Proof. Let $v_h : [0; 2h^\alpha]^2 \to \mathbb{R}^2$ be a vector-valued function defined by (7) and

$$v_2(x_1, x_2) = v_1(x_2, x_1).$$

We have that

$$\nabla v(x) \in \left\{W_1', W_2', W_3, W_4'\right\} \text{ a.e. in } \Omega.$$

Let us denote

$$Q_1 = [0; \frac{4}{3}h^\alpha] \times [0; \frac{4}{3}h^\alpha],\ Q_2 = [0; \frac{4}{3}h^\alpha] \times [\frac{4}{3}h^\alpha; 2h^\alpha] \text{ and } Q_4 = [\frac{4}{3}h^\alpha; 2h^\alpha] \times [0; \frac{4}{3}h^\alpha]$$

We proceed in several steps. The following equalities give us the successive steps of modifications of the affine function v:

(i) $W_2' = \frac{3}{4}W_2 + \frac{1}{4}W_3$;
(ii) $W_1' = \frac{1}{2}W_1 + \frac{1}{2}W_2'$;
(iii) $W_4' = \frac{2}{5}W_1' + \frac{3}{5}W_4$.

Then one proceeds as follows:

Step 1: Second lamination

On Q_2 a lamination with W_2, W_3 replaces W_2';
On Q_1 a lamination with W_1, W_2' replaces W_1';
On Q_4 a lamination with W_1', W_4 replaces W_4';

Step 2: *Third lamination*
On Q_1 a lamination with W_2, W_3 replaces W_2';
On Q_4 a lamination with W_1, W_2' replaces W_1';

Step 3: *Fourth lamination*
On Q_4 a lamination with W_2, W_3 replaces W_2';

For the reader convenience the scheme is illustrated by the pictures below.

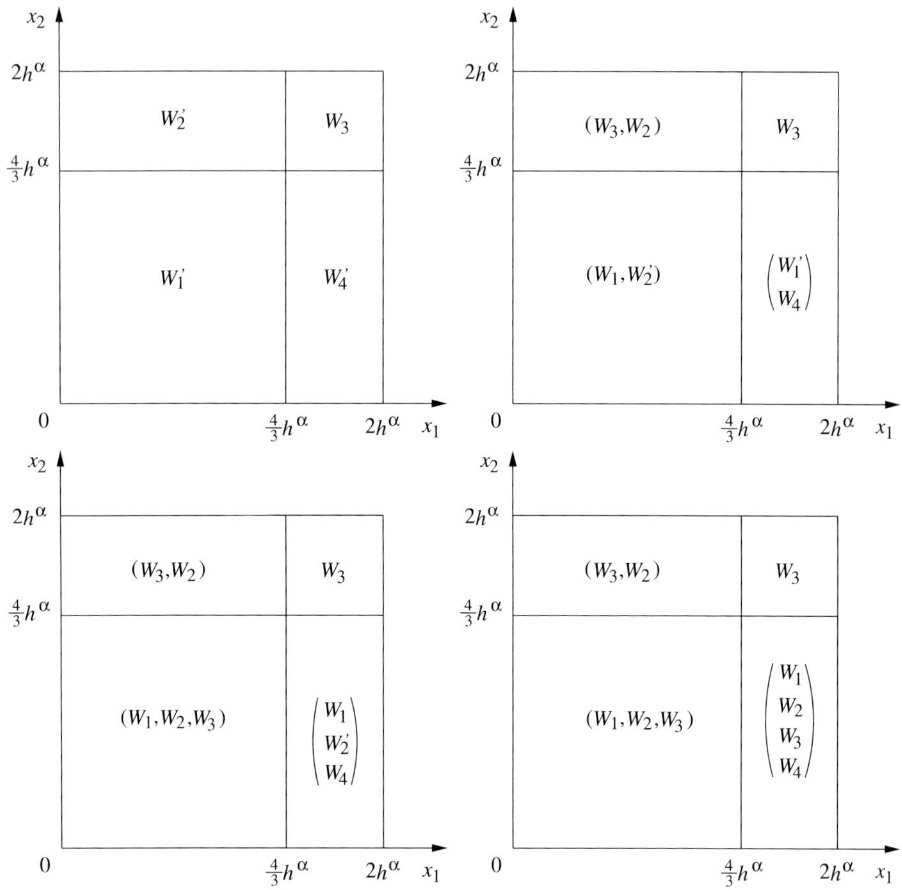

FIGURE 4

Let $k, l, m \in \mathbb{N}^*$ to be chosen later on.
Step 1: *Second lamination.*

1) On Q_2 a lamination with W_2, W_3 replaces W_2'. Set

$$\omega_1(x) = \omega_1(x_1) = \int_0^{x_1} \Big(2\chi_{\frac{3}{4}}\big(\frac{3kt}{4h^\alpha}\big) - 1\Big)\,dt.$$

Let $\partial_2 Q_2 \subset \partial Q_2$ be defined by

$$\partial_2 Q_2 = \Big\{(x_1, x_2) \in Q_2 : x_2 = \frac{4}{3}h^\alpha \text{ or } x_2 = 2h^\alpha\Big\}.$$

One sets for every $x \in Q_2$

$$u_1(x) = \omega_1(x) \wedge (v_1(x) + \operatorname{dist}(x, \partial_2 Q_2)) \text{ and,}$$
$$w(x) = (u_1(x), v_2(x)).$$

Then $\nabla w \in \{W_2, W_3\}$ a.e. in Q_2 except on a small neighborhood of $\partial_2 Q_2$, when

$$\operatorname{dist}(x, \partial_2 Q_2) < \frac{1}{2}\frac{h^\alpha}{k}.$$

2) On Q_1 a lamination with W_1, W_2' replaces W_1'. Set

$$\omega_2(x) = \omega_2(x_2) = \int_0^{x_2} \Big(3\chi_{\frac{1}{2}}(\frac{3kt}{4h^\alpha}) - 1\Big)dt.$$

Let $\partial_1 Q_1 \subset \partial Q_1$ be defined by

$$\partial_1 Q_1 = \Big\{(x_1, x_2) \in Q_1 : x_1 = 0 \text{ or } x_1 = \frac{4}{3}h^\alpha\Big\}.$$

One sets for every $x \in Q_1$

$$u_2(x) = \omega_2(x) \wedge (v_2(x) + \operatorname{dist}(x, \partial_1 Q_1)) \text{ and,}$$
$$w(x) = (v_1(x), u_2(x)).$$

Then $w = (v_1, u_2)$ satisfies

$$\nabla w(x) \in \{W_1, W_2'\} \text{ a.e. on } Q_1$$

except on a small neighborhood of $\partial_1 Q_1$, when

$$\operatorname{dist}(x, \partial_1 Q_1) < \frac{3}{4}\frac{h^\alpha}{k}.$$

3) On Q_4 a lamination with W_1', W_4 replaces W_4'. Set

$$\omega_{1,4}(x) = \omega_{1,4}(x_1) = \frac{2}{3}h^\alpha\Big(1 + \int_0^{\frac{3}{2}\frac{x_1}{h^\alpha} - 2} \frac{5}{2}\chi_{\frac{2}{5}}(kt) - 2dt\Big).$$

Let $\partial_2 Q_4 \subset \partial Q_4$ be defined by

$$\partial_2 Q_4 = \Big\{(x_1, x_2) \in Q_4 : x_2 = 0 \text{ or } x_2 = \frac{4}{3}h^\alpha\Big\}.$$

One sets for every $x \in Q_4$

$$u_1(x) = \omega_{1,4}(x) \wedge (v_1(x) + \operatorname{dist}(x, \partial_2 Q_4)) \text{ and,}$$
$$w(x) = (u_1(x), v_2(x)).$$

Then $w = (u_1, v_2)$ satisfies $\nabla w \in \{W_4, W_1'\}$ a.e. in Q_4 except on a small neighborhood of $\partial_2 Q_4$, when

$$\operatorname{dist}(x, \partial_2 Q_4) < \frac{3}{4}\frac{h^\alpha}{k}.$$

Step 2: Third lamination.

1) On Q_1 a lamination with W_2, W_3 replaces W_2';
We change the function w on the strips Q_1' defined by

$$Q_1' = \bigcup_{p=0}^{k-1} \Sigma_p \text{ where } \Sigma_p = [0; \frac{4}{3}h^\alpha] \times \Big\{ [\frac{4}{6}\frac{h^\alpha}{k}; \frac{4}{3}\frac{h^\alpha}{k}] + \{p\frac{4}{3}\frac{h^\alpha}{k}\} \Big\}.$$

Set

$$\omega_{1,1}(x) = \int_0^{x_1} \Big(2\chi_{\frac{3}{4}}(\frac{3lt}{4h^\alpha}) - 1 \Big) dt.$$

For each $p \in \{0, \ldots, k-1\}$ one sets $\partial_2 \Sigma_p \subset \partial Q_1'$ given by

$$\partial_2 \Sigma_p = \Big\{ (x_1, x_2) \in Q_1' : x_2 = \frac{4}{6}\frac{h^\alpha}{k} + p\frac{4}{3}\frac{h^\alpha}{k} \text{ or } x_2 = \frac{4}{3}\frac{h^\alpha}{k} + p\frac{4}{3}\frac{h^\alpha}{k} \Big\}.$$

We define for every $p = 0, \ldots, k-1$

$$u_1^p(x) = \omega_{1,1}(x) \wedge (v_1(x) + \operatorname{dist}(x, \partial_2\Sigma_p) \quad \text{for } x \in \Sigma_p,$$

and we change w like

$$w(x) = \begin{cases} (u_1^p(x), w_2(x)) & \text{if } x \in \Sigma_p \\ (v_1(x), w_2(x)) & \text{if } x \in Q_1 \setminus Q_1'. \end{cases}$$

Then $\nabla w \in \{W_2, W_3\}$ a.e. in Q_2 except on a small neighborhood of $\partial_2\Sigma_p$

$$\operatorname{dist}(x, \partial_2\Sigma_p) < \frac{1}{2}\frac{h^\alpha}{l}$$

for each $p = 0, \ldots, l-1$.

2) On Q_4 a lamination with W_1, W_2' replaces W_1';
We change the function w on the strips Q_4' defined by

$$Q_4' = \bigcup_{p=0}^{k-1} \Delta_p \text{ where } \Delta_p = \Big\{ [\frac{4}{3}h^\alpha; \frac{4}{3}h^\alpha + \frac{4}{15}\frac{h^\alpha}{k}] + \{p\frac{2}{3}\frac{h^\alpha}{k}\} \Big\} \times [\frac{4}{3}h^\alpha; 2h^\alpha].$$

We do l laminations on each strips Δ_p with a similar function as Step 1. 2) to produce a function $\omega_{2,4}$ which satisfies

$$\nabla\omega_{2,4} = (0, -1) \text{ or } (0, 2) \text{ on } \Delta_p.$$

For each $p \in \{0, \ldots, k-1\}$ one sets $\partial_1\Delta_p \subset \partial Q_1'$ given by

$$\partial_1\Delta_p = \Big\{ (x_1, x_2) \in Q_4' : x_1 = \frac{4}{3}h^\alpha + p\frac{2}{3}\frac{h^\alpha}{k} \text{ or } x_1 = \frac{4}{3}h^\alpha + \frac{4}{15}\frac{h^\alpha}{k} + p\frac{2}{3}\frac{h^\alpha}{k} \Big\}.$$

We define for every $p = 0, \ldots, k-1$

$$u_2^p(x) = \omega_{2,4}(x) \wedge (v_2(x) + \operatorname{dist}(x, \partial_1\Delta_p) \quad \text{for } x \in \Delta_p,$$

and we change w like

$$w(x) = \begin{cases} (u_1(x), u_2^p(x)) & \text{if } x \in \Delta_p, \\ (u_1(x), v_2(x)) & \text{if } x \in Q_4 \setminus Q_4'. \end{cases}$$

Then $\nabla w \in \{W_1, W_2', W_4\}$ a.e. in Q_4 except on a small neighborhood of $\partial_1 \Delta_p$

$$\operatorname{dist}(x, \partial_1 \Delta_p) < \frac{3}{4}\frac{h^\alpha}{l}$$

for each $p = 0, \dots, k-1$.

Step 3: Fourth lamination
On Q_4 a lamination with W_2, W_3 replaces W_2';
We change the function w on the strips of the form

$$I_{p,r} = \left\{\left[\frac{4}{3}h^\alpha; \frac{4}{3}h^\alpha + \frac{4}{15}\frac{h^\alpha}{k}\right] + \left\{p\frac{2}{3}\frac{h^\alpha}{k}\right\}\right\} \times \left\{\left[\frac{2}{3}\frac{h^\alpha}{l}; \frac{4}{3}\frac{h^\alpha}{l}\right] + \left\{r\frac{4}{3}\frac{h^\alpha}{k}\right\}\right\} \subset \Delta_p,$$

where $p = 0, \dots, k-1$ and $r = 0, \dots, l-1$. We do m laminations on each rectangle $I_{p,r}$ with a similar function as in Step 1. 1) to obtain a function $\omega_{1,4}^{p,r}$ which satisfies

$$\nabla \omega_{1,4}^{p,r} = (-1, 0) \text{ or } (1, 0) \text{ on } I_{p,r}$$

For each $p \in \{0, \dots, k-1\}$ and $r \in \{0, \dots, l-1\}$ one sets $\partial_2 I_{p,r} \subset \partial \Delta_p$ given by

$$\partial_2 I_{p,r} = \left\{(x_1, x_2) \in \partial I_{p,r} : x_2 = \frac{2}{3}\frac{h^\alpha}{l} + r\frac{4}{3}\frac{h^\alpha}{k} \text{ or } x_2 = \frac{4}{3}\frac{h^\alpha}{l} + r\frac{4}{3}\frac{h^\alpha}{k}\right\}.$$

We define for every $p = 0, \dots, k-1$ and $r = 0, \dots, l-1$

$$u_1^{p,r}(x) = \omega_{1,4}^{p,r}(x) \wedge (u_1(x) + \operatorname{dist}(x, \partial_2 I_{p,r}) \quad \text{for } x \in I_{p,r},$$

and we change w like

$$w(x) = \begin{cases} (u_1^{p,r}(x), u_2^p(x)) & \text{if } x \in I_{p,r}, \\ (u_1(x), u_2^p(x)) & \text{otherwise in } \Delta_p. \end{cases}$$

Then $\nabla w \in \{W_1, W_2, W_3\}$ a.e. in Δ_p except on a small neighborhood of $\partial_2 I_{p,r}$ for which it holds

$$\operatorname{dist}(x, \partial_2 I_{p,r}) < \frac{1}{2}\frac{h^\alpha}{mk}$$

for each $p = 0, \dots, k-1$ and $r = 0, \dots, l-1$.

We extend then w by periodicity of period $2h^\alpha$ in both directions x_1, x_2. In order to match the 0 boundary conditions we set

$$\hat{w} = (w_1 \wedge \operatorname{dist}(., \Gamma), w_2 \wedge \operatorname{dist}(., \Gamma))$$

and finally one defines

$$\tilde{w} = \tilde{w}_h \text{ the interpolant of } \hat{w} \text{ on } \tau_h.$$

Since $\nabla\tilde{w}$ is bounded independently of h, then

$$\int_\Omega \varphi(\nabla\tilde{w})dx \leq C|\{x \in \Omega : \nabla\tilde{w}(x) \notin K\}|. \tag{12}$$

Let us evaluate in terms of h the right-hand side of (12). The gradient of $\tilde{w}_h$ does possibly not belong to K on different subsets (see the previous section). As before we have:

$$|S_1| \leq Ch^\alpha, \quad |S_2| \leq Ch^{1-\alpha}.$$

Let us now examine more closely our lamination process. Suppose that we are laminating a cell of size $a \times b$, say $[x_0; x_0 + a] \times [y_0; y_0 + b]$, in the x_1 direction. If we split this cell in k the gradient of the laminate does not belong to the good set:

- on a domain near the lateral border of size $a.\frac{a}{k}$,
- on strips of width h, i.e., on a domain of size khb.

The size of the total domain is of the order of magnitude:

$$\Big(a.\frac{a}{k} + khb\Big) \times \# \text{ of cells}$$

(i.e., # of cells that we are laminating). In the process above we have laminated cells of size $h^\alpha \times h^\alpha$ (in fact $c_1h^\alpha \times c_2h^\alpha$ but these constants are playing no rôle here.) Thus the set of bad gradients which will always remain after the first lamination is of size

$$\frac{C}{h^{2\alpha}}\Big(h^\alpha.\frac{h^\alpha}{k} + khh^\alpha\Big) = C\Big(\frac{1}{k} + kh^{1-\alpha}\Big).$$

We then laminate in l, k cells of size $h^\alpha \times \frac{h^\alpha}{k}$. This generates an unremovable set of bad gradients of size:

$$\frac{C}{h^{2\alpha}}k\Big(h^\alpha.\frac{h^\alpha}{l} + lh\frac{h^\alpha}{k}\Big) = C\Big(\frac{k}{l} + lh^{1-\alpha}\Big).$$

Finally we laminate lk cells of size $\frac{h^\alpha}{k} \times \frac{h^\alpha}{l}$ in m and this produces a set of bad gradients of size:

$$\frac{C}{h^{2\alpha}}lk\Big(\frac{h^\alpha}{k}.\frac{h^\alpha}{km} + mh\frac{h^\alpha}{l}\Big) = C\Big(\frac{l}{km} + mkh^{1-\alpha}\Big).$$

We deduce that

$$\int_\Omega \varphi(\nabla\tilde{w})dx \leq C\Big\{h^\alpha + h^{1-\alpha} + \frac{1}{k} + \frac{k}{l} + \frac{l}{mk} + (k + l + mk)h^{1-\alpha}\Big\}.$$

The function $m \mapsto mkh^{1-\alpha} + \frac{l}{mk}$ reaches its minimum for $m = l^{\frac{1}{2}}k^{-1}h^{-\frac{1-\alpha}{2}}$. It follows that

$$\int_\Omega \varphi(\nabla\tilde{w})dx \leq C\Big\{h^\alpha + \frac{1}{k} + \frac{k}{l} + kh^{1-\alpha} + lh^{1-\alpha} + (lh^{1-\alpha})^{\frac{1}{2}}\Big\}.$$

Since $h \ll 1$ if l is chosen such that $lh^{1-\alpha} < (lh^{1-\alpha})^{\frac{1}{2}}$ (see below) we get

$$\int_\Omega \varphi(\nabla\tilde{w})dx \leq C\Big\{h^\alpha + \frac{1}{k} + \frac{k}{l} + kh^{1-\alpha} + (lh^{1-\alpha})^{\frac{1}{2}}\Big\}.$$

The function $l \mapsto (lh^{1-\alpha})^{\frac{1}{2}} + \frac{k}{l}$ is minimum for $l = (2k)^{\frac{2}{3}} h^{-\frac{1-\alpha}{3}}$. It follows

$$\int_\Omega \varphi(\nabla \tilde{w})dx \leq C\Big\{h^\alpha + \frac{1}{k} + k^{\frac{1}{3}} h^{\frac{1-\alpha}{3}}\Big\}.$$

The function $k \mapsto k^{\frac{1}{3}} h^{\frac{1-\alpha}{3}} + \frac{1}{k}$ is minimum for $k = 3^{\frac{3}{4}} h^{-\frac{1-\alpha}{4}}$ and we obtain

$$\int_\Omega \varphi(\nabla \tilde{w})dx \leq C\Big\{h^\alpha + h^{\frac{1-\alpha}{4}}\Big\}.$$

The best constant α is $\alpha = \frac{1-\alpha}{4}$ i.e., $\alpha = \frac{1}{5}$. This completes the proof. □

Remark 3.1. *It would be interesting to check the sharpness of these estimates. For results in this direction see* [3], [6].

Remark 3.2. *Our estimates give direct proofs of the Propositions* 2.1 *and* 3.1.

Acknowledgements

This paper was written while the first author was visiting the Institute of Mathematics of the University of Zürich. The financial support provided by the Institute, as well as the support of the programme HMS2000, is gratefully acknowledged. The second author is supported by the Swiss National Science Foundation under the contracts # 20-67618.02 and 20-103300/1.

References

[1] M. Chipot, The appearance of microstructures in problems with incompatible wells and their numerical approach, Numerische Math. **83** (1999): 325–352.

[2] M. Chipot, *Elements of nonlinear analysis*, Birkhäuser Advanced Texts 2000.

[3] M. Chipot, S. Müller, *Sharp Energy Estimates For Finite Element Approximations of Non-convex Problems.* Proceedings of a IUTAM Symposium on Variations of Domains and Free-Boundary Problems in Solid Mechanics, P. Argoul, M. Frémond, Q.S. Nguyen Eds., Kluwer, (1999), p. 317–325.

[4] B. Dacorogna, *Direct methods in the calculus of variations*, Springer, 1989.

[5] G. Dolzmann, *Variational Methods for Crystalline Microstructure-Analysis and Computation*, Springer Lecture Notes in Mathematics # 1803, Springer Berlin, 2003.

[6] A. Lorent, An optimal scaling law for finite element approximations of a variational problem with non-trivial microstructure, M2AN. **5** (2001): 921–934.

O. Anza Hafsa and M. Chipot
University of Zurich
Angewandte Mathematik
Winterthurerstr. 190
CH-8057 Zurich

Progress in Nonlinear Differential Equations
and Their Applications, Vol. 64, 227–242

Nonlinear Singular Elliptic Problems: Recent Results and Open Problems

Jesús Hernández, Francisco J. Mancebo and José M. Vega

To Herbert Amann, with admiration, acknowledgement and friendship

1. Introduction

Reaction-diffusion equations and systems have been studied very extensively in the last years as models for many problems arising in applications as, e.g., population dynamics (Lotka-Volterra systems) combustion, chemically reacting systems, etc. A typical example is

$$\frac{\partial u}{\partial t} - \Delta u = f(x,u) \quad \text{in } \Omega\times]0,+\infty[\tag{1.1}$$

where Ω is a bounded domain in $\mathbb{R}^n$, Δ is the Laplacian, which models diffusion and the nonlinearity f represents the reaction term. This evolution equation must be supplemented with appropriate boundary and initial conditions. Of course, this equation admits much more general versions, but it still retains most of the interesting features arising along this theory. The reader may consult the references in [32]. Stationary solutions of this parabolic equation are very important, since they correspond to equilibria solutions, and its existence and multiplicity have been considered in detail as well.

Very often only positive solutions are the relevant ones, a fact which is very closely related with the meaning of the unknowns (concentrations, populations, etc.). Usually the nonlinearity is a smooth (actually C^1 or at least locally Lipschitz) function such that $f(0) \geq 0$ but functions like $f(u) = u^\alpha$ with $0 < \alpha < 1$ are considered in the literature, since they arise in fractional order chemical reactions or nonlinear diffusion models, see the book [15] and the references therein. Also nonlinearities defined for $u \in]0,+\infty[$ but going to infinity when u goes to zero are of interest in situations as chemical heterogeneous catalysts or non-Newtonian fluids, besides its intrinsic mathematical interest. A model example of this kind is

$$\begin{aligned} -\Delta u &= \frac{\lambda}{u^\alpha} \quad \text{in } \Omega \\ u &= 0 \quad \text{on } \partial\Omega, \end{aligned} \tag{1.2}$$

where $\alpha > 0$ and $\lambda > 0$ is a real parameter. Problems of this type, for monotone decreasing nonlinearities, were treated by Fulks and Maybee in the pioneering paper [18], a fact which is really remarkable if one takes into account that the considerable development in the study of reaction-diffusion equations and systems only started in the late sixties. They proved existence and uniqueness for the parabolic problem by using a fixed point argument for the associated integral equation and also existence and uniqueness for the stationary problem; moreover solutions of the parabolic problem tend to the unique solution of the elliptic one. As usual, the one-dimensional case was studied by using ordinary differential equations methods that cannot be extended to the case of a general bounded domain (see, e.g., [19]). It is clear that, due to the singular character of the problem, only positive solutions are to be considered.

The general case was studied later by different authors by replacing the original problem by an approximate one that is regular, where the usual techniques (sub- and super-solutions, bifurcation) can be applied and then go to the limit to obtain the solution to the singular problem. This was done first by Stuart in [33] by replacing the homogeneous Dirichlet boundary condition by (positive) nonhomogeneous ones.

Then Crandall, Rabinowitz and Tartar in [12] proved several interesting results. First, they obtain very general existence results (valid for any $\alpha > 0$ in the case of the model example (1.2) and for differential operators not necessarily in divergence form) by an approximation argument using either sub- and super-solutions in the monotone case or global bifurcation (and a more sophisticated convergence argument for continua) in the general one. They only look for classical solutions in the space $C^2(\Omega) \cap C(\overline{\Omega})$ and do not try to get optimal regularity results. However it is shown that if $\alpha > 1$ the solutions are Hölder continuous (more precisely in $C^{\frac{2}{1+\alpha}}(\overline{\Omega})$) and a necessary and sufficient condition for solutions to be Lipschitz up to the boundary (which amounts to $0 < \alpha < 1$ in the case model) is obtained in [12] Theorem 3.25. Simpler proofs of some of these results were obtained by Lazer and McKenna in [28], where they studied the behavior of solutions close to the boundary and also showed that classical solutions are in the Sobolev space $H_0^1(\Omega)$ if and only if $0 < \alpha < 3$.

Several improvements concerning both existence and regularity were obtained by Gomes [21] and del Pino [14]. And some general regularity results were obtained by Gui and Hua-Lin [22]. We also mention existence results by using approximation and then sub- and super-solutions due to Coclite ([9]–[11]), see also [30], [31]. Existence of weak solutions in the Sobolev space $H_0^1(\Omega)$ has been studied in [35]. Finally Ghergu and Radulescu [20] and Aranda and Lami-Dozo [3] use asymptotic bifurcation to prove existence, and the latter get some (partial) uniqueness as well.

Other existence results may be found in [34]. Multiplicity results can be found in [27], [35] and [23]. In a recent interesting paper Dávila and Montenegro [13] study both positive and dead core solutions for the singular problem in Application 4 below. Apparently the techniques used there can be applied elsewhere.

Recently the authors of this paper have extended the method of sub- and super-solutions to these problems (see [24]). First, they proved a general existence theorem for maximal and minimal positive solutions in the interval between ordered sub- and super-solutions. We intended to get a theorem as similar as possible to the well-known theorems in the regular case (see, e.g., [1]) which could be applied directly avoiding any kind of approximation procedure. To obtain this generalization, we have needed some additional conditions are imposed to sub- and super-solutions (see Remark 2.1). We allow operators not necessarily in divergence form, where even some coefficients may be singular. In order to implement this program it was necessary to have suitable existence, uniqueness and regularity results for linear problems with singular data (regularity is important in order to get compactness of the solution operator) and to extend the Strong Maximum Principle when the coefficients are singular. These results are still useful for studying the existence of principal eigenvalues for singular linear eigenvalue problems (now in the sense of having unbounded coefficients close to the boundary) extending to the case of general operators the results proved by Berstsch and Rostamian in [5] in the variational case by using variational arguments, weighted Sobolev spaces and the Hardy-Sobolev inequality. We also obtained as corollaries of the same general theorems the differentiability of the solution (or Green's) operator in the interior of the positive cone (i.e., functions $u > 0$ in Ω and $\dfrac{\partial u}{\partial n} < 0$ on $\partial\Omega$) and a linearized stability theorem again as in the regular case. In a subsequent paper we apply the results in [24] to many examples, extending and unifying in this way most of the results in the literature. Most of the above-mentioned results will be reviewed in the survey paper [26].

The paper is organized as follows. After this introduction, we collect in Section 2 the main results in [24], which will be applied in Section 3. Full proofs and many related results and comments can be found in [24], [26] and its references. In Section 3 we consider a few selected applications exhibiting some interesting features (see [25], [26] for more details).

2. Main general theorems

In this section we collect, usually without proof, the main general results which will be applied to the problems considered in Section 3. Most of them are taken from [24], where the reader will also find complements and comments. We have also included here several results which are more or less easy corollaries of the main theorems in [24] we consider of independent interest in order to be used in Section 3 and elsewhere.

We deal with the semilinear elliptic problem

$$Lu \equiv -\sum_{i,j=1}^{n} a_{ij}(x)\frac{\partial^2 u}{\partial x_i \partial x_j} + \sum b_i(x)\frac{\partial u}{\partial x_i} = f(x,u) \quad \text{in } \Omega \tag{2.1}$$

$$u = 0 \qquad \text{on} \quad \partial\Omega \tag{2.2}$$

under the following assumptions, which hold for some fixed α such that $-1 < \alpha < 1$.

(H.1) $\Omega \subset \mathbb{R}^n$ is a bounded domain, with a $C^{3,\gamma}$ boundary, for some $\gamma > 0$ if $n > 1$. Note that the distance from $x \in \Omega$ to $\partial\Omega$, $d(x)$, defines a function $d \in C^{2,\gamma}(\bar{\Omega}_1)$, with $\Omega_1 = \{x \in \Omega : d(x) < \rho_1\}$ for some $\rho_1 > 0$.

(H.2) *The second-order part of the operator $-L$ is uniformly, strongly elliptic in Ω. Also, for all $i, j, k = 1, \ldots, n$, $a_{ij} = a_{ji} \in C^3(\Omega) \cap C(\bar{\Omega})$, $b_i \in C^2(\Omega)$, and there is a constant K such that $|\partial a_{ij}/\partial x_k| + |b_i| < K[1 + d(x)^\alpha]$ and $|\partial^2 a_{ij}/\partial x_i \partial x_j| + |\partial b_i/\partial x_j| < Kd(x)^{\alpha-1}$ for all $x \in \Omega$.* As a consequence, the functions a_{ij}, $x \to d(x)\partial a_{ij}(x)/\partial x_k$ and $x \to d(x)b_i(x)$ are in $C^{0,\delta}(\bar{\Omega})$ whenever $0 < \delta < \min\{\alpha + 1, \gamma\}$.

(H.3) *There is an integer $m > 0$ such that f, $\partial^j f/\partial u^j$, $\partial^j f/\partial u^{j-1}\partial x_k \in C(\Omega \times]0, \infty[)$ for all $k = 1, \ldots, n$ and all $j = 1, \ldots, m+1$. And if $u : \Omega \to \mathbb{R}$ is such that $0 < k_1 d(x) < u(x) < k_2 d(x)$ for all $x \in \Omega$ and some constants k_1 and k_2, then $|f(x, u(x))| < K_0[1 + d(x)^\alpha]$ and $|\partial^j f(x, u(x))/\partial u^j| + \sum_{k=1}^{n} |\partial^j f(x, u(x))/\partial u^{j-1}\partial x_k| < K_j d(x)^{\alpha-j}$ for all $x \in \Omega$, all $k = 1, \ldots, n$ and all $j = 1, \ldots, m+1$, where K_j (can depend on k_1 and k_2 but) is independent of u .*

As the reader can see, (H.2) allows some singularity for the coefficients b_i at the boundary. We also require the coefficients in L to be such as the formal adjoint operator L^* defined as

$$L^* u \equiv -\sum \frac{\partial}{\partial x_i}\left(a_{ij}\frac{\partial u}{\partial x_j}\right) - \sum \frac{\partial}{\partial x_i}\left[\left(b_i - \sum \frac{\partial a_{ij}}{\partial x_j}\right)u\right]$$

also satisfies (H.2). Since all sums apply to the values $1, \ldots, n$ of the involved indexes, the limits 1 and n are omitted hereafter in the symbol $\sum$. Note that assumption (H.3) is satisfied, for example, by the usual power-law nonlinearities $f(x, u) = g(x)u^{\alpha_1}$ if $\alpha_1 > -1$ and $g \in C^1(\bar{\Omega})$ or, more generally, for $g \in C^1(\Omega)$ satisfying $|g(x)| \le kd(x)^{\alpha_2}$ for some $k > 0$ and some α_2 such that $|\alpha_1 + \alpha_2| < 1$.

Our main existence theorem is the following (Theorem 4.1 in [24]).

Theorem 2.1. *Under the assumptions* (H.1)–(H.3), *let us assume that* (2.1), (2.2) *has a sub-solution u_0 and a super-solution u^0 such that u_0, $u^0 \in C^2(\Omega) \cap C_0^{1,\delta}(\bar{\Omega})$ for some $\delta > 0$ and $0 < kd(x) < u_0(x) \le u^0(x)$ for all $x \in \Omega$. Then* (2.1), (2.2) *possesses a minimal and a maximal solution in the interval $[u_0, u^0]$, u_* and u^*, which are such that $u_*, u^* \in C^2(\Omega) \cap C_0^{1,\delta}(\bar{\Omega})$ whenever $0 < \delta < \delta_0 = \min\{\gamma, \alpha+1\}$ and $u_0 \le u_* \le u^* \le u^0$ in Ω. Also, u_* (resp. u^*) is the $C_0^{1,\delta}(\Omega)$-limit from below* *(resp. from above) of a monotone sequence of sub-solutions (resp. super-solutions) of* (2.1) (2.2). *Moreover, we have* $\frac{\partial u^*}{\partial n} < 0$, $\frac{\partial u_*}{\partial n} < 0$ *on* $\partial\Omega$.

As it was pointed out after Theorem 4.1 in [24], the condition $u_0 > kd(x) > 0$ is often too stringent for applications. The Lemma 4.2 in [24] is instrumental in order to extend the applicability of Theorem 2.1 above.

Remark 2.1. *Theorem* 2.1 *provides an existence result for positive solutions in the interval between ordered sub- and super-solutions. In this sense it is similar to well-known results for nonsingular elliptic problems (see, e.g.,* [1] *and its references). However, the singularity forces us to deal only with positive solutions. Moreover, the additional assumption* $u_0 = u^0 = 0$ *on* $\partial\Omega$ *is made here, and also we impose* $u_0 > kd(x) > 0$ *in* Ω.

Remark 2.2. *Concerning the regularity of solutions given by Theorem* 2.1, *we have proved that they are in* $C_0^{1,\delta}(\bar{\Omega})$ *with* $\delta < \delta_0 = \min\{\gamma, \alpha+1\}$, *a result that coincides with the one obtained in* [22] *for a particular nonlinearity. Other (sometimes nonoptimal) regularity results were given in* [12] *(where the case* $\alpha < -1$ *was also studied),* [21], [28], [14] *and* [9], [10], [35] *and* [34]. *Notice that we work from the beginning in these spaces, instead of getting some kind of weak solutions and then improve their smoothness by using some regularity techniques. In particular by the Strong Maximum principle in Appendix B in* [24] $\frac{\partial u}{\partial n} < 0$ *on* $\partial\Omega$. *But examples due to Ouyang, Shi and Yao* [29] *show that there are solutions such that* $\frac{\partial u}{\partial n} = 0$ *on* $\partial\Omega$. *The above references provide more information concerning the boundary behavior of solutions.*

The proof of Theorem 2.1 follows the usual pattern for results of this type whose main ingredients are:

i) an existence and uniqueness result for the associated linear problem giving the iterates converging to the (maximal and minimal) solutions;
ii) the compactness of the solution operator for the linear problem in i);
iii) a comparison result (following from a Maximum Principle) for the same problem.

In our case, we have stated an existence and uniqueness theorem for the linear problem

$$\begin{aligned} Lu &= M(x)v \quad \text{in } \Omega \\ u &= 0 \qquad\quad\ \ \text{on } \partial\Omega, \end{aligned} \tag{2.3}$$

where $M(x)$is a weight function satisfying the natural assumption

(H.3′) $M \in C^1(\Omega)$ and for all $k = 1, \ldots, n$, $d(x)^{2-\alpha}|\partial M(x)/\partial x_k|$ is bounded in Ω. As a consequence, the function $x \to d(x)^2 M(x)$ is in $C^{0,\delta}(\bar{\Omega})$ whenever $0 < \delta < \min\{\gamma, 1+\alpha\}$ and $d(x)^{1-\alpha}M(x)$ is bounded in Ω.

The natural character of assumption (H.3′) comes out from the fact that if we consider the linearized problem along the positive solution u we obtain

$$\begin{aligned} LU - M(x)U &= \lambda U && \text{in } \Omega \\ U &= 0 && \text{on } \partial\Omega, \end{aligned} \tag{2.4}$$

where $M(x) = f_u(x, u(x))$.

Proposition 2.1. (Proposition 2.3 in [24]) *Let Ω, $\mathcal{L}$ and M satisfy assumptions* (H.1)–(H.2) *and* (H.3′). *If $v \in C_0^{0,1}(\bar{\Omega})$ then* (2.3) *has a unique solution $u \in C^2(\Omega) \cap C^{1,\delta}(\bar{\Omega})$ for all δ such that $0 < \delta < \delta_0 = \min\{\gamma, \alpha + 1\}$ and there is a constant $K > 0$, which* (can depend on δ but) *is independent of v, such that*

$$\|u\|_{C^{1,\delta}(\bar{\Omega})} \leq K \|v\|_{C^{0,1}(\bar{\Omega})}. \tag{2.5}$$

Proof. See Appendix A in [24].

We also add a special instance of Proposition 2.1 which will be useful in Section 3.

Corollary 2.1. *Under the assumptions of Proposition* 2.1, *if $0 < \beta < 1$, there exists a unique solution $u \in C^2(\Omega) \cap C^{1,\delta}(\bar{\Omega})$ for any $0 < \delta < \delta_0 = \min\{\gamma, 1 - \beta\}$ of the problem*

$$\begin{aligned} Lu &= \frac{1}{d(x)^\beta} && \text{in } \Omega \\ u &= 0 && \text{on } \partial\Omega. \end{aligned} \tag{2.6}$$

Moreover, we have $\dfrac{\partial u}{\partial n} < 0 \quad$ *on* $\partial\Omega$.

The necessary extension of the Strong Maximum Principle to our singular problem was obtained in the Appendix B in [24]; we will come back to this point below. And the compactness of the solution operator to (2.3) follows readily from estimate (2.5) (See the proof of Theorem 2.6 in [24].) This property is also instrumental for the extension of classical results concerning existence and properties of principal eigenvalues to the case of singular coefficients. We actually consider the more general linear eigenvalue problem

$$\begin{aligned} LU - M(x)U &= \lambda N(x)U && \text{in } \Omega \\ U &= 0 && \text{on } \partial\Omega, \end{aligned} \tag{2.7}$$

where N satisfies

(H.4) N is (strictly) positive in Ω and satisfies (H.3′).

Theorem 2.2. (Theorem 2.6 in [24]) *Under the assumptions* (H.1)–(H.2), (H.3′) *and* (H.4), *the spectrum of the linear eigenvalue problem* (2.7), *with $U \in C^2(\Omega) \cap C^{0,1}(\bar{\Omega})$, is such that*

(i) *It consists, at most, of an infinite, countable set of eigenvalues which are isolated, and the eigenfunctions are in $C^{1,\delta}(\bar{\Omega})$ for all δ such that $0 < \delta < \delta_0 = \min\{\gamma, \alpha + 1\}$.*

(ii) *It contains a unique principal eigenvalue (i.e., a real eigenvalue with an associated eigenfunction in the interior of the positive cone of* $C_0^1(\bar{\Omega})$*, namely, such that* $U > 0$ *in* Ω *and* $\partial U/\partial n < 0$ *on* $\partial\Omega$*), which is simple, and it is the only eigenvalue with this property.*

Uniqueness of the positive solution of (2.1)–(2.2) can be proved if the nonlinearity $f(x,u)$ satisfies a concavity assumption. This is well known in the regular case (see, e.g., [1]) and we extend the result to the present situation. The idea in the proof goes back as far to Krasnoselskii and we have adapted the version given by Brezis and Kamin in Method III in the Appendix in [6] to obtain the following theorem.

Theorem 2.3. *Under the assumptions of Theorem* 2.1 *and moreover*

(H.5) *For any* $x \in \Omega$*, the mapping* $\dfrac{f(x,u)}{u}$ *is (strictly) decreasing for* $u > 0$*, there is a unique positive solution to problem* (2.1)–(2.2).

Proof. See [6] and [25].

For more uniqueness results see [30] (where an argument involving integration by parts in [2] or [6] is used) or [31].

We have also studied in [24] the linearization of the semilinear problem (2.1)–(2.2) around a smooth solution such that $u > 0$ in Ω, $\dfrac{\partial u}{\partial n} < 0$ on $\partial\Omega$. With this aim we rewrite the problem in the integral form

$$\mathcal{F}(u) \equiv u - G(f(\cdot,u)) = 0, \tag{2.8}$$

where $G : C_0^1(\bar{\Omega}) \to C_0^1(\bar{\Omega})$ is the Green operator corresponding to the linear equation

$$LU = V \quad \text{in} \quad \Omega, \quad U = 0 \quad \text{on} \quad \partial\Omega, \tag{2.9}$$

(defined as $G(V) = U$). Then we have

Theorem 2.4. ([24], Theorem 3.1) *Under the assumptions* (H.1)–(H.3)*, the operator* $\mathcal{F} : C_0^1(\bar{\Omega}) \to C_0^1(\bar{\Omega})$ *defined in* (2.8) *is of class* C^m *in the positive cone of* $C_0^1(\bar{\Omega})$ *(that is, the set of those functions of* $C_0^1(\bar{\Omega})$ *that satisfy* $u > 0$ *in* Ω *and* $\partial u/\partial n < 0$ *on* $\partial\Omega$*), where* $m \geq 1$ *is as defined in Assumption* (H.3) *and the linear operator* $\mathcal{F}'(u) : C_0^1(\bar{\Omega}) \to C_0^1(\bar{\Omega})$ *is given by*

$$\mathcal{F}'(u)v = v - G(f_u(\cdot,u)v). \tag{2.10}$$

If $m > 1$ *and* $1 < j \leq m$ *then the* j*-linear operator ,* $\partial^j\mathcal{F}(u)/\partial u^j \equiv \mathcal{F}^{(j)}(u) : [C_0^1(\bar{\Omega})]^j \to C_0^1(\bar{\Omega})$*, is given by*

$$\mathcal{F}^{(j)}(u)(v_1,\ldots,v_j) = -G((\partial^j f(\cdot,u)/\partial u^j)v_1\cdots v_j). \tag{2.11}$$

Lemma 2.1. *If $0 < \alpha < 1$ and $0 < p < 1$, then there exists $\tilde{\lambda} > 0$ such that for any $\lambda \geq \tilde{\lambda}$ the problem*

$$\begin{aligned} Lu + \frac{1}{u^\alpha} &= \lambda u^p \quad \text{in } \Omega \\ u &= 0 \quad \text{on } \partial\Omega, \end{aligned} \tag{2.12}$$

has a positive solution.

Proof. Consider, for $\lambda > 0$ fixed, the associated problem

$$\begin{aligned} Lu + \frac{\delta}{u^\alpha} &= \lambda u^p \quad \text{in } \Omega \\ u &= 0 \quad \text{on } \partial\Omega, \end{aligned} \tag{2.13}$$

and the associated equation $u - G(-\frac{\delta}{u^\alpha} + \lambda u^p) = 0$ where G is the Green's function. For $\delta = 0$ (2.13) has a unique positive solution $\bar{u}$ and it is easy to see by using Lemma 2.2 below with $f(u) = u^p$ that $H(u) = (1-p)u^p > 0$ and $\lambda_1(L - \lambda p\bar{u}^{p-1}) > 0$. Hence by the Implicit Function Theorem for $(\delta, u) = (0, \bar{u})$, there exists $\delta_0 > 0$ such that for any $0 < \delta < \delta_0$ (2.13) has a positive solution. Now we write $u = \mu v$ and hence

$$\begin{aligned} Lv + \frac{\delta}{\mu^{\alpha+1} v^\alpha} &= \lambda \mu^{p-1} v^p \quad \text{in } \Omega \\ v &= 0 \quad \text{on } \partial\Omega. \end{aligned} \tag{2.14}$$

If we pick $\delta = \mu^{1+\alpha}$ ($\delta \to 0$ if and only if $\mu \to 0$) we obtain that

$$\begin{aligned} Lv + \frac{1}{v^\alpha} &= \frac{\lambda}{\mu^{1-p}} v^p \quad \text{in } \Omega \\ v &= 0 \quad \text{on } \partial\Omega. \end{aligned} \tag{2.15}$$

has a positive solution for $\lambda > 0$ and $\mu > 0$ "small", and this ends the proof. (Notice that $1 - p > 0$).

In [24] we have also extended standard linearized stability results for regular semilinear parabolic problems to solutions that are in the positive cone of $C_0^1(\bar{\Omega})$ as steady states of

$$\frac{\partial u}{\partial t} + Lu = f(x,u) \quad \text{in} \quad \Omega, \quad u = 0 \quad \text{on} \quad \partial\Omega, \quad u(\cdot, 0) = u_0 \quad \text{in} \quad \Omega, \tag{2.16}$$

by using the fact that for $\alpha > -\dfrac{1}{n}$ the operator L is sectorial in $L_q(\Omega)$ for any $q > n$. Notice that if the linear differential operator has, e.g., constant coefficients, then we can get rid of the limitation $\alpha > -\dfrac{1}{n}$.

Theorem 2.5. ([24], Theorem 4.6) *In addition to the assumptions* (H.1)–(H.3), *let us assume that $\alpha > -1/n$, let $u_s \in C^2(\Omega) \cap \mathcal{C}$ (= the positive cone of $C_0^1(\bar{\Omega})$) be a solution of* (2.1)–(2.2), *and let $M \equiv f_u(\cdot, u_s)$. If the principal eigenvalue of* (2.4) *is strictly positive* (resp., *strictly negative) then u_s is an exponentially stable* (resp., *unstable) steady state of* (2.16) *in the Lyapunov sense, with the norm of $C^1(\bar{\Omega})$. Also, if $u_0 \in \mathcal{C}$ then the problem* (2.16) *has a unique solution, $t \to u(\cdot, t) \in \mathcal{C}$, in a* *maximal existence interval,* $0 \leq t < T \leq \infty$, *and if* $T < \infty$ *then there is a sequence* $\{t_m\}$ *such that* $t_m \nearrow T$ *and either* $\max\{u(x,t_m) : x \in \Omega\} \to \infty$, *or* $u(x,t_m) \searrow 0$ *for some* $x \in \Omega$, *or* $\min\{\partial u(x,t_m)/\partial n : x \in \partial\Omega\} \to 0$ *as* $m \to \infty$.

The following lemma provides a simple, rather general and useful, result in order to check that the principal eigenvalue of (2.4) is strictly positive. This property will be useful when applying the Implicit Function Theorem.

Lemma 2.2. *Suppose that* $u \in C^2(\Omega) \cap C_0^1(\bar{\Omega})$ *is a solution of* (2.1)–(2.2) *such that* $u > 0$ *in* Ω *and* $\frac{\partial u}{\partial n} < 0$ *on* $\partial\Omega$ *and let* μ_1 *be the principal eigenvalue of the linearized problem*

$$Lw - f_u(x,u)w = \mu\, w \quad in \quad \Omega, \qquad w = 0 \quad on \quad \partial\Omega, \tag{2.17}$$

around the solution u. *Then if* $f(x,u)$ *satisfies the assumption*

(H.6) $$H(x,u) = f(x,u) - u f_u(x,u) > 0, \tag{2.18}$$

we have $\mu_1 > 0$.

3. Applications

In this section we consider a few selected applications, already studied in the literature, which exhibit some interesting features or specially relevant ones in one way or another. Additional applications with detailed proofs can be found in [25] or in [26] and its references.

We assume all along this section that Ω is a bounded domain in $\mathbb{R}^n$ with the smoothness given by (H.1) and the differential operator L satisfies assumptions (H.2).

Application 1: The model case
For the model problem

$$\begin{aligned} Lu &= \frac{\lambda K(x)}{u^\alpha} && \text{in } \Omega \\ u &= 0 && \text{on } \partial\Omega, \end{aligned} \tag{3.1}$$

we prove (see [24]) the following

Theorem 3.1. *If* $0 < \alpha < 1$, *and* $K(x)$ *is a function such that* $0 < K(x) \leq k_1 d(x)^{-\beta}$ *where* $\beta > 0$ *is such that* $\alpha + \beta < 1$ *and* $f(x,u) = K(x)u^{-\alpha}$ *satisfies* (H.3), *then for any* $\lambda > 0$ *there exists a unique positive solution* $z(\lambda) > 0$ *of* (3.1) *with* $z(\lambda) \in C^2(\Omega) \cap C_0^{1,\delta}(\overline{\Omega})$, *where* $\delta < \delta_0 = \min\{\gamma, \alpha+1\}$. *Moreover, the mapping* $\lambda \to z(\lambda)$ *from* $(0,\infty)$ *into* $C_0^{1,\delta}(\overline{\Omega})$ *is* C^∞ *and* $z(\lambda)$ *is asymptotically stable.*

Proof. As a sub-solution we try $u_0 \equiv ce$, where e is the unique solution of

$$\begin{aligned} Le &= K(x) && \text{in } \Omega \\ e &= 0 && \text{on } \partial\Omega, \end{aligned} \tag{3.2}$$

given by Corollary 2.1 . We have

$$Lu_0 - \frac{\lambda K(x)}{(u_0)^\alpha} = cK(x) - \frac{\lambda K(x)}{c^\alpha e^\alpha} = \frac{(c^{1+\alpha}e^\alpha - \lambda)K(x)}{c^\alpha e^\alpha}$$

and u_0 will be a sub-solution if, for $\lambda > 0$ given, $c^{1+\alpha}\|e\|^\alpha_{L^\infty} \leq \lambda$, which holds for $c > 0$ small enough.

In order to find a super-solution, we start by recalling that it follows from Theorem 2.2 that there exist $\lambda_1 > 0$ eigenvalue and $\varphi_1 > 0$ eigenfunction to

$$\begin{aligned} L\varphi_1 &= \lambda_1\varphi_1 \quad \text{in } \Omega \\ \varphi_1 &= 0 \qquad\quad \text{on } \partial\Omega, \end{aligned} \tag{3.3}$$

normalized by $\|\varphi_1\|_{L^\infty} = 1$.

Then there is (again by Corollary 2.1, notice that $0 < \alpha + \beta < 1$) a unique solution $u_1 > 0$ to

$$\begin{aligned} Lu_1 &= \frac{K(x)}{\varphi_1^\alpha} \quad \text{in } \Omega \\ u_1 &= 0 \qquad\quad \text{on } \partial\Omega, \end{aligned} \tag{3.4}$$

and we look for a super-solution $u^0 \equiv \mu u_1$ for a suitable μ. We have for $\lambda > 0$ given

$$Lu^0 - \frac{\lambda K(x)}{(u^0)^\alpha} = \frac{\mu K(x)}{\varphi_1^\alpha} - \frac{\lambda K(x)}{\mu^\alpha u_1^\alpha} = \frac{(\mu^{\alpha+1}u_1^\alpha - \lambda\varphi_1^\alpha)K(x)}{\mu^\alpha\varphi_1^\alpha u_1^\alpha}$$

and it follows from the smoothness of u_1 and φ_1 that $\mu u_1 > \varphi_1$ for $\mu > 0$ large enough, and hence

$$Lu^0 - \frac{\lambda K(x)}{(u^0)^\alpha} \geq \frac{(\mu - \lambda)K(x)\varphi_1^\alpha}{\mu^\alpha\varphi_1^\alpha u_1^\alpha} = \frac{(\mu - \lambda)K(x)}{\mu^\alpha u_1^\alpha} > 0$$

for μ large enough.

It is clear that for μ large enough we have $u_0 \leq u^0$ and then Theorem 2.1 gives existence. Uniqueness follows immediately from Theorem 2.3 and the rest of the statement from Theorem 2.4 for $f(x,u) = K(x)u^{-\alpha}$, $H(x,u) = K(x)u^{-\alpha} + \alpha K(x)u^{-\alpha} = (1+\alpha)\, K(x)u^{-\alpha} > 0$, Theorem 2.5 and Lemma 2.2.

Theorem 3.1 extends in different ways existence and uniqueness results, with a simpler proof, in [12], [28]. The smoothness of the branch $z(\lambda)$ and its stability seem to be new in the literature.

Application 2: A "mixed" nonlinearity

We consider now the problem

$$\begin{aligned} Lu &= \frac{\lambda}{u^\alpha} + u^p \quad \text{in } \Omega \\ u &= 0 \qquad\qquad \text{on } \partial\Omega. \end{aligned} \tag{3.5}$$

Theorem 3.2. *If $0 < \alpha < 1$ and $p > 1$, then there is a $\bar{\lambda} > 0$ such that for $0 < \lambda < \bar{\lambda}$ there is a positive solution to (3.5).*

Proof. It is clear that $u_0 \equiv z(\lambda) > 0$, $z(\lambda)$ being the unique solution of 3.1 for $K(x) \equiv 1$ is a sub-solution. Now we look for a super-solution of the form $u^0 \equiv Az(\lambda)$, $z(\lambda) > 0$ solution to (3.1). We have

$$\begin{aligned} Lu^0 - \frac{\lambda}{(u^0)^\alpha} - (u^0)^p =& \frac{\lambda A}{z^\alpha} - \frac{\lambda}{A^\alpha z^\alpha} - A^p z^p \\ =& \frac{\lambda(A^{\alpha+1} - 1) - A^{p+\alpha} z^{p+\alpha}}{A^\alpha z^\alpha} \end{aligned}$$

which will be positive if

$$\lambda(A^{1+\alpha} - 1) \geq A^{p+\alpha} \|z(\lambda)\|_\infty^{p+\alpha}.$$

But since $z(\lambda) = \lambda^{\frac{1}{1+\alpha}} z(1)$, it is enough to have

$$\lambda(A^{1+\alpha} - 1) \geq A^{p+\alpha} \|z(1)\|_\infty^{p+\alpha} \lambda^{\frac{p+\alpha}{1+\alpha}}.$$

For $A > 1$ fixed, this inequality can be written as

$$\lambda \geq C(A) \lambda^{\frac{p+\alpha}{1+\alpha}}$$

and since $\dfrac{p+\alpha}{1+\alpha} > 1$ it turns out that u^0 will be a super-solution for $\lambda > 0$ small enough.

We always have $u_0 \leq u^0$ and existence follows from Theorem 2.1.

Problem (3.5) for $L = -\Delta$ was studied by Coclite and Palmieri [11] using a much more complicated approximation argument involving sub- and super-solutions. Later the case $-1 < \alpha < 0$, corresponding to the sum of a concave and a convex nonlinearities was considered in [2], where the existence of a minimal positive solution was shown using sub- and super-solutions; the authors also proved that the solutions small enough (in the $L^\infty(\Omega)$ norm) are unique. Existence of a second positive solution was then proved by an application of the Mountain Pass Theorem. A sharp result showing that there are exactly two positive solutions for $N = 1$ was obtained in [7]. In the singular case this second solution was obtained by using different (and rather more complicated) variational arguments (see [23] [35] and [27]).

Application 3: Open multiplicity problems

We analyze now the problem

$$\begin{aligned} Lu + \frac{1}{u^\alpha} &= \lambda f(x) \quad \text{in } \Omega \\ u &= 0 \qquad\quad\ \text{on } \partial\Omega, \end{aligned} \tag{3.6}$$

where $0 < \alpha < 1$, $f \geq 0$ and $\lambda > 0$ is a real parameter.

Theorem 3.3. *If $0 < \alpha < 1$, c is a strictly positive constant and f is such that $0 \leq f(x) \leq \dfrac{c}{d(x)^\alpha}$ in Ω, then there exists $\bar{\lambda} > 0$ such that if $\lambda > \bar{\lambda}$ there exists at least a positive solution to (3.6), where $d(x) = dist(x, \partial\Omega)$.*

Proof. We pick as a super-solution the only solution to $Lw = \lambda f(x)$ in Ω, $w = 0$ on $\partial\Omega$. By the Maximum Principle, $w > 0$ on Ω.

We obtain a sub-solution by reasoning as in the proof of Lemma 2.1. Consider the auxiliary problem

$$\begin{aligned} Lu + \frac{\delta}{u^\alpha} &= f(x) \quad \text{in } \Omega \\ u &= 0 \quad\quad\; \text{on } \partial\Omega. \end{aligned} \tag{3.7}$$

Proceeding as in the proof of Lemma 2.1 it is obtained that the problem

$$Lv + \frac{1}{v^\alpha} = \frac{1}{\mu} f(x)$$

has one solution for $\frac{1}{\mu}$ large enough.

Define $A = \{\lambda > 0 / \exists \text{ solution } u > 0\}$. Again A is nonempty and $0 \notin A$. If $\bar{\lambda} = \inf A$ and $\lambda > \bar{\lambda}$ then there exists $\tilde{\lambda} \in A$ such that $\bar{\lambda} < \tilde{\lambda} < \lambda$. Then if $\tilde{u}$ is a solution of (3.6) for $\lambda = \tilde{\lambda}$ we get

$$L\tilde{u} + \frac{1}{\tilde{u}^\alpha} - \lambda f(x) = (\tilde{\lambda} - \lambda) f(x) \le 0$$

and we have a sub-solution.

Remark 3.1. *Díaz, Morel and Oswald* [17] *proved the same result in Theorem* 3.3 *for* $L = -\Delta$ *and* $f \in L^\infty(\Omega)$. *Uniqueness raises interesting questions, see the results and examples given in* [13] *and* [8] *for* $n = 1$.

Application 4: "Stopping" branches
Let us consider the problem

$$\begin{aligned} Lu + \frac{1}{u^\alpha} &= \lambda u^p \quad \text{in } \Omega \\ u &= 0 \quad\quad \text{on } \partial\Omega \end{aligned} \tag{3.8}$$

with $0 < \alpha < 1$ and $0 < p < 1$.

Theorem 3.4. *If* $0 < \alpha < 1$ *and* $0 < p < 1$, *there is a* $\bar{\lambda} > 0$ *such that for any* $\lambda > \bar{\lambda}$ *there exists at least a positive solution of* (3.8). *Let* $\tilde{\varepsilon} > 0$ *be a fixed number small enough, then for* λ *large enough there is a unique solution* u *of* (3.8) *such that* $\lambda^{\frac{1}{1-p}} \tilde{\varepsilon} \varphi_1 < u$, *where* φ_1 *is defined by* (3.3).

Proof. It is not difficult to find a super-solution. Indeed, it suffices to pick the unique positive solution of problem $-\Delta u = \lambda u^p$ in Ω, $u = 0$ on $\partial\Omega$.

In order to find a sub-solution, we use Lemma 2.1 which provides existence of a solution for a $\lambda > 0$ large enough. If we define

$$A = \{\lambda \in \mathbb{R} / \exists u > 0 \text{ solution to (3.8)}\}$$

then A is nonempty and $0 \notin A$. It is easy to see that if $a \in A$ and $b \ge a$, then $b \in A$. Indeed, if u is a solution for $\lambda = a$ then it is a sub-solution for $\lambda = b$, $b \in A$.

Now, let $\lambda_* = \inf A$. If $\lambda > \lambda^*$, there exists $\lambda_* < \tilde{\lambda} < \lambda$ with $\tilde{\lambda} \in A$ and if $\tilde{u} > 0$ is a solution for $\lambda = \tilde{\lambda}$, we have

$$L\tilde{u} + \frac{1}{\tilde{u}^\alpha} - \lambda(\tilde{u})^p = (\tilde{\lambda} - \lambda)\tilde{u}^p \leq 0$$

and $\tilde{u}$ is a sub-solution.

Remark 3.2. *The above result extends Theorem* 1.2 *in* [29] *and Lemma* 4.11 *in* [13]. *The stability of the maximal positive solution is also proved in* [13] *for* $L = -\Delta$. *Moreover, the radial case is studied in* [29] *where the exact number of positive solutions is obtained; besides the maximal (stable) positive solution for* $\lambda > \lambda_0$ *we obtain in the general case by using sub- and super-solutions, existence of a second "lower" positive solution* $\underline{u}(\lambda)$ *is proved for a bounded interval* $]\lambda_0, \lambda_1[$ *of* λ; *moreover, these solutions form a smooth branch and* $\underline{u}(\lambda_1) = \dfrac{\partial \underline{u}(\lambda_1)}{\partial n} = 0$ *on* ∂B, *where* B *is the unit ball. This situation strongly recalls the transition from positive to nonnegative dead core solutions described in detail for the problem* $-\Delta u + u^\beta = \lambda u^\alpha$ *in* Ω, $u = 0$ *on* $\partial\Omega$ *for* $0 < \beta < \alpha < 1$, *in the case* $\Omega =]0, 1[$ *in* [16]. *This suggests to deal with the singular problem by using a similar approach improving in this way the results in* [29] *and looking for an answer for the conjecture on the behavior near the boundary made there. The analogous singular problem* $-1 < \beta < \alpha < 0$ *could be treated by similar arguments.*

Application 5: Positive versus dead core solutions in mathematical biology
Our last example is the problem

$$\begin{aligned} Lu &= \lambda M(x) u^\alpha && \text{in} \quad \Omega, \\ u &= 0 && \text{on} \quad \partial\Omega. \end{aligned} \tag{3.9}$$

This problem arises in population dynamics when looking for stationary solutions to the usual logistic equation with nonlinear diffusion. After a change of unknown we get the problem (3.9) with $0 < \alpha < 1$.

It is well known (and easy to show) that if $M \equiv 1$ for each $\lambda > 0$ there is a nontrivial positive solution which is asymptotically stable. If, however, $M(x)$ changes sign over Ω, the situation becomes more delicate. A very careful and detailed study was carried out in [4]. Since $0 < \alpha < 1$, the nonlinearity is not locally Lipschitz near the origin and "dead core" solutions (i.e., solutions identically zero on subdomains of Ω with positive measure) may appear, see [15]. Existence of these dead core solutions was obtained in [4] by using sub- and super-solutions and uniqueness of solutions satisfying some additional conditions was proved as well. All these results were closely related with the structure and properties of the subsets where $M(x)$ is positive.

But no result concerning existence of positive solutions was given in [4]. In [24] the authors gave a sufficient condition on $M(x)$ for existence of positive solutions. More precisely, they actually prove the following result, under the condition that the unique solution to the linear problem

$$\begin{aligned} LU &= M(x) \quad \text{in} \quad \Omega \\ U &= 0 \qquad\quad \text{on} \quad \partial\Omega, \end{aligned} \tag{3.10}$$

is such that $U > 0$ on Ω and $\dfrac{\partial U}{\partial n} < 0$ on $\partial\Omega$. This condition is satisfied if $M(x) \geq 0$ but there are also changing sign functions $M(x)$ that satisfy this property.

Theorem 3.5. *If the operator L and the domain Ω satisfy assumptions* (H.1) *and* (H.2), *let α be such that $0 < \alpha < 1$ and $M \in C^1(\Omega)$ be such that $|M(x)| \leq K_1 d(x)^{\alpha_2}$ for some K_1 and some α_2 such that $-1 < \alpha_2 < 1 - \alpha$. Then there is a unique positive solution $u(\lambda)$ of* (3.9) *for any $\lambda > 0$. Moreover the mapping $\lambda \to u(\lambda)$ from $(0, +\infty)$ into $C_0^{1,\delta}(\bar{\Omega})$ is C^∞ and $u(\lambda)$ is asymptotically stable.*

Proof. We only sketch it, details can be found in [24]. Let Ψ be the unique solution of $L\Psi = |M(x)|$ in Ω, $\Psi = 0$ on $\partial\Omega$ then $u^0 = C\Psi$ is a super-solution of (3.9) for C large enough. Reasoning as in [24] it is obtained that $u_0 = (\lambda(1-\alpha)\Psi)^{\frac{1}{1-\alpha}}$ is a sub-solution. This gives existence. Uniqueness follows from an argument which seems due to Spruck (see [24], [4], [6]). The last part is proved by means of Theorem 2.5 and Lemma 2.2.

Acknowledgement

Research partially supported by the Spanish Ministry of Education Grant MTM2004-03808 and REN 2003-0223 and by the European Community HPRN-CT-2002-00274.

References

[1] H. Amann. Fixed point equations and nonlinear eigenvalue problems in ordered Banach spaces. *SIAM Review*, 18:620–709, 1976.

[2] A. Ambrosetti, H. Brezis, and G. Cerami. Combined effects of concave and convex nonlinearities in some elliptic problems. *J. Funct. Anal*, 122:519–543, 1994.

[3] C. Aranda and E. Lami-Dozo. Multiple solutions to a singular nonlinear Dirichlet problem. *to appear*, 00:1–2, 2002.

[4] C. Bandle, A. Pozio, and A. Tesei. The asymptotic behavior of the solutions of degenerate parabolic equations. *Trans. Amer. Math. Soc*, 303:487–501, 1987.

[5] M. Bertsch and R. Rostamian. The principle of linearized stability for a class of degenerate diffusion equations. *J. Differ. Equat*, 57:373–405, 1985.

[6] H. Brezis and S. Kamin. Sublinear elliptic equations in $\mathbb{R}^n$. *Manus. Math.*, 74:87–106, 1992.

[7] T. Cazenave, M. Escobedo, and A. Pozio. Some stability properties for minimal solutions of $-\Delta u = \lambda g(u)$. *Portugaliae Mathematica*, 59:373–391, 2002.

[8] Y.S. Choi, A.C. Lazer, and P.J. McKenna. Some remarks on a singular elliptic boundary value problem. *Nonlinear Anal., TMA*, 32:305–314, 1998.

[9] M.M. Coclite. On a singular nonlinear Dirichlet problem II. *Boll. U. Mat. Ital*, 5(B):955–975, 1991.

[10] M.M. Coclite. On a singular nonlinear Dirichlet problem IV. *Nonlinear Anal., TMA*, 23:925–936, 1994.

[11] M.M. Coclite and G. Palmieri. On a singular nonlinear Dirichlet problem. *Comm. Partial Differ. Equat*, 14:1315–1327, 1989.

[12] M.G. Crandall, P.H. Rabinowitz, and L. Tartar. On a Dirichlet problem with singular nonlinearity. *Comm. Part. Diff. Equat*, 2:193–222, 1977.

[13] J. Dávila and M. Montenegro. Positive versus free boundary solutions to a singular equation. *Jour. d'Analyse Math.*, 90:303–305, 2003.

[14] M. del Pino. A global estimate for the gradient in a singular elliptic boundary value problem. *Proc. Royal Soc. Edinburgh*, 122(A):341–352, 1992.

[15] J.I. Díaz. *Nonlinear Partial Differential Equations and Free Boundaries.* Pitman, 1985.

[16] J.I. Díaz and J. Hernández. Global bifurcation and continua of nonnegative solutions for a quasilinear elliptic problem. *C.R. Acad. Sci. Paris, Serie I*, 329:587–592, 1999.

[17] J.I. Díaz, J.M. Morel, and L. Oswald. An elliptic equation with singular nonlinearity. *Comm. Part. Diff. Equat*, 12:1333–1344, 1987.

[18] W. Fulks and J.S. Maybee. A singular nonlinear equation. *Osaka J. Math*, 12:1–19, 1960.

[19] J.A. Gatica, V. Oliker, and P. Waltman. Singular nonlinear boundary value problems for second-order ordinary differential equations. *J. Diff. Equat.*, 79:62–78, 1989.

[20] M. Ghergu and V.D. Radulescu. Bifurcation and asymptotics for the Lane-Emden equation. *C.R. Acad. Sci. Paris, Serie I*, 337:259–264, 2003.

[21] S.N. Gomes. On a singular nonlinear elliptic problem. *SIAM J. Math. Anal*, 17:1359–1369, 1986.

[22] C. Gui and F. Hua Lin. Regularity of an elliptic problem with a singular nonlinearity. *Proc. Roy. Soc. Edinburgh*, 123(A):1021–1029, 1993.

[23] Y. Haitao. Multiplicity and asymptotic behavior of positive solutions for a singular semilinear elliptic problem. *J. Differ. Equat.*, 189:487–512, 2003.

[24] J. Hernández, F.J. Mancebo, and J.M. Vega. On the linearization of some singular nonlinear elliptic problems and applications. *Annales I.H.P., Anal. non linéaire*, 19(6):777–813, 2002.

[25] J. Hernández, F.J. Mancebo, and J.M. Vega. Positive solutions for singular nonlinear elliptic equations. *Preprint*, 2005.

[26] J. Hernández, F.J. Mancebo, and J.M. Vega. Survey in preparation. In M. Chipot and P. Quittner, editors, *Handbook of Differential Equations*, Elsevier, Amsterdam, 2006.

[27] N. Hirano, C. Saccon, and N. Shioji. Existence of multiple positive solutions for singular elliptic problems with concave and convex nonlinearities. *Adv. Differ. Eqs.*, 9:197–220, 2004.

[28] A.C. Lazer and P.J. McKenna. On a singular nonlinear elliptic boundary value problem. *Proc. Amer. Math. Soc*, 111:721–730, 1991.

[29] T. Ouyang, J. Shi, and M. Yao. Exact Multiplicity and bifurcation of solutions of a singular equation. *Preprint*, 1996.

[30] J. Shi and M. Yao. Positive solutions of elliptic equations with singular nonlinearity. *Preprint*, 1996.

[31] J. Shi and M. Yao. On a singular nonlinear semilinear elliptic problem. *Proc. Royal Soc. Edinburgh*, 138(A):1389–1401, 1998.

[32] J. Smoller. *Shock Waves and Reaction-Diffusion Equations.* Springer-Verlag, 1985.

[33] C.A. Stuart. Existence and approximation of solutions of nonlinear elliptic equations. *Math. Z.*, 147:53–63, 1976.

[34] M. Wiegner. A Degenerate Diffusion Equation with a Nonlinear Source Term. *Nonlinear Anal., TMA*, 28:1977–1995, 1997.

[35] S. Yijing, W. Shaoping, and L. Yiming. Combined effects of singular and superlinear nonlinearities in some singular boundary value problems. *J. Differ. Equat.*, 176:511–531, 2001.

Jesús Hernández
Departamento de Matemáticas
Universidad Autónoma de Madrid
E-28049 Madrid

Francisco J. Mancebo and José M. Vega
E.T.S.I. Aeronáuticos
Universidad Politécnica de Madrid
Plaza del Cardenal Cisneros 3
E-28040 Madrid

Progress in Nonlinear Differential Equations
and Their Applications, Vol. 64, 243–251

The Navier-Stokes Flow in the Exterior of Rotating Obstacles

Matthias Hieber

Dedicated to Prof. H. Amann on the Occasion of his 65th Birthday

Abstract. In this note we describe recent results on the equations of Navier-Stokes in the exterior of a rotating domain. After rewriting the problem on a fixed exterior domain Ω in $\mathbb{R}^n$, it is shown that for initial data $u_0 \in L^p_\sigma(\Omega)$ with $p \geq n$ and which are satisfying a certain compatibility condition there exists a unique local mild solution to the Navier-Stokes problem. In the case of the whole space of $\mathbb{R}^n$, this local mild solution is even analytic in the space variable x.

1. Introduction

In this paper we describe recent developments in the mathematical analysis of the Navier-Stokes flow past rotation obstacles. Observe that whereas the description of the Navier-Stokes flow in the exterior of a fixed obstacle can be regarded as fairly well understood (see, e.g., [26], [31], [12], [28], [30], [1], [8], [19], [18]), this is not the case for rotating or moving obstacles. It is the aim of this paper to discuss and present recent results in the latter direction.

Let us start with the following situation: we denote by O denotes a three-dimensional rigid body, the obstacle, and assume that the exterior domain $\Omega = \mathbb{R}^3 \backslash O$ is filled with a viscous, incompressible fluid. Assume that the obstacle is rotating about the x_3-axis with constant angular velocity $\omega = (0,0,1)^T$. We set

$$\Omega(t) := \{y(t) = e^{tR}x, x \in \Omega\} \text{ and } \Gamma(t) := \{y(t) = e^{tR}x, x \in \Gamma\}$$

where $R = \begin{pmatrix} 1 & -1 & 0 \\ 1 & 1 & 0 \\ 0 & 0 & 1 \end{pmatrix}$ is the 3×3-matrix describing the above rotation. Then the motion past the rotating obstacle O is governed by the equations of

Navier-Stokes given by

$$\begin{aligned} \partial_t w - \Delta w + w \cdot \nabla w + \nabla q &= 0, & y \in \Omega(t), \quad t > 0, \\ \nabla \cdot w &= 0, & y \in \Omega(t), \quad t > 0, \\ w &= Ry, & y \in \Gamma(t), \quad t > 0, \\ w(y,0) &= w_0(y), & y \in \Omega. \end{aligned} \tag{1.1}$$

Here $w = w(y,t)$ and $q(y,t)$ denote the velocity and the pressure of the fluid, respectively. The boundary condition on $\Gamma(t)$ is the usual no-slip boundary condition.

By a suitable change of variables we are led to the following set of equations defined on the fixed domain Ω:

$$\begin{aligned} \partial_t u - \Delta u + u \cdot \nabla u - Rx \cdot \nabla u + Ru + \nabla p = 0, & \quad x \in \Omega, \quad t > 0, \\ \nabla \cdot u = 0, & \quad x \in \Omega, \quad t > 0, \\ u = Rx, & \quad x \in \Gamma, \quad t > 0, \\ u(x,0) = w_0(x), & \quad x \in \Omega. \end{aligned} \tag{1.2}$$

Observe first that the term $Ru = (-u_2, u_1, 0)^T$, the Coriolis force, can be considered as a lower-order perturbation of the Laplacian. On the other hand, the first-order term

$$Rx \cdot \nabla u = -x_2 \partial_1 u + x_1 \partial_2 u$$

cannot be regarded as subordinate to the Laplacian.

The above equation (1.2) was first considered by Borchers [6] and by Chen and Miyakawa [9] in the framework of weak solutions. Later, Hishida [23], [24] proved, that the solution of the linearized equation is governed by a strongly continuous semigroup on $L^2_\sigma(\Omega)$. Moreover, he proved the existence of a unique local mild solution in $L^2_\sigma(\Omega)$ to problem (1.2) provided the initial data w_0 belongs to suitable subset of $L^2(\Omega)$.

For the case of $\Omega = \mathbb{R}^n$ and $1 < p < \infty$, Babin, Mahalov and Nicolaenko [3], [4] proved the existence of a local mild solution to (1.2) provided w_0 lies in $L^p_\sigma(\mathbb{R}^n)$ or w_0 is a periodic function satisfying certain properties.

The stationary problem, i.e., the time independent problem, was considered before by Galdi [13] and by Farwig, Hishida and Müller [11] on $\mathbb{R}^3$. They proved a priori L^p-estimates for the elliptic problem using Littlewood-Paley theory.

Recently, the original time dependent problem (1.2) was considered for $\Omega = \mathbb{R}^n$ and for arbitrary $n \times n$ matrices M by Sawada and Hieber in [22] and for exterior domains by Geissert, Heck and Hieber in [14]. The main technique in [22] and [14] was to show that the operator $-A$ defined as

$$Au = P(-\Delta - Mx \cdot \nabla)u + Mu$$

and equipped with a suitable domain generates a C_0-semigroup T on $L^p_\sigma(\mathbb{R}^n)$ or $L^p_\sigma(\Omega)$, respectively. The semigroup T satisfies in addition L^p-L^q smoothing estimates as well as gradient estimates similar to the ones known for the usual Stokes operator. By adjusting Kato's iteration procedure (see [25], [17], [19], [21]) to the given situation one obtains a local mild solution to the problem on $\mathbb{R}^n$.

Note that in contrast to the situation of the classical Navier-Stokes equation, the semigroup T on $L^p_\sigma(\Omega)$ is not expected to be analytic, even for $p = 2$.

2. The Navier-Stokes flow with undounded drift terms: the case of $\mathbb{R}^n$

We start this section by considering the equation

$$\begin{aligned} \partial_t u - \Delta u + u \cdot \nabla u - Mx \cdot \nabla u + Mu + \nabla p &= 0 \quad \text{in } \mathbb{R}^n \times (0,T), \\ \nabla \cdot u &= 0 \quad \text{in } \mathbb{R}^n \times (0,T), \\ u(0) &= u_0 \quad \text{in } \mathbb{R}^n, \end{aligned} \tag{2.1}$$

where $u_0 \in L^p_\sigma(\mathbb{R}^n)^n$ for some p satisfying $1 < p < \infty$. Here M denotes an $n \times n$-matrix with real coefficients.

We define the operator A in $L^p_\sigma(\mathbb{R}^n)^n$ as

$$Au := -\Delta u - Mx \cdot \nabla u + Mu \tag{2.2}$$

with domain $D(A) := \{u \in W^{2,p}(\mathbb{R}^n)^n \cap L^p_\sigma(\mathbb{R}^n)^n; Mx \cdot \nabla u \in L^p(\mathbb{R}^n)^n\}$.

Modifying techniques developed in [10], [27] and [29] it was proved in [22] that $-A$ generates a C_0-semigroup e^{-tA} on $L^p_\sigma(\mathbb{R}^n)^n$, which is not analytic.

Applying the Helmholtz projection $\mathbb{P}$ to (2.1) we may rewrite (2.1) as

$$\left\{ \begin{aligned} u'(t) + Au + \mathbb{P}u \cdot \nabla u - 2\mathbb{P}Mu &= 0 \\ u(0) &= u_0. \end{aligned} \right. \tag{2.3}$$

In the given situation of $\mathbb{R}^n$, the Helmholtz projection $\mathbb{P}$ can be expressed explicitly by $\mathbb{P} := (\delta_{ij} + R_i R_j)_{1 \le i,j \le n}$, where δ_{ij} denotes the Kronecker's delta, and R_i is the ith Riesz transform on $\mathbb{R}^n$ defined by $R_i := \partial_i(-\Delta)^{-1/2}$, $i = 1, \dots, n$. Note that A and $\mathbb{P}$ commute, since $\nabla \cdot Au = 0$ if $\nabla \cdot u = 0$. For $T > 0$ we call a function $u \in C([0,T); L^p_\sigma(\mathbb{R}^n))$ a *mild solution* of (2.3) if u satisfies the integral equation

$$u(t) = e^{-tA}u_0 - \int_0^t e^{-(t-s)A}\mathbb{P}u(s) \cdot \nabla u(s) ds + 2\int_0^t e^{-(t-s)A}\mathbb{P}Mu(s) ds, \qquad t > 0. \tag{2.4}$$

and $u(0) = u_0$.

We now state the local existence and uniqueness result for mild solutions of (2.3) in L^p spaces which was proved recently in [22]. Note that the underlying semigroup e^{-tA} is *not* analytic; hence it is a priori not obvious that the classical iteration procedure due to Kato is applicable in the given situation.

Theorem 2.1 ([22]). **(Local Existence and Uniqueness)** *Let $n \ge 2$, $p \in [n, \infty)$ and $q \in [p, \infty]$. Let M be an $n \times n$-matrix with real coefficients and assume that $u_0 \in L^p_\sigma(\mathbb{R}^n)$. Then there exist $T_0 > 0$ and a unique mild solution u of* (2.3) *such that*

$$t^{\frac{n}{2}(\frac{1}{p}-\frac{1}{q})}u \in C([0,T_0); L^q_\sigma(\mathbb{R}^n))$$
$$t^{\frac{n}{2}(\frac{1}{p}-\frac{1}{q})+\frac{1}{2}}\nabla u \in C([0,T_0); L^q(\mathbb{R}^n)).$$

Regularity questions of mild solution of (2.3) are of particular interest. The estimates for higher-order derivatives given in the following result imply that the mild solution to (2.3) is even analytic in the space variable x. In fact, the following result holds true.

Theorem 2.2 ([22]). *Let $n \geq 2, u_0 \in L^n_\sigma(\mathbb{R}^n), r \in (n,\infty)$ and $q \in [n,\infty]$. Assume that $\|e^{tM}\| \leq 1$ for all $t \geq 0$. Let u be the local mild solution of (2.3) for some $T > 0$. Assume further that there exist constants $M_1, M_2 \geq 0$ such that*

$$\sup_{0<t<T} \|u(t)\|_n \leq M_1 < \infty \qquad \text{and} \sup_{0<t<T} t^{\frac{n}{2}(\frac{1}{n}-\frac{1}{r})}\|u(t)\|_r \leq M_2 < \infty.$$

Then there exist constants K_1 and K_2 such that

$$\|\nabla^m u(t)\|_q \leq K_1(K_2 m)^m t^{-\frac{m}{2}-\frac{n}{2}(\frac{1}{n}-\frac{1}{q})}, \qquad t \in (0,T),\ m \in \mathbb{N}_0. \tag{2.5}$$

It follows from the above Theorem 2.2 that the mild solution u of (2.3) is analytic in x. More precisely, we have the following estimate on the radius of analyticity of u.

Corollary 2.3. *Let u be a mild solution of (2.3) satisfying the assumptions of Theorem 2.2. Then u is analytic in x. Moreover, there exists a constant $C > 0$ such that the radius $\rho(t)$ of analyticity is given by*

$$\rho(t) \geq \limsup_{m\to\infty} \left(\frac{\|\nabla^m u(t)\|_\infty}{m!}\right)^{-1/m} \geq C\sqrt{t}.$$

In the following section we deal with the corresponding situation in exterior domains.

3. The Stokes semigroup for unbounded drift terms: exterior domains

Let $\Omega \subset \mathbb{R}^n$ be a domain in $\mathbb{R}^n$ with $C^{1,1}$-boundary and let $1 < p < \infty$. We then set

$$\mathcal{L}u := (\Delta + Mx \cdot \nabla)u - Mu$$

and define its realization L_Ω in $L^p(\Omega)^n$ by

$$\begin{aligned} L_\Omega u &:= \mathcal{L}u \\ D(L_\Omega) &:= \{u \in W^{2,p}(\Omega)^n \cap W^{1,p}_0(\Omega)^n : Mx \cdot \nabla u \in L^p(\Omega)^n\}. \end{aligned}$$

Of course, if Ω is bounded, then $D(L_\Omega) = W^{2,p}(\Omega)^n \cap W^{1,p}_0(\Omega)^n$. In the following we will not distinguish between $L^p(\Omega)^n$ and $L^p(\Omega)$ and often write

$$L_\Omega u := \Delta u + Mx \cdot \nabla u - Mu.$$

Moreover, we define the operator A_Ω in $L^p_\sigma(\Omega)^n$ by

$$A_\Omega u := P_\Omega L_\Omega u.$$

with domain $D(A_\Omega) := \{u \in W^{2,p}(\Omega)^n \cap W^{1,p}_0(\Omega)^n \cap L^p_\sigma(\Omega)^n : Mx \cdot \nabla u \in L^p(\Omega)^n\}$. Here, $P_\Omega : L^p(\Omega) \to L^p_\sigma(\Omega)$ denotes the Helmholtz projection. For the proof of the local existence result in exterior domains (see Theorem 4.1) we need to consider perturbations of L_Ω and A_Ω of the following form: for $b \in C_c^\infty(\overline{\Omega})^n$ define $\mathcal{L}_b$ by

$$\mathcal{L}_b u := \mathcal{L}u + b \cdot \nabla u + u \cdot \nabla b$$

and set

$$L_{\Omega,b} u := \mathcal{L}_b u, \quad D(L_{\Omega,b}) := D(L_\Omega)$$

as well as

$$A_{\Omega,b} u := A_\Omega u + P_\Omega(b \cdot \nabla u + u \cdot \nabla b), \quad D(A_{\Omega,b}) := D(A_\Omega).$$

Here $u = (u_1, \ldots, u_n) \in L^p(\mathbb{R}^n)^n$.

The following result states that the operator $A_{\Omega,b}$ generates a C_0-semigroup $T_{\Omega,b}$ on $L^p_\sigma(\Omega)$ which satisfies so-called L^p-L^q smoothing properties. They will be the key ingredient in the proof of the local existence Theorem 4.1.

Proposition 3.1 ([14]). *Let $1 < p < \infty$ and let $\Omega \subset \mathbb{R}^n$ be an exterior domain with $C^{1,1}$-boundary. Then the operator $A_{\Omega,b}$ generates a C_0-semigroup $T_{\Omega,b}$ on $L^p_\sigma(\Omega)$. In particular, A_Ω generates a C_0-semigroup T_Ω on $L^p_\sigma(\Omega)$. Moreover, let $1 < p < q < \infty$. Then there exist constants $C > 0, \omega \geq 0$ such that for $f \in L^p_\sigma(\Omega)$*

(a) $\|T_{\Omega,b}(t)f\|_{L^q_\sigma(\Omega)} \leq Ct^{-\frac{n}{2}\left(\frac{1}{p}-\frac{1}{q}\right)} e^{\omega t}\|f\|_{L^p_\sigma(\Omega)}, \quad t > 0,$

(b) $\|\nabla T_{\Omega,b}(t)f\|_{L^p(\Omega)} \leq Ct^{-\frac{1}{2}} e^{\omega t}\|f\|_{L^p_\sigma(\Omega)}, \quad t > 0.$

Moreover, for $f \in L^p_\sigma(\Omega)$

$$\|t^{\frac{n}{2}\left(\frac{1}{p}-\frac{1}{q}\right)} T_{\Omega,b}(t)f\|_{L^q_\sigma(\Omega)} + \|t^{\frac{1}{2}} \nabla T_{\Omega,b}(t)f\|_{L^p(\Omega)} \to 0, \quad \text{for} \quad t \to 0.$$

Recall that the first assertion of the above proposition was proved first by Hishida [23], [24] for the special case $p = 2$. Observe also that the corresponding results for the classical Stokes operator can be found in [16], [7] and [20].

The proof of Proposition 3.1 relies on techniques on Laplace Transforms [2], sums of convolution operators and on properties of solutions of the equation $\operatorname{div} u = f$ in suitable function spaces. For Sobolev spaces of nonnegative order there is a well-known answer to the latter problem due to Bogovskiĭ [5].

Proposition 3.2 (Bogovskiĭ, [5]). *Let $\Omega \subset \mathbb{R}^n$, $n \geq 2$, be a bounded Lipschitz domain, $1 < p < \infty$ and $k \in \mathbb{N}_0$. Then there exists a continuous operator $V : W_0^{k,p}(\Omega) \to (W_0^{k+1,p}(\Omega))^n$ such that*

$$\operatorname{div} Vf = f \tag{3.1}$$

for all $f \in W_0^{k,p}(\Omega)$ satisfying $\int_\Omega f = 0$.

For an extension of this result to Sobolev spaces of negative order, see the monograph of Galdi [12] or [15].

The result on iterated convolutions (for a proof see [14]) needed for the proof of Proposition 3.1 reads as follows.

Lemma 3.3. *Let X, Y be Banach spaces and let $T : (0,\infty) \to \mathcal{L}(Y,X)$ and $S : (0,\infty) \to \mathcal{L}(Y)$ be strongly continuous functions. Assume that*

$$\|T(t)\|_{\mathcal{L}(Y,X)} \le C_0 t^\alpha e^{\omega t}, \quad \|S(t)\|_{\mathcal{L}(Y)} \le C_0 t^\beta e^{\omega t}, \quad t>0,$$

for some $C_0, \omega > 0$ and $\alpha, \beta > -1$. For $f \in Y$ set $T_0(t)f := T(t)f$ and

$$T_n(t)f := \int_0^t T_{n-1}(t-s)S(s)f \,ds, \quad n\in\mathbb{N},\ t>0.$$

Then there exist $C, \tilde\omega > 0$ such that

$$\sum_{n=0}^\infty \|T_n(t)f\|_X \le Ct^\alpha e^{\tilde\omega t}\|f\|_Y, \quad t>0.$$

In the following section we modify the well-known iteration procedure due to Kato [25] to our situation. Of central importance are the $L^p - L^q$-smoothing properties as well as the gradient estimates for $T_{\Omega,b}$ developed in Proposition 3.1.

4. The Navier-Stokes flow in exterior domains

In this section we are considering the Navier-Stokes equations in the exterior domain $\Omega(t)$ as defined in Section 1. Recall that after rewriting the problem to a fixed domain, the Navier-Stokes equations read as follows

$$\begin{array}{rcllr} \partial_t u - \Delta u + u\cdot\nabla u - Mx\cdot\nabla u + Mu + \nabla p &=& 0, & x\in\Omega, & t>0,\\ \nabla\cdot u &=& 0, & x\in\Omega, & t>0,\\ u &=& Mx, & x\in\Gamma, & t>0,\\ u(x,0) &=& u_0(x), & x\in\Omega, & \end{array} \tag{4.1}$$

where $\operatorname{div} u_0 = 0$ and the matrix M satisfies $\operatorname{tr} M = 0$. Furthermore, assume that the initial value u_0 satisfies the compatibility condition

$$u_0\cdot n = Mx\cdot n \quad \text{on} \quad \partial\Omega,$$

where n denotes the outer normal.

Let $\zeta \in C_c^\infty(\mathbb{R}^n)$ be a cut-off function with $0\le\zeta\le 1$ and $\zeta = 1$ near Γ. Further let $\operatorname{supp}\nabla\zeta =: K$. We then define $b : \mathbb{R}^n \to \mathbb{R}^n$ by

$$b(x) := \zeta Mx - B_K((\nabla\zeta)Mx),$$

where B is the operator defined in Proposition 3.2. Then $\operatorname{div} b = 0$ and $b(x) = Mx$ on Γ. Setting $\tilde u := u - b$ it follows that $\tilde u$ satisfies

$$\begin{array}{rl} \partial_t u - \Delta u + u\cdot\nabla u - Mx\cdot\nabla u & \\ +Mu + b\cdot\nabla u + u\cdot\nabla b + \nabla p = F_2 & \text{in } \Omega\times(0,T),\\ \nabla\cdot u = 0 & \text{in } \Omega\times(0,T),\\ u|_\Gamma = 0, & \\ u(0) = u_0 - b, & \end{array} \tag{4.2}$$

with $\nabla \cdot (u_0 - b) = 0$ in Ω and $F_2 = -\Delta b - Mx \cdot \nabla b + Mb + b \cdot \nabla b$, provided u satisfies (4.1). Applying the Helmholtz projection P_Ω to (4.2), we may rewrite (4.2) as an evolution equation in $L^p_\sigma(\Omega)$ as

$$\begin{aligned} u' - A_{\Omega,b}u + P_\Omega(u \cdot \nabla u) &= P_\Omega F_2, \\ u(0) &= u_0 - b. \end{aligned} \tag{4.3}$$

Note that the compatibility condition on u_0 implies $u_0 - b \in L^p_\sigma(\Omega)$. In the following, given $0 < T < \infty$, we call a function $u \in C([0,T); L^p_\sigma(\Omega))$ a *mild solution* of (4.3) if u satisfies the integral equation

$$\begin{aligned} u(t) = T_{\Omega,b}(t)(u_0 - b) &- \int_0^t T_{\Omega,b}(t-s)P_\Omega(u \cdot \nabla u)(s)\,\mathrm{d}s \\ &+ \int_0^t T_{\Omega,b}(t-s)P_\Omega F_2(s)\,\mathrm{d}s, \quad 0 < t < T. \end{aligned}$$

Our main result concerning the Navier-Stokes flow past rotating obstacles is then the following theorem.

Theorem 4.1 ([14])**.** *Let $n \geq 2$, $n \leq p \leq q < \infty$ and $u_0 - b \in L^p_\sigma(\Omega)$. Then there exists $T_0 > 0$ and a unique mild solution u of (4.3) such that*

$$\begin{aligned} t \mapsto t^{\frac{n}{2}\left(\frac{1}{p}-\frac{1}{q}\right)}u(t) &\in C\left([0,T_0]\,; L^q_\sigma(\Omega)\right), \\ t \mapsto t^{\frac{n}{2}\left(\frac{1}{p}-\frac{1}{q}\right)+\frac{1}{2}}\nabla u(t) &\in C\left([0,T_0]\,; L^q(\Omega)\right). \end{aligned}$$

References

[1] H. Amann, *On the strong solvability of the Navier-Stokes equations.* J. Math. Fluid Mech., **2** (2000), 16–98.

[2] W. Arendt, Ch. Batty, M. Hieber, and F. Neubrander, *Vector-valued Laplace Transforms and Cauchy Problems.* Birkhäuser, Basel, 2001.

[3] A. Banin, A. Mahalov, and B. Nicolaenko, *Global regularity of 3D rotating Navier-Stokes equations for resonant domains.* Indiana Univ. Math. J. **48** (1999), 1133–1176.

[4] A. Banin, A. Mahalov and B. Nicolaenko, *3D Navier-Stokes and Euler equations with initial data characterized by uniformly large vorticity.* Indiana Univ. Math. J. **50** (2001), 1–35.

[5] M.E. Bogovskiĭ, *Solution of the first boundary value problem for an equation of continuity of an incompressible medium.* Dokl. Akad. Nauk SSSR **248** (1979), 1037–1040

[6] W. Borchers, *Zur Stabilität und Faktorisierungsmethode für die Navier-Stokes-Gleichungen inkompressibler viskoser Flüssigkeiten*, Habilitationschrift Universität Paderborn, 1992.

[7] W. Borchers, and H. Sohr, *On the semigroup of the Stokes operator for exterior domains in L^q-spaces.* Math. Z. **196**, (1987), 415–425.

[8] M. Cannone, *Harmonic analysis tools for solving the incompressible Navier-Stokes equations.* Handbook of Mathematical Fluid Dynamics, vol 3 (eds. S. Friedlander, D. Serre), Elsevier, 2003.

[9] Z. Chen, and T. Miyakawa, *Decay properties of weak solutions to a perturbed Navier-Stokes system in* $\mathbb{R}^n$. Adv. Math. Sci. Appl. **7** (1997), 741–770.

[10] G. Da Prato, and A. Lunardi, *On the Ornstein-Uhlenbeck operator in spaces of continuous functions.* J. Func. Anal. **131** (1995), 94–114.

[11] R. Farwig, T. Hishida, and D. Müller, L^q*-theory of a singular 'winding' integral operator arising from fluid dynamics*, Preprint, TU Darmstadt, 2003.

[12] G.P. Galdi, *An Introduction to the Mathematical Theory of the Navier-Stokes equations. Vol. I*, Springer Tracts in Natural Philosophy, Vol. 38, Springer 1998.

[13] G.P. Galdi, *Steady flow of a Navier-Stokes fluid around a rotating obstacle*, J. Elasticity **71** (2003), no. 1-3, 1–31.

[14] M. Geissert, H. Heck, and M. Hieber, L^p*-Theory of the Navier-Stokes flow past rotating or moving obstacles*, Preprint (2004).

[15] M. Geissert, H. Heck, and M. Hieber, *Remarks on Bogovskii's operator in Sobolev spaces of negative order*, Preprint (2004).

[16] Y. Giga, *Analyticity of the semigroup generated by the Stokes operator in* L_r *spaces*, Math. Z. **178** (1981), no. 3, 297–329.

[17] Y. Giga, *Solutions for semilinear parabolic equations in* L^p *and regularity of weak solutions of the Navier-Stokes system.* J. Differential Equations, **62** (1986), 186–212.

[18] Y. Giga, S. Matsui and O. Sawada, *Global existence of two-dimensional Navier-Stokes flow with nondecaying initial velocity.* J. Math. Fluid Mech., **3** (2001), 302–315.

[19] Y. Giga, and T. Miyakawa, *Solutions in* L^r *of the Navier-Stokes initial value problem.* Arch. Rational Mech. Anal., **89** (1985), 267–281.

[20] Y. Giga, H. Sohr, *On the Stokes operator in exterior domains.* J. Fac. Sci. Univ. Tokyo, **36**, (1989), 313–333.

[21] Y. Giga and O. Sawada, *On regularizing-decay rate estimate for solutions to the Navier-Stokes initial value problem.* Nonlinear Anal. and Appl., (2003), to appear.

[22] M. Hieber, and O. Sawada, *The Navier-Stokes equations in* $\mathbb{R}^n$ *with linearly growing initial data.* Arch. Rational Mech. Anal., to appear 2004.

[23] T. Hishida, *An existence theorem for the Navier-Stokes flow in the exterior of a rotating obstacle.* Arch. Rat. Mech. Anal., **150** (1999), 307–348.

[24] T. Hishida, *The Stokes operator with rotation effect in exterior domains.* Analysis, **19** (1999), 51–67.

[25] T. Kato, *Strong* L^p*-solutions of Navier-Stokes equations in* $\mathbb{R}^n$ *with applications to weak solutions.* Math. Z., **187** (1984), 471–480.

[26] O. Layzhenskaya, *The Mathematical Theory of Viscous Incompressible Flow.* Gordon and Breach, New York, 1969.

[27] A. Lunardi, *On the Ornstein-Uhlenbeck operator in* L^2*-spaces with respect to invariant measures.* Trans. Amer. Math. Soc. **349** (1997), 155–169.

[28] A. Majda, *Vorticity and the mathematical theory of incompressible fluid flow.* Comm. Pure Appl. Math., **34** (1986), 187–220.

[29] G. Metafune, J. Prüss, A. Rhandi, and R. Schnaubelt, *The domain of the Ornstein-Uhlenbeck operator on an L^p-space with invariant measure.* Ann. Sc. Norm. Super. Pisa Cl. Sci., **1** (2002), 471–485.

[30] H. Sohr, *The Navier-Stokes equations*, Birkhäuser Advanced Texts: Birkhäuser, Basel, 2001.

[31] V.A. Solonnikov, *Estimates for solutions of nonstationary Navier-Stokes equations.* J. Soviet Math., **8** (1977), 467–529.

Matthias Hieber
Technische Universität Darmstadt
Fachbereich Mathematik
Schlossgartenstr. 7
D-64289 Darmstadt, Germany
e-mail: `hieber@mathematik.tu-darmstadt.de`

Progress in Nonlinear Differential Equations
and Their Applications, Vol. 64, 253–264

A Global Bifurcation Result for Variational Inequalities

Milan Kučera, Jan Eisner and Lutz Recke

Dedicated to Professor Herbert Amann in the occasion of his 65th birthday

1. Introduction

Let H be a real Hilbert space with the inner product $\langle \cdot, \cdot \rangle$ and the norm $\| \cdot \|$, let K be a closed convex cone with the vertex at the origin in H. We will consider variational inequalities

$$\lambda \in \mathbb{R}, \; u \in K: \quad \langle u - \lambda Lu - G(\lambda, u), \varphi - u \rangle \geq 0 \text{ for all } \varphi \in K \tag{1.1}$$

with a real bifurcation parameter λ. We assume that

$$\begin{aligned} &L : H \to H \text{ is a linear compact mapping,} \\ &G : \mathbb{R} \times H \to H \text{ is a continuous nonlinear compact mapping,} \\ &\lim_{\|u\| \to 0} \tfrac{\|G(\lambda, u)\|}{\|u\|} = 0 \text{ uniformly on compact } \lambda\text{-intervals.} \end{aligned} \tag{1.2}$$

Denote by

$$C_I := \overline{\{(\lambda, u) \in \mathbb{R} \times H \, : \, (1.1) \text{ is fulfilled}, \; \|u\| \neq 0\}}$$

the closure of the set of all nontrivial solutions to (1.1). The aim of this paper is to show that for a given couple (λ_0, u_0) satisfying the "linearized variational inequality" and certain particular assumptions, there is a component of C_I bifurcating at $(\lambda_0, 0)$ from the set of trivial solutions in the direction u_0. This component is global in the sense that it is unbounded or reaches a point (λ, u) satisfying certain conditions determined by u_0. An application is given to a simple reaction-diffusion system with a unilateral nonlocal boundary condition.

The variational inequality (1.1) is equivalent to the equation $u = P_K(\lambda Lu + G(\lambda, u))$, where P_K is the projection onto the cone K. If the Leray-Schauder index of the mapping $I - P_K(\lambda L)$ changes between two values λ_1 and λ_2 then there is a global bifurcation between λ_1 and λ_2 (see [7]). However, $I - P_K(\lambda L)$ is a nonlinear

The research has been supported by the grant A 1019202 of the Grant Agency of the Academy of Sciences of the Czech Republic and by Deutsche Forschungsgemeinschaft (Sfb 555).

nonsmooth mapping and its index can be calculated only in particular situations (see [7, 9]). We consider here a situation when there need not be a jump of the index for this nonlinear mapping but there is a jump of the index of the linear mapping $I - \lambda PL$, where P is the orthogonal projection onto a certain subspace determined by u_0. Using Dancer's global bifurcation theorem [1] for the equation

$$u = P(\lambda Lu + G(\lambda, u)) \tag{1.3}$$

and some relationship between the sets of solutions to (1.1) and (1.3), we obtain our result.

2. Main result

It is well known that the cone K can be written in the form

$$K = \{u \in H : \langle u, v_\alpha \rangle \geq 0 \text{ for all } \alpha \in \mathcal{A}\}, \tag{2.1}$$

where $\mathcal{A}$ is a certain set of indices and $\{v_\alpha\}_{\alpha \in \mathcal{A}}$ is a family of vectors in H, $\|v_\alpha\| = 1$ (see [2]). We assume that $\mathcal{A} \neq \emptyset$, i.e., $K \neq H$. We will always consider a fixed solution $(\lambda_0, u_0) \in \mathbb{R} \times H$ of the "linearized" variational inequality (more precisely, the variational inequality with the linearized operator)

$$\lambda \in \mathbb{R}, \ u \in K: \ \langle u - \lambda Lu, \varphi - u \rangle \geq 0 \text{ for all } \varphi \in K. \tag{2.2}$$

We assume that $\|u_0\| = 1$ and

$$\mathcal{A}_0 := \{\alpha \in \mathcal{A} : \langle u_0, v_\alpha \rangle = 0\} \neq \emptyset. \tag{2.3}$$

Let us introduce the closed subspace

$$H_0 := \{u \in H : \ \langle u, v_\alpha \rangle = 0 \text{ for all } \alpha \in \mathcal{A}_0\}$$

and the orthogonal projection P of H onto H_0. We assume that there is $c_0 > 0$ such that

$$\langle u_0, v_\alpha \rangle \geq c_0 \ \text{ for all } \alpha \in \mathcal{A} \setminus \mathcal{A}_0, \tag{2.4}$$

$$\lambda_0 \langle Lu_0, (I-P)\varphi \rangle \leq -c_0 \|(I-P)\varphi\| \ \text{ for all } \varphi \in K, \tag{2.5}$$

$$\lambda_0 \text{ is an (algebraically) simple characteristic value of the operator } PL, \tag{2.6}$$

i.e., $(\lambda_0)^{-1}$ is a simple eigenvalue of PL. In what follows, solutions (λ, u) to (1.1) satisfying the following conditions will play a basic role:

$$\langle u, v_\alpha \rangle > 0 \text{ for all } \alpha \in \mathcal{A} \setminus \mathcal{A}_0, \tag{2.7}$$

$$\langle \lambda Lu + G(\lambda, u), (I-P)\varphi \rangle < 0 \text{ for all } \varphi \in K \text{ with } (I-P)\varphi \neq 0. \tag{2.8}$$

We denote

$$C_I(u_0) := \text{the component of } \overline{\{(\lambda, u) \in C_I \ : \ u \in H_0, \ (2.7), (2.8) \ \text{ hold }\}} \text{ containing } (\lambda_0, 0).$$

Remark 2.1. *The assumptions* (2.4), (2.5) *ensure that the couple* (λ_0, u_0) *satisfies* (2.2) *if and only if it satisfies* $u_0 = PLu_0$ *(see Lemma* 2.5 *and Remark* 2.9 *below). The assumption* (2.6) *guaranties that* u_0 *is the only normalized solution to* (2.2) *satisfying* $u \in H_0$, (2.7)*. Hence, the continuum* $C_I(u_0)$ *(if it is nonempty) bifurcates at* $(\lambda_0, 0)$ *in the direction* u_0 *(see Remark* 2.10 *for details). Let us remark that there could exist solutions* (λ_0, u_1) *to* (2.2) *with* $\|u_1\| = 1$ *and* $u_1 \neq u_0$ *which satisfy* (2.3)–(2.6)*, but with another set* $\mathcal{A}_0$*, another subspace* H_0 *and another projection* P *and, hence, another correspondingly defined component* $C_I(u_1)$ *of solutions to* (1.1).

Remark 2.2. *In many applications the variational inequality* (1.1) *can be interpreted as a principle of virtual work. Then, for a solution* (λ, u) *of* (1.1) *with* $u \in H_0$*, the condition* (2.8) *says the following. If there is a nonactive constraint for a given "virtual displacement"* $\varphi - u$ *among those active for* u_0 *(i.e.,* $\langle \varphi, v_\alpha \rangle > 0$ *for a certain* $\alpha \in \mathcal{A}_0$*), then the "virtual work", carried out by the generalized force on such a displacement, should be negative (i.e.,* $\langle \lambda Lu + G(\lambda, u), \varphi - u \rangle < 0$*), cf.* [5, Remark 2.2] *for more details. In the application discussed in Section* 3*, the condition* (2.8) *is equivalent to* (3.9)*, which means a nontrivial flux through an "active" part of the boundary.*

Theorem 2.3. *Let* (λ_0, u_0) *satisfy* (2.4)–(2.6)*. Then at least one of the following conditions is fulfilled:*

(i) $C_I(u_0)$ *is unbounded;*

(ii) $C_I(u_0)$ *contains a point* $(\lambda, u) \neq (\lambda_0, 0)$ *for which* $\langle u, v_\alpha \rangle = 0$ *for some* $\alpha \in \mathcal{A} \setminus \mathcal{A}_0$ *or* $\langle \lambda Lu + G(\lambda, u), (I - P)\varphi \rangle = 0$ *for some* $\varphi \in K$ *with* $(I - P)\varphi \neq 0$.

Let us note that the point (λ, u) from (ii) can be of the type $(\lambda, 0)$, $\lambda \neq \lambda_0$. Of course, in this case (2.2) with this λ has a nontrivial solution. Further properties of the set $C_I(u_0)$ will be mentioned on the end of this section.

Remark 2.4. *Let* C_I^0 *be the component of* C_I *containing* $C_I(u_0)$*. Our considerations do not exclude the case when* C_I^0 *(starting in the direction* u_0*) can return to* $(\lambda_0, 0)$ *in the direction of an other solution* $\tilde{u}_0 \neq u_0$ *of* (2.2) *(see Remark* 2.1*), e.g., that* C_I^0 *"contains a closed loop".*

Lemma 2.5. *If* (λ, u) *satisfies* (1.3) *and the weakened conditions* (2.7), (2.8) *where the sharp sign* $>$ *is replaced by* $\geq$*, then* (λ, u) *fulfills also* (1.1)

Proof. We have $u \in H_0$ for any solution of (1.3) and this together with the weakened condition (2.7) gives $u \in K$. Further, $\langle u - \lambda Lu - G(\lambda, u), P(\varphi - u) \rangle = \langle P(u - \lambda Lu - G(\lambda, u)), \varphi - u \rangle = 0$ and $\langle u - \lambda Lu - G(\lambda, u), (I - P)(\varphi - u) \rangle = -\langle \lambda Lu + G(\lambda, u), (I - P)\varphi \rangle \geq 0$ for all $\varphi \in K$ by the weakened condition (2.8). Adding the expressions with P and $I - P$ we get (1.1). □

For $\varepsilon > 0$, $R > 0$, we denote

$$K_\varepsilon(u_0) = \left\{ (\lambda, u) \in \mathbb{R} \times H : \left\| \frac{u}{\|u\|} - u_0 \right\| < \varepsilon \right\},$$
$$B_R(\lambda_0, 0) = \{ (\lambda, u) \in \mathbb{R} \times H : |\lambda - \lambda_0| + \|u\| \leq R \}.$$

Lemma 2.6. *Let (λ_0, u_0) satisfy (2.4), (2.5). If $(\lambda, u) \in K_{c_0}(u_0)$ with c_0 from (2.4) then $\langle u, v_\alpha \rangle \geq c(u) := \|u\|(c_0 - \|\frac{u}{\|u\|} - u_0\|) > 0$ for all $\alpha \in \mathcal{A} \setminus \mathcal{A}_0$. There is $\varepsilon > 0$ such that if $(\lambda, u) \in B_\varepsilon(\lambda_0, 0) \cap K_\varepsilon(u_0)$ then the condition (2.8) holds.*

Proof. The first assertion follows from a direct calculus. Assume that there is no $\varepsilon > 0$ such that (2.8) holds for all (λ, u) considered. Then there are (λ_n, u_n) such that $(\lambda_n, u_n) \to (\lambda_0, 0)$, $\frac{u_n}{\|u_n\|} \to u_0$ and $\langle \lambda_n L u_n + G(\lambda_n, u_n), (I-P)\varphi_n \rangle \geq 0$ for some $\varphi_n \in K$, $(I-P)\varphi_n \neq 0$. Due to the assumptions (2.5), (1.2), we have

$$c_0 \leq \left\langle \frac{\lambda_n L u_n + G(\lambda_n, u_n)}{\|u_n\|} - \lambda_0 L u_0, \frac{(I-P)\varphi_n}{\|(I-P)\varphi_n\|} \right\rangle \to 0,$$

which is a contradiction. □

Remark 2.7. *Let us recall the Dancer's global bifurcation theorem [1]. Consider a real Banach space E with the norm $\|\cdot\|$. Let $\tilde{L} : E \to E$ and $N : \mathbb{R} \times E \to E$ be a linear compact mapping and a continuous nonlinear compact mapping such that $\lim_{\|u\| \to 0} \frac{\|N(\lambda,u)\|}{\|u\|} = 0$ uniformly on compact λ-intervals. Let λ_0 be a simple characteristic value of the operator $\tilde{L}$, let $u_0 \in E$ and $u_0^* \in E^*$ be the corresponding eigenvector of $\tilde{L}$ and of the adjoint operator $\tilde{L}^*$, respectively, such that $\langle u_0^*, u_0 \rangle = 1$ (here $\langle \cdot, \cdot \rangle$ denotes the duality pairing between E and E^*). Set*

$$C = \overline{\{(\lambda, u) \in \mathbb{R} \times E : \|u\| \neq 0, \ u - \lambda \tilde{L} u - N(\lambda, u) = 0\}}$$

and let C_0 be the component of C containing $(\lambda_0, 0)$. Following [1], let us define subcontinua C_0^+ and C_0^- of C_0 starting at $(\lambda_0, 0)$ in the direction of u_0 and $-u_0$, respectively. For $\eta \in (0,1)$, set

$$K_\eta = \{(\lambda, u) \in \mathbb{R} \times E : |\langle u, u_0^* \rangle| > \eta \|u\|\}$$
$$K_\eta^+ = \{(\lambda, u) \in K_\eta : \langle u, u_0^* \rangle > 0\}, \ K_\eta^- = \{(\lambda, u) \in K_\eta : \langle u, u_0^* \rangle < 0\}.$$

By Lemma 1.24 in [10],

$$\text{there exists } R > 0 \text{ such that } (C \setminus \{(\lambda_0, 0)\}) \cap B_R(\lambda_0, 0) \subset K_\eta. \tag{2.9}$$

For each $r \in (0, R]$ denote by $D_r^\pm$ the components of the sets $\{(\lambda_0, 0)\} \cup (C \cap B_r(\lambda_0, 0) \cap K_\eta^\pm)$, respectively, containing $(\lambda_0, 0)$. Denote by $C_{0,r}^+$ and $C_{0,r}^-$ the components of $\overline{C_0 \setminus D_r^-}$ and $\overline{C_0 \setminus D_r^+}$, respectively, containing $(\lambda_0, 0)$. Set

$$C_0^+ = \overline{\bigcup_{0<r\leq R} C_{0,r}^+}, \qquad C_0^- = \overline{\bigcup_{0<r\leq R} C_{0,r}^-}.$$

It follows from (2.9) that the definitions of C_0^+ and C_0^- are independent of the choice of $\eta \in (0,1)$, both C_0^+ and C_0^- are connected and $C_0 = C_0^+ \cup C_0^-$ (see also [1]). The Dancer's global bifurcation theorem [1] states that

$$\text{either } C_0^+ \cap C_0^- \neq \{(\lambda_0, 0)\} \text{ or both } C_0^+ \text{ and } C_0^- \text{ are unbounded.} \tag{2.10}$$

Proof of Theorem 2.3. Set $E = H$, $\tilde{L}u = PLu$, $N(\lambda, u) = PG(\lambda, u)$ and consider the sets C, C_0^+ and C_0^- from Remark 2.7 corresponding to this choice. Hence, C_0^+ and C_0^- is the half-branch of nontrivial solutions to (1.3) bifurcating at $(\lambda_0, 0)$ in the direction u_0 and $-u_0$, respectively. Let us denote

C_A^+ – the component of the set $\overline{\{(\lambda, u) \in C_0^+ : (2.7), (2.8) \text{ hold}\}}$ containing $(\lambda_0, 0)$.

Let ε be from Lemma 2.6. Standard considerations using (1.2) imply that there is $R > 0$ such that

$$(C \cap B_R(\lambda_0, 0)) \setminus \{(\lambda_0, 0)\} \subset K_\varepsilon(u_0) \cup (-K_\varepsilon(u_0)) \tag{2.11}$$

(see [10], Lemma 1.24). Let us choose R such that (2.11) holds, $0 < R < \varepsilon$. We know from Lemma 2.6 that (2.7), (2.8) hold for all $(\lambda, u) \in B_\varepsilon(\lambda_0, 0) \cap K_\varepsilon(u_0)$, and if $(\lambda, u) \in -K_\varepsilon(u_0)$ then clearly (2.7) is not fulfilled. It follows by using (2.11) and then realizing the definition of C_A^+ and (2.10), that

$$\left(C_A^+ \cap B_R(\lambda_0, 0)\right) \setminus \{(\lambda_0, 0)\} = C_0^+ \cap B_R(\lambda_0, 0) \cap K_\varepsilon(u_0) \neq \emptyset. \tag{2.12}$$

Any $(\lambda, u) \in C_A^+$ satisfies (1.3) and the weakened conditions (2.7), (2.8) where the sharp sign $>$ is replaced by $\geq$. Hence, Lemma 2.5 implies that any $(\lambda, u) \in C_A^+$ satisfies (1.1). The set C_A^+ is connected and contains $(\lambda_0, 0)$ and therefore it follows

$$C_A^+ \subset C_I(u_0). \tag{2.13}$$

Now, let us assume that (ii) does not hold. First, let us show that then $C_A^+ = C_0^+$. If $C_A^+ \neq C_0^+$ then there is $(\lambda, u) \in C_0^+$ such that at least one of the conditions $\langle u, v_\alpha \rangle \leq 0$ for some $\alpha \in \mathcal{A} \setminus \mathcal{A}_0$ or $\langle \lambda Lu + G(\lambda, u), (I - P)\varphi \rangle \geq 0$ for some $\varphi \in K$ with $\|(I - P)\varphi\| \neq 0$ holds. We know from (2.12) and Lemma 2.6 that $C_A^+ \setminus \{(\lambda_0, 0)\}$ contains points (λ, u) satisfying (2.7), (2.8), and it follows by using the connectedness of C_0^+, (2.13) and the definition of C_A^+ that one of the cases mentioned in (ii) must occur. Further, let us show that $C_0^+ \cap C_0^- = \{(\lambda_0, 0)\}$. Indeed, if $C_0^+ \cap C_0^- \neq \{(\lambda_0, 0)\}$ then C_A^+ contains a closed loop and, according to the definition of C_0^+, C_0^- and the assumption (2.4) (i.e., $\langle -u_0, v_\alpha \rangle < 0$ for all $\alpha \in \mathcal{A} \setminus \mathcal{A}_0$), C_A^+ must contain a point (λ, u) with $\langle u, v_\alpha \rangle < 0$ for some $\alpha \in \mathcal{A} \setminus \mathcal{A}_0$, which is not true. Hence, $C_A^+ = C_0^+$ must be unbounded by (2.10). Now, (i) is a consequence of (2.13). □

Remark 2.8. (*see* [5, Lemma 2.3]) *Let us suppose that $\mathcal{A}_0$ is a finite set and assume that the vectors $\{v_\alpha\}_{\alpha \in \mathcal{A}_0}$ are linearly independent. Then there exists a basis $\{v_\alpha^*\}_{\alpha \in \mathcal{A}_0}$ in* span $\{v_\alpha : \alpha \in \mathcal{A}_0\}$ *such that $\langle v_\alpha^*, v_\beta \rangle = \delta_{\alpha\beta}$ for $\alpha, \beta \in \mathcal{A}_0$ and $(I - P)u = \sum_{\alpha \in \mathcal{A}_0} \langle u, v_\alpha^* \rangle v_\alpha = \sum_{\alpha \in \mathcal{A}_0} \langle u, v_\alpha \rangle v_\alpha^*$ for $u \in H$. Then the condition* (2.5) *holds if and only if $\langle \lambda_0 L u_0, v_\alpha^* \rangle < 0$ for all $\alpha \in \mathcal{A}_0$. Moreover, we have $\langle \lambda Lu + G(\lambda, u), (I - P)\varphi \rangle = 0$ for some $\varphi \in K$ with $(I - P)\varphi \neq 0$ (the condition from (ii) in Theorem* 2.3) *if and only if $\langle \lambda Lu + G(\lambda, u), v_\alpha^* \rangle = 0$ for some $\alpha \in \mathcal{A}_0$.*

Remark 2.9. *If $(\lambda, u) \in \mathbb{R} \times H_0$ is a solution to* (1.1) *such that there is $c > 0$ with $\langle u, v_\alpha \rangle \geq c$ for all $\alpha \in \mathcal{A} \setminus \mathcal{A}_0$, then (λ, u) is also a solution to* (1.3) (*see* [5, Lemma

2.2]). *Consequently, Lemmas* 2.5 *and* 2.6 *imply that the problems* (1.1) *and* (1.3) *are equivalent in* $B_R(\lambda_0, 0) \cap K_\varepsilon(u_0)$ *for sufficiently small* R *and* ε.

Analogously, if $(\lambda, u) \in \mathbb{R} \times H_0$ *is a solution to* (2.2) *with* $\langle u, v_\alpha \rangle \geq c$ *for all* $\alpha \in \mathcal{A} \setminus \mathcal{A}_0$, *then* $u = \lambda PLu$. *Hence,* (2.6) *implies that* u_0 *is an isolated normalized solution to* (2.2) *for* $\lambda = \lambda_0$.

Remark 2.10. *In the proof of Theorem* 2.3, ε *could be taken arbitrarily small and* R *was such that* (2.11) *holds. Hence, the definitions of* $C_I(u_0)$ *and* C_A^+ *and Remark* 2.9 *imply that for any* $\varepsilon > 0$ *there is* $R > 0$ *such that* $(C_I(u_0) \cap B_R(\lambda_0, 0)) \setminus \{(\lambda_0, 0)\} = (C_A^+ \cap B_R(\lambda_0, 0)) \setminus \{(\lambda_0, 0)\}$.

Therefore (2.12) *ensures that* $(C_I(u_0) \cap B_R(\lambda_0, 0)) \setminus \{(\lambda_0, 0)\} \subset K_\varepsilon(u_0)$ *and in this sense one can say that* $C_I(u_0)$ *bifurcates in the direction* u_0.

3. Application

Let Ω be a bounded domain in $\mathbb{R}^n$, $n \geq 2$, with a Lipschitzian boundary $\partial\Omega$. Let Γ_D and Γ_N be open subsets of $\partial\Omega$, $\operatorname{meas}\Gamma_D > 0$, $\operatorname{meas}(\partial\Omega \setminus (\Gamma_D \cup \Gamma_N)) = 0$. Furthermore, let Γ_U be a subset of Γ_N. We will consider the problem

$$\left.\begin{aligned} &d_1^0 \Delta u_1 + \lambda(b_{11}u_1 + b_{12}u_2 + g(u_1,u_2)u_1) = 0 \\ &d_2^0 \Delta u_2 + \lambda(b_{21}u_1 + b_{22}u_2) = 0 \end{aligned}\right. \quad \text{in } \Omega, \tag{3.1}$$

$$\begin{aligned} &u_1 = u_2 = 0 \text{ on } \Gamma_D, \quad \frac{\partial u_1}{\partial n} = 0 \text{ on } \Gamma_N, \quad \frac{\partial u_2}{\partial n} = 0 \text{ on } \Gamma_N \setminus \Gamma_U, \\ &\int_{\Gamma_U} u_2 \,\mathrm{d}\Gamma \geq 0, \ \frac{\partial u_2}{\partial n} \geq 0, \ \frac{\partial u_2}{\partial n} \text{ is constant}, \ \frac{\partial u_2}{\partial n} \cdot \int_{\Gamma_U} u_2 \,\mathrm{d}\Gamma = 0 \text{ on } \Gamma_U. \end{aligned} \tag{3.2}$$

Here $d^0 := (d_1^0, d_2^0)$ is a couple of positive numbers (diffusion coefficients), $b_{ij} \in \mathbb{R}$ $(i, j = 1, 2)$, g is a real differentiable function on $\mathbb{R}^2$, $g(0,0) = 0$, $\lambda \in \mathbb{R}$ is the bifurcation parameter.

Set $H_D := \{\varphi \in W^{1,2}(\Omega) : \varphi = 0 \text{ on } \Gamma_D \text{ in the sense of traces}\}$, $H := H_D \times H_D$. The elements of H will be denoted by $u = (u_1, u_2)$. Introduce the inner product in H_D by

$$\langle \varphi, \psi \rangle := \int_\Omega \nabla\varphi \cdot \nabla\psi \,\mathrm{d}x \quad \text{for all } \varphi, \psi \in H_D$$

and denote by $\|\cdot\|$ the corresponding norm in H_D. The same symbols will be used for the corresponding inner product and norm in H. Further, consider the closed convex cone

$$K_2 := \left\{ \varphi \in H_D : \int_{\Gamma_U} \varphi \,\mathrm{d}\Gamma \geq 0 \right\}$$

with the vertex at the origin in H_D. A weak solution to (3.1), (3.2) is a couple $u = (u_1, u_2) \in H$ satisfying the variational inequality

$$\begin{aligned} &u_1 \in H_D : \ d_1^0 u_1 - \lambda(b_{11}Au_1 + b_{12}Au_2 + G_1(u_1,u_2)u_1) = 0, \\ &u_2 \in K_2 : \ \langle d_2^0 u_2 - \lambda(b_{21}Au_1 + b_{22}Au_2), \psi - u_2 \rangle \geq 0 \ \text{ for all } \psi \in K_2 \end{aligned} \tag{3.3}$$

where the operators $A : H_D \to H_D$, $G_1 : H \to \mathcal{L}(H_D)$ (the space of all linear mappings in H_D) are defined by $\langle A\varphi, \psi \rangle = \int_\Omega \varphi\psi \,\mathrm{d}x$, $\langle G_1(u_1,u_2)\varphi, \psi \rangle =$

$\int_\Omega g(u_1,u_2)\varphi\psi\,\mathrm{d}x$ for all $u_1,u_2,\varphi,\psi \in H_D$. Clearly, A is a linear, completely continuous, positive symmetric operator. We will assume $g(\xi,\eta) \le c(1+|\xi|^{q-2}+|\eta|^{q-2})$, q arbitrary for $n=2$, $q<\frac{2n}{n-2}$ for $n>2$, so that G_1 is well defined, that the map $(u_1,u_2)\in H \mapsto G_1(u_1,u_2)u_1 \in H_D$ is completely continuous and $\|G_1(u_1,u_2)u_1\|/\|(u_1,u_2)\| \to 0$ for $\|(u_1,u_2)\|\to 0$.

We need also the "linearized" variational inequality (which is a nonlinear problem again)

$$\begin{aligned} &u_1 \in H_D:\ d_1^0 u_1 - \lambda(b_{11}Au_1 + b_{12}Au_2) = 0\\ &u_2 \in K_2:\ \langle d_2^0 u_2 - \lambda(b_{21}Au_1 + b_{22}Au_2), \psi - u_2\rangle \ge 0 \ \text{ for all } \psi \in K_2. \end{aligned} \tag{3.4}$$

This is a weak formulation of the problem

$$\begin{aligned} d_1^0\Delta u_1 + \lambda(b_{11}u_1 + b_{12}u_2) = 0\\ d_2^0\Delta u_2 + \lambda(b_{21}u_1 + b_{22}u_2) = 0 \end{aligned} \quad \text{in } \Omega \tag{3.5}$$

with (3.2).

The problem (3.3) can be written in the form (1.1) with the cone $K := H_D \times K_2$, which is of the form (2.1) with $\mathcal{A} := \{0\}$, $v_0 = (0, v_2^0) \in H$ defined by $\langle v_0, u\rangle = \int_{\Gamma_U} u_2\,\mathrm{d}\Gamma$ for all $u=(u_1,u_2)\in H$, and with the operators $Lu := ((b_{11}Au_1 + b_{12}Au_2)/d_1^0, (b_{21}Au_1 + b_{22}Au_2)/d_2^0)$, $G(\lambda,u) := \big(\lambda G_1(u_1,u_2)u_1/d_1^0, 0\big)$ for all $\lambda\in\mathbb{R}$, $u=(u_1,u_2)\in H$. The conditions (1.2) follow from our growth assumptions.

Remark 3.1. *If $v\in H_D$, $\Delta v \in L_2(\Omega)$ then the normal derivative $\frac{\partial v}{\partial n}$ can be defined as a linear bounded functional on the space H_D by $\big(\frac{\partial v}{\partial n},\varphi\big) = \int_\Omega \Delta v\varphi + \nabla v \nabla\varphi dx$ for all $\varphi\in H_D$ where $(\cdot,\cdot)$ denotes the dual pairing. All equalities and inequalities for the normal derivative used can be understood in the sense of this functional (see* [5, Remark 5.2] *for details). It is easy to see that $u=(u_1,u_2)\in H$ is a solution of* (3.3) *or of* (3.4) *(i.e., a weak solution of* (3.1) *or* (3.5) *with* (3.2)*) if and only if $\Delta u_1, \Delta u_2 \in L^2(\Omega)$,* (3.1) *or* (3.5)*, respectively, holds a.e. in Ω and the boundary conditions* (3.2) *are fulfilled in the sense mentioned. Analogously for the other problems considered. See* [5, Observation 5.2].

By a solution of any boundary value problem we will always mean its weak solution. Further, we will assume that

$$b_{11}>0,\ b_{12}<0,\ b_{21}>0,\ b_{22}<0,\ b_{11}+b_{22}<0,\ \det b_{ij}>0. \tag{3.6}$$

We will need the following notation (see also Fig. 1A):

$$\begin{aligned} h_j &:= \left\{d=[d_1,d_2]\in\mathbb{R}_+^2 : d_2 = \tfrac{b_{12}b_{21}/\kappa_j^2}{d_1 - b_{11}/\kappa_j} + \tfrac{b_{22}}{\kappa_j}\right\},\ j=1,2,\dots,\\ D_S &:= \left\{d\in\mathbb{R}_+^2 : d \text{ is to the right from all } h_j,\ j=1,2,\dots\right\},\\ D_U &:= \left\{d\in\mathbb{R}_+^2 : d \text{ is to the left from at least one } h_j\right\},\\ \omega &:= \overline{D_S}\cap\overline{D_U}, \end{aligned}$$

where κ_j are eigenvalues of $\Delta v + \kappa v = 0$ on Ω with $v=0$ on Γ_D, $\frac{\partial v}{\partial n}=0$ on Γ_N. Under the assumption (3.6), this notation has the following interpretation related to the problem

$$\begin{aligned} d_1\Delta u_1 + b_{11}u_1 + b_{12}u_2 = 0 \\ d_2\Delta u_2 + b_{21}u_1 + b_{22}u_2 = 0 \end{aligned} \quad \text{in } \Omega \tag{3.7}$$

$$u_1 = u_2 = 0 \text{ on } \Gamma_D, \quad \frac{\partial u_1}{\partial n} = \frac{\partial u_2}{\partial n} = 0 \text{ on } \Gamma_N \tag{3.8}$$

with parameters $(d_1, d_2) \in \mathbb{R}^2_+$. The trivial solution of the parabolic system corresponding to (3.7), (3.8) is stable for $d \in D_S$ (the domain of stability) and unstable for $d \in D_U$ (the domain of instability). Simultaneously, the trivial solution of ODE obtained for $d_1 = d_2 = 0$ is stable. Hence, the "diffusion driven instability" occurs, which plays an essential role in many mathematical models in biology (see, e.g., [6]).

Remark 3.2. *Let* (3.6) *hold. Then the problem* (3.7), (3.8) *has a nontrivial solution if and only if* $d \in h_j$ *for some* $j = 1, 2, \ldots$. *See, e.g.,* [8] *for details.*

Lemma 3.3. *Let* (3.6) *hold. Let* (λ, u) *satisfy* (3.4) *with* $d^0/\lambda \in D_S$, $\|u\| \neq 0$. *Then*

$$\frac{\partial u_2}{\partial n} > 0 \ \text{on } \Gamma_U. \tag{3.9}$$

Proof. If $\frac{\partial u_2}{\partial n} = 0$ on Γ_U then u is simultaneously a weak solution to the problem (3.5), (3.8) (see also Remark 3.1). Hence, it cannot be $d^0/\lambda \in D_S$ due to Remark 3.2. □

Now we will distinguish two cases:

$$d^0/\lambda \notin h_1 \text{ for all } \lambda > 0 \tag{3.10}$$

$$d^0/\lambda_c \in h_1 \text{ for some } \lambda_c > 0, \quad d^0/\lambda \in D_S \text{ for all } 0 < \lambda < \lambda_c. \tag{3.11}$$

Remark 3.4. *Since* h_j *have the joint tangent intersecting the origin,* (3.10) *means that the straight line* $\sigma := \{d^0/\lambda : \lambda > 0\}$ *does not intersect the border* ω *between* D_S *and* D_U, *while* (3.11) *occurs always if* σ *intersects* ω, d^0/λ_c *being the first intersection point.*

Further, $C_I(u_0)$ will be the set from Section 1 for our particular H, K, L, G, H_0, P. We will assume that

$$g(u_1, u_2) \leq 0 \text{ for all } (u_1, u_2) \in \mathbb{R}^2. \tag{3.12}$$

Theorem 3.5. *Let* (3.6), (3.12) *hold. Let us assume that* $(\lambda_0, u_0) = (\lambda_0, u_1^0, u_2^0)$ *satisfies* (3.4), $d^0/\lambda_0 \in D_S$, $\|u_0\| > 0$ *and*

$$\|u_1^0\|^2 + \frac{d_2^0}{d_1^0}\frac{b_{12}}{b_{21}}\|u_2^0\|^2 \neq 0. \tag{3.13}$$

(a) *If* (3.10) *holds and there is no* $(\lambda, 0) \in C_I(u_0)$ *with* $\lambda \neq \lambda_0$ *then* $C_I(u_0)$ *is unbounded.*
(b) *If* (3.11) *holds and there is no* $(\lambda, 0) \in C_I(u_0)$ *with* $\lambda \neq \lambda_0$, $\lambda < \lambda_c$, *then* $C_I(u_0)$ *is unbounded or for any* $\lambda \in (\lambda_0, \lambda_c)$ *there is* $u \in H$ *such that* $(\lambda, u) \in C_I(u_0)$, $\|u\| \neq 0$.

The last assertion in (b) means that the set $C_I(u_0)$ emanating at $(\lambda_0, 0)$ reaches at least λ_c, i.e., d^0/λ reaches the border between D_S and D_U (see Fig. 1). The existence of the points (λ_0, u_0) satisfying our assumptions will be discussed on the end of the paper.

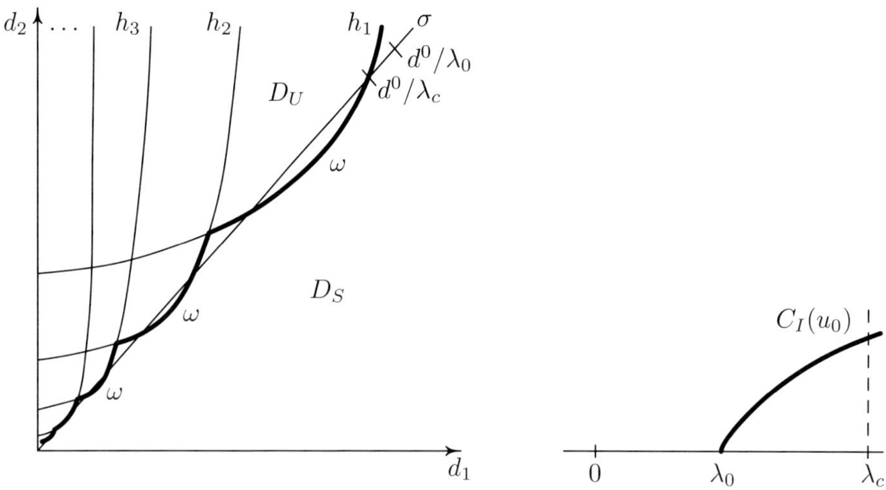

FIGURE 1. Left: The system of hyperbolas h_j, their envelope ω, domain of stability D_S (to the right from ω) and instability D_U (to the left from ω), the straight line $\sigma = \{d^0/\lambda : \lambda > 0\}$.
Right: A possible bifurcation continuum $C_I(u_0)$ emanating from $(\lambda_0, 0)$ in the case (3.11).

Remark 3.6. *Let* (3.6) *hold. The weak formulation of* (3.7), (3.8) *is*

$$\begin{aligned} d_1u_1 - b_{11}Au_1 - b_{12}Au_2 &= 0 \\ d_2u_2 - b_{21}Au_1 - b_{22}Au_2 &= 0. \end{aligned} \tag{3.14}$$

Expressing u_2 from the second equation and substituting to the first one, we see that (3.14) *is equivalent to*

$$\begin{aligned} &u_2 = b_{21}(d_2I - b_{22}A)^{-1}Au_1 \\ &d_1u_1 - b_{11}Au_1 - b_{12}b_{21}A(d_2I - b_{22}A)^{-1}Au_1 = 0. \end{aligned} \tag{3.15}$$

It follows from Remark 3.2 *that for any fixed $d_2 > 0$, the greatest eigenvalue $d_{\max}(d_2)$ of the linear operator $S_{d_2} := b_{11}A + b_{12}b_{21}A(d_2I - b_{22}A)^{-1}A$ coincides with the maximal d_1 such that $(d_1, d_2) \in \omega$, and therefore it is positive. The operator S_{d_2} is compact and symmetric, and it follows from the variational characterization of its greatest positive eigenvalue that*

$$d_{\max}(d_2) = \max_{w \in H_D, \|w\|=1} \left\langle b_{11}Aw + b_{12}b_{21}A(d_2I - b_{22}A)^{-1}Aw, w \right\rangle.$$

Further, it follows from the definition of D_U and D_S that $(d_1, d_2) \in D_U$ if and only if $d_1 < d_{\max}(d_2)$, and $(d_1, d_2) \in D_S$ if and only if $d_1 > d_{\max}(d_2)$.

For the proof of Theorem 3.5, we will compare the problems (3.7), (3.8) and

$$\begin{aligned} d_1\Delta u_1 + b(x)u_1 + b_{12}u_2 = 0 \\ d_2\Delta u_2 + b_{21}u_1 + b_{22}u_2 = 0 \end{aligned} \quad \text{in } \Omega \tag{3.16}$$

with (3.8), the parameters $d_1, d_2 \in \mathbb{R}_+$ and $b \in L^r(\Omega)$ with r arbitrary for $n = 2$, $r < \frac{n}{2}$ for $n > 2$. The weak formulation of (3.16), (3.8) is

$$\begin{aligned} d_1u_1 - \tilde{A}u_1 - b_{12}Au_2 = 0, \\ d_2u_2 - b_{21}Au_1 - b_{22}Au_2 = 0 \end{aligned} \tag{3.17}$$

where the operator $\tilde{A} : H_D \to H_D$ is defined by $\langle \tilde{A}v, \varphi \rangle = \int_\Omega b(x)v\varphi \, dx$ for all $v, \varphi \in H_D$.

Lemma 3.7. *Let (3.6) hold, $b \in L^r(\Omega)$ with r as above, $b(x) \le b_{11}$. If (3.16), (3.8) has a nontrivial solution for some $(d_1, d_2) \in \mathbb{R}^2_+$ then $(d_1, d_2) \in \overline{D_U}$. If the nonlinear problem (3.1), (3.8) has a nontrivial solution for some $\lambda > 0$ and (3.12) holds then $d^0/\lambda \in \overline{D_U}$.*

Proof. Applying considerations from Remark 3.6 to (3.17) and using the assumption $b(x) \le b_{11}$, for any solution (u_1, u_2) to (3.16) and (3.8) we obtain

$$d_1 = \left\langle \tilde{A}u_1 + b_{12}b_{21}A(d_2I - b_{22}A)^{-1}Au_1, u_1 \right\rangle \|u_1\|^{-2} \le d_{\max}(d_2).$$

Hence, Remark 3.6 yields $(d_1, d_2) \in \overline{D_U}$. If (u_1, u_2) is a solution to (3.1), (3.8) then it is a solution to (3.16), (3.8) with $d_i := d_i^0/\lambda$, $i = 1, 2$, and $b(x) := b_{11} + g(u_1(x), u_2(x)) \le b_{11}$ by (3.12). Hence, the last assertion of our Lemma is a consequence of the first one. □

Proof of Theorem 3.5. We will apply Theorem 2.3 with $\mathcal{A}_0 = \{0\}$, $H_0 = \{(\varphi, \psi) \in H : \int_{\Gamma_U} \psi \, d\Gamma = 0\}$. Lemma 3.3 and (3.2) yield that

$$\frac{\partial u_2^0}{\partial n} > 0 \text{ on } \Gamma_U, \quad \int_{\Gamma_U} u_2^0 \, d\Gamma = 0 \tag{3.18}$$

under our assumptions. Hence, $u_0 \in H_0$ and (2.4) is clearly fulfilled. In the notation of Remark 2.8 we have $v_0^* = v_0$ and we obtain $\langle \lambda_0 Lu_0, v_0^* \rangle = \int_\Omega \frac{\lambda_0}{d_2^0}(b_{21}u_1^0 + b_{22}u_2^0)v_2^0 \, dx = -\int_\Omega \Delta u_2^0 v_2^0 \, dx = \langle u_2^0, v_2^0 \rangle - \int_{\Gamma_U} \frac{\partial u_2^0}{\partial n} v_2^0 \, d\Gamma = \langle u_0, v_0 \rangle - c\langle v_0, v_0 \rangle = -c$, where $c := \frac{\partial u_2^0}{\partial n}$ on Γ_U. Hence, (2.5) follows from (3.18) by using Remark 2.8.

Due to Remark 2.1, for the proof of (2.6), we need to show that

$$\ker(I - \lambda_0 PL) = \text{span}\,\{(u_0)\}, \ \langle u_0, u_0^* \rangle \ne 0 \tag{3.19}$$

where u_0^* is an eigenvector corresponding to λ_0 of the adjoint operator L^*P. The couple $u = (u_1, u_2) \in H$ satisfies $u = \lambda PLu$ if and only if $\Delta u_1, \Delta u_2 \in L^2(\Omega)$, (3.5) holds a.e. in Ω and the boundary conditions

$$\begin{aligned} &u_1 = u_2 = 0 \text{ on } \Gamma_D, \quad \frac{\partial u_1}{\partial n} = 0 \text{ on } \Gamma_N, \quad \frac{\partial u_2}{\partial n} = 0 \text{ on } \Gamma_N \setminus \Gamma_U, \\ &\int_{\Gamma_U} u_2 \,\mathrm{d}\Gamma = 0, \ \frac{\partial u_2}{\partial n} \text{ is constant on } \Gamma_U \end{aligned} \tag{3.20}$$

are fulfilled (cf. Remark 3.1). Clearly, (λ_0, u_0) satisfies (3.5), (3.20). If there were an other normalized solution $u = (u_1, u_2) \neq \pm u_0$ to (3.5), (3.20) then we could find their linear combination $\tilde{u} := (\tilde{u}_1, \tilde{u}_2)$ such that $\frac{\partial \tilde{u}_2}{\partial n} = 0$ on Γ_U. Hence, $\tilde{u}$ would be a solution of (3.5), (3.8), which would contradict Remark 3.2 because we have $d^0/\lambda_0 \in D_S$. Hence, the first condition in (3.19) is proved. The second condition in (3.19) follows from (3.13) because we can choose $u_0^* = (u_1^*, u_2^*)$, $u_1^* = u_1^0$, $u_2^* = \frac{d_2^0 b_{12}}{d_1^0 b_{21}} u_2^0$.

If (ii) from Theorem 2.3 does not hold then $C_I(u_0)$ is unbounded.

Further, let the condition (ii) from Theorem 2.3 be fulfilled. Assume that there is no $(\lambda, 0) \in C_I(u_0)$, $\lambda \neq \lambda_0$, as in the assumptions of (a), (b) in Theorem 3.5. We have $\mathcal{A} \setminus \mathcal{A}_0 = \emptyset$ and therefore there is $(\lambda, u) = (\lambda, u_1, u_2) \in C_I(u_0)$ such that $\|u\| \neq 0$, $\langle \lambda Lu + G(\lambda, u), (I - P)\varphi \rangle = 0$ for some $\varphi \in K$ with $(I - P)\varphi \neq 0$. Because of Remark 2.8 and the fact $\mathcal{A} = \mathcal{A}_0 = \{0\}$, we have $\langle \lambda Lu + G(\lambda, u), v_0^* \rangle = 0$. Similarly as above we obtain $\langle \lambda Lu + G(\lambda, u), v_0^* \rangle = \int_\Omega \frac{\lambda}{d_2^0}(b_{21}u_1 + b_{22}u_2)v_2^0 \,\mathrm{d}x = -c$ where $c := \frac{\partial u_2}{\partial n}$ on Γ_U. This implies $\frac{\partial u_2}{\partial n} = 0$ on Γ_U. Hence (u_1, u_2) is a solution of (3.16), (3.8) with $b(x) = b_{11} + g(u_1, u_2)$, $d = d^0/\lambda$. Lemma 3.7 implies that $d^0/\lambda \in \overline{D_U}$. In the case (3.10), this is excluded. In the case (3.11), $d^0/\lambda \in \overline{D_U}$ implies $\lambda \geq \lambda_c$ and the connectedness of $C_I(u_0)$ implies that for any $\lambda \in (\lambda_0, \lambda_c)$ there is $u \in H$, $\|u\| \neq 0$, $(\lambda, u) \in C_I(u_0)$. □

Let us note that in the case (3.11), the existence of at least one (λ_0, u_0) satisfying all assumptions of Theorem 3.5 with the exception of (3.13) is guaranteed by the following Theorem 3.8, which follows from [4, Theorem 2.1]. It can be shown by the methods used in [3, 4, 9] that such points exist also in the case (3.10) when the straight line σ does not intersect ω but is sufficiently close to it. The assumption (3.13) must be checked in concrete examples.

Theorem 3.8. *Let* (3.6) *and* (3.11) *hold. Then there is a bifurcation point* $\lambda_I < \lambda_c$ *of the problem* (3.1), (3.2). *In particular,* $d^I := (d_1^0/\lambda_I, d_2^0/\lambda_I) \in D_S$ *and there is a nontrivial solution to* (3.3). *(See Fig. 1.)*

References

[1] E.N. Dancer, On the structure of solutions of non-linear eigenvalue problems. Indiana Univ. Math. J., Vol. 23, No. 11 (1974), 1069–1076.

[2] E. DiBenedetto, Real Analysis, Birkhäuser Advanced Texts, Birkhäuser, Boston, 2002.

[3] P. Drábek, M. Kučera, M. Míková, Bifurcation points of reaction-diffusion systems with unilateral conditions. Czechoslovak Math. J., 35 (1985), 639–660.

[4] J. Eisner, M. Kučera, Spatial patterning in reaction-diffusion systems with nonstandard boundary conditions, Fields Institute Communications 25 (2000), 239–256.

[5] J. Eisner, M. Kučera, L. Recke, Smooth continuation of solutions and eigenvalues for variational inequalities based on the implicit function theorem. J. Math. Anal. Appl. 274, No. 1 (2002), 159–180.

[6] J.D. Murray, Mathematical Biology, 2nd ed., Biomathematics Texts, Vol. 19, (1993), Springer-Verlag Berlin, Heidelberg.

[7] V.K. Le, K. Schmitt, Global Bifurcation in Variational Inequalities. Application to Obstacle and Unilateral Problems. Springer-Verlag, New York, 1997.

[8] M. Mimura, Y. Nishiura, M. Yamaguti, Some diffusive prey and predator systems and their bifurcation problems. Ann. N.Y. Acad. Sci. 316 (1979), 490–521.

[9] P. Quittner, Bifurcation points and eigenvalues of inequalities of reaction-diffusion type. J. Reine Angew. Math., 380 (1987), 1–13.

[10] P.H. Rabinowitz, Some global results for non-linear eigenvalue problems. J. Functional Analysis 7 (1971), 487–513.

[11] L. Recke, J. Eisner, M. Kučera, Smooth bifurcation for variational inequalities based on the implicit function theorem. J. Math. Anal. Appl. 275 No. 2 (2002), 615–641.

Milan Kučera and Jan Eisner
Mathematical Institute of the Academy
of Sciences of the Czech Republic
Žitná 25
115 67 Prague 1

Lutz Recke
Humboldt University
Unter den Linden 6
D-10099 Berlin

Progress in Nonlinear Differential Equations
and Their Applications, Vol. 64, 265–272

On Elliptic Non-divergence Operators with Measurable Coefficients

Peer Christian Kunstmann

Abstract. We study properties of the coefficient matrices of non-divergence operators on $\mathbb{R}^n$ aiming at sectoriality and R-sectoriality of these operators. In particular, we present results on approximation, scaling, and the behavior in the L_p-scale.

1. Introduction

Let $m \in \mathbb{N}$. In this paper we study sectoriality and R-sectoriality of elliptic operators of order $2m$ in $L_p(\mathbb{R}^n, \mathbb{C})$, $p \in (1, \infty)$, which have the following form

$$A = \sum_{|\alpha|=2m} a_\alpha(x) D^\alpha, \quad D(A) = W_p^{2m}(\mathbb{R}^n, \mathbb{C}), \tag{1}$$

where $a_\alpha : \mathbb{R}^n \to \mathbb{C}$ are bounded measurable functions. Here we use the usual multi-index notation $\alpha = (\alpha_1, \dots, \alpha_n) \in \mathbb{N}_0^n$, $|\alpha| = \alpha_1 + \dots + \alpha_n$, and $D^\alpha = D_1^{\alpha_1} \dots D_n^{\alpha_n}$ where $D_j := -i\partial/\partial x_j$ for $j = 1, \dots, n$.

The classical three step approach to show sectoriality of non-divergence operators can be described as follows: First study operators with constant coefficients. Second study small perturbations of constant coefficient operators. In the last step use localization and patch together the localized operators (we refer in particular to the localization scheme in [1] for this last step). The approach has also been successfully used to establish a number of further properties such as maximal L_p-regularity, R-sectoriality, boundedness of imaginary powers (BIP), or boundedness of an H^∞-functional calculus. For the second and third step, it seems to be unavoidable to assume some kind of smoothness for the highest-order coefficients, e.g., uniform continuity ([1], [9], [13], [14], [5]) or VMO ([10]) for the properties of sectoriality and R-sectoriality, and Hölder continuity for BIP ([16]) or a bounded H^∞-calculus ([1], [6]) (at least, Hölder continuity is needed for boundary value problems; on $\mathbb{R}^n$, however, it is known that uniform continuity ([7]) or even VMO

([8]) is sufficient). It is also very well known that for all these properties lower-order terms $\sum_{|\beta|<2m} a_\beta(x)D^\beta$ with bounded and measurable a_β may be studied by perturbation methods. Hence we may neglect lower-order terms in our study.

In this paper we somehow change the point of view. We make no assumptions on the smoothness of the coefficients and we do not use localization. Instead, we investigate what can be said in the general case by looking not on a particular operator in a particular L_p, but on sets of operators in a particular L_p or on a particular operator in a scale of L_p-spaces. On the technical side we limit ourselves to abstract nonsense and hand-waving. Consequently, our observations are not of an *absolute* (certain regularities imply certain properties) but rather of a *relative* nature. Nevertheless, they give some useful information.

The paper is organized as follows: in Section 2 we study approximation of coefficients and prove a result (Theorem 1) on approximation in the *strong* operator topology (not in *uniform* operator topology, the latter being, of course, very well known and usually used for bounded uniformly continuous coefficients). This result on sectoriality also holds for R-sectoriality and makes clear how a precise control on the constants may be exploited. The result on R-sectoriality is new, and even the result on sectoriality has, to our knowledge, not been explicitly stated before. We recall that, for operators in divergence form, scaling may sometimes be used to *improve* constants in certain estimates (cf. [2]). In Section 3 we concentrate on non-divergence operators with coefficients that are *invariant* under scaling. On the one hand, this additional assumption makes life more easy and, in fact, the results are somehow analogous to those for operators in divergence form. On the other hand, however, famous counterexamples in elliptic theory belong to this class. In Section 4, the last one, we use a result from abstract interpolation theory to investigate the behavior of a given operator in a *scale* of L_p-spaces. These results are also new.

Notation: We denote tuples $(a_\alpha)_{|\alpha|=2m}$ of functions $a_\alpha \in L_\infty(\mathbb{R}^n,\mathbb{C})$ by $\mathfrak{a}$ and the set of all such $\mathfrak{a}$ by $\mathscr{A}^{2m}$. For a given tuple of coefficients $\mathfrak{a} \in \mathscr{A}^{2m}$ and $p \in (1,\infty)$ fixed we define the associated operator in $L_p(\mathbb{R}^n,\mathbb{C})$ by

$$(\mathrm{Op}(\mathfrak{a})u)(x) := \sum_{|\alpha|=2m} a_\alpha(x)(D^\alpha u)(x), \quad x \in \mathbb{R}^n.$$

where $u \in D(\mathrm{Op}(\mathfrak{a})) := W_p^{2m}(\mathbb{R}^n,\mathbb{C})$. We define

$$\mathscr{S}_p^{2m} := \{\mathfrak{a} \in \mathscr{A}^{2m} : \mathrm{Op}(\mathfrak{a}) \text{ is quasi-sectorial in } L_p(\mathbb{R}^n,\mathbb{C})\}.$$

Recall that an operator A in a Banach space X is called *sectorial* if, for some $\theta_0 \in (0,\pi)$, the spectrum $\sigma(A)$ of A is contained in the closure of the open sector $\Sigma_{\theta_0} := \{z \in \mathbb{C}\setminus\{0\} : |\arg z| < \theta_0\}$ and, for all $\theta \in (0,\pi-\theta_0)$, the set $\{\lambda(\lambda+A)^{-1} : \lambda \in \Sigma_\theta\}$ is bounded. The operator A is called *quasi-sectorial* if $\nu + A$ is sectorial for some $\nu \geq 0$. If, in these definitions, the set $\{\lambda(\lambda+A)^{-1} : \lambda \in \Sigma_\theta\}$ is even R-bounded, then the operator A is called *R-sectorial* or *quasi-R-sectorial*, respectively. We denote the set of coefficients $\mathfrak{a}$ such that $\mathrm{Op}(\mathfrak{a})$ is quasi-R-sectorial in $L_p(\mathbb{R}^n,\mathbb{C})$ by $\mathscr{RS}_p^{2m}$. Recall that a set $\mathbb{T}$ of linear operators $X \to Y$ is called

R-bounded, if there is a constant $C > 0$ such that, for all $k \in \mathbb{N}$, $T_1, \ldots, T_k \in \mathbb{T}$, and $x_1, \ldots, x_k \in X$, one has

$$\mathbb{E}\|\sum_{j=1}^{k} \varepsilon_j T_j x_j\|_Y \leq C\mathbb{E}\|\sum_{j=1}^{k} \varepsilon_j x_j\|_X,$$

where (ε_j) denotes a sequence of independent and symmetric, $\{\pm 1\}$-valued random variables on a probability space, e.g., the Rademachers. The infimum of all such C is called the R-bound of $\mathbb{T}$. For the notions of R-boundedness, R-sectoriality, and their connection with maximal L_p-regularity we refer to [14].

Since all functions are complex-valued on $\mathbb{R}^n$ we drop $\mathbb{R}^n$ and $\mathbb{C}$ in notation and write only L_p and W_p^{2m}.

2. Sets of coefficients and approximation

On $\mathscr{A}^{2m}$ we consider the norm $\|\mathfrak{a}\|_\infty := \sum_{|\alpha|=2m} \|a_\alpha\|_\infty$. For any $q \in [1, \infty]$ we denote by τ_q the topology induced by the strong operator topology when considering the elements a_α of $\mathfrak{a}$ as pointwise multipliers on $L^q(\mathbb{R}^n)$. Then $\tau_1 \subset \tau_q \subset \tau_r \subset \tau_\infty$ for $1 \leq q \leq r \leq \infty$ and τ_∞ is induced by $\|\cdot\|_\infty$. Note that $\mathfrak{a}_k \to \mathfrak{a}\,(k \to \infty)$ in τ_q for $q \in [1, \infty)$ if $(\mathfrak{a}_k)$ is a sequence such that $\sup_k \|\mathfrak{a}_k\|_\infty < \infty$, and $\mathfrak{a}_k \to \mathfrak{a}\,(k \to \infty)$ point- and componentwise a.e.

In the following $p \in (1, \infty)$ is fixed. For a tuple $\mathfrak{a}$ of coefficients and $A = \mathrm{Op}(\mathfrak{a})$ we introduce the following notation for the constants appearing in a priori estimates, where $\nu \geq 0$ and $\theta \in (0, \pi)$:

$$\begin{aligned} P_{\theta,\nu}(\mathfrak{a}) := \inf\{M > 0 : \forall \lambda \in \Sigma_\theta, u \in W_p^{2m} \\ : \|\lambda u\|_p + \|(-\Delta)^m u\|_p \leq M\|(\lambda + \nu + A)u\|_p\}. \end{aligned}$$

We use the convention $\inf \emptyset := \infty$. For $\nu = 0$ we just write $P_\theta(\mathfrak{a}) = P_{\theta,0}(\mathfrak{a})$. We also shall use the constants in resolvent estimates:

$$M_{\theta,\nu}(\mathfrak{a}) := \sup\{\|\lambda(\lambda + \nu + A)^{-1}\|_{p\to p} + \|(-\Delta)^m(\lambda + \nu + A)^{-1}\|_{p\to p} : \lambda \in \Sigma_\theta\}.$$

Again we let $M_\theta(\mathfrak{a}) := M_{\theta,0}(\mathfrak{a})$ and understand $M_{\theta,\nu}(\mathfrak{a}) = \infty$ if $-\Sigma_\theta \not\subset \rho(\nu + A)$. Since we are also interested in R-sectoriality, we introduce the corresponding R-boundedness counterparts. We denote by $P_{\theta,\nu}^R(\mathfrak{a})$ the infimum of all constants $C > 0$ such that, for all $k \in \mathbb{N}$, $\lambda_1, \ldots, \lambda_k \in \Sigma_\theta$, and $u_1, \ldots, u_k \in W_p^{2m}$ we have

$$\mathbb{E}\|\sum_{j=1}^{k} \varepsilon_j \lambda_j u_j\|_p + \mathbb{E}\|\sum_{j=1}^{k} \varepsilon_j (-\Delta)^m u_j\|_p \leq C\mathbb{E}\|\sum_{j=1}^{k} \varepsilon_j (\lambda_j + \nu + A)u_j\|_p, \qquad (2)$$

and we denote by $R_{\theta,\nu}(\mathfrak{a})$ the R-bound of the set

$$\{(\lambda(\lambda + \nu + A)^{-1}, (-\Delta)^m(\lambda + \nu + A)^{-1}) : \lambda \in \Sigma_\theta\},$$

considered as operators from L_p into $L_p \times L_p$. Recall that p was fixed. If we want to indicate the special L^p-space to which we refer then we use p as an additional subscript, e.g., $M_{\theta,\nu,p}(\mathfrak{a})$ etc.

We have the following theorem on approximation of coefficients.

Theorem 1. *Let* $p \in (1,\infty)$ *and* $\mathscr{K} \subset \mathscr{S}_p^{2m}$ *be convex and* $\|\cdot\|_\infty$*-bounded. Suppose that* $M,\nu \geq 0$ *and* $\theta \in (0,\pi)$ *are such that* $M_{\theta,\nu} \leq M$ *on* $\mathscr{K}$*. Then* $\overline{\mathscr{K}}^{\tau_p} \subset \mathscr{S}_p^{2m}$ *and* $M_{\theta,\nu} \leq M$ *on* $\overline{\mathscr{K}}^{\tau_p}$*.*

If, in addition, $\mathscr{K} \subset \mathscr{R}\mathscr{S}_p^{2m}$ *and* $R_{\theta,\nu} \leq M$ *on* $\mathscr{K}$*, then* $\overline{\mathscr{K}}^{\tau_p} \subset \mathscr{R}\mathscr{S}_p^{2m}$ *and* $R_{\theta,\nu} \leq M$ *on* $\overline{\mathscr{K}}^{\tau_p}$*.*

Proof. First we notice that $M_{\theta,\nu} \leq M$ on $\mathscr{K}$ implies $P_{\theta,\nu} \leq M$ on $\mathscr{K}$. Then we observe that $P_{\theta,\nu}$ is lower semi-continuous for τ_p on $\mathscr{A}^{2m}$ which is immediate from the definitions. Hence $P_{\theta,\nu} \leq M$ on $\overline{\mathscr{K}}^{\tau_p}$. For $\mathfrak{a} \in \overline{\mathscr{K}}^{\tau_p}$ we choose $\mathfrak{a}_0 \in \mathscr{K}$ and consider $\mathfrak{a}_\mu := (1-\mu)\mathfrak{a}_0 + \mu\mathfrak{a}$ for $\mu \in [0,1]$. Since $\overline{\mathscr{K}}^{\tau_p}$ is again convex, we have $P_{\theta,\nu}(\mathfrak{a}_\mu) \leq M$ for all $\mu \in [0,1]$. Since $\mathfrak{a}_0 \in \mathscr{K} \subset \mathscr{S}_p^{2m}$, the continuity method yields the first assertion.

Since $\overline{\mathscr{K}}^{\tau_p} \subset \mathscr{S}_p^{2m}$, the assertion $R_{\theta,\nu} \leq M$ on $\overline{\mathscr{K}}^{\tau_p}$ is equivalent to $P^R_{\theta,\nu} \leq M$ on $\overline{\mathscr{K}}^{\tau_p}$. Now we notice that

$$\mathbb{E}\|\sum_{j=1}^k \varepsilon_j f_j\|_p = \frac{1}{2^k}\sum_{\sigma_j=\pm1}\|\sum_{j=1}^k \sigma_j f_j\|_p, \tag{3}$$

and the RHS hereof is just an average over choices of signs in the sum. Hence, if we let $\mathfrak{a}_m \to \mathfrak{a}$ in τ_p, then (3) shows that (2) for $\mathrm{Op}(\mathfrak{a}_m)$ in place of A will converge to (2) for $\mathrm{Op}(\mathfrak{a})$ in place of A. Thus $P^R_{\theta,\nu} \leq M$ on $\mathscr{K}$ implies $P^R_{\theta,\nu} \leq M$ on $\overline{\mathscr{K}}^{\tau_p}$, which in turn implies $R_{\theta,\nu} \leq M$ on $\overline{\mathscr{K}}^{\tau_p}$. □

Remark 2. The proof shows that the convexity assumption may be weakened. It is sufficient that $\overline{\mathscr{K}}^{\tau_p}$ is arc-wise connected for τ_∞.

For applicability of the above results we want to mention the following.

Remark 3. For a tuple $\mathfrak{a} = (a_\alpha)_{|\alpha|=2m} \in \mathscr{A}^{2m}$ we denote the associated symbol by

$$\mathrm{Sy}(\mathfrak{a},x,\xi) := \sum_{|\alpha|=2m} a_\alpha(x)\xi^\alpha, \quad x,\xi \in \mathbb{R}^n,$$

where, as usual, $\xi^\alpha := \xi_1^{\alpha_1}\cdot\ldots\cdot\xi_n^{\alpha_n}$ for $\xi = (\xi_1,\ldots,\xi_n) \in \mathbb{R}^n$ and $\alpha = (\alpha_1,\ldots,\alpha_n) \in \mathbb{N}_0^n$. If $K \subset \{z \in \mathbb{C} : \mathrm{Re}\, z > 0\}$ is compact and convex then

$$\mathscr{A}_K^{2m} := \{\mathfrak{a} \in \mathscr{A}^{2m} : \text{ for a.e. } x \in \mathbb{R}^n\text{: } \forall|\xi| = 1 : \mathrm{Sy}(\mathfrak{a},x,\xi) \in K\}$$

is convex. Let $\mathfrak{a} = (a_\alpha) \in \mathscr{A}_K^{2m}$. Taking a standard mollifier sequence $(\rho_k) = (k^n\rho(k\cdot))$ we have that all $\mathfrak{a}_k := (\rho_k * a_\alpha)_{|\alpha|=2m}$, $k \in \mathbb{N}$, belong to $\mathscr{A}_K^{2m}$, and that $\mathfrak{a}_k \to \mathfrak{a}\,(k \to \infty)$ in τ_q for any $q \in [1,\infty)$. Hence via Theorem 1 a good control on sectoriality [R-sectoriality] constants for smooth coefficients yields sectoriality [R-sectoriality] also for non-smooth coefficients.

3. Scaling and scaling invariant coefficients

We define, for $\mu > 0$, the dilation operator $S_\mu f := f(\mu\cdot)$ and extend it to tuples $\mathfrak{a} = (a_\alpha)_{|\alpha|=m} \in \mathscr{A}^{2m}$ via $S_\mu \mathfrak{a} = (S_\mu a_\alpha)_{|\alpha|=2m}$. Observe that $\|S_\mu f\|_p = \mu^{-n/p}\|f\|_p$ for $f \in L_p(\mathbb{R}^n)$ and $p \in (1,\infty)$. For $\mu > 0$ and $\mathfrak{a} \in \mathscr{A}^{2m}$ we have

$$S_{1/\mu}\mathrm{Op}(\mathfrak{a})S_\mu u = \mu^{2m}\mathrm{Op}(S_{1/\mu}\mathfrak{a})u \tag{4}$$

for $u \in W_p^{2m}$. Sine S_μ is an isomorphism of L_p with inverse $S_{1/\mu}$, we conclude that S_μ leaves also $\mathscr{S}_p^{2m}$ invariant.

Lemma 4. *Let $p \in (1,\infty)$, $\mathfrak{a} \in \mathscr{A}^{2m}$ and $\theta \in (0,\pi)$, $\nu \geq 0$. For any $\mu > 0$ we have*

$$K_{\theta,\nu}(\mathfrak{a}) = K_{\theta,\nu\mu^{-2m}}(S_{1/\mu}\mathfrak{a}) \quad \textit{where } K \in \{P, M, R, P^R\}. \tag{5}$$

Proof. The other cases being similar we only study $K = P$. In

$$\|\lambda u\|_p + \|(-\Delta)^m u\|_p \leq M\|(\lambda + \nu + \mathrm{Op}(\mathfrak{a}))u\|_p$$

we write $u = S_\mu v$ and $\|S_{1/\mu}\cdot\|_p$ in place of $\|\cdot\|_p$. By (4) we obtain

$$\|\lambda u\|_p + \|\mu^{2m}(-\Delta)^m u\|_p \leq M\|(\lambda + \nu + \mu^{2m}\mathrm{Op}(S_{1/\mu}\mathfrak{a}))u\|_p.$$

Dividing by μ^{2m} we obtain the assertion since $\mu^{-2m}\Sigma_\theta = \Sigma_\theta$. □

Remark 5. If we have $\nu = 0$ in the preceding lemma we see that the constants cannot depend on, e.g., the modulus of continuity or the Hölder norm of the a_α in $\mathfrak{a}$. This makes finiteness of the constants for $\nu = 0$ the most interesting case.

We call an $\mathfrak{a} \in \mathscr{A}^{2m}$ *scaling-invariant* if $S_\mu\mathfrak{a} = \mathfrak{a}$ for all $\mu > 0$. For scaling-invariant coefficients we obtain the following.

Proposition 6. *Let $p \in (1,\infty)$ and $\mathfrak{a} \in \mathscr{A}^{2m}$ be scaling-invariant. For any $\theta \in (0,\pi)$ and $\nu > 0$ we have*

$$K_{\theta,\nu}(\mathfrak{a}) = K_\theta(\mathfrak{a}) \quad \textit{where } K \in \{P, M, R, P^R\}. \tag{6}$$

In particular, $P_{\theta,\nu}(\mathfrak{a}) < \infty$ for some θ, ν implies the Gårding inequality

$$\|Op(\mathfrak{a})u\|_p \geq (P_\theta(\mathfrak{a}))^{-1}\|u\|_p, \quad u \in W_p^{2m}. \tag{7}$$

Moreover, if $(0,\infty) \cap \rho(-Op(\mathfrak{a})) \neq \emptyset$ then $Op(\mathfrak{a}) \in \mathscr{S}_p^{2m}$.

Proof. For (6) we use (5) in Lemma 4 and let $\mu \to \infty$ (where we may use Theorem 1). We obtain (7) by letting $\lambda \to 0$ in the a priori estimate. To prove the last assertion let $\lambda_0 > 0$ be in $\rho(-\mathrm{Op}(\mathfrak{a}))$ and $M := \|(\lambda_0 + \mathrm{Op}(\mathfrak{a}))^{-1}\|_{p\to p}$. By (4) and scaling-invariance we have, for any $\mu > 0$ and $A = \mathrm{Op}(\mathfrak{a})$,

$$\begin{aligned} M = \|S_{1/mu}(\lambda_0 + A)^{-1}S_\mu\|_{p\to p} &= \|(\lambda_0 + \mu^{2m}A)^{-1}\|_{p\to p} \\ &= \|\mu^{-2m}(\lambda_0\mu^{-2m} + A)^{-1}\|_{p\to p}, \end{aligned}$$

which means $\|\lambda(\lambda + A)^{-1}\|_{p\to p} = M\lambda_0$ for any $\lambda > 0$. □

For general $\mathfrak{a} \in \mathscr{S}^{2m}$ we would obtain (7) only for $\nu + \mathrm{Op}(\mathfrak{a})$ and some $\nu > 0$ in place of $\mathrm{Op}(\mathfrak{a})$.

Examples of scaling-invariant operators are, of course, constant coefficient operators, for which the technique is well known. Other examples are given by operators

$$-(\Delta + c \sum_{j,k=1}^{n} \frac{x_j x_k}{|x|^2} \partial_j \partial_k), \quad c \geq 0.$$

For $n \geq 3$, operators of this type provide well-known counterexamples in the theory of elliptic boundary value problems (we refer to [15, Sect. 1.1]). It seems, however, not to be clear, if they also serve as counterexamples to sectoriality on $\mathbb{R}^n$.

4. Extrapolation via interpolation

In this section we exploit Sneiberg's Lemma (cf. [2, Lem. 23, p. 53], cf. also [12]):

Lemma 7 (Sneiberg's Lemma). *Let (X_θ) and (Y_θ) complex interpolation scales of reflexive spaces where $\theta \in [0,1]$. If S is a linear operator which is bounded $X_\theta \to Y_\theta$ for each $\theta \in [0,1]$. Then the following subsets of $(0,1)$ are open:*

$$\{\theta : \exists \eta > 0 : \|Sx\|_{Y_\theta} \geq \eta \|x\|_{X_\theta}\}, \qquad \{\theta : S : X_\theta \to Y_\theta \text{ is an isomorphism}\}.$$

The second statement follows from the first if we also consider the dual operator S^* in the dual scales (Y_θ^*) and (X_θ^*).

Theorem 8. *For fixed $\mathfrak{a} \in \mathscr{A}^{2m}$, $\theta \in (0,\pi)$, the following subsets of $(1,\infty)$ are open:*

$$\{p \in (1,\infty) : K_{\theta,0,p} < \infty\} \quad \text{where } K \in \{P, M, R, P^R\}.$$

For $\nu > 0$ also the subsets

$$\{p \in (1,\infty) : K_{\theta,\nu,p} < \infty \text{ and } \nu + Op(\mathfrak{a}) \text{ is invertible in } L_p\}$$
$$\text{where } K \in \{P, M, R, P^R\}$$

are open.

Proof. We choose a dense sequence (λ_j) in Σ_θ and let $A := \mathrm{Op}(\mathfrak{a})$. We first treat the case $\nu = 0$ and start with $K \in \{P, M\}$. We shall apply Lemma 7 to

$$X_p := \{(u_j) \in (W_p^{2m})^{\mathbb{N}} : \|(u_j)\|_{X_p} := (\sum_j (\|\lambda_j u_j\|_p + \|(-\Delta)^m u_j\|_p)^p)^{1/p} < \infty\}$$

and $Y_p := l_p(L_p)$ and the operator $S : (u_j) \mapsto ((\lambda_j + A)u_j)$. Observe that

$$\|(\lambda_j + A)u_j\|_p \leq \|\lambda_j u_j\|_p + \|Au_j\|_p \leq \|\lambda_j u_j\|_p + C\|(-\Delta)^m u_j\|_p$$

where C depends on p and $\|\mathfrak{a}\|_\infty$. This yields boundedness of $S : X_p \to Y_p$ for any $p \in (1,\infty)$. Now Lemma 7 proves the claim.

For $K \in \{P^R, R\}$ we use the same operator S but we use

$$X_p := \{(u_j) \in (W_p^{2m})^{\mathbb{N}} : \|(u_j)\|_{X_p} := \mathbb{E}\|\sum_j \varepsilon_j \lambda_j u_j\|_p + \mathbb{E}\|\sum_j \varepsilon_j (-\Delta)^m u_j\|_p < \infty\}$$

and $Y_p := \mathrm{Rad}(L_p)$ with norm $\|(f_j)\|_{Y_p} := \mathbb{E}\|\sum_j \varepsilon_j f_j\|_p$. For elementary properties of $\mathrm{Rad}(X)$ we refer to [14]. Since L_p-spaces have non-trivial type we have $[\mathrm{Rad}(L_p), \mathrm{Rad}(L_q)]_\theta = \mathrm{Rad}(L_r)$ for $1/r = (1-\theta)/p + \theta/q$ (cf. [11]), and (Y_p) is a complex interpolation scale. A similar argument shows that (X_p) is a complex interpolation scale. Now we observe

$$\begin{aligned}\mathbb{E}\|\sum_j \varepsilon_j(\lambda_j + A)u_j\|_p &\leq \mathbb{E}\|\sum_j \varepsilon_j \lambda_j u_j\|_p + \mathbb{E}\|A\sum_j \varepsilon_j u_j\|_p \\ &\leq \mathbb{E}\|\sum_j \varepsilon_j \lambda_j u_j\|_p + C\mathbb{E}\|\sum_j (-\Delta)^m u_j\|_p\end{aligned}$$

where C is the same constant as before. Thus Lemma 7 yields the claim for $K \in \{P^R, R\}$.

The case $\nu > 0$ is treated very similar. We just replace A by $\nu + A$ and $(-\Delta)^m$ by $1 + (-\Delta)^m$. Observe, e.g., that $P_{\theta,\nu,p}(\mathfrak{a}) < \infty$ and $\nu + A$ has a bounded inverse $L_p \to L_p$ if and only if there is a $C > 0$ such that

$$\|\lambda u\|_p + \|(1 + (-\Delta)^m)u\|_p \leq C\|(\lambda + \nu + A)u\|_p, \quad u \in W_p^{2m}.$$

For $k \in \{M, R, P^R\}$ the situation is similar. □

References

[1] H. Amann, M. Hieber, G. Simonett, Bounded H_∞ calculus for elliptic operators, Differential Integral Equations 7, 613–653 (1994).

[2] P. Auscher, Ph. Tchamitchian, Square root problem for divergence operators and related topics, Astérisque 249, Paris, Société Mathématique de France (1998).

[3] S. Blunck, P.C. Kunstmann, Weighted norm estimates and maximal regularity, Adv. Differ. Equ. 7, 1513–1532 (2002).

[4] E.B. Davies, Uniformly elliptic operators with measurable coefficients, J. Funct. Anal. 132, 141–169 (1995).

[5] R. Denk, M. Hieber, J. Prüss, $\mathcal{R}$-boundedness, Fourier multipliers and problems of elliptic and parabolic type, Mem. Am. Math. Soc. 788 (2003).

[6] R. Denk, G. Dore, M. Hieber, J. Prüss, A. Venni, New thoughts on old results of R. T. Seeley, Math. Ann. 328, 545–583 (2004).

[7] X.T. Duong, G. Simonett, H_∞-calculus for elliptic operators with non-smooth coefficients, Differ. Integral Equ. 10, 201–217 (1997).

[8] X.T. Duong, Li Xin Yan, Bounded holomorphic functional calculus for non-divergence form differential operators, Differ. Integral Equ. 15, 709–730 (2002).

[9] R. Haller, H. Heck, M. Hieber, Muckenhoupt weights and maximal L^p-regularity, Arch. Math. 81, 422–430 (2003).

[10] H. Heck, M. Hieber, Maximal L^p-regularity for elliptic operators with VMO-coefficients, J. Evol. Equ. 3, 332–359 (2003).

[11] N.J. Kalton, P.C. Kunstmann, L. Weis, Perturbation and interpolation theorems for the H^∞-calculus with applications to differential operators, to appear in Math. Ann.

[12] N.J. Kalton, M. Mitrea, Stability results on interpolation scales of quasi-Banach spaces and applications, Trans. Am. Math. Soc. 350, 3903–3922 (1998).

[13] P.C. Kunstmann, Maximal L^p-regularity for second-order elliptic operators with uniformly continuous coefficients on domains, in Iannelli, Mimmo (ed.) et al., Evolution equations: applications to physics, industry, life sciences and economics, Basel, Birkhäuser. Prog. Nonlinear Differ. Equ. Appl. 55, 293–305 (2003).

[14] P.C. Kunstmann, L. Weis, Maximal L_p-regularity for parabolic equations, Fourier multiplier theorems and H^∞-functional calculus, in M. Iannelli, R. Nagel, S. Piazzera (eds.), *Functional Analytic Methods for Evolution Equations*, Springer Lecture Notes Math. 1855, 65–311 (2004).

[15] A. Maugeri, D.K. Palagachev, L.G. Softova, Elliptic and Parabolic Equations with Discontinuous Coefficients, Mathematical Research Vol. 109, Wiley-VCH, Berlin, 2000.

[16] J. Prüss, H. Sohr, Imaginary powers of elliptic second-order differential operators in L^p-spaces, Hiroshima Math. J. 23, 161–192 (1993).

Peer Christian Kunstmann
Mathematisches Institut I
Universität Karlsruhe
Englerstr. 2
D-76128 Karlsruhe, Germany
e-mail: peer.kunstmann@math.uni-karlsruhe.de

Progress in Nonlinear Differential Equations
and Their Applications, Vol. 64, 273–290

A Chemotaxis Model with Threshold Density and Degenerate Diffusion

Philippe Laurençot and Dariusz Wrzosek*

Abstract. A quasilinear degenerate parabolic system modelling the chemotactic movement of cells is studied. The system under consideration has a similar structure as the classical Keller-Segel model, but with the following features: there is a threshold value which the density of cells cannot exceed and the flux of cells vanishes when the density of cells reaches this threshold value. Existence and uniqueness of weak solutions are proved. In the one-dimensional case, flat-hump-shaped stationary solutions are constructed.

Mathematics Subject Classification (2000). 35K55, 35K65, 34B15, 34C25.

Keywords. Chemotaxis equations, degenerate diffusion, quasilinear parabolic equation, compactness method.

1. Introduction

We investigate the well-posedness and the properties of steady states for a degenerate parabolic system describing the aggregation phase of the chemosensitive motion of cells. This system is actually a particular case of the general system introduced by Patlak [7] and Keller and Segel [4] and describes the space and time evolution of the cell density u and the chemoattractant density v (see also [3] and the references therein). More precisely, it reads

$$\partial_t u = \nabla \cdot (D(u)\, \nabla u - h(u)\, \nabla v) \quad \text{in} \quad (0,+\infty) \times \Omega\,, \tag{1}$$

$$\partial_t v = \Delta v + g(u,v) \quad \text{in} \quad (0,+\infty) \times \Omega\,, \tag{2}$$

$$(D(u)\, \nabla u - h(u)\, \nabla v)\, \dot{\nu} = 0 \quad \text{on} \quad (0,+\infty) \times \partial\Omega\,, \tag{3}$$

$$(u,v)(0) = (u_0, v_0) \quad \text{in} \quad \Omega\,, \tag{4}$$

where Ω is a bounded open subset of $\mathbb{R}^N$ with smooth boundary $\partial\Omega$ and ν denotes the outward normal unit vector field to $\partial\Omega$. Here, $D(u)$ denotes the density-dependent diffusion coefficient and $h(u)$ is usually written in the form

* The second author was supported by Polish KBN grant 2 P03A 03022.

$h(u) = u\ h_1(u)$ where h_1 is commonly referred to as the chemotactic sensitivity function. The function g describes the rates of production and degradation of the chemoattractant v, and the classical Keller-Segel model corresponds to the case where D and h_1 are constants.

The basic assumption made here is the existence of a threshold value for the cell density which corresponds to a tight packing state. In other words, cells cannot accumulate without bound at a given point of Ω and the corresponding mathematical assumption is that the function h vanishes identically when u crosses a given value, say $u \geq 1$. The effect of a threshold cell density or a volume-filling effect has been taken into account in the modelling of chemotaxis phenomenon in [2, 6], and the well-posedness and large time behavior of solutions to chemotaxis systems incorporating the volume-filling effect were studied in [9, 10]. A particular feature of the models studied in the above-mentioned papers is that there is a relationship between D and h: more precisely, there is a non-increasing function q such that $q(0) = 1$, $q(1) = 0$,

$$D(r) = q(r) - r\ q'(r) \quad \text{and} \quad h(r) = r\ q(r)\,, \quad r \in [0,1]\,. \tag{5}$$

Furthermore, q is assumed to be such that D is bounded from below by a positive constant, so that (1)–(4) is a non-degenerate parabolic system.

Now, since a threshold density for u has been introduced in the model, it is natural to wonder whether this threshold density is attained by u during the evolution. Numerical experiments performed in [6] suggest that the threshold density may be reached by the u-component of stationary states. To the contrary, it turns out that the u-component of steady states to (1)–(3) never attains the threshold value 1 when (5) and the non-degeneracy assumption on D are fulfilled (unless $u \equiv 1$). This was shown in [10, Proposition 1.4] and we extend it in Proposition 8 below. Furthermore, as we will see below, removing the non-degeneracy assumption on D is not sufficient to warrant that the u-component of steady states could take the value 1.

The purpose of this paper is thus twofold: on the one hand, we establish the well-posedness of (1)–(4) without the condition (5) and allow the diffusion coefficient to vanish at the threshold value ($D(1) = 0$). From a biological viewpoint, this means that the diffusivity of cells decreases gradually to zero as u approaches values close to the threshold. On the other hand, we exhibit a condition on D and h which guarantees that, at least in the one-dimensional case, there are stationary states (u, v) to (1)–(3) such that u reaches the threshold density either on large sets (*flat-hump-shaped solutions*) or at one point (*spike-shaped solutions*), see Theorem 10 and Remark 11 below.

Let us now state the assumptions on the data we will use in the sequel, together with the main results we obtain in this paper. We assume that $D \in \mathcal{C}^2(\mathbb{R})$, $h \in \mathcal{C}^2(\mathbb{R})$ satisfy

$$D(1) = h(0) = h(1) = 0 \quad \text{and} \quad \begin{cases} D(r) > 0 & \text{for} \quad r \in [0,1)\,, \\ h(r) > 0 & \text{for} \quad r \in (0,1)\,, \end{cases} \tag{6}$$

and that the reaction term $g \in \mathcal{C}^2(\mathbb{R}^2)$ satisfies the (one-sided) growth condition

$$g(r,0) \geq 0 \quad \text{and} \quad \partial_s g(r,s) \leq \kappa \quad \text{for} \quad (r,s) \in [0,1] \times [0,+\infty)\,, \tag{7}$$

for some $\kappa > 0$.

As for the initial data, we assume that

$$(u_0, v_0) \in L^\infty(\Omega;\mathbb{R}^2) \quad \text{with} \quad 0 \leq u_0 \leq 1 \quad \text{and} \quad 0 \leq v_0 \quad \text{a.e. in} \quad \Omega\,. \tag{8}$$

Further assumptions on D, h and g will be needed to ensure the uniqueness of a solution to (1)–(4) and read: there are $C_0 > 0$, $C_1 > 0$ and functions $g_1 \in \mathcal{C}^2(\mathbb{R})$ and $g_2 \in \mathcal{C}^2(\mathbb{R})$ such that $g_1(0) \geq 0$, $g_2(0) \geq 0$,

$$(h(r) - h(s))^2 \leq C_0\ (r-s)\ (\mathcal{D}(r) - \mathcal{D}(s))\ , \tag{9}$$

$$g(r,s) = g_1(s) + r\ g_2(s) \quad \text{and} \quad \max\{g_1'(s), g_2'(s)\} \leq C_1\,, \tag{10}$$

for $(r,s) \in [0,1] \times [0,+\infty)$, where

$$\mathcal{D}(r) := \int_0^r D(s)\ ds\,, \quad r \in [0,1]\,.$$

Definition 1. *Let $(u_0, v_0) \in L^\infty(\Omega;\mathbb{R}^2)$ be such that (8) holds true. A weak solution to (1)–(4) is a couple (u,v) of functions such that, for each $T > 0$,*

$$0 \leq u(t,x) \leq 1\,, \quad 0 \leq v(t,x) \quad \textit{a.e. in} \quad (0,T) \times \Omega\,, \tag{11}$$

$$u \in L^\infty((0,T)\times\Omega) \cap \mathcal{C}_w([0,T];L^2(\Omega))\,,$$
$$\mathcal{D}(u) \in L^2(0,T;H^1(\Omega))\,, \quad u(0) = u_0\,, \tag{12}$$

$$v \in L^\infty((0,T)\times\Omega) \cap H^1(0,T;L^2(\Omega)) \cap L^2(0,T;H^2(\Omega))\,, \quad v(0) = v_0\,, \tag{13}$$

and (u,v) satisfy

$$\int_0^T \langle \partial_t u, \psi\rangle\ dt \quad + \quad \int_0^T \int_\Omega (\nabla\mathcal{D}(u) - h(u)\ \nabla v)\,(t,x) \cdot \nabla\psi(t,x)\ dxdt = 0$$

$$\partial_t v \quad = \quad \Delta v + g(u,v) \quad \textit{a.e. in} \quad (0,T)\times\Omega\,,$$

$$\partial_\nu v \quad = \quad 0 \quad \textit{a.e. on} \quad (0,T)\times\partial\Omega\,,$$

for each $t \in [0,T]$ and $\psi \in L^2(0,T;H^1(\Omega))$. Here, $\langle .,.\rangle$ denotes the duality pairing between $H^1(\Omega)$ and $H^1(\Omega)'$.

Here, $\mathcal{C}_w([0,T];L^2(\Omega))$ denotes the set of functions from $[0,T]$ in $L^2(\Omega)$ which are continuous for the weak topology of $L^2(\Omega)$.

Theorem 2. *Consider $(u_0,v_0) \in L^\infty(\Omega;\mathbb{R}^2)$ such that (8) holds true. Under the assumptions (6) and (7) on D, h and g, there exists at least a weak solution (u,v) to (1)–(4) in the sense of Definition 1 such that*

$$\int_\Omega u(t,x)\ dx = \int_\Omega u_0(x)\ dx \quad \textit{for each} \quad t \geq 0\,. \tag{14}$$

In addition, if D, h and g also fulfil (9) and (10), then (1)–(4) has a unique weak solution.

As already mentioned, the well-posedness of (1)–(4) has been established in [2, 9] when D and h are related by (5) and $D(r) \geq \delta > 0$ for $r \in [0,1]$. In particular, the case where $q(1) = q'(1) = 0$ is not covered by the analysis performed in [9] and Theorem 2 thus provides an existence theorem for that case. As for uniqueness, we may use Proposition 4 to check that D and h fulfil the condition (9) if $q'(r) \sim -(1-r)^k$ as $r \to 1$ for some $k > 0$. Indeed, in that case, we have $h'(r) \sim -(1-r)^k$ and $D(r) \sim (1-r)^k$ as $r \to 1$.

The proof of Theorem 2 is given in Section 2.1. In the last section, we state some properties of steady states to (1)–(3) and, in the one-dimensional case, construct stationary solutions (u,v) to (1)–(3) such that $u \equiv 1$ on a non-empty subinterval of Ω.

2. Well-posedness

2.1. Existence

The existence statement in Theorem 2 is proved by a compactness method. Owing to the degeneracy of D for $u = 1$, we first study a non-degenerate regularized problem for which the well-posedness is deduced from the abstract theory developed by Amann [1]. More precisely, we have the following result:

Theorem 3. *Let $p > N$ and consider $(u_0, v_0) \in W^{1,p}(\Omega;\mathbb{R}^2)$ such that (8) holds true. Let D, h and g be functions satisfying (6) and (7), and assume further that there is $\eta > 0$ such that*

$$D(r) \geq \eta \quad \textit{for} \quad r \in [0,1]\,. \tag{15}$$

Then there is a unique classical solution

$$(u,v) \in \mathcal{C}([0,+\infty) \times \bar{\Omega};\mathbb{R}^2) \cap \mathcal{C}^{1,2}((0,+\infty) \times \bar{\Omega};\mathbb{R}^2)$$

to (1)–(4) such that $(u(t,x), v(t,x)) \in [0,1] \times [0,+\infty)$ for each $(t,x) \in [0,+\infty) \times \bar{\Omega}$. In addition,

$$\int_\Omega u(t,x)\, dx = \int_\Omega u_0(x)\, dx \quad \textit{for each} \quad t \geq 0\,. \tag{16}$$

Proof. Owing to (15) and the continuity of D, there exists $\delta > 0$ such that

$$D \geq \eta/2 \quad \text{on} \quad (-\delta, 1+\delta)\,. \tag{17}$$

We put $D_0 := (-\delta, +\infty) \times (-\delta, 1+\delta)$ and define $a \in \mathcal{C}^2(D_0; \mathcal{M}_2(\mathbb{R}))$ by

$$a(y) = (a^{m,n})_{1 \leq m,n \leq 2} := \begin{pmatrix} 1 & 0 \\ -h(y_2) & D(y_2) \end{pmatrix} \quad \text{for} \quad y = (y_1, y_2) \in D_0\,.$$

We next set $a_{j,k}(y) := a(y)\,\delta_{j,k}$ for $1 \le j,k \le N$ and $y \in D_0$ and introduce the operators

$$\mathcal{A}(y)z := -\sum_{j,k=1}^{N} \partial_j \left(a_{j,k}(y)\,\partial_k z\right),$$

$$\mathcal{B}(y)z := \sum_{j,k=1}^{N} \nu_j \cdot a_{j,k}(y)\,\partial_k z,$$

and the function

$$f(y) = (f^m)_{1\le m\le 2} := \begin{pmatrix} g(y_2,y_1) \\ 0 \end{pmatrix},$$

with $z = (z_1, z_2)$. With these notations, an abstract formulation of (1)–(4) reads

$$\begin{aligned} \partial_t z + \mathcal{A}(z)z &= f(z), \\ \mathcal{B}(z)z &= 0, \\ z(0) &= (v_0, u_0), \end{aligned}$$

with $z = (v,u)$. Owing to (17), the eigenvalues of the matrix $a(y)$ are positive for each $y \in D_0$, and the boundary-value operator $(\mathcal{A},\mathcal{B})$ is of separated divergence form in the sense of [1, Example 4.3 (e)]. Consequently, the boundary-value operator $(\mathcal{A},\mathcal{B})$ is normally elliptic by [1, Section 4]. We are then in a position to apply [1, Theorems 14.4 & 14.6] to conclude that (1)–(4) has a unique maximal classical solution

$$z := (v,u) \in \mathcal{C}([0,T_\star)\times\bar\Omega;\mathbb{R}^2) \cap \mathcal{C}^{1,2}((0,T_\star)\times\bar\Omega;\mathbb{R}^2),$$

where $T_\star \in (0,+\infty]$. In addition, thanks to (6) and (7), we have $a^{1,2}(0,y_2) = 0$, $a^{2,1}(y_1,0) = -h(0) = 0$ and $f^1(0,y_2) = g(y_2,0) \ge 0$. These properties along with (8) and [1, Theorem 15.1] ensure that $v(t,x) \ge 0$ and $u(t,x) \ge 0$ for $(t,x) \in [0,T_\star)\times\bar\Omega$. We next observe that $\tilde z := (v, 1-u)$ is a solution to a similar initial-boundary value problem with

$$\tilde a(y) = (\tilde a^{m,n})_{1\le m,n\le 2} := \begin{pmatrix} 1 & 0 \\ -h(1-y_2) & D(1-y_2) \end{pmatrix} \quad \text{for} \quad y = (y_1,y_2) \in D_0.$$

Since $\tilde a^{2,1}(y_1,0) = -h(1) = 0$, a further application of [1, Theorem 15.1] and (8) yield that $1 - u(t,x) \ge 0$ for $(t,x) \in [0,T_\star)\times\bar\Omega$. So far, we have thus proved that

$$0 \le u(t,x) \le 1 \quad \text{and} \quad 0 \le v(t,x) \tag{18}$$

for $(t,x) \in [0,T_\star)\times\bar\Omega$. Also, integrating (1) over $(0,t)\times\Omega$ implies that (16) holds true for $t \in [0,T_\star)$. We finally infer from (2), (7) and (18) that

$$\partial_t v - \Delta v \le g(u,0) + \kappa\, v \le C\,(1+v)$$

with $C := \|g(.,0)\|_{L^\infty(0,1)} + \kappa$, from which we conclude that

$$v(t,x) \le (1 + \|v_0\|_{L^\infty})\, e^{Ct}, \quad (t,x) \in [0,T_\star)\times\bar\Omega.$$

We are then in a position to use [1, Theorem 15.5] to conclude that $T_\star = +\infty$. Recalling (18), the proof of Theorem 3 is complete. □

Proof of Theorem 2 (Existence). For $\varepsilon \in (0,1)$ and $r \in \mathbb{R}$, we put $D^\varepsilon(r) := D(r)+\varepsilon$. Also, there exist $(u_0^\varepsilon, v_0^\varepsilon) \in W^{1,N+1}(\Omega;\mathbb{R}^2)$ satisfying (8) and

$$\|v_0^\varepsilon\|_{L^\infty} \le 2\,\|v_0\|_{L^\infty} \quad \text{and} \quad \|u_0^\varepsilon - u_0\|_{L^2} + \|v_0^\varepsilon - v_0\|_{L^2} \le \varepsilon\,. \tag{19}$$

By Theorem 3, the initial-boundary value problem (1)–(4) with D^ε instead of D and $(u_0^\varepsilon, v_0^\varepsilon)$ instead of (u_0, v_0) has a unique classical solution $(u^\varepsilon, v^\varepsilon)$ such that

$$0 \le u^\varepsilon(t,x) \le 1 \quad \text{and} \quad 0 \le v^\varepsilon(t,x)\,, \quad (t,x) \in [0,+\infty)\times\bar{\Omega}\,. \tag{20}$$

In the following we denote by C_i, $i \ge 1$, any positive constant which does not depend on $\varepsilon \in (0,1)$. The dependence of the C_i's upon additional parameters will be indicated explicitly.

We next proceed as in the last step of the proof of Theorem 3 to show that, for each $T > 0$,

$$0 \le v^\varepsilon(t,x) \le C_1(T)\,, \quad (t,x) \in [0,T]\times\bar{\Omega}\,. \tag{21}$$

Consequently, the right-hand side of (2) is bounded on finite time intervals and classical parabolic estimates warrant that, for each $T > 0$ and $p \in (1,\infty)$, there is a positive constant $C_2(p,T)$ such that

$$\|v^\varepsilon\|_{L^p(0,T;W^{2,p}(\Omega))} + \|\partial_t v^\varepsilon\|_{L^p((0,T)\times\Omega)} \le C_2(p,T)\,. \tag{22}$$

We next introduce the functions $\mathcal{D}^\varepsilon$ and $\tilde{\mathcal{D}}^\varepsilon$ defined by

$$\frac{d^2\tilde{\mathcal{D}}^\varepsilon}{dr^2} = \frac{d\mathcal{D}^\varepsilon}{dr} = D^\varepsilon \quad \text{and} \quad \tilde{\mathcal{D}}^\varepsilon(0) = \mathcal{D}^\varepsilon(0) = 0\,.$$

We multiply (1) by $\mathcal{D}^\varepsilon(u^\varepsilon)$ and integrate over Ω to obtain

$$\begin{aligned}
\frac{d}{dt}\int_\Omega \tilde{\mathcal{D}}^\varepsilon(u^\varepsilon)\,dx &= -\int_\Omega \nabla\mathcal{D}^\varepsilon(u^\varepsilon)\cdot\left[\nabla\mathcal{D}^\varepsilon(u^\varepsilon) - h(u^\varepsilon)\,\nabla v^\varepsilon\right]\,dx \\
&\le -\|\nabla\mathcal{D}^\varepsilon(u^\varepsilon)\|_{L^2}^2 + \|h\|_{L^\infty(0,1)}\,\|\nabla\mathcal{D}^\varepsilon(u^\varepsilon)\|_{L^2}\,\|\nabla v^\varepsilon\|_{L^2} \\
&\le -\frac{1}{2}\,\|\nabla\mathcal{D}^\varepsilon(u^\varepsilon)\|_{L^2}^2 + \frac{\|h\|_{L^\infty(0,1)}^2}{2}\,\|\nabla v^\varepsilon\|_{L^2}^2\,,
\end{aligned}$$

whence

$$\int_0^T\int_\Omega |\nabla\mathcal{D}^\varepsilon(u^\varepsilon)|^2\,dxdt \le C_3(T) \tag{23}$$

by (22). Recalling (1), we deduce from (6), (20), (22) and (23) that

$$\int_0^T \|\partial_t u^\varepsilon\|_{H^1(\Omega)'}^2\,dt \le C_4(T)\,. \tag{24}$$

Also, the non-negativity of D implies that $(d\mathcal{D}^\varepsilon/dr) \ge D$, which, together with (23), implies that

$$\int_0^T\int_\Omega |\nabla\mathcal{D}(u^\varepsilon)|^2\,dxdt \le C_3(T)\,. \tag{25}$$

We now define the function $P \in \mathcal{C}^2(\mathbb{R})$ by $P(0) = 0$ and $dP/dr := D^2$ and claim that

$$(P(u^\varepsilon)) \quad \text{is bounded in} \quad \left\{ w \in L^2(0,T;H^1(\Omega)) \,,\ \ \partial_t w \in L^1(0,T;W^{1,N+1}(\Omega)') \right\} \tag{26}$$

for each $T > 0$. Indeed, on the one hand, it follows from (6), (20) and (25) that

$$\|P(u^\varepsilon)\|_{L^\infty} \le \|P\|_{L^\infty(0,1)} \,,$$

and

$$\int_0^T \|\nabla P(u^\varepsilon)\|_{L^2}^2 \ dt \le \|D\|_{L^\infty(0,1)}^2 \int_0^T \|\nabla \mathcal{D}(u^\varepsilon)\|_{L^2}^2 \ dt \le C_5(T) \,.$$

On the other hand, we infer from (6) and the continuous embedding of $W^{1,N+1}(\Omega)$ in $L^\infty(\Omega)$ that, for $\varphi \in W^{1,N+1}(\Omega)$ and $t \in (0,T)$,

$$\begin{aligned}
\left| \int_\Omega \varphi \ \partial_t P(u^\varepsilon)(t) \ dx \right| &\le \|\partial_t u^\varepsilon(t)\|_{H^1(\Omega)'} \ \left\| D(u^\varepsilon(t))^2 \ \varphi \right\|_{H^1(\Omega)} \\
&\le \|\partial_t u^\varepsilon(t)\|_{H^1(\Omega)'} \ \|D\|_{L^\infty(0,1)}^2 \ \|\varphi\|_{H^1} \\
&\quad + 2 \ \|\partial_t u^\varepsilon(t)\|_{H^1(\Omega)'} \ \left\| \frac{dD}{dr} \right\|_{L^\infty(0,1)} \ \|\nabla \mathcal{D}(u^\varepsilon(t))\|_{L^2} \ \|\varphi\|_{L^\infty} \\
&\le C_6 \ \|\partial_t u^\varepsilon(t)\|_{H^1(\Omega)'} \ (1 + \|\nabla \mathcal{D}(u^\varepsilon(t))\|_{L^2}) \ \|\varphi\|_{W^{1,N+1}} \,.
\end{aligned}$$

and the right-hand side of the above estimate belongs to $L^1(0,T)$ by (24) and (25). Thanks to (26), we may apply a classical compactness result [5, Corollary 4] to conclude that $(P(u^\varepsilon))$ is relatively compact in $L^2((0,T)\times\Omega)$ for each $T > 0$. Since P is increasing, the relative compactness of $(P(u^\varepsilon))$ in $L^2((0,T)\times\Omega)$ and the bound (20) allow us to deduce that (u^ε) is relatively compact in $L^p((0,T)\times\Omega)$ for each $p \in [1,\infty)$ and $T > 0$. Also, (20), (24) and a classical compactness result [5, Théorème 1.12.1] ensure that (u^ε) is relatively compact in $\mathcal{C}([0,T];H^1(\Omega)')$ for each $T > 0$. Recalling (22), we conclude from the above analysis that there are functions $(u,v) \in L^\infty((0,T)\times\Omega;\mathbb{R}^2)$ and a subsequence of $(u^\varepsilon, v^\varepsilon)$ (not relabeled) such that

$$(u^\varepsilon, v^\varepsilon) \longrightarrow (u,v) \quad \text{in} \quad L^p((0,T)\times\Omega;\mathbb{R}^2) \,, \tag{27}$$

and

$$(u^\varepsilon, v^\varepsilon) \longrightarrow (u,v) \quad \text{in} \quad \mathcal{C}([0,T];H^1(\Omega)') \times \mathcal{C}([0,T];L^p(\Omega)) \,, \tag{28}$$

for each $p \in [1,\infty)$ and $T > 0$.

Furthermore, (20) and (27) warrant that $(u(t,x), v(t,x)) \in [0,1] \times [0,+\infty)$ for almost every $(t,x) \in [0,+\infty) \times \Omega$, while (25) ensures that $\mathcal{D}(u)$ belongs to $L^2(0,T;H^1(\Omega))$ for each $T > 0$. Recalling (19) and the definition of D^ε, the convergences (27), (28) and classical arguments allow us to show that (u,v) is actually a solution to (1)–(4) in the sense of Definition 1 and complete the proof of the existence part of Theorem 2. □

2.2. Uniqueness

In this section, we prove the uniqueness statement of Theorem 2 under the additional assumptions (9) and (10) on D, h and g. The proof relies on a classical duality technique: introducing the subset $L_0^2(\Omega)$ of $L^2(\Omega)$ defined by

$$L_0^2(\Omega) := \left\{ w \in L^2(\Omega) \,, \quad \int_\Omega w(x)\ dx = 0 \right\} ,$$

we denote by $\mathcal{N}w$ the unique solution to

$$-\Delta \mathcal{N}w = w \quad \text{in} \quad \Omega \,, \tag{29}$$

$$\partial_\nu \mathcal{N}w = 0 \quad \text{on} \quad \partial\Omega \,, \tag{30}$$

in $H^2(\Omega) \cap L_0^2(\Omega)$.

Proof of Theorem 2 (Uniqueness). Let (u,v) and $(\hat{u},\hat{v})$ be two solutions to (1)–(4) in the sense of Definition 1 and fix $T>0$. We put

$$U(t,x) := u(t,x) - \hat{u}(t,x) \,, \qquad V(t,x) := v(t,x) - \hat{v}(t,x) \,,$$

for $(t,x) \in [0,T] \times \Omega$. Then $U(t) \in L_0^2(\Omega)$ for each $t \in [0,T]$ and it follows from Definition 1 and (14) that

$$\begin{aligned} &\int_0^t \langle \partial_t U, \psi \rangle\ ds \\ &= - \int_0^t \int_\Omega \left[\nabla \left(\mathcal{D}(u) - \mathcal{D}(\hat{u}) \right) - \left(h(u) - h(\hat{u}) \right)\ \nabla v - h(\hat{u})\ \nabla V \right] \cdot \nabla \psi\ dxds \end{aligned} \tag{31}$$

for each $\psi \in L^2(0,t;H^1(\Omega))$. We may then take $\psi = \mathcal{N}U$ in (31) and obtain, for each $\delta \in (0,1)$,

$$\begin{aligned} \int_\Omega |\nabla \mathcal{N}U(t)|^2\ dx &= \int_\Omega |\nabla \mathcal{N}U(0)|^2\ dx + 2 \int_0^t \langle \partial_t U, \mathcal{N}U \rangle ds \\ &= 2 \int_0^t \int_\Omega \Delta \mathcal{N}U\ \left(\mathcal{D}(u) - \mathcal{D}(\hat{u}) \right)\ dxds \\ &\quad + 2 \int_0^t \int_\Omega \left(h(u) - h(\hat{u}) \right)\ \nabla \mathcal{N}U \cdot \nabla v\ dxds \\ &\quad + 2 \int_0^t \int_\Omega h(\hat{u})\ \nabla \mathcal{N}U \cdot \nabla V\ dxds \\ &\le -2 \int_0^t \int_\Omega (u - \hat{u})\ \left(\mathcal{D}(u) - \mathcal{D}(\hat{u}) \right)\ dxds \\ &\quad + 2 \int_0^t \|\nabla v\|_{L^\infty}\ \|\nabla \mathcal{N}U\|_{L^2}\ \|h(u) - h(\hat{u})\|_{L^2}\ ds \\ &\quad + 2\ \|h\|_{L^\infty(0,1)} \int_0^t \|\nabla \mathcal{N}U\|_{L^2}\ \|\nabla V\|_{L^2}\ ds \end{aligned}$$

$$
\begin{aligned}
&\leq -2 \int_0^t \int_\Omega (u-\hat{u})\ (\mathcal{D}(u)-\mathcal{D}(\hat{u}))\ dxds \\
&\quad + \delta \int_0^t \int_\Omega (h(u)-h(\hat{u}))^2\ dxds + \frac{1}{\delta} \int_0^t \|\nabla v\|_{L^\infty}^2\ \|\nabla \mathcal{N} U\|_{L^2}^2\ ds \\
&\quad + \delta \int_0^t \|\nabla V\|_{L^2}^2\ ds + \frac{C}{\delta} \int_0^t \|\nabla \mathcal{N} U\|_{L^2}^2\ ds\,.
\end{aligned}
$$

It next follows from (2), (10) and the boundedness of $\hat{u}$, v and $\hat{v}$ on $(0,T)\times\Omega$ that

$$
\begin{aligned}
&\frac{1}{2}\ \frac{d}{dt}\|V\|_{L^2}^2 + \|\nabla V\|_{L^2}^2 \\
&\quad = \int_\Omega V\ \left[g_1(v)-g_1(\hat{v}) + \hat{u}\ (g_2(v)-g_2(\hat{v}))\right]\ dx + \int_\Omega g_2(v)\ U\ V\ dx \\
&\quad \leq C(T)\ \|V\|_{L^2}^2 + \int_\Omega \nabla \mathcal{N} U \cdot \nabla\left(g_2(v)\ V\right)\ dx \\
&\quad \leq C(T)\ \left(\|V\|_{L^2}^2 + \|\nabla V\|_{L^2}\ \|\nabla \mathcal{N} U\|_{L^2} + \|\nabla v\|_{L^\infty}\ \|V\|_{L^2}\ \|\nabla \mathcal{N} U\|_{L^2}\right) \\
&\quad \leq \frac{1}{2}\ \|\nabla V\|_{L^2}^2 + C(T)\ \|V\|_{L^2}^2 + C(T)\ \left(1+\|\nabla v\|_{L^\infty}^2\right)\ \|\nabla \mathcal{N} U\|_{L^2}^2\,,
\end{aligned}
$$

whence

$$
\begin{aligned}
&\|V(t)\|_{L^2}^2 + \int_0^t \|\nabla V\|_{L^2}^2\ ds \\
&\qquad \leq C(T) \int_0^t \|V\|_{L^2}^2\ ds + C(T) \int_0^t \left(1+\|\nabla v\|_{L^\infty}^2\right)\ \|\nabla \mathcal{N} U\|_{L^2}^2\ ds\,.
\end{aligned}
$$

We now sum the previous two inequalities with $\delta \leq \min\{1/4, 1/C_0\}$ fixed and use (9) to deduce that

$$
\|V(t)\|_{L^2}^2 + \|\nabla \mathcal{N} U(t)\|_{L^2}^2 \leq C(T) \int_0^t \left(1+\|\nabla v\|_{L^\infty}^2\right)\ \left(\|V\|_{L^2}^2 + \|\nabla \mathcal{N} U\|_{L^2}^2\right)\ ds\,. \tag{32}
$$

As u and v are bounded in $(0,T)\times\Omega$, classical parabolic regularity results and (2) imply that $v \in L^p(0,T;W^{2,p}(\Omega))$ for each $p \in (1,\infty)$. By the Sobolev embedding, there is $p \geq 2$ large enough such that $\nabla v \in L^2(0,T;L^\infty(\Omega))$. The Gronwall lemma and (32) then entail that $V(t) = \nabla \mathcal{N} U(t) = 0$ for every $t \in [0,T]$, which completes the proof. □

In the final result of this section, we provide a sufficient condition on h and D which guarantees that (9) holds true.

Proposition 4. *If $h'\ D^{-1/2} \in L^\infty(0,1)$, then* (9) *holds true.*

Proof. For $0 \leq r \leq r' \leq 1$, we have

$$\begin{aligned} |h(r') - h(r)| &\leq \int_r^{r'} |h'(s)|\ ds \leq C \int_r^{r'} D^{1/2}(s)\ ds \\ &\leq C \left(\int_r^{r'} D(s)\ ds \right)^{1/2} |r - r'|^{1/2} \\ &\leq C\ |r - r'|^{1/2}\ |\mathcal{D}(r) - \mathcal{D}(r')|^{1/2}\ , \end{aligned}$$

whence the claim. □

In particular, $D(r) = (1-r)^m$ and $h(r) = r\ (1-r)^n$ fulfil (6) and (9) for $m \geq (n+2)/2$ and $n \geq 2$.

3. Stationary states

In this section, we assume that

$$g(r,s) := \gamma\ r - \beta\ s \tag{33}$$

for some $\beta > 0$ and $\gamma > 0$.

Definition 5.

(i) *A stationary solution to* (1)–(3) *is a couple* (u, v) *of functions in* $L^\infty(\Omega; \mathbb{R}^2)$ *such that* $\mathcal{D}(u) \in H^1(\Omega)$, $v \in H^2(\Omega)$,

$$(u(x), v(x)) \in [0,1] \times [0, +\infty) \quad a.e.\ in \quad \Omega\ , \tag{34}$$

$$\nabla \mathcal{D}(u) - h(u)\ \nabla v = 0 \quad a.e.\ in \quad \Omega\ , \tag{35}$$

$$-\Delta v + \beta\ v = \gamma\ u \quad a.e.\ in \quad \Omega\ , \tag{36}$$

$$\partial_\nu v = 0 \quad a.e.\ on \quad \partial\Omega\ . \tag{37}$$

(ii) *If* $M \in [0, |\Omega|]$, *a stationary solution to* (1)–(3) *with mass* $(M, \gamma M/\beta)$ *is a stationary solution* (u, v) *to* (1)–(3) *satisfying in addition*

$$\int_\Omega u(x)\ dx = \frac{\beta}{\gamma} \int_\Omega v(x)\ dx = M\ . \tag{38}$$

Obviously, $(u, v) = (0, 0)$ and $(u, v) = (1, \gamma/\beta)$ are the only stationary solutions to (1)–(3) with mass $(0, 0)$ and $(|\Omega|, \gamma\ |\Omega|/\beta)$, respectively. Also, if $m \in (0, 1)$, $(m, \gamma m/\beta)$ is a stationary solution to (1)–(3) with mass $(m|\Omega|, \gamma m|\Omega|/\beta)$.

We next describe some properties of stationary solutions to (1)–(3) with mass $(M, \gamma M/\beta)$ for $M \in (0, |\Omega|)$. For further use, we introduce the auxiliary functions

$$j(r) := \int_{1/2}^r \frac{D(s)}{h(s)}\ ds \quad \text{and} \quad J(r) := \int_{1/2}^r j(s)\ ds\ , \quad r \in (0,1)\ ,$$

which enjoy the following property.

Lemma 6. *The function j is an increasing function from $(0,1)$ onto $(-\infty, j_1)$ where*

$$j_1 := \int_{1/2}^1 \frac{D(s)}{h(s)}\, ds \in (0,+\infty]\,.$$

Proof. The strict monotonicity of j follows at once from (6). We also infer from (6) that, for $s \in (0,1/2]$,

$$\frac{D(s)}{h(s)} \geq \frac{\min\limits_{[0,1/2]} D}{\|h'\|_{L^\infty(0,1)}}\, \frac{1}{s} \geq \frac{C}{s}$$

for some $C > 0$, so that D/h does not belong to $L^1(0,1/2)$. □

Proposition 7. *Let $M \in (0,|\Omega|)$ and consider a stationary solution (u,v) to (1)–(3) with mass $(M,\gamma M/\beta)$. Then $\Omega_u := \{x \in \Omega\,, u(x) \in [0,1)\}$ is a non-empty open subset of Ω, $j(u) - v$ is constant on each connected component of Ω_u and $u \in \mathcal{C}(\bar{\Omega}) \cap \mathcal{C}^1(\Omega_u)$. In addition, $u(x) > 0$ for $x \in \bar{\Omega}$ and*

$$v(x) \in \left[0, \frac{\gamma}{\beta}\right]\,, \qquad x \in \bar{\Omega}\,. \tag{39}$$

Proof. Since $u \in L^\infty(\Omega)$, we infer from (36) and (37) that $v \in W^{2,p}(\Omega)$ for every $p \in (1,\infty)$. Choosing p large enough, we realize that $\nabla v \in \mathcal{C}(\bar{\Omega})$, which, together with (34) and (35), implies that $\mathcal{D}(u) \in W^{1,\infty}(\Omega)$. Consequently, $\mathcal{D}(u) \in \mathcal{C}(\bar{\Omega})$ and, since $\mathcal{D}$ is increasing, we have also $u \in \mathcal{C}(\bar{\Omega})$. A further use of (35) then yields that $\mathcal{D}(u) \in \mathcal{C}^1(\bar{\Omega})$. Therefore, recalling that $u \geq 0$, we deduce that Ω_u is an open subset of Ω (which is non-empty by (34) since $M < |\Omega|$) and (6) ensures that $u \in \mathcal{C}^1(\Omega_u)$ with $\nabla \mathcal{D}(u) = D(u)\, \nabla u$ on Ω_u.

Now, let Λ be a connected component of $\{x \in \Omega\,,\ u(x) \in (0,1)\}$. For $x \in \Lambda$, (35) also reads $\nabla\,(j(u) - v)\,(x) = 0$. Therefore, there is $\lambda \in \mathbb{R}$ such that $j(u(x)) = v(x) + \lambda$ for each $x \in \Lambda$. Thanks to the monotonicity of j, we have in fact $u(x) = j^{-1}(v(x) + \lambda) \geq j^{-1}(\lambda) > 0$ for $x \in \Lambda$. The continuity of u then entails that $u(x) \geq j^{-1}(\lambda) > 0$ for $x \in \bar{\Lambda}$. Consequently, $u(x) > 0$ for $x \in \bar{\Omega}$ and $\Omega_u = \{x \in \Omega\,,\ u(x) \in (0,1)\}$. It then follows from (35) that $\nabla\,(j(u) - v) = 0$ on Ω_u.

Finally, the bounds (39) are straightforward consequence of (36) and (37) by the comparison principle. □

3.1. The non-degenerate case: $D/h \notin L^1(1/2,1)$

We next derive some upper and lower bounds for stationary solutions which generalize [10, Proposition 1.4].

Proposition 8. *Let $M \in (0,|\Omega|)$ and consider a stationary solution (u,v) to (1)–(3) with mass $(M,\gamma M/\beta)$. Assume further that $j_1 = +\infty$. Then*

$$j^{-1}\left(j\left(\frac{M}{|\Omega|}\right) - \frac{\gamma}{\beta}\right) \leq u(x) \leq j^{-1}\left(j\left(\frac{M}{|\Omega|}\right) + \frac{\gamma}{\beta}\right)\,, \qquad x \in \bar{\Omega}\,.$$

In particular, $u(x) \in (0,1)$ for $x \in \bar{\Omega}$.

Proof. Let Λ be a connected component of Ω_u. By Proposition 7, there is $\lambda \in \mathbb{R}$ such that

$$j(u(x)) = v(x) + \lambda \quad \text{for} \quad x \in \Lambda\,. \tag{40}$$

Since $j_1 = +\infty$, we infer from (39) that $u(x) \le j^{-1}((\gamma/\beta) + \lambda) < 1$, from which we conclude that $\Lambda = \Omega_u = \Omega$ by the continuity of u. Therefore,

$$M = \int_\Omega u(x)\; dx = \int_\Omega j^{-1}(v(x) + \lambda)\; dx \in \left[|\Omega|\; j^{-1}(\lambda), |\Omega|\; j^{-1}\left(\frac{\gamma}{\beta} + \lambda\right)\right] ,$$

whence

$$\lambda \in \left[j\left(\frac{M}{|\Omega|}\right) - \frac{\gamma}{\beta}, j\left(\frac{M}{|\Omega|}\right)\right] .$$

The claimed bounds then follow from (39), (40) and the above estimates for λ. □

Remark 9. *When D and h are related by (5), we have $j(r) = \ln(2r) - \ln(q(r)) + \ln(q(1/2))$ and $D/h \notin L^1(1/2, 1)$.*

3.2. The degenerate case: $D/h \in L^1(1/2, 1)$

In that case, $j_1 < +\infty$. In contrast to the previous case where we have shown that the first component u of any steady state to (1)–(3) cannot reach the threshold value 1 (unless $u \equiv 1$), we shall prove in this subsection that there are stationary solutions (u, v) to (1)–(3) such that $u = 1$ on some subset of Ω with non-empty interior, so that the graph of u is flat-hump-shaped. To this end, we recall that, by Proposition 7, $j(u) - v$ is constant on each connected component of $\Omega_u = \{x \in \Omega\,, u(x) \in [0, 1)\}$ and look for a stationary solution (u, v) such that the constant does not depend on the connected component of Ω_u. Introducing $f := j^{-1}$ and $f_\lambda(.) := f(. + \lambda)$ for $\lambda \in \mathbb{R}$, we thus look for a stationary solution (u, v) to (1)–(3) such that there is $\lambda \in \mathbb{R}$ for which $u(x) = f_\lambda(v(x))$ for $x \in \Omega_u$. Since $u(x) = 1$ for $x \in \Omega \setminus \Omega_u$, we actually have

$$u(x) = F_\lambda(v(x)) \quad \text{for} \;\; x \in \bar{\Omega}\,, \tag{41}$$

where

$$F_\lambda(y) = \left\{ \begin{array}{ccl} f_\lambda(y) & \text{for} & y \le v_\lambda := j_1 - \lambda\,, \\ 1 & \text{for} & y > v_\lambda\,. \end{array} \right. \tag{42}$$

Next, (36) and (41) yield that

$$-\Delta v = \gamma F_\lambda(v) - \beta v \quad \text{in} \;\; \Omega \tag{43}$$

with no-flux boundary conditions on $\partial\Omega$. Observe that the continuity of v established in Proposition 7 and (43) imply that

$$-\Delta v \;=\; \gamma - \beta v \quad \text{in} \;\; \Omega_0 := \Omega \setminus \bar{\Omega}_u\,, \tag{44}$$

$$v \;=\; v_\lambda \quad \text{on} \;\; \partial\Omega_0\,. \tag{45}$$

We finally recall that (39) warrants that

$$v_\lambda \le \gamma/\beta\,. \tag{46}$$

From now on, we restrict ourselves to the one-dimensional case $\Omega = (0, l)$ and provide a sufficient condition for the existence of a flat-hump-shaped stationary solution to (1)–(3). To this end, we say that the triple $(\lambda, \gamma, \beta) \in \mathbb{R} \times (0, +\infty)^2$ satisfies the condition **(C)** if

$$
\textbf{(C)} \quad \begin{cases} \text{There are } \varrho_{-1}, \varrho_0, \varrho_1 = \gamma/\beta \text{ such that } \varrho_{-1} < \varrho_0 < v_\lambda < \varrho_1 \text{ and} \\ F_\lambda(\varrho_i) = \dfrac{\beta}{\gamma}\varrho_i, \quad i \in \{-1, 0, 1\}, \\ F_\lambda(y) > \dfrac{\beta}{\gamma} y, \quad y \in (\varrho_0, \varrho_1), \\ F_\lambda(y) < \dfrac{\beta}{\gamma} y, \quad y \in (\varrho_{-1}, \varrho_0), \end{cases}
$$

Theorem 10. *Suppose that there is a triple (λ, γ, β) which satisfies the condition* **(C)** *together with*

$$
\frac{1}{v_\lambda - \varrho_{-1}} \int_{\varrho_{-1}}^{v_\lambda} f_\lambda(y)\, dy < \frac{\beta}{\gamma} \frac{v_\lambda + \varrho_{-1}}{2}. \tag{47}
$$

Then, for l large enough, there exists a flat-hump-shaped stationary solution (u, v) to (1)–(3) in $\Omega = (0, l)$. More precisely, there exists $x_1 \in (0, l/2)$ such that $u(x) \in [0, 1)$ for $x \in [0, x_1) \cup (l - x_1, l]$, $u(x) = 1$ for $x \in [x_1, l - x_1]$ and $v(x) \geq v_\lambda$ for $x \in [x_1, l - x_1]$.

If additionally

$$
f' \in L^2(j_1 - \delta, j_1) \quad \text{for some } \delta > 0, \tag{48}
$$

then $u \in H^1(0, l)$.

Proof. Step 1. Let us first consider the one-dimensional boundary-value problem

$$
v'' = \Phi(v) \quad \text{in } (0, l), \quad v'(0) = v'(l) = 0, \tag{49}
$$

$$
\Phi(v) = -\gamma\, F_\lambda(v) + \beta\, v. \tag{50}
$$

Notice that the previous second-order ordinary differential equation is equivalent to the hamiltonian system

$$
v' = w, \quad w' = \Phi(v), \tag{51}
$$

with "energy" E and "potential" $\mathcal{U}$ given by

$$
E(v, w) = \frac{1}{2} w^2 + \mathcal{U}(v), \quad \mathcal{U}(v) = -\int_{\varrho_0}^{v} \Phi(y) dy. \tag{52}
$$

We denote by $x \longmapsto \varphi(x; v_0, w_0) = (\varphi^1(x; v_0, w_0), \varphi^2(x; v_0, w_0))$ the solution to (51) starting from (v_0, w_0) at $x = 0$ and restrict the analysis of the phase portrait of (51) to the strip $P = \{(v, w) : v \in [\varrho_{-1}, \varrho_1]\}$. It follows from **(C)** that there are three equilibria $A_i := (\varrho_i, 0)$, $i \in \{-1, 0, 1\}$ in P. Since $\Phi(v) > 0$ for $v \in (\varrho_{-1}, \varrho_0)$ and $\Phi(v) < 0$ for $v \in (\varrho_0, \varrho_1)$ by **(C)**, we realize that $\mathcal{U}$ has a local minimum at ϱ_0 and $\mathcal{U}(\varrho_0) = 0$. Consequently, for $c \in \mathcal{I} := (0, \min\{\mathcal{U}(\varrho_{-1}), \mathcal{U}(\varrho_1)\})$, the level sets $E_c = \{(v, w) : E(v, w) = c\}$ of E are closed curves, from which we deduce that any orbit starting from $(v_0, 0)$ such that $\mathcal{U}(v_0) \in \mathcal{I}$ is periodic and surrounds the equilibrium A_0.

Step 2. In order to satisfy the boundary conditions we seek a periodic solution (v, w) to (51) such that $v(0) \in (\varrho_{-1}, \varrho_0)$, $w(0) = 0$, and which crosses the line $l_\lambda := \{(v, w) : v = v_\lambda\}$. Suppose that

$$\mathcal{U}(\varrho_{-1}) > \mathcal{U}(v_\lambda), \tag{53}$$

and notice that $\mathcal{U}(v_\lambda) < \mathcal{U}(\varrho_1)$ since $v_\lambda \in (\varrho_0, \varrho_1)$ and $\mathcal{U}$ is increasing on (ϱ_0, ϱ_1). Therefore, $\mathcal{U}(v_\lambda) \in \mathcal{I}$ and there exists $v_0 \in (\varrho_{-1}, \varrho_0)$ such that $\mathcal{U}(v_0) \in \mathcal{I}$ and $\mathcal{U}(v_0) > \mathcal{U}(v_\lambda)$. Owing to Step 1, the solution $\varphi(.; v_0, 0)$ to (51) is periodic with period $T(v_0)$ and reaches l_λ for some $x = x_1$. We then set $l = T(v_0)$ and define

$$v(x) = \varphi^1(x; v_0, 0) \quad \text{for } x \in [0, l], \tag{54}$$

$$u(x) = f_\lambda(v(x)) \quad \text{for } x \in [0, x_1) \cup (l - x_1, l], \tag{55}$$

$$u(x) = 1 \quad \text{for } x \in [x_1, l - x_1]. \tag{56}$$

It is then straightforward to check that the couple (u, v) thus constructed is a stationary solution to (1)–(3) in the sense of Definition 5 and that $v \in \mathcal{C}^2([0, l])$ and $u \in \mathcal{C}([0, l])$. These two facts actually imply that $\mathcal{D}(u) \in \mathcal{C}^1([0, l])$. Notice further that v satisfies

$$\begin{aligned} v_{xx}(x) &= -\gamma + \beta v(x) \quad \text{for } x \in (x_1, l - x_1), \\ v(x_1) &= v(l - x_1) = v_\lambda. \end{aligned}$$

It remains to check (53). It follows from (50) that (53) also reads

$$\int_{\varrho_0}^{v_\lambda} F_\lambda(y)dy - \frac{\beta}{2\gamma}(v_\lambda^2 - \varrho_0^2) < \int_{\varrho_0}^{\varrho_{-1}} F_\lambda(y)dy - \frac{\beta}{2\gamma}(\varrho_{-1}^2 - \varrho_0^2),$$

which is equivalent to (47).

We finally assume that f' enjoys the additional property (48). Owing to (54), (55) and (56), to show that $u_x \in L^2(0, l)$ it is sufficient to prove that

$$u_x \in L^2(x_1 - \varepsilon, x_1) \quad \text{and} \quad u_x \in L^2(l - x_1 - \varepsilon, l - x_1) \tag{57}$$

for ε small enough. Since $v_x(x) > 0$ for $x \in [x_1 - \delta, x_1]$ we infer from (55) that

$$\begin{aligned} \int_{x_1-\varepsilon}^{x_1} |u_x(x)|^2 \, dx &= \int_{x_1-\varepsilon}^{x_1} |f'(v(x) + \lambda)|^2 \, (v_x(x))^2 \, dx \\ &= \int_{v(x_1-\varepsilon)+\lambda}^{v(x_1)+\lambda} |f'(z)|^2 \, v_x(v^{-1}(z - \lambda)) \, dz, \end{aligned}$$

and the first assertion of (57) readily follows from (48). A similar argument provides the second assertion of (57) and completes the proof. □

Remark 11.

(i) *It clearly follows from the proof that, for* λ, γ, β *satisfying the assumptions of Theorem* 10, *there exists a unique (singular) solution such that* u *is spike-shaped (it suffices to choose the unique* $v_0 \in (\varrho_{-1}, \varrho_0)$ *such that* $\mathcal{U}(v_0) = \mathcal{U}(v_\lambda)$ *in the second step of the proof of Theorem* 10*). In that case* $\Omega_u = (0, l) \setminus \{l/2\}$.

(ii) *It is possible to merge a finite number of flat-hump-shaped stationary solutions to obtain a multi-hump solution on a suitably enlarged interval.*

(iii) *It is possible to modify the proof of Theorem* 10 *to construct a stationary hump-shaped solution which is not symmetric with respect to* $l/2$. *To this end one should choose two different values of* λ *on each connected component of the set* Ω_u. *Proceeding as in the proof of Theorem* 10 *one can construct separately a part of the solution defined on a subinterval containing* 0 *and another part on a subinterval containing* l. *Then one can merge both in such a way that, at a merging point* x_0, *we have*

$$\lim_{x \to x_0^-} v_x(x) = 0 = \lim_{x \to x_0^+} v_x(x)$$

and there is a possible jump discontinuity of v_{xx} *at* x_0 .

We now specify conditions under which Theorem 10 can be applied. We first show that the condition **(C)** is valid under a simple condition on j' or f' near j_1.

Proposition 12. *If*

$$\lim_{y \to j_1^-} f'(y) > \frac{\beta}{\gamma}, \tag{58}$$

then there is $\varepsilon > 0$ *such that the triple* (λ, β, γ) *satisfies the condition* **(C)** *for any* $\lambda \in (j_1 - \gamma/\beta, j_1 + \varepsilon - \gamma/\beta)$.

Proof. We put $\lambda' = j_1 - \gamma/\beta$, so that $v_{\lambda'} = \gamma/\beta$. Since

$$\lim_{y \to -\infty} f(y) = 0, \quad f(0) = \frac{1}{2},$$

and $f_{\lambda'}(v_{\lambda'}) = 1$, the graph of $y \mapsto F_{\lambda'}(y)$ has at least one intersection point $S = (\varrho_{\lambda'}, F_{\lambda'}(\varrho_{\lambda'}))$ satisfying $\varrho_{\lambda'} < v_{\lambda'}$ with the straight line $y \mapsto (\beta/\gamma)y$. Suppose that S is the closest intersection point with respect to $R = (\gamma/\beta, 1)$. If S is a transversal intersection point it is sufficient to shift the graph of $y \mapsto F_\lambda(y)$ to the left by a small $\lambda'' \in (0, \varepsilon)$ and then (λ, γ, β) satisfies **(C)** with $\lambda \in (\lambda', \lambda' + \varepsilon)$. If S is not a transversal intersection point, using the fact that $f_\lambda \in \mathcal{C}^1(-\infty, j_1)$, a small shift of the graph to the right creates a transversal intersection in the vicinity of S. Then one proceeds as in the previous case. □

Proposition 13. *Consider* $(\beta, \gamma) \in (0, +\infty) \times (0, +\infty)$. *Assume that* D/h *has a limit* j_2 *in* $[0, +\infty]$ *as* $r \to 1$ *such that*

$$j_2 > \frac{\gamma}{\beta}, \tag{59}$$

and that $(D/h)''(r) < 0$ *for every* $r \in [0,1)$ *such that* $(D/h)'(r) = 0$. *Introducing*

$$r_0 := \sup\left\{r \in (0,1)\,, j'(r) = \frac{\gamma}{\beta}\right\}, \tag{60}$$

the triple (λ, γ, β) *satisfies the condition* **(C)** *if*

$$j_1 - \frac{\gamma}{\beta} < \lambda < j(r_0) - \frac{\gamma}{\beta}\, r_0\,. \tag{61}$$

Proof. For $r \in (0,1)$ and $\lambda \in \mathbb{R}$, we put

$$\psi(r) := \frac{\gamma}{\beta}\, r - j(r) \quad \text{and} \quad \psi_\lambda(r) := \psi(r) + \lambda\,.$$

Since $\psi'(r) = (\gamma/\beta) - (D/h)(r)$, the assumptions (6) and (59) guarantee that

$$\lim_{r\to 1} \psi'(r) = \frac{\gamma}{\beta} - j_2 > 0 \quad \text{and} \quad \lim_{r\to 0} \psi'(r) = -\infty\,.$$

Consequently, ψ' has a least one zero in $(0,1)$ and we denote by r_0 the largest zero of ψ'. Owing to (59), we clearly have $r_0 < 1$ and $\psi'(r) > 0$ for $r \in (r_0, 1)$, which implies that $\psi''(r_0) \geq 0$. Either $\psi''(r_0) > 0$ and there is $\varepsilon > 0$ such that $\psi'(r) < 0$ for $r \in (r_0 - \varepsilon, r_0)$. Or $\psi''(r_0) = 0$ and the assumption in Proposition 13 ensures that $\psi'''(r_0) > 0$. We therefore conclude that, in this case also, there is $\varepsilon > 0$ such that $\psi'(r) < 0$ for $r \in (r_0 - \varepsilon, r_0)$.

Consider now $\lambda \in (-\psi(1), -\psi(r_0))$. Then $\psi_\lambda(1) > 0 > \psi_\lambda(r_0)$. Recalling that $\psi_\lambda' = \psi'$ is positive in $(r_0, 1)$, there is a unique $s_0 \in (r_0, 1)$ such that $\psi_\lambda(s_0) = 0$, $\psi_\lambda(r) > 0$ for $r \in (s_0, 1)$ and $\psi_\lambda(r) < 0$ for $r \in [r_0, s_0)$. Next, $\psi_\lambda(r) \to +\infty$ as $r \to 0$ by Lemma 6. Consequently, ψ_λ has at least one zero in $(0, r_0)$ and we denote by $s_{-1} \in (0, r_0)$ the largest zero of ψ_λ in $(0, r_0)$. Then, $\psi_\lambda(r) < 0$ for $r \in (s_{-1}, s_0)$.

Introducing $\varrho_i := (\gamma\, s_i)/\beta$, $i = -1, 0$, and observing that

$$f_\lambda(y) = f\left(\psi_\lambda\left(\frac{\beta}{\gamma}\, r\right) + j\left(\frac{\beta}{\gamma}\, r\right)\right),$$

we easily see that the condition **(C)** is fulfilled. □

We now check that there are situations in which Theorem 10 can be applied. More precisely, we assume that $h(r) = r\, D(r)$ for $r \in [0,1]$. In that particular case, we have $j(r) = \ln 2r$ for $r \in (0,1)$, $j_1 = \ln 2$, $j_2 = 1$ and $r_0 = \beta/\gamma$. On the one hand, by Proposition 13, the triple (λ, γ, β) fulfils the condition **(C)** if

$$\frac{\gamma}{\beta} > 1 \quad \text{and} \quad \ln 2 - \frac{\gamma}{\beta} < \lambda < \ln 2 - \ln\left(\frac{\gamma}{\beta}\right) - 1\,. \tag{62}$$

In addition, when (λ, γ, β) fulfils (62), a further study of the function $y \longmapsto f(y + \lambda) - (\beta y)/\gamma$ on $(-\infty, v_\lambda)$ shows that

$$0 < \varrho_{-1} < \ln 2 - \ln\left(\frac{\gamma}{\beta}\right) - \lambda\,. \tag{63}$$

On the other hand, since $h(r) = r\ D(r)$ for $r \in [0,1]$, we have

$$\int_{\varrho_{-1}}^{v_\lambda} f_\lambda(y)\ dy = \int_{(\beta\varrho_{-1})/\gamma}^{1} r\ j'(r)\ dr = 1 - \frac{\beta}{\gamma}\ \varrho_{-1}\,.$$

Now assume that

$$\frac{\gamma}{\beta} > 2 \quad \text{and} \quad \lambda = \ln 2 - \frac{\gamma}{\beta} + \delta\,,$$

for some $\delta \in (0, (\gamma/\beta) - 2)$. Then $v_\lambda > 2$ and we infer from (63) that

$$\begin{aligned}
\nu &= \frac{\beta}{2\gamma}\ \left(v_\lambda^2 - \varrho_{-1}^2\right) - \int_{\varrho_{-1}}^{v_\lambda} f_\lambda(y)\ dy \\
&= \frac{\beta}{\gamma}\ \left[\frac{v_\lambda^2 - \varrho_{-1}^2}{2} - \frac{\gamma}{\beta} + \varrho_{-1}\right] \\
&\geq \frac{\beta}{\gamma}\ \left[\frac{v_\lambda - 2}{2}\ (v_\lambda - \varrho_{-1}) - \frac{\gamma}{\beta} + v_\lambda\right] \\
&\geq \frac{\beta}{\gamma}\ \left[\frac{v_\lambda - 2}{2}\ \left(v_\lambda - \ln 2 + \ln\left(\frac{\gamma}{\beta}\right) + \lambda\right) - \frac{\gamma}{\beta} + v_\lambda\right] \\
&\geq \frac{\beta}{\gamma}\ \left[\frac{1}{2}\ \left(\frac{\gamma}{\beta} - \delta - 2\right)\ \ln\left(\frac{\gamma}{\beta}\right) - \delta\right] \\
&> 0
\end{aligned}$$

for δ small enough. Observe that the larger γ/β is, the larger δ can be.

As a final remark, let us point out that, in the case $h(r) = r\ D(r)$, a flat-hump-shaped stationary solution (u,v) to (1)–(3) coincides with a stationary solution to the so-called minimal chemotaxis model in which $D \equiv 1$ and $h(r) = r$ (cf. [3]) on the set $\{x : u(x) < 1\}$.

Acknowledgments

Partial support from the EU financed network HYKE no. HPRN-CT-2002-00282 and the Université de Toulouse I is gratefully acknowledged. Also, part of this work was done while both authors enjoyed the friendly atmosphere of the workshop “Nonlinear Elliptic and Parabolic Problems: A Special Tribute to the Work of Herbert Amann” held in Zürich from June 28 to June 30, 2004.

References

[1] H. Amann, *Nonhomogeneous linear and quasilinear elliptic and parabolic boundary value problems*, in “Function Spaces, Differential Operators and Nonlinear Analysis”, H. Triebel, H.J. Schmeisser (eds.), Teubner-Texte Math. **133**, Teubner, Stuttgart, 1993, pp. 9–126.

[2] T. Hillen and K. Painter, *Global existence for a parabolic chemotaxis model with prevention of overcrowding*, Adv. Appl. Math. **26** (2001), 280–301.

[3] D. Horstmann, *From 1970 until present: the Keller-Segel model in chemotaxis and its consequences. I.*, Jahresber. Deutsch. Math.-Verein. **105** (2003) 103–165.

[4] E. Keller and L. Segel, *Initiation of slime mold aggregation viewed as an instability*, J. Theor. Biol. **26** (1970), 399–415.

[5] J.-L. Lions, *Quelques Méthodes de Résolution des Problèmes aux Limites non Linéaires*, Dunod, Paris, 1969.

[6] K. Painter and T. Hillen, *Volume-filling and quorum-sensing in models for chemosensitive movement*, Canadian Appl. Math. Q. **10** (2002), 501–543.

[7] C.S. Patlak, *Random walk with persistence and external bias*, Bull. Math. Biol. Biophys. **15** (1953), 311–338.

[8] J. Simon, *Compact sets in the space* $L^p(0,T;B)$, Ann. Mat. Pura Appl. **146** (1987), 65–96.

[9] D. Wrzosek, *Global attractor for a chemotaxis model with prevention of overcrowding*, Nonlinear Anal. **59** (2004), 1293–1310.

[10] D. Wrzosek, *Long time behaviour of solutions to a chemotaxis model with volume filling effect*, Proc. Roy. Soc. Edinburgh Sect. A, to appear.

Philippe Laurençot
Mathématiques pour l'Industrie et la Physique
CNRS UMR 5640
Université Paul Sabatier–Toulouse 3
118 route de Narbonne
F–31062 Toulouse cedex 4, France
e-mail: laurenco@mip.ups-tlse.fr

Dariusz Wrzosek
Institute of Applied Mathematics and Mechanics
Warsaw University
Banacha 2
02-097 Warszawa, Poland
e-mail: darekw@mimuw.edu.pl

Progress in Nonlinear Differential Equations
and Their Applications, Vol. 64, 291–327

In the Blink of an Eye

Julián López-Gómez and Marcela Molina-Meyer

Dedicated to H. Amann, with our deepest admiration,
at the occasion of his retirement from Zürich University

Abstract. The mathematical analysis of this paper shows how the effects of strategic local symbiosis provide an exceptional mechanism to increase productivity in highly competitive environments. The most striking consequence from our analysis is that productivity can blow-up in cooperation areas, though some of the species might become extinct elsewhere, as a result of the aggressions received from competitors. As a by-product, it is realized why strategic symbiosis effects help to avoid massive extinctions of populations, or industrial and financial companies. Going beyond, it has been numerically observed that, in the presence of local strategic symbiosis, high level aggressions might provoke a substantial increment of the complexity of the system; a mechanism that might explain the extraordinary bio-diversity of Earth's biosphere, as well as the complexity of Global Economy.

Mathematics Subject Classification (2000). 34A34, 34C23, 92B05.

Keywords. Symbiosis in competitive environments, coexistence states, complexity, biodiversity.

1. Introduction

544 millions years ago there were 3 animal phyla with their corresponding variety of external forms, though at 538 millions years ago there were 38, the same number that exists today, except for one or two extinctions, Parker [44]. Consequently, the vast existing diversity of body architectures has emerged during a five-million-year interval beginning 543 millions year ago. Such a time interval is referred to in the specialized literature as the *Cambrian explosion* – the first fossils from that period of time where found in the Cambrian Hills in Wales by the recognized geologist Sedgwick, who named it as the *Cambrian*. Cambrian explosion supports that the history of life has been governed by long time intervals of *micro-evolution* – gradual evolution – combined with short *macro-evolution* periods which have been the most prolific generators of biodiversity, in contrast with the classical paradigm of

Darwin and Wallace, according with it evolution occurred gradually. Parker [44] defends the theory that Cambrian explosion took place as a result of a revolutionary event occurred in between 544 and 543 million-years-ago: in one-million-year interval most of skin energy receptors specialized into eyes; in the blink of an eye vision was born. Consequently, it seems that Cambrian explosion was extraordinarily facilitated by the explosion of all kind of mutual interactions between Precambrian animals as a natural consequence from vision. Quite strikingly, it seems that, indeed, predating and competition mechanisms are in the origin of the extraordinary diversity of the Earth biosphere.

A reminiscent phenomenology has been observed as well in studying the diversity and productivity of grass-lands, Tilman [46]. Quite surprisingly, some species in competition have better productivity than any individual species in monoculture; a phenomenon usually referred to as *overyielding* which is far from well understood. Cambrian explosion might be regarded as a genuine overyielding phenomenon originated by the extraordinary increment of animal interactions as a result of eyes formation. Consequently, understanding overyielding seems to be imperative in order to explain biodiversity, complexity and productivity.

This paper provides a possible mathematical approach to the study of these crucial problems by analyzing the effects of local strategic symbiosis in the dynamics of competing species systems; a mechanism also adopted by commercial and industrial companies to avoid extinction and improve productivity, simultaneously generating complex commercial networks and global market diversity.

Competition, as most ecologists employ the word, means the active demand by a number of individuals of the same species – intraspecific competition – or members of a number of species at the same trophic level – interspecific competition – for a common resource or requirement that is actually, or potentially, limiting, Wilson [47], Begon et al. [4]. It is commonly agreed that this definition is consistent with the general assumptions of the Lotka-Volterra equations, which still conform the basis of the mathematical theory of competition. Though non-spatial models are far from realistic, the spatial ones allowed to establish the range of validity of the competitive exclusion principle, [25], [8], one of the greatest paradigms of evolution, not well understood yet. Rather than extinction, in nature one observes segregation when the level of the aggressions blows-up, and, after appropriate periods of time, species differentiation, as a result of spatial and temporal heterogeneities.

Our mathematical analysis reveals that overyielding is a direct consequence from the super-linear character of competition in the presence of strategic symbiosis, which might explain diversity through the existence of multiple stable coexistence states. These multiplicities might allow the species to evolve adopting different profiles according to their inhabiting territories, though a most rigorous analysis of this issue would require a strong additional multidisciplinary effort.

Focusing attention into the mathematical contains of this paper, more than in the eventual scientific significance of the underlying ideas, this paper is distributed as follows. In Section 2 we introduce the mathematical model. It consists of a generalized Lotka-Volterra system where the interaction coefficients are assumed to change sign; so, it is far from being classified within the classical families of Lotka-Volterra models, where the interactions have constant sign. Section 3 analyzes the qualitative behavior of the semi-trivial positive solutions of the model. Section 4 ascertains the shape of the curves of neutral stability of the semi-trivial positive solutions. Section 5 recalls a series of fundamental results for the classical competition Lotka-Volterra model. Collecting those results is imperative to realize the significance of the new mathematical results of this paper. Section 6 gives a general existence result concerning the coexistence states of the competition model in the presence of local symbiosis. Some of these results are reminiscent of well-known results for general superlinear indefinite problems, while others are completely new. It is worth-mentioning that our model is a sort of superlinear indefinite system and, hence, all known properties for those problems have a counterpart for the system. Section 7 analyzes the behavior of one of the species, say u, when the intrinsic growth rate of the competitor, say v, blows-up. Its main result shows that the species u must entirely segregate into the regions where it is free from the aggressions of v, arbitrarily increasing within the cooperation areas. As the classical Lotka-Volterra competition model with diffusion entails the extinction of u, our result shows how strategic symbiosis is an extraordinary mechanism to avoid extinction, increasing simultaneously productivity, even when the level of the aggressions within competition areas is arbitrarily high. This result might be a breakthrough from the point of view of the applications of the underlying theory to Population Dynamics and Macro-Economy. Finally, in Section 8 we provide an example where our numerical experiments show the existence of multiple stable coexistence states. Such kind of phenomenology is rather striking, since the single equation admits, at most, a unique stable positive solution. Consequently, it seems that competition is an extraordinary mechanism to generate multiple stable positive solutions. This feature should position to many multidisciplinary scientists in the right direction to realize why predating and competition are on the basis of the great diversity of Earth Biosphere, as well as on the foundations of the great complexity of Global Economy.

Although we are dealing with general systems, without any kind of variational structure, where no *global* comparison techniques are available, the maximum principle and the nice properties derived from it, whose study was systematically carried out from the seminal results of Amann (cf. [1] and the list of references therein), still are applicable to make *local* comparisons which are on the basis of the mathematical analysis carried out overall this honoring paper. Consequently, comparison techniques might be imperative to explain why, in the blink of an eye, Biosphere diversity blown-up, and why, Markets globalization, facilitated by the development of new communications and technologies, is originating a drastic increment of Markets complexity, as well in the blink of an eye, or course!

2. The mathematical model

Throughout this paper, we will work with two species, u and v, interacting in a certain habitat Ω, which is modeled by a smooth bounded domain of $\mathbb{R}^N$, $N \geq 1$, which is assumed to be divided into three regions, Ω_+, Ω_- and Ω_0, in such a way that the species u and v compete within Ω_+, cooperate within Ω_-, and are free from the action of the other within Ω_0. Figure 1 shows one among the multiple admissible configurations. Subsequently, to simplify the exposition, we shall always assume that, like in Figure 1,

$$\bar{\Omega}_0 \cup \bar{\Omega}_- \subset \Omega \quad \text{and} \quad \bar{\Omega}_0 \cap \bar{\Omega}_- = \varnothing\,,$$

thought this condition is far from necessary for the validity of the analysis carried out in this paper. Also, though it is everywhere assumed that $\Omega_- \neq \varnothing \neq \Omega_0$, some of these regions might be empty.

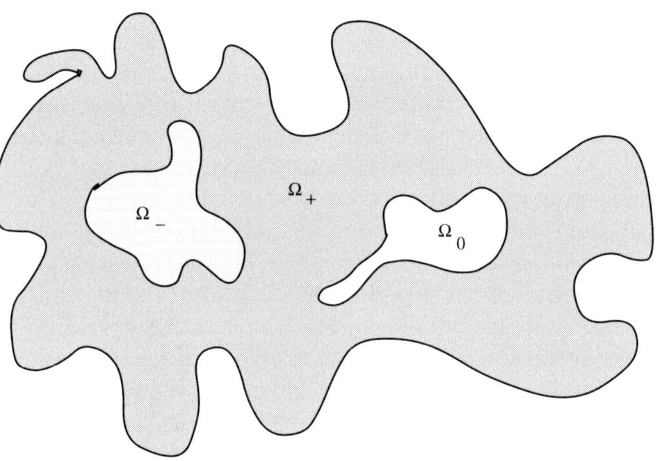

FIGURE 1. An admissible habitat.

Then, our spatial heterogeneous model for analyzing the dynamics of the species u and v is the following:

$$\begin{cases} \left.\begin{aligned} \frac{\partial u}{\partial t} - \Delta u = \lambda u - u^2 - buv \\ \frac{\partial v}{\partial t} - \Delta v = \mu v - v^2 - cuv \end{aligned}\right\} & \text{in } \Omega \times \mathbb{R}_+\,, \\ u = v = 0 & \text{on } \partial\Omega \times \mathbb{R}_+\,, \\ u(\cdot,0) = u_0 \geq 0\,, \qquad v(\cdot,0) = v_0 \geq 0\,, & \text{in } \Omega\,, \end{cases} \tag{2.1}$$

where Ω is a bounded domain of $\mathbb{R}^N$, $N \geq 1$, with boundary $\partial\Omega$, of class $\mathcal{C}^{2+\nu}$ for some $\nu \in (0,1)$, Δ is the Laplacian of $\mathbb{R}^N$, $\mathbb{R}_+ := (0,\infty)$, $\lambda, \mu \in \mathbb{R}$, b, $c \in \mathcal{C}^\nu(\bar{\Omega})$ are two functions changing sign in Ω, and $u_0, v_0 \in W_0^{2,p}(\Omega)$, for some $p > N/2$. Typically, Ω represents the inhabiting area, u and v measure the densities of the species at $x \in \Omega$ after time $t > 0$, the Laplacian measures the dispersion rates of u

and v overall Ω – they have been normalized to unity –, λ and μ are the intrinsic growth rates of u and v, respectively, and $b(x)$, $c(x)$ model the interaction patterns, and other spatial effects, between u and v. The regions where b are c are positive (resp. negative) are the areas where competition (resp. cooperation) takes place, the regions where $bc < 0$ are the areas where some of the species preys on the other, and the regions where some of these coefficients vanishes provide us with the areas where the corresponding species is free from the action of the other. Finally, u_0 and v_0 are the initial population densities. Our boundary conditions entail the habitat Ω to be surrounded by completely hostile areas. As we will think of (2.1) as a model for competition in the presence of strategic symbiosis, we will suppose

$$\begin{aligned}\Omega_0 &= \operatorname{Int} b^{-1}(0) = \operatorname{Int} c^{-1}(0)\,,\\ \Omega_+ &= b^{-1}(\mathbb{R}_+) = c^{-1}(\mathbb{R}_+)\,,\\ \Omega_- &= b^{-1}(-\mathbb{R}_+) = c^{-1}(-\mathbb{R}_+)\,.\end{aligned}$$

Consequently, Ω_- is the region where cooperation takes place, Ω_+ is the region where competition takes place, and Ω_0 is the region where both species are free from the action of the other. It should be noted that, under our general assumptions, (2.1) *is not a Lotka-Volterra model.* Now, setting

$$h^+ := \max\{h, 0\}\,, \quad h^- := h^+ - h\,, \qquad h \in \mathcal{C}(\bar\Omega)\,,$$

we have that

$$b = b^+ - b^-\,, \qquad c = c^+ - c^-\,,$$

$$\bar\Omega_+ = \operatorname{supp} b^+ = \operatorname{supp} c^+\,, \qquad \bar\Omega_- = \operatorname{supp} b^- = \operatorname{supp} c^-\,,$$

and $\Omega_0 = \operatorname{Int}[\Omega \setminus (\Omega_+ \cup \Omega_-)]$. Moreover, $b^+ > 0$, $b^- > 0$, $c^+ > 0$ and $c^- > 0$. Throughout this paper, for any $h \in \mathcal{C}(\bar\Omega)$ it is said that $h > 0$ if $h \geq 0$ and $h \neq 0$. Also, given $h \in \mathcal{C}^1(\bar\Omega)$, it is said that $h \gg 0$ if it lies in the interior of the cone of positive functions of $\mathcal{C}^1(\bar\Omega)$. Under these assumptions, for each $p > N/2$ and u_0, $v_0 \in W_0^{2,p}(\Omega)$, there exists a maximal *existence time*

$$T = T(u_0, v_0, \lambda, \mu, b, c) \in (0, \infty]$$

such that (2.1) possesses a unique solution

$$(u, v) := (u(x, t; u_0, v_0, \lambda, \mu, b, c), v(x, t; u_0, v_0, \lambda, \mu, b, c))\,,$$

satisfying

$$u\,,\ v \in \mathcal{C}(\bar\Omega \times [0, T)) \cap \mathcal{C}^{2+\nu, 1+\frac{\nu}{2}}(\bar\Omega \times (0, T))$$

and

$$\lim_{t \uparrow T} \left(\|u(\cdot, t)\|_{\mathcal{C}(\bar\Omega)} + \|v(\cdot, t)\|_{\mathcal{C}(\bar\Omega)}\right) = \infty \qquad \text{if} \quad T < \infty\,.$$

Moreover, by the parabolic maximum principle, for each $t \in (0, T)$, $u(\cdot, t) \gg 0$ if $u_0 > 0$, while $u = 0$ if $u_0 = 0$. By symmetry, v satisfies the same property. In general, $T < \infty$, since b and c change sign in Ω. Consequently, in the absence of the other, each species grows according to the logistic low

$$\begin{cases} \frac{\partial w}{\partial t} - \Delta w = \gamma w - w^2 & \text{in } \Omega \times (0,\infty)\,, \\ w = 0 & \text{on } \partial\Omega \times (0,\infty)\,, \\ w(\cdot,0) = w_0 \geq 0\,, & \text{in } \Omega\,, \end{cases} \tag{2.2}$$

and, hence, $\lim_{t\uparrow\infty}(u,v)$ is determined by the non-negative steady states of (2.1) with a vanishing component. Note that the steady states of (2.1) are the non-negative solutions (u,v) of the system

$$\begin{cases} -\Delta u = \lambda u - u^2 - b(x)uv & \\ -\Delta v = \mu v - v^2 - c(x)uv & \text{in } \Omega\,, \\ u = v = 0 & \text{on } \partial\Omega\,. \end{cases} \tag{2.3}$$

whose structure might be extraordinarily involved, since it depends upon the geometry of the habitat Ω and the spatial dimension, as well as on b and c, because, in a number of regimes, (2.3) exhibits a genuine super-linear behavior.

As in the classical spatially homogeneous Lotka-Volterra models,

$$\begin{cases} u' = \lambda u - u^2 - buv\,, \\ v' = \mu v - v^2 - cuv\,, \end{cases} \tag{2.4}$$

where $' = \frac{d}{dt}$ and b, $c \in \mathbb{R}$, in order to ascertain the dynamics of (2.1), it is imperative to analyze the existence, multiplicity and stability properties of all non-negative solutions of (2.3). Consequently, at our present early stage, most of our effort will be focused to analyze (2.3). Although under some appropriate circumstances the limiting profiles of the solutions of (2.1) are given by the *metasolutions* of (2.3) (e.g., [30]), here we restrict ourselves to study the classical solutions of (2.3).

Besides $(0,0)$, Problem (2.3) exhibits three types of non-negative solutions. Those of the form $(u,0)$, or $(0,v)$, often refereed to as the *semi-trivial* positive solutions, and the so called *coexistence states*, which are the solutions of the form (u,v) with both components positive. In any circumstance, as a consequence from the strong maximum principle, $u \gg 0$ if $u > 0$, and $v \gg 0$ if $v > 0$.

3. The curves of change of stability of the semi-trivial states

Subsequently, for each smooth subdomain $D \subset \Omega$ and $V \in L^\infty(D)$, we shall denote by $\sigma[-\Delta + V; D]$ the *lowest eigenvalue* of the linear boundary value problem

$$\begin{cases} (-\Delta + V)w = \tau w & \text{in } D\,, \\ w|_{\partial D} = 0\,, \end{cases} \tag{3.1}$$

referred to as the *principal eigenvalue* of (3.1); it is algebraically simple, and its associated eigenfunction, φ, can be chosen to satisfy $\varphi \gg 0$ in D – the *principal eigenfunction* –; $\sigma[-\Delta + V; D]$ is the unique eigenvalue of (3.1) to a positive eigenfunction (cf. Amann [2]). Subsequently, we set

$$\sigma := \sigma[-\Delta; \Omega]\,.$$

It is folklore that (2.3) possesses a semi-trivial positive solution of the form $(u, 0)$ if, and only if, $\lambda > \sigma$. Moreover, if $\lambda > \sigma$ and we denote by θ_γ the unique positive solution of

$$\begin{cases} -\Delta w = \gamma w - w^2 & \text{in } \Omega\,, \\ w|_{\partial\Omega} = 0\,, \end{cases} \tag{3.2}$$

then, $(\theta_\lambda, 0)$ is the unique semi-trivial positive solution of the form $(u, 0)$. Similarly, (2.3) possesses a semi-trivial positive solution of the form $(0, v)$ if, and only if, $\mu > \sigma$, and, in such case, $(0, \theta_\mu)$ is the unique one. Condition $\lambda > \sigma$ measures the facilities (among them, the size and natural resources) that Ω should offer to each of the species to maintain it in the absence of the other; e.g., in the absence of v, u is driven to extinction if $\lambda \leq \sigma$.

In this section we study the stability of the semi-trivial positive solutions of (2.3). To show the significance of this analysis we will first concentrate into the spatially homogeneous model (2.4) with $\lambda, \mu, b, c \in \mathbb{R}_+$, which is the classical competition Lotka-Volterra model. In such case, the semi-trivial positive solutions are $(\lambda, 0)$ and $(0, \mu)$ and the coexistence state should be given by

$$(u, v) = \left(\frac{\lambda - b\mu}{1 - bc}, \frac{\mu - c\lambda}{1 - bc} \right) \tag{3.3}$$

if it exists and, in addition, $bc \neq 1$. Regarding applications, the case $bc = 1$ is not of interest, since, generically, $bc \neq 1$. So, suppose $bc \neq 1$. Then, two different situations must be distinguished, according to the size of bc:

- $0 < bc < 1$, referred to as the *low level competition* case.
- $bc > 1$, refereed to as the *high level competition* case.

In case $bc < 1$, (3.3) provides us with a coexistence state of (2.4) if, and only if,

$$\lambda > b\mu \quad \text{and} \quad \mu > c\lambda\,, \tag{3.4}$$

while, in case $bc > 1$, (3.3) is a coexistence state if, and only if,

$$\lambda < b\mu \quad \text{and} \quad \mu < c\lambda\,. \tag{3.5}$$

As the matrices of the linearizations of (2.4) at $(\lambda, 0)$ and $(0, \mu)$ are

$$\begin{pmatrix} -\lambda & -b\lambda \\ 0 & \mu - c\lambda \end{pmatrix} \quad \text{and} \quad \begin{pmatrix} \lambda - b\mu & 0 \\ -c\mu & -\mu \end{pmatrix},$$

respectively, it is obvious that the straight lines

$$\mu = c\lambda \quad \text{and} \quad \lambda = b\mu \tag{3.6}$$

provide us with the curves of change of stability of the semi-trivial states $(\lambda, 0)$ and $(0, \mu)$ in the (λ, μ)-parameter space. Indeed, $(\lambda, 0)$ is a stable node if $\mu < c\lambda$, while it is a saddle point if $\mu > c\lambda$. Similarly, $(0, \mu)$ is stable if $\lambda < b\mu$, while it is unstable if $\lambda > b\mu$.

Therefore:

- In case $bc < 1$, (2.4) has a coexistence state (unique) if, and only if, $(\lambda, 0)$ and $(0, \mu)$ are saddle points.
- In case $bc > 1$, (2.4) has a coexistence state (unique) if, and only if, $(\lambda, 0)$ and $(0, \mu)$ are stable nodes.

Actually, it is folklore that in case $bc < 1$ the coexistence state is a global attractor for all positive solutions of (2.4), while, in case $bc > 1$, the coexistence state is a saddle point and, generically, there is extinction of some of the species, according to the relative size of the initial populations. Moreover, when one of the semi-trivial states is linearly stable and the other linearly unstable, then one of the species becomes extinct. This elementary analysis makes apparent the significance of the curves of neutral stability of the semi-trivial positive solutions in ascertaining the dynamics of the positive solutions of (2.4).

Now, we analyze the linearized stability of the semi-trivial positive solutions $(\theta_\lambda, 0)$ and $(0, \theta_\mu)$ of (2.3) as steady-states of (2.1). Suppose $\lambda > \sigma$. Then, the local attractive character of $(\theta_\lambda, 0)$ is given through the sign of the real parts of the eigenvalues τ of the linear eigenvalue problem

$$\begin{cases} -\Delta u = \lambda u - 2\theta_\lambda u - b\theta_\lambda v + \tau u \\ -\Delta v = \mu v - c\theta_\lambda v + \tau v \end{cases} \quad \text{in } \Omega\,, \qquad u = v = 0 \quad \text{on } \partial\Omega\,. \tag{3.7}$$

By the linearized stability principle, $(\theta_\lambda, 0)$ is exponentially asymptotically stable if all those real parts are positive, while it is unstable if it has an eigenvalue with negative real part. Suppose τ is an eigenvalue of (3.7) with associated eigenfunction $(u, v) \neq (0, 0)$. If $v = 0$, then

$$(-\Delta + 2\theta_\lambda - \lambda)u = \tau u \qquad \text{in } \Omega\,, \qquad u|_{\partial\Omega} = 0\,,$$

and, since $\sigma[-\Delta + 2\theta_\lambda - \lambda; \Omega] > \sigma[-\Delta + \theta_\lambda - \lambda; \Omega] = 0$, necessarily $\tau > 0$. Therefore, such eigenvalues do not affect the local behavior of $(\theta_\lambda, 0)$. If $v \neq 0$, then τ must be an eigenvalue of $-\Delta + c\theta_\lambda - \mu$, and, hence, $\tau \in \mathbb{R}$ and

$$\tau \geq \sigma[-\Delta + c\theta_\lambda - \mu; \Omega] = \sigma[-\Delta + c\theta_\lambda; \Omega] - \mu\,.$$

Moreover, since $\sigma[-\Delta + c\theta_\lambda; \Omega] - \mu$ is the lowest one among these eigenvalues, the curve $\mu = \sigma[-\Delta + c\theta_\lambda; \Omega]$ [in the (λ, μ)-plane] provides us with the geometrical locus of the set of values (λ, μ) where $(\theta_\lambda, 0)$ is neutrally stable. Most precisely, $(\theta_\lambda, 0)$ is linearly stable (l.s.) if, and only if, $\mu < \sigma[-\Delta + c\theta_\lambda; \Omega]$, while $(\theta_\lambda, 0)$ is linearly unstable (l.u.) if, and only if, $\mu > \sigma[-\Delta + c\theta_\lambda; \Omega]$. By symmetry, $(0, \theta_\mu)$ is l.s. if, and only if, $\lambda < \sigma[-\Delta + b\theta_\mu; \Omega]$, while it is l.u. if, and only if, $\lambda > \sigma[-\Delta + b\theta_\mu; \Omega]$. Consequently, the curves

$$\mu = \sigma[-\Delta + c\theta_\lambda; \Omega] \quad \text{and} \quad \lambda = \sigma[-\Delta + b\theta_\mu; \Omega] \tag{3.8}$$

will play a similar role, with respect to (2.1), as the curves $\mu = c\lambda$ and $\lambda = b\mu$ for its spatially homogeneous counterpart (2.4). These curves will be refereed to as the *curves of change of stability*, or neutral stability, of the semi-trivial solutions.

4. Limiting behaviors of the curves of change of stability

The following result will provide us with the limiting behavior at $(\lambda, \mu) = (\sigma, \sigma)$ and at infinity of the curves (3.8).

Proposition 4.1. *Suppose $a \in \mathcal{C}^\nu(\bar\Omega)$ and set, for each $\gamma > \sigma$,*

$$f(\gamma) := \sigma[-\Delta + a(x)\theta_\gamma; \Omega].$$

Then, f is real analytic in (σ, ∞), and it admits the asymptotic expansion

$$f(\gamma) = \sigma + (\gamma - \sigma) \int_\Omega a\varphi^3 + O(|\gamma - \sigma|^2) \qquad \text{as} \quad \gamma \downarrow \sigma\,, \tag{4.1}$$

where $\varphi \gg 0$ is the unique eigenfunction of σ satisfying $\int_\Omega \varphi^3 = 1$. Moreover,

$$\lim_{\gamma\uparrow\infty} \frac{f(\gamma)}{\gamma} = \min_{\bar\Omega} a\,. \tag{4.2}$$

In particular, if we define $f(\sigma) := \sigma$, then $f \in \mathcal{C}([\sigma, \infty))$. Furthermore, setting

$$\Omega_0^a := \operatorname{Int} a^{-1}(0)\,, \qquad \Omega_-^a := \operatorname{Int} a^{-1}(-\mathbb{R}_+)\,,$$

one has that

$$f(\gamma) \le \sigma[-\Delta; \Omega_0^a \cup \Omega_-^a]\,, \qquad \gamma \ge \sigma\,, \tag{4.3}$$

and, actually,

$$\lim_{\gamma\uparrow\infty} f(\gamma) = \sigma[-\Delta; \Omega_0^a] \tag{4.4}$$

if $a \ge 0$ (i.e., $\Omega_-^a = \emptyset$) and Ω_0^a is a nice smooth open subset of Ω.

Proof. The fact that the map $(\sigma, \infty) \to \mathcal{C}^\nu(\bar\Omega)$, $\gamma \mapsto \theta_\gamma$, is real analytic follows from the implicit function theorem, since θ_γ is the unique positive solution of (3.2) and it is non-degenerate. By the simplicity of the principal eigenvalue it readily follows that f is real analytic in (σ, ∞). Moreover, since $\theta_\gamma \to 0$ as $\gamma \downarrow \sigma$ in $\mathcal{C}^\nu(\bar\Omega)$, $\lim_{\gamma\downarrow\sigma} f(\sigma) = \sigma$, and, hence, $f \in \mathcal{C}([\sigma, \infty))$ if we define $f(\sigma) := \sigma$. The asymptotic expansion (4.1) follows from the main theorem of Crandall & Rabinowitz [9] by adapting the argument of the proof of [25, Lemma 3.6] (cf. [15, Lemma 4.3] as well). Now, note that for each $\gamma > \sigma$,

$$\frac{f(\gamma)}{\gamma} = \sigma[-\frac{1}{\gamma}\Delta + a(x)\vartheta_\gamma; \Omega]\,,$$

where $\vartheta_\gamma := \theta_\gamma/\gamma$; ϑ_γ is the unique positive solution of

$$-\frac{1}{\gamma}\Delta w = w - w^2 \qquad \text{in} \quad \Omega\,, \qquad w|_{\partial\Omega} = 0\,,$$

and, hence, thanks to, e.g., [20, Theorem 3.4], $\lim_{\gamma\uparrow\infty} \vartheta_\gamma = 1$, uniformly in compact subsets of Ω. Therefore, adapting the proof of [20, Lemma 3.1], (4.2) holds. Estimate (4.3) follows from the fact that $a \le 0$ in $\Omega_0^a \cup \Omega_-^a$. Indeed, for each $\gamma \ge \sigma$,

$$f(\gamma) = \sigma[-\Delta + a\theta_\gamma; \Omega] \le \sigma[-\Delta + a\theta_\gamma; \Omega_0^a \cup \Omega_-^a] \le \sigma[-\Delta; \Omega_0^a \cup \Omega_-^a]\,.$$

Finally, by [25, Theorem 3.4], (4.4) holds if $a \ge 0$. $\square$

In analyzing the behavior of the curves (3.8) by means of Proposition 4.1 one must take $a \in \{b, c\}$. In such case, a changes sign, and, hence, $\min_{\bar\Omega} a < 0$, and the sign of the limiting slope $\lim_{\gamma \downarrow \sigma} f'(\sigma) = \int_\Omega a\varphi^3$ at $\gamma = \sigma$ depends upon the relative sizes of a^+ and a^- versus φ^3. In fact, there exists $\varepsilon > 0$ such that $f(\gamma)$ is increasing in $[\sigma, \sigma + \varepsilon]$ if $\int_\Omega a\varphi^3 > 0$, while it is decreasing if $\int_\Omega a\varphi^3 < 0$. Thus, the local behavior of $f(\gamma)$ at $\gamma = \sigma$ depends, among other things, upon the sign of $\int_\Omega a\varphi^3$. Also, by (4.2), the map $\gamma \mapsto f(\gamma)$ is asymptotic, at $\gamma = \infty$, to the straight line $\gamma \mapsto \min_{\bar\Omega} a \cdot \gamma$. Moreover, due to the estimate (4.3), within the quadrant

$$\min\{\lambda, \mu\} > \sigma_1 := \sigma[-\Delta; \Omega_0 \cup \Omega_-], \tag{4.5}$$

$(\theta_\lambda, 0)$ and $(0, \theta_\mu)$ are linearly unstable.

5. The classical competition model

It is model (2.1) under the very special requirement that

$$b(x) > 0 \quad \text{and} \quad c(x) > 0 \quad \text{for each} \quad x \in \bar\Omega. \tag{5.1}$$

In particular, $\Omega_- = \Omega_0 = \varnothing$. This model has been extensively analyzed in [5], [10], [11], [39], [40], [12], [23] [41], [17], [19], [43], [27], [20], [13], [29], and the list of references therein. This section gives a short summary of some of the most important findings of these references; to illustrate the most striking differences between the classical model and (2.1). Suppose (5.1) and $a \in \{b, c\}$. Then, the function $f(\lambda)$, $\gamma \geq \sigma$, defined in Proposition 4.1, in increasing in γ. Moreover,

$$\gamma \mapsto \sigma + (\gamma - \sigma) \int_\Omega a\varphi^3 \quad \text{and} \quad \gamma \mapsto \min_{\bar\Omega} a \cdot \gamma \tag{5.2}$$

provide us with the tangents of $f(\gamma)$ at $\gamma = \sigma$ and at infinity, respectively. The shape of the regions enclosed by the curves (3.8), where the attractiveness character of $(\theta_\lambda, 0)$ and $(0, \theta_\mu)$ is the same, is strongly dependent on the intensity of the competition, measured by the function bc.

5.1. Low level of aggressions

According to the theory developed in Furter & López-Gómez [19] (going back to [39] and [41]), there exists a constant $K_1 \in (0, 1)$ such that the curves (3.8) cannot meet within the quadrant $\min\{\lambda, \mu\} > \sigma$ if

$$\|b\|_\infty < 1, \quad \|c\|_\infty < 1, \qquad \|bc\|_\infty < K_1. \tag{5.3}$$

Moreover, the region enclosed by the curves (3.8) consists of the values of (λ, μ) for which $(\theta_\lambda, 0)$ and $(0, \theta_\mu)$ are linearly unstable. In Figure 2 we have represented the curves (3.8) in the special case when b and c are positive constants. Except for the values of the slopes of the limiting tangents (5.2), these curves, in the general case when b and c are arbitrary functions satisfying (5.3), look very similar to those shown in Figure 2.

As illustrated by Figure 2, the curves (3.8) divide the quadrant $\min\{\lambda, \mu\} > \sigma$ in three regions. In Region 1, $(\theta_\lambda, 0)$ is l.u. and $(0, \theta_\mu)$ is l.s.; in Region 2, $(\theta_\lambda, 0)$

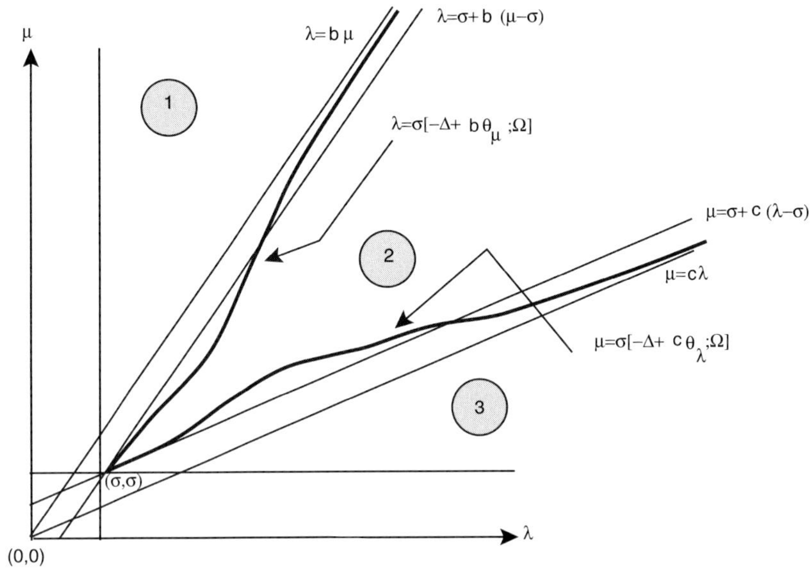

FIGURE 2. The curves (3.8) under condition (5.3).

and $(0,\theta_\mu)$ are, simultaneously, l.u.; and, in Region 3, $(\theta_\lambda,0)$ is l.s. and $(0,\theta_\mu)$ is linearly unstable. By a result of Blat & Brown [5], (2.3) possesses a coexistence state for each (λ,μ) in Region 2. Actually, by a result of Hess & Lazer [23], (2.1) is *compressive*, or *permanent*, in Region 2, i.e., the ω-limit of any positive solution of (2.1) lies in the interior of the positive quadrant; with both components bounded away from zero. Further, by a result of López-Gómez & Sabina [43], for each (λ,μ) in Region 2, (2.3) possesses a unique coexistence state if $\|bc\|_\infty$ is sufficiently small, which is a global attractor for all positive solutions of (2.1).

As an effect from dispersion, in strong contrast with the behavior exhibited by the spatially homogeneous model (2.4), (2.3) can exhibit a coexistence state within Regions 1 and 3. Actually, the set of values of (λ,μ) for which (2.3) possesses a coexistence can be strictly larger than the region enclosed by the curves (3.8). Figure 3 shows an admissible configuration where such a situation occurs (see Eilbeck et al. [17] and Furter & López-Gómez [19], [20], for further details).

In Figure 3, Region 1 is the region where $(\theta_\lambda,0)$ is l.u., $(0,\theta_\mu)$ is l.s., and (2.3) does not exhibit a coexistence state. Similarly, Region 5 is the region where $(\theta_\lambda,0)$ is l.s., $(0,\theta_\mu)$ is l.u., and (2.3) does not exhibit a coexistence state. Thanks to a result attributable to Dancer [12], in Region 1 $(0,\theta_\mu)$ is a global attractor for all positive solutions of (2.1), while in Region 5 $(\theta_\lambda,0)$ attracts to all positive solutions of (2.1). For each pair of values (λ,μ) in the interior of Regions 2 and 4, (2.3) possesses at least two coexistence states. Actually, as a result of the theory of monotone dynamical systems, (2.3) possesses a minimal and a maximal coexistence

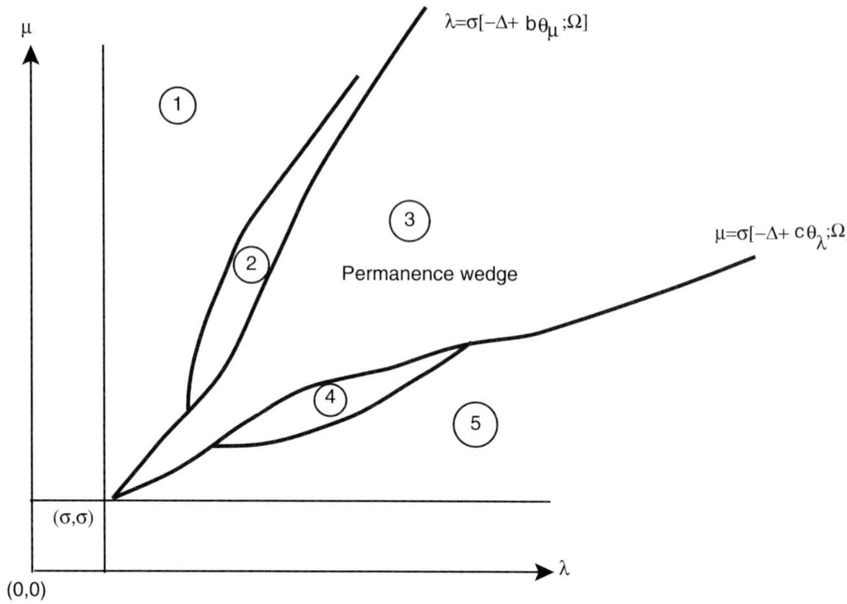

FIGURE 3. Most relevant regions of the quadrant $\min\{\lambda, \mu\} > \sigma$.

state; one of them stable and the other unstable. (cf. Dancer [12] and López-Gómez [30]). Finally, Region 3 in Figure 3 stands for the region where (2.1) is permanent.

Further, by the a priori bounds of Furter & López-Gómez [19] and López-Gómez & Sabina [43], for each $\lambda > \sigma$, there exists $\mu_2 = \mu_2(\lambda)$ such that (2.3) does not admit a coexistence state if $\mu > \mu_2$. Therefore,

$$(0, \theta_\mu) \text{ attracts to all positive solutions of (2.1) if } \mu > \mu_2 \,. \tag{5.4}$$

By symmetry, for each $\mu > \sigma$, there exists $\lambda_2 = \lambda_2(\mu)$ such that (2.3) does not admit a coexistence state if $\lambda > \lambda_2$, and, hence,

$$(\theta_\lambda, 0) \text{ attracts to all positive solutions of (2.1) if } \lambda > \lambda_2 \,. \tag{5.5}$$

Consequently, u (resp. v) is driven to extinction by v (resp. u) if μ (resp. λ) is sufficiently large, which is a rather natural result from the point of view of Population Dynamics, in complete agreement with the behavior exhibited by the classical non-spatial model (2.4).

Figure 4 shows a genuine bifurcation diagram of non-negative solutions of (2.3), where $\lambda > \sigma$ has been fixed and μ has been regarded as the main bifurcation parameter. By the celebrated local bifurcation theorem of Crandall & Rabinowitz [9], the curves (3.8) provide us with the set of values of (λ, μ) for which a curve of coexistence states bifurcates from the semi-trivial positive solutions.

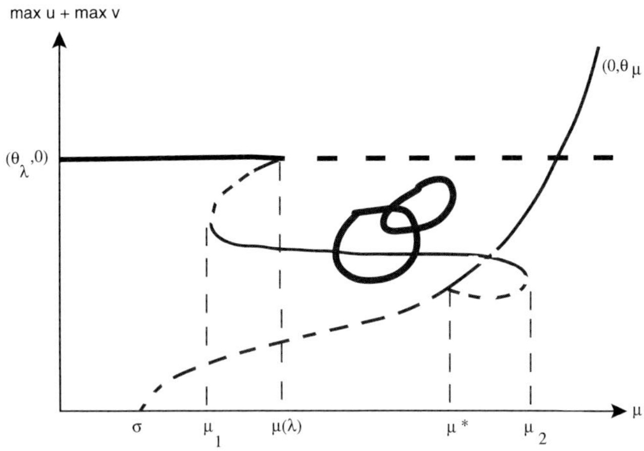

FIGURE 4. A genuine bifurcation diagram of non-negative steady-states.

In Figure 4, we have denoted

$$\mu(\lambda) := \sigma[-\Delta + c\theta_\lambda; \Omega] , \tag{5.6}$$

the value of μ where the stability of $(\theta_\lambda, 0)$ is lost, while μ^* stands for the unique value of μ for which

$$\lambda = \sigma[-\Delta + b\theta_{\mu^*}; \Omega] , \tag{5.7}$$

where $(0, \theta_\mu)$ becomes stable. Thanks to López-Gómez [29, Theorem 7.2.2], (2.3) possesses a continuum of coexistence states linking the states $(\mu(\lambda), \theta_\lambda, 0)$ and $(\mu^*, 0, \theta_{\mu^*})$. Incidentally, it should be noted that the unilateral global theorems of Rabinowitz [45] cannot be applied to get these results as they are wrong in the way they are stated (cf. the recent counterexample of Dancer [14]; prompted from the discussion carried out in López-Gómez [29]). In Figure 4, dashed lines represent curves filled in by unstable solutions. If we assume that Figure 4 shows all the non-negative solutions of (2.3), then, it is apparent that the following occurs:

- $(\theta_\lambda, 0)$ is a global attractor for all positive solutions of (2.1) if $\mu < \mu_1$, where $\mu_1 \leq \mu(\lambda)$ is the minimal value of μ for which (2.3) has a coexistence state.
- For each $\mu \in (\mu_1, \mu(\lambda)) \cup (\mu^*, \mu_2)$ – it might be empty –, (2.3) possesses two coexistence states, at least, one of them stable and another unstable, though the fine topological structure of the continuum of coexistence states might be quite involved.
- For each $\mu \in (\mu(\lambda), \mu^*)$, (2.3) has a coexistence state and (2.1) is permanent.
- For each $\mu > \mu_2$, $(0, \theta_\mu)$ attracts to all positive solutions of (2.1).

5.2. Intermediate level of aggressions

As b and c grow and bc crosses the critical value 1, then the problem of analyzing the fine structure of the region enclosed by the curves (3.8), and, in particular, finding out the cardinal of its components, becomes extraordinarily more involved, since these curves meet in the interior of the quadrant $\min\{\lambda,\mu\} > \sigma$. Indeed, Eilbeck et al. [17] found that, if b and c are positive constants such that

$$bc > 1 \quad \text{and} \quad bc - 1 \sim 0\,, \tag{5.8}$$

then the curves (3.8) cross, in dramatic contrast with the situation exhibited by the model under low level of aggressions. Figure 5 shows one admissible situation. The crossing point near (σ,σ) perturbs from (σ,σ) as $bc-1>0$ separates from 0. The other crossing point perturbs from infinity as a result of the change of the relative positions between the curves (3.8) (cf. [27]).

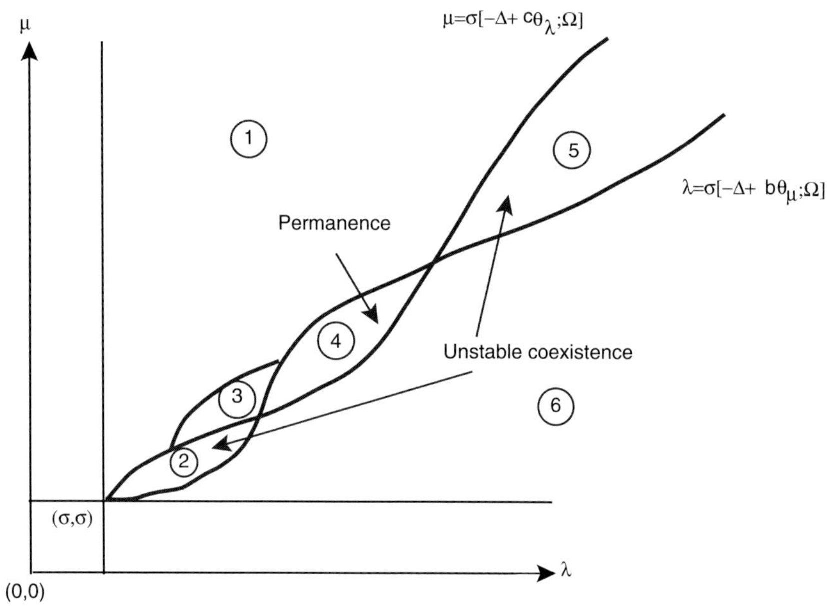

FIGURE 5. Relevant (λ,μ)-regions for an intermediate level of aggressions.

Now, in Region 1 $(0,\theta_\mu)$ is a global attractor for the positive solutions of (2.1). In Regions 2 and 5, $(\theta_\lambda,0)$ and $(0,\theta_\mu)$ are simultaneously l.s. and (2.1) admits an unstable coexistence state. In Region 3, $(0,\theta_\mu)$ is l.s., $(\theta_\lambda,0)$ is l.u., and the model exhibits, at least, two coexistence states, one stable and another unstable. In Region 4, (2.3) has a coexistence state and (2.1) is permanent. Finally, in Region 6, $(\theta_\lambda,0)$ is a global attractor for the positive solutions of (2.1).

At the crossing points of the curves (3.8), where both semi-trivial positive solutions are linearly neutrally stable, (2.3) might not admit a coexistence state as it has been proven by Dancer [13], by sharpening the theory developed by Eilbeck et al. [17] and Furter & López-Gómez [19]. Dancer's example has shown that a number of published results in the literature are completely false, as a consequence from a wrong use of the topological degree (cf. [13] for further details).

In the presence of spatial heterogeneities, i.e., in the general case when b and c are arbitrary positive functions satisfying (5.8), Furter & López-Gómez [19] developed a number of strategies to construct examples where the curves (3.8) cross at an arbitrarily large number of points, by using singularity theory techniques at the co-dimension two singularity (σ, σ).

5.3. Severe aggressions

This is the case when bc is sufficiently large, separated away from 1.

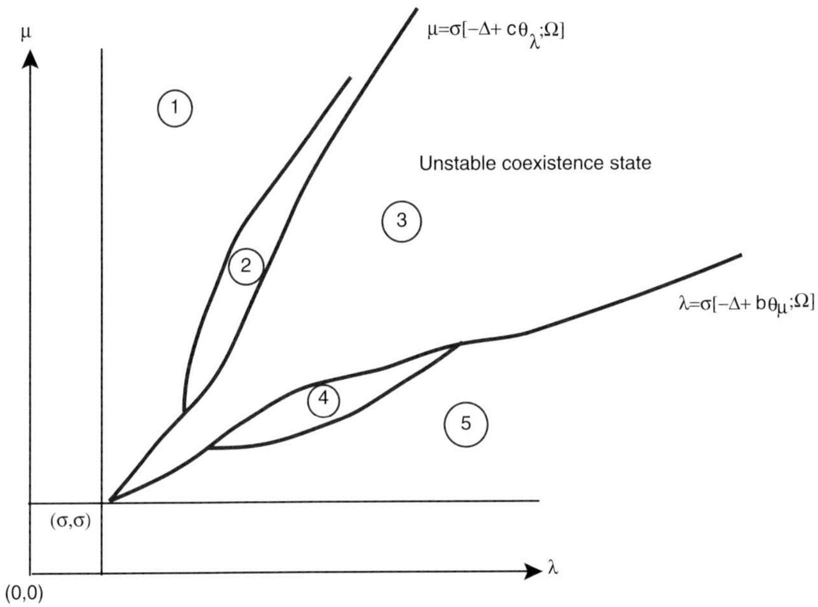

FIGURE 6. Relevant (λ, μ) regions for severe aggressions.

Now, the qualitative behavior of (2.1) changes drastically, as in Region 3, instead of permanence, (2.1) may exhibit extinction of some of the species, u or v, according to the relative size of each of the initial populations.

Note that, in all circumstances analyzed in this section, (5.4) and (5.5) remain valid. Therefore, for any λ, u is driven to extinction by v if μ – the intrinsic growth rate of v – is sufficiently large, while, for any μ, v is driven o extinction by u if λ – the intrinsic growth rate of u – is sufficiently large. As a result of local symbiosis effects, our model does not always exhibit such a behavior, as it will become apparent later.

6. The general competitive-symbiotic prototype model

Throughout this section we work under the general assumptions of Section 2. In particular, Ω_+, Ω_- and Ω_0 are non-empty and b, c satisfy the sign requirements of Section 2. Subsequently, for any $a \in \mathcal{C}(\bar\Omega)$ we denote

$$a_L := \min_{\bar\Omega} a\,, \qquad a_M := \max_{\bar\Omega} a\,. \tag{6.1}$$

Then, thanks to Proposition 4.1, the curves (3.8) look like shows Figure 7. Although they might cross as many times as we wish in a neighborhood of (σ,σ), this is far from being important in our subsequent analysis. Note that if $b^- \downarrow 0$ and $c^- \downarrow 0$, then, on any compact subset of the quadrant $\min\{\lambda,\mu\} \geq \sigma$, (3.8) must approximate uniformly to the curves of change of stability of the semi-trivial positive solutions of the associated competition model

$$\begin{cases} -\Delta u = \lambda u - u^2 - b^+(x)uv & \\ -\Delta v = \mu v - v^2 - c^+(x)uv & \text{in } \Omega\,, \\ u = v = 0 & \text{on } \partial\Omega\,, \end{cases} \tag{6.2}$$

as a result of the continuous dependence of the principal eigenvalues with respect to their potentials. This suggests that the symbiotic effects of (2.1) should play their most significant role for sufficiently large intrinsic growth rates. Indeed, by (4.2), $\mu = c_L\lambda$ and $\lambda = b_L\mu$ are the tangents at infinity of the curves $\mu = \sigma[-\Delta + c\theta_\lambda;\Omega]$ and $\lambda = \sigma[-\Delta + b\theta_\mu;\Omega]$, respectively, and, due to (4.3), the whole quadrant

$$\min\{\lambda,\mu\} > \sigma_1 := \sigma[-\Delta;\Omega_0 \cup \Omega_-] \tag{6.3}$$

entirely consists of values of (λ,μ) where $(\theta_\lambda,0)$ and $(0,\theta_\mu)$ are l.u.; this is a completely new phenomenology, inherent to the model we are dealing with, which cannot be covered by the classical competition model revisited in Section 5.

Note that $\lim_{\max\{|\Omega_0|,|\Omega_-|\}\downarrow 0}\sigma_1 = \infty$ (cf. [26]), where $|\cdot|$ stands for the Lebesgue measure of $\mathbb{R}^N$. Subsequently, we set

$$\lambda_s := \max_{\mu>\sigma} \sigma[-\Delta + b\theta_\mu;\Omega]\,, \qquad \mu_s := \max_{\lambda>\sigma} \sigma[-\Delta + c\theta_\lambda;\Omega]\,. \tag{6.4}$$

Note that $\max\{\lambda_s,\mu_s\} < \sigma_1$. As it will become clear later, (2.3) exhibits a genuine *competitive behavior* within the box delimited by (σ,σ) and (λ_s,μ_s), while it inherits a *symbiotic behavior* within $\lambda > \lambda_s$, $\mu > \mu_s$. Indeed, thanks to [29, Theorem 7.2.2], if we fix $\lambda > \lambda_s$ and regard to μ as the main bifurcation parameter, then there is a component $\mathfrak{C}_{(\mu,\theta_\lambda,0)}$ of the set of coexistence states of (2.3) that emanates from the semi-trivial state $(\theta_\lambda,0)$ at $\mu = \sigma[-\Delta + c\theta_\lambda;\Omega]$. Moreover, $\mathfrak{C}_{(\mu,\theta_\lambda,0)}$ satisfies some of the following alternatives: either $\mathfrak{C}_{(\mu,\theta_\lambda,0)}$ is unbounded in $\mathbb{R}\times\mathcal{C}_0(\bar\Omega)\times\mathcal{C}_0(\bar\Omega)$, or $(\mu^*,0,\theta_{\mu^*}) \in \bar{\mathfrak{C}}_{(\mu,\theta_\lambda,0)}$, for some μ^* such that $\lambda = \sigma[-\Delta + b\theta_{\mu^*};\Omega]$. Obviously, the last alternative cannot occur if $\lambda > \lambda_s$, and, hence, in this case, $\mathfrak{C}_{(\mu,\theta_\lambda,0)}$ is unbounded in $\mathbb{R}\times\mathcal{C}_0(\bar\Omega)\times\mathcal{C}_0(\bar\Omega)$. Now, we will study some numerical one-dimensional examples to illustrate the different global

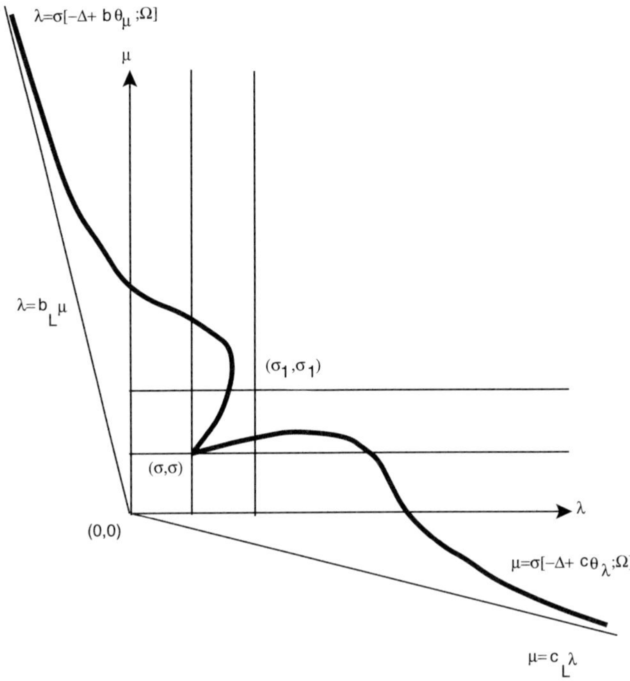

FIGURE 7. Relevant (λ, μ)-regions for the general model.

behaviors that the component $\mathfrak{C}_{(\mu,\theta_\lambda,0)}$ can have. Precisely, we consider

$$\begin{cases} -u'' = \lambda u - u^2 - b(x)uv & \text{in } (0,1)\,, \\ -v'' = \mu v - v^2 - b(x)uv & \\ u(0) = u(1) = v(0) = v(1) = 0\,, \end{cases} \tag{6.5}$$

where

$$b(x) := \begin{cases} b^+ \sin\left(\frac{\pi x}{0.4}\right) & \text{if } 0.0 \le x \le 0.4\,, \\ -b^- \sin\left[\frac{\pi}{0.2}(x-0.4)\right] & \text{if } 0.4 \le x \le 0.6\,, \\ b^+ \sin\left[\frac{\pi}{0.4}(x-0.6)\right] & \text{if } 0.6 \le x \le 1.0\,. \end{cases} \tag{6.6}$$

for some positive constants $b^+ > 0$ and $b^- > 0$. In this example, $\Omega_0 = \varnothing$, $\Omega_- = (0.4, 0.6)$, and $\Omega_+ = (0.0, 0.4) \cup (0.6, 1.0)$. Figure 8 shows a plot of the curves (3.8) for the special choice

$$b^+ = 2\,, \qquad b^- = 0.4\,. \tag{6.7}$$

They have been computed by combining a path-following code for approximating θ_λ and θ_μ, through centered divided differences, with the inverse power method to compute the associated principal eigenvalues. Note that, thanks to Proposition 4.1,

$$\lim_{\lambda\uparrow\infty} \frac{\sigma[-\frac{d^2}{dx^2} + b\theta_\lambda; (0,1)]}{\lambda} = \lim_{\mu\uparrow\infty} \frac{\sigma[-\frac{d^2}{dx^2} + b\theta_\mu; (0,1)]}{\mu} = b_L = -0.4\,,$$

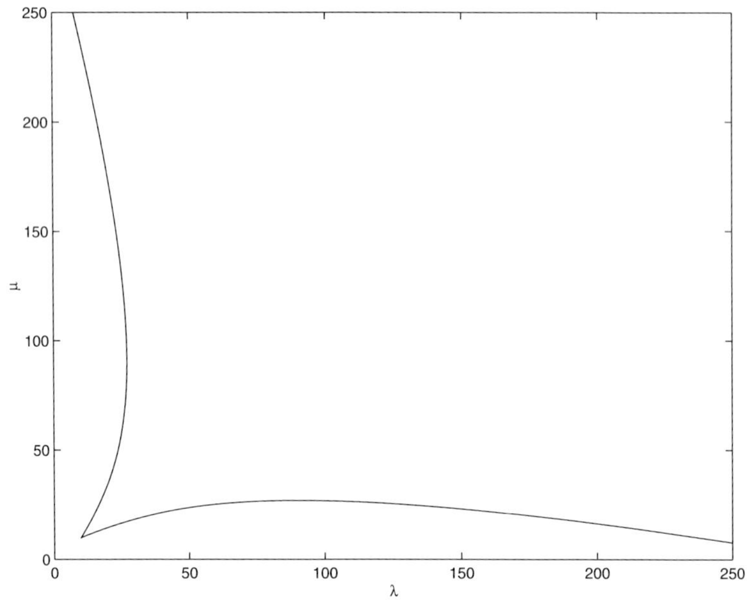

FIGURE 8. The curves (3.8) for the model (6.5), (6.6), (6.7).

provide us with the slopes at infinity of the curves (3.8). Figure 9 shows a plot of the component $\mathfrak{C}_{(\mu,\theta_\lambda,0)}$ for $\lambda = 48.98 > \lambda_s$. To compute it we have fixed $\lambda = 48.98$ and used μ as the main bifurcation parameter by coupling a pure spectral method with collocation and a path global continuation solver (cf. [24], [35], [8], and [38] for any further detail required).

In Figure 9 the component $\mathfrak{C}_{(\mu,\theta_{48.98},0)}$ emanates from $(\theta_{48.98}, 0)$ at the critical value of the parameter

$$\mu(48.98) := \sigma[-\frac{d^2}{dx^2} + b\theta_{48.98}; (0,1)] \sim 23.51$$

and it is actually defined for all further value of μ for which we tried to compute it. All computed coexistence states are exponentially asymptotically stable and, hence, they are local attractors for the positive solutions of (2.1). Figure 9 shows four significant curves; stable solutions are indicated by solid lines, unstable by dashed lines. We are plotting the value of μ against the norm

$$|u| + |v| := \max_{0\leq x\leq 1} u(x) + \max_{0\leq x\leq 1} v(x)\,.$$

The two horizontal lines describe the trivial state $(0,0)$, and the semi-trivial state $(\theta_{48.98}, 0)$; $|\theta_{48.98}| \sim 44.02$. Then, Figure 9 shows the curve of semi-trivial positive solutions $(0, \theta_\mu)$. This curve emanates from $(0,0)$ at $\mu = \pi^2$ and it grows with μ for any further value of the parameter. Finally, Figure 9 shows the curve of coexistence states. It is the curve emanating super-critically from the state $(\theta_{48.98}, 0)$

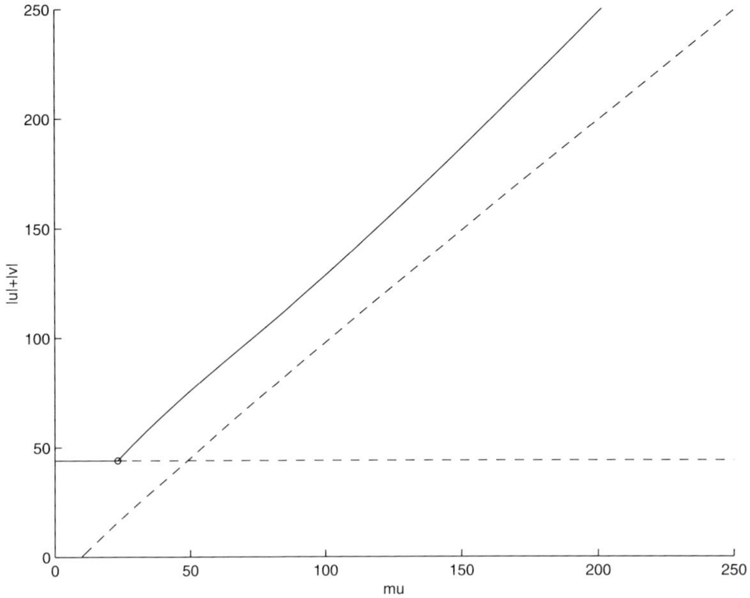

FIGURE 9. A plot of the component $\mathfrak{C}_{(\mu,\theta_{48.98},0)}$ of (6.5), (6.6), (6.7).

at the value of μ where $(\theta_\lambda, 0)$ loses stability. Strongly motivated by our numerical computations, we conjecture that in this example $\mathcal{P}_\mu \mathfrak{C}_{(\mu,\theta_{48.98},0)} = (\pi^2, \infty)$, where $\mathcal{P}_\mu$ stands for the μ–projection operator. If, instead of (6.7), we make the choice

$$b^+ = 2\,, \qquad b^- = 1.2\,, \tag{6.8}$$

then, $b_L = -1.2$ and, thanks again to Proposition 4.1,

$$\lim_{\lambda\uparrow\infty} \frac{\sigma[-\frac{d^2}{dx^2} + b\theta_\lambda; (0,1)]}{\lambda} = \lim_{\mu\uparrow\infty} \frac{\sigma[-\frac{d^2}{dx^2} + b\theta_\mu; (0,1)]}{\mu} = -1.2\,,$$

which provides us with the slopes at infinity of the curves (3.8), whose plots are shown in Figure 10. Rather naturally, as bigger is b^- as wider is the region where both semi-trivial states are unstable, as a result of strengthening symbiotic effects if b^- grows, which explains why the region enclosed by the curves (3.8) is larger in Figure 10 than in Figure 9.

Figure 11 shows a plot of $\mathfrak{C}_{(\mu,\theta_{48.98},0)}$ for the special choice (6.8). Now, the component bifurcates from $(\theta_{48.98}, 0)$ at $\mu \sim 8.46$ and it exhibits a sub-critical turning point at $\mu \sim 165.1353$, where it turns backward. The coexistence states along the lower-half curve are exponentially asymptotically stable, while the solutions along the upper-half curve are linearly unstable. In this example, based upon our numerical experiments, a reasonable conjecture is $\mathcal{P}_\mu \mathfrak{C}_{(\mu,\theta_{48.98},0)} = (-\infty, 165.1353]$, though a proof of this fact will not be given here in.

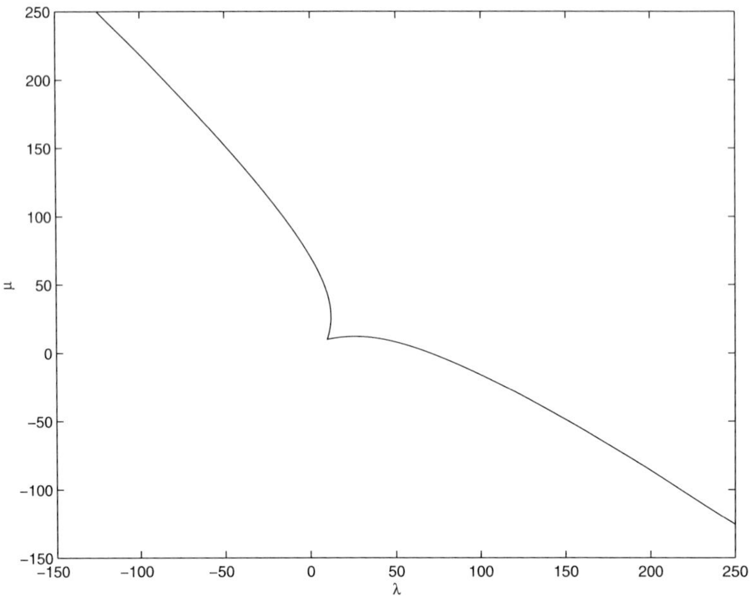

FIGURE 10. The curves (3.8) for the model (6.5), (6.6), (6.8).

In general, the following result is satisfied.

Theorem 6.1. *Suppose* $\lambda > \lambda_s$,

$$b_M^- < 1\,, \qquad c_M^- < 1\,, \tag{6.9}$$

and (2.3) *possesses a coexistence state. Then,*

$$\lambda > (1 - b_M^- c_M^-)\,\sigma - b_M^-\,\mu\,, \qquad \mu > (1 - b_M^- c_M^-)\,\sigma - c_M^-\,\lambda\,. \tag{6.10}$$

Moreover, there exists $\mu_0(\lambda) \in \mathbb{R}$ *such that*

$$(1 - b_M^- c_M^-)\sigma - c_M^-\lambda < \mu_0(\lambda) \leq \sigma[-\Delta + c\theta_\lambda; \Omega] \tag{6.11}$$

and

$$\mathcal{P}_\mu \mathfrak{C}_{(\mu,\theta_\lambda,0)} \in \{\, [\mu_0(\lambda), \infty)\,,\, (\sigma[-\Delta + c\theta_\lambda; \Omega], \infty)\,\}\,. \tag{6.12}$$

In particular, (2.3) *possesses a coexistence state for each* $\mu > \sigma[-\Delta + c\theta_\lambda; \Omega]$.

Note that $b_M^- = c_M^-$ in example (6.5), (6.6). Thus, (6.9) holds for the choice (6.7), while it fails for (6.8). Therefore, by simply looking at Figures 9 and 11, it seems that some kind of condition (6.9) is really necessary for the validity of the theorem. Most crucially, thanks to Theorem 6.1, it is obvious that the *principle of competitive exclusion* fails to be true for model (2.1), because if $\Omega_0 \cup \Omega_-$ can maintain to u in the absence of v, i.e., if $\lambda > \sigma_1$, then $\lambda > \lambda_s$ and, for any $\mu > \mu_0(\lambda)$, (2.3) possesses a coexistence state. Consequently, u can avoid extinction, even in the most dramatic case when the intrinsic growth rate of its competitor v, measured

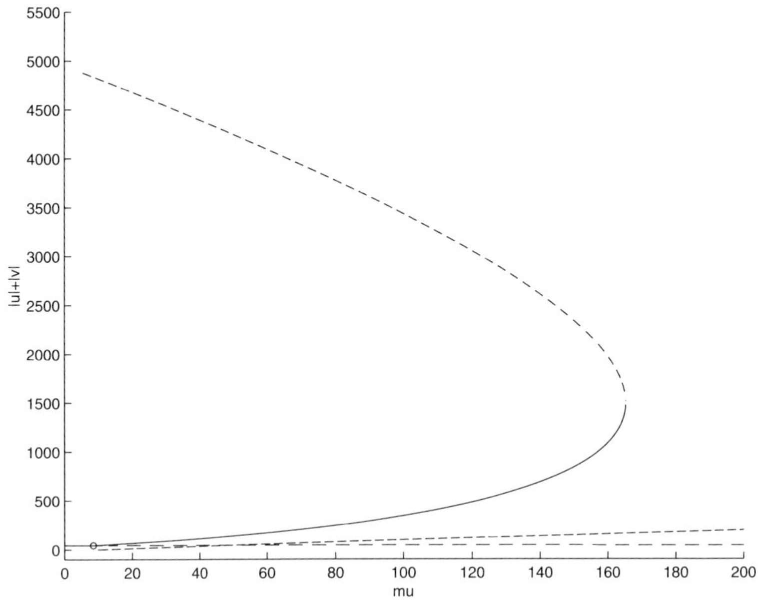

FIGURE 11. The component $\mathfrak{C}_{(\mu,\theta_{48.98},0)}$ of (6.5), (6.6), (6.8).

by μ, blows-up. In Section 7 we shall show that, rather naturally, u must pay the price of segregation within $\Omega_0 \cup \Omega_-$ if $\mu \uparrow \infty$. Therefore, strategic symbiosis seem to be imperative to avid extinction in the presence of good competitors; an strategy adopted by most of world-leading companies, of course !

Proof of Theorem 6.1. Suppose (u_0, v_0) is a coexistence state of (2.3). Then,

$$\begin{cases} -\Delta u_0 = \lambda u_0 - u_0^2 - b u_0 v_0 \leq \lambda u_0 - u_0^2 + b^- u_0 v_0 \leq \lambda u_0 - u_0^2 + b_M^- u_0 v_0 \,, \\ -\Delta v_0 = \mu v_0 - v_0^2 - c u_0 v_0 \leq \mu v_0 - v_0^2 + c^- u_0 v_0 \leq \mu v_0 - v_0^2 + c_M^- u_0 v_0 \,, \end{cases}$$

in Ω, while $u_0 = v_0 = 0$ on $\partial\Omega$. Thus, (u_0, v_0) is a positive sub-solution of the pure symbiotic model

$$\begin{cases} -\Delta u = \lambda u - u^2 + b_M^- uv & \text{in } \Omega \,, \\ -\Delta v = \mu v - v^2 + c_M^- uv & \\ u = v = 0 \,, & \text{on } \partial\Omega \,. \end{cases} \tag{6.13}$$

Due to (6.9), for each constant $N \geq \max\left\{ \frac{\lambda}{1-b_M^-}, \frac{\mu}{1-c_M^-} \right\}$, the pair (N, N) is a super-solution of (6.13), and – if necessary – enlarging N one has that $(u_0, v_0) \leq (N, N)$. Consequently, thanks to [15, Theorem 8.7], (6.13) has a coexistence state $(\hat{u}_0, \hat{v}_0)$ such that

$$u_0 \leq \hat{u}_0 \leq N \,, \qquad v_0 \leq \hat{v}_0 \leq N \,. \tag{6.14}$$

Therefore, due [15, Lemma 6.1], (λ, μ) satisfies (6.10). Moreover,

$$\hat{u}_0 \le \frac{\lambda + \mu b_M^-}{1 - b_M^- c_M^-}\,, \qquad \hat{v}_0 \le \frac{\mu + \lambda c_M^-}{1 - b_M^- c_M^-}\,. \tag{6.15}$$

In particular, (6.14) and (6.15) imply

$$u_0 \le \frac{\lambda + \mu b_M^-}{1 - b_M^- c_M^-}\,, \qquad v_0 \le \frac{\mu + \lambda c_M^-}{1 - b_M^- c_M^-}\,, \tag{6.16}$$

for any coexistence state (u_0, v_0) of (2.3). Note that, due to (6.10), $\lambda + \mu b_M^- > 0$ and $\mu + \lambda c_M^- > 0$. Consequently, the coexistence states of (2.3) possess uniform a priori bounds in L^∞ on any compact subset of $(\lambda, \mu) \in \mathbb{R}^2$. The remaining assertions of the theorem follow straight ahead from the fact that $\mathfrak{C}_{(\mu,\theta_\lambda,0)}$ is an unbounded closed and connected set. □

When, instead of (6.9), the following conditions are satisfied

$$b_M^- > 1\,, \qquad c_M^- > 1\,, \tag{6.17}$$

then (2.3) might not exhibit coexistence states for sufficiently large μ, as strongly supported by Figure 11. But, from this fact, it should not concluded that the principle of competitive exclusion will hold as in the classical competition model. Actually, as larger are the symbiotic effects between u and v, measured by b^- and c^-, as greater should be the possibilities of avoiding extinction for each of the species. Consequently, the absence of coexistence states when b^- and c^- grow should entail some kind of Malthusian growth for the species. To realize what's going on, we will focus our attention into the simplest case when

$$\lambda = \mu \quad \text{and} \quad b = c\,. \tag{6.18}$$

Then, all pairs of the form $(u, v) = (w, w)$, where w is a positive solution of

$$\begin{cases} -\Delta w = \lambda w - (1 + b) w^2 & \text{in } \Omega\,, \\ w|_{\partial\Omega} = 0\,, \end{cases} \tag{6.19}$$

are coexistence states of (2.3). The solutions of (6.19) exhibit a number of different behaviors according to the size of b^-. If $b_M^- < 1$, then $1 + b(x) > 0$ for each $x \in \bar{\Omega}$ and, hence, (6.19) has a positive solution (unique) if, and only if, $\lambda > \sigma$; denoted by w_λ. Moreover, if $\lambda > \sigma$, then w_λ is a global attractor for

$$\begin{cases} \frac{\partial w}{\partial t} - \Delta w = \lambda w - (1 + b) w^2 & \text{in } \Omega \times (0, \infty)\,, \\ w = 0 & \text{on } \partial\Omega \times (0, \infty)\,, \\ w(\cdot, 0) = w_0 > 0 & \text{in } \Omega\,. \end{cases} \tag{6.20}$$

Consequently, (w_λ, w_λ) attracts to all solutions of (2.1) with $u_0 = v_0 > 0$.

If $b_M^- = 1$ in the closure of some appropriate smooth subdomain $D_0 \subset \Omega_-$, then $1 + b = 0$ in $\bar{D}_0$ and $1 + b(x) > 0$ for each $x \in \Omega \setminus \bar{D}_0$. Thus, thanks to the theory of Fraile et al. [18], (6.19) has a positive solution (unique) if, and only if, $\sigma < \lambda < \sigma_0 := \sigma[-\Delta; D_0]$, say w_λ. Moreover, as in the previous case, w_λ is a global attractor for (6.20), and, therefore, (w_λ, w_λ) is a global attractor for all solutions of (2.1) with $u_0 = v_0 > 0$ if $\sigma < \lambda < \sigma_0$. Furthermore, thanks to the theory of

García-Melián et al. [21], $L_{\sigma_0} := \lim_{\lambda\uparrow\sigma_0} w_\lambda$ provides us with the minimal positive solution of the singular problem

$$\begin{cases} -\Delta w = \sigma_0 w - (1+b)w^2 & \text{in } \Omega\setminus\bar{D}_0\,, \\ w|_{\partial D_0} = \infty\,, \end{cases} \tag{6.21}$$

in $\Omega\setminus\bar{D}_0$, while $\lim_{\lambda\uparrow\sigma_0} w_\lambda = \infty$ in $\bar{D}_0$. For each $\lambda \geq \sigma_0$, the dynamics of (6.20) is regulated by the minimal *metasolutions* of (6.19) supported in $\Omega\setminus\bar{D}_0$. By a metasolution it is meant the extension by infinity of the solutions of (6.21) (cf. [28], [31] and [33]). Therefore, the limit as $t\uparrow\infty$ of the solutions of (2.1) with $\lambda\geq\sigma_0$ and $u_0 = v_0 > 0$ is regulated by the metasolutions of (6.19), like in [30]. In such case, (2.3) cannot admit a coexistence state.

If $b_M^- > 1$, then the function $1+b(x)$ changes sign in Ω and, in such case, the dynamics of (6.20) becomes much more involved, since (6.20) is a superlinear indefinite problem. Among the main results available in this case, it is known that there exists $\lambda^* > \sigma$ such that (6.19) cannot have a positive solution if $\lambda > \lambda^*$ (cf. Amann & López-Gómez [3]). Moreover, if $\lambda^* > \sigma$ and $\sigma < \lambda < \lambda^*$, then the minimal positive solution is the unique non-negative stable solution of (6.20) (cf. Gómez-Reñasco and López-Gómez [22]), though, quite strikingly, as a result of the superlinear character of the model, (6.19) might exhibit an arbitrarily large number of positive – multiple spike – solutions; all of them necessarily unstable; the uniqueness of the stable solution in this general class of superlinear indefinite problems showing the extraordinary depth beneath the celebrated theorem of Amann concerning the stability of the minimal positive solution, [1]. As in the previous case, if $\lambda > \lambda^*$, then the dynamics of (6.20) are regulated by the metasolutions of (6.19), which can be reached in a finite, or infinity, time according to the size of b_M^-, but this – rather sharp – analysis is outside the general scope of this work (cf. López-Gómez and Quittner [42] and [34] for complete details).

The previous discussion, besides showing the necessity of condition (6.9) for the validity of Theorem 6.1, supports the thesis that, in the absence of coexistence states, the species will exhibit a genuine Malthusian growth within cooperation areas, up to blow-up in a finite time if b^-, or c^-, are sufficiently large.

A model that might avoid the previous blow-up phenomenology, simultaneously maintaining arbitrarily large the population levels as time passes by is the following *fast dispersion* model

$$\begin{cases} \frac{\partial u}{\partial t} - \Delta(u^m) = \lambda u - u^2 - b(x)uv & \\ \frac{\partial v}{\partial t} - \Delta(v^n) = \mu v - v^2 - c(x)uv & \text{in } \Omega\times\mathbb{R}_+\,, \\ u = v = 0 & \text{on } \partial\Omega\times\mathbb{R}_+\,, \\ u(\cdot,0) = u_0 \geq 0\,, \qquad v(\cdot,0) = v_0 \geq 0\,, & \text{in } \Omega\,, \end{cases}$$

where $n > 1$ and $m > 1$ are sufficiently close to 1, but we refrain of giving more details here (cf. M. Delgado et al. [16] for the case of one single equation).

7. Segregation within cooperation areas

The next result provides us with the profiles of the coexistence states of (2.3) when the intrinsic growth rate of the species v, μ, reaches high levels. It explains the ultimate reasons why the principle of competition exclusion fails in our model (2.1). As the level of the aggressions from the competitor blows-up, the species segregates within regions where cooperation takes place, so avoiding extinction, and facilitating diversity through further differentiation.

Theorem 7.1. *Suppose $\lambda > \lambda_s$ (see (6.4)),*

$$b_M^- < 1\,, \qquad c_M^- < 1\,, \qquad c_M^+ < \frac{1 - b_M^- c_M^-}{b_M^-}\,, \tag{7.1}$$

and let $\{(\mu_n, u_n, v_n)\}_{n\geq 1}$ be a sequence of coexistence states of (2.3) such that $\mu_n \uparrow \infty$ as $n \uparrow \infty$. Then,

$$\lim_{n\to\infty} v_n = \infty \quad \textit{uniformly in compact subsets of } \Omega\,, \tag{7.2}$$

and

$$\lim_{n\to\infty} u_n = \begin{cases} 0 & \textit{in } \Omega_+ \\ \theta_{[\lambda,\Omega_0]} & \textit{in } \Omega_0 \\ \infty & \textit{in } \Omega_- \end{cases} \tag{7.3}$$

where $\theta_{[\lambda,\Omega_0]}$ is the maximal non-negative solution of

$$\begin{cases} -\Delta u = \lambda u - u^2 & \textit{in } \Omega_0\,, \\ u = 0 & \textit{on } \partial\Omega_0\,. \end{cases}$$

Note that $\theta_{[\lambda,\Omega_0]} = 0$ if $\lambda \leq \sigma[-\Delta;\Omega_0]$, while $\theta_{[\lambda,\Omega_0]} \gg 0$ if $\lambda > \sigma[-\Delta;\Omega_0]$. Also, note that the third condition of (7.1) imposes a control on the size of c_M^+, which measures the stress suffered by v from its interaction with u. So, (7.1) seems rather natural for getting (7.2). By Theorem 6.1, the first two estimates of (7.1) guarantee the existence of a coexistence state of (2.3) for any sufficiently large μ. Note that the model (6.5), (6.6), (6.7) fits within the setting of Theorem 7.1, since

$$2 = c_M^+ < \frac{1 - b_M^- c_M^-}{b_M^-} = \frac{1-(0.4)^2}{0.4} = 2.1\,.$$

Figure 12 shows a series of plots of u_n and v_n for a sequence of coexistence states (μ_n, u_n, v_n) of (6.5), (6.6), (6.7), with $\lambda = 48.98$ and μ_n increasing. Actually, it shows the components of a series of coexistence states along the curve of coexistence states of Figure 9. Their point-wise behaviors, as μ_n increases, completely agrees with the predictions made from Theorem 7.1. Indeed, v_n grows to infinity all over $(0,1)$ as $\mu_n \uparrow \infty$, while $u_n \uparrow \infty$ as $\mu_n \uparrow \infty$ in $\Omega_- = (0.4, 0.6)$, and $u_n \downarrow 0$ as $\mu_n \uparrow \infty$ in $\Omega_+ = (0.0, 0.4) \cup (0.6, 1.0)$. Rather naturally, $v_n \uparrow \infty$ at a higher rate within Ω_+ than in Ω_-, as an effect of symbiosis with u within Ω_+.

The profiles of u_n and v_n clearly illustrate the permanence of both species as $\mu \uparrow \infty$ at the cost of the segregation of u within the cooperation area. A rather new phenomenology that cannot occur in the classical competing species

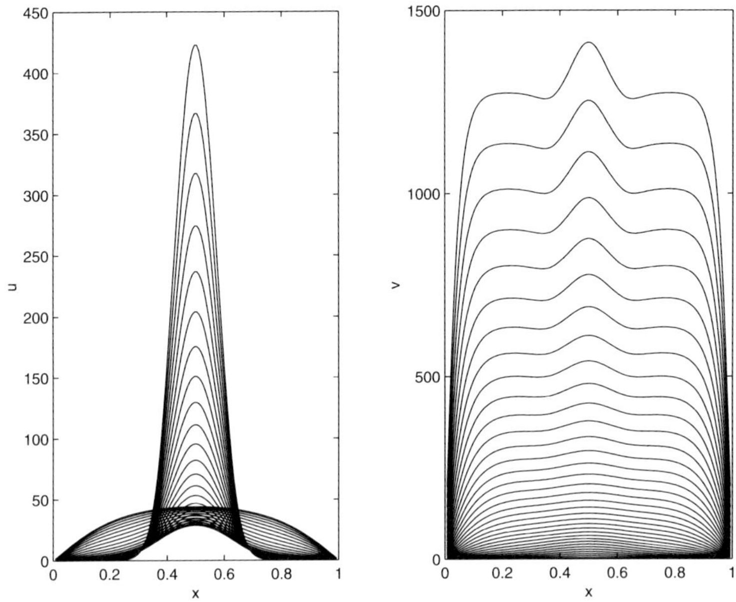

FIGURE 12. A series of coexistence states along $\mathfrak{C}_{(\mu,\theta_{48.98},0)}$ for (6.5), (6.6), (6.7).

models revisited in Section 5, where the principle of competitive exclusion predicts extinction of u.

Proof of Theorem 7.1*:* Fix $\lambda > \lambda_s$ and a sequence $\{(\mu_n, u_n, v_n)\}_{n\geq 1}$ of coexistence states of (2.3) with $\lim_{n\to\infty} \mu_n = \infty$, whose existence is guaranteed by Theorem 6.1. Without lost of generality we can assume that $\mu_n \geq 0$ for each $n \geq 1$. Thanks to (6.16), we already know that

$$u_n \leq \frac{\lambda + \mu_n b_M^-}{1 - b_M^- c_M^-}, \qquad n \geq 1,$$

and, hence,

$$\begin{aligned} -\Delta v_n &= \mu_n v_n - v_n^2 - c^+ u_n v_n + c^- u_n v_n \geq \mu_n v_n - v_n^2 - c^+ u_n v_n \\ &\geq \left(\mu_n - c^+ \frac{\lambda + \mu_n b_M^-}{1 - b_M^- c_M^-} \right) v_n - v_n^2 \\ &\geq \left[\left(1 - \frac{b_M^- c_M^+}{1 - b_M^- c_M^-} \right) \mu_n - \frac{\lambda\, c_M^+}{1 - b_M^- c_M^-} \right] v_n - v_n^2 . \end{aligned}$$

Thus, setting

$$\gamma_n := \left(1 - \frac{b_M^- c_M^+}{1 - b_M^- c_M^-} \right) \mu_n - \frac{\lambda\, c_M^+}{1 - b_M^- c_M^-}, \qquad n \geq 1,$$

we find that, for each $n \geq 1$, v_n is a positive super-solution of

$$\begin{cases} -\Delta w = \gamma_n w - w^2 & \text{in } \Omega\,, \\ w = 0 & \text{on } \partial\Omega\,. \end{cases} \tag{7.4}$$

Due to (7.1), $\lim_{n\to\infty} \gamma_n = \infty$ and, in particular, there is $n_0 \geq 1$ such that $\gamma_n > \sigma$ for each $n \geq n_0$. For such values of n, (7.4) has a (unique) positive solution, say w_n. Thus, $v_n \geq w_n$ if $n \geq n_0$. Now, by [20, Theorem 3.4], $\lim_{n\to\infty} w_n = \infty$ uniformly in compact subsets of Ω. Therefore, $\lim_{n\to\infty} v_n = \infty$.

Once proven (7.2), the validity of (7.3) will be obtained from the u_n-problem

$$\begin{cases} (-\Delta + bv_n)\, u_n = \lambda u_n - u_n^2 & \text{in } \Omega\,, \\ u_n = 0 & \text{on } \partial\Omega\,, \end{cases} \qquad n \geq 1\,.$$

Subsequently, we will prove each of the parts of (7.3) separately.

Part 1: Demographic explosion in Ω_-. It suffices to show that for each $x_0 \in \Omega_-$ there exists $\rho > 0$ such that $\bar{B}_\rho(x_0) \subset \Omega_-$ and $\lim_{n\to\infty} u_n = \infty$ uniformly in $\bar{B}_\rho(x_0)$. Indeed, let $x_0 \in \Omega_-$ and $R > 0$ with $\bar{B}_R(x_0) \subset \Omega_-$, and consider the auxiliary problems

$$\begin{cases} \left(-\Delta - b^- \min\limits_{\bar{B}_R(x_0)} v_n\right) U = \lambda U - U^2 & \text{in } B_R(x_0)\,, \\ U = 0 & \text{on } \partial B_R(x_0)\,. \end{cases} \tag{7.5}$$

For each $n \geq 1$, we have that

$$\sigma[-\Delta - b^- \min_{\bar{B}_R(x_0)} v_n; B_R(x_0)] \leq \sigma[-\Delta; B_R(x_0)] - \min_{\bar{B}_R(x_0)} b^- \cdot \min_{\bar{B}_R(x_0)} v_n$$

and, hence,

$$\lim_{n\to\infty} \sigma[-\Delta - b^- \min_{\bar{B}_R(x_0)} v_n; B_R(x_0)] = -\infty\,,$$

since $v_n \to \infty$ uniformly in $\bar{B}_R(x_0)$. In particular, there exists $n_0 \geq 1$ such that, for each $n \geq n_0$,

$$\lambda > \sigma[-\Delta - b^- \min_{\bar{B}_R(x_0)} v_n; B_R(x_0)]\,.$$

Consequently, (7.5) possesses a unique positive solution, say U_n. As u_n is a positive super-solution of (7.5), for each $n \geq n_0$, $u_n|_{B_R(x_0)} \geq U_n$ in $B_R(x_0)$ and, hence, to obtain the limiting behavior of u_n in Ω_- it suffices to prove that

$$\lim_{n\to\infty} U_n = \infty \qquad \text{uniformly in} \quad \bar{B}_{\frac{R}{2}}(x_0)\,. \tag{7.6}$$

To show (7.6), we perform the change of variable $U = \varepsilon_n^{-1} Z$ where $\varepsilon_n^{-1} := \min_{\bar{B}_R(x_0)} v_n$, to transform (7.5) into

$$\begin{cases} -\varepsilon_n \Delta Z = (\lambda\varepsilon_n + b^-)\, Z - Z^2 & \text{in } B_R(x_0)\,, \\ Z = 0 & \text{on } \partial B_R(x_0)\,. \end{cases} \tag{7.7}$$

Thanks to [20, Theorem 3.6], one has that $\lim_{n\to\infty} Z_n = b^-$ uniformly in $\bar{B}_\rho(x_0)$, for each $\rho \in (0, R)$, where Z_n is the unique positive solution of (7.7). This shows

(7.6). Consequently, $\lim_{n\to\infty} u_n = \infty$ uniformly in compact subsets of Ω_-, which concludes the proof of this part.

Part 2: Demographic extinction in Ω_+. Let $x_0 \in \Omega_+$ and $R > 0$ such that $\bar{B}_R(x_0) \subset \Omega_+$, and consider the singular auxiliary problems

$$\begin{cases} (-\Delta + \alpha_n) U = \lambda U - U^2 & \text{in } B_R(x_0)\,, \\ U = \infty & \text{on } \partial B_R(x_0)\,, \end{cases} \tag{7.8}$$

where

$$\alpha_n := \min_{\bar{B}_R(x_0)} b^+ \cdot \min_{\bar{B}_R(x_0)} v_n\,.$$

We already know that $\lim_{n\to\infty} \alpha_n = \infty$. Moreover, for each $n \geq 1$, (7.8) possesses a unique solution, denoted by L_n (cf., e.g., [33]), and, since u_n is a positive subsolution of (7.8), $u_n \leq L_n$ for each $n \geq 1$. Therefore, to show that $u_n \to 0$ uniformly on compact subsets of Ω_+ it suffices to prove that $\lim_{n\to\infty} L_n = 0$ uniformly in $\bar{B}_{\frac{R}{2}}(x_0)$. Indeed, set $D := B_{R/2}(x_0)$ and let $\varphi \gg 0$ be a principal eigenfunction associated to $\sigma[-\Delta; D]$. Then, multiplying the differential equation of L_n by φ, integrating in D and applying the formula of integration by parts, gives

$$(\alpha_n - \lambda + \sigma[-\Delta; D]) \int_D L_n\varphi = -\int_{\partial D} L_n \frac{\partial \varphi}{\partial \nu} - \int_D L_n^2 \varphi \leq -\int_{\partial D} L_n \frac{\partial \varphi}{\partial \nu}\,,$$

where ν stands for the outward unit normal of D. Let L_0 be the unique large solution of $-\Delta U = \lambda U - U^2$ in D. As, for each $n \geq 1$, $\lambda - \alpha_n \leq \lambda$, we have that $L_n \leq L_0$, $n \geq 1$, and, hence, since φ has negative flux on ∂D, we obtain that

$$(\alpha_n - \lambda + \sigma[-\Delta; D]) \int_D L_n\varphi \leq -\int_{\partial D} L_0 \frac{\partial \varphi}{\partial \nu}\,, \qquad n \geq 1\,.$$

Thus, since $\lim_{n\to\infty}(\alpha_n - \lambda + \sigma[-\Delta; D]) = \infty$, necessarily $\lim_{n\to\infty} \int_D L_n\varphi = 0$. On the other hand, thanks to the results of [28], for each $n \geq 1$, L_n is radially symmetric and, for n sufficiently large, $\min_{\bar{D}} L_n = L_n(x_0)$, since $-\Delta L_n \leq 0$. Necessarily, $\lim_{n\to\infty} L_n(x_0) = 0$, and, therefore, $\lim_{n\to\infty} u_n(x_0) = 0$, since $u_n \leq L_n$. As this holds for each $x_0 \in \Omega_+$, the functions u_n approximate to zero in Ω_+ as $n \to \infty$. To show the uniformity of that convergence in $\bar{D}$ one can argue as follows. For each $x \in D - B_{\frac{R}{2}}(x_0)$, let L_n^x denote the large solution of

$$(-\Delta + \alpha_n) U = \lambda U - U^2 \quad \text{in} \quad B_{\frac{R}{2}}(x)\,, \qquad U|_{\partial B_{\frac{R}{2}}(x)} = \infty\,.$$

Since $L_n \leq L_n^x = L_n^{x_0}(x_0 - x + \cdot)$ in $B_{\frac{R}{2}}(x)$, we have that $L_n(x) \leq L_n^{x_0}(x_0)$ for each $x \in D$, which implies $\lim_{n\to\infty} L_n = 0$ uniformly in $\bar{D}$ and concludes the second part of the proof.

Part 3: Demographic stabilization in Ω_0. First, suppose that

$$\lambda > \sigma[-\Delta; \Omega_0] \tag{7.9}$$

and denote by θ_λ the unique positive solution of

$$-\Delta u = \lambda u - u^2 \quad \text{in} \quad \Omega_0\,, \qquad u|_{\partial\Omega_0} = 0\,. \tag{7.10}$$

As, for each $n \geq 1$, u_n is a positive super-solution of (7.10), we find that

$$u_n \geq \theta_\lambda \qquad \text{in} \quad \bar{\Omega}_0\,, \qquad n \geq 1\,. \tag{7.11}$$

Now, for each sufficiently small $\delta > 0$, consider the δ-neighborhood of Ω_0

$$\Omega_\delta := \{\, x \in \Omega \ : \ \operatorname{dist}(x, \Omega_0) < \delta \,\}\,;$$

δ is assumed to be chosen so that $\partial\Omega_\delta \subset \Omega_+$ (see Figure 1). By monotonicity, (7.9) implies

$$\lambda > \sigma[-\Delta; \Omega_0] > \sigma[-\Delta; \Omega_\delta]\,, \tag{7.12}$$

since $\bar{\Omega}_0 \subset \Omega_\delta$. Thus, the problem

$$-\Delta u = \lambda u - u^2 \qquad \text{in} \quad \Omega_\delta\,, \qquad u|_{\partial\Omega_\delta} = 0\,, \tag{7.13}$$

possesses a unique positive solution. Let denote it by $\theta_{[\lambda,\delta]}$ and consider a smooth function of the following type

$$\Phi_\delta := \begin{cases} \theta_{[\lambda,\delta]} & \text{in } \Omega_{\frac{\delta}{2}} \\ \psi_\delta & \text{in } \bar{\Omega}_\delta \setminus \Omega_{\frac{\delta}{2}} \end{cases}$$

where ψ_δ is any positive extension (bounded away from zero) of $\theta_{[\lambda,\delta]}|_{\Omega_{\frac{\delta}{2}}}$ to the totality of Ω_δ. We now show that for sufficiently large n the function Φ_δ provides us with a positive super-solution of the following problem

$$\begin{cases} (-\Delta + bv_n)\, U = \lambda U - U^2 & \text{in} \quad \Omega_\delta\,, \\ U = u_n & \text{on} \quad \partial\Omega_\delta\,, \end{cases} \tag{7.14}$$

whose unique positive solution is u_n. Since ψ_δ is bounded away from zero in $\partial\Omega_\delta$ and, thanks to the result already obtained in Part 2, we have that

$$\lim_{n\to\infty} u_n = 0 \qquad \text{uniformly in} \quad \partial\Omega_\delta \subset \Omega_+\,.$$

Thus, there exists $n_0 \geq 1$ such that for each $n \geq n_0$ one has that $\Phi_\delta \geq u_n$ in $\partial\Omega_\delta$. Moreover, in $\Omega_{\frac{\delta}{2}}$ one has that $\Phi_\delta = \theta_{[\lambda,\delta]}$ and, hence,

$$(-\Delta + bv_n)\, \Phi_\delta = (-\Delta + bv_n)\, \theta_{[\lambda,\delta]} > -\Delta\theta_{[\lambda,\delta]} = \lambda\theta_{[\lambda,\delta]} - \theta^2_{[\lambda,\delta]}\,,$$

whereas in $\Omega_\delta \setminus \Omega_{\frac{\delta}{2}}$, we have $\Phi_\delta = \psi_\delta$ and, so,

$$(-\Delta + bv_n)\, \Phi_\delta = (-\Delta + bv_n)\, \psi_\delta \geq \lambda\psi_\delta - \psi_\delta^2\,,$$

if n is sufficiently large, since $\lim_{n\to\infty} \min_{\bar{\Omega}_\delta \setminus \Omega_{\delta/2}} (bv_n) = \infty$. Summarizing, there exists $n_0 \geq 1$ such that, for each $n \geq n_0$, Φ_δ is a positive strict super-solution of (7.14), whose unique solution is u_n. Consequently, $\Phi_\delta \geq u_n$, $n \geq n_0$. In particular, $\limsup u_n|_{\Omega_0} \leq \theta_{[\lambda,\delta]}$ for each sufficiently small $\delta > 0$. Therefore,

$$\theta_\lambda \leq \liminf_{n\to\infty} u_n|_{\Omega_0} \leq \limsup_{n\to\infty} u_n|_{\Omega_0} \leq \theta_{[\lambda,\delta]}\,.$$

Now, thanks to theory of Cano-Casanova & López-Gómez [7], $\lim_{\delta\downarrow 0} \theta_{[\lambda,\delta]} = \theta_\lambda$ uniformly in $\bar{\Omega}_0$, and, consequently, $\lim_{n\to\infty} u_n|_{\bar{\Omega}_0} = \theta_\lambda$ uniformly in $\bar{\Omega}_0$, which concludes the proof of the theorem under condition (7.9).

Now, suppose that $\lambda \leq \sigma[-\Delta;\Omega_0]$. Then, for each $\varepsilon > 0$ the function u_n is a positive sub-solution of

$$\begin{cases} (-\Delta + bv_n)U = (\sigma[-\Delta;\Omega_0]+\varepsilon)U - U^2 & \text{in } \Omega\,, \\ U = 0 & \text{on } \partial\Omega\,, \end{cases} \tag{7.15}$$

and, hence,

$$u_n \leq u_{[n,\varepsilon]} \qquad \text{in } \Omega\,, \tag{7.16}$$

where $u_{[n,\varepsilon]}$ stands for the unique positive solution of (7.15); it exists for sufficiently large n. As

$$\rho_\varepsilon := \sigma[-\Delta;\Omega_0] + \varepsilon > \sigma[-\Delta;\Omega_0],$$

by the result that we have just obtained, we know that

$$\lim_{n\to\infty} u_{[n,\varepsilon]} = \theta_{\rho_\varepsilon} \qquad \text{uniformly in } \bar{\Omega}_0\,, \tag{7.17}$$

where $\theta_{\rho_\varepsilon}$ is the unique positive solution of

$$\begin{cases} -\Delta U = \rho_\varepsilon U - U^2 & \text{in } \Omega_0\,, \\ U = 0 & \text{on } \partial\Omega_0\,. \end{cases}$$

Therefore, thanks to (7.16) and (7.17), it is apparent that for each $\varepsilon > 0$

$$\limsup_{n\to\infty} u_n \leq \theta_{\rho_\varepsilon}\,. \tag{7.18}$$

On the other hand, as a consequence from the celebrated theorem of Crandall & Rabinowitz [9], it is rather clear that $\lim_{\varepsilon\downarrow 0}\theta_{\rho_\varepsilon} = 0$, uniformly in $\bar{\Omega}_0$. Consequently, (7.18) implies $\lim_{n\to\infty} u_n = 0$ uniformly in $\bar{\Omega}_0$, which concludes the proof. □

8. Generating biological and economical complexity

In this section we consider the prototype model (6.5), (6.6) with

$$b(x) := \begin{cases} -0.4\sin(5\pi x)\,, & 0.0 \leq x \leq 0.2, \\ 20\sin[5\pi(x-0.2)]\,, & 0.2 < x \leq 0.4, \\ -0.4\sin[5\pi(x-0.4)]\,, & 0.4 < x \leq 0.6, \\ 40\sin[5\pi(x-0.6)]\,, & 0.6 < x \leq 0.8, \\ -0.4\sin[5\pi(x-0.8)]\,, & 0.8 < x \leq 1.0. \end{cases} \tag{8.1}$$

For this choice, u and v compete in $\Omega_+ = (0.2, 0.4)\cup(0.6, 0.8)$, while they cooperate in $\Omega_- = (0, 0.2)\cup(0.4, 0.6)\cup(0.8, 1)$. Note that the level of the aggressions between u and v has been chosen relatively high, 20 and 40, in comparison with the level of the cooperative effects within symbiotic areas, 0.4. Consequently, at first glance, symbiotic interactions are relatively weak. For this choice, $\sigma_1 = 25\pi^2 \sim 246.74$ and, hence, in the quadrant $\min\{\lambda,\mu\} > 246.74$ both semi-trivial states are linearly unstable, which is clearly illustrated by Figure 13.

Thanks to Proposition 4.1, the tangents at infinity of the curves (3.8) are given by $\mu = -0.4\times\lambda$ and $\lambda = -0.4\times\mu$, respectively, which complete agrees with the numerical computations. Figure 14 shows the bifurcation diagram of non-negative solutions for a $\lambda \sim 140$; it shows three significant curves for describing

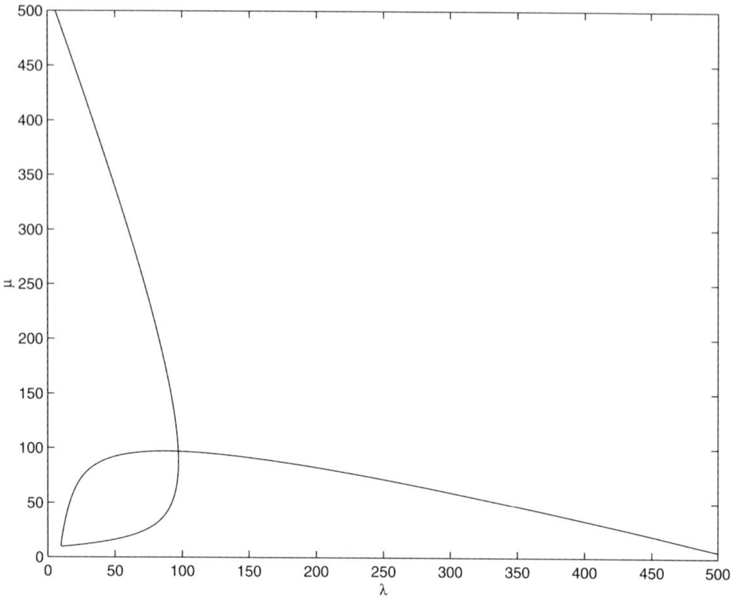

FIGURE 13. The curves (3.8) for the choice (6.5), (6.6) and (8.1).

the dynamics of (2.1); stable solutions are indicated by solid lines, unstable by dashed lines. As $(0,0)$ has been represented in the μ–axis, its character cannot be distinguished in the plot, though it is always unstable. The horizontal line describes the semi-trivial state $(\theta_{140}, 0)$; $|\theta_{140}| \sim 116.57$. Then, Figure 14 shows the curve of semi-trivial solutions $(0, \theta_\mu)$, which emanates from $(0,0)$ at $\mu = \pi^2$ and it grows for all further values of μ. All solutions along this curve are unstable. The component of coexistence states $\mathfrak{C}_{(\mu,\theta_{140},0)}$ emanates, super-critically, from $(\theta_{140}, 0)$ at

$$\mu = \sigma[-\frac{d^2}{dx^2} + b\theta_{140}; (0,1)] \sim 93.40\,,$$

where the previous stability of $(\theta_{140}, 0)$ is lost. All coexistence states are stable until $\mathfrak{C}_{(\mu,\theta_{140},0)}$ reaches its first turning point, at $\mu_1 \sim 171.0$, where it turns backwards. Once crossed the first turning point all coexistence states become unstable, with *Morse index* one, until $\mathfrak{C}_{(\mu,\theta_{140},0)}$ reached its second turning point, super-critical, at $\mu_2 = 126.47$. The Morse index counts the dimension of the unstable manifold of the steady-states. And so on... until the eight turning point exhibited by $\mathfrak{C}_{(\mu,\theta_{140},0)}$ is crossed, super-critically, at $\mu_8 := 100.7$, where all solutions become stable for all further values of μ.

Table 1 collects the parameter values of each of the turning points exhibited by $\mathfrak{C}_{(\mu,\theta_{140},0)}$ as well as the Morse indices of the coexistence states along each of the associated arcs of curve, $c_0, c_1, \ldots, c_8$; c_0 stands for the arc of curve of coexistence states emanating from the semitrivial positive solution $(\theta_{140}, 0)$ and, for each $j \geq 1$,

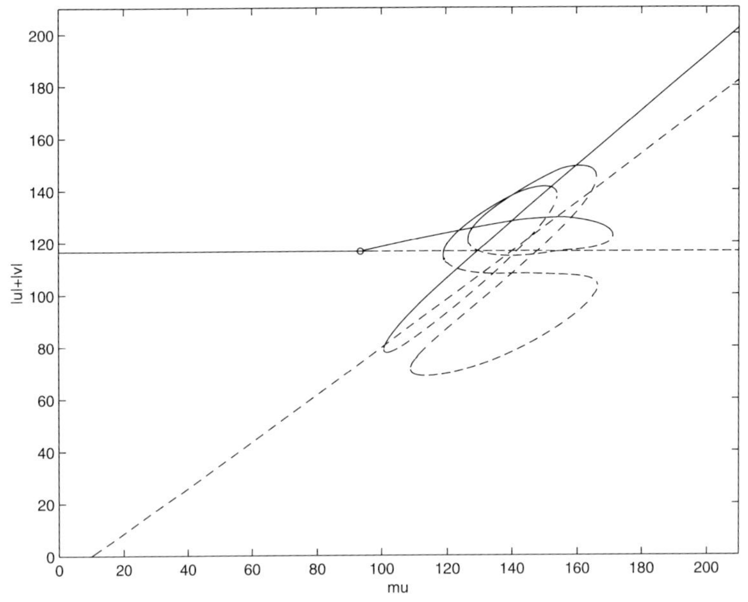

FIGURE 14. The component $\mathfrak{C}_{(\mu,\theta_{140},0)}$ of (6.6), (6.7), (8.1).

c_j stands for the arc of curve in between the turning points corresponding to μ_j and μ_{j+1}. Consequently, for each $\mu \in (126.47, 153.93)$, (6.5), (6.6), (8.1) possesses 9 coexistence states: 4 among them are asymptotically stable, and the remaining 5 are unstable. The Morse index changed by ± 1 whenever crossed a turning point, as predicted by well-known results of classical bifurcation theory. Figure 15 shows a series of plots of representative coexistence states along c_0. We have plotted the profiles of the u and the v-components of the coexistence states for a series of values of μ in between the bifurcation point from $(\theta_{140}, 0)$ and the first turning point at $\mu_1 = 171.0$.

arc	Morse index	Turning point
c_0	0	171.0
c_1	1	126.47
c_2	0	165.9
c_3	1	109.0
c_4	2	166.2
c_5	1	118.97
c_6	0	153.93
c_7	1	100.7
c_8	0	*none*

Table 1. The turning points and the Morse index along the arcs c_j.

As the value of μ separates from the bifurcation value $\mu = 93.40$, u decreases from θ_{140}, whereas v increases from zero. In Figure 15 and in all subsequent plots of coexistence states, the arrows in each of the components of $(0,1)$ where b has a constant sign, indicate (very roughly) if the corresponding species grows, $\uparrow$, or, conversely, decays, $\downarrow$, as μ moves between two consecutive turning points.

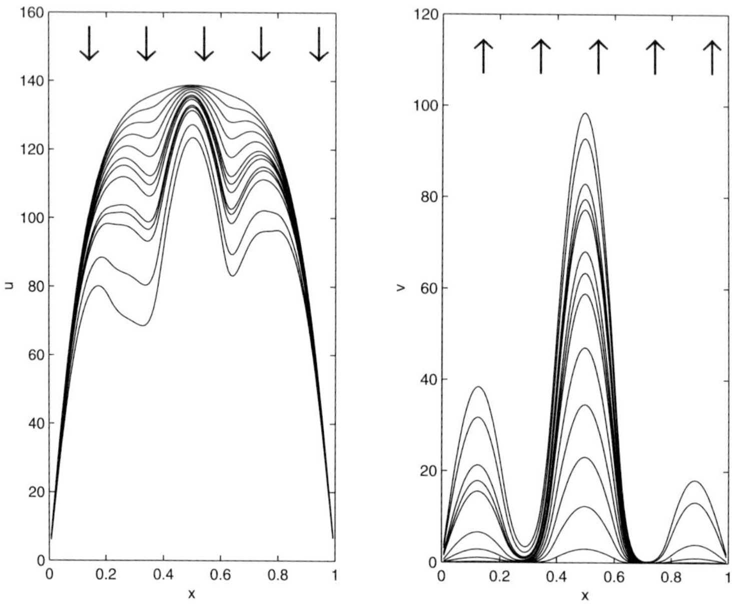

FIGURE 15. Representative coexistence states along c_0.

Rather naturally, as the intrinsic growth rate of v, measured by μ, increases, the species v grows. Moreover, it grows at the fastest rate within the symbiotic regions, $b < 0$, where cooperation with u takes place, than in the competitive domains, $b > 0$, where u and v compete. Not surprisingly, u decays as μ increases, and its fastest decay rates occur within regions where competition takes place.

Similarly, Figure 16 shows the plots of a series of representative coexistence states along the arc of curve c_2. As in Figure 15, we have plotted the profiles of the u and the v-components of the computed coexistence states for a series of values of μ in between the turning points $\mu_2 = 126.47$ and $\mu_3 = 165.9$. As along c_0, the coexistence states along c_2 exhibit a genuine *sub-linear behavior*, in the sense that u is point-wise decreasing and v is increasing in $(0,1)$ for all values of μ. Such a nice monotonicity seems to be closely related to the attractive character of these coexistence states, as it is lost along arcs where solutions are unstable.

Once crossed the last turning point exhibited by $\mathfrak{C}_{(\mu,\theta_{140},0)}$, at $\mu_8 = 100.7$, the coexistence states become stable for all further value of μ. In Figure 17 we have represented the plots of a series of representative coexistence states along c_8.

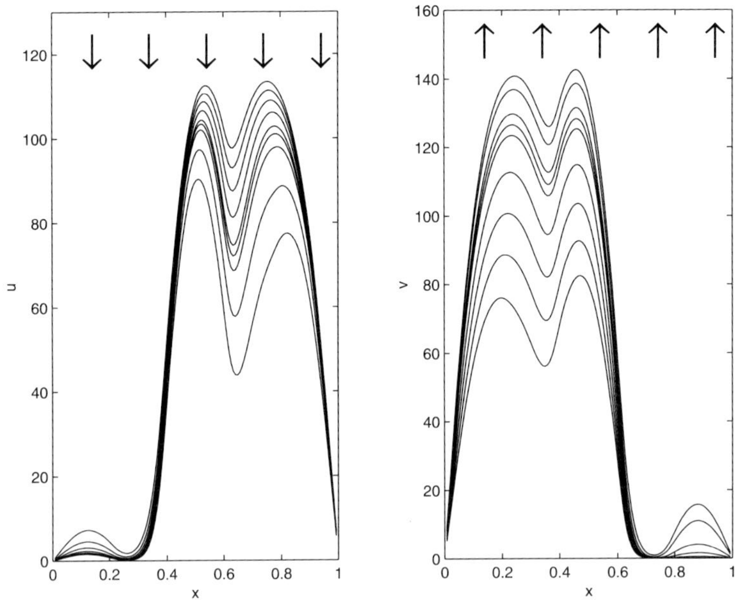

FIGURE 16. Representative coexistence states along c_2.

As illustrated by Figure 17, the distribution of the population v grows everywhere in $(0,1)$ as μ increases, reaching arbitrarily large values, while the species u grows arbitrarily in the regions where cooperation with v takes place, $(0,0.2)$, $(0.4,0.6)$ and $(0.8,1)$, though it naturally decays up to extinction within the regions where u is stressed by v, $(0.2,0.4)$ and $(0.6,0.8)$, in complete agreement with Theorem 7.1. It should be noted that, for the choice (8.1), one has

$$b_M^- = c_M^- = 0.4 < 1\,, \qquad c_M^+ = 40 > \frac{1 - b_M^- c_M^-}{b_M^-} = \frac{1-(0.4)^2}{0.4} = 2.1\,,$$

and, consequently, the last estimate of (7.1) fails to be true in this example, which suggests that the range of validity of Theorem 7.1 might be substantially wider.

This example shows that, although (6.19) possesses at most a unique stable positive solution (by the main theorem of Gómez-Reñasco & López-Gómez [22]), (2.3) might have an arbitrarily large number of stable coexistence states for some appropriate range of values of the parameters involved in the setting of the model, so, strongly suggesting that competition combined with some strategic symbiotic effects might result into a substantial growth of the dynamical complexity of the system, a sort of principle corroborating that vision indeed extraordinarily facilitated the Cambrian explosion.

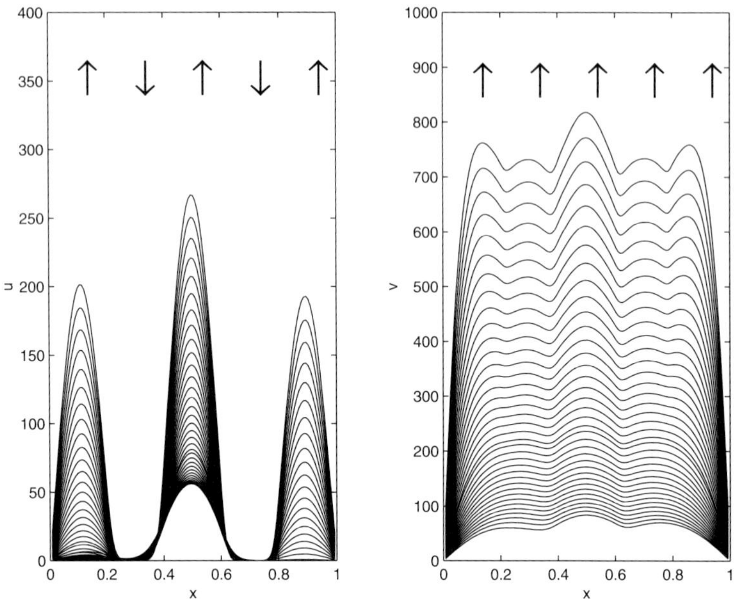

FIGURE 17. Representative coexistence states along c_8.

Acknowledgements

The authors thank to the Spanish Ministry of Education and Science under grant REN2003-00707.

References

[1] H. Amann, Fixed point equations and nonlinear eigenvalue problems in ordered Banach spaces, *SIAM Rev.* **18** (1976), 620–709.

[2] H. Amann, Maximum principles and principal eigenvalues, in *Ten Mathematical Essays on Approximation in Analysis and Topology* (J. Ferrera, J. López-Gómez and F.R. Ruíz del Portal, eds.), pages 1–60, Elsevier, Amsterdam, 2005.

[3] H. Amann and J. López-Gómez, A priori bounds and multiple solutions for superlinear indefinite elliptic problems, *J. Diff. Eqns.* **146** (1998), 336–374.

[4] M. Begon, J.L. Harper, and C.R. Townsend, *Ecology, Individual, Populations and Communities*, Blackwell Scientific Publications, Cambridge, Massachusetts, 1990.

[5] J. Blat and K.J. Brown, Bifurcation of steady-state solutions in predator-prey and competition systems, *Proc. Roy. Soc. Edinburgh* **97A** (1984), 21–34.

[6] S. Cano-Casanova and J. López-Gómez, Permanence under strong aggression is possible, *An. Inst. H. Poincaré, Anal. non Lin.* **20** (2003), 999–1041.

[7] S. Cano-Casanova and J. López-Gómez, Varying domains in a general class of sublinear elliptic problems, *Elec. J. Diff. Eqns.* **74** (2004), 1–41.

[8] S. Cano-Casanova, J. López-Gómez, and M. Molina-Meyer, Permanence through spatial segregation in heterogeneous competition, in *Proc. 9th IEEE International Conference on Methods and Models in Automation and Robotics* (R. Kaszyński, Ed.), pages 123–130, Szczecin, 2003.

[9] M.G. Crandall and P.H. Rabinowitz, Bifurcation from simple eigenvalues, *J. Funct. Anal.* **8** (1971), 321–340.

[10] E.N. Dancer, On positive solutions of some pairs of differential equations, I, *Trans. Amer. Math. Soc.* **284** (1984), 729–743.

[11] E.N. Dancer, On positive solutions of some pairs of differential equations, II, *J. Diff. Eqns.* **60** (1985), 236–258.

[12] E.N. Dancer, The existence and the uniqueness of positive solutions for competing species models with diffusion, *Trans. Amer. Math. Soc.* **326** (1991), 829–859.

[13] E.N. Dancer, A counterexample of competing species equations, *Diff. Int. Eqns.* **9** (1996), 239–246.

[14] E.N. Dancer, Bifurcation from simple eigenvalues and eigenvalues of geometric multiplicity one, *Bull. London Math. Soc.* **34** (2002), 533–538.

[15] M. Delgado, J. López-Gómez and A. Suárez, On the symbiotic Lotka-Volterra model with diffusion and transport effects, *J. Diff. Eqns.* 160 (2000), 175–262.

[16] M. Delgado, J. López-Gómez and A. Suárez, Non-linear versus linear diffusion from classical solutions to metasolutions, *Adv. Diff. Eqns.* **7** (2002), 1101–1124.

[17] J.C. Eilbeck, J.E. Furter and J. López-Gómez, Coexistence in the competition model with diffusion, *J. Diff. Eqns.* **107** (1994), 96–139.

[18] J.M. Fraile, P. Koch, J. López-Gómez and S. Merino, Elliptic eigenvalue problems and unbounded continua of positive solutions of a nonlinear elliptic problem, *J. Diff. Eqns.* **127** (1996), 295–319.

[19] J.E. Furter and J. López-Gómez, On the existence and uniqueness of coexistence states for the Lotka-Volterra competition model with diffusion and spatially dependent coefficients, *Nonl. Anal. TMA* **25** (1995), 363–398.

[20] J.E. Furter and J. López-Gómez, Diffusion-mediated permanence problem for a heterogeneous Lotka-Volterra competition model, *Proc. Roy. Soc. Edinburgh* **127A** (1997), 281–336.

[21] J. García Molián, R. Gómez-Reñasco, J. López-Gómez and J.C. Sabina de Lis, Pointwise growth and uniqueness of positive solutions for a class of sublinear elliptic problems where bifurcation from infinity occurs, *Arch. Rat. Mech. Anal.* **145** (1998), 261–289.

[22] R. Gómez-Reñasco and J. López-Gómez, The effects of varying coefficients on the dynamics of a class of superlinear indefinite reaction-diffusion equations, *J. Diff. Eqns.* **167** (2000), 36–72.

[23] P. Hess and A. Lazer, On an abstract competition model and applications, *Nonl. Anal. Th. Meth. Appns.* **16** (1991), 917–940.

[24] J. López-Gómez, *Estabilidad y Bifurcación Estática. Aplicaciones y Métodos Numéricos*, Cuadernos de Matemática y Mecánica, Cursos y Seminarios 4, CONICET, Santa Fe, 1988

[25] J. López-Gómez, Permanence under strong competition, in *Dynamical Systems and Applications*, pages 473–488, World Scientific Series in Applied Analysis 4, Singapore, 1995.

[26] J. López-Gómez, The maximum principle and the existence of principal eigenvalues for some linear weighted boundary value problems, *J. Diff. Eqns.* **127** (1996), 263–294.

[27] J. López-Gómez, On the structure of the permanence region for competing species models with general diffusivities and transport effects, *Disc. Cont. Dyn. Sys.* **2** (1996), 525–542.

[28] J. López-Gómez, Large solutions, metasolutions, and asymptotic behavior of the regular positive solutions of a class of sublinear parabolic problems, *Elec. J. Diff. Eqns.* **Conf. 05** (2000), 135–171.

[29] J. López-Gómez, *Spectral Theory and Nonlinear Functional Analysis*, Research Notes in Mathematics 426, CRC Press, Boca Raton (FL), 2001.

[30] J. López-Gómez, Coexistence and meta-coexistence for competing species, *Houston J. of Maths.*, 29 (2003), 483–536.

[31] J. López-Gómez, Dynamics of parabolic equations. From classical solutions to metasolutions, *Diff. Int. Eqns.* **16** (2003), 813–828.

[32] J. López-Gómez, The boundary blow-up rate of large solutions, *J. Diff. Eqns.* **195** (2003), 25–45.

[33] J. López-Gómez, Metasolutions: Malthus versus Verhulst in population dynamics. A dream of Volterra, in *Handbook of Differential Equations "Stationary Partial Differential Equations"* (M. Chipot and P. Quittner, Eds.), Elsevier Science B. V., North Holland, to appear.

[34] J. López-Gómez, Global existence versus blow-up in superlinear indefinite parabolic problems, *Sci. Mat. Japonica* **61** (2005), 493–517.

[35] J. López-Gómez, J.C. Eilbeck, M. Molina and K.N. Duncan, Structure of solution manifolds in a strongly coupled elliptic system, *IMA J. Numer. Anal.* **12** (1992), 405–428.

[36] J. López-Gómez and M. Molina-Meyer, Symbiosis in competitive environments, in *Variational Methods and Related Topics*, Vol. 12, pages 118–141, RIMS KYOTO 1347, Kyoto, 2003.

[37] J. López-Gómez and M. Molina-Meyer, Singular perturbations in Economy and Ecology. The effect of strategic symbiosis in random competitive environments, *Adv. Math. Sci. Appns.* **14** (2004), 87–107.

[38] J. López-Gómez and M. Molina-Meyer, Biodiversity through competition, in *Proc. 10th IEEE International Conference on Methods and Models in Automation and Robotics* (R. Kaszyński, Ed.), pages 39–44, Szczecin, 2004.

[39] J. López Gómez, M. Molina-Meyer and M. Villarreal, Existence and numerical simulation of coexistence states for some elliptic Lotka-Volterra models, *Quadernos del IAC Mauro Picone* **27** (1989), 119–133.

[40] J. López Gómez, M. Molina-Meyer and M. Villarreal, Numerical simulation of coexistence states, *SIAM J. Numer. Anal.* **29** (1992), 1074–1092.

[41] J. López-Gómez and R.M. Pardo, Coexistence regions in Lotka-Volterra models with Diffusion, *Nonl. Anal. Th. Meth. Appns.* **19** (1992), 11–28.

[42] J. López-Gómez and P. Quittner, Complete and energy blow-up in indefinite superlinear parabolic problems, *Disc. Cont. Dyn. Syst.*, to appear.

[43] J. López-Gómez and J.C. Sabina de Lis, Coexistence states and global attractivity for some convective diffusive competing species models, *Trans. Amer. Math. Soc.* **347** (1995), 3797–3833.

[44] A. Parker, *In the blink of an eye: the cause of the most dramatic event in the history of life*, Free Press, London, 2003.

[45] P.H. Rabinowitz, Some global results for nonlinear eigenvalue problems, *J. Funct. Anal.* **7** (1971), 487–513.

[46] D. Tilman, Diversity and Production in European Grasslands, *Science* **285** (1999), 1099–1100.

[47] E.O. Wilson, *Sociobiology*, Harvard University Press, Cambridge, Massachusetts, 1980.

Julián López-Gómez
Dpto. de Matemática Aplicada
Univ. Complutense de Madrid
E-28040 Madrid, Spain
e-mail: Lopez_Gomez@mat.ucm.es

Marcela Molina-Meyer
Dpto. de Matemáticas
Univ. Carlos III de Madrid
E-28911 Leganés, Madrid, Spain
e-mail: mmolinam@math.uc3m.es

Progress in Nonlinear Differential Equations
and Their Applications, Vol. 64, 329–349

Generalized Minimal Cardinal of the λ-slices of the Semi-bounded Components Arising in Global Bifurcation Theory

Julián López-Gómez and Carlos Mora-Corral

Dedicated to H. Amann, with our deepest admiration

Abstract. We use the degree introduced by P. Benevieri and M. Furi [3] to obtain the generalized minimal cardinal of the number of solutions of the λ-sections of those components of the set of nontrivial solutions of an abstract equation of the form $\mathfrak{F}(\lambda, x) = 0$ that are compact in one direction of the parameter; here $\mathfrak{F}$ is a $\mathcal{C}^1$ Fredholm map of index 1 such that $\mathfrak{F}(\lambda, 0) = 0$ for all $\lambda \in \mathbb{R}$. These bounds are given in terms of the parity of the linearized Fredholm family $D_2\mathfrak{F}(\cdot, 0)$. The parity is a local invariant measuring the change of the orientation of $D_2\mathfrak{F}(\lambda, 0)$ as λ crosses an interval. The set of eigenvalues of $D_2\mathfrak{F}(\cdot, 0)$ is not assumed to be discrete. Therefore, as regards applications, the theory developed in this paper should be extremely versatile.

Mathematics Subject Classification (2000). 47J15, 58C30, 47H11, 58C40.

Keywords. Bifurcation theory. Counting the number of solutions of λ-sections.

1. Introduction

In this paper, if E and F are two real Banach spaces, the set $\mathcal{L}(E, F)$ denotes the space of linear bounded operators from E to F, and $\Phi_0(E, F)$ stands for the subset of $\mathcal{L}(E, F)$ consisting of all Fredholm operators of index zero. Throughout this work,

$$\mathfrak{F} : \mathbb{R} \times E \to F$$

is a continuous map satisfying the following properties:

F1. For each $\lambda \in \mathbb{R}$, the partial map $\mathfrak{F}(\lambda, \cdot) \in \mathcal{C}^1(E, F)$ and $D_2\mathfrak{F}(\lambda, x) \in \Phi_0(E, F)$ for all $x \in E$, where D_2 stands for derivation with respect to $x \in E$.

F2. The map $D_2\mathfrak{F} : \mathbb{R} \times E \to \Phi_0(E, F)$ is continuous.

F3. $\mathfrak{F}(\lambda, 0) = 0$ for each $\lambda \in \mathbb{R}$.

Our main goal is to analyze the global topological behavior of the *compact components* of the set of nontrivial solutions of the equation

$$\mathfrak{F}(\lambda, x) = 0, \qquad \lambda \in J, \tag{1.1}$$

where, for some fixed $\Lambda \in \mathbb{R}$,

$$J \in \{\mathbb{R}, (-\infty, \Lambda], [\Lambda, \infty)\}.$$

As usual, by a *component* it is meant a maximal (for the inclusion) connected subset. Under assumption F3, $\mathbb{R} \times \{0\}$ provides us with a solution curve of (1.1) in $\mathbb{R} \times E$. A *bifurcation value* of $\mathfrak{F} = 0$ from $\mathbb{R} \times \{0\}$ is any $\lambda_0 \in \mathbb{R}$ for which $(\lambda_0, 0)$ belongs to the closure of

$$S := \mathfrak{F}^{-1}(0) \cap [\mathbb{R} \times (E \setminus \{0\})].$$

The set of nontrivial solutions of $\mathfrak{F} = 0$ is then defined as the union of S with $B \times \{0\}$, where $B \subset \mathbb{R}$ is a set of possible bifurcation values; the precise concept will be given in Section 4.

In P. Benevieri and M. Furi [5] some bifurcation results concerning (1.1) for the special case where $J = \mathbb{R}$ were derived: a local one, [5, Theorem 3.1], in the spirit of M.A. Krasnosel'skii [16]; and two global ones, [5, Theorems 3.2 and 3.6], in the spirit of P.H. Rabinowitz [25]. The aim of this paper is twofold:

- To show that the *sign jump* (a key concept ensuring bifurcation, going back within our setting to [5, Theorem 3.1] and reminiscent of the concept of *crossing number* attributable to H. Poincaré) can be detected by means of the algebraic multiplicity of $D_2\mathfrak{F}(\cdot, 0)$ (cf. [18], [23], and the list of references therein).
- To ascertain the *generalized minimal number of solutions* of the λ-slices of the compact components of the set of nontrivial solutions of (1.1), for an unbounded interval J, through the *signature* of the component; this shares the spirit of [20] and [21], where the special case when the map $(\lambda, x) \mapsto \mathfrak{F}(\lambda, x) - x$ is completely continuous was dealt with.

Although in this paper $\mathfrak{F}$ is slightly more regular than in [20] and [21], where $\mathfrak{F}$ was imposed to be completely continuous, it should be noted that here $\mathfrak{F}$ is allowed to be a general Fredholm operator of index 1, not necessary a compact perturbation of the identity; essentially, to avoid compactness, one must impose $\mathcal{C}^1$ regularity. Consequently, we are substantially enlarging the range of applicability of the abstract theory developed therein.

Fig. 3 in [21] illustrates how a compact component of (1.1) in case $J \neq \mathbb{R}$ might be a subset of a larger unbounded component of the set of nontrivial solutions of $\mathfrak{F}$ in the whole space $\mathbb{R} \times E$, for which no available result in the literature applies, except those coming from [20] and [21]. Consequently, the methodology adopted here might result into a breakthrough in global bifurcation theory. Actually, the classic results by P.H. Rabinowitz [25], E.N. Dancer [7], R.J. Magnus [22] and J. Ize [13] exclusively treated the case $J = \mathbb{R}$, as well as most of the work originated from them, obtaining or refining the so-called *global alternative* of

P.H. Rabinowitz, which reads as follows. If $J = \mathbb{R}$ and $\mathfrak{C}$ is a component of (1.1) emanating from $\mathbb{R} \times \{0\}$ at a nonlinear eigenvalue $\lambda_0 \in \mathbb{R}$ – bifurcation value based on the linear part –, then some of the following (non-exclusive) options occurs:

1. The component $\mathfrak{C}$ is unbounded in $\mathbb{R} \times E$.
2. There is another nonlinear eigenvalue $\lambda_1 \in \mathbb{R} \setminus \{\lambda_0\}$ such that $(\lambda_1, 0) \in \mathfrak{C}$.

This result is certainly seminal. Its importance and popularity is due to its huge number of applications in wide areas of nonlinear differential equations, as it provides a universal property (very appealing and easily remembered) satisfied by the compact components of the set of non-trivial solutions; in many circumstances, it can be used to conclude that $\mathfrak{C}$ is unbounded. Thus, that result provides one of the most powerful techniques to get general existence results in nonlinear eigenvalue problems. This is the case, for instance, if $\mathfrak{C}$ consists of non-negative solutions of a nonlinear equation satisfying a generalized maximum principle in an ordered Banach space (cf. [19] and the list of references therein).

Almost simultaneously with the publication of the seminal theorems of P.H. Rabinowitz [25] and the sharpened versions of E.N. Dancer [7], H. Amann and S.A. Weiss obtained their celebrated uniqueness theorem of the Leray-Schauder topological degree, which appeared in [2] after a rather lengthy elaboration process. The theorem of Amann and Weiss axiomatized the topological degree, which has shown to be the most powerful topological tool for nonlinear analysis developed in the 20th century. At that time, H. Amann had already familiarized himself with the use of the topological degree in obtaining a number of *multiplicity results* within the context of abstract fixed point equations in ordered Banach spaces and its applications to the search of positive solutions in nonlinear elliptic eigenvalues problems. Undoubtedly, those results, collected by himself in one of the most influential references on nonlinear analysis of the second half of the 20th century, [1], have shown to be a milestone in the development of modern nonlinear analysis.

The spirit of H. Amann [1] has also impregnated the present honoring paper, where the topological degree for Fredholm mappings is used to count the minimal number of solutions of the λ-slices of all bounded components of (1.1), so giving the topological degree the original use for which it was introduced by L.E.J. Brouwer [6] and further extended by J. Leray and J. Schauder [17]. Consequently, the spirit of this paper is extremely classic, strongly influenced by P.H. Rabinowitz and H. Amann, where some new mathematical challenges are addressed. Among them, the development, or eventual adaptation, of techniques from topological shape theory for ascertaining the minimal "topological complexity" of the compact components of (1.1) – a rather natural question once one knows the minimal number of solutions of the λ-slices of $\mathfrak{C}$ in terms of its *signature*, which concept we will introduce in Section 7.

The distribution of this paper is as follows. In Section 2 we recall the concepts and properties of orientation and topological degree for Fredholm maps. In Section 3 it is shown that the algebraic multiplicity detects every change of orientation of any Φ_0-valued family of real variable at an algebraic eigenvalue; as a by-product, it provides us with a sufficient condition for a real number to be a bifurcation value from $\mathbb{R} \times \{0\}$. The change of orientation is, therefore, the counterpart of the change of the Leray-Schauder index. In Section 4 we introduce the concepts of admissible families and associated parities. Roughly speaking, an admissible family is a locally finite family of disjoint open intervals such that $D_2\mathfrak{F}(\lambda, 0)$ is an isomorphism for each λ in each of those intervals, and the parity is a local invariant measuring the change of orientation of $D_2\mathfrak{F}(\lambda, 0)$ as λ crosses one of those intervals. Section 5 gives a formula for the degree of $\mathfrak{F}$ in an appropriate open set containing a compact component of the set of nontrivial solutions of (1.1) by means of the associated parities; this is the main result of this paper. Section 6 applies the main theorem to establish some (optimal) lower bounds for the number of solutions of the λ-sections (obtained by fixing the value of the parameter λ) of any compact component of the set of nontrivial solutions of (1.1). Finally, in Section 7 the concepts of *signature* and *generalized minimal cardinal* are introduced. Then, the reader is sent to the final sections of [21] for a detailed discussion illustrating the profoundness of the abstract theory developed here.

Throughout this paper, the subset of $\mathcal{L}(E, F)$ consisting of the topological isomorphisms is denoted by $\mathrm{Iso}(E, F)$, and I_E stands for the identity map in E.

2. Topological degree for Fredholm maps

This section collects the concepts of orientation and topological degree for $\mathcal{C}^1$ Fredholm maps given in P. Benevieri and M. Furi [3, 4, 5], which sharpen those going back to P.M. Fitzpatrick and J. Pejsachowicz [9] (cf. [4] for a comparison with other related constructions).

Let E, F, G be three real Banach spaces, and, for each $L \in \Phi_0(E, F)$, let $\mathcal{C}(L)$ denote the (non-empty) set of finite-rank operators $A \in \mathcal{L}(E, F)$ for which $L + A \in \mathrm{Iso}(E, F)$. In $\mathcal{C}(L)$ an equivalence relation can be defined by declaring that $A, B \in \mathcal{C}(L)$ are equivalent, $A \sim_L B$, if

$$\det\left[(L+B)^{-1}(L+A)\right] > 0\,.$$

Note that

$$(L+B)^{-1}(L+A) = I_E + (L+B)^{-1}(A-B)$$

is a finite-rank perturbation of the identity and, hence, its determinant is well defined (see, e.g., T. Kato [14, Section III.§4.3]). The relation $\sim_L$ has exactly two equivalence classes. Each of these equivalence classes is called an *orientation* of L. The operator L is said to be oriented once an orientation has been chosen. In such a case, the orientation given to L will be denoted by $\mathcal{C}_+(L)$, and we will call $\mathcal{C}_-(L) := \mathcal{C}(L) \setminus \mathcal{C}_+(L)$. Once an operator $L \in \mathrm{Iso}(E, F)$ has been oriented, its sign, denoted by $\operatorname{sgn} L$, is defined through

$$\operatorname{sgn} L := \begin{cases} 1\,, & \text{if } 0 \in \mathcal{C}_+(L)\,, \\ -1\,, & \text{if } 0 \in \mathcal{C}_-(L)\,. \end{cases}$$

This notation should not be confused with the concept of sign of a real number

$$\operatorname{sign} a = a/|a|\,, \qquad a \in \mathbb{R} \setminus \{0\}\,.$$

For each pair of oriented operators $L_1 \in \Phi_0(E,F)$ and $L_2 \in \Phi_0(F,G)$, their *oriented composition* is defined as the operator L_2L_1 equipped with the product orientation $\mathcal{C}_+(L_2L_1)$, which is taken as the orientation containing

$$L_2A_1 + A_2A_1 + A_2L_1,$$

where $A_1 \in \mathcal{C}_+(L_1)$ and $A_2 \in \mathcal{C}_+(L_2)$. This concept is consistent, as $\mathcal{C}_+(L_2L_1)$ is independent of the choice of A_1 and A_2. The following result is satisfied.

Lemma 2.1. *Let $L_1 \in \mathrm{Iso}(E,F)$ and $L_2 \in \mathrm{Iso}(F,G)$ be two oriented isomorphisms and consider the oriented composition L_2L_1. Then,*

$$\operatorname{sgn}(L_2L_1) = \operatorname{sgn} L_2 \operatorname{sgn} L_1\,.$$

Proof. For each $i \in \{1,2\}$, let $A_i \in \mathcal{C}_+(L_i)$. Then, thanks to the properties of the determinant (cf., e.g., T. Kato [14]),

$$\begin{aligned}
\det\big[(L_2L_1 + L_2A_1 \; &+ A_2A_1 + A_2L_1)^{-1}(L_2L_1)\big]\\
&= \det\left[(L_1+A_1)^{-1}(L_2+A_2)^{-1}L_2L_1\right]\\
&= \det\left[(L_1+A_1)^{-1}L_1L_1^{-1}(L_2+A_2)^{-1}L_2L_1\right]\\
&= \det\left[(L_1+A_1)^{-1}L_1\right]\,\det\left[L_1^{-1}(L_2+A_2)^{-1}L_2L_1\right]\\
&= \det\left[(L_1+A_1)^{-1}L_1\right]\,\det\left[(L_2+A_2)^{-1}L_2\right],
\end{aligned}$$

which readily concludes the proof. □

Now, let X be a topological space and $\mathfrak{L} \in \mathcal{C}(X,\Phi_0(E,F))$. An *orientation* of $\mathfrak{L}$ is a map $X \ni x \mapsto \alpha(x)$ such that $\alpha(x)$ is an orientation of $\mathfrak{L}(x)$ for all $x \in X$, and, for each $x_0 \in X$ and $A \in \alpha(x_0)$, there exists a neighborhood $\mathcal{U}$ of x_0 in X such that

$$A \in \bigcap_{x\in\mathcal{U}} \alpha(x).$$

Although $\mathfrak{L}$ might not admit an orientation, the following result holds from P. Benevieri and M. Furi [3, 4, 5], where we send for further details.

Proposition 2.2. *Suppose X is a simply-connected topological space. Then, every*

$$\mathfrak{L} \in \mathcal{C}(X,\Phi_0(E,F))$$

possesses exactly two orientations, say $\mathcal{C}_+(\mathfrak{L})$ and $\mathcal{C}_-(\mathfrak{L})$, and each of them is uniquely determined by the orientation of $\mathfrak{L}(x)$ for an arbitrary $x \in X$. More precisely, for every $x \in X$ and every orientation $\alpha(x)$ of $\mathfrak{L}(x)$, there exists a unique $\varepsilon \in \{+,-\}$ such that $\mathcal{C}_\varepsilon(\mathfrak{L})(x) = \alpha(x)$.

In this paper the space X will always be simply-connected, since $X \in \{I, E, I\times E\}$ for some interval $I \subset \mathbb{R}$. Consequently, throughout the remaining of this paper X will be assumed to be simply connected, and every family $\mathfrak{L} \in \mathcal{C}(X,\Phi_0(E,F))$

will be thought of as oriented by $\mathcal{C}_+(\mathfrak{L})$. Moreover, if $g \in \mathcal{C}^1(E,F)$ satisfies $Dg(x) \in \Phi_0(E,F)$ for each $x \in E$, then it will be said that g is oriented once an orientation $\mathcal{C}_+(Dg)$ for Dg has been chosen. Similarly, any operator $\mathfrak{F} \in \mathcal{C}(\mathbb{R}\times E, F)$ satisfying F1 and F2 of Section 1 will be assumed to be oriented, and this means that an orientation $\mathcal{C}_+(D_2\mathfrak{F})$ has been chosen for $D_2\mathfrak{F}$.

Subsequently, $\mathcal{A}$ stands for the set of pairs $(g,\mathcal{U})$ consisting of an oriented function $g \in \mathcal{C}^1(E,F)$ satisfying $Dg(x) \in \Phi_0(E,F)$ for all $x \in E$, and an open set $\mathcal{U} \subset E$ for which $g^{-1}(0) \cap \mathcal{U}$ is compact. According to P. Benevieri and M. Furi [3], a $\mathbb{Z}$-valued topological degree is defined in $\mathcal{A}$ and satisfies the usual properties of normalization, additivity and homotopy invariance. In this paper, we denote by $\operatorname{Deg}(g,\mathcal{U})$ the degree of g in $\mathcal{U}$ with respect to the origin 0, according to [3]. The following result collects some of the main properties of this topological degree (cf. P. Benevieri and M. Furi [3, 5] for a proof). Throughout the remaining of this paper, given any subset $S \subset \mathbb{R}\times E$ and $\lambda \in \mathbb{R}$ we will define

$$S_\lambda := \{\, x \in E \;:\; (\lambda, x) \in S \,\}\,.$$

Proposition 2.3. *Suppose $g \in \mathcal{C}^1(E,F)$. Then, the following properties hold:*

1. *(Additivity) If $\mathcal{U}_1, \mathcal{U}_2$ are two disjoint open subsets of an open set $W \subset E$ such that $(g,W) \in \mathcal{A}$ and $g^{-1}(0) \cap W \subset \mathcal{U}_1 \cup \mathcal{U}_2$, then*

$$\operatorname{Deg}(g,W) = \operatorname{Deg}(g,\mathcal{U}_1) + \operatorname{Deg}(g,\mathcal{U}_2)\,.$$

2. *(Regular-value formula) If $(g,\mathcal{U}) \in \mathcal{A}$, and $Dg(x) \in \operatorname{Iso}(E,F)$ for every $x \in g^{-1}(0) \cap \mathcal{U}$, then*

$$\operatorname{Deg}(g,\mathcal{U}) = \sum_{x \in g^{-1}(0)\cap\mathcal{U}} \operatorname{sgn} Dg(x).$$

3. *(Homotopy invariance) Let Ω be an open subset of $\mathbb{R}\times E$, two real numbers $a \le b$, and $\mathfrak{F} : \mathbb{R}\times E \to F$ an oriented continuous function satisfying F1 and F2 of Section 1 such that*

$$\{(\lambda,x) \in \Omega \;:\; \mathfrak{F}(\lambda,x) = 0,\ \lambda \in [a,b]\} \quad \textit{is compact.}$$

Then, $\operatorname{Deg}(\mathfrak{F}(\lambda,\cdot),\Omega_\lambda)$ is constant for each $\lambda \in [a,b]$.

Let $g \in \mathcal{C}^1(E,F)$ and suppose that for some open subsets $\mathcal{U}_1, \mathcal{U}_2$ of E and some $x \in E$,

$$g^{-1}(0) \cap [\mathcal{U}_1 \cup \mathcal{U}_2] \subset \{x\}\,.$$

Then, by the additivity property of Proposition 2.3,

$$\operatorname{Deg}(g,\mathcal{U}_1) = \operatorname{Deg}(g,\mathcal{U}_2)\,.$$

This common value will be denoted by $\operatorname{Ind}(x,g)$ and called the *index* of g at x.

3. Change of orientation through the algebraic multiplicity

In many circumstances, a generalized concept of algebraic multiplicity has been defined for Φ_0-valued families of a real or complex variable (e.g., I.C. Gohberg and E.I. Sigal [12], P. Sarreither [26], R.J. Magnus [22], J. Ize [13], P.J. Rabier [24], or [18], and the references therein). Actually, it was shown in [23] that there is a unique possible concept of algebraic multiplicity. In this section we show that this algebraic multiplicity detects every change of orientation of an operator family at any *algebraic eigenvalue*, a concept going back to [18, Definition 4.3.1]. Our result extremely sharpens the corresponding theorems of P.M. Fitzpatrick and J. Pejsachowicz [8]. The following definition fixes the concept of algebraic eigenvalue.

Definition 3.1. *Let Ω be an open interval of $\mathbb{R}$, and $\mathfrak{L} \in \mathcal{C}^r(\Omega, \Phi_0(E,F))$, for some integer $r \geq 0$. A value $\lambda_0 \in \Omega$ is said to be an algebraic eigenvalue of $\mathfrak{L}$ if there exist $C, \delta > 0$, and an integer $0 \leq k \leq r$ such that $\mathfrak{L}(\lambda) \in \mathrm{Iso}(E,F)$ for $0 < |\lambda - \lambda_0| < \delta$ and*

$$\|\mathfrak{L}(\lambda)^{-1}\|_{\mathcal{L}(F,E)} \leq C|\lambda - \lambda_0|^{-k}, \qquad 0 < |\lambda - \lambda_0| < \delta\,. \tag{3.1}$$

If k is the minimum integer number for which (3.1) *holds for some $C > 0$ and $\delta > 0$, then λ_0 is said to be a k-algebraic eigenvalue.*

Note that λ_0 is a 0-algebraic eigenvalue of $\mathfrak{L}$ if and only if $\mathfrak{L}(\lambda_0) \in \mathrm{Iso}(E,F)$. If λ_0 is an algebraic eigenvalue of $\mathfrak{L}$, then it was shown in [18] that the algebraic multiplicity $\chi[\mathfrak{L};\lambda_0]$ of $\mathfrak{L}$ at λ_0 can be defined and is a natural number. In [18, Chapter 5] the theories of [18, Chapter 4] and R.J. Magnus [22] are combined to obtain the following result.

Proposition 3.2. *Let Ω be an open interval, and $\lambda_0 \in \Omega$ an algebraic eigenvalue of*

$$\mathfrak{L} \in \mathcal{C}^r(\Omega, \Phi_0(E,F)).$$

Then, one of the following exclusive alternatives occurs:

(a) *$\mathfrak{L}(\lambda_0) \in \mathrm{Iso}(E,F)$ and $\chi[\mathfrak{L};\lambda_0] = 0$.*

(b) *λ_0 is a k-algebraic eigenvalue of $\mathfrak{L}$ for some $1 \leq k \leq r$, and there exist k finite-rank projections $P_1, \ldots, P_k \in \mathcal{L}(E,E) \setminus \{0\}$ such that*

$$\mathfrak{L}(\lambda) = \mathfrak{M}(\lambda)\,[(\lambda-\lambda_0)P_1 + I_E - P_1]\cdots[(\lambda-\lambda_0)P_k + I_E - P_k]\,, \quad \lambda \in \Omega,$$

for a certain $\mathfrak{M} \in \mathcal{C}^{r-k}(\Omega, \Phi_0(E,F))$ satisfying $\mathfrak{M}(\lambda_0) \in \mathrm{Iso}(E,F)$. Moreover,

$$\chi[\mathfrak{L};\lambda_0] = \sum_{i=1}^{k} \mathrm{rank}\, P_i\,,$$

independently of the projections P_i chosen.

The main theorem in this section establishes that the algebraic multiplicity $\chi[\mathfrak{L};\lambda_0]$ detects every *sign jump* (as discussed by P. Benevieri and M. Furi [5]) of $\mathfrak{L}$ at an algebraic eigenvalue λ_0. It reads as follows.

Theorem 3.3. *Let Ω be an open interval, and $\lambda_0 \in \Omega$ a k-algebraic eigenvalue of the oriented family $\mathfrak{L} \in \mathcal{C}^r(\Omega, \Phi_0(E,F))$, for some $0 \leq k \leq r$. Then,* $\operatorname{sgn} \mathfrak{L}(\lambda)$ *changes as λ crosses λ_0 if and only if $\chi[\mathfrak{L};\lambda_0]$ is odd.*

Proof. Suppose $\mathfrak{L}(\lambda_0) \in \operatorname{Iso}(E,F)$ and, for example, $\operatorname{sgn} \mathfrak{L}(\lambda_0) = 1$. Then, by definition of orientation, $0 \in \mathcal{C}_+(\mathfrak{L}(\lambda))$ for $\lambda \simeq \lambda_0$, and, hence, $\operatorname{sgn} \mathfrak{L}(\lambda) = 1$ for $\lambda \simeq \lambda_0$.

Suppose $\mathfrak{L}(\lambda_0)$ is not invertible. Then, due to Proposition 3.2(b), we can decompose $\mathfrak{L}$ in the form

$$\mathfrak{L}(\lambda) = \mathfrak{M}(\lambda) f_1(\lambda) \cdots f_k(\lambda), \qquad \lambda \in \Omega, \tag{3.2}$$

where the family $\mathfrak{M} \in \mathcal{C}(\Omega, \Phi_0(E,F))$ satisfies $\mathfrak{M}(\lambda_0) \in \operatorname{Iso}(E,F)$, and

$$f_i(\lambda) = (\lambda - \lambda_0)P_i + I_E - P_i, \qquad \lambda \in \Omega, \quad 1 \leq i \leq k,$$

for some finite-rank projections $P_1, \ldots, P_k \in \mathcal{L}(E,E) \setminus \{0\}$.

Clearly, the statement of the proposition is independent of the orientation chosen for $\mathfrak{L}$. We choose the following orientations: for each $1 \leq i \leq k$, the orientation of $f_i(\lambda_0)$ is defined as $P_i \in \mathcal{C}_+(f_i(\lambda_0))$, and $0 \in \mathcal{C}_+(\mathfrak{M}(\lambda_0))$. This determines the orientations $\mathcal{C}_+(f_i)$ for $1 \leq i \leq k$, and $\mathcal{C}_+(\mathfrak{M})$. The orientation of $\mathfrak{L}$ is defined as the product orientation given by (3.2). Now fix $1 \leq i \leq k$ and $\lambda \simeq \lambda_0$ with $\lambda \neq \lambda_0$. Then, $P_i \in \mathcal{C}_+(f_i(\lambda))$ and

$$\det\left[f_i(\lambda)^{-1}(f_i(\lambda) + P_i)\right] = \det\left(I_E + (\lambda - \lambda_0)^{-1} P_i\right) = \left(1 + (\lambda - \lambda_0)^{-1}\right)^{\operatorname{rank} P_i}.$$

Consequently,

$$\operatorname{sgn} f_i(\lambda) = \operatorname{sign}(\lambda - \lambda_0)^{\operatorname{rank} P_i}, \qquad \lambda \simeq \lambda_0, \quad \lambda \neq \lambda_0.$$

Therefore, by Lemma 2.1 and Proposition 3.2(b), it is apparent that

$$\operatorname{sgn} \mathfrak{L}(\lambda) = \operatorname{sign}(\lambda - \lambda_0)^{\chi[\mathfrak{L};\lambda_0]}, \quad \lambda \simeq \lambda_0, \quad \lambda \neq \lambda_0,$$

which concludes the proof. □

The relevance of Theorem 3.3 comes from the fact that, according to P. Benevieri and M. Furi [5, Theorem 3.1], a change of $\operatorname{sgn} \mathfrak{L}(\lambda)$ as λ crosses λ_0 entails λ_0 being a bifurcation value of $\mathfrak{F} = 0$ from $\mathbb{R} \times \{0\}$ for every continuous function $\mathfrak{F} : \mathbb{R} \times E \to F$ vanishing at $\Omega \times \{0\}$ and such that $D_2\mathfrak{F}(\cdot, 0) = \mathfrak{L}$. Therefore, combining [5, Theorem 3.1] with Theorem 3.3 gives rise to the next result.

Theorem 3.4. *If λ_0 is an algebraic eigenvalue of $\mathfrak{L}$ with $\chi[\mathfrak{L};\lambda_0] \in 2\mathbb{N}+1$, then λ_0 is a bifurcation value from $\mathbb{R} \times \{0\}$ for every continuous function $\mathfrak{F} : \mathbb{R} \times E \to F$ satisfying* F1, F2 *and* F3 *of Section* 1 *such that $D_2\mathfrak{F}(\cdot, 0) = \mathfrak{L}$.*

Actually, $\mathfrak{F}$ needs only to be defined in a neighborhood of $(\lambda_0, 0)$. Also, condition $\chi[\mathfrak{L};\lambda_0] \in 2\mathbb{N}+1$ is optimal even in the classic setting where the map $(\lambda, x) \mapsto \mathfrak{F}(\lambda, x) - x$ is compact; see [18, Theorem 4.3.4] for the precise statement and proof of this fact.

4. Admissible families and associated parities

To introduce the concepts of *admissible families* and *associated parities*, it is convenient to recall some elementary facts about locally finite families. A family $\mathcal{A}$ whose elements are subsets of a topological space X is said to be *locally finite* if for each $x \in X$ there exists a neighborhood Ω of x such that the set $\{A \in \mathcal{A} : A \cap \Omega \neq \varnothing\}$ is finite. An elementary property of locally finite families is the fact that for every compact K of X and every locally finite family $\mathcal{A}$, the set $\{A \in \mathcal{A} : A \cap K \neq \varnothing\}$ is finite. The next lemma shows that every locally finite family of disjoint open intervals can be *ordered*; these are the families with which we will work. Given A, B two non-empty subsets of $\mathbb{R}$, we will write $A < B$ when $a < b$ for all $a \in A$ and $b \in B$. If $A = \{a\}$ (respectively, $B = \{b\}$) the notation $a < B$ (respectively, $A < b$) will also be used.

Lemma 4.1. *Let $\mathcal{A}$ be a non-empty, locally finite family of disjoint, non-empty open intervals of $\mathbb{R}$. Then, $\mathcal{A}$ is finite or countable, and there exist $r, s \in \mathbb{Z} \cup \{-\infty, \infty\}$ with $r \leq s$ such that $\mathcal{A}$ can be expressed as $\mathcal{A} = \{A_i\}_{i=r}^{s}$, where $A_{i-1} < A_i$ for all $i \in \mathbb{Z} \cap [r+1, s]$.*

Proof. For each $n \in \mathbb{N}$ define $\mathcal{A}_n := \{A \in \mathcal{A} : [-n, n] \cap A \neq \varnothing\}$, which is a finite set because $\mathcal{A}$ is locally finite. As $\mathcal{A} = \bigcup_{n=1}^{\infty} \mathcal{A}_n$, then $\mathcal{A}$ is countable. Fix $n \in \mathbb{N}$ such that $\mathcal{A}_n$ is non-empty; as it is formed by a finite number of disjoint, non-empty open intervals, then there exist $p_0, q_0 \in \mathbb{Z}$ with $p_0 \leq q_0$ such that $\mathcal{A}_n = \{A_{p_0}, \dots, A_{q_0}\}$ with $A_{p_0} < \cdots < A_{q_0}$. It is apparent that $\mathcal{A}_{n+1} = \{A_{p_1}, \dots, A_{q_1}\}$ for some integers p_1, q_1 with $p_1 \leq p_0 \leq q_0 \leq q_1$, and $A_{p_1} < \cdots < A_{q_1}$. An inductive process concludes the proof. ☐

Definition 4.2. *Let $r, s \in \mathbb{Z} \cup \{-\infty, \infty\}$ be with $r \leq s$. The family $\{J_i\}_{i=r}^{s}$ is said to be an admissible family for a given set $\Sigma \subset \mathbb{R}$ (or just an admissible family, if Σ is clear from the context) when it is a locally finite family of open and disjoint intervals $J_i \subset \mathbb{R}$ such that $J_i \neq \varnothing$ and $J_i \cap \Sigma = \varnothing$ for all $i \in \mathbb{Z} \cap [r, s]$. In such a case, according to Lemma 4.1, we can (and will) suppose that $J_{i-1} < J_i$ for all $i \in \mathbb{Z} \cap [r+1, s]$.*

Now, let $\mathfrak{L} \in \mathcal{C}(\mathbb{R}, \Phi_0(E, F))$ be an oriented family, consider the closed set

$$\Sigma := \{\lambda \in \mathbb{R} : \mathfrak{L}(\lambda) \notin \mathrm{Iso}(E, F)\} \tag{4.1}$$

and an admissible family $\{J_i\}_{i=r}^{s}$ for Σ. The map

$$\bigcup_{i=r}^{s} J_i \ni \lambda \mapsto \operatorname{sgn} \mathfrak{L}(\lambda) \in \{-1, 1\}$$

is continuous, and the proof of this fact is essentially the first paragraph of the proof of Theorem 3.3. Therefore, for each $i \in \mathbb{Z} \cap [r, s]$ there exists $a_i \in \{-1, 1\}$ such that $\operatorname{sgn} \mathfrak{L}(\lambda) = a_i$ for every $\lambda \in J_i$. We define the *parity map* P associated with $\mathfrak{L}$ and with the admissible family $\{J_i\}_{i=r}^{s}$ as

$$P : \mathbb{Z} \cap [r+1, s] \to \{-1, 0, 1\}, \qquad P(i) = \frac{a_i - a_{i-1}}{2}. \tag{4.2}$$

Note that, when one sets

$$\Gamma_0 := \{i \in \mathbb{Z} \cap [r+1, s] : a_{i-1} = a_i\}, \qquad \Gamma_1 := \{i \in \mathbb{Z} \cap [r+1, s] : a_{i-1} \neq a_i\},$$

the parity P satisfies the following properties:

- $P(i) = 0$ if $i \in \Gamma_0$.
- $P(i) \in \{-1, 1\}$ if $i \in \Gamma_1$.
- $P(i)P(j) = -1$ if $i, j \in \Gamma_1$ with $i < j$ and $(i, j) \cap \Gamma_1 = \varnothing$.

Moreover, every map defined in $\mathbb{Z} \cap [r+1, s]$ satisfying these three properties must equal P or $-P$. Thus, either set, Γ_0 and Γ_1, determines P up to a sign switch. The knowledge of P up to a sign switch suffices in order to apply the results in the next sections. This fact also increases the relevance of Theorem 3.3, since it helps to compute the sets Γ_0 and Γ_1.

The family of compact intervals $\{I_i\}_{i=r+1}^{s}$ associated with the admissible family $\{J_i\}_{i=r}^{s}$ is defined by

$$I_i := [\sup J_{i-1}, \inf J_i], \qquad i \in \mathbb{Z} \cap [r+1, s]. \tag{4.3}$$

Each of these I_i is non-empty, but it may consist of a single point. Also, note that $I_i < I_{i+1}$ for $i \in \mathbb{Z} \cap [r+1, s-1]$. The proof of following lemma, being elementary, is omitted here.

Lemma 4.3. *The family $\{I_i\}_{i=r+1}^{s}$ is locally finite.*

As a consequence of Lemma 4.3, if K is a compact subset of $\mathbb{R} \times E$, then the set

$$\{i \in \mathbb{Z} \cap [r+1, s] : (I_i \times \{0\}) \cap K \neq \varnothing\}$$

is finite, since $\{I_i \times \{0\}\}_{i=r+1}^{s}$ is locally finite in $\mathbb{R} \times E$.

Throughout the remaining of this paper, we consider a continuous map $\mathfrak{F} : \mathbb{R} \times E \to F$ satisfying F1, F2 and F3 of Section 1 with $D_2\mathfrak{F}(\cdot, 0) = \mathfrak{L}$, and also an admissible family $\{J_i\}_{i=r}^{s}$ for the set Σ defined in (4.1); we then define

$$S := \mathfrak{F}^{-1}(0) \cap [\mathbb{R} \times (E \setminus \{0\})], \qquad \mathfrak{S} := S \cup \left[\left(\mathbb{R} \setminus \bigcup_{i=r}^{s} J_i \right) \times \{0\} \right]. \tag{4.4}$$

The set $\mathfrak{S}$ will be subsequently referred to as the *set of nontrivial solutions* of $\mathfrak{F} = 0$ with respect to the admissible family $\{J_i\}_{i=r}^{s}$. According to the next lemma, $\mathfrak{S}$ is closed.

Lemma 4.4. *The set $\mathfrak{S}$ is closed in $\mathbb{R} \times E$.*

Proof. It suffices to show that $\bar{S} \subset \mathfrak{S}$. Let $\{(\lambda_n, x_n)\}_{n \in \mathbb{N}}$ be a sequence in S with limit $(\lambda, x) \in \mathbb{R} \times E$. By continuity, we have $\mathfrak{F}(\lambda, x) = 0$. If $x \neq 0$, then $(\lambda, x) \in S$. If $x = 0$, then λ is a bifurcation value from $\mathbb{R} \times \{0\}$ and, due to the Implicit Function Theorem, $\lambda \in \Sigma$; thus, $(\lambda, 0) \in (\mathbb{R} \setminus \bigcup_{i=r}^{s} J_i) \times \{0\}$. Although $\mathfrak{F}$ is not of class $\mathcal{C}^1$ in its two variables, it has enough regularity to apply the Implicit Function Theorem (e.g., S.G. Krantz and H.R. Parks [15, Theorem 3.4.10]). □

The most common situation arising in the literature occurs when Σ is discrete. Then, the optimal choice of an admissible family consists in taking the components (open intervals) of $\mathbb{R} \setminus \Sigma$. In this case, every I_i consists of a single point, and

$$\Sigma = \bigcup_{i=r+1}^{s} I_i .$$

5. The topological degree of $\mathfrak{F}$ in terms of the parity map

Subsequently, we consider an oriented continuous map $\mathfrak{F} : \mathbb{R} \times E \to F$ satisfying F1, F2 and F3 of Section 1 with $\mathfrak{L} := D_2\mathfrak{F}(\cdot, 0)$, an admissible family $\{J_i\}_{i=r}^{s}$ for the set Σ defined in (4.1), the parity P associated with $\mathfrak{L}$ and with $\{J_i\}_{i=r}^{s}$, and the associated family $\{I_i\}_{i=r+1}^{s}$ defined through (4.3). Also, this section supposes that, for some $\Lambda \in \mathbb{R}$,

$$J \in \big\{\mathbb{R}\,, (-\infty, \Lambda]\,, [\Lambda, \infty)\big\}, \tag{5.1}$$

and that $\mathfrak{C}$ is a compact component of $\mathfrak{S} \cap (J \times E)$. Moreover, we define

$$\mathcal{B} := \{ i \in \mathbb{Z} \cap [r+1, s] : (I_i \times \{0\}) \cap \mathfrak{C} \neq \varnothing \} .$$

The set $\mathcal{B}$ is finite, since $\mathfrak{C}$ is compact and the family $\{I_i \times \{0\}\}_{i=r+1}^{s}$ is locally finite. Note that

$$(J \cap \bigcup_{i \in \mathcal{B}} I_i) \times \{0\} \subset \mathfrak{C},$$

because $\mathfrak{C}$ is a component of $\mathfrak{S} \cap (J \times E)$. Also, every $\lambda \in \mathbb{R} \setminus \bigcup_{i=r}^{s} J_i$ with $(\lambda, 0) \in \mathfrak{C}$ satisfies $\lambda \in \bigcup_{i=r+1}^{s} I_i$, since $\mathfrak{C}$ is bounded. Therefore,

$$\mathfrak{C} \cap (\mathbb{R} \times \{0\}) = \Big(J \cap \bigcup_{i \in \mathcal{B}} I_i \Big) \times \{0\} . \tag{5.2}$$

The proof of the following topological property can be found in M. Furi and M.P. Pera [10, 11]. This lemma will be very useful in the proof of Proposition 5.3.

Lemma 5.1. *Suppose Y is a locally compact Hausdorff topological space and $Y_0 \subset Y$ is compact. Then, some of the following alternatives occurs:*

1. *There exists a compact set in Y containing Y_0 and having empty boundary.*
2. *There exists a connected set C in Y contained in $Y \setminus Y_0$ such that $\bar{C}$ is not compact and $\bar{C} \cap Y_0 \neq \varnothing$.*

The following concept will play a crucial role in the subsequent analysis. The ball of radius $R > 0$ centred at $(\lambda, x) \in \mathbb{R} \times E$ is denoted by $B_R(\lambda, x)$.

Definition 5.2. *Let $\beta > 0$. An open set $\Omega \subset \mathbb{R} \times E$ is said to be an open β-isolating neighborhood of $\mathfrak{C}$ in $J \times E$ if*

$$\mathfrak{C} \subset \Omega \subset \mathfrak{C} + B_\beta(0,0)$$

and the set

$$\Omega \cap \Big[S \cup \Big(\bigcup_{i \in \mathcal{B}} I_i \times \{0\} \Big) \Big] \cap (J \times E)$$

is compact, where S is the set defined in (4.4). *An open set $\Omega \subset \mathbb{R} \times E$ is said to be an open isolating neighborhood of $\mathfrak{C}$ in $J \times E$ if it is an open β-isolating neighborhood of $\mathfrak{C}$ in $J \times E$ for some $\beta > 0$.*

The following result establishes the existence of open β-isolating neighborhoods.

Proposition 5.3. *Suppose* (5.1) *and $\mathfrak{C}$ is a compact component of $\mathfrak{S} \cap (J \times E)$. Then, for each $\beta > 0$, the set $\mathfrak{C}$ possesses an open β-isolating neighborhood in $J \times E$.*

Proof. Let $\beta > 0$ and consider

$$Y_0 := \mathfrak{C}, \qquad D := S \cup \Big(\bigcup_{i \in \mathcal{B}} I_i \times \{0\} \Big), \qquad Y := D \cap \big[\mathfrak{C} + B_\beta(0,0) \big] \cap (J \times E),$$

where S is defined in (4.4). Let us prove that D is locally compact. Let $(\lambda, x) \in D$; in order to construct a compact neighborhood of (λ, x) in D we distinguish two cases.

First, we suppose that $x \neq 0$; let $r > 0$ be such that $B_r(\lambda, x) \cap (\mathbb{R} \times \{0\}) = \varnothing$. As P. Benevieri and M. Furi [5, Lemma 3.5] showed, $\mathfrak{F}$ is locally proper and, hence, $\mathfrak{F}^{-1}(0)$ is locally compact. Thus, there exists a compact neighborhood K of (λ, x) in $\mathfrak{F}^{-1}(0)$ such that $K \subset B_r(\lambda, x)$. Then, K is a compact neighborhood of (λ, x) in D.

Secondly, suppose that $x = 0$. Then, $\lambda \in \bigcup_{i \in \mathcal{B}} I_i$; as $\bigcup_{i=r}^{s} J_i \cup \bigcup_{i \in \mathcal{B}} I_i$ is open in $\mathbb{R}$, there exists $r > 0$ such that

$$(\lambda - r, \lambda + r) \subset \bigcup_{i=r}^{s} J_i \cup \bigcup_{i \in \mathcal{B}} I_i.$$

As $\mathfrak{F}^{-1}(0)$ is locally compact, there exists a compact neighborhood K of $(\lambda, 0)$ in $\mathfrak{F}^{-1}(0)$ such that $K \subset B_r(\lambda, 0)$. Then,

$$K' := K \setminus [(\mathbb{R} \setminus \bigcup_{i \in \mathcal{B}} I_i) \times \{0\}]$$

is a compact neighborhood of $(\lambda, 0)$ in D. Indeed, the only not immediate fact is that K' is compact, and this will we proved as soon as we show that K' is closed in K. So let $\{(\lambda_n, x_n)\}_{n \in \mathbb{N}}$ be a sequence in K' with limit $(\lambda_0, 0)$ such that $\lambda_0 \in \mathbb{R} \setminus \bigcup_{i \in \mathcal{B}} I_i$. We shall derive a contradiction. By the election of r we have that $\lambda_0 \in \bigcup_{i=r}^{s} J_i$, so λ_0 is not a bifurcation value. Then, there exists $n_0 \in \mathbb{N}$ such that $x_n = 0$ for all $n \geq n_0$. Therefore, $\lambda_n \in \bigcup_{i \in \mathcal{B}} I_i$ for all $n \geq n_0$, a contradiction.

So D is locally compact. Therefore, Y is a locally closed subset of the locally compact space D, so Y is locally compact. The set Y_0 is compact and, due to (5.2), it is contained in Y. Now we prove that Alternative 2 of Lemma 5.1 cannot happen. Let C be a connected subset of $Y \setminus Y_0$ such that $\bar{C} \cap Y_0 \neq \varnothing$, where $\bar{C}$ stands for the closure of C in $\mathbb{R} \times E$. Then $\bar{C}$ is a connected set of $\mathfrak{S} \cap (J \times E)$ with non-empty intersection with $\mathfrak{C}$, and, hence, $\bar{C} \subset \mathfrak{C}$. Thus, $\bar{C} \cap Y = \bar{C}$ is compact. Therefore, Alternative 1 of Lemma 5.1 must occur. According to it, there exists a compact set K with empty boundary in Y and such that $Y_0 \subset K \subset Y$. Necessarily, K is open in Y and consequently $K = \mathcal{U} \cap Y$ for some $\mathcal{U}$ open set in $\mathbb{R} \times E$. The set $\mathcal{U} \cap [\mathfrak{C} + B_\beta(0,0)]$ constitutes an open β-isolating neighborhood of $\mathfrak{C}$ in $J \times E$. $\square$

Subsequently, B_R denotes the ball of radius R centred at the origin in E. Also, recall that for a subset $S \subset \mathbb{R} \times E$ and $\lambda \in \mathbb{R}$, the set S_λ is formed by all $x \in E$ such that $(\lambda, x) \in E$, and, as a usual convention, a summation over the empty set is taken as zero. The main result of this section reads as follows.

Theorem 5.4. *Suppose* (5.1) *and* $\mathfrak{C}$ *is a compact component of* $\mathfrak{S} \cap (J \times E)$. *Let* Ω *be an open isolating neighborhood of* $\mathfrak{C}$ *in* $J \times E$. *Then, for each* $\lambda^* \in J \setminus \bigcup_{i \in \mathcal{B}} I_i$ *there exists* $\rho^* > 0$ *such that for all* $0 < \rho \leq \rho^*$,

$$\operatorname{Deg}\left(\mathfrak{F}(\lambda^*, \cdot), \Omega_{\lambda^*} \setminus \bar{B}_\rho\right) = 2 \operatorname{sign} J^* \sum_{i \in J^*} P(i)\,, \tag{5.3}$$

where

$$J^* := \begin{cases} \{i \in \mathcal{B} : I_i > \lambda^*\} & \text{if } J = [\Lambda, \infty), \\ \{i \in \mathcal{B} : I_i < \lambda^*\} & \text{if } J = (-\infty, \Lambda], \end{cases}$$

$$\operatorname{sign} J^* := \begin{cases} 1 & \text{if } J^* = \{i \in \mathcal{B} : I_i > \lambda^*\}, \\ -1 & \text{if } J^* = \{i \in \mathcal{B} : I_i < \lambda^*\}, \end{cases}$$

and $J^* \in \{\{i \in \mathcal{B} : I_i > \lambda^*\}, \{i \in \mathcal{B} : I_i < \lambda^*\}\}$ *if* $J = \mathbb{R}$.

Proof. Suppose $J = [\Lambda, \infty)$. Fix an orientation for $\mathfrak{F} : \mathbb{R} \times E \to F$. By Definition 5.2, the set

$$K := \Omega \cap \Big[S \cup \Big(\bigcup_{i \in \mathcal{B}} I_i \times \{0\}\Big)\Big] \cap (J \times E)$$

is compact. Now, we distinguish two different cases. Suppose

$$J^* = \varnothing\,. \tag{5.4}$$

Consider $b > \lambda^*$ such that $\Omega \subset (-\infty, b) \times E$. The set K is disjoint from $[\lambda^*, b] \times \{0\}$, so there exists $\rho^* > 0$ such that

$$K \cap ([\lambda^*, b] \times \bar{B}_{\rho^*}) = \varnothing$$

This implies that for each $0 < \rho \leq \rho^*$,

$$\big\{(\lambda, x) \in \Omega \setminus (\mathbb{R} \times \bar{B}_\rho) : \mathfrak{F}(\lambda, x) = 0, \lambda \in [\lambda^*, b]\big\} = K \cap \big([\lambda^*, b] \times (E \setminus B_\rho)\big),$$

which is compact. Therefore, the homotopy invariance of Proposition 2.3 shows that

$$\operatorname{Deg}\left(\mathfrak{F}(\lambda^*, \cdot), \Omega_{\lambda^*} \setminus \bar{B}_\rho\right) = \operatorname{Deg}\left(\mathfrak{F}(b, \cdot), \Omega_b \setminus \bar{B}_\rho\right) = 0,$$

because $\Omega_b = \varnothing$. This concludes the proof of the theorem under condition (5.4). Now, suppose

$$J^* = \{i_1, \ldots, i_M\}\,, \tag{5.5}$$

instead of (5.4), where $i_1 < \cdots < i_M$. Take $b \in \mathbb{R}$ such that $\Omega \subset (-\infty, b) \times E$. Let $\delta > 0$ be such that

$$\bigcup_{i \in \mathcal{B}} \left[(J \cap I_i) \times \{0\} \right] + B_\delta(0,0) \subset \Omega,$$

the intervals $I_i + (-\delta, \delta)$ are disjoint for $i \in \mathcal{B}$, and

$$\lambda^* \notin \bigcup_{i \in \mathcal{B}} I_i + (-\delta/2, \delta/2).$$

Also, $\delta > 0$ can be chosen so that, when we denote

$$s_0^+ := \lambda^*, \quad s_{M+1}^- := b, \qquad s_j^- := \inf I_{i_j} - \delta/2, \quad s_j^+ := \sup I_{i_j} + \delta/2, \quad 1 \le j \le M,$$

we have that $s_j^- \in J_{i_j - 1}$ and $s_j^+ \in J_{i_j}$ for $1 \le j \le M$. We observe that for all $1 \le j \le M$,

$$\left\{ (\lambda, x) \in \Omega : \mathfrak{F}(\lambda, x) = 0, \lambda \in [s_j^-, s_j^+] \right\} = \left[K \cap \left([s_j^-, s_j^+] \times E \right) \right] \cup \left([s_j^-, s_j^+] \times \{0\} \right)$$

is compact. Therefore, by the homotopy invariance of Proposition 2.3,

$$\operatorname{Deg}(\mathfrak{F}(s_j^-, \cdot), \Omega_{s_j^-}) = \operatorname{Deg}(\mathfrak{F}(s_j^+, \cdot), \Omega_{s_j^+}), \qquad 1 \le j \le M. \tag{5.6}$$

Now, since K is disjoint from $\bigcup_{j=0}^M ([s_j^+, s_{j+1}^-] \times \{0\})$, there exists $\rho^* > 0$ such that

$$K \cap \bigcup_{j=0}^{M} \left([s_j^+, s_{j+1}^-] \times \bar{B}_{\rho^*} \right) = \varnothing.$$

This implies that for each $0 < \rho \le \rho^*$ and $0 \le j \le M$, the set

$$\left\{ (\lambda, x) \in \Omega \setminus (\mathbb{R} \times \bar{B}_\rho) : \mathfrak{F}(\lambda, x) = 0, \lambda \in [s_j^+, s_{j+1}^-] \right\} = K \cap \left([s_j^+, s_{j+1}^-] \times (E \setminus B_\rho) \right)$$

is compact. By the homotopy invariance of Proposition 2.3, there exist $d_0, \ldots, d_M \in \mathbb{Z}$ such that

$$\operatorname{Deg}\left(\mathfrak{F}(\lambda, \cdot), \Omega_\lambda \setminus \bar{B}_\rho \right) = d_j, \quad \lambda \in [s_j^+, s_{j+1}^-], \quad 0 \le j \le M.$$

Moreover, $d_M = 0$, because $\Omega_b = \varnothing$.

By the additivity and the regular-value formula of Proposition 2.3, for each $\rho > 0$ small enough and each $1 \le j \le M$, we have

$$\begin{aligned} \operatorname{Deg}\left(\mathfrak{F}(s_j^-, \cdot), \Omega_{s_j^-} \right) &= d_{j-1} + \operatorname{sgn} D_2 \mathfrak{F}(s_j^-, 0), \\ \operatorname{Deg}\left(\mathfrak{F}(s_j^+, \cdot), \Omega_{s_j^+} \right) &= d_j + \operatorname{sgn} D_2 \mathfrak{F}(s_j^+, 0). \end{aligned}$$

These equalities together with (5.6) and (4.2) give

$$d_{j-1} - d_j = 2P(i_j), \qquad 1 \le j \le M.$$

Thus, adding them and taking into account that $d_M = 0$, we get

$$d_0 = 2 \sum_{j=1}^{M} P(i_j)$$

and, therefore,

$$\mathrm{Deg}\left(\mathfrak{F}(\lambda^*,\cdot),\Omega_{\lambda^*}\setminus\bar{B}_\rho\right)=2\sum_{j=1}^{M}P(i_j)\,. \tag{5.7}$$

Thanks to (5.5), equation (5.7) provides us with (5.3) when $J=[\Lambda,\infty)$. The case when $J=(-\infty,\Lambda]$ can be proved similarly. Consequently, we omit the details of its proof.

Finally, suppose $J=\mathbb{R}$, let $\mathfrak{C}$ be a compact component of $\mathfrak{S}$ and consider an open isolating neighborhood Ω of $\mathfrak{C}$ in $\mathbb{R}\times E$, and $\lambda^*\in\mathbb{R}\setminus\bigcup_{i\in\mathcal{B}}I_i$. Take a $\Lambda\leq\lambda^*$ such that $\Omega\subset[\Lambda,\infty)\times E$. Then, $\mathfrak{C}$ is a component of $\mathfrak{S}\cap([\Lambda,\infty)\times E)$, and Ω is an open isolating neighborhood of $\mathfrak{C}$ in $[\Lambda,\infty)\times E$. Therefore, thanks to the first part of the proof,

$$\mathrm{Deg}\left(\mathfrak{F}(\lambda^*,\cdot),\Omega_{\lambda^*}\setminus\bar{B}_\rho\right)=2\sum_{i\in\mathcal{B},\,I_i>\lambda^*}P(i).$$

Similarly,

$$\mathrm{Deg}\left(\mathfrak{F}(\lambda^*,\cdot),\Omega_{\lambda^*}\setminus\bar{B}_\rho\right)=-2\sum_{i\in\mathcal{B},\,I_i<\lambda^*}P(i)\,,$$

which completes the proof. □

Theorem 5.4 provides us with a valuable information about the structure of the compact components of $\mathfrak{S}$. The following result should be compared with R. J. Magnus [22, Theorem 2.10], E. N. Dancer [7, Corollary 2] and P. Benevieri and M. Furi [5, Theorem 3.2]. It constitutes a substantial extension of the main results in these references. Actually, it is the result from which the global alternative of P.H. Rabinowitz [25] readily follows.

Corollary 5.5. *Suppose $\mathfrak{C}$ is a compact component of $\mathfrak{S}$. Then,*

$$\sum_{i\in\mathcal{B}}P(i)=0\,.$$

6. Lower estimates of $\mathrm{Card}\,\mathfrak{C}_\lambda$

In this section we use Theorem 5.4 to estimate the number of elements of $\mathfrak{C}_\lambda$. The notation J^* introduced in the statement of Theorem 5.4 will be subsequently maintained.

Theorem 6.1. *Suppose* (5.1) *and $\mathfrak{C}$ is a compact component of $\mathfrak{S}\cap(J\times E)$. Suppose, in addition, that there exist $\lambda^*\in J\setminus\bigcup_{i\in\mathcal{B}}I_i$ and $n\in\mathbb{N}$ different isolated zeros of $\mathfrak{F}(\lambda^*,\cdot)$, say $x_1,\ldots,x_n\in\mathfrak{C}_{\lambda^*}$, such that*

$$\mathrm{Ind}(x_j,\mathfrak{F}(\lambda^*,\cdot))\in\{-1,0,1\}\,,\qquad 1\leq j\leq n\,. \tag{6.1}$$

Set

$$n_\pm:=\mathrm{Card}\left\{j\in\{1,\ldots,n\}\,:\,\mathrm{Ind}\left(x_j,\mathfrak{F}(\lambda^*,\cdot)\right)=\pm1\right\}. \tag{6.2}$$

Then,

$$\operatorname{Card}\mathfrak{C}_{\lambda^*} \geq n+1 \quad \textit{whenever} \quad 2\Big|\sum_{i\in J^*} P(i)\Big| \neq |n_+ - n_-|\,. \tag{6.3}$$

In particular, $\operatorname{Card}\mathfrak{C}_{\lambda^*} \geq n+1$ *if* $n_+ + n_- \in 2\mathbb{N}+1$.

Proof. Suppose $\mathfrak{C}_{\lambda^*} = \{x_1, \dots, x_n\}$. Necessarily, $x_j \neq 0$ for each $1 \leq j \leq n$, since $\lambda^* \in J \setminus \bigcup_{i\in\mathcal{B}} I_i$ and (5.2). Thus, $\mathfrak{C}_{\lambda^*} \cap \bar{B}_\rho = \varnothing$ for sufficiently small $\rho > 0$. The set $\mathfrak{C}$ is compact and has empty intersection with

$$Z := \{\lambda^*\} \times \big\{x \in E \setminus \mathfrak{C}_{\lambda^*} : \mathfrak{F}(\lambda^*, x) = 0\big\},$$

which is closed since x_j is an isolated zero of $\mathfrak{F}(\lambda^*, \cdot)$ for each $1 \leq j \leq n$. Therefore, there exists $\beta > 0$ such that $(\mathfrak{C} + B_\beta(0,0)) \cap Z = \varnothing$. By Proposition 5.3, there exists Ω an open isolating neighborhood of $\mathfrak{C}$ in $J \times E$ such that $\Omega \cap Z = \varnothing$. Then,

$$\mathfrak{C}_{\lambda^*} = \big\{x \in \Omega_{\lambda^*} \setminus \bar{B}_\rho \,:\, \mathfrak{F}(\lambda^*, x) = 0\big\}$$

and, according to the additivity property of Proposition 2.3, we find that

$$\operatorname{Deg}\big(\mathfrak{F}(\lambda^*, \cdot), \Omega_{\lambda^*} \setminus \bar{B}_\rho\big) = \sum_{j=1}^{n} \operatorname{Ind}\big(x_j, \mathfrak{F}(\lambda^*, \cdot)\big)\,.$$

Therefore, it follows from Theorem 5.4 that

$$2\Big|\sum_{i\in J^*} P(i)\Big| = |n_+ - n_-|\,,$$

which completes the proof of (6.3). Finally, if $n_+ + n_-$ is odd, then $n_+ - n_-$ is odd as well, which concludes the proof. □

Theorem 6.2. *Suppose* (5.1), $\mathfrak{C}$ *is a compact component of* $\mathfrak{S} \cap (J \times E)$, *and* $\lambda^* \in J \setminus \bigcup_{i\in\mathcal{B}} I_i$ *is a parameter value for which* $\mathfrak{C}_{\lambda^*}$ *consists of exactly* $n \in \mathbb{N}$ *isolated zeros of* $\mathfrak{F}(\lambda^*, \cdot)$, *say* $x_1, \dots, x_n$, *satisfying* (6.1). *Then,*

$$\operatorname{Card}\mathfrak{C}_{\lambda^*} \geq 2\Big|\sum_{i\in J^*} P(i)\Big|\,. \tag{6.4}$$

Moreover, $\operatorname{Card}\mathfrak{C}_{\lambda^*} \in 2\mathbb{N}$ *if* $n - n_+ - n_- \in 2\mathbb{N}$, *where* n_+ *and* n_- *are defined in* (6.2).

Proof. We have $0 \notin \mathfrak{C}_{\lambda^*}$, because of (5.2). Moreover, an adaptation of the argument given in the proof of Theorem 6.1 gives

$$\begin{aligned} 2\Big|\sum_{i\in J^*} P(i)\Big| &= \Big|\sum_{j=1}^{n} \operatorname{Ind}\big(x_j, \mathfrak{F}(\lambda^*, \cdot)\big)\Big| \\ &\leq \sum_{j=1}^{n} |\operatorname{Ind}(x_j, \mathfrak{F}(\lambda^*, \cdot))| = n_+ + n_- \leq \operatorname{Card}\mathfrak{C}_{\lambda^*} \end{aligned}$$

and, therefore, (6.4) holds.

Further, suppose $n - n_+ - n_- \in 2\mathbb{N}$. Then,

$$2\left|\sum_{i\in J^*} P(i)\right| = \left|\sum_{j=1}^{n} \operatorname{Ind}\left(x_j, \mathfrak{F}(\lambda^*, \cdot)\right)\right| = |n_+ - n_-|,$$

and, hence, $n_+ + n_-$ and n are even. This concludes the proof. □

7. Concepts of Signature and Generalized Minimal Cardinal

Strongly motivated by Theorem 6.2, the following concepts are rather natural.

Definition 7.1. *Suppose* (5.1) *and* $\mathfrak{C}$ *is a compact component of* $\mathfrak{S}\cap(J\times E)$*; assume* $\mathfrak{C}_\Lambda \neq \emptyset$ *and* $\Lambda \notin \bigcup_{i\in\mathcal{B}} I_i$ *if* $J \in \{(-\infty, \Lambda], [\Lambda; \infty)\}$*. Then:*

(a) *When* $\mathcal{B} = \{i_1, \dots, i_N\}$ *with* $i_1 < \dots < i_N$*, the signature of* $\mathfrak{C}$ *in* $J \times E$ *is defined by*

$$\operatorname{Signature}[\mathfrak{C}; J\times E] := \begin{pmatrix} i_1 & \cdots & i_N \\ P(i_1) & \cdots & P(i_N) \end{pmatrix},$$

whereas the signature of $\mathfrak{C}$ *in* $J \times E$ *is said to be empty if* $\mathcal{B} = \emptyset$*.*

(b) *When* $\mathcal{B} = \{i_1, \dots, i_N\}$ *with* $i_1 < \dots < i_N$*, the generalized minimal cardinal of* $\mathfrak{C}$ *in* $J \times E$ *is the map*

$$\mathrm{GMC} := \mathrm{GMC}_{[\mathfrak{C};J\times E]} \;:\; J \setminus \bigcup_{i\in\mathcal{B}} I_i \to \mathbb{N}$$

defined by

$$\mathrm{GMC}(\lambda) := \begin{cases} 2\max\left\{1, \left|\displaystyle\sum_{j=1}^{N} P(i_j)\right|\right\}, & \text{if } \lambda \in [\Lambda, \inf I_{i_1}), \\ 2\max\left\{1, \left|\displaystyle\sum_{j=k+1}^{N} P(i_j)\right|\right\}, & \text{if } \lambda \in (\sup I_{i_k}, \inf I_{i_{k+1}}),\ 1 \le k \le N-1, \\ 0, & \text{if } \lambda \in (\sup I_{i_N}, \infty), \end{cases}$$

if $J = [\Lambda, \infty)$*, by*

$$\mathrm{GMC}(\lambda) := \begin{cases} 2\max\left\{1, \left|\displaystyle\sum_{j=1}^{N} P(i_j)\right|\right\}, & \text{if } \lambda \in (\sup I_{i_N}, \Lambda], \\ 2\max\left\{1, \left|\displaystyle\sum_{j=1}^{k} P(i_j)\right|\right\}, & \text{if } \lambda \in (\sup I_{i_k}, \inf I_{i_{k+1}}),\ 1 \le k \le N-1, \\ 0, & \text{if } \lambda \in (-\infty, \inf I_{i_1}), \end{cases}$$

if $J = (-\infty, \Lambda]$, *and by*

$$\mathrm{GMC}(\lambda) := \begin{cases} 0\,, & \text{if } \lambda \in (-\infty, \inf I_{i_1}) \cup (\sup I_{i_N}, \infty)\,, \\ 2\max\left\{1, \left|\sum_{j=1}^{k} P(i_j)\right|\right\}, & \text{if } \lambda \in (\sup I_{i_k}, \inf I_{i_{k+1}})\,, \quad 1 \le k \le N-1\,, \end{cases}$$

if $J = \mathbb{R}$, *whereas we take*

$$\mathrm{GMC} := 0 \quad \text{if} \quad \mathcal{B} = \varnothing\,.$$

Note that $\mathrm{GMC}_{[\mathfrak{C}; J\times E]}$ is uniquely determined from the Signature of $\mathfrak{C}$ in $J \times E$, once the admissible family $\mathcal{A}$ for Σ is fixed. With these concepts in mind, Theorem 6.2 may be rephrased as follows:

> $\mathrm{GMC}_{[\mathfrak{C}; J\times E]}(\lambda)$ is a lower bound of $\operatorname{Card} \mathfrak{C}_\lambda$ for every value of $\lambda \in J \setminus \bigcup_{i\in\mathcal{B}} I_i$ for which $\mathfrak{C}_\lambda$ consists of a finite number of isolated zeros of $\mathfrak{F}(\lambda, \cdot)$ satisfying (6.1).

Subsequently we will assume that $J = [\Lambda, \infty)$. In the special case that the signature of $\mathfrak{C}$ in $J \times E$ is empty we have defined $\mathrm{GMC}_{[\mathfrak{C}; J\times E]} = 0$ and it should be noted that, although $\mathfrak{C}_\lambda$ might have an arbitrarily large number of solutions for some set of values of the parameter $\lambda > \Lambda$, it might have none for each $\lambda > \Lambda$. Actually, this occurs if $\mathcal{P}_\lambda \mathfrak{C} = \{\Lambda\}$, where $\mathcal{P}_\lambda$ stands for the λ-projection operator. In the case when

$$\mathrm{Signature}[\mathfrak{C}; J \times E] \in \left\{ \begin{pmatrix} i_1 & i_2 \\ -1 & -1 \end{pmatrix}, \begin{pmatrix} i_1 & i_2 \\ 1 & 1 \end{pmatrix} \right\}, \tag{7.1}$$

according to Definition 7.1 we have that

$$\mathrm{GMC}_{[\mathfrak{C}; J\times E]}(\lambda) = \begin{cases} 4\,, & \text{if } \lambda \in [\Lambda, \inf I_{i_1})\,, \\ 2\,, & \text{if } \lambda \in (\sup I_{i_1}, \inf I_{i_2})\,, \\ 0\,, & \text{if } \lambda > \sup I_{i_2}\,. \end{cases}$$

Figure 1 shows some admissible components with signature (7.1) with the following additional assumptions: every $\lambda \in [\Lambda, \infty) \setminus (I_{i_1} \cup I_{i_2})$ is a *strongly regular* parameter value of $\mathfrak{C}$, $\operatorname{Card} \mathfrak{C}_\Lambda = 4$, $I_{i_1} = \{\sigma_1\}$, and $I_{i_2} = \{\sigma_2\}$. By a strongly regular parameter value it is meant any value of λ for which $\mathfrak{C}_\lambda$ consists of exactly $n \in \mathbb{N}$ isolated zeros of $\mathfrak{F}(\lambda, \cdot)$, say $x_1, \dots, x_n$, satisfying

$$\mathrm{Ind}(x_j, \mathfrak{F}(\lambda, \cdot)) \in \{-1, 1\}\,, \qquad 1 \le j \le n\,.$$

It should be noted that, thanks to Theorem 6.1, $\mathfrak{C}_\lambda$ must possess an even number of solutions for each $\lambda \in [\Lambda, \infty) \setminus (I_{i_1} \cup I_{i_2})$. The analysis carried out in the last sections of [21] showed that all situations illustrated in Figure 1 are admissible. The five cases shown by Figure 1 correspond with each of the possible cases accordingly to the number of arcs of $\mathfrak{C}$ connecting $\{\Lambda\} \times \mathfrak{C}_\Lambda$ to $I_{i_1} \times \{0\}$.

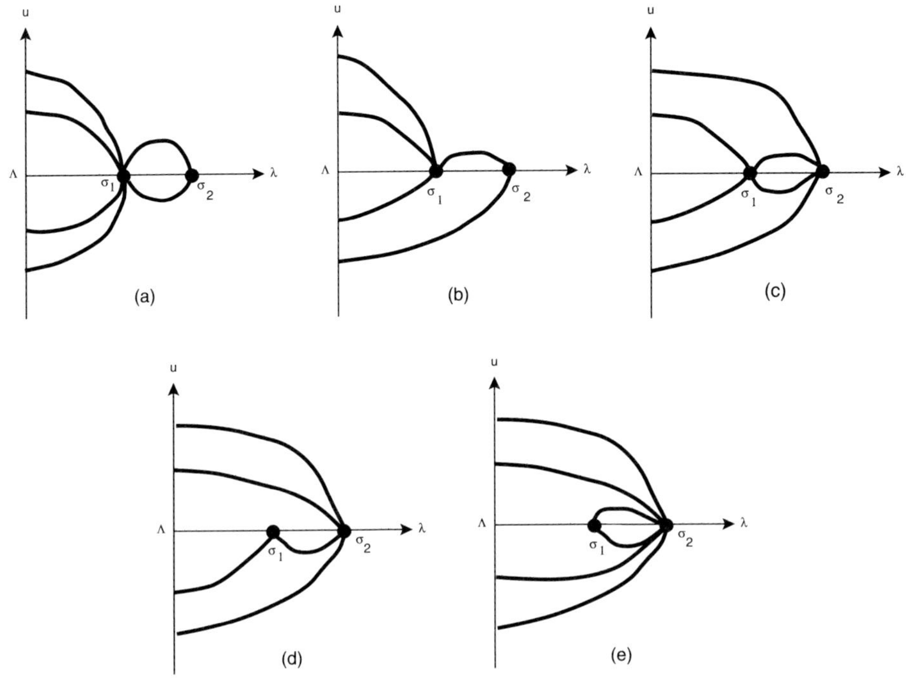

FIGURE 1. Some admissible components with $\operatorname{Card}\mathfrak{C}_\Lambda = 4$.

Now, suppose

$$\text{Signature}[\mathfrak{C}; J \times E] \in \left\{ \begin{pmatrix} i_1 & i_2 \\ \pm 1 & \mp 1 \end{pmatrix}, \begin{pmatrix} i_1 & i_2 \\ \pm 1 & 0 \end{pmatrix}, \begin{pmatrix} i_1 & i_2 \\ 0 & \pm 1 \end{pmatrix}, \begin{pmatrix} i_1 & i_2 \\ 0 & 0 \end{pmatrix} \right\}. \tag{7.2}$$

Then,

$$\text{GMC}_{[\mathfrak{C}; J\times E]}(\lambda) = \begin{cases} 2, & \text{if } \lambda \in [\Lambda, \inf I_{i_1}) \cup (\sup I_{i_1}, \inf I_{i_2}), \\ 0, & \text{if } \lambda > \sup I_{i_2}. \end{cases} \tag{7.3}$$

Figure 2 shows some admissible components with signature in the set (7.2) in the special case that every $\lambda \in [\Lambda, \infty) \setminus (I_{i_1} \cup I_{i_2})$ is a strong regular parameter value of $\mathfrak{C}$, $\operatorname{Card}\mathfrak{C}_\Lambda = 2$, $I_{i_1} = \{\sigma_1\}$ and $I_{i_2} = \{\sigma_2\}$. Again, (7.3) provides us with the minimal number of solutions in all these cases. When some degenerate points appear along any of these curves of $\mathfrak{C}$, the number of solutions might drastically increase, or decrease, of course, but it seems the previous diagrams provide us with the minimal topological patterns that $\mathfrak{C}$ should contain.

The analysis already carried out in [21, Sections 6,7], where we send for further details, justifies why GMC(λ) is indeed a generalized minimal cardinal of $\mathfrak{C}$.

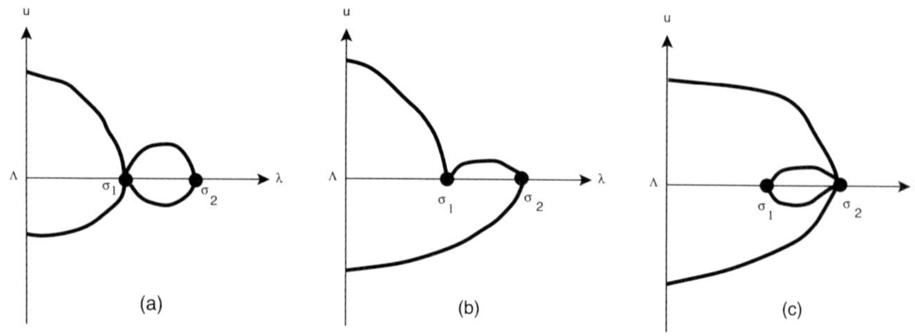

FIGURE 2. Some admissible components with Card $\mathfrak{C}_\Lambda = 2$.

Acknowledgements

The authors thank the Ministry of Education and Science of Spain for research support under Grant REN2003-00707. The second author also thanks the Spanish MECD for partial support, through a predoctoral fellowship under the supervision of J. López-Gómez.

References

[1] H. Amann, Fixed point equations and nonlinear eigenvalue problems in ordered Banach spaces, *SIAM Rev.* **18** (1976), 620–709.

[2] H. Amann and S.A. Weiss, On the uniqueness of the topological degree, *Math. Z.* **130** (1973), 39–54.

[3] P. Benevieri and M. Furi, A simple notion of orientability for Fredholm maps of index zero between Banach manifolds and degree theory, *Ann. Sci. Math. Québec* **22** (1998), 131–148.

[4] P. Benevieri and M. Furi, On the concept of orientability for Fredholm maps between real Banach manifolds, *Topol. Meth. Nonl. Anal.* **16** (2000), 279–306.

[5] P. Benevieri and M. Furi, Bifurcation results for families of Fredholm maps of index zero between Banach spaces. Nonlinear analysis and its applications (St. John's, NF, 1999), *Nonlinear Anal. Forum* **6** (2001), 35–47.

[6] L.E.J. Brouwer, Über Abbildung von Mannigfaltigkeiten, *Math. Ann.* **71** (1912), 97–115.

[7] E.N. Dancer, On the structure of solutions of non-linear eigenvalue problems, *Ind. Univ. Math. J.* **23** (1974), 1069–1076.

[8] P.M. Fitzpatrick and J. Pejsachowicz, Parity and generalized multiplicity, *Trans. Amer. Math. Soc.* **326** (1991), 281–305.

[9] P.M. Fitzpatrick and J. Pejsachowicz, Orientation and the Leray-Schauder theory for fully nonlinear elliptic boundary value problems, Memoirs of the American Mathematical Society 483, Providence, R. I., 1993.

[10] M. Furi and M.P. Pera, A continuation principle for periodic solutions of forced motion equations on manifolds and applications to bifurcation theory, *Pacific J. Math.* **160** (1993), 219–244.

[11] M. Furi and M.P. Pera, Global Bifurcation of Fixed Points and the Poincaré Translation Operator on Manifolds, *Annali di Mat. Pura ed Appl. (IV)* **173** (1997), 313–331.

[12] I.C. Gohberg and E.I. Sigal, An Operator Generalization of the Logarithmic Residue Theorem and the Theorem of Rouché, *Math. Sbornik* 84(126) (1971), 607–629. English Trans.: *Math. USSR Sbor.* **13** (1971), 603–625.

[13] J. Ize, Bifurcation theory for Fredholm operators, Memoirs of the American Mathematical Society 174, Providence, R. I., 1976.

[14] T. Kato, *Perturbation Theory for Linear Operators*, Grundlehren der mathematischen Wissenschaften 132, Springer, Berlin 1980.

[15] S.G. Krantz and H.R. Parks, *The implicit function theorem. History, theory, and applications,* Birkhäuser, Boston, MA, 2002.

[16] M.A. Krasnosel'skii, *Topological Methods in the Theory of Nonlinear Integral Equations,* Gos. Izdat. Tehn.-Teor. Lit., Moscow, 1956.

[17] J. Leray and J. Schauder, Topologie et équations functionelles, *Ann. Sci. École Norm. Sup. Sér. 3* **51** (1934), 45–78.

[18] J. López-Gómez, *Spectral Theory and Nonlinear Functional Analysis,* Chapman & Hall / CRC. Research Notes in Mathematics 426, Boca Raton, 2001.

[19] J. López-Gómez and M. Molina-Meyer, Bounded components of positive solutions of abstract fixed point equations: mushrooms, loops and isolas, *J. Diff. Eqns.* **209** (2005), 416–441.

[20] J. López-Gómez and C. Mora-Corral, Counting solutions of nonlinear abstract equations, *Top. Meth. Nonl. Anal.* **24** (2004), 307–335.

[21] J. López-Gómez and C. Mora-Corral, Minimal complexity of semi-bounded components in bifurcation theory, *Nonl. Anal. T.M.A.* **58** (2004), 749–777.

[22] R.J. Magnus, A Generalization of Multiplicity and the Problem of Bifurcation, *Proc. London Math. Soc.* **32** (1976), 251–278.

[23] C. Mora-Corral, On the Uniqueness of the Algebraic Multiplicity, *J. London Math. Soc.* **69** (2004), 231–242.

[24] P.J. Rabier, Generalized Jordan chains and two bifurcation theorems of Krasnoselskii, *Nonl. Anal. T.M.A.* **13** (1989), 903–934.

[25] P.H. Rabinowitz, Some global results for nonlinear eigenvalue problems, *J. Funct. Anal.* **7** (1971), 487–513.

[26] P. Sarreither, Transformationseigenschaften endlicher Ketten und allgemeine Verzweigungsaussagen, *Math. Scand.* **35** (1974), 115–128.

Julián López-Gómez and Carlos Mora-Corral
Facultad de Ciencias Matemáticas
Departamento de Matemática Aplicada
Universidad Complutense de Madrid
E-28040 Madrid, Spain
e-mail: Lopez_Gomez@mat.ucm.es
e-mail: Carlos_Mora@mat.ucm.es

Progress in Nonlinear Differential Equations
and Their Applications, Vol. 64, 351–356

Blow-up of Solutions of a Semilinear Heat Equation with a Visco-elastic Term

Salim A. Messaoudi

Abstract. In this work we consider an initial boundary value problem related to the equation

$$u_t - \Delta u + \int_0^t g(t-s)\Delta u(x,s)ds = |u|^{p-2}u$$

and prove, under suitable conditions on g and p, a blow-up result for solutions with negative or vanishing initial energy. This result improves an earlier one by the author.

Mathematics Subject Classification (2000). 35K05 - 35K65.

Keywords. Blow-up, finite time, viscoelastic, relaxation function, vanishing energy.

1. Introduction

In this work we study the finite-time blow-up of solutions for the following initial boundary value problem

$$\begin{aligned} u_t - \Delta u + \int_0^t g(t-s)\Delta u(x,s)ds &= |u|^{p-2}u, & x \in \Omega, \quad & t > 0 \\ u(x,t) &= 0, & x \in \partial\Omega, \quad & t \geq 0 \\ u(x,0) &= u_0(x), & x \in \Omega, \quad & \end{aligned} \tag{1.1}$$

where $g : R_+ \to R_+$ is a bounded C^1 function, $p > 2$, and Ω is a bounded domain of $\mathbb{R}^n$ $(n \geq 1)$, with a smooth boundary $\partial\Omega$.

This type of equations arises from a variety of mathematical models in engineering and physical sciences. For example, in the study of heat conduction in materials with memory, the classical Fourier's law of heat flux is replaced by the following form

$$q = -d\nabla u - \int_{-\infty}^t \nabla\left[k(x,t)u(x,\tau)\right] d\tau, \tag{1.2}$$

where u is the temperature, d the diffusion coefficient and the integral term represents the memory effect in the material. The study of this type of equations has drawn a considerable attention see [3], [4], [12] , [14] [15]. From a mathematical point of view, one would expect the integral term to be dominated by the leading term in the equation. Therefore, the theory of parabolic equations applies to this type of equations.

In the absence of the memory term ($g = 0$) problem (1.1) has been studied by various authors and several results concerning global and nonglobal existence have been established. For instance, in the early 1970's Levine [8] introduced the concavity method and showed that solutions with negative energy blow up in finite time. Later, this method was improved by Kalantarov and Ladyzhenskaya [7] to accommodate more general situations. Ball [2] also studied (1.1) with $f(u, \nabla u)$ instead of $|u|^{p-2}u$ and established a nonglobal existence result in bounded domains. This result had been extended to unbounded domains by Alfonsi and Weissler [1].

For the quasilinear case, Junning [6] studied

$$\begin{aligned} u_t - \operatorname{div}(|\nabla u|^{m-2}\nabla u) &= f(u), \quad && x \in \Omega, \quad t > 0 \\ u(x,t) &= 0, && x \in \partial\Omega \quad t \geq 0 \\ u(x,0) &= u_0(x), && x \in \Omega, \end{aligned} \tag{1.3}$$

and established a global existence result. He also proved a nonglobal existence result under the condition

$$\frac{1}{m}\int_\Omega |\nabla u_0(x)|^m dx - \int_\Omega F(u_0(x))dx \leq -\frac{4(m-1)}{mT(m-2)^2}\int_\Omega u_0^2(x)dx \tag{1.4}$$

where $F(u) = \int_0^u f(s)ds$. More precisely he showed that if there exists $T > 0$, for which (1.4) holds then the solution blows up in a time less than T. This type of results have been extensively generalized and improved by Levine, Park, and Serrin in a paper [9], where the authors proved some global, as well as nonglobal, existence theorems. Their result, when applied to problem (1.3), requires that

$$\frac{1}{m}\int_\Omega |\nabla u_0(x)|^m dx - \int_\Omega F(u_0(x))dx < 0. \tag{1.5}$$

We note that the inequality (1.5) implies (1.4). In a note, Messaoudi [10] extended the blow-up result to solution with initial datum satisfying

$$\frac{1}{m}\int_\Omega |\nabla u_0(x)|^m dx - \int_\Omega F(u_0(x))dx \leq 0. \tag{1.6}$$

In the present work, we consider (1.1) and prove, for suitable conditions on p and g, a blow-up result for solutions with negative or vanishing initial energy. This result improves an earlier one in [10].

2. Blow-up

In order to state and prove our result we introduce the "modified" energy functional

$$E(t) = \frac{1}{2}(g \diamond \nabla u)(t) + \frac{1}{2}(1 - \int_0^t g(s)ds) \|\nabla u(t)\|_2^2 - \frac{1}{p}\|u(t)\|_p^p \tag{2.1}$$

where

$$(g \diamond v)(t) = \int_0^t g(t-\tau)\|v(t) - v(\tau)\|_2^2 \, d\tau. \tag{2.2}$$

For the relaxation function g and p, we assume that

$$g(s) \geq 0, \qquad g'(s) \leq 0, \qquad 1 - \int_0^\infty g(s)ds = l > 0 \tag{2.3}$$

and

$$2 < p \leq \frac{2(n-1)}{n-2}, \; n > 2, \qquad p > 2, \quad n = 1, 2. \tag{2.4}$$

By multiplying the equation in (1.1) by u_t, integrating over Ω we get, after some manipulations, see [11],

$$\frac{d}{dt}E(t) = -\left(\frac{1}{2}g(t)\|\nabla u(t)\|_2^2 - \frac{1}{2}(g' \diamond \nabla u)(t) + \int_\Omega |u_t|^2 u_t dx\right) \leq 0, \tag{2.5}$$

for regular solutions. The same result can be established, for weaker solutions and for almost every t, by a simple density argument.

Similarly to [13], we give a definition for a strong solution of (1.1).

Definition: A strong solution of (1.1) is a function $u \in C([0,T); H_0^1(\Omega)) \cap C^1([0,T); L^2(\Omega))$, satisfying (2.5) and

$$\int_0^t \int_\Omega \left(\nabla u.\nabla\phi - \int_0^s \nabla u(\tau).\nabla\phi(s)d\tau + u_t\phi - |u|^{p-2} u\phi\right) dxds = 0,$$

for all t in $[0,T)$ and all ϕ in $C([0,T), H_0^1(\Omega))$.

Remark 2.1: Condition (2.4) is needed so that $|u|^{p-2}u \in L^2(\Omega)$; hence $\int_\Omega |u|^{p-2} u\phi dx$ makes sense. Condition $1 - \int_0^\infty g(s)ds = l > 0$ is necessary to guarantee the parabolicity of the system (1.1).

Theorem. *Assume that* (2.3) *and* (2.4) *hold. Given* $u_0 \in H_0^1(\Omega)$ *satisfying* $E(0) \leq 0$. *If*

$$\int_0^\infty g(s)ds < \frac{p-2}{p-3/2} \tag{2.6}$$

then any strong solution of (1.1) *blows up in finite time.*

Proof. We define

$$\mathcal{L}(t) = \frac{1}{2}\int_\Omega u^2(x,t)dx$$

and differentiate with respect to t to get

$$\begin{aligned}
\mathcal{L}'(t) &= \int_\Omega u u_t(x,t)dx \\
&= \int_\Omega u\Delta u dx - \int_\Omega u(x,t)\int_0^t g(t-s)\Delta u(x,s)dsdx + \int_\Omega |u|^p dx \\
&= -\int_\Omega |\nabla u|^2 dx + \int_\Omega\int_0^t g(t-s)\nabla u(x,t).\nabla u(x,s)dsdx + \int_\Omega |u|^p dx \\
&\geq -\int_\Omega |\nabla u|^2 dx + \int_0^t g(t-s)\|\nabla u(t)\|_2^2 d\tau + \int_\Omega |u|^p dx \qquad (2.7)\\
&\quad -\int_0^t g(t-s)\int_\Omega |\nabla u(t).[\nabla u(s)-\nabla u(t)]|dxd\tau.
\end{aligned}$$

By using the Schwarz inequality, (2.7) takes the form

$$\begin{aligned}
\mathcal{L}'(t) &\geq \int_\Omega |u|^p dx - (1-\int_0^t g(s)ds)\|\nabla u(t)\|_2^2 \qquad (2.8)\\
&\quad -\int_0^t g(t-\tau)\|\nabla u(t)\|_2\|\nabla u(\tau)-\nabla u(t)\|_2 d\tau
\end{aligned}$$

By applying Young's inequality to the last term of (2.8) we arrive at

$$\mathcal{L}'(t) \geq \int_\Omega |u|^p dx - \left[1-\frac{3}{4}\int_0^t g(s)ds\right]\|\nabla u(t)\|_2^2 - (g\diamond\nabla u)(t) \qquad (2.9)$$

We then substitute for $\|\nabla u(t)\|_2^2$ from (2.1); hence (2.9) becomes

$$\begin{aligned}
\mathcal{L}'(t) &\geq \int_\Omega |u|^p dx + 2\frac{\left[1-\frac{3}{4}\int_0^t g(s)ds\right]}{(1-\int_0^t g(s)ds)}H(t) \\
&\quad + \left(\frac{1-\frac{3}{4}\int_0^t g(s)ds}{(1-\int_0^t g(s)ds)} - 1\right)(g\diamond\nabla u)(t) \qquad (2.10)\\
&\quad -\frac{2}{p}\frac{1-\frac{3}{4}\int_0^t g(s)ds}{(1-\int_0^t g(s)ds)}\int_\Omega |u|^p dx \geq \gamma\int_\Omega |u|^p dx
\end{aligned}$$

where

$$\gamma = 1-\frac{2}{p}\frac{1-\frac{3}{4}\int_0^\infty g(s)ds}{(1-\int_0^\infty g(s)ds)} > 0$$

because of (2.6). Next we have, by the embedding of the L^q spaces,

$$\mathcal{L}^{p/2}(t) \leq C\|u\|_p^p. \qquad (2.11)$$

By combining (2.10) and (2.11) we get

$$\mathcal{L}'(t) \geq \Gamma L^{p/2}(t), \qquad (2.12)$$

A direct integration of (2.12) then yields

$$\mathcal{L}^{p/2-1}(t) \geq \frac{1}{\mathcal{L}^{1-p/2}(0) - \Gamma t}.$$

Therefore $\mathcal{L}$ blows up in a time

$$t^* \leq \frac{1}{\Gamma \mathcal{L}^{(p/2)-1}(0)}. = \frac{1}{\Gamma}\left(\frac{1}{2}\int_\Omega u_0^2(x)dx\right)^{1-p/2}$$

Acknowledgment

The author would like to express their sincere thanks to King Fahd University of Petroleum and Minerals for its support. This work has been funded by KFUPM under Project # MS/VISCO ELASTIC/270.

References

[1] Alfonsi L. and F. Weissler, Blow-up in $\mathbb{R}^n$ for a parabolic equation with a damping nonlinear gradient term, *Progress in nonlinear differential equations and their applications* **7** (1992), 1–20.

[2] Ball J., Remarks on blow-up and nonexistence theorems for nonlinear evolution equations, *Quart. J. Math. Oxford Ser.* **28** (1977), 473–486

[3] Da Prato G. and M. Iannelli, Existence and regularity for a class of integro-differential equations of parabolic type, *J. Math. Anal. Appl.*, **112** (1985), 36–55.

[4] A. Friedman, Mathematics in Industrial Problems, Springer-Verlag, New York, 1992.

[5] Friedman A., Partial differential equations of parabolic type, *Prentice-Hall* Englewood N. J. 1964.

[6] Junning Z., Existence and nonexistence of solutions for $u_t = div(|\nabla u|^{p-2}\nabla u) + f(\nabla u, u, x, t)$, *J. Math. Anal. Appl.* **172** (1993), 130–146.

[7] Kalantarov V. K. and O. A. Ladyzhenskaya, The occurrence of collapse for quasilinear equations of parabolic and hyperbolic type, *J. Soviet Math.* **10** (1978), 53–70.

[8] Levine H., Some nonexistence and instability theorems for solutions of formally parabolic equations of the form $Pu_t = -Au + F(u)$, *Archive Rat. Mech. Anal.* **51** (1973), 371–386.

[9] Levine H., S. Park, and J. Serrin, Global existence and nonexistence theorems for quasilinear evolution equations of formally parabolic type *J. Diff. Eqns* **142** (1998), 212–229.

[10] Messaoudi S.A., A note on blow-up of solutions of a quasilinear heat equation with vanishing initial energy, *J. Math. Anal. Appl.* **273** (2002), 243–247.

[11] Messaoudi S.A., Blow-up and global existence in a nonlinear viscoelastic equation, *Mathematische Nachrichten* vol. **260** (2003), 58–66.

[12] Nohel J.A., Nonlinear Volterra equations for the heat flow in materials with memory, Integral and functional differential equations, Lecture notes in pure and applied mathematics, edited by T.L. Herdman, S.M. Rankin, III, H.W. Stech, Marcel Dekker Inc., 1981.

[13] Pucci P. and J. Serrin, asymptotic stability for nonlinear parabolic systems, *Energy methods in continuum mechanics*, (Oviedo, 1994), 66–74, Kluwer Acad. Publ., Dordrecht, 1996.

[14] Yin H.M., On parabolic Volterra equations in several space dimensions, *SIAM J. Mathematical Analysis* **22** (1991), 1723–1737.

[15] Yin H. M., Weak and classical solutions of some Volterra integro-differential equations, *Communication in P.D.E* **17** (1992), 1369–1385.

Salim A. Messaoudi
1. Mathematical Sciences Department
KFUPM, Dhahran 31261
Saudi Arabia
e-mail: messaoud@kfupm.edu.sa

Progress in Nonlinear Differential Equations
and Their Applications, Vol. 64, 357–371

Finite Element Methods for Investigating the Moving Boundary Problem in Biological Development

Cornel Marius Murea and George Hentschel

Abstract. We describe two finite element algorithms which can be used to study organogenesis or organ development during biological development. Such growth can often be reduced to a free boundary problem with similarities to two-fluid flow in the presence of surface tension, though material is added at a constant growth rate to the developing organ. We use the specific case of avian limb development to discuss our algorithms.

1. Introduction

Biological development involves both growth and changes of form which can often involves free moving boundaries [22]. Such moving boundary problems are similar in some respects to two fluid flow interfaces such as the Hele-Shaw problem with surface tension also called Mullins-Sekerka problem. In general, however, in contrast to incompressible flows, growth due to mitosis and nutrients ensure that material is constantly being added (and sometimes removed when cell death or apoptosis occurs). In addition, specific boundary conditions (in general different for each organ or cell type considered) will result in a more complex boundary value problem than those studied in Hele Shaw cells. In this paper to be specific we shall consider avian limb development, though we believe that similar finite element algorithms described here will be useful for other problems of organogenesis or organ morphogenesis and biological development.

In this paper we will consider the evolution of two-dimensional moving domains. The more realistic but more complex case of three-dimensional domains separated by two-dimensional interfaces will be described in a future publication. In the case of the avian limb, the ventral-dorsal length scale (back of limb to palm) is normally small compared to the proximal-distal (tip of finger to point at which the limb joins the main body of the organism) or the posterior-anterior distance (from thumb to little finger) and therefore two-dimensional simulations are quite informative. In addition, at the developmental stage we are interested in, namely the embryo the whole limb which is only of a millimeter in scale, has approximately the shape of an ellipse with boundary $\Gamma_1(t)$ and a boundary $\Gamma_2(t)$ grafted to the trunk of the organism. In general only the growth velocity of $\Gamma_1(t)$ parallel to the gradient of a pressure, while the growth of $\Gamma_2(t)$ can be described by the motion of the joining vertex with the main trunck (in the more complex three-dimensional case this single point becomes a closed one-dimensional contour). The pressure in the limb whose gradient describes the rate of growth of the limb is the solution of a Poisson problem with Dirichlet boundary conditions depending on the curvature of the boundary.

In Section 2 we give a brief description of some relevant aspects of avian limb development. Then in Section 3 we present a mathematical formulation of the resulting free boundary problem.

In Section 4 two algorithms are described to solve numerically integrate the resulting equations of motion and find the dynamical evolution of the interface.

In the first algorithm the boundary of the domain is approached by a polygon and the pressure is computed by a Finite Element Method. The computed pressure is a piecewise linear function, globally continuous. The curvature is computed as the inverse of the ray of the circle passing through three consecutive vertices of the boundary. The gradient of the pressure is then computed in a vertex of the mesh, as a weighted mean of the gradients in the neighborhood triangles. For the time discretization, we use the forward finite difference Euler's scheme. A dynamic mesh technique is used in order to generate a triangular mesh at each time step. Starting from the mesh at the precedent time step and knowing the boundary at the current time step, we generate a mesh by redistributing the interior vertices using an optimization algorithm. The number of the interior vertices are constant. Also the connections of all meshes are the same, i.e., if i, j, k are the vertices of a triangle in the mesh at the precedent time step, these points are the vertices of a triangle in the current mesh.

While in the second algorithm, the boundary is approached by cubic spline interpolation which gives a curve twice continuously differentiable. The curvature is computed using the parametrization of the splines. Again at each time step, a new mesh is generated, but this time, the generation of the current mesh is independent from the previous one.

In Section 5 we describe the numerical tests of our algorithms. Finally in Section 6, we give a brief discussion of the potential of these approaches for studies of organogenesis and biological development.

2. Early avian limb development

Early avian limb development presents a beautiful example of organogenesis and biological pattern formation: Well-defined developmental axes exist which need to be understood. Limb growth changes the size and shape of internal domains in which biochemical processes occur. Cartilege formation via mesenchymal cell condensation occurs which will later differentiate into bone and form the skeletal limb. Many of these features appear to be robust: if comparisons are made anatomically with such an apparently different vertabrates as chicken and mouse, it is remarkable the extent to which the gross features of patterning observed during development are conserved by evolution. All of which suggests that universal physical mechanisms controlling development exist.

The embryo produces the raw materials (e.g., proteins, polysaccharides, RNAs) for its development from the available nutrients, according to rules embodied in the genetic code; diffusion, spreading, differential adhesion and chemotaxis transport these materials to specific regions of the organism. The mechanical or chemical changes which may take place in the course of the transport (change of concentration, cell shape, adhesiveness and cohesiveness) are signals that often affect the production of the building material itself, that is, gene activity.

In the course of these events cells differentiate and become more specialized. Differentiation involves regulated gene expression, but elaborate interactions among cells determine where and when new genes are expressed. In addition, morphogenetic changes require coordinated cell movement. The formation of the avian limb requires the establishment of proximal-distal positional gradients and transverse periodic modulations of morphogens to control the formation of individual and multiple parallel skeletal elements. These morphogen patterns act on limb mesenchyme to promote the formation of precartilage condensations, and ultimately the chondrocytes that will give rise to the cartilaginous primordia of the limb skeleton, which ultimately are replaced by bone.

In this paper we wish to investigate only one aspect of this morphogenesis. Namely what type of overall shape is to be expected as a result of growth of the developing embryonic limb. To investigate this problem we consider a minimal model which incorporates the key features of this biological growth. Key is the addition of material at a rate S to the extracellular matrix in which the cells move (more generally the rate of growth will be $S(\mathbf{x}, t)$ as it could be both spatially varying and have a temporal dependence due to genetic switching mechanisms). This means that that the material flow in the limb will obey

$$\nabla \cdot \mathbf{v} = S. \tag{1}$$

We treat growth of the limb as due to a creeping flow because of the very low Reynolds numbers involved [8]. Therefore we can expect the flow to obey Darcy's Law

$$\mathbf{v} = -\alpha \nabla p \tag{2}$$

where p is a pseudo pressure field, which obeys $\Delta p = -S/\alpha$ in the limb domain.

Finally we need biologically reasonable boundary conditions. As it appear there is no flow of material into the main body of the organism at $\Gamma_2(t)$ we shall take slip boundary condition here

$$\mathbf{v} \cdot \nu = 0, \tag{3}$$

where ν is the outer unit normal vector to the boundary, while the elastic properties of the epithelial layer of cells forming the skin layer at $\Gamma_1(t)$ will result a pressure at this boundary obeying

$$p = \gamma\kappa, \tag{4}$$

where γ is the effective surface tension of the limb [9] while κ is the limb curvature.

The equation of the normal velocity of the boundary $\Gamma_1(t)$ is

$$V_\nu = \mathbf{v} \cdot \nu. \tag{5}$$

The above condition requires that the boundary $\Gamma_1(t)$ moves with the fluid.

It is the integration of this free boundary value problem that we study below. This mathematical model agrees favorably with the analysis presented in [3] based on biological experiments where the limb is considered as a homogeneous and highly hydrated core embedded in an dense envelope.

3. The free boundary problem

We study the evolution of a bounded connected open domain $\Omega(t)$ of $\mathbb{R}^2$ with boundary $\partial\Omega(t) = \Gamma_1(t) \cup \Gamma_2(t)$, where $\Gamma_1(t)$ and $\Gamma_2(t)$ are two non-empty subsets of $\partial\Omega(t)$. Here $t \geq 0$ is the time. We assume that $\Gamma_1(t)$ is a non-closed curve of class C^2 and its ends evolve on the Oy ax. The boundary $\Gamma_2(t)$ is the segment which has the same ends as $\Gamma_1(t)$. Let ν denote the outer unit normal vector to the boundary.

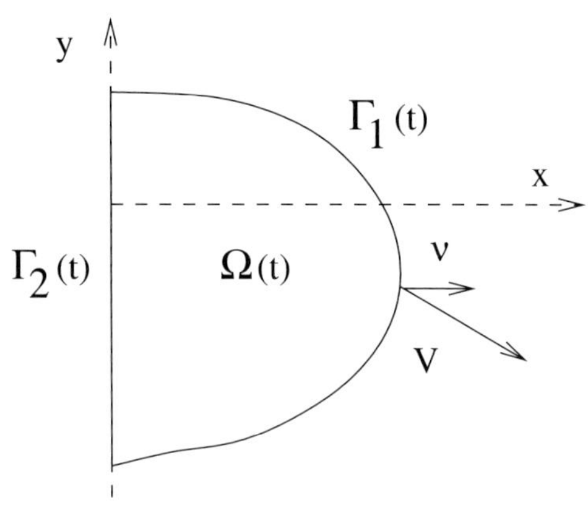

FIGURE 1. Schematic illustration of the free boundary problem

From the equations (1)–(5), we can eliminate $\mathbf{v}$ and we obtain a system in p only. In the moving domain $\Omega(t)$, we have to find the pressure $p(x,y,t) : \Omega(t) \to \mathbb{R}$, such that

$$\begin{cases} -\Delta p &= S/\alpha & \text{in } \Omega(t) \\ p &= \gamma\kappa & \text{on } \Gamma_1(t) \\ \dfrac{\partial p}{\partial \nu} &= 0 & \text{on } \Gamma_2(t) \end{cases} \tag{6}$$

where α, γ, S are positive real constants and κ is the curvature of $\Gamma_1(t)$. We use the sign convention that convex domains have positive curvature of the boundary.

The boundary $\Gamma_1(t)$ evolves according to the law

$$V_\nu = -\alpha \frac{\partial p}{\partial \nu} \tag{7}$$

where V_ν is the normal velocity of $\Gamma_1(t)$.

We know the initial domain

$$\Omega(0) = \Omega^0. \tag{8}$$

We consider the problem (6)–(8) of determining the evolution of $\Omega(t)$ and to find the pressure $p(x,y,t)$ for $t \in [0,T]$, where $T > 0$ is a given real constant.

This problem is similar to the Hele-Shaw problem with surface tension, but in our case the pressure is no longer harmonic ($\Delta p \neq 0$).

Let $\overline{p}(x,y,t) = \frac{S}{4\alpha}\left(x^2+y^2\right)$. We set $P = p + \overline{p}$ and we obtain from (6)–(7) the following problem: in the moving domain $\Omega(t)$, we have to find the pressure $P(x,y,t) : \Omega(t) \to \mathbb{R}$, such that

$$\begin{cases} \Delta P &= 0 & \text{in } \Omega(t) \\ P &= \gamma\kappa + \overline{p} & \text{on } \Gamma_1(t) \\ \dfrac{\partial P}{\partial \nu} &= 0 & \text{on } \Gamma_2(t) \end{cases} \tag{9}$$

and the normal velocity of the boundary $\Gamma_1(t)$ is

$$V_\nu = -\alpha\left(\frac{\partial P}{\partial \nu} - \frac{\partial \overline{p}}{\partial \nu}\right). \tag{10}$$

Though we cannot prove the existence and uniqueness of our solution, the existencc and uniqueness of classical solution for the Hele-Shaw with surface tension problem was proved in [7], suggests that this is the case here also. It is possible that, in order to obtain the existence and uniqueness of solution, we have to prescribe the angles between the boundaries $\Gamma_1(t)$ and $\Gamma_2(t)$. The problem of existence and uniqueness is now under active investigation by G. Simonett.

As it was shown in [12], the shape of the moving domain is determined solely by its normal velocity. In other words, if the velocity of $\Gamma_1(t)$ has the form

$$\mathbf{V} = V_\nu \cdot \nu + V_\tau \cdot \tau$$

where τ is the unit tangent vector to the boundary and V_ν is given by (7), then the movement of the domain and the pressure are the same as in the case $V_\tau = 0$. We set $V_\tau = -\alpha\left(\nabla p \cdot \tau\right)$ and then $\mathbf{V} = -\alpha \nabla p$. The advantage of this choice is that we do not need to evaluate the normal vector to the boundary when we compute the velocity of the boundary.

It is convenient to describe the curve $\Gamma_1(t)$ by the parametric coordinates

$$\begin{array}{rcl} x & = & r_1(\theta,t), \\ y & = & r_2(\theta,t), \end{array} \qquad \theta \in [a,b] .$$

Let us introduce the following generalized cylinder:

$$\Omega_T = \bigcup_{t\in]0,T[} (\Omega(t) \times \{t\}) .$$

The problem (6), (7) and (8) is equivalent to the following: find $r = (r_1, r_2) : [a,b] \times [0,T] \to \mathbb{R}^2$ and $p : \overline{\Omega}_T \to \mathbb{R}$, such that

$$\begin{array}{rcll} \dfrac{\partial r_1}{\partial t}(\theta,t) & = & -\alpha \dfrac{\partial p}{\partial x}\left(r_1(\theta,t), r_2(\theta,t), t\right), & \forall \theta \in [a,b], \forall t \in]0,T[\\ \dfrac{\partial r_2}{\partial t}(\theta,t) & = & -\alpha \dfrac{\partial p}{\partial y}\left(r_1(\theta,t), r_2(\theta,t), t\right), & \forall \theta \in [a,b], \forall t \in]0,T[\\ r(\theta,0) & = & \left(r_1^0(\theta), r_2^0(\theta)\right), & \forall \theta \in [a,b] \end{array}$$

where $r^0 = \left(r_1^0, r_2^0\right)$ is a parametric representation of $\Gamma_1(0)$ and $p(x,y,t)$ is the solution of (6).

Since $\frac{\partial p}{\partial \nu} = 0$ and $\nu = (-1,0)^T$ on $\Gamma_2(t)$, we have $\frac{\partial p}{\partial x} = 0$ on $\Gamma_2(t)$. If we suppose that $p \in \mathcal{C}^2(\Omega(t)) \cap \mathcal{C}^1\left(\overline{\Omega}(t)\right)$, we obtain that $\frac{\partial p}{\partial x} = 0$ at the ends of $\Gamma_2(t)$, also

$$\begin{array}{l} \frac{\partial p}{\partial x}\left(r_1(a,t), r_2(a,t), t\right) = 0, \forall t \in]0,T[\\ \frac{\partial p}{\partial x}\left(r_1(b,t), r_2(b,t), t\right) = 0, \forall t \in]0,T[. \end{array}$$

Then $\frac{\partial r_1}{\partial t}(a,t) = \frac{\partial r_1}{\partial t}(b,t) = 0$ and consequently $r_1(a,t) = r_1(b,t) = 0$, $\forall t \in]0,T[$.

The boundary could be parametrized in multiple ways, but the solution must be independent of parametrization.

4. Numerical methods

The free boundary problem (6)–(8) is similar to the Hele-Shaw problem with surface tension also called Mullins-Sekerka problem.

To solve numerically the Hele-Shaw problem with surface tension, there exists an efficient approach named $\theta - L$ introduced in [12]. The variables are the tangent angle θ to the moving boundary and its arc length L. This framework makes the application of an implicit method for time integration easy and it permits to study the problem in a long time interval [13], [4]. In this approach a Fredholm like boundary integral has solved and the integral representation of a harmonic function is used. This is specific to some linear problems with constant coefficients. This method is not appropriate if we replace the linear Darcy's law (2) by the non-linear Navier-Stokes equation with a volume source located at a certain point in the domain

$$\rho_0\left(\frac{\partial \mathbf{v}}{\partial t}+(\mathbf{v}\cdot\nabla)\,\mathbf{v}\right)-\mu\Delta\mathbf{v}+\nabla p=\mathbf{f}+\frac{\mu}{3}\nabla\left(\nabla\cdot\mathbf{v}\right)$$

where $\rho_0 > 0$ is the density of the fluid and $\mu > 0$ its viscosity.

A frequented framework used for Navier-Stokes equation in moving domain is Arbitrary Lagrangian Eulerian together with the Finite Element Method [14].

Other approaches are Time-Space Finite Elements [1], Level Set Method [21] and Immersed Boundary Methods [17]. The last one was employed to study the avian limb development in [6]. One of the disadvantage of the continuum models is the complex implementation required to handle the moving boundary of the domain where we have to solve a system of PDEs.

In [15] the software *CompuCell* is presented, where a purely continuum approach for morphogenesis is used in combination with a discrete cellular automata. One of the part of *CompuCell* is based on the cellular Potts model (CMP). A criticism of this formalism is that it neglects simple force balance between cells.

In this paper we present two algorithms which belong to the general framework called "front-tracking methods" [5]. The numerical results were produced for the Darcy's law, but these algorithms could be used for the steady Navier-Stokes equation also.

4.1. The first algorithm

For the time discretization, we use the forward finite differences Euler's scheme. We denote by Δt the time step and by $N = T/\Delta t$ the number of time steps. We approximate $\Gamma_1(n\Delta t)$ by a polygonal line Γ_1^n of vertices (x_i^n, y_i^n) for $i = 0, \ldots, M$. We have $x_0^n = x_M^n = 0$, for all n. We denote by Ω^n the polygonal domain bounded by Γ_1^n and the Oy ax. For each vertex (x_i, y_i) of Γ_1^n, we compute the discrete curvature $\kappa^n(x_i, y_i)$ as the inverse of the ray of the circle passing through the three points (x_{i-1}^n, y_{i-1}^n), (x_i^n, y_i^n) and (x_{i+1}^n, y_{i+1}^n).

The problem (6) is solved numerically by the Finite Element Method. The computed pressure p^n is approached by P_1 function, globally continuous. We follow [18] for computing the discrete gradient of p^n. Let A be a vertex of Γ_1^n. We denote by $star(A)$ the set of all triangles T of the mesh such that A is a vertex of T. We compute the discrete gradient of p^n at the point A as following:

$$\frac{\sum_{T\in star(A)} Area(T)\cdot\nabla p^n{}_{|T}}{\sum_{T\in star(A)} Area(T)},$$

where $p^n{}_{|T}$ is the linear function representing the restriction of the function p^n on the triangle T.

Algorithm 1

Generate a triangular mesh for Ω^0 using **freefem+** [2].
for each n **from** 0 **to** $N-1$ **do**

Step 1: Compute the discrete curvature $\kappa^n(x_i, y_i)$ at each vertex (x_i, y_i) of Γ_1^n.

Step 2: Compute p^n by the Finite Element Method

$$\begin{cases} -\Delta p^n(x,y) &= S/\alpha, & \text{in } \Omega^n \\ p^n(x_i,y_i) &= \gamma\,\kappa^n(x_i,y_i), & \forall (x_i,y_i) \text{ vertex of } \Gamma_1^n \\ \dfrac{\partial p^n}{\partial \nu}(x,y) &= 0, & \text{on } \Gamma_2^n. \end{cases}$$

Step 3: Compute the discrete gradient of p^n at each vertex of Γ_1^n.

Step 4: Compute the vertices of Γ_1^{n+1}

$$x_i^{n+1} = x_i^n - \alpha\Delta t \frac{\partial p^n}{\partial x}(x_i^n, y_i^n)$$

$$y_i^{n+1} = y_i^n - \alpha\Delta t \frac{\partial p^n}{\partial y}(x_i^n, y_i^n)$$

for each vertex $(x_i^n, y_i^n) \in \Gamma_1^n$, $i = 0, 1, \ldots, M$, then set $x_0^{n+1} = x_M^{n+1} = 0$.

Step 5: The boundary Γ_2^{n+1} is the segment of ends $\left(x_M^{n+1}, y_M^{n+1}\right)$ and $\left(x_0^{n+1}, y_0^{n+1}\right)$.

Step 6: Compute a triangular dynamic mesh for Ω^{n+1}, where $\partial\Omega^{n+1} = \Gamma_1^{n+1} \cup \Gamma_2^{n+1}$ using the algorithm [16].

endfor;

At the **Step 6**, we start from the mesh at the precedent time step and knowing the boundary at the current time step, we generate a mesh by redistributing the interior vertices using an optimization algorithm. The number of the interior vertices are constant. Also, the connections of all meshes are the same, i.e., if i, j, k are the vertices of a triangle in the mesh at the precedent time step, these points are the vertices of a triangle in the current mesh.

4.2. The second algorithm

Let (x_i^n, y_i^n) for $i = 0, \ldots, M$ be points on $\Gamma_1(n\Delta t)$. Let $\{a = s_0 < s_1 < \cdots < s_M = b\}$ be a partition of an interval $[a,b]$. We will compute the interpolating cubic spline functions $\Gamma_1^n = \{(x(s), y(s))\,,\ s \in [a,b]\}$ with the properties:

- $x(s)$ and $y(s)$ are twice continuously differentiable on $[a,b]$,
- $x(s)$ and $y(s)$ coincide on every subinterval $[s_i, s_{i+1}]$, $i = 0, \ldots, M-1$ with polynomials of degree three,
- $x(s_i) = x_i^n$ and $y(s_i) = y_i^n$ for $i = 0, \ldots, M$,
- $x''(a) = x''(b) = 0$ and $y'(a) = y'(b) = 0$.

For the numerical tests, we have chosen $s_i = i$, for $i = 0, \ldots, M$.

In order to prevent the oscillations, we could choose $a = 0$ and

$$s_{i+1} - s_i = \sqrt{\left(x_i^n - x_{i+1}^n\right)^2 + \left(y_i^n - y_{i+1}^n\right)^2}.$$

The curve $(x(s), y(s))$, $s \in [a, b]$ has a continuous curvature given by

$$k(s) = \frac{x'(s)y''(s) - x''(s)y'(s)}{\left((x'(s))^2 + (y'(s))^2\right)^{3/2}}. \tag{11}$$

Algorithm 2

Let (x_i^0, y_i^0) for $i = 0, \ldots, M$ be points on $\Gamma_1(0)$.
for each n **from** 0 **to** $N-1$ **do**

Step 1: Compute the cubic spline functions

$$\Gamma_1^n = \{(x(s), y(s)), \; s \in [a, b]\}$$

Step 2: Compute the curvature $\kappa^n(x_i, y_i)$ at each vertex (x_i^n, y_i^n) using (11).

Step 3: Generate a triangular mesh for Ω^n using **freefem+** [2], where $\partial\Omega^n = \Gamma_1^n \cup \Gamma_2^n$ and Γ_2^n is the segment of ends (x_M^n, y_M^n) and (x_0^n, y_0^n).

Step 4: Step 5: Step 6: Compute p^n, ∇p^n and (x_i^{n+1}, y_i^{n+1}) as in the Algorithm 1.

endfor;

At he step **Step 3**, the generation of the current mesh is independent from the previous one.

We shall now describe some numerical tests of the efficacy of these algorithms in studies of organogenesis.

5. Numerical tests

5.1. The initial domain is a semicircle

First let consider the case where the initial domain is a semicircle of ray R_0. Then, if we set the parametric representation of $\Gamma_1(0)$ as

$$\begin{aligned} r_1^0(\theta) &= R_0\cos(\theta), \\ r_2^0(\theta) &= R_0\sin(\theta), \end{aligned} \qquad \theta \in \left[-\frac{\pi}{2}, \frac{\pi}{2}\right]$$

the evolution of the boundary $\Gamma_1(t)$ is described by

$$\begin{aligned} r_1(\theta, t) &= R_0\, e^{\frac{St}{2}}\cos(\theta), \\ r_2(\theta, t) &= R_0\, e^{\frac{St}{2}}\sin(\theta), \end{aligned} \qquad \theta \in \left[-\frac{\pi}{2}, \frac{\pi}{2}\right], \quad t > 0.$$

The pressure has the form

$$p(x, y, t) = \frac{S}{4\alpha}\left(R_0^2\, e^{St} - x^2 - y^2\right) + \frac{\gamma}{R_0\, e^{\frac{St}{2}}}.$$

The algorithms have been implemented using the programming language C++ and the Finite Element classes of F. Hecht [10]. The numerical results were displayed using `gnuplot`.

The first simulation was performed using the **Algorithm 1** for $R_0 = 1$, $S = 2$, $\gamma = 1$, $\alpha = 0.5$, $\Delta t = 0.05$. The number of time steps is $N = 10$.

A dynamic mesh technique is used in order to generate a triangular mesh at each time step. We have used the algorithm described in [16] for the mesh generation. The initial mesh has: 208 vertices, 362 triangles, $M = 32$ (the number of vertices on the boundary Γ_1), $h = 0.155178$ (the mesh size).

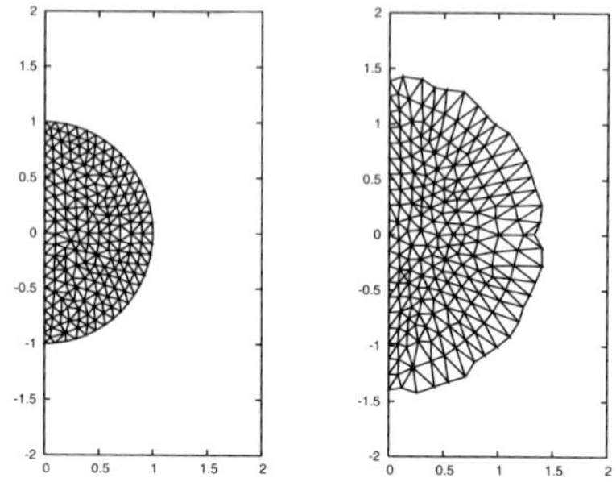

FIGURE 2. The initial mesh (left) and the mesh after 10 time steps (right)

The number of vertices, triangles, boundary edges are the same for the first 10 meshes. In the below table, we see the evolution of the mesh size h.

Time step (n)	1	2	3	4	5
Mesh size (h)	0.155	0.171	0.187	0.202	0.216

Time step (n)	6	7	8	9	10
Mesh size (h)	0.231	0.244	0.256	0.266	0.276

After 10 time steps, the domain might be a semicircle of ray $e^{0.5} \approx 1.648721$.

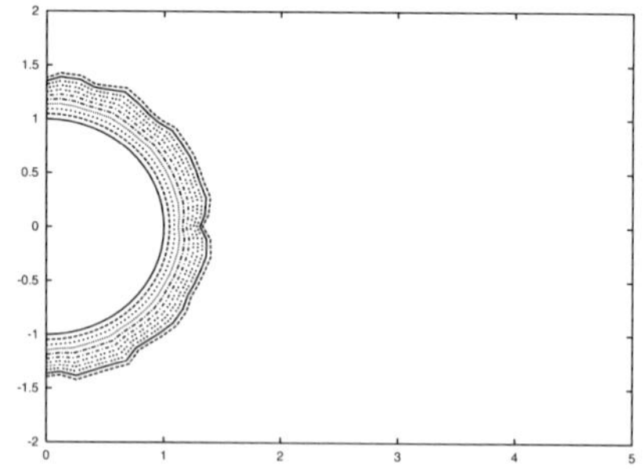

FIGURE 3. The evolution of the moving boundary (0–10 time steps)

The second simulation was performed using the **Algorithm 2** for $\Delta t = 0.0005$ and $N = 1000$ (the number of time steps). The others parameters are the same as in the first simulation.

The software `freefem+` [2] was used to generate a triangular mesh at each time step. The numbers of the vertices and of the triangles are not the same for the all meshes. For example, the mesh after 1000 time steps has 205 vertices and 356 triangles.

After 1000 time steps, the domain might be a semicircle of ray $e^{0.5} \approx 1.648721$.

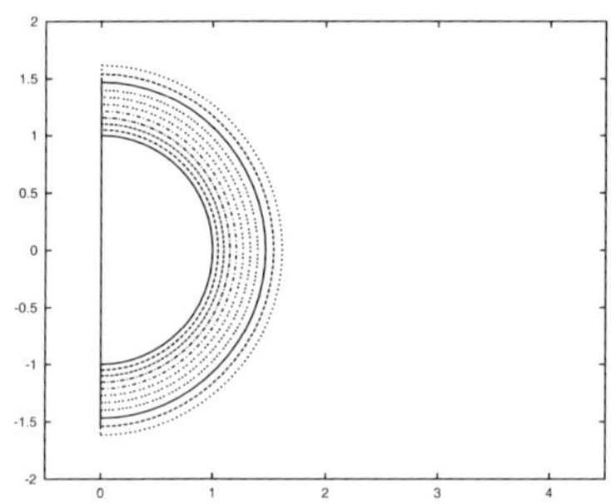

FIGURE 4. The boundary after 0, 100, ..., 1000 time steps

5.2. A non-convex initial domain

Let now consider a case when the initial domain is non-convex as in Figure 5. The boundary $\Gamma_1(0)$ has two flat parts on the bottom, on the top and three semicircles of rays $r1 = 0.6$, $r2 = 0.2$ and $r3 = 0.2$ respectively.

The simulations ware performed for: $S = 2$, $\alpha = 0.5$, $\gamma = 0.5$, $\Delta t = 0.0001$ and $N = 180$ (the number of time steps).

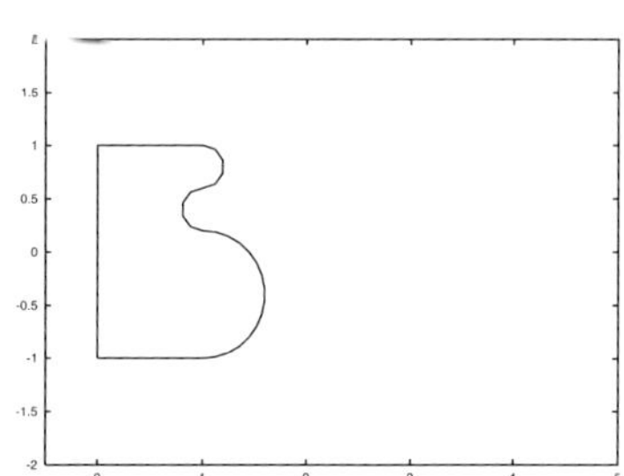

FIGURE 5. The initial domain

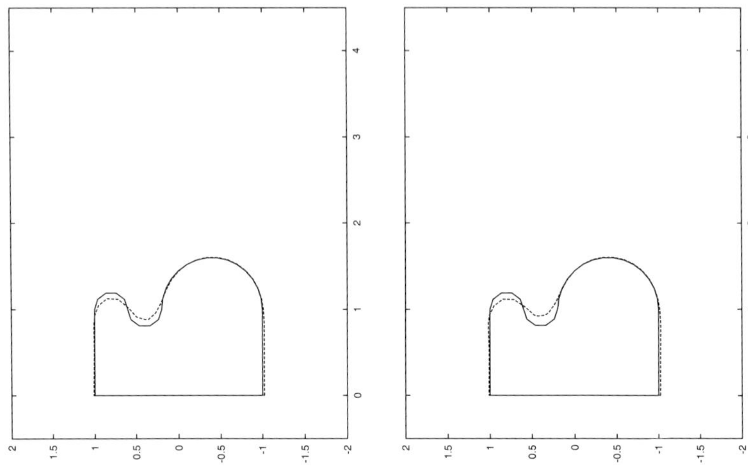

FIGURE 6. The initial (continuous) and the final (dashed) boundaries. Algorithms 1 (left) and 2 (right)

We have observed that the pressure is almost constant near the two flat parts of $\Gamma_1(t)$ and near the largest semicircle. Consequently, these parts of boundary don't move.

We obtain boundary with self-intersection (like the 8) after 182 time steps using the Algorithm 2 and after 230 time steps using the Algorithm 1.

5.3. Concluding remarks

The second algorithm is superior to the first one due in principal to a better approximation of the curvature and a better mesh. We can improve the results by moving the boundary along the normal velocity which preserves a reasonable distribution of the vertices on the boundary. The velocity of the boundary could be computed more accurate by using P2 Finite Element for the pressure. Also adapting mesh techniques could be employed for improving the quality of the mesh.

In the first numerical test, we have solved the free boundary problem until time $t = 0.5$ and in the second, until $t = 0.018$.

We have to use implicit in time algorithms in order to study this kind of free boundary problem in a long time interval. The Arbitrary Lagrangian Eulerian framework together with the Finite Element Method will be employed in a future paper. These methods could be employed for the Navier-Stokes equation with surface tension.

6. Discussion

The algorithms described above could form the basis for many important investigations of organogenesis and biological development in general. Of course they will need to be extended in several directions to give a quantitative picture of how growth and form develops. Obviously highly computationally intensive three-dimensional simulations are necessary. While genetic switching mechanisms will need to be incorporated in order to understand the temporal properties of biological development. But in addition to these questions many other lines of investigation need to be developed.

For example we know that cell condensation and bone development depend on reaction diffusion mechanisms in a progress zone of undifferentiated cells at the tip of the limb. The size of this progress zone can be expected to have a significant impact on the resulting prepattern created in the limb [19, 20, 11]. This is because even at the most basic level the number of standing waves of a heterogeneous distribution of a chemical species formed by a reaction-diffusion mechanism depends both on the scale of the basic pattern (set by the magnitude of the biological parameters) and on the space available for this pattern to develop (set by the domain size). Thus a very question is how the size of this progress zone changes with time? In order to answer this question studies of limb growth will form a vital ingredient.

Another question involves how the skeletal elements themselves, once formed, would influence growth of the developing limb? Fairly significant changes in internal domain organisation occur between early development when the stylopod and zeugopod are created, and later on when the digits appear. This question will require the development of algorithms for complex connected domains in which the skeletal elements create internal boundaries to growth. In addition, the existence of such internal domains will in turn influence reaction-diffusion mechanisms in the interdigital regions. Such reaction-diffusion mechanisms might be relevant to properly controlled cell death leading ultimately to digit formation. The influence will be very dependent on the relative scale of the patterning compared to the size of the interdigital domains.

Clearly these algorithms need to be developed in several directions for studies of biological development and these will be reported in future publications.

Acknowledgments

The authors are very grateful to Professor Stuart A. Newman from the New York Medical College whose questions and comments have contributed to improve the contents of this paper.

References

[1] Bansch E., Finite element discretization of the Navier-Stokes equations with a free capillary surface. *Numer. Math.* **88** (2001) no. 2, 203–235.

[2] Bernardi D., Hecht F., Ohtsuka K., Pironneau O., A finite element software for PDE: freefem+. `http://www-rocq.inria.fr/Frederic.Hecht`

[3] Borkhvardt V.G., Growth and Shaping of the Fin and Limb Buds. *Russian Journal of Developmental Biology*, **31** (2000) no. 3, 154–161.

[4] Ceniceros H.D., Hou T.Y., Si H., Numerical study of Hele-Shaw flow with suction. *Phys. Fluids* **11** (1999) no. 9, 2471–2486.

[5] Crank J., Free and moving boundary problems. The Clarendon Press, Oxford University Press, New York, 1987.

[6] Dillon R., Othmer H.G., A mathematical Model for outgrowth and spatial pattering of the vertebrate limb bud. *J. Theor. Biol.*, **197** (1999) 295–330.

[7] Escher J., Simonett G., Classical solutions for Hele-Shaw models with surface tension. *Adv. Differential Equations* **2** (1997) no. 4, 619–642.

[8] Forgacs G., Foty R.A., Shafrir Y., Steinberg M.S., Viscoelastic properties of living tissues: a quantitative study. *Biophysical Journal* **74** (1998) 2227–2234.

[9] Foty R.A., Pfleger C.M., Forgacs G., Steinberg M.S., Surface tensions of embryonic tissues predict their mutual envelopment behavior. *Development* **122** (1996) 1611–1620.

[10] Hecht F., C++ et Eléments Finis, Cours DEA, Université Paris VI, `http://www-rocq.inria.fr/Frederic.Hecht`

[11] Hentschel H.G.E., Glimm T., Glazier J.A., Newman S.A., Dynamical Mechanisms for Skeletal Pattern Formation in the Avian Limb, *Proc. Royal Soc. B.* **271** (2004) 1713–1722.

[12] Hou T.Y., Lowengrub J.S., Shelley M.J., Removing the stiffness from interfacial flows with surface tension. *J. Comput. Phys.* **114** (1994) 312–328.

[13] Hou T. Y., Lowengrub J. S., Shelley M. J., The long-time motion of vortex sheets with surface tension. *Phys. Fluids* **9** (1997) no. 7, 1933–1954.

[14] Hughes T.J.R., Liu W.K., Zimmermann T.K., Lagrangian-Eulerian finite element formulation for incompressible viscous flows. *Comput. Methods Appl. Mech. Engrg.* **29** (1981) no. 3, 329–349.

[15] Izaguirre J.A., Chaturvedi R., Huang C., Cickovski T., Cof J., Thomas G., Forgac G., Alber M., Hentschel G., Newman S.A., Glazier J.A., COMPUCELL, a multi-model framework for simulation of morphogenesis, *Bioinformatics* **20** (2004) no. 7, 1129–1137.

[16] Murea C.M., Dynamic meshes generation using the relaxation method with applications to fluid-structure interaction problems. *An. Univ. Bucuresti Mat.* **47** (1998) no. 2, 177–186.

[17] Peskin C.S., The immersed boundary method. *Acta Numer.* **11** (2002), 479–517.

[18] Pironneau O., Hecht F., Introduction au Calcul Scientifique en C++, Cours Matrîse de Mathématiques, Ingénierie Mathématique. Université Paris VI, 2000. `http://www-rocq.inria.fr/Frederic.Hecht`

[19] Newman S.A., Frisch H.L., Dynamics of skeletal pattern formation in developing chick limb. *Science* **205** (1979), 662–668.

[20] Newman S.A., Frisch H.L., Percus J.K. On the stationary state analysis of reaction-diffusion mechanisms for biological pattern formation. *J. Theor. Biol.* **134**, (1988) 183–197.

[21] Sethian J. A., *Level set methods: evolving interfaces in geometry, fluid mechanics, computer vision and material science*, Cambridge University Press, 1996.

[22] Wolpert L., Beddington R., Brockes J., Jessell T., Lawrence P., Meyerowitz E., *Principles of Development.* Oxford, New York, Tokyo. Oxford University Press, 1998.

Cornel Marius Murea
Laboratoire de Mathématiques et Applications
Université de Haute-Alsace
4, rue des Frères Lumière
F-68093 Mulhouse Cedex, France
e-mail: c.murea@uha.fr

George Hentschel
Emory University
Department of Physics
Atlanta, GA 30322 USA
e-mail: phshgeh@physics.emory.edu

Progress in Nonlinear Differential Equations
and Their Applications, Vol. 64, 373–390

Existence of Weak Solutions to the Equations of Stationary Motion of Heat-conducting Incompressible Viscous Fluids

J. Naumann

Dedicated to Professor H. Amann on the occasion of his 65th birthday

1. Introduction

The stationary motion of a heat-conducting incompressible viscous fluid in a domain $\Omega \subset \mathbb{R}^n$ ($n = 2$ resp. $n = 3$) is governed by the following system of conservation laws:

$$\nabla \cdot \mathbf{u} = 0 \qquad \text{mass,} \tag{1.1}$$

$$(\mathbf{u} \cdot \nabla)\mathbf{u} = \nabla \cdot S - \nabla p + f \qquad \text{momentum,} \tag{1.2}$$

$$\gamma(\theta)(\mathbf{u} \cdot \nabla\theta) = \mathrm{tr}(SD) - \nabla \cdot q \qquad \text{energy,} \tag{1.3}$$

where

$$\mathbf{u} = (u_1, \ldots, u_n) \quad \text{velocity,}$$

$$S = \{S_{ij}\} \quad \text{extra stress}^{1)}, \ p = \text{ pressure}, \quad f = (f_1, \ldots, f_n) \ \text{ external force,}$$

$$\theta = \text{ temperature}, \ \gamma(\theta) = \text{ specific heat,}$$

$$D = \{D_{ij}\}, \ D_{ij} = D_{ij}(\mathbf{u}) = \frac{1}{2}\left(\frac{\partial u_i}{\partial x_j} + \frac{\partial u_j}{\partial x_i}\right) \quad \text{rate of strain}$$

$$\mathrm{tr}(SD) = S_{ij}D_{ij} \quad \text{trace of } SD, \quad q = \text{ heat flux}$$

(cf., e.g., Landau/Lifschitz [9], Müller [15], Serrin [19], Spurk [21]). The first term on the right-hand side of (1.3) represents the production of heat due to viscosity, while the second term represents the heat due to the heat flux.

---

1) In what follows, Latin indices take the values $1, \ldots, n$ ($n = 2$ resp. $n = 3$). Repeated indices imply summation over $1, \ldots, n$.

We specify the constitutive assumptions for the extra stress tensor S and the vector of heat flux q:

$$\begin{cases} S = \mu(\theta)D & \text{Newton's law,} \\ q = -\kappa(\theta)\nabla\theta & \text{Fourier's law} \end{cases} \tag{1.4}$$

where

$$\mu(\theta) = \text{viscosity}\ , \quad \kappa(\theta) = \text{heat capacity}$$

(cf. [15], [21] for a discussion of these constitutive assumptions). □

We consider system (1.1)–(1.4) in a bounded domain $\Omega \subset \mathbb{R}^n$ ($n = 2$ resp. $n = 3$) whose boundary $\partial\Omega$ is assumed to satisfy the following conditions:

$$\partial\Omega \text{ is Lipschitzian,}$$
$$\partial\Omega = \Gamma_0 \cup \Gamma_1 \ \ (\Gamma_0 \cap \Gamma_1 = \emptyset, \ \ \Gamma_0 \text{ relatively open}).$$

With these assumptions, the boundary conditions upon $\mathbf{u}$ and θ we are going to consider, are as follows:

$$\mathbf{u} = 0 \qquad \text{on } \partial\Omega, \tag{1.5}$$

$$\theta = \theta_0 \qquad \text{on } \Gamma_0, \tag{1.6}$$

$$q \cdot \nu = 0 \qquad \text{on } \Gamma_1. \tag{1.7}$$

Here θ_0 is a given function, while $\nu = (\nu_1, \dots, \nu_n)$ denotes the unit outward normal with respect to Ω along $\partial\Omega$.

Boundary condition (1.5) means that the fluid adheres to the boundary $\partial\Omega$; conditions (1.6), (1.7) model the situation where the temperature of the fluid is prescribed on Γ_0, while there is no heat flux through Γ_1. □

Various modifications of problem (1.1)–(1.7) have been studied in recent time. First, Morimoto [12], [13] proved the existence of weak solutions to the stationary Boussinesq equations (with $\mu = \text{const} > 0, \kappa = \text{const} > 0$). These equations can be formally obtained by replacing f in (1.2) by a term depending linearly on θ, and by neglecting $\operatorname{tr}(SD) = \mu D_{ij}(\mathbf{u})D_{ij}(\mathbf{u})$ in (1.3). This modification of (1.3) simplifies considerably proving the existence of weak solutions to (1.1)–(1.7). General outflow boundary conditions for the stationary Boussinesq equations are studied by the same author in [14]. A uniqueness result for weak solutions to the stationary Boussinesq equations with temperature dependent viscosity $\mu = \mu(\theta)$ has been proved by Goncharova [8].

With $\mu = \text{const} > 0$, $\kappa = \text{const} > 0$, f in (1.2) replaced by a nonlinear term involving θ, and $\Gamma_1 = \emptyset$, Łukaszewicz [11] proved the existence of a strong solution to (1.1)–(1.6). The proof is based on estimates for strong solutions to the Stokes system.

In [2], Baranger and Mikelič studied (1.1)–(1.3) with constitutive assumptions

$$S = \mu(\theta, |D|)D, \quad q = -\kappa\nabla\theta \quad (\kappa = \text{const} > 0),$$

where the conditions upon $\mu(\theta, \cdot)$ make S to include the well-known power-law model of non-Newtonian fluid mechanics as special case. The authors present a proof of existence of a weak solution to (1.1)–(1.3), (1.5), (1.6) (i.e., $\Gamma_1 = \emptyset$) for the case when the term $(\mathbf{u} \cdot \nabla)\mathbf{u}$ is neglected in (1.2). An important step of their proof consists of solving (1.3), (1.6) by decomposing $\theta = \theta_1 + \theta_2$ where θ_1 is determined by a linear elliptic problem, whereas θ_2 is determined by a weakly nonlinear elliptic problem.

The constitutive assumptions

$$S = \tau(D), \quad q = -\kappa \nabla\theta \quad (\kappa = \text{ const } > 0)$$

are considered in Roubíček [18], where the conditions upon the tensor-valued function τ are motivated by the power law model. In that paper, source terms depending on θ, are added to the right-hand side of (1.2) and (1.3), respectively. An existence theorem for weak solutions to (1.1)–(1.3) under rather general constitutive assumptions on S and q has been established by Consiglieri [5].

In [2], [11] and [18] the Schauder fixed point theorem is used for proving the existence of a weak solution to the boundary value problem under consideration. All the papers mentioned above, are concerned with the special case $\gamma \equiv 1$.

Under the constitutive assumptions

$$S = \mu(\theta)a(D), \quad q = -\kappa(\theta)b(\nabla\theta),$$

Consiglieri and Shilkin [6] proved the existence of a weak solution $\{\mathbf{u}, \theta\}$ to (1.1)–(1.3) where $\mathbf{u}$ possesses locally integrable second-order derivatives. From this result the local Hölder continuity of $\nabla\mathbf{u}$ and $\nabla\theta$ is then deduced in case $n = 2$ (note that [12] contains an interior differentiability result for weak solutions to the stationary Boussinesq equations). Finally, with

$$S = \mu(\theta)D, \quad q = -\nabla\theta,$$

Shilkin [20] proved a result on the partial Hölder continuity of weak solutions to (1.1)–(1.3) for the case $n = 3$. □

The aim of the present paper is to prove the existence of a weak solution $\{\mathbf{u}, \theta\}$ to (1.1)–(1.7) for variable κ and under very mild conditions upon the specific heat γ. Our method of proof is based on an approximation argument which does not require any uniqueness result for weak solutions to auxiliary problems. In addition, we obtain an estimate on the integral

$$\int\limits_{\Omega} \frac{|\nabla\theta|^2}{(1 + \theta - \theta_0)^{1+\delta}} dx \qquad (\delta > 0)$$

and a result on the higher integrability of $\nabla\mathbf{u}$ and $\nabla\theta$.

The paper is arranged as follows:

2. Statement of the main result
3. Proof of the main result
   3.1 Existence of an approximate solution $\{\mathbf{u}_\varepsilon, \theta_\varepsilon\}$

2. Statement of the main result

We consider the following conditions upon the data in (1.4), (1.3) and (1.2):

$$\mu \in C(\mathbb{R}),\ 0 < \mu_1 \le \mu(\xi) \le \mu_2 < +\infty \quad \forall \xi \in \mathbb{R} \quad (\mu_1, \mu_2 = \text{const}), \tag{2.1}$$

$$\kappa \in C(\mathbb{R}),\ 0 < \kappa_1 \le \kappa(\xi) \le \kappa_2 < +\infty \quad \forall \xi \in \mathbb{R} \quad (\kappa_1, \kappa_2 = \text{const}), \tag{2.2}$$

$$\gamma \in C(\mathbb{R}),\ |\gamma(\xi)| \le \gamma_0 < +\infty \quad \forall \xi \in \mathbb{R} \quad (\gamma_0 = \text{const}), \tag{2.3}$$

$$f \in [L^s(\Omega)]^n \quad (s > 1 \text{ if } n = 2,\ s = \frac{6}{5} \text{ if } n = 3). \tag{2.4}$$

Next, let $W^{1,p}(\Omega)$ $(1 \le p < +\infty)$ denote the usual Sobolev space. Define

$$W_0^{1,p}(\Omega) := \{\varphi \in W^{1,p}(\Omega) \mid \varphi = 0 \text{ a.e. on } \partial\Omega\},$$
$$\mathbf{D}_0^{1,p}(\Omega) := \{\mathbf{v} \in [W_0^{1,p}(\Omega)]^n \mid \operatorname{div} \mathbf{v} = 0 \text{ a.e. in } \Omega\}$$

and

$$W_{\Gamma_0}^{1,p}(\Omega) := \{\varphi \in W^{1,p}(\Omega) \mid \varphi = 0 \text{ a.e. on } \Gamma_0\}.$$

Finally, for $\mathbf{u}, \mathbf{v} \in [W^{1,2}(\Omega)]^n$ we define

$$D(\mathbf{u}, \mathbf{v}) := D_{ij}(\mathbf{u}) D_{ij}(\mathbf{v}) = \frac{1}{4}\left(\frac{\partial u_i}{\partial x_j} + \frac{\partial u_j}{\partial x_i}\right)\left(\frac{\partial v_i}{\partial x_j} + \frac{\partial v_j}{\partial x_i}\right)$$

Definition. *Assume* (2.1)–(2.4), *and* $\theta_0 \in W^{1,2}(\Omega)$. *The pair* $\{\mathbf{u}, \theta\} \in \mathbf{D}_0^{1,2}(\Omega) \times W^{1,p}(\Omega)$
$(1 < p < 2$ if $n = 2$, $\frac{6}{5} \le p < \frac{3}{2}$ if $n = 3)$ *is called a weak solution to* (1.1) - (1.7) *if*

$$\int_\Omega \mu(\theta) D(\mathbf{u}, \mathbf{v}) dx + \int_\Omega u_i \frac{\partial u_j}{\partial x_i} v_j dx = \int_\Omega f_i v_i dx \quad \forall\, \mathbf{v} \in \mathbf{D}_0^{1,2}(\Omega), \tag{2.5}$$

$$\left\{\begin{array}{l} \displaystyle\int_\Omega \kappa(\theta) \nabla\theta \cdot \nabla\varphi dx + \int_\Omega \gamma(\theta) u_i \frac{\partial \theta}{\partial x_i} \varphi dx = \int_\Omega \mu(\theta) D(\mathbf{u}, \mathbf{u}) \varphi dx \\ \forall\, \varphi \in W_{\Gamma_0}^{1,q}(\Omega), \end{array}\right. \tag{2.6}$$

$$\theta = \theta_0 \quad \textit{a.e. on} \quad \Gamma_0. \tag{2.7}$$

The main result of our paper is the following

Theorem 1. *Assume* (2.1)–(2.4), *and*

$$\theta_0 = \text{const}\ ^{1)}.$$

Then there exists $\{\mathbf{u}, \theta\} \in \mathbf{D}_0^{1,2}(\Omega) \times \Big(\bigcap_{p\in[1,\frac{n}{n-1}[} W^{1,p}(\Omega)\Big)$ *which satisfies* (2.5)–(2.7). *In addition, there holds*

$$\theta \geq \theta_0 \quad a.e.\ in \quad \Omega, \tag{2.8}$$

$$\int_\Omega |\nabla \mathbf{u}|^2 dx \leq c_1 \|f\|^2_{[L^s(\Omega)]^n}, \tag{2.9}$$

$$\int_\Omega \frac{|\nabla\theta|^2}{(1+\theta-\theta_0)^{1+\delta}} dx \leq \frac{c_2}{\delta}\|f\|^2_{[L^s(\Omega)]^n} \quad \forall\delta > 0, \tag{2.10}$$

$$\begin{cases} \displaystyle\int_\Omega |\nabla\theta|^p dx \leq \Big(\|f\|^p_{[L^s(\Omega)]^n} + \|f\|^{2p/(1-\delta)}_{[L^s(\Omega)]^n}\Big) \\ \forall\, p\in[1,2[, \quad \forall\delta\in]0,1[\qquad \text{if} \quad n=2, \\ \forall\, p\in[1,\frac{3}{2}[, \quad \delta = \dfrac{3-2p}{3-p} \qquad \text{if} \quad n=3 \end{cases} \tag{2.11}$$

$(c_i = \text{const}; i = 1,2,3)$.

Remark 1. The existence of weak solutions in

$$\bigcap_{p\in[1,\frac{n}{n-1}[} W^{1,p}(\Omega) \quad (n\geq 2)$$

to linear, uniformly elliptic equations with L^∞-coefficients, right-hand side in L^1 and zero boundary conditions has been proved in Stampacchia [22]. A very general theory of linear and semi-linear elliptic boundary value problems involving measures as right-hand sides, has been developed by Amann and Quittner [1]. Existence results for very large classes of nonlinear elliptic equations with right-hand side in L^1, are proved by Rakotoson [17] and Boccardo and Gallouët [3], [4].

The coefficient $\kappa(\theta)$ and the presence of the term $\gamma(\theta)u_i\dfrac{\partial\theta}{\partial x_i}$ in (1.3) prevent proving the existence of a weak solution to this equation (cf. (2.6)) by appealing to the existence results in [22], [1] and [3], [4], [17]. Nevertheless, using techniques developed in [3], [4], [17] we are able to establish integral estimates on $\nabla\theta$. For this, the key observation is that the integral

$$\int_\Omega \gamma(\theta)u_i\frac{\partial\theta}{\partial x_i}\varphi dx$$

vanishes for appropriate functions φ. □

---

1) This assumption we made for the sake of technical simplicity of the calculations in Sect. 3.3.

3. Proof of the main result

3.1. *Existence of an Approximate Solution* $\{\mathbf{u}_\varepsilon, \theta_\varepsilon\}$

We begin with proving the following existence result.

Proposition 1. *Suppose that the conditions of Theorem 1 hold. For every $\varepsilon > 0$, there exists $\{\mathbf{u}_\varepsilon, \theta_\varepsilon\} \in \mathbf{D}_0^{1,2}(\Omega) \times W^{1,2}(\Omega)$ such that*

$$\int_\Omega \mu(\theta_\varepsilon) D(\mathbf{u}_\varepsilon, \mathbf{v})dx + \int_\Omega u_{\varepsilon i} \frac{\partial u_{\varepsilon j}}{\partial x_i} v_j dx = \int_\Omega f_i v_i dx \quad \forall\, \mathbf{v} \in \mathbf{D}_0^{1,2}(\Omega), \tag{3.1}$$

$$\begin{cases} \displaystyle\int_\Omega \kappa(\theta_\varepsilon)\nabla\theta_\varepsilon \cdot \nabla\varphi dx + \int_\Omega \gamma(\theta_\varepsilon) u_{\varepsilon i} \frac{\partial \theta_\varepsilon}{\partial x_i}\varphi dx = \int_\Omega \mu(\theta_\varepsilon) \frac{D(\mathbf{u}_\varepsilon, \mathbf{u}_\varepsilon)}{1 + \varepsilon D(\mathbf{u}_\varepsilon, \mathbf{u}_\varepsilon)} \varphi dx \\ \forall\, \varphi \in W_{\Gamma_0}^{1,2}(\Omega), \end{cases} \tag{3.2}$$

$$\theta_\varepsilon = \theta_0 \quad a.e.\ on\ \Gamma_0. \tag{3.3}$$

Proof. By Korn's inequality, $[W_0^{1,2}(\Omega)]^n$ is a Hilbert space with respect to the scalar product

$$\mathbf{u}, \mathbf{v} \mapsto \int_\Omega D(\mathbf{u}, \mathbf{v})dx.$$

We define $\mathcal{H} := \mathbf{D}_0^{1,2}(\Omega) \times W_{\Gamma_0}^{1,2}(\Omega)$. By $\{\mathbf{u}, \theta\}, \{\mathbf{v}, \varphi\}, \dots$ we denote the elements of $\mathcal{H}$. Clearly, $\mathcal{H}$ is a Hilbert space with respect to the scalar product

$$(\{\mathbf{u}, \theta\}, \{\mathbf{v}, \varphi\}) := \int_\Omega (D(\mathbf{u}, \mathbf{v}) + \nabla\theta \cdot \nabla\varphi)dx.$$

The associated norm is

$$\|\{\mathbf{u}, \theta\}\| = \left(\int_\Omega (D(\mathbf{u}, \mathbf{u}) + |\nabla\theta|^2)dx \right)^{1/2}.$$

We furnish $(\mathbf{D}_0^{1,2}(\Omega))^* \times (W_{\Gamma_0}^{1,2}(\Omega))^*$ [2)] with the norm

$$\left(\|\mathbf{u}^*\|^2_{(\mathbf{D}_0^{1,2}(\Omega))^*} + \|\theta^*\|^2_{(W_{\Gamma_0}^{1,2}(\Omega))^*} \right)^{1/2}.$$

Then $\mathcal{H}^*$ is linearly isometric to $(\mathbf{D}_0^{1,2}(\Omega))^* \times (W_{\Gamma_0}^{1,2}(\Omega))^*$. Identifying both spaces, then for $\{\mathbf{u}^*, \theta^*\} \in \mathcal{H}^*$ and $\{\mathbf{u}, \theta\} \in \mathcal{H}$ we may write

$$\langle \{\mathbf{u}^*, \theta^*\}, \{\mathbf{u}, \theta\} \rangle = \langle \mathbf{u}^*, \mathbf{u} \rangle + \langle \theta^*, \theta \rangle.$$

---

2) Let X be a normed linear space. We denote by X^* the dual of X, by $\|\cdot\|_{X^*}$ the dual norm on X^* and by $\langle x^*, x \rangle$ the value of $x^* \in X^*$ at $x \in X$.

Let $\varepsilon > 0$. The assumptions (2.1)–(2.3) permit to define a nonlinear mapping $\mathcal{A} : \mathcal{H} \to \mathcal{H}^*$ by

$$\begin{aligned}\langle \mathcal{A}(\{\mathbf{u},\theta\}), \{\mathbf{v},\varphi\}\rangle \quad := \quad & \int\limits_\Omega \mu(\theta+\theta_0) D(\mathbf{u},\mathbf{v}) dx + \int\limits_\Omega u_i \frac{\partial u_j}{\partial x_i} v_j dx \\ & + \int\limits_\Omega \kappa(\theta+\theta_0)\nabla\theta\cdot\nabla\varphi dx + \int\limits_\Omega \gamma(\theta+\theta_0) u_i \frac{\partial \theta}{\partial x_i}\varphi dx \\ & - \int\limits_\Omega \mu(\theta+\theta_0)\frac{D(\mathbf{u},\mathbf{u})}{1+\varepsilon D(\mathbf{u},\mathbf{u})}\varphi dx, \ \{\mathbf{u},\theta\},\{\mathbf{v},\phi\}\in\mathcal{H}.\end{aligned}$$

We show that $\mathcal{A}$ is *coercive*, i.e.,

$$\frac{1}{\|\{\mathbf{u},\theta\}\|}\langle \mathcal{A}(\{\mathbf{u},\theta\}), \{\mathbf{u},\theta\}\rangle \mapsto +\infty \quad \text{as} \quad \|\{\mathbf{u},\theta\}\| \to \infty. \tag{3.4}$$

Indeed, we have

$$\int\limits_\Omega u_i \frac{\partial u_j}{\partial x_i} u_j dx = 0 \quad \forall \mathbf{u} \in \mathbf{D}_0^{1,2}(\Omega), \tag{3.5$_1$}$$

$$\int\limits_\Omega \gamma(\theta+\theta_0) u_i \frac{\partial\theta}{\partial x_i}\theta dx = 0 \quad \forall \mathbf{u} \in \mathbf{D}_0^{1,2}(\Omega), \quad \forall\, \theta \in W^{1,2}(\Omega). \tag{3.5$_2$}$$

Here (3.5$_1$) is readily seen by integration by parts. To verify (3.5$_2$), define

$$\Phi(t) := \int\limits_0^t s\gamma(s+\theta_0) ds, \quad t\in\mathbb{R}.$$

Then

$$\Phi \in C^1(\mathbb{R}), \quad |\Phi'(t)| \le \gamma_0 |t| \quad \forall t\in\mathbb{R}$$

(cf. (2.3)), and for every $\theta \in W^{1,2}(\Omega)$

$$\frac{\partial}{\partial x_i}\Phi(\theta) = \theta\gamma(\theta+\theta_0)\frac{\partial\theta}{\partial x_i} \quad \text{a.e. in } \ \Omega \quad (i = 1,\dots,n),$$

$$\Phi(\theta) \in W^{1,r}(\Omega) \quad (r \in [1,2[\ \text{ if } \ n=2, \ r = \frac{3}{2} \ \text{ if } \ n = 3).$$

Thus, by integration by parts,

$$\int\limits_\Omega \gamma(\theta+\theta_0) u_i \frac{\partial\theta}{\partial x_i}\theta dx = \int\limits_\Omega u_i \frac{\partial}{\partial x_i}\Phi(\theta) dx = 0.$$

We obtain

$$\langle \mathcal{A}(\{\mathbf{u},\theta\}), \{\mathbf{u},\theta\}\rangle \ge \mu_1 \int\limits_\Omega D(\mathbf{u},\mathbf{u}) dx + \kappa_1 \int\limits_\Omega |\nabla\theta|^2 dx - \frac{\mu_2}{\varepsilon}\int\limits_\Omega |\theta| dx$$

for all $\{\mathbf{u},\theta\} \in \mathcal{H}$. Whence (3.4).

Next, we show that $\mathcal{A}$ is *pseudo-monotone, i.e.,*

1° *$\mathcal{A}$ maps bounded sets in $\mathcal{H}$ into bounded sets in $\mathcal{H}^*$,*

2° *for every sequence $(\{\mathbf{u}_k, \theta_k\}) \subset \mathcal{H}$ with*

$$\{\mathbf{u}_k, \theta_k\} \to \{\mathbf{u}, \theta\} \text{ weakly in } \mathcal{H},$$
$$\limsup \langle \mathcal{A}(\{\mathbf{u}_k, \theta_k\}), \{\mathbf{u}_k, \theta_k\} - \{\mathbf{u}, \theta\} \rangle \leq 0$$

there exists a subsequence (not relabelled) such that

$$\liminf \langle \mathcal{A}(\{\mathbf{u}_k, \theta_k\}), \{\mathbf{u}_k, \theta_k\} - \{\mathbf{v}, \varphi\} \rangle \geq \langle \mathcal{A}(\{\mathbf{u}, \theta\}), \{\mathbf{u}, \theta\} - \{\mathbf{v}, \varphi\} \rangle$$

for all $\{\mathbf{v}, \varphi\} \in \mathcal{H}$.

Indeed, 1° is easily deduced from the definition of $\mathcal{A}$. To establish 2°, we note that there exists a subsequence of $(\{\mathbf{u}_k, \theta_k\})$ (not relabelled) such that

$$u_{ki} \to u_i, \ \theta_k \to \theta \ \text{ strongly in } \ L^4(\Omega),$$
$$\theta_k \to \theta \ \text{ a.e. in } \ \Omega$$

as $k \to +\infty$ $(i = 1, \ldots, n)$.

Observing that

$$\begin{aligned}
&\int_\Omega \mu(\theta_k + \theta_0) D(\mathbf{u}_k - \mathbf{u}, \mathbf{u}_k - \mathbf{u}) dx + \int_\Omega \kappa(\theta_k + \theta_0) |\nabla(\theta_k - \theta)|^2 dx \\
&\quad = \langle \mathcal{A}(\{\mathbf{u}_k, \theta_k\}), \{\mathbf{u}_k, \theta_k\} - \{\mathbf{u}, \theta\} \rangle \\
&\qquad - \int_\Omega \mu(\theta_k + \theta_0) D(\mathbf{u}, \mathbf{u}_k - \mathbf{u}) dx - \int_\Omega \kappa(\theta_k + \theta_0) \nabla\theta \cdot \nabla(\theta_k - \theta) dx \\
&\qquad - \int_\Omega u_{ki} \frac{\partial u_{kj}}{\partial x_i} \left(u_{kj} - u_j \right) dx - \int_\Omega \gamma(\theta_k + \theta_0) u_{ki} \frac{\partial \theta_k}{\partial x_i} (\theta_k - \theta) dx \\
&\qquad + \int_\Omega \mu(\theta_k + \theta_0) \frac{D(\mathbf{u}_k, \mathbf{u}_k)}{1 + \varepsilon D(\mathbf{u}_k, \mathbf{u}_k)} (\theta_k - \theta) dx,
\end{aligned}$$

we find

$$\begin{aligned}
&\min\{\mu_1, \kappa_1\} \limsup \int_\Omega (D(\mathbf{u}_k - \mathbf{u}, \mathbf{u}_k - \mathbf{u}) + |\nabla(\theta_k - \theta)|^2) dx \\
&\quad \leq \limsup \langle \mathcal{A}(\{\mathbf{u}_k, \theta_k\}), \{\mathbf{u}_k, \theta_k\} - \{\mathbf{u}, \theta\} \rangle \\
&\quad \leq 0.
\end{aligned}$$

Hence

$$\mathbf{u}_k \to \mathbf{u} \text{ strongly in } [W^{1,2}(\Omega)]^n, \ \theta_k \to \theta \text{ strongly in } W^{1,2}(\Omega),$$

and (again by passing to a subsequence if necessary)

$$D(\mathbf{u}_k, \mathbf{u}_k) \to D(\mathbf{u}, \mathbf{u}) \quad \text{a.e. in} \quad \Omega$$

as $k \to \infty$. Thus, for every $\varphi \in W^{1,2}(\Omega)$,

$$\lim \int_\Omega \mu(\theta_k + \theta_0) \frac{D(\mathbf{u}_k, \mathbf{u}_k)}{1 + \varepsilon D(\mathbf{u}_k, \mathbf{u}_k)} (\theta_k - \varphi) dx = \int_\Omega \mu(\theta + \theta_0) \frac{D(\mathbf{u}, \mathbf{u})}{1 + \varepsilon D(\mathbf{u}, \mathbf{u})} (\theta - \varphi) dx.$$

It follows

$$\liminf \langle \mathcal{A}(\{\mathbf{u}_k, \theta_k\}), \{\mathbf{u}_k, \theta_k\} - \{\mathbf{v}, \varphi\} \rangle \geq \langle \mathcal{A}(\{\mathbf{u}, \theta\}), \{\mathbf{u}, \theta\} - \{\mathbf{v}, \varphi\} \rangle$$

for all $\{\mathbf{v}, \varphi\} \in \mathcal{H}$. Whence 2°.

From the theory of pseudo-monotone operators we obtain: for every $\varepsilon > 0$ there exists $\{\mathbf{u}_\varepsilon, \hat{\theta}_\varepsilon\} \in \mathcal{H}$ such that

$$\mathcal{A}(\{\mathbf{u}_\varepsilon, \hat{\theta}_\varepsilon\}) = \{f, 0\}$$

(cf., e.g., Lions [10]).

The pair $\{\mathbf{u}_\varepsilon, \theta_\varepsilon\}$, $\theta_\varepsilon := \hat{\theta}_\varepsilon + \theta_0$, satisfies (3.1)–(3.3). □

3.2. *A priori-estimates on* $\nabla \mathbf{u}_\varepsilon$

We insert $\mathbf{v} = \mathbf{u}_\varepsilon$ in (3.1). By (3.5_1),

$$\int_\Omega \mu(\theta_\varepsilon) D(\mathbf{u}_\varepsilon, \mathbf{u}_\varepsilon) dx = \int_\Omega f_i u_{\varepsilon i} dx.$$

Thus

$$\int_\Omega |\nabla \mathbf{u}_\varepsilon|^2 dx \leq c_1 \|f\|^2_{[L^s(\Omega)]^n} \quad \forall \varepsilon > 0. \tag{3.6}$$

□

3.3. *A priori estimates on* $\nabla \theta_\varepsilon$

For notational simplicity, throughout this section we write $\mathbf{u}$ and θ in place of $\mathbf{u}_\varepsilon$ and θ_ε, respectively.

Inserting $\varphi = (\theta_\varepsilon - \theta_0)^-$ ($\xi^- := \min\{\xi, 0\}$, $\xi \in \mathbb{R}$) in (3.2) gives

$$\begin{aligned} &\int_\Omega \kappa(\theta) |\nabla(\theta - \theta_0)^-|^2 dx + \int_\Omega \gamma(\theta) u_i \frac{\partial \theta}{\partial x_i} (\theta - \theta_0)^- dx \\ &\quad = \int_\Omega \mu(\theta) \frac{D(\mathbf{u}, \mathbf{u})}{1 + \varepsilon D(\mathbf{u}, \mathbf{u})} (\theta - \theta_0)^- dx \\ &\quad \leq 0. \end{aligned}$$

Recalling that $\theta = \hat{\theta} + \theta_0$, we have

$$\begin{aligned}\int_\Omega \gamma(\theta) u_i \frac{\partial \theta}{\partial x_i}(\theta - \theta_0)^- dx &= \int_{\{x|\hat{\theta}(x)<0\}} \gamma(\hat{\theta}+\theta_0) u_i \frac{\partial \hat{\theta}}{\partial x_i} \hat{\theta} dx \\ &= \int_{\{x|\hat{\theta}(x)<0\}} u_i \frac{\partial}{\partial x_i} \Phi(\hat{\theta}) dx \\ &= \int_\Omega u_i \frac{\partial}{\partial x_i} \Phi(\hat{\theta}^-) dx \\ &= 0\end{aligned}$$

(cf. the proof of (3.5_2)). It follows

$$\int_\Omega |\nabla(\theta - \theta_0)^-|^2 dx = 0.$$

Thus

$$\theta \geq \theta_0 \quad \text{a.e. in } \Omega. \tag{3.7}$$

To prove estimates on $\nabla\theta$, we make use of arguments of Rakotoson [17]. Similar techniques are developed in Boccardo and Gallouët [3], [4].

Let $\delta > 0$. The function $\varphi = 1 - \dfrac{1}{(1+\theta-\theta_0)^\delta}$ is admissible in (3.2) (observe (3.7)). We obtain

$$\begin{aligned}\delta \int_\Omega \kappa(\theta) \frac{|\nabla\theta|^2}{(1+\theta-\theta_0)^{1+\delta}} dx + \int_\Omega \gamma(\theta) u_i \frac{\partial\theta}{\partial x_i}\left(1 - \frac{1}{(1+\theta-\theta_0)^\delta}\right) dx \\ = \int_\Omega \mu(\theta) \frac{D(\mathbf{u},\mathbf{u})}{1+\varepsilon D(\mathbf{u},\mathbf{u})}\left(1 - \frac{1}{(1+\theta-\theta_0)^\delta}\right) dx.\end{aligned}$$

Here the second integral on the left-hand side vanishes. Indeed, defining

$$\Psi(t) := \int_0^t \gamma(s+\theta_0)\Big(1 - \frac{1}{(1+s)^\delta}\Big) ds, \quad t \in [0, +\infty[,$$

we find

$$\int_\Omega \gamma(\theta) u_i \frac{\partial\theta}{\partial x_i}\left(1 - \frac{1}{(1+\theta-\theta_0)^\delta}\right) dx = \int_\Omega u_i \frac{\partial}{\partial x_i} \Psi(\theta - \theta_0) dx = 0$$

(cf. the proof of (3.5_2)). Observing (3.6) it follows

$$\int_\Omega \frac{|\nabla\theta|^2}{(1+\theta-\theta_0)^{1+\delta}}dx \le \frac{c_2}{\delta}\|f\|^2_{[L^s(\Omega)]^n}, \tag{3.8}$$

where the constant c_2 is independent of both ε and δ.

Now, let

$$\begin{aligned} &p \in]1,2[,\ \delta \in]0,1[&& \text{if} \quad n=2,\\ &p \in]1,\tfrac{3}{2}[,\ \delta = \frac{3-2p}{3-p} && \text{if} \quad n=3. \end{aligned}$$

Then

$$\begin{aligned} &\frac{p(1+\delta)}{2-p} \in]1,+\infty[&& \text{if} \quad n=2,\\ &\frac{p(1+\delta)}{2-p} = \frac{3p}{3-p} && \text{if} \quad n=3. \end{aligned}$$

Combining Hölder's inequality, estimate (3.8) and Sobolev's embedding theorem we obtain

$$\begin{aligned} \int_\Omega |\nabla\theta|^p dx &\le \left(\int_\Omega \frac{|\nabla\theta|^2}{(1+\theta-\theta_0)^{1+\delta}}dx\right)^{p/2}\left(\int_\Omega (1+\theta-\theta_0)^{p(1+\delta)/(2-p)}dx\right)^{(2-p)/2}\\ &\le \left(\frac{c_2}{\delta}\right)^{p/2}\|f\|^p_{[L^s(\Omega)]^n}\left(\int_\Omega (1+\theta-\theta_0)^{p(1+\delta)/(2-p)}dx\right)^{(2-p)/2}\\ &\le 2^{(2-p)/2}\left(\frac{c_2}{\delta}\right)^{p/2}\|f\|^p_{[L^s(\Omega)]^n}\times\\ &\quad\times\left\{(\text{meas }\Omega)^{(2-p)/2} + C_0\left(\int_\Omega |\nabla\theta|^p dx\right)^{(1+\delta)/2}\right\} \end{aligned}$$

(here C_0 denotes the constant of the embedding $W^{1,p}_{\Gamma_0}(\Omega) \subset L^{p(1+\delta)/(2-p)}(\Omega)$; note that if $n=2$, then $C_0 \to +\infty$ as $p \to 2$). Since $\dfrac{1+\delta}{2} < 1$, we finally obtain by the aid of Young's inequality

$$\int_\Omega |\nabla\theta|^p dx \le c_3\left(\|f\|^p_{[L^2(\Omega)]^n} + \|f\|^{2p/(1-\delta)}_{[L^s(\Omega)]^n}\right) \tag{3.9}$$

with $c_3 = \text{const}$ independent of ε. □

3.4. *Passage to limit* $\varepsilon \to 0$

By passing to a subsequence, from (3.6) and (3.9) we conclude

$$\mathbf{u}_\varepsilon \to \mathbf{u} \text{ weakly in } [W^{1,2}(\Omega)]^n,\ \theta_\varepsilon \to \theta \text{ weakly in } W^{1,p}(\Omega), \tag{3.10}$$

and (again by passing to a subsequence if necessary)

$$\mathbf{u}_\varepsilon \to \mathbf{u} \text{ strongly in } [L^4(\Omega)]^n, \ \theta_\varepsilon \to \theta \text{ strongly in } L^1(\Omega), \tag{3.11}$$

$$\theta_\varepsilon \to \theta \text{ a.e. in } \Omega \tag{3.12}$$

as $\varepsilon \to 0$.

Clearly, $\mathbf{u} \in \mathbf{D}_0^{1,2}(\Omega)$, $\theta = \theta_0$ a.e. on Γ_0 and $\theta \geq \theta_0$ a.e. in Ω. Using (3.10), (3.11) and (3.12), the passage to limit $\varepsilon \to 0$ in (3.1) easily gives (2.5). Estimates (2.9) and (2.11) are readily deduced from (3.6) and (3.9), respectively.

To carry out the passage to limit $\varepsilon \to 0$ in (3.2), we first note that from (3.1) it follows that

$$\begin{aligned}\int_\Omega \mu(\theta_\varepsilon) D(\mathbf{u}_\varepsilon - \mathbf{u}, \mathbf{u}_\varepsilon - \mathbf{u}) dx &= -\int_\Omega \mu(\theta_\varepsilon) D(\mathbf{u}, \mathbf{u}_\varepsilon - \mathbf{u}) dx \\ &\quad - \int_\Omega u_{\varepsilon i} \frac{\partial u_{\varepsilon j}}{\partial x_i}(u_{\varepsilon j} - u_j) dx + \int_\Omega f_i(u_{\varepsilon i} - u_i) dx.\end{aligned}$$

Hence, by (3.10), (3.11) and (3.12),

$$\lim_{\varepsilon \to 0} \int_\Omega \mu(\theta_\varepsilon) D(\mathbf{u}_\varepsilon - \mathbf{u}, \mathbf{u}_\varepsilon - \mathbf{u}) dx = 0. \tag{3.13}$$

We obtain $\mathbf{u}_\varepsilon \to \mathbf{u}$ strongly in $[W^{1,2}(\Omega)]^n$, and then (for a subsequence)

$$D(\mathbf{u}_\varepsilon, \mathbf{u}_\varepsilon) \to D(\mathbf{u}, \mathbf{u}) \text{ a.e. in } \Omega \tag{3.14}$$

as $\varepsilon \to 0$.

Taking into account the elementary identity

$$\frac{|a|^2}{1 + \varepsilon |a|^2} - |b|^2 = \frac{(a - b, a + b)}{1 + \varepsilon |a|^2} - \frac{\varepsilon |a|^2 |b|^2}{1 + \varepsilon |a|^2}, \ a, b \in \mathbb{R}^n,$$

we obtain by the aid of (3.12), (3.13) and (3.14)

$$\lim_{\varepsilon \to 0} \int_\Omega \left| \mu(\theta_\varepsilon) \frac{D(\mathbf{u}_\varepsilon, \mathbf{u}_\varepsilon)}{1 + \varepsilon D(\mathbf{u}_\varepsilon, \mathbf{u}_\varepsilon)} - \mu(\theta) D(\mathbf{u}, \mathbf{u}) \right| dx = 0.$$

Now the passage to limit $\varepsilon \to 0$ in (3.2) with $\varphi \in W^{1,p'}_{\Gamma_0}(\Omega)$ $\left(1 < p < \frac{n}{n-1}\right)$ is readily carried out. This gives (2.6).

Finally, from (3.8), (3.10) and (3.12) we conclude

$$\frac{\dfrac{\partial \theta_\varepsilon}{\partial x_i}}{(1 + \theta_\varepsilon - \theta_0)^{(1+\delta)/2}} \longrightarrow \frac{\dfrac{\partial \theta}{\partial x_i}}{(1 + \theta - \theta_0)^{(1+\delta)/2}} \quad \text{weakly in } L^2(\Omega)$$

as $\varepsilon \to 0$ $(i = 1, \dots, n)$. It follows

$$\int\limits_{\Omega} \frac{|\nabla\theta|^2}{(1+\theta-\theta_0)^{1+\delta}} dx \leq \liminf \int\limits_{\Omega} \frac{|\nabla\theta_\varepsilon|^2}{(1+\theta_\varepsilon-\theta_0)^{1+\delta}} dx$$

$$\leq \frac{c_2}{\delta} \|f\|^2_{[L^s(\Omega)]^n}. \qquad \square$$

4. Higher integrability

In this section, we consider system (1.1)–(1.7) for the case $n = 3$. For what follows, let

$$f \in [L^s(\Omega)]^3 \quad \left(s > \frac{6}{5}\right) \tag{4.1}$$

(cf. (2.4)).

4.1. *Higher integrability of* $\nabla \mathbf{u}_\varepsilon$

Let (2.1)–(2.3) and (4.1) be satisfied, and let $\theta_0 \in W^{1,2}(\Omega)$. Let $\{\mathbf{u}, \theta\} \in \mathbf{D}^{1,2}(\Omega) \times W^{1,p}(\Omega) \left(\frac{6}{5} \leq p < \frac{3}{2}\right)$ be a weak solution to (1.1)–(1.7) (cf. Sect. 2).

From Wolff [23] it follows:

$$\exists\, t > 2 : \nabla\mathbf{u} \in [L^t(\Omega)]^9, \ \|\nabla\mathbf{u}\|_{[L^t(\Omega)]^9} \leq K_0\Big(\|f\|_{[L^s(\Omega)]^3}\Big), \tag{4.2}$$

where K_0 is a continuous increasing function defined on $[0, +\infty[$. This result which holds for dimensions $n = 2$ and $n = 3$, is proved by the method of higher integrability by reverse Hölder inequality (cf. Giaquinta [7] for a presentation of this method for weak solutions to nonlinear elliptic systems). We note that this method does not provide an information about the size of t. $\square$

Let (2.1)–(2.3) and (4.1) be satisfied, and let $\theta_0 =$ const. Let $\{\mathbf{u}_\varepsilon, \theta_\varepsilon\} \in \mathbf{D}_0^{1,2}(\Omega) \times W^{1,2}(\Omega)$ $(\varepsilon > 0)$ be as in Prop. 1 (cf. Sect. 3.1).

Again appealing to [23] we obtain:

$$\exists\, t > 2 : \nabla\mathbf{u}_\varepsilon \in [L^t(\Omega)]^9, \ \|\nabla\mathbf{u}_\varepsilon\|_{[L^t(\Omega)]^9} \leq K_0\Big(\|f\|_{[L^s(\Omega)]^3}\Big), \tag{4.2'}$$

with the same function K_0 as above in (4.2). Clearly, neither t nor K_0 depend on ε[3)].

We shall use (4.2′) to establish a result on the higher integrability of $\nabla\theta_\varepsilon$.

[3)] Note that the estimates on the integrals providing (4.2′) depend on μ_1 and μ_2 (cf. (2.1)), but not on $\mu(\theta_\varepsilon)$ itself.

4.2. *Higher integrability of* $\nabla\theta_\varepsilon$

We continue to assume (2.1)–(2.3), (4.1) and $\theta_0 = \text{const}$. Let $\{\mathbf{u}_\varepsilon, \theta_\varepsilon\} \in \mathbf{D}_0^{1,2}(\Omega) \times W^{1,2}(\Omega)$ $(\varepsilon > 0)$ be as in Prop. 1 (cf. Sect. 3.1). The function θ_ε represents a weak solution to the nonlinear elliptic boundary value problem

$$-\frac{\partial}{\partial x_i}\left(\kappa(\theta)\frac{\partial\theta}{\partial x_i}\right) + \gamma(\theta)u_{\varepsilon i}\frac{\partial\theta}{\partial x_i} - \mu(\theta)\frac{D(\mathbf{u}_\varepsilon, \mathbf{u}_\varepsilon)}{1+\varepsilon D(\mathbf{u}_\varepsilon, \mathbf{u}_\varepsilon)} = 0 \text{ in } \Omega,$$

$$\theta = \theta_0 \text{ on } \Gamma_0, \quad \frac{\partial\theta}{\partial\nu} = 0 \text{ on } \Gamma_1,$$

where $\mathbf{u}_\varepsilon$ can be viewed as a given function (cf. (3.2)). From (4.2′) it follows

$$\begin{aligned}\int_\Omega \left(\mu(\theta_\varepsilon)\frac{D(\mathbf{u}_\varepsilon, \mathbf{u}_\varepsilon)}{1+\varepsilon D(\mathbf{u}_\varepsilon, \mathbf{u}_\varepsilon)}\right)^{t/2} dx &\le \mu_2^{t/2}\int_\Omega (D(\mathbf{u}_\varepsilon, \mathbf{u}_\varepsilon))^{t/2}dx \\ &\le \hat{K}_0\left(\|f\|_{[L^s(\Omega)]^3}\right),\end{aligned} \tag{4.3}$$

where $\hat{K}_0$ is a continuous increasing function defined on $[0, +\infty[$.

Proving the higher integrability of $\nabla\theta_\varepsilon$ by the method of reverse Hölder inequality requires $\frac{t}{2} > \frac{6}{5}$ in (4.3) (cf. [7] for local gradient estimates, and Naumann and Wolff [16] for global gradient estimates). However, this method does not provide the inequality $t > \frac{12}{5}$ in (4.2′). Moreover, establishing a global higher integrability result for $\nabla\theta_\varepsilon$ requires additional assumptions on $\bar{\Gamma}_0 \cap \Gamma_1$ (cf. [18]). □

We now prove a higher integrability result for $\nabla\theta_\varepsilon$ by a variant of the gradient estimate technique developed in Sect. 3.3.

Let t be as in (4.2′). We define

$$p_0 := \frac{6(t-1)}{3t-2}.$$

Clearly,

$$\frac{3}{2} < p_0 < 2.$$

We then have the following

Proposition 2. *There holds*

$$\nabla\theta_\varepsilon \in [L^{p_0}(\Omega)]^3, \ \|\nabla\theta_\varepsilon\|_{[L^{p_0}(\Omega)]^3} \le \frac{c_0 t}{t-2}\left(K_0\left(\|f\|_{[L^2(\Omega)]^3}\right)\right)^2,$$

where

c_0 = const *independent of* ε *and* t,
K_0 *according to* (4.2').

Proof. Let $\tau > 0$. The function

$$\varphi := \frac{\theta_\varepsilon - \theta_0}{\tau + (\theta_\varepsilon - \theta_0)^{2/t}}\ ^{4)}$$

is admissible in (3.2). We obtain

$$\int\limits_\Omega \frac{\kappa(\theta_\varepsilon)}{(\tau + (\theta_\varepsilon - \theta_0)^{2/t})^2} \nabla\theta_\varepsilon \cdot \Big(\tau\nabla(\theta_\varepsilon - \theta_0) + \Big(1 - \frac{2}{t}\Big)(\theta_\varepsilon - \theta_0)^{2/t}\nabla(\theta_\varepsilon - \theta_0)\Big)dx$$
$$+ \int\limits_\Omega \gamma(\theta_\varepsilon) u_{\varepsilon i} \frac{\partial\theta_\varepsilon}{\partial x_i} \ \frac{\theta_\varepsilon - \theta_0}{\tau + (\theta_\varepsilon - \theta_0)^{2/t}} dx$$
$$= \int\limits_\Omega \mu(\theta_\varepsilon) \frac{D(\mathbf{u}_\varepsilon, \mathbf{u}_\varepsilon)}{1 + \varepsilon D(\mathbf{u}_\varepsilon, \mathbf{u}_\varepsilon)} \ \frac{\theta_\varepsilon - \theta_0}{\tau + (\theta_\varepsilon - \theta_0)^{2/t}} \ dx. \tag{4.4}$$

Define

$$\Psi_0(\sigma) := \int\limits_0^\sigma \gamma(\eta + \theta_0)\frac{\eta}{\tau + \eta^{2/t}} d\eta, \quad \sigma \in [0, +\infty[$$

(cf. (3.5_2) and Sect. 3.3). Then, by integration by parts,

$$\int\limits_\Omega \gamma(\theta_\varepsilon) u_{\varepsilon i} \frac{\partial\theta_\varepsilon}{\partial x_i} \ \frac{\theta_\varepsilon - \theta_0}{\tau + (\theta_\varepsilon - \theta_0)^{2/t}} \ dx = \int\limits_\Omega u_{\varepsilon i} \frac{\partial}{\partial x_i} \Psi_0(\theta_\varepsilon - \theta_0) dx$$
$$= 0.$$

With the help of (4.2′) from (4.4) it follows

$$\kappa_1\Big(1 - \frac{2}{t}\Big)\int\limits_\Omega \frac{|\nabla\theta_\varepsilon|^2}{\tau + (\theta_\varepsilon - \theta_0)^{2/t}} dx \le \int\limits_\Omega \mu(\theta_\varepsilon)\frac{D(\mathbf{u}_\varepsilon, \mathbf{u}_\varepsilon)}{1 + \varepsilon D(\mathbf{u}_\varepsilon, \mathbf{u}_\varepsilon)} \ \frac{\theta_\varepsilon - \theta_0}{\tau + (\theta_\varepsilon - \theta_0)^{2/t}} \ dx$$
$$\le C_1 \int\limits_\Omega |\nabla \mathbf{u}_\varepsilon|^2 (\theta_\varepsilon - \theta_0)^{(t-2)/t} dx$$
$$\le C_1\Big(K_0\Big(\|f\|_{[L^2(\Omega)]^3}\Big)\Big)^2 \Big(\int\limits_\Omega (\theta_\varepsilon - \theta_0) dx\Big)^{(t-2)/t}$$

---

4) The use of this function in (3.2) has been suggested by J. Wolf.

with $C_1 = \text{const}$ independent of ε. Thus, by Hölder's inequality,

$$\int_\Omega |\nabla\theta_\varepsilon|^{p_0}\,dx \le$$

$$\le \left(\int_\Omega \frac{|\nabla\theta_\varepsilon|^2}{\tau+(\theta_\varepsilon-\theta_0)^{2/t}}\,dx\right)^{p_0/2}\left(\int_\Omega (\tau+(\theta_\varepsilon-\theta_0)^{2/t})^{p_0/(2-p_0)}dx\right)^{(2-p_0)/2}$$

$$\le 2\left(\frac{C_1 t}{\kappa_1(t-2)}\right)^{p_2/2}\left(K_0\Big(\|f\|_{[L^s(\Omega)]^3}\Big)\right)^2\left(\int_\Omega(\theta_\varepsilon-\theta_0)dx\right)^{p_0(t-2)/2t} \tag{4.5}$$

$$\times\left\{\tau(\text{mes }\Omega)^{(2-p_0)/2}+\left(\int_\Omega(\theta_\varepsilon-\theta_0)^{2p_0/t(2-p_0)}dx\right)^{(2-p_0)/2}\right\}.$$

By the definition of p_0,

$$\frac{2p_0}{t(2-p_0)}=\frac{3p_0}{3-p_0}=p_0^* \quad \text{(Sobolev embedding exponent)}.$$

We obtain

$$\int_\Omega(\theta_\varepsilon-\theta_0)dx \le C_2\left(\int_\Omega|\nabla\theta_\varepsilon|^{p_0}dx\right)^{1/p_0},$$

$$\int_\Omega(\theta_\varepsilon-\theta_0)^{p_0^*}dx \le C_3\left(\int_\Omega|\nabla\theta_\varepsilon|^{p_0}dx\right)^{p_0^*/p_0}$$

($C_1, C_2 = \text{const}$). Inserting both estimates into (4.5) and carrying out the passage to limit $\tau\to 0$ gives

$$\int_\Omega|\nabla\theta_\varepsilon|^{p_0}dx$$

$$\le C_4\left(\frac{t}{t-2}\right)^{p_0/2}\left(K_0\Big(\|f\|_{L^s(\Omega)]^3}\Big)\right)^{p_0}\left(\int_\Omega|\nabla\theta_\varepsilon|^{p_0}dx\right)^{(t-2)/2t+p_0^*(2-p_0)/2p_0}.$$

Finally, observing that

$$\frac{t-2}{2t}+\frac{p_0^*(2-p_0)}{2p_0}=\frac{1}{2},$$

we obtain the result claimed. □

The estimates on $\nabla\mathbf{u}_\varepsilon$ and $\nabla\theta_\varepsilon$ obtained in Sect. 3.2 and 3.3, and those in (4.2′) and Prop. 2 provide the following

Theorem 2. *Assume* (2.1)–(2.3), (4.1) *and* $\theta_0 = const.$ *Then there exists* $t > 2$ *such that: There exists a pair* $\{\mathbf{u}, \theta\} \in \mathbf{D}_0^{1,t}(\Omega) \times W^{1,p_0}(\Omega)$ $\left(p_0 = \dfrac{6(t-1)}{3t-2}\right)$ *satisfying* (2.5)–(2.11), *and*

$$\begin{aligned} \|\nabla \mathbf{u}\|_{[L^t(\Omega)]^9} &\leq K_0\Big(\|f\|_{[L^s(\Omega)]^3}\Big), \\ \|\nabla \theta\|_{[L^{p_0}(\Omega)]^3} &\leq \frac{c_0 t}{t-2}\Big(K_0\Big(\|f\|_{[L^s(\Omega)]^3}\Big)\Big)^2. \end{aligned}$$

Remark 2. The higher integrability of $\nabla\theta$ in the case $n = 2$ will be discussed in a forthcoming paper. □

Literatur

[1] AMANN, H.; QUITTNER, P., *Elliptic boundary value problems involving measures: existence, regularity, and multiplicity.* Advances Diff. Equs. 3 (1998), 753–813.

[2] BARANGER, J.; MIKELIČ, A., *Stationary solutions to a quasi-Newtonian flow with viscous heating.* Math. Models Meth. Appl. Sci. 5 (1995), 725–738.

[3] BOCCARDO, L.; GALLOUËT, T., *Nonlinear elliptic equations with right-hand side measures.* Commun. Partial Diff. Equs. 17 (1992), 641–655.

[4] —— ; —— , *Strongly nonlinear elliptic equations having natural growth terms and* L^1 *data.* Nonl. Analysis, Theory, Meth. & Appl. 19 (1992), 572–579.

[5] CONSIGLIERI, L., *Stationary weak solutions for a class of non-Newtonian fluids with energy transfer.* Int. J. Non-Linear Mech. 32 (1997), 961–972.

[6] CONSIGLIERI, L.; SHILKIN, T., *Regularity to stationary weak solutions in the theory of generalized Newtonian fluids with energy transfer.* Zapiski Nauchn. Seminarov POMI 271 (2000), 122–150.

[7] GIAQUINTA, M., **Multiple integrals in the calculus of variations and nonlinear elliptic systems.** Ann. Math. Studies 105, Princeton Univ. Press, Princeton, New Jersey 1983.

[8] GONCHAROVA, O.N., *On the uniqueness of the solution of a stationary problem for equations of free convection* (Russian). Dinamika Sploshn. Sredy. Sibirsk. Otdel. AN SSSR, Novosibirsk, no. 97 (1990), 22–28.

[9] LANDAU, L.D.; LIFSCHITZ, E.M., **Lehrbuch der theoretischen Physik; Band VI: Hydrodynamik**. 3. Aufl., Akademie-Verlag, Berlin 1978.

[10] LIONS, J.L., *Quelques méthodes de résolution des problèmes aux limites non linéaires.* Dunod, Gauthier-Villars, Paris 1969.

[11] ŁUKASZEWICZ, G., *On the stationary flows of viscous, incompressible and heat-conducting fluids.* Math. Methods Appl. Sci. 10 (1988), 329–337.

[12] MORIMOTO, H., *On the existence of weak solutions of equation of natural convection.* J. Fac. Sci. Univ. Tokyo, Sect. IA, Math. 36 (1989), 87–102.

[13] —— , *On the existence and uniqueness of the stationary solution to the equations of natural convection.* Tokyo J. Math. 14 (1991), 217–226.

[14] ——, *A note on the existence of solutions to stationary Boussinesq equations under general outflow conditions.* In: H. Amann et al. (eds.): Navier-Stokes equations and related nonlinear problems. Utrecht, VSP/TEV (1998); pp. 183–193.

[15] Müller, I., **Grundzüge der Thermodynamik**. 3. Aufl., Springer-Verlag, Berlin 2001.

[16] Naumann, J.; Wolff, M., *On a global L^q-gradient estimate on weak solutions of nonlinear elliptic systems.* Dip. Mat. Univ. Catania (rapporto interno, 18. 02. 1991) (19 pp.).

[17] Rakotoson, J.M., *Quasilinear elliptic problems with measures as data.* Diff. Integral Equs. 4 (1991), 449–457.

[18] Roubíček, T., *Steady-state bouyancy-driven viscous flow with measure data.* Math. Bohemica 126 (2001), 493–504.

[19] Serrin, J., *Mathematical principles of classical fluid mechanics.* In: **Handbuch der Physik** (Hrsg. S. Flügge), Bd. VIII/1: Strömungsmechanik I, Springer-Verlag, Berlin 1959; pp. 125–263.

[20] Shilkin, T., *Partial regularity of weak solutions of the stationary 3D-Boussinesq system.* Zapiski Nauchn. Seminarov POMI 288 (2002), 256–270.

[21] Spurk, J.H., **Strömungslehre**. 4. Aufl., Springer-Verlag, Berlin 1996.

[22] Stampacchia, G., *Le problème de Dirichlet pour les équations elliptiques du second ordre à coefficients discontinus.* Ann. Inst. Fourier Grenoble 15 (1965), 189–258.

[23] Wolff, M., *A global L^κ-gradient estimate on weak solutions to nonlinear stationary Navier-Stokes equations under mixed boundary conditions.* Inst. f. Math., Humboldt-Univ. Berlin, preprint nr. 96-3 (1996).

J. Naumann
Institut für Mathematik (Sitz: Rudower Chaussee 25)
Humboldt-Universität zu Berlin
Unter den Linden 6
D-10099 Berlin, BRD

Progress in Nonlinear Differential Equations
and Their Applications, Vol. 64, 391–402

Liouville Type Theorems and Complete Blow-up for Indefinite Superlinear Parabolic Equations

Peter Poláčik and Pavol Quittner

Dedicated to Herbert Amann

1. Introduction

In this paper we study nonnegative bounded solutions of parabolic equations of the form

$$u_t - \Delta u = h(x_1)f(u), \qquad x = (x_1, x_2, \dots, x_N) \in \mathbb{R}^N,\ t \in \mathbb{R}, \tag{1.1}$$

where h, f are nondecreasing continuous functions,

$$h(0) = 0, \quad h \text{ is strictly increasing for } x_1 > 0, \qquad \lim_{x_1 \to +\infty} h(x_1) = +\infty. \tag{1.2}$$

We are particularly interested in solutions defined for all $t \in \mathbb{R}$. Below we refer to such solutions as *entire solutions*. Our main goal is to prove a Liouville type result to the effect that there are no positive bounded entire solutions, see Theorem 1.1 below. If, for example, $f(u) = u^p$, $p > 1$, the theorem guarantees that (1.1) does not possess any such solutions. Using this fact we obtain a new result on complete blow-up for the problem

$$\left.\begin{aligned} u_t - \Delta u &= \lambda u + a(x)u^p, && x \in \Omega,\ t > 0,\\ u &= 0, && x \in \partial\Omega,\ t > 0,\\ u(x,0) &= u_0(x) \geq 0, && x \in \Omega, \end{aligned}\right\} \tag{1.3}$$

where Ω is a smooth bounded domain in $\mathbb{R}^N$, $\lambda \in \mathbb{R}$, $a \in C(\bar{\Omega})$ changes sign and $p > 1$ is subcritical, see Theorem 1.2 below. More specifically, our nonexistence result for (1.1) together with known nonexistence results for the equation

$$u_t - \Delta u = u^p, \qquad x \in \mathbb{R}^N,\ t \in \mathbb{R},$$

The first author was supported in part by NSF grant DMS-0400702.
The second author was supported in part by VEGA grant 1/0259/03.

guarantee suitable a priori bounds for solutions of (1.3) and these bounds can be efficiently used in the study of blow-up.

Let us explain the relation between a priori bounds and Liouville type theorems in more detail.

First consider steady states of (1.3) with $a > 0$. In order to show an a priori bound, we proceed by contradiction assuming that there exist positive equilibria u_k, $k = 1, 2, \dots$, of (1.3) satisfying

$$M_k := \max_{x\in\Omega} u_k(x) = u_k(x_k) \to \infty.$$

Set

$$v_k(y) := \lambda_k^{2/(p-1)} u_k(x_k + \lambda_k y), \qquad \lambda_k := M_k^{-(p-1)/2}.$$

Then a subsequence of (suitable transforms of) v_k converges to a positive bounded solution of one of the following limit problem

$$-\Delta v = v^p \quad \text{in } \mathbb{R}^N, \tag{1.4}$$

$$-\Delta v = v^p \quad \text{in } \mathbb{H}^N, \qquad v = 0 \quad \text{on } \partial\mathbb{H}^N, \tag{1.5}$$

where $\mathbb{H}^N := \{x \in \mathbb{R}^N : x_1 > 0\}$. However, if

$$p < p_S := \frac{N+2}{(N-2)_+} \tag{1.6}$$

then problems (1.4) and (1.5) do not possess such solutions, see [6] and [11], respectively. Consequently, (1.6) guarantees a uniform bound for positive equilibria of (1.3) if $a > 0$.

If a changes sign then in addition to the limit problems mentioned above one has to deal with equations of the form

$$-\Delta v = h(x_1)v^p \quad \text{in } \mathbb{R}^N, \tag{1.7}$$

(or the corresponding problem in a half-space), where h depends on the behavior of a close to its zero set (for example, $h(x_1) = x_1$ if zero is a regular value of a). The solvability of such equations and related a priori estimates have been intensively studied, see [4, 2, 7, 8] and references therein.

A priori bounds play an important role in the study of equilibria. For example, they appear as crucial ingredients in proofs of the existence of positive equilibria. Similarly, in the study of time dependent solutions of (1.3), suitable a priori bounds are often crucial tools of analysis. These bounds have either the form

$$\|u(\cdot,t)\|_\infty \le C(\|u_0\|_\infty, \delta, \tau) \qquad \text{for any } t \in (0, \min(\tau, T(u_0) - \delta)), \tag{1.8}$$

where $\|\cdot\|_\infty$ denotes the norm in $L^\infty(\Omega)$, $T(u_0)$ denotes the maximal existence time of the solution of (1.3) and $\delta, \tau > 0$ are arbitrary, or

$$\|u(\cdot,t)\|_\infty \le C(\|u_0\|_\infty) \qquad \text{for any } t > 0 \ \text{ provided } T(u_0) = \infty. \tag{1.9}$$

In fact, estimate (1.8) can be used in the study of blow-up and estimate (1.9) is useful in the study of equilibria and connecting orbits.

A connection between Liouville theorems and the a priori bounds is furnished, similarly as in the case the equilibria, by a scaling argument. Assume that estimates

(1.8) or (1.9) fail and consider solutions u_k of (1.3) with uniformly bounded initial data $u_{0,k}$ such that $\sup_{t\in I_k} \|u(\cdot,t)\|_\infty \to \infty$, where $I_k = [0, \min(\tau, T(u_{0,k}) - \delta))$ in the case of (1.8), $I_k = [0, \infty)$ in the case of (1.9). Then, for a suitable subsequence of u_k, we can find $x_k \in \Omega$ and $t_k \in I_k$ such that $M_k := u_k(x_k, t_k) \to \infty$ and the functions u_k remain bounded by $2M_k$ on the time intervals $(t_k - kM_k^{1-p}, t_k + kM_k^{1-p})$, see [21, 20]. Set

$$v_k(y,s) := \lambda_k^{2/(p-1)} u_k(x_k + \lambda_k y, t_k + \lambda_k^2 s), \qquad \lambda_k := M_k^{-(p-1)/2}.$$

Standard parabolic estimates imply that, along a subsequence, v_k converges to a positive bounded entire solution of a parabolic problem on the whole space or on a half-space. In case $a > 0$, the limit problem has one of the following forms

$$v_t - \Delta v = v^p, \quad x \in \mathbb{R}^N,\ t \in \mathbb{R}, \tag{1.10}$$

$$v_t - \Delta v = v^p, \quad x \in \mathbb{H}^N,\ t \in \mathbb{R}, \qquad v = 0, \quad x \in \partial\mathbb{H}^N,\ t \in \mathbb{R}. \tag{1.11}$$

In addition, still with $a > 0$, one can obtain a uniform bound for the corresponding energy functionals

$$E(u_k(\cdot,t)) := \frac{1}{2}\int_\Omega |\nabla u_k|^2\,dx - \frac{\lambda}{2}\int_\Omega u_k^2\,dx - \frac{1}{p+1}\int_\Omega a u_k^{p+1}\,dx, \tag{1.12}$$

for $t \in I_k$ and this bound guarantees a uniform estimate for the time derivatives of u_k in $L^2(\Omega \times I_k)$. Using this estimate it is easy to see that the limit function v does not depend on t, hence the limit problems (1.10) and (1.11) reduce to (1.4) and (1.5), respectively. Consequently, the nonexistence results mentioned above imply a priori bounds (1.8) and (1.9) for solutions of (1.3) if (1.6) holds and $a > 0$.

If, on the other hand, a changes sign (or if we considered problem (1.3) with some lower-order perturbations destroying the variational structure, see [21]) then the energy bound for (1.12) is not known and one has to guarantee the nonexistence of positive bounded solutions for the time-dependent problems (1.10), (1.11). In addition, for sign-changing functions a we also have to show nonexistence results for limit problems of the form

$$v_t - \Delta v = h(x_1)v^p, \quad x \in \mathbb{R}^N,\ t \in \mathbb{R}. \tag{1.13}$$

Unlike the nonexistence results for the elliptic problems (1.4), (1.5) and (1.7), Liouville type theorems for the parabolic counterparts (1.10), (1.11) and (1.13) do not seem to be known in the full subcritical range. Optimal nonexistence results are well known for these parabolic problems if we replace the time interval $t \in \mathbb{R}$ with $t \in (0, \infty)$. However, these Fujita type results provide just sufficient conditions for the nonexistence of entire solutions. For example, the problem

$$v_t - \Delta v = v^p, \quad x \in \mathbb{R}^N,\ t \in (0,\infty),$$

does not possess positive global solutions if and only if $p \le 1 + 2/N$. This restriction on p is certainly not optimal for positive entire solutions of problem (1.10). In fact, the nonexistence of such solutions follows from [5] for any $p < N(N+2)/(N-1)^2$ and the nonexistence of radial positive bounded entire solutions of (1.10) for any $p < p_S$ is proved in [17] (see also [16] for such result on radially decreasing solutions in dimensions $N \leq 3$ or [15, 18] for related results mentioned below). Similarly, the Fujita type result for (1.13) with $h(x_1) = x_1$ requires $p \leq 1+3/(N+1)$ (see [20]), but the nonexistence result for entire positive bounded solutions of this problem is true for any $p > 1$. This follows from the following theorem which is the main result of this paper.

Theorem 1.1. *Suppose that $h \in C(\mathbb{R})$ and $f \in C^1(\mathbb{R}^+, \mathbb{R}^+)$ are nondecreasing functions satisfying* (1.2) *and $f(0) = f'(0) = 0$, $f(u) > 0$ for $u > 0$. Then the only nonnegative bounded classical solution of the equation*

$$u_t - \Delta u = h(x_1)f(u), \qquad x \in \mathbb{R}^N,\ t \in \mathbb{R}, \tag{1.14}$$

is the trivial solution $u \equiv 0$.

The proof of this theorem is based on the method of moving planes and it was inspired by the proof of the corresponding elliptic Liouville theorem in [8]. The elliptic result now becomes a special case of Theorem 1.1 dealing with steady states. We remark that our assumptions on the functions h and f are weaker than those in [8], so we have a slight improvement of the theorem in [8] even in this special case.

Nonexistence results for (1.13) and (1.10) were used in [20] in order to prove a priori estimates and complete blow-up results for (1.3). Since the nonexistence for (1.13) was proved only for $p \leq 1 + 3/(N+1)$ in [20], the main result in [20] required this restriction on the exponent p. Here we can use our new nonexistence result for (1.13) (valid for any $p > 1$) and the nonexistence result for (1.10) from [5] to justify that the condition $p \leq 1 + 3/(N+1)$ in [20] can be replaced with $p < N(N+2)/(N-1)^2$. In order to formulate the corresponding statement, let us first recall the notion of complete blow-up.

Assume $T(u_0) < \infty$ and let u_k be solutions of the approximation problems

$$\left.\begin{aligned} u_t - \Delta u &= f_k(x,u), && x \in \Omega,\ t > 0,\\ u &= 0, && x \in \partial\Omega,\ t > 0,\\ u(x,0) &= u_0(x), && x \in \Omega, \end{aligned}\right\} \tag{1.15}$$

where

$$f_k(x,u) := \begin{cases} \lambda u + a(x)\min(u^p, k), & \text{if } a(x) \geq 0,\\ \lambda u + a(x)u^p, & \text{if } a(x) < 0, \end{cases} \qquad k = 1, 2, 3, \dots.$$

The solutions u_k exist globally and $u_{k+1} \geq u_k$, $k = 1, 2, \dots$. Set

$$\bar{u}(x,t) := \lim_{k\to\infty} u_k(x,t) \in [0,\infty], \qquad x \in \overline{\Omega},\ t \in [0,\infty).$$

Then $\bar{u}(x,t) = u(x,t)$ for any $x \in \overline{\Omega}$ and $t < T(u_0)$ and $\bar{u}$ remains finite in $\Omega^- := \{x \in \Omega : a(x) < 0\}$ for any $t > 0$. Let $D \subset \Omega$. We say that u *blows up completely in D at $t = T$* if $\bar{u}(x,t) = \infty$ for any $x \in D$ and any $t > T$. The following theorem is an immediate consequence of results in [20] and Theorem 1.1.

Theorem 1.2. *Assume that* $p < N(N+2)/(N-1)^2$, $\Omega^+ := \{x \in \Omega : a(x) > 0\}$ *is connected and*

$$a \in C^2(\overline{\Omega}), \quad a < 0 \ \text{ on } \partial\Omega, \quad \nabla a(x) \neq 0 \ \text{ whenever } a(x) = 0. \tag{1.16}$$

If $T(u_0) < \infty$ *then the solution* u *of* (1.3) *blows up completely in* Ω^+ *at* $t = T(u_0)$. *In addition, the function*

$$T : \{u_0 \in L^\infty(\Omega) : u_0 \geq 0\} \to (0, \infty] : u_0 \mapsto T(u_0)$$

is continuous.

If $a \equiv 1$ then a complete blow-up result was shown in [3] for the full subcritical range $p \in (1, p_S)$. On the other hand, it is known that solutions of (1.3) need not blow up completely if $p > p_S$, see [10] or [9] and the references therein. Similarly, the function T need not be continuous if $p > p_S$.

Liouville type theorems have many other applications than those discussed above. As an illustration we mention two related results on the problem

$$\left.\begin{aligned} u_t - \Delta u &= u^p, && x \in \mathbb{R}^N,\ t > 0, \\ u(x,0) &= u_0(x) \geq 0, && x \in \mathbb{R}^N, \end{aligned}\right\} \tag{1.17}$$

where $u_0 \in L^\infty(\mathbb{R}^N)$.

The first one concerns positive solutions of the equation

$$u_t - \Delta u = u^p, \qquad x \in \mathbb{R}^N,\ t \in (-\infty, T), \tag{1.18}$$

satisfying the estimate

$$u(x,t) \leq C(T-t)^{-\frac{1}{p-1}} \quad (x \in \mathbb{R}^N, t < T), \tag{1.19}$$

where C is a constant. Note that there is a "trivial" family of such solutions: $u(x,t) = \kappa(T_1 - t)^{-1/(p-1)}$, $T_1 \geq T$, where $\kappa = (1/(p-1))^{1/(p-1)}$. A theorem of [15] asserts that this family contains all positive solutions of (1.18) satisfying (1.19) if $p < p_S$. This result is used in a theorem on the blow-up rate of positive unbounded solutions of (1.17).

The second result concerns a different class of solutions of (1.18), namely solutions satisfying

$$\varphi_1(x) \leq u(x,t) \leq \varphi_2(x) \quad (x \in \mathbb{R}^N, t < T), \tag{1.20}$$

where $\varphi_1 < \varphi_2$ are positive radial steady states of (1.17) (φ_2 can be singular). It is well known (see [14]) that the related steady states exist if and only if $N > 10$ and $p \geq p_{JL}$, where

$$p_{JL} := \frac{(N-2)^2 - 4N + 8\sqrt{N-1}}{(N-2)(N-10)}.$$

In fact, there is an increasing curve of steady states, which constitutes a "trivial" set of solutions satisfying (1.20). According to a theorem of [18], each solution of (1.18) satisfying (1.20) is a steady state if $p > p_{JL}$. This result is used in the proof of quasiconvergence (convergence to a set of steady states) of positive solutions of (1.17) bounded above by the singular steady state.

Finally let us mention that a priori bounds (1.8) and (1.9) can also be obtained for sign-changing solutions. In that case we cannot use rescaling but one can use energy and interpolation estimates, see [19, 1, 13] and the references therein.

2. Proof of Theorem 1.1

Assume on the contrary that (1.14) possesses a positive bounded classical solution u. Denote $Lu := u_t - \Delta u$, $M := \sup u$. Given $\lambda \in \mathbb{R}$, set

$$\begin{aligned}\Sigma_\lambda &:= \{x \in \mathbb{R}^N : x_1 < \lambda\},\\ x^\lambda &= (x_1^\lambda, x_2, \dots, x_N) := (2\lambda - x_1, x_2, \dots, x_N),\\ w_\lambda(x,t) &:= u(x^\lambda, t) - u(x,t), \quad (x,t) \in \bar{\Sigma}_\lambda \times \mathbb{R},\\ \lambda(t) &:= \sup\{\tilde\lambda : w_\lambda(x,t) \geq 0 \text{ for all } x \in \Sigma_\lambda \text{ and } \lambda < \tilde\lambda\},\\ \lambda^* &:= \inf\{\lambda(t) : t \in \mathbb{R}\}.\end{aligned}$$

We will prove the following assertions.

Claim 1. If $x \in \Sigma_{\min(\lambda,0)}$, $t \in \mathbb{R}$ and $w_\lambda(x,t) \leq 0$ then $Lw_\lambda(x,t) \geq 0$.

Claim 2. $\lambda(t) \geq 0$ for all $t \in \mathbb{R}$.

Claim 3. The mapping $t \mapsto \lambda(t)$ is nondecreasing.

Claim 4. $\lambda^* = \infty$.

Claim 5. $u \equiv 0$.

Proof of Claim 1. Let $x_1 < \min(\lambda, 0)$ and $w_\lambda(x,t) \leq 0$. If $x_1^\lambda \geq 0$ then $h(x_1^\lambda) \geq 0 \geq h(x_1)$ and

$$Lw_\lambda(x,t) = h(x_1^\lambda) f(u(x^\lambda,t)) - h(x_1) f(u(x,t)) \geq 0.$$

If $x_1^\lambda < 0$ then $h(x_1) \leq h(x_1^\lambda) \leq 0$ and, since f is nondecreasing,

$$\begin{aligned}Lw_\lambda(x,t) &= h(x_1^\lambda) f(u(x^\lambda,t)) - h(x_1) f(u(x,t))\\ &\geq h(x_1^\lambda)\big(f(u(x^\lambda,t)) - f(u(x,t))\big) \geq 0.\end{aligned}$$

Proof of Claim 2. Let $\lambda \leq 0$. We prove $w_\lambda(x,t) \geq 0$ for all $x \in \Sigma_\lambda$ and $t \in \mathbb{R}$. Assume on the contrary that

$$\inf_{x \in \Sigma_\lambda} w_\lambda(x, t_0) < 0 \quad \text{for some } t_0 \in \mathbb{R}. \tag{2.1}$$

Set

$$g(y) := \ln(\lambda + 1 - y) + 1, \ y \leq \lambda, \qquad \text{and} \qquad \bar w_\lambda(x,t) := \frac{w_\lambda(x,t)}{g(x_1)}.$$

Notice that $g(y) \geq g(\lambda) = 1$, $g''(y) = -(\lambda + 1 - y)^{-2} < 0$ and

$$g(x_1) L\bar w_\lambda(x,t) = Lw_\lambda(x,t) + 2(\partial_{x_1} \bar w_\lambda(x,t)) g'(x_1) + \bar w_\lambda(x,t) g''(x_1). \tag{2.2}$$

Assume $x \in \Sigma_\lambda$ and $w_\lambda(x,t) \leq 0$. Then $Lw_\lambda(x,t) \geq 0$ due to Claim 1, hence

$$L\bar{w}_\lambda(x,t) \geq 2(\partial_{x_1}\bar{w}_\lambda)\frac{g'}{g} + \bar{w}_\lambda\frac{g''}{g}.$$

We compare $\bar{w}_\lambda$ with the solution of the problem

$$z_t - z_{yy} = F(y,z,z_y),\ y \in \mathbb{R},\ t > 0, \qquad z(y,0) = -M,\ y \in \mathbb{R},$$

where

$$F(y,z,z_y) := \begin{cases} 2z_y g'/g, & y < \lambda - 1, \\ 2z_y g'/g - az, & y \in [\lambda - 1, \lambda], \\ 0, & y > \lambda, \end{cases}$$

and $a := -g''(\lambda-1)/g(\lambda-1) > 0$. Note that, by the maximum principle, $z(y,t) < 0$ for all $y \in \mathbb{R}$, $t \geq 0$, and, since $F(y,-M,0)) \geq 0$ with the strict inequality in $(\lambda-1,\lambda)$, z is increasing in t.

In view of the obvious inequalities

$$\begin{aligned} F(x_1,z,z_{x_1}) &\leq 2z_{x_1}\frac{g'}{g} + z\frac{g''}{g} && (x_1 < \lambda), \\ \bar{w}_\lambda(x,t_0 - T) &\geq z(x_1,0) && (x \in \Sigma_\lambda), \\ w_\lambda(x,t) &= 0 > z(x_1,t) && (x \in \partial\Sigma_\lambda), \end{aligned}$$

the maximum principle implies $\bar{w}_\lambda(x,t_0) \geq z(x_1,T)$ for any $T > 0$ and $x \in \Sigma_\lambda$. We now show that $z(x_1,T) \to 0$ as $T \to \infty$, which yields a contradiction to (2.1).

First off, z being increasing in t and negative implies that the limit $\bar{z}(y) := \lim_{t\to+\infty} z(y,t)$ is a solution of $-\bar{z}'' = F(y,\bar{z},\bar{z}')$ satisfying $-M \leq \bar{z} \leq 0$. Since $F(y,\bar{z},\bar{z}') = 0$ for $y > \lambda$, the function $\bar{z}$ is constant on $[\lambda,\infty)$. One can also easily show that all bounded solutions of $-\bar{z}'' = 2\bar{z}'g'/g$ on $(-\infty,\lambda-1)$ are constant, hence $\bar{z}$ is constant on $(-\infty,\lambda-1]$. Finally, on $[\lambda-1,\lambda]$, $\bar{z}$ satisfies the following equation, with matching conditions at the boundary points:

$$-\bar{z}'' = 2\bar{z}'g'/g - a\bar{z}, \qquad \bar{z}'(\lambda-1) = \bar{z}'(\lambda) = 0.$$

Assume that $\bar{z}$ is not identically zero in $[\lambda-1,\lambda]$ and let $\bar{z}$ attain its negative minimum in this interval at $y_0 \in [\lambda-1,\lambda]$ Then $\bar{z}'(y_0) = 0$, hence $\bar{z}''(y_0) = a\bar{z}(y_0) < 0$ which yields a contradiction. Consequently, $\bar{z} \equiv 0$ in $\mathbb{R}$, as claimed.

Proof of Claim 3. Let $t_0 \in \mathbb{R}$ and $\lambda \leq \lambda(t_0)$. Then $w_\lambda(x,t_0) \geq 0$ for any $x \in \Sigma_\lambda$. If $x \in \Sigma_\lambda$ then $x_1^\lambda > x_1$ and

$$\begin{aligned} Lw_\lambda(x,t) &= h(x_1^\lambda)f(u(x^\lambda,t)) - h(x_1)f(u(x,t)) \\ &\geq h(x_1^\lambda)\big(f(u(x^\lambda,t)) - f(u(x,t))\big) \\ &= h(x_1^\lambda)f'(\theta(x,t))w_\lambda(x,t) \end{aligned}$$

for some $\theta(x,t)$ between $u(x,t)$ and $u(x^\lambda,t)$. Since $w_\lambda(x,t) = 0$ for $x \in \partial\Sigma_\lambda$, the maximum principle implies $w_\lambda(x,t) \geq 0$, hence $\lambda(t) \geq \lambda(t_0)$, for all $t \geq t_0$. This proves that the function $t \mapsto \lambda(t)$ is nondecreasing.

Proof of Claim 4. Claim 2 guarantees $\lambda^* \geq 0$. Assume $\lambda^* < \infty$ and choose $t_k \to -\infty$. Then $\lambda(t_k) \to \lambda^*$ hence there exist $\lambda_k \searrow \lambda^*$ such that

$$\inf_{x \in \Sigma_{\lambda_k}} w_\lambda(x, t_k) < 0. \tag{2.3}$$

There exists $C_f > 0$ such that $f' \leq C_f$ on $[0, M]$. Set $b_2 := h(\lambda^* + 1)C_f$, fix $\delta > 0$ such that

$$2\delta^{-2} \geq 3^3(2b_2 + 1) \tag{2.4}$$

and $\eta > 0$ such that

$$f' \leq \frac{\delta}{h(\lambda^* + 1)(\lambda^* + 1)^3} \quad \text{on } [0, \eta]. \tag{2.5}$$

Let $\varepsilon \in (0, \delta)$ be sufficiently small (as specified below). Fix k such that $\lambda_k < \lambda^* + \varepsilon$ and set $\lambda := \lambda_k$,

$$g(y) := 2 - \frac{\delta}{\delta + \lambda - y}, \; y \in (-\infty, \lambda], \qquad \bar{w}_\lambda(x, t) := \frac{w_\lambda(x, t)}{g(x_1)}.$$

We next derive an inequality satisfied by $\bar{w}_\lambda$ in the set where it is negative.

Let $x \in \Sigma_\lambda$ and $w_\lambda(x, t) \leq 0$. First we treat the case $x_1 < \lambda^* - \delta$. If $Lw_\lambda(x, t) \geq 0$, then (2.2) gives

$$L\bar{w}_\lambda(x, t) \geq 2(\partial_{x_1} \bar{w}_\lambda)\frac{g'}{g}. \tag{2.6}$$

Assume $Lw_\lambda(x, t) < 0$. Then Claim 1 guarantees that $x_1 > 0$. Standard local parabolic estimates give

$$|\nabla u(x, t)| \leq C_\lambda \quad \text{whenever } x_1 \in [0, 2\lambda].$$

Since

$$u(x^{\lambda^*}, t) \geq u(x, t) \geq u(x^\lambda, t), \tag{2.7}$$

$h(x_1) \geq 0$, and

$$|x^{\lambda^*} - x^\lambda| = |x_1^{\lambda^*} - x_1^\lambda| = 2(\lambda - \lambda^*) \leq 2\varepsilon,$$

we have

$$\begin{aligned} 0 > Lw_\lambda(x, t) &= h(x_1^\lambda)f(u(x^\lambda, t)) - h(x_1)f(u(x, t)) \\ &\geq \big(h(x_1^\lambda) - h(x_1)\big)f(u(x^{\lambda^*}, t)) + h(x_1^\lambda)\big(f(u(x^\lambda, t)) - f(u(x^{\lambda^*}, t))\big) \\ &\geq \big(h(\lambda^*) - h(\lambda^* - \delta)\big)f(u(x^{\lambda^*}, t)) - h(2\lambda^* + 1)C_f\big(u(x^{\lambda^*}, t) - u(x^\lambda, t)\big) \\ &\geq \big(h(\lambda^*) - h(\lambda^* - \delta)\big)f(u(x^{\lambda^*}, t)) - 2h(2\lambda^* + 1)C_f C_\lambda \varepsilon. \end{aligned}$$

Consequently,

$$f(u(x^{\lambda^*}, t)) \leq \frac{2h(2\lambda^* + 1)C_f C_\lambda}{h(\lambda^*) - h(\lambda^* - \delta)}\varepsilon$$

and this bound guarantees $u(x^{\lambda^*},t) \le \eta$ if ε is small enough. Now (2.7) and (2.5) imply

$$\begin{aligned} Lw_\lambda(x,t) &\ge h(x_1)\big(f(u(x^\lambda,t)) - f(u(x,t))\big) \\ &\ge h(\lambda^*)\frac{\delta}{h(\lambda^*+1)(\lambda^*+1)^3} w_\lambda(x,t) \ge b_1 w_\lambda(x,t), \end{aligned}$$

where

$$b_1 := \frac{\delta}{(\lambda^*+1)^3} \le \frac{\delta}{(\delta+\lambda-y)^3} = -\frac{g''(y)}{2} \le -\frac{g''(y)}{g(y)} \quad \text{for } y \in [0,\lambda^*].$$

Hence we infer from (2.2) that (2.6) holds in this case as well. We conclude that

$$L\bar{w}_\lambda(x,t) \ge 2(\partial_{x_1}\bar{w}_\lambda)\frac{g'}{g} \qquad \text{whenever } w_\lambda(x,t) \le 0 \text{ and } x_1 < \lambda^* - \delta.$$

Next assume $x_1 \in [\lambda^* - \delta, \lambda]$. Then

$$\begin{aligned} Lw_\lambda(x,t) &\ge h(x_1)\big(f(u(x^\lambda,t)) - f(u(x,t))\big) \\ &\ge h(\lambda^*+1)C_f w_\lambda(x,t) = b_2 w_\lambda(x,t). \end{aligned}$$

Notice that

$$-g''(y) = \frac{2\delta}{(\delta+\lambda-y)^3} \ge 2b_2 + 1 \ge g(y)b_2 + 1 \quad \text{for } y \in [\lambda^* - \delta, \lambda],$$

due to (2.4), hence

$$L\bar{w}_\lambda(x,t) \ge 2(\partial_{x_1}\bar{w}_\lambda)\frac{g'}{g} - \frac{\bar{w}_\lambda}{g} \qquad \text{whenever } w_\lambda(x,t) \le 0 \text{ and } x_1 \in [\lambda^* - \delta, \lambda].$$

We employ the above inequalities in a comparison argument, similarly as in the proof of Claim 2. By the maximum principle, we have $\bar{w}_\lambda(x,t_k) \ge z(x_1,T)$ for any $T > 0$ and $x \in \Sigma_\lambda$, where z is the solution of the problem

$$z_t - z_{yy} = F(y,z,z_y), \; y \in \mathbb{R}, t > 0, \qquad z(y,0) = -M, \; y \in \mathbb{R},$$

where

$$F(y,z,z_y) := \begin{cases} 2z_y g'/g, & y < \lambda^* - \delta, \\ 2z_y g'/g - az, & y \in (\lambda^* - \delta, \lambda), \\ 0, & y > \lambda, \end{cases}$$

and $a := \inf_{[\lambda^*-\delta,\lambda]} 1/g = 1/g(\lambda^* - \delta)$. One shows, by the same arguments as in the proof of Claim 2, that $z(x_1,T) \to 0$ as $T \to \infty$, hence $\bar{w}_\lambda(x,t_k) \ge 0$, which contradicts (2.3).

Proof of Claim 5. Set

$$\Omega_\sigma := \{x = (x_1,\tilde{x}) \in \mathbb{R} \times \mathbb{R}^{N-1} : \sigma < x_1 < \sigma + 1, \; |\tilde{x}| < \frac{1}{2}\}, \quad \sigma \ge 0.$$

Let μ_0 be the first eigenvalue of the problem

$$-\Delta\varphi = \mu\varphi \;\text{ in } \Omega_0, \qquad \varphi = 0 \;\text{ on } \partial\Omega_0,$$

and let φ_0 be the corresponding positive eigenfunction satisfying $\max_{\Omega_0}\varphi_0 = 1$. Set

$$\varphi_\sigma(x) := \varphi_0(x_1 - \sigma, \tilde{x})$$

and

$$\psi_\sigma(t) := \int_{\Omega_\sigma} u(x,t)\varphi_\sigma(x)\,dx.$$

Then

$$\psi_\sigma(0) \geq c_0 := \psi_0(0)$$

since u is nondecreasing in x_1. Denote

$$\Omega_\sigma(t) := \{x \in \Omega_\sigma : u(x,t)\varphi_\sigma(x) \geq c_0/2\}.$$

If $t \geq 0$ is such that $\psi_\sigma(t) \geq c_0$ then $|\Omega_\sigma(t)| \geq \xi := c_0/(2M)$. Indeed,

$$c_0 \leq \int_{\Omega_\sigma} u(x,t)\varphi_\sigma(x)\,dx \leq |\Omega_\sigma(t)| \cdot M + \frac{c_0}{2},$$

and the stated inequality follows. Notice that $\xi \in (0,1)$ and

$$\int_{\Omega_\sigma(t)} u(x,t)\varphi_\sigma(x)\,dx \geq \xi\frac{c_0}{2} \geq \xi\int_{\Omega_\sigma\setminus\Omega_\sigma(t)} u(x,t)\varphi_\sigma(x)\,dx,$$

hence

$$\int_{\Omega_\sigma(t)} u(x,t)\varphi_\sigma(x)\,dx \geq \frac{\xi}{1+\xi}\int_{\Omega_\sigma} u(x,t)\varphi_\sigma(x)\,dx = \frac{c_0}{2M+c_0}\psi_\sigma(t).$$

We conclude that, as long as $\psi_\sigma(t) \geq c_0$, we have

$$\begin{aligned}
\frac{d}{dt}\psi_\sigma(t) &= \int_{\Omega_\sigma} \Delta u(x,t)\varphi_\sigma(x)\,dx + \int_{\Omega_\sigma} h(x_1)f(u(x,t))\varphi_\sigma(x)\,dx \\
&\geq \int_{\Omega_\sigma} u(x,t)\Delta\varphi_\sigma(x)\,dx + h(\sigma)\int_{\Omega_\sigma(t)} f(u(x,t))\varphi_\sigma(x)\,dx \\
&\geq \int_{\Omega_\sigma} u(x,t)\Delta\varphi_\sigma(x)\,dx + h(\sigma)\int_{\Omega_\sigma(t)} \frac{f(u(x,t))}{M}u(x,t)\varphi_\sigma(x)\,dx \\
&\geq -\mu_0\psi_\sigma(t) + h(\sigma)f\Big(\frac{c_0}{2}\Big)\frac{1}{M}\int_{\Omega_\sigma(t)} u(x,t)\varphi_\sigma(x)\,dx \\
&\geq \psi_\sigma(t)\Big[-\mu_0 + h(\sigma)f\Big(\frac{c_0}{2}\Big)\frac{1}{M}\frac{c_0}{2M+c_0}\Big].
\end{aligned}$$

Taking σ large, the expression in the square brackets becomes greater than one, hence $\psi'_\sigma(t) \geq \psi_\sigma(t)$. Since $\psi_\sigma(0) \geq c_0$, we deduce that the inequality is valid for all $t \geq t_0$. Consequently, $\psi_\sigma(t) \to +\infty$ as $t \to \infty$ which contradicts the boundedness of u.

Acknowledgment

This research was done while the second author was visiting University of Minnesota. He thanks the School of Mathematics for the hospitality. The authors also thank Yihong Du for correcting a choice of the function g in the proof of Claim 4.

References

[1] N. Ackermann, T. Bartsch, P. Kaplický and P. Quittner, A priori bounds, nodal equilibria and connecting orbits in indefinite superlinear parabolic problems, preprint.

[2] H. Amann, J. López-Gómez, A priori bounds and multiple solutions for superlinear indefinite elliptic problems, *J. Differential Equations* **146** (1998), 336–374.

[3] P. Baras and L. Cohen, Complete blow-up ater T_{max} for the solution of a semilinear heat equation, *J. Funct. Anal.* **71** (1987), 142–174.

[4] H. Berestycki, I. Capuzzo-Dolcetta and L. Nirenberg, Superlinear indefinite elliptic problems and nonlinear Liouville theorems, *Topol. Methods Nonlinear Anal.* **4** (1994), 59–78.

[5] M.-F. Bidaut-Véron, Initial blow-up for the solutions of a semilinear parabolic equation with source term, In: *Equations aux dérivées partielles et applications, articles dédiés à Jacques-Louis Lions*, Gauthier-Villars, Paris (1998), 189–198.

[6] W. Chen and C. Li, Classification of solutions of some nonlinear elliptic equations, *Duke Math. J.* **63** (1991), 615–622.

[7] W. Chen and C. Li, Indefinite elliptic problems in a domain, *Discrete Contin. Dyn. Syst.* **3** (1997), 333–340.

[8] Y. Du and S. Li, Nonlinear Liouville theorems and a priori estimates for indefinite superlinear elliptic equations, *Adv. Differential Equations* **10** (2005), 841–860.

[9] M. Fila, H. Matano and P. Poláčik, Immediate regularization after blow-up, *SIAM J. Math. Analysis*, to appear.

[10] V. Galaktionov and J.L. Vázquez, Continuation of blow-up solutions of nonlinear heat equations in several space dimensions, *Comm. Pure Appl. Math.* **50** (1997), 1–67.

[11] B. Gidas and J. Spruck, A priori bounds for positive solutions of nonlinear elliptic equations, *Comm. Partial Differential Equations* **6** (1981), 883–901.

[12] Y. Giga, A bound for global solutions of semilinear heat equations, *Comm. Math. Phys.* **103** (1986), 415–421.

[13] Y. Giga, S. Matsui and S. Sasayama, Blow up rate for semilinear heat equation with subcritical nonlinearity, *Indiana Univ. Math. J.* **53** (2004), 483–514.

[14] C. Gui, W.-M. Ni and X. Wang, On the stability and instability of positive steady states of a semilinear heat equation in $\mathbb{R}^n$, *Comm. Pure Appl. Math.* **45** (1992), 1153–1181.

[15] F. Merle and H. Zaag, Optimal estimates for blowup rate and behavior for nonlinear heat equations, *Comm. Pure Appl. Math.* **51** (1998), 139–196.

[16] J. Matos and Ph. Souplet, Universal blow-up rates for a semilinear heat equation and applications, *Adv. Differential Equations* **8** (2003), 615–639.

[17] P. Poláčik and P. Quittner, A Liouville-type theorem and the decay of radial solutions of a semilinear heat equation, *Nonlinear Analysis TMA*, to appear.

[18] P. Poláčik and E. Yanagida, A Liouville property and quasiconvergence for a semilinear heat equation, *J. Differential Equations* **208** (2005), 194–214.

[19] P. Quittner, Continuity of the blow-up time and a priori bounds for solutions in superlinear parabolic problems, *Houston J. Math.* **29** (2003), 757–799.

[20] P. Quittner and F. Simondon, A priori bounds and complete blow-up of positive solutions of indefinite superlinear parabolic problems, *J. Math. Anal. Appl.* **304** (2005), 614–631.

[21] P. Quittner and Ph. Souplet, A priori estimates of global solutions of superlinear parabolic problems without variational structure, *Discrete Contin. Dyn. Syst.* **9** (2003), 1277–1292.

Peter Poláčik
School of Mathematics
University of Minnesota
Minneapolis
MN 55455, USA

Pavol Quittner
Department of Applied Mathematics and Statistics
Comenius University
Mlynská dolina
84248 Bratislava, Slovakia

Progress in Nonlinear Differential Equations
and Their Applications, Vol. 64, 403–411

On Reducing the 2d Navier-Stokes Equations to a System of Delayed ODEs

Dalibor Pražák

Abstract. We give a simple proof that projecting the 2d Navier-Stokes equations to sufficiently many eigenfunctions of the Stokes operator leads to a system of delayed ODEs. The proof is based on the repeated use of the so-called squeezing property. The reduced system is uniquely solvable and dissipative. Moreover, the solutions on the attractor to the full NSEs are in one-to-one correspondence to the solutions on a compact, invariant subset to a global attractor of the reduced system.

Mathematics Subject Classification (2000). 37D45, 75D05.

Keywords. Navier-Stokes equations, global attractor, equations with delay.

0. Introduction

Is the behavior of a given dissipative PDE finite-dimensional for large times? There is vast literature dealing with this question, see [10], [9], [1] and the references therein. Generally speaking, the answer is yes, yet it comes in a variety of results.

The "simplest" possible type of theorems state the existence of a global attractor with finite (fractal or Hausdorff) dimension. This means, all solutions are attracted to a set that is finite-dimensional. However, the convergence to the attractor can be very slow, and the distance is still measured in a infinite-dimensional space. Hence, we would like to be more specific in expressing the ultimate finite-dimensionality.

One possible approach is the following. Assume that our PDE has an abstract form

$$u_t + Au = F(u) \tag{1}$$

where $u = u(t)$ lives in a suitable (infinite-dimensional) space, A is a linear operator and F is the lower-order nonlinearity. Let P be a projection to first N eigenfunctions of A and set $p = Pu$. We would like to show that for N large enough, the behavior of u is dominated by p.

Research supported by GACR grant 201/01/D054 and MSM grant 1132 00007.

Here come the theorems on the number of determining modes. In many situations one indeed proves that if $p_1 - p_2 \to 0$, then $u_1 - u_2 \to 0$, meaning for $t \to \infty$. See [8], for example.

But let us proceed even further. The projection P gives an equation for p

$$p_t + Ap = PF(u) \tag{2}$$

and one could ask whether we can 'decouple' this equation from (1) so that it would be possible to solve for p independently. More precisely, we seek a function $\Phi : \mathcal{R}(P) \to \mathcal{R}(I-P)$ such that if p solves the reduced equation

$$p_t + Ap = PF(p + \Phi(p)) \tag{3}$$

then $p + \Phi(p)$ solves (1). Such a decoupling function (suitably smooth) is of course nothing else than the so-called inertial manifold.

There is a well-known method of Lyapunov-Perron for constructing such a manifold, see, e.g., [6]. In this approach, the function Φ is sought by means of a Banach contraction theorem in a suitable space of functions from $\mathcal{R}(P)$ to $\mathcal{R}(I-P)$.

The key assumption around which the contraction argument evolves is that the gap between the spectra of AP and $A(I-P)$ is sufficiently large to dominate the nonlinearity F. Unfortunately, there are important examples of PDEs for which this 'gap condition' fails. The 2d Navier-Stokes equations are probably the most prominent example.

At any rate, the spectral gap condition is rather restrictive. Several attempts were made to extend the validity of Lyapunov-Perron construction to more general situations.

In [3] Debussche and Temam showed that, regardless the spectral gap, one can use the fixed point argument to obtain a function $\Phi : \mathcal{R}(P) \times \mathcal{R}(I-P) \to \mathcal{R}(I-P)$ such that $\Phi(p(t), q(t-T)) = q(t)$, where $q = (I-P)u$. That means, one can find $q(t)$ provided we know $p(t)$ and also q at some earlier time $t - T$. Thus Φ is, as the authors called it, an inertial manifold with delay.

Another generalization of Lyapunov-Perron method was presented by Hale and Raugel in [7]. Here the authors proved that the fixed point argument will work – once again without the gap condition – to obtain a function Φ from the space of $\mathcal{R}(P)$-valued functions to the space of $\mathcal{R}(I-P)$-valued functions. In other words, one can find the trajectory $\{q(s)\}_{s\in\mathbb{R}}$ provided we know the trajectory $\{p(s)\}_{s\in\mathbb{R}}$.

We would also like to mention a different approach of [4]. Here the authors started directly from the fact that the attractor has finite dimension. Hence by Mañé's theorem there always exists a projection P that is – when restricted to the attractor – invertible. Thus one can write

$$p_t + Ap = PF(P^{-1}(p)) \,. \tag{4}$$

There are certain disadvantages to this approach, namely (i) one doesn't know P explicitly and (ii) in general, P^{-1} is only Hölder continuous, hence (4) is not uniquely solvable.

But let us return to the papers [3] and [7]. Both of them obtain an interesting corollary. Although the function Φ would not reduce (1) to a system of ODEs, it can still reduce it to a system of ODEs with an infinite delay. Moreover, the properties of Φ show that the delay has at least Lipschitz dependence with exponentially fading memory.

The aim of our paper is to give an alternative and perhaps simpler proof of this reduction. Without recalling to a fixed point argument we show in an elementary way (see Lemma 2.1) that the value $u(t)$ can be computed from the past values of $p(t)$. Moreover, this functional dependence is Lipschitz continuous. This reduces the equation to a system of delayed ODEs, first only on the global attractor. However, since the domain of a Lipschitz function can easily be extended to the whole space, (cf. Lemma 1.4), we finally embed the original equation into a full system of delayed ODEs (12).

We then briefly study the properties of the reduced system. We prove existence, uniqueness (Theorem 3.1), dissipativity and existence of global attractor (Theorem 3.2). Finally, in Theorem 3.3 we summarize the correspondence between the solutions of the original equation (5) and the reduced system (12).

Our paper deals exclusively with the 2d Navier-Stokes system on torus. Yet the argument depends mainly on the squeezing property (9) that holds quite generally, cf. [4]. Hence a similar construction certainly can be carried out for other dissipative equations.

Last, but not least, one is also lead to several questions of a more general character:

- How larger is $\mathcal{A}_r$ (the attractor to the reduced system) than $\mathcal{A}$ (the attractor to the full Navier-Stokes system)? In particular, has $\mathcal{A}_r$ finite dimension?
- Does the reduced system describe also the behavior outside the attractor? For example, does it have some kind of tracking property?
- Can the infinite delay be replaced by the bounded one?

We would like to address these issues in the future research.

Notation. If p is a function on $(-\infty, t]$ then $[p^t](s) = p(t+s)$ is a function on $(-\infty, 0]$ Less formally we use a superscript p^0 to emphasize that also the past values are taken into account, e.g., $\mathcal{A}^0$ denotes the set of all backward trajectories in $\mathcal{A}$.

1. Preliminaries

Consider the Navier-Stokes system

$$\begin{aligned} u_t + (u\cdot\nabla)\,u - \nu\triangle u + \nabla p &= f \\ \operatorname{div} u &= 0 \\ u|_{t=0} &= u_0 \end{aligned} \tag{5}$$

where $u(x,t) : \Omega \times [0,\infty) \to \mathbb{R}^n$ and $p(x,t) : \Omega \times [0,\infty) \to R^n$ are the velocity and pressure and $f(x) : \Omega \to R^n$ are given external forces. We restrict ourselves to the two-dimensional spatially-periodic setting, i.e., $\Omega = (0,\omega)^2$ and u, p are with zero mean and periodic boundary conditions.

We set

$$\mathcal{V} = \left\{ u : \mathbb{R}^2 \to \mathbb{R}^2 : \ \operatorname{div} u = 0, \ \int_\Omega u = 0, \ u \text{ is periodic} \right\},$$
$$H^k = \operatorname{cl}_{W^{k,2}(\Omega)} \mathcal{V} \qquad \text{with } \|u\|_{H^k} = \|\nabla^k u\|_2 .$$

Fix $f \in H^1$. It is well known (see, e.g., [2]) that for any $u_0 \in H^0$, there exists unique $u \in C([0,T]; H^0) \cap L^2(0,T; H^1)$ a weak solution to (5). Hence one can define $S(t) : H^0 \to H^0$ the solution operators to (5).

The dynamical system $(S(t), H^0)$ is dissipative - there exists a bounded $\mathcal{B} \subset H^0$ which is uniformly absorbing and positively invariant for $S(t)$. Due to the regularizing effect one can have additional smoothness of $\mathcal{B}$. We will need that $\mathcal{B}$ is bounded in H^2 and closed in H^0. See, e.g., [4] for the proof.

The set

$$\mathcal{A} = \bigcap_{t>0} \operatorname{cl}_{H^0} \left\{ \bigcup_{s>t} S(s)\mathcal{B} \right\}$$

is the so-called global attractor. By definition it is a compact set that is uniformly attracting and invariant for the solutions, cf. [10]. Note that $\mathcal{A} \subset \mathcal{B}$.

The following standard results will be used later.

Lemma 1.1. *The convective term $R(u) = (u \cdot \nabla)\, u$ is on $\mathcal{B}$ Lipschitz continuous from H^1 to H^0.*

Proof. Set $w = u - v$. The embedding $W^{2,2} \subset W^{1,4} \subset L^\infty$ gives

$$\begin{aligned} \|(u \cdot \nabla)\, u - (v \cdot \nabla)\, v\|_2 &\le \|u \cdot \nabla w\|_2 + \|w \cdot \nabla v\|_2 \\ &\le c_1 \big(\|u\|_\infty \|\nabla w\|_2 + \|w\|_4 \|\nabla v\|_4 \big) \\ &\le c_2 \|\nabla w\|_2 \big(\|u\|_{2,2} + \|v\|_{2,2} \big) \end{aligned}$$

Since $\mathcal{B}$ is bounded in H^2, the conclusion follows. □

Lemma 1.2. *For fixed $\tau > 0$, the mapping $S(\tau)$ is on $\mathcal{B}$ Lipschitz continuous from H^0 to H^1.*

Proof. Subtract equations for u, v and test by $w = u - v$ to get

$$\frac{1}{2} \frac{d}{dt} \|w\|_2^2 + \nu \|\nabla w\|_2^2 \le \int_\Omega \big| [R(u) - R(v)] \cdot w \big| = I .$$

By previous lemma

$$I \le \|R(u) - R(v)\|_2 \|w\|_2 \le c_1 \|\nabla w\|_2 \|w\|_2 \le \frac{\nu}{2} \|\nabla w\|_2^2 + \frac{c_2}{2} \|w\|_2^2 .$$

Hence

$$\frac{d}{dt}\|w\|_2^2 + \nu\|\nabla w\|_2^2 \leq c_2\|w\|_2^2 . \tag{6}$$

By Gronwall's lemma

$$\|w(t)\|_2^2 \leq e^{c_2 t}\|w(0)\|_2^2 , \qquad t \geq 0 . \tag{7}$$

Further, integrating (6) over $t \in (0,\tau)$ and using (7) gives

$$\nu\int_0^\tau \|\nabla w\|_2^2 \leq c_2\int_0^\tau \|w\|_2^2 + \|w(0)\|_2^2 \leq \|w(0)\|_2^2\big(1 + c_2\tau e^{c_2\tau}\big) . \tag{8}$$

Next, we test the equation for w by $-\triangle w$ to get

$$\frac{1}{2}\frac{d}{dt}\|\nabla w\|_2^2 + \nu\|\nabla^2 w\|_2^2 \leq \int_\Omega \big|[R(u) - R(v)]\cdot\triangle w\big| = J .$$

Here

$$J \leq \|R(u) - R(v)\|_2\|\triangle w\|_2 \leq c_3\|\nabla w\|_2\|\nabla^2 w\|_2 \leq \nu\|\nabla^2 w\|_2^2 + \frac{c_4}{2}\|\nabla w\|_2^2 .$$

Hence

$$\frac{d}{dt}\|\nabla w\|_2^2 \leq c_4\|\nabla w\|_2^2 .$$

Multiply by a smooth function ξ such that $\xi(0) = 0$ and $\xi(\tau) = 1$. Integrating over $t \in (0,\tau)$ (by parts on the left) gives

$$\|\nabla w(\tau)\|_2^2 \leq \int_0^\tau \big(c_4|\xi| + |\xi'|\big)\|\nabla w\|_2^2 .$$

Combining with (8) gives the conclusion. □

Lemma 1.3. *Let P be a projection to the first N eigenfunctions of the Stokes operator (i.e., $-\triangle$ restricted to H^0.) For N large enough there exist $t_* > 0$ and $\theta \in (0,1/4)$ such that $S = S(t_*)$ satisfies on for any $u, v \in \mathcal{B}$ the so-called squeezing property:*

$$\|Su - Sv\|_2 \geq \sqrt{2}\|P(Su - Sv)\|_2 \implies \|Su - Sv\|_2 < \theta\|u - v\|_2 . \tag{9}$$

Proof. For the proof including the estimate

$$N = \operatorname{rank} P \leq c\,G^2\Big(\ln G^4\nu^2\lambda_1 + 1\Big) \tag{10}$$

see [4, pp. 57–60]. Here λ_1 the first eigenvalue of the Stokes operator and $G = \|f\|_2/\nu^2\lambda_1$ is the Grashof number. We will need a weaker form of squeezing property:

$$\|Su - Sv\|_2 \leq \sqrt{2}\|P(Su - Sv)\|_2 + \theta\|u - v\|_2 . \tag{11}$$

□

Remark. One can obtain better Lipschitz constants in Lemmas 1.1, 1.2 using more delicate estimates of the convective term. On the other hand, (9) or (11) can be deduced from the 'smoothing property' of Lemma 1.2. However, the estimate of N $(= \operatorname{rank} P)$ would be worse.

Finally, we will need the following abstract result.

Lemma 1.4. *Let X be a normed space. Let $B \subset X$ be bounded and $\widehat{F} : B \to \mathbb{R}$ be a Lipschitz mapping. Then there exists $F : X \to \mathbb{R}$ such that $F|_B = \widehat{F}$ and F is Lipschitz, bounded and has a bounded support.*

Proof. See, e.g., [5].

2. The reduced system

Set

$$\begin{aligned} \mathcal{A}^0 &= \{ u(t) : (-\infty, 0] \to H^0 \ \text{ is a solution to (5) with } u(0) \in \mathcal{A} \} \\ \mathcal{T} &= \{ Pu : \ u \in \mathcal{A}^0 \} \end{aligned}$$

Observe that $\mathcal{A}^0$ is bounded in $C((-\infty,0];H^0) \cap L^\infty(-\infty,0;H^2)$ and $\mathcal{T}$ is bounded in $C((-\infty,0];\mathbb{R}^N)$ and compact in $\mathcal{X} = C_\rho((-\infty,0];\mathbb{R}^N)$ with $\rho = e^{\gamma t}$, $\gamma > 0$ where

$$\|\chi\|_{\mathcal{X}} = \sup_{t \le 0} |\chi(t)| e^{\gamma t} .$$

To prove the compactness: given $\chi_n \in \mathcal{T}$ one finds corresponding $u_n \in \mathcal{A}^0$. Using the regularity of $\mathcal{A}$ and the equation, one obtains a subsequence converging to $\overline{u}$ in (say) $L^2_{\mathrm{loc}}(-\infty,0;H^1)$. This is enough to conclude that $\overline{u} \in \mathcal{A}^0$ and $\chi_n = Pu_n \to P\overline{u}$ in $\mathcal{X}$.

The following lemma is a key step of our construction.

Lemma 2.1. *There exists a mapping $E : \mathcal{T} \to H^1$ such that $E\{Pu^0\} = u(0)$ for $\forall u \in \mathcal{A}^0$. Moreover, E is Lipschitz continuous from $\mathcal{X}$ to H^1 provided that $\gamma > 0$ is small enough.*

Proof. Set $w = u - v$ where u, $v \in \mathcal{A}^0$. Using Lemma 1.2 and then (11) repeatedly one has

$$\begin{aligned} c_1^{-1} \|\nabla w(0)\|_2 &\le \|w(-t_*)\|_2 \le \sqrt{2} \|Pw(-t_*)\|_2 + \theta \|w(-2t_*)\|_2 \\ &\le \sqrt{2}\|Pw(-t_*)\|_2 + \sqrt{2}\theta \|Pw(-2t_*)\|_2 + \theta^2 \|w(-3t_*)\|_2 \\ &\dots \\ &\le \sqrt{2} \sum_{k=1}^{n} \theta^{k-1} \|Pw(-kt_*)\|_2 + \theta^n \|w(-(n+1)t_*)\|_2 . \end{aligned}$$

Letting $n \to \infty$ and using the boundedness of $w(t)$ one finds

$$\|\nabla w(0)\|_2 \le c_1\sqrt{2}\sum_{k=1}^{\infty}\theta^{k-1}\|Pw(-kt_*)\|_2 \le c_2\|Pw^0\|_{\mathcal{X}}\sum_{k=1}^{\infty}\theta^k e^{\gamma k t_*}\,.$$

In the second inequality we estimated $\|Pw(t)\| \le \|Pw^0\|_{\mathcal{X}}e^{-\gamma t}$. If $\ln\theta + \gamma t_* < 0$, the last sum converges and we are done. □

Now we can deal with the projection of the nonlinear term.

Lemma 2.2. *There exists a mapping $B : \mathcal{X} \to \mathbb{R}^N$ such that if $u \in \mathcal{A}^0$ then $B(Pu^0) = PR(u(0))$. Moreover, B is Lipschitz, bounded and has bounded support.*

Proof. We define $\widehat{B} : \mathcal{T} \to \mathbb{R}^N$ by $\widehat{B}(\chi) = PR(E(\chi))$. By Lemmas 2.1, 1.1, $\widehat{B}$ is Lipschitz and $\widehat{B}(Pu^0) = PR(u(0))$ if $u \in \mathcal{A}^0$. Applying Lemma 1.4 to each of the components $\widehat{B}_i$, $i = 1,\dots,N$ gives the desired mapping B. □

Let now u be a solution to (5) with values in $\mathcal{A}$. We can assume that u is defined for all $t \in \mathbb{R}$. Applying P to (5) gives (denoting $p(t) = Pu(t)$)

$$p_t + Ap + B(p^t) = g\,. \tag{12}$$

Here $A = \nu\,\mathrm{diag}\,(\lambda_1,\dots,\lambda_N)$, $g = Pf$ and B comes from Lemma 2.2.

3. Properties of the reduced system

We first establish the existence and uniqueness of solutions.

Theorem 3.1. *For any $\chi \in \mathcal{X}$ and $T > 0$ there exist unique $p : (-\infty, T] \to \mathbb{R}^N$ a solution to (12) with $p^0 = \chi$.*

Proof. Assume p_1, p_2 are two solutions to (12). Multiplying the equation for $z = p_1 - p_2$ by z gives

$$\frac{1}{2}\frac{d}{dt}|z|^2 + \nu\lambda_1|z|^2 \le |B(p_1^t) - B(p_2^t)|\,|z| = I\,.$$

By Lemma 2.2

$$|I| \le c_1\|z^t\|_{\mathcal{X}}|z| \le \nu\lambda_1|z|^2 + \frac{c_2}{2}\|z^t\|_{\mathcal{X}}^2\,.$$

Hence

$$\frac{d}{dt}|z|^2 \le c_2\sup_{s\le t}|z(s)|^2 e^{2\gamma(s-t)}\,.$$

Integrating over $s \in (0,\tau)$ (where $\tau \in (0,T)$ is fixed) gives

$$|z(\tau)|^2 \le |z(0)|^2 + c_2 T\sup_{s\le 0}|z(s)|e^{2\gamma s} + c_2 T\sup_{s\in[0,T]}|z(s)|^2\,.$$

Coming to $\sup_{\tau\in[0,T]}$ on the left gives (as $|z(0)|^2 \le \|z^0\|_{\mathcal{X}}^2$)

$$\|z\|_{C([0,T])}^2 \le (c_2T+1)\|z^0\|_{\mathcal{X}}^2 + c_2T\|z\|_{C([0,T])}^2\,.$$

Thus if $c_2 T < 1$ one has a local existence and uniqueness by a fixed point argument. As c_2 is independent of the data, the results extend globally. We omit further details. □

Next we show that the reduced system remains dissipative.

Theorem 3.2. *The system* (12) *has a global attractor* $\mathcal{A}_r \subset \mathcal{X}$ *and* $\mathcal{A}_r$ *is bounded in* $C((-\infty, 0]; \mathbb{R}^N)$.

Proof. Multiply (12) by p to obtain

$$\frac{1}{2}\frac{d}{dt}|p|^2 + \nu\lambda_1|p|^2 + B(p^t)\,p \le |f|\,|p|\,.$$

Let $R_0 > 0$ be such that $B(\chi) = 0$ if $\|\chi\|_{\mathcal{X}} \ge R_0$. We can assume $|f| - \nu\lambda_1 R_0 \le -1$. Since $\|p^t\|_{\mathcal{X}} \ge |p(t)|$, on has

$$\frac{1}{2}\frac{d}{dt}|p|^2 \le |p|\big[|f| - \nu\lambda_1|p|\big] \le -|p|$$

whenever $|p(t)| \ge R_0$. One concludes that $|p(t)| \le R_0$ for $t \ge t_0$.

Set $\mathcal{B}_r = \{\chi \in \mathcal{X} : \ |\chi(t)| \le R_0 \text{ for } \forall t \le 0\}$. The auxiliary function $q = q(t)$ is defined as $p(t)$ if $t \ge t_0$ and $p(t_0)$ if $t < t_0$. Clearly $q^t \in \mathcal{B}_r$, hence

$$\begin{aligned}\operatorname{dist}_{\mathcal{X}}(p^t, \mathcal{B}_r) &\le \|q^t - p^t\|_{\mathcal{X}} = \sup_{s \le t_0} |p(s) - q(s)| e^{\gamma(s-t)} \\ &\le R_0 e^{\gamma(t_0 - t)} + \|p^{t_0}\|_{\mathcal{X}}\, e^{\gamma(t_0-t)} \to 0 \quad \text{as} \quad t \to \infty\,.\end{aligned}$$

Thus the set $\mathcal{B}_r$ is uniformly attracting. Hence it is enough to find an attractor to solutions starting from $\mathcal{B}_r$. Due to [10, Remark 1.4] it suffices to verify the asymptotic compactness: if $\chi_n = p_n^0 \in \mathcal{B}_r$ and $t_n \to \infty$, the sequence $p_n^{t_n}$ has a converging subsequence in $\mathcal{X}$ (cf. .)

However, $p_n^{t_n}$ are uniformly bounded on $(-\infty, 0]$ and its derivatives are uniformly bounded on $[-t_n, 0]$ from the equation. Thus there is a subsequence converging locally uniformly in $(-\infty, 0]$ and hence in $\mathcal{X}$. □

Now we can prove our main theorem that relates the large time behavior of (5) and (12). Recall that $\mathcal{A}$, $\mathcal{A}_r$ are global attractors to (5), (12) respectively, $\mathcal{A}^0$ are backward trajectories on $\mathcal{A}$ and $\mathcal{T} = P\mathcal{A}^0$. Here P is a projection to first N eigenfunctions of the Stokes operator and N is estimated in (10). The mapping E was introduced in Lemma 1.2.

Theorem 3.3.

(i) *Let u be a solution to* (5) *with an initial condition $u_0 \in \mathcal{A}$. Then $p(t) = Pu(t)$ is a solution to* (12) *with an initial condition $\chi = Pu^0 \in \mathcal{T}$.*

(ii) *Conversely, let p be a solution to* (12) *with an initial condition $\chi \in \mathcal{T}$. Then $u(t) = E(p^t)$ is a solution to* (5) *with an initial condition $u_0 = E(\chi) \in \mathcal{A}$.*

(iii) *The set $\mathcal{T} \subset \mathcal{A}_r$ is compact and positively invariant with respect to* (12).

Proof. Ad (i). Assume u is defined for all $t \in \mathbb{R}$. (One can prove that the backward continuation is unique, but we do not need that.) Then apply P to find that $p = Pu$ solves (12), see after Lemma 2.2.

Ad (ii). Take $u^0 \in \mathcal{A}^0$ such that $Pu^0 = \chi$ and extend u to all $t \geq 0$. By (i), $q = Pu$ solves (12) with an initial condition χ. However, due to the uniqueness (Theorem 3.1), $p = q$. In particular, by Lemma 2.1, $u(t) = E(p^t)$ for all $t \geq 0$.

Ad (iii). For the compactness of $\mathcal{T}$ see before Lemma 2.1. The invariance follows easily from (i), (ii) and the invariance of $\mathcal{A}$ with respect to (5). □

References

[1] Babin A.V., Vishik, M.I., "Attractors of evolution equations," Studies in Mathematics and its Applications 25 North-Holland, Amsterdam, 1992.

[2] Constantin P., Foias C., "Navier-Stokes equations," University of Chicago Press, Chicago, 1988.

[3] Debussche A., Temam R., *Some new generalizations of inertial manifolds,* Discrete Contin. Dyn. Syst. (4) 2 (1996), 543–558.

[4] Eden A., Foias C., Nicolaenko B., Temam R., "Exponential attractors for dissipative evolution equations," Masson, Paris, 1994.

[5] Engelking R., "General Topology," Heldermann Verlag, Berlin, 1989.

[6] Foias C., Sell G.R., Temam R., *Inertial manifolds for nonlinear evolutionary equations,* J. Differ. Equations (2) 73 (1988), 309–353.

[7] Hale J.K., Raugel G., *Regularity, determining modes and Galerkin methods,* J. Math. Pures Appl. (9) 82 (2003), 1075–1136.

[8] Jones D.A., Titi E.S., *Upper bounds on the number of determining modes, nodes, and volume elements for the Navier-Stokes equations,* Indiana Univ. Math. J. (3) 42 (1993), 875–887.

[9] Sell G.R., You Y., "Dynamics of evolutionary equations," Applied Mathematical Sciences 143, Springer, New York, 2002.

[10] Temam R., "Infinite-dimensional dynamical systems in mechanics and physics," Springer Verlag, New York, 1997.

Dalibor Pražák
Department of Mathematical Analysis
Charles University Prague
Sokolovska 83
186 75 Prague 8, Czech Republic
e-mail: prazak@karlin.mff.cuni.cz

Progress in Nonlinear Differential Equations
and Their Applications, Vol. 64, 413–419

Quasilinear Parabolic Equations in L^p

Joachim Rehberg

Abstract. The paper contains a local existence and uniqueness result for quasilinear parabolic equations on a three-dimensional domain including mixed boundary conditions and discontinuous coefficients.

Mathematics Subject Classification (2000). 35B45, 35D05/10, 35J05/10/20/60/65, 35Q40, 47A55/60/75, 47H05/10/15.

Keywords. Partial differential equations, nonsmooth domains, mixed boundary conditions, discontinuous coefficients.

1. Introduction

In modelling of real world problems one often is confronted with partial differential equations which act on nonsmooth domains, possess discontinuous coefficients or/and are complemented by mixed boundary conditions [1], [25] [4] [17], [2], [7], [12], [9], [5]. In particular, in simulation of the current flow in real semiconductor devices this is unavoidable, or the model is meaningless [25]. In this sense, the focus of this paper is on quasilinear equations of the (formal) type

$$\frac{\partial u}{\partial t} - \operatorname{div}\big(F(t,u)\mu \operatorname{grad} u\big) = G(t,u,\operatorname{grad} u), \quad u(T_0) = u_0 \tag{1}$$

including Neumannn, Dirichlet or mixed type boundary conditions. The aim is to regard these equations in a suitably chosen L^p space; only this justifies the application of Gauss' theorem which allows to re-establish the first principle formulation (cf. [26])

$$\frac{\partial}{\partial t}\int_\Xi u\,dx - \int_{\partial\Xi} \nu\cdot j\,d\sigma = \int_\Xi G(t,u,\operatorname{grad} u)\,dx \tag{2}$$

(j being the species current belonging to the particle species represented by u and Ξ any suitable subdomain of Ω). In fact, only the knowledge that $\operatorname{div} j$ is a (locally integrable) function allows to define the normal component of the current across boundaries of subdomains. Let us mention that this also has important consequences for the numerical treatment of the equations, see [8] and [21]. It turns out that – among other things – an isomorphism property for the associated elliptic operators

$$-\operatorname{div}\mu\operatorname{grad}+1 : H^{1,q}_\Gamma(\Omega) \longmapsto H^{-1,q}_\Gamma(\Omega) \tag{3}$$

(Γ being the Neumann boundary part) is of great use if q is greater than the dimension of the domain Ω (see [18] or [23]). The problem is that for arbitrary (elliptic, L^∞) coefficients the occurring $q's$ do exceed 2 (cf. [15]), but in general only arbitrarily little (cf.[24]). This suffices for the treatment of two-dimensional problems (cf. [18] or [19]), while in the meanwhile the necessity grows to regard not only two-dimensional problems (mostly as cuts of the originally three-dimensional ones) but the three-dimensional models themselves. In particular, this is true in device modelling (cf. [10], [22]) and in many of the above-mentioned applications. This makes it desirable to find a class of (three-dimensional) domains Ω, coefficient functions μ and Neumann parts Γ of the boundary, such that the operator (3) provides a topological isomorphism for a $q > 3$. This we will do first within this paper. Afterwards we establish some functional analytic tools with which we finally solve (1).

2. Notations/General assumptions

1. Assumption. We assume $\Omega \subset \mathbb{R}^3$ to be a bounded three-dimensional Lipschitz domain. Γ is an open boundary part satisfying the following condition: For any $x \in \partial\Omega$ there is an open neighborhood $\mathcal{U}$ of x and a bi-Lipschitz mapping L_x from $\mathcal{U}$ onto an open set $\mathcal{V}$ such that $L_x(\mathcal{U} \cap (\Omega \cup \Gamma))$ is either

- the open unit cube
- the open unit cube combined with its ground plate
- the open unit cube combined with half of its ground plate.

For $s \in [0, 1]$ and $q \in]1, \infty[$ we denote by $H^{s,q}$ the space of Bessel potentials with differentiability index s and integrability index q on Ω. By $H^{s,q}_\Gamma$ we denote the closure of

$$\{\psi|_\Omega : \psi \in C_0^\infty(\mathbb{R}^3), supp\psi \cap (\partial\Omega \setminus \Gamma) = \emptyset\}$$

in $H^{s,q}$. If $s \in [-1, 0[$ then $H^{s,q}_\Gamma$ shall denote the dual of $H^{-s,q'}_\Gamma$, provided $\frac{1}{q} + \frac{1}{q'} = 1$.

2. Assumption. We will always assume that μ is an L^∞ function on Ω taking its values in the set of **real**, symmetric 3×3 matrices. Additionally, μ shall satisfy the usual ellipticity condition

$$\underline{\mu} := ess \inf_{x\in\Omega} \inf_{\|y\|_{\mathbb{R}^3}=1} (\mu(x)y) \cdot y > 0. \tag{4}$$

The operator

$$-\operatorname{div}\mu\operatorname{grad} : H^{1,2}_\Gamma \longmapsto H^{-1,2}_\Gamma \tag{5}$$

is defined by

$$\langle -\operatorname{div}\mu\operatorname{grad}\psi_1, \psi_2\rangle = \int_\Omega (\mu\operatorname{grad}\psi_1)\cdot\operatorname{grad}\psi_2\, dx. \tag{6}$$

3. The elliptic regularity result

3. Proposition. *If the domain Ω, the Neumann boundary part Γ and the coefficient function μ satisfy one of the following conditions I–V then there is a number $\hat{q} = \hat{q}(\Omega, \Gamma, \mu) > 3$ such that (3) provides a topological isomorphism for every $q \in [2, \hat{q}]$.*

Condition I Ω is Lipschitzian, $\Gamma = \emptyset$, $\mu \equiv 1$ cf. [16].

Condition II Ω is Lipschitzian, $\Gamma = \partial\Omega$, $\mu \equiv 1$, cf. [27].

Condition III Ω is a polyhedral Lipschitz domain, $\Gamma = \emptyset$, μ is constant on (finitely many) polyhedral subdomains and the resulting analytic operator pencils satisfy a specific condition on the edges, cf. [23].

Condition IV Ω is a convex polyhedron, $\bar{\Gamma} \cap (\partial\Omega \setminus \Gamma)$ is a finite union of line segments, $\mu \equiv 1$, cf. [3].

Condition V Ω is Lipschitzian, $\Omega = \Omega_1 \cup \partial\Omega_1 \cup (\Omega \setminus \bar{\Omega}_1)$, Ω_1 is C^1, μ is uniformly continuous on Ω_1 and on $\Omega \setminus \bar{\Omega}_1$, cf. [6].

4. Proposition. *(Gröger's localization principle, cf. [15] and [6]) The assertion of the preceding proposition remains true if there is a (finite) localization of Ω and Γ such that the resulting local objects satisfy one of the Conditions I–V.*

4. Functional analytic instruments

Having the goal in mind to treat (1) in an L^p space, we give the following

5. Definition. If $p \in [\frac{6}{5}, \infty[$ we denote by $-\nabla \cdot \mu\nabla$ the maximal restriction of the operator $-\operatorname{div} \mu \operatorname{grad}$ to the space L^p. The corresponding domain is denoted by $\mathcal{D}$.

6. Proposition. *Let U be a positive, essentially bounded function on Ω with an upper bound $\hat{U}$ and a strictly positive lower bound $\underline{U}$. Then the operator $-U\nabla \cdot \mu\nabla$ is densely defined and satisfies the resolvent estimate for z with $\Re z \geq 0$:*

$$\|(-U\nabla \cdot \mu\nabla + 1 + z)^{-1}\|_{\mathcal{B}(L^p)} \leq \Lambda\Big(\frac{\|\mu\|_{L^\infty}}{\underline{\mu}}, p\Big)\frac{\hat{U}}{\underline{U}} \cdot \frac{1}{1+|z|}. \tag{7}$$

Hereby, the function $\Lambda : [1, \infty[\times]1, \infty[\longmapsto]0, \infty[$ is continuous.

7. Proposition. *Assume*

$$\theta \in]0, 1[,\ q \in]1, \infty[,\ s_0, s_1 \in [-1, 1] \setminus \{1/q, -1/q'\}$$

and $s = (1 - \theta)s_0 + \theta s_1$. Then

$$[H_\Gamma^{s_0,q}, H_\Gamma^{s_1,q}]_\theta = H_\Gamma^{s,q}. \tag{8}$$

5. Result

At first we will now fix the Banach space, namely we put $p := \frac{q}{2}$.

8. Remark. Generically, q cannot be expected to exceed 4 and, hence, $p \leq 2$.

9. Theorem. *Assume* $3 < q \leq 4$. *For* $\alpha \in]\frac{1}{2} + \frac{3}{2q}, 1]$ *the space* $[L^p, \mathcal{D}]_\alpha$ *continuously embeds into* $H^{1,q}_\Gamma$.

10. Lemma. *Assume* $s \in [0,1[$ *and* $\tau \in]s, 1]$ *and set* $\varrho = \tau - s$. *Then the operator* $(-\operatorname{div}\mu \operatorname{grad} + 1)^{\varrho/2}$ *maps* $H^{-s,2}_\Gamma$ *isomorphically onto* $H^{-\tau,2}_\Gamma$.

Proof. The operator $-\operatorname{div}\mu\operatorname{grad}+1$ is positive and generates analytic semigroups on both, $H^{-1,2}_\Gamma$ and L^2. Thus, powers of $-\operatorname{div}\mu\operatorname{grad}+1$ and its restriction to the range space L^2 are well defined and the usual rules for calculus hold. In this sense, we consider the operators

$$B \stackrel{\text{def}}{=} (-\operatorname{div}\mu\operatorname{grad}+1)^{1/2} : H^{1,2}_\Gamma \mapsto L^2\,,$$

$$C \stackrel{\text{def}}{=} (-\operatorname{div}\mu\operatorname{grad}+1)^{1/2} : L^2 \mapsto H^{-1,2}_\Gamma\,.$$

Clearly, we have

$$(B^*)^\varrho \;=\; (B^\varrho)^* \;=\; C^\varrho.$$

By a well-known theorem B^ϱ maps the domain of B^τ isomorphically onto the domain of B^s. On the other hand, B is positive and selfadjoint, thus

$$\operatorname{dom}(B^\gamma) = [L^2, \operatorname{dom}(B)]_\gamma = [L^2, H^{1,2}_\Gamma]_\gamma\ ,\ \gamma \in]0,1[$$

holds. Because these interpolation spaces are identical with $H^{\gamma,2}_\Gamma$, B^ϱ provides a topological isomorphism between $H^{\tau,2}_\Gamma$ and $H^{s,2}_\Gamma$. Hence, by duality, $C^\varrho = (-\operatorname{div}\mu\operatorname{grad})^{\varrho/2}$ maps $H^{-s,2}_\Gamma$ isomorphically onto $H^{-\tau,2}_\Gamma$. □

We now give the proof of Theorem 9:

Assume $s = 3(\frac{1}{p} - \frac{1}{2})$ and $\tau = \frac{3}{q} - \frac{1}{2}$ and set $\varrho = \tau - s$. The definition of s and τ yields the (continuous) embeddings

$$L^p \hookrightarrow H^{-s,2}_\Gamma \quad \text{and} \quad H^{-\tau,2}_\Gamma \hookrightarrow H^{-1,q}_\Gamma.$$

By definition, we have $\alpha > \frac{1}{2} + \frac{3}{2q} = 1 - \rho/2$. Thus, one gets

$$[L^p, \mathcal{D}]_\alpha \hookrightarrow \operatorname{dom}\big((-\nabla\cdot\mu\nabla + 1)^{1-\rho/2}\big).$$

Hence, it suffices to prove

$$\operatorname{dom}\big((-\nabla\cdot\mu\nabla + 1)^{1-\rho/2}\big) \hookrightarrow H^{1,q}_\Gamma,$$

or, what is the same, $(-\nabla\cdot\mu\nabla+1)^{\rho/2-1} \in \mathcal{B}(L^p; H^{1,q}_\Gamma)$.

Denoting by ι and κ the embedding constants between the corresponding spaces, we may estimate

$$\begin{aligned}
&\|(-\nabla\cdot\mu\nabla+1)^{\varrho/2-1}\|_{\mathcal{B}(L^p;H^{1,q}_\Gamma)}\\
&\le \|(-\operatorname{div}\mu\operatorname{grad}+1)^{-1}\|_{\mathcal{B}(H^{-\tau,2}_\Gamma;H^{1,q}_\Gamma)}\|(-\operatorname{div}\mu\operatorname{grad}+1)^{\varrho/2}\|_{\mathcal{B}(L^p;H^{-\tau,2}_\Gamma)}\\
&\le \iota\kappa\|(-\operatorname{div}\mu\operatorname{grad}+1)^{-1}\|_{\mathcal{B}(H^{-1,q}_\Gamma;H^{1,q}_\Gamma)}\|(-\operatorname{div}\mu\operatorname{grad}+1)^{\varrho/2}\|_{\mathcal{B}(H^{-s,2}_\Gamma;H^{-\tau,2}_\Gamma)}.
\end{aligned}$$

11. Theorem. *Assume that Ω,Γ,μ form a constellation which is described in Section 3. Suppose, additionally, that the function F in* (1) *depends Hölder continuously from t and locally Lipschitzian from the second argument. Finally, let the dependence of G in* (1) *from t also be Hölder continuous, from the second argument locally Lipschitzian and from ∇u not be stronger than quadratic. Then, if the (Dirichlet homogenized) initial value belongs to the space $(L^p,\mathcal{D})^0_{\alpha,\infty}$ with $\alpha\in]\frac{1}{2}+\frac{3}{2q},1]$, then the equation* (1) *has locally in time a unique solution which belongs to*

$$C^1(]T_0,T];L^p)\cap C(]T_0,T];\mathcal{D})\cap C([T_0,T];(L^p,\mathcal{D})^0_{\alpha,\infty}).$$

Proof. The proof results by applying the local existence result from [20]. The suppositions of that theorem are assured by Proposition 6 and Theorem 9 on one hand and our suppositions made on the mappings F and G on the other. □

Acknowledgement

The author thanks J. Elschner, J.A. Griepentrog, H.-Chr. Kaiser, and G. Schmidt from the Weierstrass Institute for Applied Analysis and Stochastics, Berlin, for scientific collaboration on the topic.

References

[1] H. Amann, *Nonhomogeneous Linear and Quasilinear Elliptic and Parabolic Boundary Value Problems*, Schmeisser, Hans-Jürgen et al (Eds.), Function spaces, differential operators and nonlinear analysis. Survey articles and communications of the international conference held in Friedrichsroda, Germany, September 20–26, 1992. Stuttgart: B. G. Teubner Verlagsgesellschaft.

[2] S.N. Antontsev, M. Chipot, *The thermistor problem: Existence, smoothness, uniqueness, blowup*, SIAM J. Math. Anal. **25** (1994), 1128–1156.

[3] M. Dauge, *Neumann and mixed problems on curvilinear polyhedra*, Integral Equations Oper. Theory **15**, (1992), 227–261 (1992).

[4] P. Degond, S. Genieys, A. Jüngel, *A steady-state system in non-equilibrium thermodynamics including thermal and electrical effects*, Math. Meth. Appl. Sci **21** XV (1998) 1399–1413.

[5] F. Duderstadt, D. Hömberg, A.M. Kludnev, *A mathematical model for impulsive resistance welding*, Math. Meth. Appl. Sci. **26** 717–737.

[6] J. Elschner, H.-Chr. Kaiser, J. Rehberg, G. Schmidt, *Optimal regularity for elliptic operators occurring in real world problems*, in preparation.

[7] P. Fabrie, T. Gallouet, *Modelling wells in porous media flow*, Math. Mod. Meth. Appl. Sci. **10** V (2000) 673–70.

[8] J. Fuhrmann, H. Langmach, *Stability and existence of solutions of time-implicit finite volume schemes for viscous nonlinear conservation laws*, Appl. Num. Math. **37** (2001) 201–230.

[9] H. Gajewski, K. Gröger, *Initial boundary value problems modelling heterogeneous semiconductor devices*, in: Surveys on Analysis, Geometry and Mathematical Physics, B. W. Schulze, H. Triebel (Eds.) Teubner-Texte zur Mathematik, Leipzig, 1990.

[10] H. Gajewski, H.-Chr. Kaiser, H. Langmach, R. Nürnberg, R.H. Richter, *Mathematical Modelling and Numerical Simulation of Semiconductor Detectors*, in: MATHEMATICS: Key Technology for the Future, W. Jäger, H.-J. Krebs (Eds.) Springer (2002) 355–364.

[11] H. Gajewski, K. Gröger, *Reaction-diffusion processes of electrically charged species*, Math. Nachr. **177** (1996), 109–130.

[12] A. Glitzky, R. Hünlich, *Global estimates and asymptotics for electro-reaction-diffusion equations*, Appl. Anal. **66** (1997) 205–226.

[13] J.A. Griepentrog, H.-Chr. Kaiser, J. Rehberg, *Heat kernel and resolvent properties for second-order elliptic differential operators with general boundary conditions on* L^p, Adv. Math. Sci. Appl. **11** (1), (2001) 87–112.

[14] J.A. Griepentrog, K. Gröger, H.-Chr. Kaiser, J. Rehberg, *Interpolation for function spaces related to mixed boundary value problems*, Math. Nachr. **241** (2002) 110–120.

[15] K. Gröger, *A* $W^{1,p}$*-estimate for solutions to mixed boundary value problems for second-order elliptic differential equations*, Math. Ann. **283** (1989), 679–687.

[16] D. Jerison, C. Kenig, *The inhomogeneous Dirichlet problem in Lipschitz domains*, J. Funct. Anal. **130**, (1995), 161–219.

[17] A. Jüngel, *Regularity and uniqueness of solutions to a parabolic system in nonequilibrium thermodynamics*, Nonlinear analysis: theor. appl. **41** V-VI, (2000) 669–688.

[18] H.-Chr. Kaiser, H. Neidhardt, J. Rehberg, *Classical solutions of quasilinear parabolic systems on two-dimensional domains*, to appear in NoDEA.

[19] H.-Chr. Kaiser, H. Neidhardt, J. Rehberg, *Classical solutions of quasilinear parabolic systems on two-dimensional domains*, WIAS-Preprint No. 765 (2002).

[20] A. Lunardi, *Abstract quasilinear parabolic equations*, Math. Ann. **267** (1984) 395–415.

[21] R. Luce, S. Perez, *A finite volume scheme for an elliptic equation with heterogeneous coefficients. Application to a homogenization problem* Appl. Num. Math. **38** (2001) 427–444.

[22] G. Lutz, R.H. Richter, L. Strüder, *DEPMOS-Arrays for X-ray imaging*, XRAY Optics, Instruments and Missions III, J. Trümper, B. Aschenbach (Eds.), Proceedings of SPIE, vol. 4012, (2000), 249–256.

[23] W. Maz'ya, J. Elschner, J. Rehberg, G. Schmidt, *Solutions for quasilinear evolution systems in* L^p, Arch. Rat. Mech. Anal. **171** (2004) 219–262.

[24] N.G. Meyers, *An L^p-estimate for the gradient of solutions of second-elliptic divergence equations*, Ann. Scuola Norm. Sup. Pisa **17**, (1963), 189–206.

[25] S. Selberherr, *Analysis and simulation of semiconductor devices*, Springer-Verlag, Wien-New York, 1984.

[26] A. Sommerfeld, *Thermodynamics and statistical mechanics*, Lectures on theoretical physics, vol. V, Academic Press, New York, 1956.

[27] D. Zanger, *The inhomogeneous Neumann problem in Lipschitz domains*, Commun. Partial Differ. Equations **25** (2000), 1771–1808.

Joachim Rehberg
Weierstrass Institute for Applied Analysis and Stochastics
Mohrenstr. 39
D-10117 Berlin, Germany
e-mail: `rehberg@wias-berlin.de`

Progress in Nonlinear Differential Equations
and Their Applications, Vol. 64, 421–432

Parabolic Equations in Locally Uniform Spaces

Aníbal Rodríguez-Bernal

"Nonlinear Elliptic and Parabolic Problems: A Special Tribute to the Work of Herbert Amann", Zürich, June 2004

1. The problem

In this talk we present some recent results on linear parabolic equations of the form

$$\begin{cases} u_t - \Delta u = V(x)u & \text{in } \mathbb{R}^N, \quad t > 0 \\ u(x,0) = u_0(x) \end{cases} \tag{1.1}$$

where both the potential $V(x)$ and the initial data belong to the so-called locally uniform spaces, [2]. These spaces, defined below, allow for local singularities and impose no growth restrictions at infinity. Moreover they enjoy suitable nesting and embedding properties similar as the Lebesgue spaces in a bounded domain.

Also, we consider more general parabolic equations of the form

$$\begin{cases} u_t - \displaystyle\sum_{k,l=1}^{N} a_{kl}(x)\partial_k\partial_l u + \sum_{j=1}^{N} b_j(x)\partial_j u + c(x)u = 0, \quad x \in \mathbb{R}^N \\ u(x,0) = u_0(x), \quad x \in \mathbb{R}^N \end{cases} \tag{1.2}$$

with complex-valued coefficients and initial data in uniform spaces. In this context our goal is to extend the results in [3] to uniform spaces.

Finally some results on nonlinear equations will be also presented, see [4].

2. Uniform spaces

Recall that the uniform spaces, [9, 1, 2] consists of functions $\phi \in L^q_{\text{loc}}(\mathbb{R}^N)$, such that

$$\sup_{x \in \mathbb{R}^N} \int_{B(x,1)} |\phi(y)|^q \, dy < \infty \tag{2.1}$$

Partially supported by BFM2003-03810, Spain.

with norm

$$\|\phi\|_{L^q_U(\mathbb{R}^N)} = \sup_{x\in\mathbb{R}^N} \|\phi\|_{L^q(B(x,1))}.$$

Moreover we also consider the space $\dot{L}^q_U(\mathbb{R}^N)$ which is the subspace of $L^q_U(\mathbb{R}^N)$ such that

$$\|\tau_y\phi - \phi\|_{L^q_U(\mathbb{R}^N)} \to 0 \quad \text{as} \quad |y| \to 0$$

where $\{\tau_y, y \in \mathbb{R}^N\}$ are the translations.

Analogously one can define the uniform Sobolev spaces $W^{k,q}_U(\mathbb{R}^N)$ consisting of functions $\phi \in W^{k,q}_{\text{loc}}(\mathbb{R}^N)$ such that

$$\|\phi\|_{W^{k,q}_U(\mathbb{R}^N)} = \sup_{x\in\mathbb{R}^N} \|\phi\|_{W^{k,q}(B(x,1))} < \infty$$

for $k \in \mathbb{N}$ and $\dot{W}^{k,q}_U(\mathbb{R}^N)$ is the subspace of $W^{k,q}_U(\mathbb{R}^N)$

$$\|\tau_y\phi - \phi\|_{W^{k,q}_U(\mathbb{R}^N)} \to 0 \quad \text{as} \quad |y| \to 0$$

where $\{\tau_y, y \in \mathbb{R}^N\}$ are the translations.

Note that embeddings can be sketched in the following diagram

$$\begin{array}{ccccccc}
L^\infty(\mathbb{R}^N) = & L^\infty_U(\mathbb{R}^N) \subset & \cdots & L^q_U(\mathbb{R}^N) \subset & \cdots & L^1_U(\mathbb{R}^N) \\
 & \cup & & \cup & & \cup \\
BUC(\mathbb{R}^N) = & \dot{L}^\infty_U(\mathbb{R}^N) \subset & \cdots & \dot{L}^q_U(\mathbb{R}^N) \subset & \cdots & \dot{L}^1_U(\mathbb{R}^N) \\
 & & & \cup & & \cup \\
 & & & L^q(\mathbb{R}^N) \not\subset & & \not\supset L^1(\mathbb{R}^N)
\end{array}$$

3. The heat equation

We first consider the heat equation in $L^q_U(\mathbb{R}^N)$

$$\begin{cases} u_t - \Delta u = 0 & \text{in } \mathbb{R}^N, \quad t > 0 \\ u(x,0) = u_0(x) \in L^q_U(\mathbb{R}^N) \end{cases} \tag{3.1}$$

whose solution is given by the convolution with the heat kernel

$$u(x,t) = T(t)u_0 = (4\pi t)^{-N/2} \int_{\mathbb{R}^N} e^{-\frac{|x-y|^2}{4t}} u_0(y)\, dy. \tag{3.2}$$

Then we can show the following $L^q_U - L^p_U$ estimates, [2]

Theorem 3.1.

i) *There exists $M_0 = M_0(N)$ such that for $1 \le q \le p \le \infty$*

$$\|T(t)u_0\|_{L^p_U(\mathbb{R}^N)} \le M_0\left(t^{-\frac{N}{2}(\frac{1}{q}-\frac{1}{p})} + 1\right)\|u_0\|_{L^q_U(\mathbb{R}^N)},$$

$$\|D^\beta T(t)u_0\|_{L^p_U(\mathbb{R}^N)} \le M_0 t^{-\frac{1}{2}(k-\frac{N}{p}+\frac{N}{q})}\|u_0\|_{L^q_U(\mathbb{R}^N)},$$

for any $1 \le |\beta| = k$.

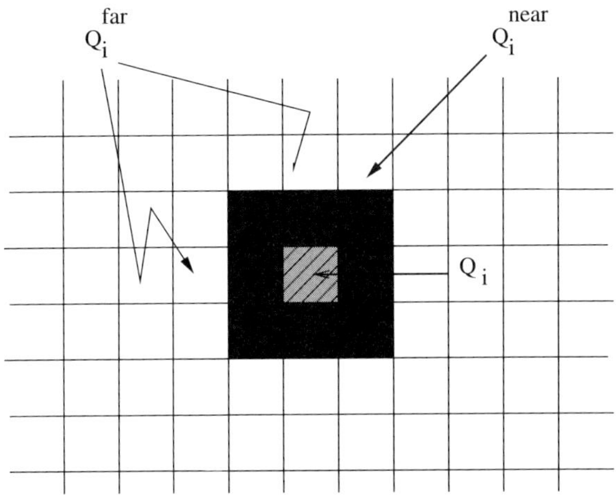

FIGURE 1. The ith cube, the near and far neighbors

ii) *For each bounded set* $B \subset \mathbb{R}^N$ *and* $u_0 \in L^q_U(\mathbb{R}^N)$

$$\|T(t)u_0 - u_0\|_{L^q(B)} \to 0, \quad \textit{as } t \to 0.$$

If $u_0 \in \dot{L}^q_U(\mathbb{R}^N)$, $1 \le q \le \infty$, *then* $T(t)u_0 \in \dot{L}^p_U(\mathbb{R}^N)$ *for* $t > 0$ *and any* $1 \le p \le \infty$ *and*

$$\|T(t)u_0 - u_0\|_{L^q_U(\mathbb{R}^N)} \to 0, \quad \textit{as } t \to 0.$$

Proof. We sketch the proof of i): Let $i \in \mathbb{Z}^N$, Q_i the open cube in $\mathbb{R}^N$ centered at i and sides parallel to the axes. Hence $Q_i \cap Q_j = \emptyset$, $i \neq j$, $\mathbb{R}^N = \cup_{i\in\mathbb{Z}^N}\overline{Q_i}$. Given $i \in \mathbb{Z}^N$, we denote $j \in N(i)$ if $\overline{Q_i} \cap \overline{Q_j} \neq \emptyset$ and then

$$Q_i^{\text{near}} = \cup_{j\in N(i)} Q_i, \qquad Q_i^{\text{far}} = \mathbb{R}^N \setminus \overline{Q_i^{\text{near}}}$$

For fixed $u_0 \in L^q_U(\mathbb{R}^N)$ and i, decompose

$$T(t)u_0 = \Big(T(t)u_0\Big)_i^{\text{near}} + \Big(T(t)u_0\Big)_i^{\text{far}}$$

with

$$\Big(T(t)u_0\Big)_i^{\text{near}} = T(t)(u_0\mathcal{X}_{Q_i^{\text{near}}}), \qquad \Big(T(t)u_0\Big)_i^{\text{far}} = T(t)(u_0\mathcal{X}_{Q_i^{\text{far}}})$$

Now the result follows from:

$$\|\Big(T(t)u_0\Big)_i^{\text{near}}\|_{L^p(Q_i)} \le (4\pi t)^{-\frac{N}{2}(\frac{1}{q}-\frac{1}{p})}\|u_0\|_{L^q(Q_i^{\text{near}})} \quad t > 0, \tag{3.3}$$

and

$$\|\Big(T(t)u_0\Big)_i^{\text{far}}\|_{L^\infty(Q_i)} \le c(t)\|u_0\|_{L^1_U(Q_i^{\text{far}})}, \quad t \ge 0 \tag{3.4}$$

with $0 \le c(t) \le Ct^{-N/2}e^{-\alpha/t}$ as $t \to 0$ and C and $\alpha > 0$ depending only on N.

In fact, using the inclusions $L^\infty(Q_i) \hookrightarrow L^p(Q_i)$ and restrictions $L^q_U(\mathbb{R}^N) \hookrightarrow L^q(Q_i^{\text{near}})$, $L^q_U(\mathbb{R}^N) \hookrightarrow L^1_U(Q_i^{\text{far}})$ from (3.4),(3.3) we get

$$\|T(t)u_0\|_{L^p(Q_i)} \le C(t^{-\frac{N}{2}(\frac{1}{q}-\frac{1}{p})} + c(t))\|u_0\|_{L^q_U(\mathbb{R}^N)}.$$

Taking the sup in i we get the result.

Now we prove (3.3) using standard estimates for the heat equation in Lebesgue spaces

$$\|\Big(T(t)u_0\Big)_i^{\text{near}}\|_{L^p(Q_i)} \le \|T(t)(u_0\mathcal{X}_{Q_i^{\text{near}}})\|_{L^p(\mathbb{R}^N)} \le (4\pi t)^{-\frac{N}{2}(\frac{1}{q}-\frac{1}{p})}\|u_0\|_{L^q(Q_i^{\text{near}})}.$$

On the other hand, to prove (3.4) we write $u_0\mathcal{X}_{Q_i^{\text{far}}} = \sum_{j\in\mathbb{Z}^N\setminus N(i)} u_0^j$ with $u_0^j = u_0\mathcal{X}_{Q_j}$. Then for each j

$$T(t)u_0^j(x) = (4\pi t)^{-N/2}\int_{Q_j} e^{-\frac{|x-y|^2}{4t}}u_0^j(y)dy$$

and denoting $d_{ij} := \inf\{\text{dist}(x,y),\ x\in Q_i, y\in Q_j\}$ we get

$$\|\Big(T(t)u_0\Big)_i^{\text{far}}\|_{L^\infty(Q_i)} \le \sum_{j\in\mathbb{Z}^N\setminus N(i)} \|T(t)u_0^j\|_{L^\infty(Q_i)}$$

$$\le (4\pi t)^{-N/2}\|u_0\|_{L^1_U(Q_i^{\text{far}})} \sum_{j\in\mathbb{Z}^N\setminus N(i)} e^{\frac{-d_{ij}^2}{4t}}.$$

Since $\#\{j\in\mathbb{Z},\ d_{ij}=k\} \le Ck^{N-1}$

$$\sum_{j\in\mathbb{Z}^N\setminus N(i)} e^{\frac{-d_{ij}^2}{4t}} \le C\sum_{k=1}^{\infty} k^{N-1}e^{\frac{-k^2}{4t}} = t^{N/2}c(t)$$

with $c(t)$ as in (3.4). □

Moreover we can also prove that the heat equation defines an analytic semigroup which is not continuous at $t=0$, as follows

Theorem 3.2. *The heat equation defines an order preserving analytic semigroup in $L^q_U(\mathbb{R}^N)$, for $1\le q\le\infty$, which is moreover continuous at $t=0$ if $u_0\in\dot{L}^q_U(\mathbb{R}^N)$. However the domain of Δ is not dense and satisfies*

$$D(\Delta)\begin{cases} \subset W^{1,p}_U & \text{for any } p<\frac{N}{N-1} \quad \text{if } q=1\\ = W^{2,q}_U & \text{for} \quad 1<q<\infty\\ \subset W^{2,p}_U & \text{for any } p<\infty \quad \text{if } q=\infty\end{cases}$$

Moreover, for $u_0\in L^q_U(\mathbb{R}^N)$, $1\le q\le\infty$

$$(0,\infty)\ni t\longmapsto T(t)u_0\in\dot{L}^\infty_U(\mathbb{R}^N) = BUC(\mathbb{R}^N)$$

is analytic.

Proof. The first part is proved by direct manipulation of (3.2). For the second part note that in particular we have proved the Calderon–Zigmud estimate: for $f \in L^q_U(\mathbb{R}^N)$, $1 < q < \infty$, we show

$$u_0(x) = \int_{\mathbb{R}^N} G(x-y) f(y)\, dy \in W^{2,q}_U(\mathbb{R}^N)$$

where $G(x)$ Green's function for $-\Delta + I$ in $\mathbb{R}^N$.

For this, fix the cube Q_i and write $f_i^{\mathrm{near}} = f\mathcal{X}_{Q_i^{\mathrm{near}}}$, $f_i^{\mathrm{far}} = f\mathcal{X}_{Q_i^{\mathrm{far}}}$. Then

$$\|(u_0)_i^{\mathrm{near}}\|_{W^{2,q}(Q_i)} \le \|(u_0)_i^{\mathrm{near}}\|_{W^{2,q}(\mathbb{R}^N)} \le C\|f\|_{L^q(Q_i^{\mathrm{near}})}$$

with C independent of i.

Using the cubes Q_j with $j \notin N(i)$, since $G \in C^\infty(\mathbb{R}^N)$ and decays exponentially, for any $k \in \mathbb{N}$

$$\|(u_0)_i^{\mathrm{far}}\|_{W^{k,\infty}(Q_i)} \le C\|f\|_{L^1_U(Q_i^{\mathrm{far}})}$$

C independent of i.

The analyticity part of the theorem is achieved by means of the following interesting Lemma, which allows to transfer analyticity to other spaces. □

Lemma 3.3. *Assume $\{S(t)\}_{t\ge 0}$ analytic semigroup in a Banach space X. Assume that for some Banach space Y and for $t > 0$,*

$$S(t) : X \to Y$$

is continuous.

Then for each $u_0 \in X$,

$$(0,\infty) \ni t \mapsto u(t) = S(t)u_0 \in Y \quad \textit{is analytic.}$$

In particular if $Y \subset X$, continuous, then $\{S(t)\}_{t\ge 0}$ is an analytic semigroup in Y.

Proof. For $t_0 > 0$, the Taylor series in X:

$$u(t) = S(t)u_0 = \sum_{m=0}^{\infty} \frac{1}{m!} u^{m)}(t_0)(t-t_0)^m, \quad \text{with} \quad u^{m)}(t_0) = S^{m)}(t_0)u_0.$$

For small enough $\epsilon > 0$, $u^{m)}(t_0) = A^m S(t_0)u_0 = S(\epsilon)A^m S(t_0-\epsilon)u_0$, where $-A$ is the infinitesimal generator in X. Hence, the series converges in Y since

$$\|u^{m)}(t_0)\|_Y \le C(\epsilon)\|A^m S(t_0-\epsilon)u_0\|_X = C(\epsilon)\|u^{m)}(t_0-\epsilon)\|_X$$

If $Y \subset X$, continuous, and $u_0 \in Y$,

$$\|S^{m)}(t_0)u_0\|_Y \le C(\epsilon)\|u^{m)}(t_0-\epsilon)\|_X \le$$

$$\le C(\epsilon)\|S^{m)}(t_0-\epsilon)\|_{\mathcal{L}(X)}\|u_0\|_X \le C(\epsilon)\|S^{m)}(t_0-\epsilon)\|_{\mathcal{L}(X)}\|u_0\|_Y.$$

Since the Taylor series converges in $\mathcal{L}(X)$ it also converges in $\mathcal{L}(Y)$. □

As the proofs above depend crucially on the Gaussian structure of the heat kernel, using [6], we can prove

Theorem 3.4. *Consider $u_t + L(u) = 0$ in $\mathbb{R}^N$ for*

$$L(u) = -\sum_{i=1}^{N} \partial_i\Big(a_{i,j}(x)\partial_j u + a_i(x)u\Big) + b_i(x)\partial_i u + c_0(x)u$$

with real coefficients $a_{i,j}, a_i, b_i, c_0 \in L^\infty(\mathbb{R}^N)$ and the ellipticity condition

$$\sum_{i,j=1}^{N} a_{i,j}(x)\xi_i\xi_j \geq \alpha_0|\xi|^2, \quad \xi \in \mathbb{R}^N, \quad \alpha_0 > 0.$$

Then the fundamental solution of the parabolic problem $u_t + Lu = 0$ in $\mathbb{R}^N$ satisfies a Gaussian bound

$$0 \leq k(x,y,t,s) \leq C(t-s)^{-N/2}e^{\omega(t-s)}e^{-c\frac{|x-y|^2}{(t-s)}}$$

for $t > s$ and $x, y \in \mathbb{R}^N$ where C, c, ω depend on the L^∞ norm of the coefficients.

Hence, the results above apply to the semigroup

$$u(t,x) = T_L(t)u_0 = \int_{\mathbb{R}^N} k(x,y,t,0)u_0(y)\,dy$$

4. Schrödinger semigroups

We consider now Schrödinger semigroups

$$\begin{cases} u_t - \Delta u = V(x)u, & x \in \mathbb{R}^N \quad t > 0, \\ u(0) = u_0 \in L^q_U(\mathbb{R}^N) \end{cases} \tag{4.1}$$

where V is in a locally uniform space. By treating this equation as a perturbation of the heat equation we get, [2]

Proposition 4.1. *Assume $V \in L^\sigma_U(\mathbb{R}^N)$, $\sigma > N/2$, then*

i) *For $1 \leq q \leq p \leq \infty$*

$$\|e^{(\Delta+V)t}u_0\|_{L^p_U(\mathbb{R}^N)} \leq Me^{at}t^{-\frac{N}{2}(\frac{1}{q}-\frac{1}{p})}\|u_0\|_{L^q_U(\mathbb{R}^N)},$$

a and M depend only on N, σ and $\|V\|_{L^\sigma_U(\mathbb{R}^N)}$.

ii) *For any bounded $B \subset \mathbb{R}^N$, $u_0 \in L^q_U(\mathbb{R}^N)$, $1 \leq q < \infty$,*

$$\|e^{(\Delta+V)t}u_0 - u_0\|_{L^q(B)} \to 0, \quad \text{as } t \to 0.$$

If moreover $u_0 \in \dot{L}^q_U(\mathbb{R}^N)$, $1 \leq q \leq \infty$, then

$$\|e^{(\Delta+V)t}u_0 - u_0\|_{L^q_U(\mathbb{R}^N)} \to 0, \quad \text{as } t \to 0.$$

Theorem 4.2. *If $V \in L^\sigma_U(\mathbb{R}^N)$, $\sigma > N/2$, then $\Delta+V$ generates an order preserving analytic semigroup in $L^q_U(\mathbb{R}^N)$, for $1 \leq q \leq \infty$ which is continuous at $t = 0$ if $u_0 \in \dot{L}^q_U(\mathbb{R}^N)$.*

Moreover for $u_0 \in L^q_U(\mathbb{R}^N)$, $1 \leq q \leq \infty$

$$(0,\infty) \ni t \longmapsto e^{(\Delta+V)t}u_0 \in \dot{L}^\infty_U(\mathbb{R}^N) = BUC(\mathbb{R}^N)$$

is analytic.

As for the decay of solutions in locally uniform spaces, we have

Lemma 4.3. *The semigroup* $e^{(\Delta+V)t}$ *decays in* $L^q_U(\mathbb{R}^N)$ *iff decays in* $L^p(\mathbb{R}^N)$ *for* $1 \leq p, q \leq \infty$

5. General second-order parabolic equations

Now we consider general second-order parabolic equations

$$\begin{cases} u_t - \sum_{k,l=1}^N a_{kl}(x)\partial_k\partial_l u + \sum_{j=1}^N b_j(x)\partial_j u + c(x)u = 0, & x \in \mathbb{R}^N \\ u(x,0) = u_0(x), \quad x \in \mathbb{R}^N \end{cases} \tag{5.1}$$

with **complex-valued** coefficients and initial data

$$u_0 \in L^q_U(\mathbb{R}^N) \quad 1 < q < \infty.$$

Note that systems can be also considered, but we stick to the scalar case for the sake of simplicity.

Hereafter we will make the following assumptions for the coefficients of the second-order differential operator A in (5.1)

- For some $0 < M$, $0 < \theta_0 < \frac{\pi}{2}$ and for all $x, \xi \in \mathbb{R}^N$ with $|\xi| = 1$

$$|A_0(x,\xi)| = |\sum_{i,j=1}^N a_{i,j}(x)\xi_i\xi_j| \geq \frac{1}{M}, \qquad |arg(A_0(x,\xi))| \leq \theta_0$$

- $a_{kl} \in BUC(\mathbb{R}^N)$, with modulus of continuity $\omega(s)$.
- $b_j \in L^{p_j}_U(\mathbb{R}^N)$ where $p_j > N$, $p_j \geq q$
- $c \in L^{p_0}_U(\mathbb{R}^N)$ where $p_0 > N/2$, $p_0 > q$
- $\sum_{k,l=1}^N \|a_{kl}\|_{BUC(\mathbb{R}^N,\omega)} + \sum_{j=1}^N \|b_j\|_{L^{p_j}_U(\mathbb{R}^N)} + \|c\|_{L^{p_0}_U(\mathbb{R}^N)} \leq M.$

Under these assumptions and considering solutions in $L^q(\mathbb{R}^N)$ the following results were proven, among other things, in [3]

Theorem 5.1. *Given* $0 < M$, $0 < \theta_0 < \theta < \pi$, A *in* $L^q(\mathbb{R}^N)$ *with domain* $W^{2,q}(\mathbb{R}^N)$, *is sectorial and for some* $\mu \in \mathbb{R}$

$$A + \mu I \in Iso(W^{2,q}(\mathbb{R}^N), L^q(\mathbb{R}^N))$$

$$\|A + \mu I\|_{\mathcal{L}(W^{2,q}(\mathbb{R}^N), L^q(\mathbb{R}^N))} + \|(A+\mu I)^{-1}\|_{\mathcal{L}(L^q(\mathbb{R}^N), W^{2,q}(\mathbb{R}^N))} \leq c.$$

Therefore $-A$, generates a C^0 analytic semigroup $\{S(t)\}$ in $L^q(\mathbb{R}^N)$ and

$$\begin{cases} u_t - \sum_{k,l=1}^N a_{kl}(x)\partial_k\partial_l u + \sum_{j=1}^N b_j(x)\partial_j u + c(x)u = 0 \\ u(0) = u_0 \in L^q(\mathbb{R}^N) \end{cases}$$

has a unique solution $u(t) = S(t)u_0$ for $t \geq 0$.

Concerning bounded imaginary powers it was shown in [3]:

Proposition 5.2. *With $0 < M$, $0 < \theta_0 < \theta < \pi$ as before if*

$$\int_0^1 \frac{\omega^{\frac{1}{3}}(t)}{t}\,dt < \infty, \tag{5.2}$$

then, $\Lambda = A + \mu I \in \mathcal{B}IP\left(L^q(\mathbb{R}^N), \mathcal{N}, \theta\right)$, i.e.,

$$\|\Lambda^{it}\|_{\mathcal{L}(L^q(\mathbb{R}^N))} \leq \mathcal{N}e^{\theta|t|}, \ t \in \mathbb{R}.$$

In particular, the fractional powers spaces satisfy

$$X_q^\alpha = [L^q(\mathbb{R}^N), W^{2,q}(\mathbb{R}^N)]_\alpha = H_q^{2\alpha}(\mathbb{R}^N)$$

which are Bessel potential spaces and $[\cdot,\cdot]_\alpha$ denotes complex interpolation.

Note that a crucial remark is that in the results above all constants $\mu, c, \mathcal{N}$, depend only on $0 < M$, θ_0, θ, ω.

In order to extend these results to locally uniform spaces we first obtain an alternative description of these spaces in terms of some weighted norms. For this consider a weight $0 < \rho \in C^2(\mathbb{R}^N) \cap L^1(\mathbb{R}^N)$ with

$$|D^\alpha \rho(x)| \leq C_\alpha \rho(x), \quad x \in \mathbb{R}^N, \quad |\alpha| \leq 2$$

and the weighted space

$$L^q_\rho(\mathbb{R}^N) = \left\{\phi \in L^q_{\text{loc}}(\mathbb{R}^N), \ \int_{\mathbb{R}^N} |\phi(x)|^q \rho(x)dx < \infty\right\},$$

with norm $\|\phi\|_{L^p_\rho(\mathbb{R}^N)} = \left(\int_{\mathbb{R}^N} |\phi(x)|^p \rho(x)dx\right)^{\frac{1}{p}}$.

If ρ does not decay too fast, see [2], then

$$L^q_U(\mathbb{R}^N) = \left\{\phi \in L^q_{\text{loc}}(\mathbb{R}^N), \ \sup_{y\in\mathbb{R}^N} \|\phi\|_{L^q_{\tau_y\rho}(\mathbb{R}^N)} < \infty\right\}$$

where $\{\tau_y, y \in \mathbb{R}^N\}$ are the translations.

Now we first obtain solutions of (5.1) in weighted spaces, [2]

Theorem 5.3. *With $0 < M$, $0 < \theta_0 < \theta < \pi$ as before, A in $L^q_\rho(\mathbb{R}^N)$ with domain $W^{2,q}_\rho(\mathbb{R}^N)$ is sectorial*

$$A + \mu I \in Iso(W^{2,q}_\rho(\mathbb{R}^N), L^q_\rho(\mathbb{R}^N))$$

$$\|A + \mu I\|_{\mathcal{L}(W^{2,q}_\rho(\mathbb{R}^N), L^q_\rho(\mathbb{R}^N))} + \|(A + \mu I)^{-1}\|_{\mathcal{L}(L^q_\rho(\mathbb{R}^N), W^{2,q}_\rho(\mathbb{R}^N))} \leq c.$$

Therefore $-A$, generates a C^0 analytic semigroup $\{S(t)\}$ in $L^q_\rho(\mathbb{R}^N)$ and

$$\begin{cases} u_t - \sum_{k,l=1}^N a_{kl}(x)\partial_k\partial_l u + \sum_{j=1}^N b_j(x)\partial_j u + c(x)u = 0 \\ u(0) = u_0 \in L^q_\rho(\mathbb{R}^N) \end{cases}$$

has a unique solution $u(t) = S(t)u_0$ for $t \geq 0$.

Moreover, if (5.2) holds, then

$$\Lambda = A + \mu I \in \mathcal{BIP}\left(L^q_\rho(\mathbb{R}^N), \mathcal{N}, \theta\right)$$

and the fractional powers spaces satisfy

$$W^{2\alpha,q}_\rho = [L^q_\rho(\mathbb{R}^N), W^{2,q}_\rho(\mathbb{R}^N)]_\alpha$$

Note that again all constants, $\mu, c, \mathcal{N}$, depend only on $0 < M$, θ_0, θ, ω and ρ, but can be taken uniform for all translations $\tau_y\rho$. Now we are ready to obtain solutions in uniform spaces

Theorem 5.4. *With $0 < M$, $0 < \theta_0 < \theta < \pi$ and ρ as before, A in $L^q_U(\mathbb{R}^N)$ with (not dense) domain $W^{2,q}_U(\mathbb{R}^N)$ is sectorial and*

$$A + \mu I \in Iso(W^{2,q}_U(\mathbb{R}^N), L^q_U(\mathbb{R}^N))$$

$$\|A + \mu I\|_{\mathcal{L}(W^{2,q}_U(\mathbb{R}^N), L^q_U(\mathbb{R}^N))} + \|(A + \mu I)^{-1}\|_{\mathcal{L}(L^q_U(\mathbb{R}^N), W^{2,q}_U(\mathbb{R}^N))} \leq c.$$

Therefore $-A$, generates a (not C^0) analytic semigroup $\{S(t)\}$ in $L^q_U(\mathbb{R}^N)$ and

$$\begin{cases} u_t - \sum_{k,l=1}^N a_{kl}(x)\partial_k\partial_l u + \sum_{j=1}^N b_j(x)\partial_j u + c(x)u = 0 \\ u(0) = u_0 \in L^q_U(\mathbb{R}^N) \end{cases}$$

has a unique solution $u(t) = S(t)u_0$ for $t \geq 0$.

Combining some results above we get

Corollary 5.5. *If $a_{kl} \in BUC(\mathbb{R}^N)$, $b_j \in L^{p_j}_U(\mathbb{R}^N)$ and $c \in L^{p_0}_U(\mathbb{R}^N)$ where $p_j > N$ and $p_0 > N/2$, for any $1 < r \leq \infty$*

$$\begin{cases} u_t - \sum_{k,l=1}^N a_{kl}(x)\partial_k\partial_l u + \sum_{j=1}^N b_j(x)\partial_j u + c(x)u = 0 \\ u(0) = u_0 \in L^r_U(\mathbb{R}^N) \end{cases}$$

has a unique solution $u(t) = S(t)u_0$ for $t \geq 0$ and

$$(0, \infty) \ni t \mapsto S(t)u_0 \in BUC(\mathbb{R}^N)$$

is analytic.

For any p, q, r the semigroup is order preserving in $L^p(\mathbb{R}^N)$ iff it is order preserving in $L^q_U(\mathbb{R}^N)$ iff it is order preserving in $L^r_\rho(\mathbb{R}^N)$.

The next result shows that under additional assumptions on the coefficients the semigroup above is continuous up to $t = 0$ in the dotted uniform spaces

Theorem 5.6. *Assume* $0 < M$, $0 < \theta_0 < \theta < \pi$, ρ *as before,* $a_{kl} \in BUC(\mathbb{R}^N)$,

$$b_j \in \dot{L}^{p_j}_U(\mathbb{R}^N) \qquad c \in \dot{L}^{p_0}_U(\mathbb{R}^N)$$

where $p_j > N$, $p_j \geq q$ *and* $p_0 > N/2$, $p_0 \geq q$.

Then A *in* $\dot{L}^q_U(\mathbb{R}^N)$ *with domain* $\dot{W}^{2,q}_U(\mathbb{R}^N)$ *is sectorial, generates a* C^0 *analytic semigroup* $\{S(t)\}$ *in* $\dot{L}^q_U(\mathbb{R}^N)$ *and*

$$\begin{cases} u_t - \sum_{k,l=1}^N a_{kl}(x)\partial_k\partial_l u + \sum_{j=1}^N b_j(x)\partial_j u + c(x)u = 0 \\ u(0) = u_0 \in \dot{L}^q_U(\mathbb{R}^N) \end{cases}$$

has a unique solution $u(t) = S(t)u_0$ *for* $t \geq 0$.

Moreover, if (5.2) *holds, then*

$$\Lambda = A + \mu I \in \mathcal{BIP}\left(\dot{L}^q_U(\mathbb{R}^N), \mathcal{N}, \theta\right)$$

and the fractional powers spaces satisfy

$$\dot{W}^{2\alpha,q}_U := [\dot{L}^q_U(\mathbb{R}^N), \dot{W}^{2,q}_U(\mathbb{R}^N)]_\alpha$$

6. Nonlinear equations

Now we present some results on nonlinear equations, see [4]

$$\begin{cases} u_t = \Delta u + f(x,u), \qquad x \in \mathbb{R}^N, \quad t > 0, \\ u(0,x) = u_0(x) \in L^q_U(\mathbb{R}^N) \end{cases} \tag{6.1}$$

where the nonlinear term satisfies

$$f(x,s) = g(x) + m(x)s + f_0(x,s)$$

$$f_0(x,0) = 0, \quad \frac{\partial}{\partial s} f_0(x,0) = 0.$$

Concerning local existence, we have

Theorem 6.1. *Assume* $1 < q < \infty$ *and*

$$g \in L^q_U(\mathbb{R}^N), \qquad m \in L^p_U(\mathbb{R}^N) \quad p > N/2.$$

Then (6.1) *is well posed for* $u_0 \in L^q_U(\mathbb{R}^N)$ *if*

$$\left|\frac{\partial}{\partial s} f_0(x,s)\right| \leq c(1 + |s|^{\rho-1}), \qquad 1 \leq \rho \leq \rho_C = 1 + \frac{2q}{N}.$$

□

Note that solutions are constructed through the variations of constants formula and that the result cover the case of the critical exponent, ρ_C, for the space $L^q_U(\mathbb{R}^N)$.

The next result ensures global existence of solutions, see [1], for similar conditions in a standard Lebesgue spaces setting,

Theorem 6.2. *Assume*

$$f(x,s)s \le C(x)|s|^2 + D(x)|s|$$

with

$$C, D \in L^p_U(\mathbb{R}^N), \quad p > N/2, \quad D \ge 0.$$

Then solutions of (6.1) are global for $u_0 \in L^q_U(\mathbb{R}^N)$*, and are bounded in* $L^\infty(\mathbb{R}^N)$ *on* $[\varepsilon, T]$*. Moreover if* $u_0 \in B \subset L^q_U(\mathbb{R}^N)$*, a bounded set, then*

$$u(t, B) \subset L^\infty(\mathbb{R}^N) \quad \textit{is bounded}$$

Proof. Using comparison and maximum principles we get

$$|u(t,x)| \le U(t,x) \qquad x \in \mathbb{R}^N$$

where

$$U_t - \Delta U = C(x)U + D(x), \qquad U(0) = |u_0|.$$

With this, $D \in L^p_U(\mathbb{R}^N)$ and $p > N/2$ implies U is bounded in $L^\infty(\mathbb{R}^N)$ on $[\varepsilon, T]$. Hence,

$$\|f_0(\cdot, u)\|_{L^q_U(\mathbb{R}^N)} \le C_0(\varepsilon, T), \quad t \in [\varepsilon, T]$$

and from the variation of constants formula, the solution is global. □

The next step is obtaining asymptotic bounds on the solutions

Theorem 6.3. *Assume the solutions of*

$$z_t - \Delta z = C(x)z$$

decay exponentially.

Let $0 \le \phi \in W^{2,p}_U(\mathbb{R}^N) \subset C^\alpha_U(\mathbb{R}^N)$ *be the unique solution of*

$$-\Delta\phi = C(x)\phi + D(x) \quad \textit{in} \quad \mathbb{R}^N.$$

Then $\limsup_{t\to\infty} |u(t,x,u_0)| \le \phi(x)$*, uniformly in* $x \in \mathbb{R}^N$*. If* $|u_0(x)| \le \phi(x)$*, then* $|u(t,x,u_0)| \le \phi(x)$ *for all* $t > 0$*.*

Proof. Now

$$|u(t,x)| \le U(t,x) \to \phi(x), \quad \text{as} \quad t \to \infty$$

and we use the variation of constants formula. □

With this we can prove

Theorem 6.4. *With the assumptions above there exists bounded invariant set* $\mathcal{A} \subset BUC(\mathbb{R}^N)$ *such that for every bounded set* $B \subset L^q_U(\mathbb{R}^N)$*,*

$$dist_{C_{\text{loc}}(\mathbb{R}^N)}\big(u(t,B), \mathcal{A}\big) \to 0, \quad \textit{as} \quad t \to \infty.$$

Example: Assume $m \in L^p_U(\mathbb{R}^N)$, with $p > N/2$ and

$$f(x,s) = m(x)s - s^3, \quad x \in \mathbb{R}^N.$$

Since $\rho = 3$ for local existence we take initial data in $L^q_U(\mathbb{R}^N)$ with $q > N$. On the other hand, global existence follows by taking $C(x) = m(x)$ and $D(x) = 0$.

Assume now that

$$m(x) = m_0(x) - m_1(x), \quad m_0, m_1 \in L^p_U(\mathbb{R}^N)$$

such that

$$e^{(\Delta - m_1(x)I)t} \quad \text{decays exponentially,}$$

e.g., m_1 can be taken as a large positive constant. Hence using Young's inequality we get

$$f(x,s)s \leq C(x)|s|^2 + D(x)|s|$$

with

$$C = -m_1 \in L^p_U(\mathbb{R}^N), \quad p > N/2, \qquad D(x) \approx |m_0|^{3/2}(x)$$

and the existence of the attractor in the local compact topology follows if $m_0 \in L^r_U(\mathbb{R}^N)$ with $r > \frac{3N}{4}$.

See [5] for related results in weighted spaces, [8] for the case of $BUC(\mathbb{R}^N)$ setting and [7] for a different construction of locally compact attractors.

References

[1] J.M. Arrieta, J.W. Cholewa, T. Dlotko, A. Rodríguez-Bernal, "Asymptotic behavior and attractors for reaction diffusion equations in unbounded domains". Nonlinear Analysis TMA. 56, 515–554 (2004).

[2] J.M. Arrieta, J.W. Cholewa, T. Dlotko, A. Rodríguez-Bernal, "Linear parabolic equations in locally uniform spaces". Math. Mod. Meth. Appl. Sci. 14, 253–293 (2004).

[3] H. Amann, M. Hieber, G. Simonett, Bounded H_∞-calculus for elliptic operators, Diff. Int. Eqns 3, 613–653 (1994).

[4] J.M. Arrieta, A.Rodríguez–Bernal, in preparation

[5] A.V. Babin, M.I. Vishik, Attractors of partial differential evolution equations in unbounded domain, Proc. Roy. Soc. Edinburgh 116A, 221–243 (1990).

[6] D. Daners, Heat kernel estimates for operators with boundary conditions, Math. Nachr. 217, 13–41 (2000).

[7] M.A. Efendiev, S.V. Zelik, The attractor for a nonlinear reaction-diffusion system in an unbounded domain, Comm. Pure Appl. Math. 54, 625–688 (2001).

[8] S. Merino, On the existence of the compact global attractor for semilinear reaction diffusion systems on $\mathbb{R}^N$, J. Diff. Eqns 132, 87–106 (1996).

[9] B. Simon, Schrödinger semigroups, Bull. Amer. Math. Soc. 7, 447–526 (1982).

Aníbal Rodríguez-Bernal
Departamento de Matemática Aplicada
Universidad Complutense de Madrid
E-28040 Madrid, Spain

Progress in Nonlinear Differential Equations
and Their Applications, Vol. 64, 433–462

Bifurcation of Traveling Waves Related to the Bénard Equations with an Exterior Force

B. Scarpellini

Abstract. We consider a time-independent Bénard equation (*) $-Aw + (\lambda_0 + \varepsilon)\mathcal{P}Mw + \mathcal{N}(u, w) + (a\partial_x + b\partial_y)w + \tau f = 0$ on the infinite layer $R^2 \times [-\frac{1}{2}, \frac{1}{2}]$. Here $f = f(z)$ is an exterior force depending only on $z \in [-\frac{1}{2}, \frac{1}{2}]$, w satisfies Dirichlet conditions in the z-direction and L_1, L_2-periodic conditions in the x, y-direction, while a, b satisfy a diophantine condition. λ_0 is the critical Rayleigh parameter. It is shown that for generic f and small τ, ε, a, b, (*) has solutions w such that $(a\partial_x + b\partial_y)w \neq 0$. These solutions give rise to periodic traveling wave solutions of the time-dependent version of (*) (with $a = b = 0$). The proof is via bifurcation methods related to Hopf bifurcation.

Mathematics Subject Classification (2000). 35B32, 37K50, 35Q35.

Keywords. Inhomogenic Bénard equation, traveling waves, Hopf bifurcation.

0. Introduction

Below we study the inhomogeneous Bénard equations with a drift term:

$$\begin{aligned} &\nu\Delta v + \lambda k\theta + a\partial_x v + b\partial_y v - (v\nabla)v + \widetilde{f} - \nabla p = 0 \\ &\mu\Delta\theta + \lambda v_3 + a\partial_x\theta + b\partial_y\theta - (v\nabla)\theta + f_4 = 0 \\ &\operatorname{div} v = 0, \quad k = (0,0,1)^t, \quad f = (\widetilde{f}, f_4), \quad \widetilde{f} = (f_1, f_2, f_3). \end{aligned} \tag{0}$$

Here $v = (v_1, v_2, v_3)$ is the velocity field, θ the temperature, p the pressure, a, b constants with $ab \neq 0$, ν the viscosity, μ the thermal conductivity and $\lambda = \sqrt{R}$ with R the Rayleigh number; the density ϱ has been set $= 1$. We consider (0) on the infinite layer $\Omega = R^2 \times (-\frac{1}{2}, \frac{1}{2})$ with v, θ subject to Dirichlet boundary conditions at $\partial\Omega$ and to L_1, L_2-periodic boundary conditions in the unbounded space directions. For $a = b = 0$, Eq. (0) reduces to the slightly rescaled original Bénard equations (see [11], Sec. (0)). The drift term $T = a\partial_x + b\partial_y$ in (0) is included since solutions v, θ of (0) give rise to traveling wave solutions

$$v(x - at, y - bt, z), \quad \theta(x - at, \; y - bt, z)$$

of the original, time-dependent Bénard equations provided with the time-dependent force $f(x-at, y-bt, z)$. The exterior force, while justified on mathematical grounds, admits also a physical interpretation. In fact, disregarding the drift term, we may look at (0) as an approximation to the magnetic Bénard problem ([1], [8], [10]) in which the induced magnetic field $h = (h_1, h_2, h_3)$ is neglected in view of its smallness and in which

$$f = ((H_0\nabla)A, \quad (H_0\nabla)B, \quad (H_0\nabla)C, 0), \quad H_0 = (A, B, C)$$

is given in terms of the impressed magnetic field H_0; for a justification of this argument, see [8], introduction.

In [10] the problem was to prove existence of solutions of (a) via Leray-Schauder degree arguments. Our interest here is in the following situation. Let the force $f = f(z)$ depend on $z \in [-\frac{1}{2}, \frac{1}{2}]$ only. In this case there is a unique solution $v_0(z)$, $\theta_0(z)$, $p_0(z)$ of Eq. (0), depending on z only. The question to be treated here is: (*) given $f = f(z)$, do there exist solutions v, θ, p of (0) depending nontrivially on x, y and such that Tv,$T\theta$ are $\not\equiv 0$, with $T = a\partial_x + b\partial_y$? Since degree arguments do not contain sufficient information to answer this question, we are forced to proceed by bifurcation arguments. Now if $ab \neq 0$ it does not seem possible to apply symmetry arguments in order to achieve reduction to the case of a simple eigenvalue. This difficulty is avoided by showing (Sect. 3, 4) that one is in a situation which is analogous to that encountered in case of Hopf bifurcation. In sect. 3 we show that if a certain linear operator L_0, arising from the linearisation of (0) with respect to the "trivial" equilibrium solution $v_0(z)$, $\theta_0(z)$, $p_0(z)$, satisfies $\sigma(L_0) \cap R = \{0\}$, then only solutions v, θ, p can bifurcate which generate trivial traveling waves, i.e., such that

$$\partial_t v(x-at, y-bt, z) = -(Tv)(x-at, y-bt, z) = 0$$

and likewise with θ, p; see Sect. 3 for details. If however $\sigma(L_0) \cap R$ contains a unique pair $\pm i\sigma$, $\sigma \neq 0$, $\sigma \in R$ of simple eigenvalues, then bifurcation of solutions v, θ with $Tv \neq 0$, $T\theta \neq 0$ is indeed possible. It is shown in Sect. 4 that under this spectral condition there are a number of conditions such that if any one of these is satisfied, bifurcation of nontrivial traveling waves occurs. A difficulty with the results in Sect. 4 is that the verification of these conditions requires a detailed knowledge of the eigenfunctions of a certain generalised Orr-Sommerfeld eigenvalue problem which is in general difficult to obtain. There is however an instance of Eq. (0) considered which, although it does not satisfy the spectral assumption of Sect. 4, is nevertheless accessible to the techniques of Sect. 4. In fact, an instance of Eq. (0) is considered in which $f = f(z)$ is small, λ close to the critical Rayleigh parameter λ_0 and where the constants a, b satisfy some diophantine inequalities, for which the existence of bifurcating solutions v, θ, p such that $Tv \neq 0$, $T\theta \neq 0$ can effectively be proved, and which thus give rise to genuine traveling wave solutions of the original, time-dependent Bénard equations with exterior force $f = f(z)$. As to related literature we refer to [14] for an extended list of references; the papers cited in [14] consider one-dimensional traveling waves in contrast to ours which are in 3D. In [15], traveling waves of the 3D-Navier Stokes equations are constructed; however this construction is not based on the standard semigroup setting and does not interfere with ours.

1. Notation

R, C denote real and complex numbers, with $\overline{\zeta}$ the complex conjugate of $\zeta \in C$. We set $\partial_j = \partial_{x_j}$, $x_1 = x$, $x_2 = y$, $x_3 = z$ and set $D^\alpha = \partial_1^{\alpha_1} \dots \partial_p^{\alpha_p}$ for any multiindex $\alpha = (\alpha_1, \dots, \alpha_p)$. For $\Omega \subseteq R^3$ a domain with segment property, $f \in \mathcal{L}^2(\Omega)$ is in $H^p(\Omega)$ iff there are $f_n \in C^p(\overline{\Omega})$, $n \geq 1$ such that $\|f_n - f\|_{L^2} \to 0$ as $n \to \infty$, and such that f_n, $n \geq 1$ is Cauchy in $H^p(\Omega)$; if $f_n \in C_0^p(\Omega)$, $n \geq 1$ then $f \in H_0^p(\Omega)$. The scalar product on $H^p(\Omega)$ is given by

$$(u,v)_p = \sum_{|\alpha| \leq p} (D^\alpha u, D^\alpha v)_0, \quad (u,v)_0 = \int_\Omega u\overline{v}\, dx^n. \tag{1}$$

We have $H^0(\Omega) = \mathcal{L}^2(\Omega)$ and set $\| \ \|_{H^p} = \| \ \|_{H^p(\Omega)}$ for Ω fixed. For vector fields $u, v \in H^p(\Omega)^m$ we set

$$\langle u, v\rangle_p = \textstyle\sum_j (u_j, v_j)_p, \quad u = (u_1, \dots, u_m), \quad v = (v_1, \dots, v_m) \tag{2}$$
$$\|u\|^2 = \|u\|^2_{L^2} = \textstyle\sum \|u_j\|^2, \|u\|^2_{H^p} = \langle u, u\rangle_p.$$

Thus $\|u\| = \|u\|_{L^2}$ and $\|u\|_{H^p}$ are interpreted according to the context. For $u = (u_1, u_2, u_3)$, $v = (v_1, \dots, v_m)$ vector fields and θ a scalar we set

$$(u\nabla)\theta = \textstyle\sum u_j \partial_j \theta, \quad (u\nabla)v = ((u\nabla)v_1, \dots, (u\nabla)v_m). \tag{3}$$

With Δ the Laplacian, ∂ a derivative and $f = (f_1, \dots, f_4)$ we put $\Delta f = (\Delta f_1, \dots, \Delta f_4)$, $\partial f = (\partial f_1, \dots, \partial f_4)$ and $\widetilde{f} = (f_1, f_2, f_3)$. For $u = (u_1, \dots, u_4)$, $v = (v_1, \dots, v_m)$ we thus stipulate

$$(u\nabla)v = (\widetilde{u}\nabla)v \tag{4}$$

in accordance with (1.3).

2. A functional setting

In order to handle the problem raised in the introduction we rephrase the system (0) in functional analytic terms along established lines. With $L_1, L_2 > 0$ fixed periods we set

$$Q_L = (0, L_1) \times (0, L_2),\ I = [-\tfrac{1}{2}, \tfrac{1}{2}],\ Q = Q_L \times I,\ \Omega = R^2 \times I \tag{0}$$

and stipulate: (a) $f \in C^p_{\text{per}}(Q)$ iff $f \in C^p(\overline{\Omega})$ and if f is L_1-periodic in x, L_2-periodic in y, (b) $f \in C^1_{\text{per},0}(Q)$ iff $f \in C^1_{\text{per}}(Q)$ and $f(x, y, \pm\frac{1}{2}) = 0$, $x, y \in R$. We then stipulate that $f \in L^2(Q)$ is in $H^p_{\text{per}}(Q)$ if there is a sequence $f_n \in C^p_{\text{per}}(Q)$, $n \geq 1$ which is Cauchy in $H^p(Q)$ and such that $\|f - f_n\|_{L^2(Q)} \to 0$ as $n \to \infty$; the space $H^1_{\text{per},0}(Q)$ is defined in an analogous way. These notions readily extend to vector fields in an obvious way; we thereby use the same notation, interpreting it according to the context. We stipulate:

$$E \subseteq \mathcal{L}^2(Q)^3 \text{ is the } L^2\text{-closure of fields } f \in H^1_{\text{per}}(Q)^2 \times H^1_{\text{per},0}(Q) \tag{1}$$

such that $\operatorname{div} f = 0$, and P is the orthogonal projection from $\mathcal{L}^2(Q)^3$ to E.

The phase space $\mathcal{E}$ of Eq. (0) is now given by

$$\begin{aligned} &\mathcal{E} = E \times I, \text{ and } \mathcal{P} \text{ is the orthogonal projection from } \mathcal{L}^2(Q)^4 \\ &\text{onto } \mathcal{E}: \text{ if } f = (\widetilde{f}, f_4) \in \mathcal{L}^2(Q)^4 \text{ then } \mathcal{P}f = (P\widetilde{f}, f_4). \end{aligned} \tag{2}$$

A Stokes operator related to the first three equations of (0) is defined via

$$\begin{aligned} &f \in \operatorname{dom}(A_S) \text{ iff } f \in (H^2_{\text{per}}(Q) \cap H^1_{\text{per},0}(Q))^3 \text{ and } \operatorname{div} f = 0, \\ &\text{and } A_S f = \nu P \Delta f \text{ for } f \in \operatorname{dom}(A_S) \end{aligned} \tag{3}$$

The Stokes operator B_S, associated with the system (0) is given by:

$$\begin{aligned} &\operatorname{dom}(B_S) = \operatorname{dom}(A_S) \times (H^2_{\text{per}}(Q) \cap H^1_{\text{per},0}(Q)); \text{ for} \\ &v \in \operatorname{dom}(A_S),\ (v,\theta) \in \operatorname{dom}(B_S), \text{ we set } B_S(v,\theta) = (A_S v, \mu\Delta\theta). \end{aligned} \tag{4}$$

One then has ([11],[12]):

Proposition 2.1: (a) A_S, B_S are selfadjoint on E, $\mathcal{E}$ respectively, (b) $A_S < 0$, $B_S < 0$, (c) A_S, B_S have compact resolvents.

For convenience we introduce the operator A and the matrix M via

$$\begin{aligned} &A = -B_S; \quad M = (a_{jk}), \quad j,k \leq 4 \text{ with } a_{34} = a_{43} = 1 \\ &\text{and } a_{jk} = 0 \text{ otherwise.} \end{aligned} \tag{5}$$

Notation: For $w \in \operatorname{dom}(A)$ we set $\|w\|_A = \|Aw\|$ and recall that $\|w\|_{H^2}$, $\|w\|_A$ define equivalent norms on $\operatorname{dom}(A)$.

In order to handle the nonlinearity in (0) we recall (1.3), (1.4). For $w_j \in \operatorname{dom}(A)$, $j = 1,2$ we set

$$\mathcal{N}(w_1, w_2) = -\mathcal{P}(w_1\nabla)w_2 = -\mathcal{P}(\widetilde{w}_1\nabla)w_2. \tag{6}$$

$\mathcal{N}(\ ,\ )$ admits a number of estimates in terms of fractional powers A^γ. Here we only need the fact that if $w_j \in H^2_{\text{per}}(Q)^4$, $j = 1,2$, then $(w_1\nabla)w_2 \in H^1_{\text{per}}(Q)^4$ and

$$\|(w_1\nabla)w_2\|_{H^1} \leq c\|w_1\|_{H^2}\|w_2\|_{H^2}. \tag{7}$$

System (0) in sect. 0 can now be put into abstract form. The pressure p is thereby eliminated with the aid of the projection $\mathcal{P}$. We assume $f \in \mathcal{E}$, i.e., $\mathcal{P}f = f$. By these stipulations, Eq. (0) assumed the forms

$$Aw = \lambda\mathcal{P}Mw + \mathcal{P}Tw + \mathcal{N}(w,w) + f, \quad T = a\partial_x + b\partial_y. \tag{8}$$

To solve (0) now amounts to seek $w \in \operatorname{dom}(A)$ such that (2.8) holds. A solution $w \in \operatorname{dom}(A)$ such that $Tw \neq 0$ is called a nontrivial traveling wave; otherwise it is called trivial.

3. Nonexistence of bifurcating traveling waves

As noted in the introduction, to find nontrivial traveling wave solutions of Eq. (0) leads to bifurcation problems whose solvability depends on certain spectral assumptions. Prior to formulate them we need some preparatory steps.

Proposition 3.1: For $f = (f_1, \ldots, f_4) \in \mathcal{E}$ we have $f_3 = 0$.

Proof. Since $\mathcal{E} = E \times I$ and $f = (\widetilde{f}, f_4)$, $\widetilde{f} = (f_1, f_2, f_3)$ we have $\widetilde{f} \in E$ by assumption. Now we recall:

$$g \in E \text{ iff } \langle g, \nabla p\rangle = 0 \text{ for all } p \in H^1(Q); \tag{1}$$

see [5] or [12], proof of Lemma 7.2 and Remarks (9.62).

For $\widetilde{f} = g$, (3.1) entails

$$\textstyle\int_I f_3(z)\partial_z p\, dz = 0, \text{ for } p \in H^1(I), \quad I = (-\frac{1}{2}, \frac{1}{2})$$

whence $f_3 = 0$. □

With $f \in \mathcal{E}$ and by Prop. 3.1, Eq. (0) admits a unique solution $v = (v_1, \ldots, v_4) = (\widetilde{v}, v_4)$, depending on z only:

$$\begin{aligned} &\nu\partial_z^2 v = f_j, \quad j = 1, 2, \quad \mu\partial_z^2 v_4 = f_4, \quad v_3 = 0, \quad \lambda\partial_z p + v_4 = 0, \\ &v_j \in H^2(I) \cap H_0^1(I), \quad p \in H^1(I). \end{aligned} \tag{2}$$

From (3.2) we read off: (a) τv is the solution associated to τf, (b) $\mathcal{N}(v, v) = 0$, (c) v does not depend on λ.

We stipulate:

$$\begin{aligned} &\text{if } \varphi \in H^1(Q) \text{ then } T\varphi = a\partial_x\varphi + b\partial_y\varphi, \\ &\text{if } w = (w_1, \ldots, w_n) \in H^1(Q)^n \text{ then } Tw = (T_{w_1}, \ldots, T_{w_n}). \end{aligned} \tag{3}$$

Proposition 3.2: If $w \in \mathrm{dom}(A)$ then $Tw \in \mathcal{E}$

Proof. By periodicity we have

$$\int_Q g\partial_s f\, dx^3 = -\int_Q f\partial_s g\, dx^3, \quad f, g \in H^1_{\mathrm{per}}(Q), \quad s \in \{x, y\} \tag{4}$$

whence

$$\langle Tf, g\rangle = -\langle f, T_g\rangle, \quad f, g \in H^1_{\mathrm{per}}(Q)^n. \tag{5}$$

Let $w = (\widetilde{w}, w_4)$ whence $\widetilde{w} \in \mathrm{dom}(A_S) \subseteq E$. Since $H^2_{\mathrm{per}}(Q)$ is H^1-dense in $H^1_{\mathrm{per}}(Q)$ and by (3.1) it suffices to show:

$$\langle T\widetilde{w}, \nabla p\rangle = 0, \quad p \in H^2_{\mathrm{per}}(Q). \tag{6}$$

(3.6) follows via (3.1), (3.5) from the equalities:

$$\langle T\widetilde{w}, \nabla p\rangle = -\langle \widetilde{w}, T\nabla p\rangle = -\langle \widetilde{w}, \nabla Tp\rangle, \ p \in H^2(Q).$$ □

By these remarks we have that τv, with $v = v(z)$ as in (3.2), is the unique x, y-independent solution of the following variant of equation (2.10):

$$-Aw_0 + \lambda\mathcal{P}Mw_0 + (1 + \chi)Tw_0 + \mathcal{N}(w_0, w_0) = \tau f, \ w_0 \in \mathrm{dom}(A). \tag{7}$$

The extra parameters χ, τ are needed later. In order to find solutions w_0 giving rise to nontrivial traveling waves we set $w_0 = \tau v + w$ in (3.7) and obtain the following equation for w:

$$(-A + \lambda \mathcal{P}M - \tau G + (1+\chi)T)w + \mathcal{N}(w,w) = 0, \; w \in \operatorname{dom}(A) \tag{8}$$

where G is the operator given by

$$Gw = -(\mathcal{N}(w,v) + \mathcal{N}(v,w)), \quad \operatorname{dom}(G) = H^2_{\text{per}}(Q)^4. \tag{9}$$

In order to handle (3.9) as a bifurcation problem we pick $\lambda_0 > 0$, $\tau_0 \geq 0$ arbitrarily for the moment, set $\lambda = \lambda_0 + \varepsilon$, $\tau = \tau_0 + \delta$ in (3.8) so as to obtain

$$(-A + \lambda_0 \mathcal{P}M - \tau_0 G + T)w + \varepsilon \mathcal{P}Mw - \delta Gw + \chi Tw + \mathcal{N}(w,w) = 0. \tag{10}$$

Given $\varepsilon, \delta, \chi$ small we look for solutions $w \in \operatorname{dom}(A)$ of (3.10) with $\|w\|_{H^2}$ small such that $\tau v + w$ is a nontrivial traveling wave. We note that a solution $w \neq 0$ of (3.10) depends nontrivially on x, y since otherwise $(\tau_0 + \delta)v + w$ and $(\tau_0 + \delta)v$ would be two different solutions of (3.8) for $\lambda = \lambda_0 + \varepsilon$, $\tau = \tau_0 + \delta$, depending on z only, what is not possible. But in order to guarantee that $\tau v + w$ is a nontrivial traveling wave, the stronger condition $Tw \neq 0$ must hold. The above bifurcation problem depends on λ_0, τ_0 which define operators $\widetilde{B}$, L_0 via:

$$\begin{aligned} \widetilde{B} &= -A + \lambda_0 \mathcal{P}M - \tau_0 G, \quad L_0 = \widetilde{B} + T, \\ \operatorname{dom}(\widetilde{B}) &= \operatorname{dom}(L_0) = \operatorname{dom}(A). \end{aligned} \tag{11}$$

Proposition 3.3: $\widetilde{B}$ and L_0 are holomorphic semigroup generators with compact resolvents.

Proof. See, e.g., [12] pg. 69, 139 for comparable situations, see also [7], Thm. 2.1, pg. 80, [4], Thm. 2.4, pg. 4.99. □

As a consequence we have

Lemma 3.1: Let $0 \notin \sigma(L_0)$. There are ε_0, δ_0, χ_0, $d_0 > 0$ as follows: if $|\varepsilon| \leq \varepsilon_0$, $|\delta| \leq \delta_0$, $|\chi| \leq \chi_0$ and if $w \in \operatorname{dom}(A)$ is a solution of (3.10) with $\|w\|_{H^2} \leq d_0$ then $w = 0$.

Proof. By Prop. 3.3 and the assumption, L_0^{-1} maps $\mathcal{E}$ boundedly onto $\operatorname{dom}(A)$, $\| \; \|_{H^2}$, thus (3.10) is equivalent to

$$w + L_0^{-1}(\varepsilon \mathcal{P}Mw - \delta Gw + \chi Tw) + L_0^{-1}\mathcal{N}(w,w) = 0. \tag{12}$$

The left-hand side of (3.12) is a real analytic mapping $\mathcal{F}(w, \varepsilon, \delta, \chi)$ from $\operatorname{dom}(A)$, $\| \; \|_{H^2} \times R^3$ into $\operatorname{dom}(A)$, $\| \; \|_{H^2}$, such that $\mathcal{F}(0, \varepsilon, \delta, \chi) = 0$ for all $\varepsilon, \delta, \chi$. Since $(\partial_w \mathcal{F}(0,0,0,0) = I$ the claim follows from the implicit function theorem. □

The situation is more involved if the assumption of the Lemma does not hold.

Theorem 1: Assume:

(a) $0 \in \sigma(L_0)$,

(b) if $\sigma \in \sigma(\widetilde{B}) \cap iR$ then $\sigma = 0$. Then there are $\varepsilon_0, \chi_0, \delta_0, d_0 > 0$ as follows: if $|\varepsilon| \le \varepsilon_0$, $|\delta| \le \delta_0$, $|\chi| \le \chi_0$ and if $w \in \mathrm{dom}(A)$ is a solution of (3.10) with $\|w\|_{H^2} \le d_0$ then $Tw = 0$.

Remarks: (a), (b) do not exclude the possibility of a small solution $w \neq 0$ of (3.10); but $w_0 = \tau v + w$ is then already a solution of

$$(-A + \lambda \mathcal{P} M) w_0 + \mathcal{N}(w_0, w_0)_\tau f = 0, \quad \lambda = \lambda_0 + \varepsilon,\ \tau = \tau_0 + \delta, \tag{13}$$

that is,

$$(\partial_t w_0)(x - at, y - bt, z) = -(Tw_0)(x - at, y - bt, z) = 0, \tag{14}$$

whence no nontrivial traveling wave is generated by $w_0 = \tau v + w$. For the proof of Thm. 1 we need some facts about Fourier series which underlie the Orr-Sommerfeld method. We recall the periods L_1, L_2 and set

$$\widehat{\alpha} = \frac{2\pi\alpha}{L_1}, \quad \widehat{\beta} = \frac{2\pi\beta}{L_2}, \quad e_{\alpha\beta} = e^{i(\widehat{\alpha}x + \widehat{\beta}y)}, \quad \alpha, \beta \in Z. \tag{15}$$

An element $w \in \mathcal{L}^2(Q)$ is then characterised by its Fourier series:

$$w = \sum_{\alpha\beta} e_{\alpha\beta} w_{\alpha\beta}, \; w_{\alpha\beta} \in \mathcal{L}^2(I), \quad \sum_{\alpha\beta} \int_I |w_{\alpha\beta}|^2 \, dz < \infty. \tag{16}$$

We then have that $w \in H^1_{\mathrm{per}}(Q)$ iff $w_{\alpha\beta} \in H^1(I)$, $\alpha, \beta \in Z$ and in addition

$$\sum_{\alpha\beta} (1 + \alpha^2 + \beta^2) \int_I |w_\alpha \beta|^2 \, dz + \sum_{\alpha\beta} \int_I |\partial_z w_{\alpha\beta}|^2 \, dz < \infty. \tag{17}$$

Likewise we have

$$w \in H^1_{\mathrm{per},0}(Q) \text{ iff } w_{\alpha\beta} \in H^1_0(I), \quad \alpha, \beta \in Z \text{ and (3.17) holds.} \tag{18}$$

Similar characterizations hold for elements $w \in H^p_{\mathrm{per}}(Q)$, $p \ge 2$; we refer to [13] for a detailed treatment of such expansions, see also [11], sec. 2. Here it is convenient to rephrase these facts in terms of suitable subspaces of $\mathcal{L}^2(Q)$, $\mathcal{L}^2(Q)^n$. We define subspaces $L_{\alpha\beta}$, $H^p_{\alpha\beta}$, $H^1_{\alpha\beta 0}$ of $\mathcal{L}^2(Q)$ via

$$\begin{aligned} L_{\alpha\beta} &= \{e_{\alpha\beta} f / f \in \mathcal{L}^2(I)\}, \quad H^p_{\alpha\beta} = \{e_{\alpha\beta} f / f \in H^p(I)\}, \\ H^1_{\alpha\beta 0} &= \{e_{\alpha\beta} f / f \in H^1_0(I)\}. \end{aligned} \tag{19}$$

Fourier expansion arguments show that one has

$$H^p_{\alpha\beta} = H^p_{\mathrm{per}}(Q) \cap L_{\alpha\beta}, \quad H^1_{\alpha\beta 0} = H^1_{\mathrm{per},0}(Q) \cap L_{\alpha\beta}. \tag{20}$$

These notions and facts extend readily to fields and product spaces such as $H^p_{\mathrm{per}}(Q)^n$, etc. We define

$$E_{\alpha\beta} = E \cap L^3_{\alpha\beta}, \quad \mathcal{E}_{\alpha\beta} = \mathcal{E} \cap L^4_{\alpha\beta} \tag{21}$$

and note that $E_{\alpha\beta}$ admits the equivalent definition:

$$\begin{aligned} &E_{\alpha\beta} \text{ is the } \mathcal{L}^2\text{-closure of fields } e_{\alpha\beta} f, \text{ with } f = (f_1, f_2, f_3) \\ &\text{in } H^1(I)^2 \times H^1_0(I) \text{ such that } i\widehat{\alpha} f_1 + i\widehat{\beta} f_2 + \partial_z f_3 = 0. \end{aligned} \tag{22}$$

From the expansion properties (3.16)–(3.18) and the stipulations (3.21) we infer the orthogonal sum relations:

$$\mathcal{L}^2(Q)^n = \sum_{\alpha\beta} \oplus L^n_{\alpha\beta}, \quad E = \sum_{\alpha\beta} \oplus E_{\alpha\beta}, \quad \mathcal{E} = \sum_{\alpha\beta} \oplus \mathcal{E}_{\alpha\beta}. \tag{23}$$

Now let $\widehat{p}_{\alpha\beta}$ be the orthogonal projection from $\mathcal{L}^2(Q)$ onto $L_{\alpha\beta}$, likewise $\widetilde{p}_{\alpha\beta}$ from $\mathcal{L}^2(Q)^3$ onto $L^3_{\alpha\beta}$ and $p_{\alpha\beta}$ from $\mathcal{L}^2(Q)^4$ onto $L^4_{\alpha\beta}$. These projections satisfy commutation relations:

Lemma 3.2:

(1) $P\widetilde{p}_{\alpha\beta} = \widetilde{p}_{\alpha\beta}P$,
(2) $\mathcal{P}p_{\alpha\beta} = p_{\alpha\beta}\mathcal{P}$,
(3) $\widetilde{p}_{\alpha\beta}A_s \subset A_s\widetilde{p}_{\alpha\beta}$,
(4) $p_{\alpha\beta}A \subseteq Ap_{\alpha\beta}$,
(5) $p_{\alpha\beta}G \subseteq Gp_{\alpha\beta}$,
(6) $p_{\alpha\beta}(\mathcal{P}M) = (\mathcal{P}M)p_{\alpha\beta}$.

Proof. As to (1) we first note:

$$\text{if } f \in L^3_{\alpha\beta} \text{ then } Pf \in L^3_{\alpha\beta}. \tag{24}$$

We first prove (3.24) for the dense subset

$$f = e_{\alpha\beta}\widetilde{f}, \quad \widetilde{f} = (f_1, f_2, f_3) \in H^1(I)^2 \times H^1_0(I). \tag{25}$$

For f in (3.25) we solve the boundary value problem:

$$(\partial_z^2 - \gamma^2)p = i\widehat{\alpha}f_1 + i\widehat{\beta}f_2 + \partial_z f_3, \quad p \in H^2(I),\ \partial_z p \in H^1_0(I) \tag{26}$$

where $\gamma^2 = \widehat{\alpha}^2 + \widehat{\beta}^2$. Now (3.26) has a unique solution if $\gamma^2 > 0$ and unique up to an additive constant if $\gamma = 0$ (since $\int_I \partial_z f_3\, dz = 0$). We thus have

$$Pf = e_{\alpha\beta}\widetilde{f} - \nabla(e_{\alpha\beta}p) \in L^3_{\alpha\beta} \cap E = E_{\alpha\beta}. \tag{27}$$

With (3.24) proved for f's of type (3.25), (3.24) follows by a density and approximation argument. Now let $f \in \mathcal{L}^2(Q)^3$ From (3.23), (3.24) and the boundedness of P we infer

$$f = \sum_{\varepsilon\delta} \widetilde{P}_{\varepsilon\delta}f, \quad Pf = \sum_{\varepsilon\delta} P\widetilde{p}_{\varepsilon\delta}f = \sum_{\varepsilon\delta} P\widetilde{p}_{\varepsilon\delta}f$$

whence

$$\widetilde{p}_{\alpha\beta}Pf = \sum_{\varepsilon\delta} \widetilde{p}_{\alpha\beta}\widetilde{p}_{\varepsilon\delta}P\widetilde{p}_{\varepsilon\delta}f = \widetilde{p}_{\alpha\beta}P\widetilde{p}_{\alpha\beta}f = P\widetilde{p}_{\alpha\beta}f$$

proving (1). Claim (2) now follows from (1) and (2.2).

In order to prove (3), (4) we consider the Dirichlet Laplacian Δ:

$$\mathrm{dom}(\Delta) = H^2_{\mathrm{per}}(Q) \cap H^1_{\mathrm{per},0}(Q), \quad \Delta f = (\partial_x^2 + \partial_y^2 + \partial_z^2)f \tag{28}$$

and show

$$\widehat{p}_{\alpha\beta}\Delta \subseteq \Delta\widehat{p}_{\alpha\beta}, \text{ that is } \Delta^{-1}\widehat{p}_{\alpha\beta} = \widehat{p}_{\alpha\beta}\Delta^{-1} \tag{29}$$

what in turn is a consequence of:

$$\text{if } f \in L_{\alpha\beta} \text{ then } \Delta^{-1} f \in L_{\alpha\beta}, \quad \alpha, \beta \in Z. \tag{30}$$

Thus let $f = e_{\alpha\beta} g$, $g \in \mathcal{L}^2(I)$ and h the solution of:

$$(\partial_z^2 - \gamma^2) h = g, \quad h \in H^2(I) \cap H_0^1(I), \quad \gamma^2 = \widehat{\alpha}^2 + \widehat{\beta}^2. \tag{31}$$

We then have

$$e_{\alpha\beta} h \in \mathrm{dom}(\Delta), \quad \Delta(e_{\alpha\beta} h) = e_{\alpha\beta} g = f, \quad \Delta^{-1} f \in L_{\alpha\beta}, \tag{32}$$

proving (3.30). Now let $f \in \mathcal{L}^2(Q)$. From (3.30) and the boundedness of Δ^{-1} we infer:

$$f = \sum_{\varepsilon\delta} \widehat{p}_{\varepsilon\delta} f, \quad \Delta^{-1} f = \sum_{\varepsilon\delta} \widehat{p}_{\varepsilon\delta} \Delta^{-1} \widehat{p}_{\varepsilon\delta} f \text{ whence}$$

$$\widehat{p}_{\alpha\beta} \Delta^{-1} f = \sum_{\varepsilon\delta} \widehat{p}_{\alpha\beta} \widehat{p}_{\varepsilon\delta} \Delta^{-1} \widehat{p}_{\varepsilon\delta} f = \widehat{p}_{\alpha\beta} \Delta^{-1} \widehat{p}_{\alpha\beta} f = \Delta^{-1} \widehat{p}_{\alpha\beta} f$$

what proves (3.29). Clauses (3), (4) now follow from (2.3)–(2.5), (3.29) and clauses (1), (2) already proved. The proofs of (5), (6) are straightforward and omitted. □

Remarks: In order to find suitable subspaces which are also invariant under $\mathcal{N}(\,,\,)$ we list a few remarks.

Proposition 3.4: Set $D(\alpha, \beta) = +i(a\widehat{\alpha} + b\widehat{\beta})$. (1) If $f \in (H^1_{\alpha\beta})^4$ then $Tf = +iD(\alpha, \beta) f$, (2) if $f \in H^1_{\mathrm{per}}(Q)^4$ then $T p_{\alpha\beta} f = +iD(\alpha, \beta) p_{\alpha\beta} f$.

Proof. Based on (3.19) we have in case (1) that $f = e_{\alpha\beta} g$ with $g \in H^1(I)^4$ and $e_{\alpha\beta}$ as in (3.15) whence

$$Tf = ((a\partial_x + b\partial_y) e_{\alpha\beta}) g = i(a\widehat{\alpha} + b\widehat{\beta}) e_{\alpha\beta} g = +iD(\alpha, \beta) f.$$

(2) follows from (1) $p_{\alpha\beta} f \in (H^1_{\alpha\beta})^4$. □

Given $\alpha_0, \beta_0 \in Z$ we consider the additive group:

$$m\{\alpha_0, \beta_0\} = \{(n\alpha_0, n\beta_0)/n \in Z\} \tag{33}$$

and note

Proposition 3.5: Let $a, b \in R$, $a \cdot b \neq 0$. Then there are $\alpha_0, \beta_0 \in Z$ such that

$$m\{\alpha_0, \beta_0\} = \{(\alpha, \beta)/a\widehat{\alpha} + b\widehat{\beta} = 0, \quad \alpha, \beta \in Z\} \qquad \bullet$$

We omit the elementary proof. Based on (3.19), (3.23), we now associate with a pair $(\alpha, \beta) \in Z^2$ the orthogonal sum

$$[\alpha, \beta]_k = \sum_n \oplus L^k_{n(\alpha,\beta)}, \quad (n(\alpha, \beta) = (n\alpha, n\beta)). \tag{34}$$

Thus $[\alpha, \beta]_k$ is the space of elements $w \in \mathcal{L}^2(Q)^k$ whose Fourier series assumes the form

$$w = \sum_n e_{n(\alpha,\beta)} w_n, \quad w_n \in \mathcal{L}^2(I)^k, \quad \sum \int_I |w_n|^2 \, dz < \infty \tag{35}$$

or likewise (for $k = 4$) the space of $w \in \mathcal{L}^2(Q)^4$ such that $p_{\varepsilon\delta} w = 0$ if $(\varepsilon, \delta) \notin m\{\alpha, \beta\}$, that is such that $w = \sum_n p_{n(\alpha\beta)} w$. Now let $\widetilde{Q}_{\alpha\beta}$ and $Q_{\alpha\beta}$ be the orthogonal projections from $\mathcal{L}^2(Q)^3$ onto $[\alpha, \beta]_3$ and from $\mathcal{L}^2(Q)^4$ onto $[\alpha, \beta]_4$ respectively. We have the commutation properties.

Lemma 3.3:

(a) $\widetilde{p}_{\alpha\beta}\widetilde{Q}_{\varepsilon\delta} = \widetilde{Q}_{\varepsilon\delta}\widetilde{p}_{\alpha\beta}$ and $p_{\alpha\beta}Q_{\varepsilon\delta} = Q_{\varepsilon\delta}p_{\alpha\beta}$,

(b) $\widetilde{Q}_{\alpha\beta}P = P\widetilde{Q}_{\alpha\beta}$ and $Q_{\alpha\beta}\mathcal{P} = \mathcal{P}Q_{\alpha\beta}$,

(c) $\widetilde{Q}_{\alpha\beta}A_S =\subseteq A_S\widetilde{Q}_{\alpha\beta}$ and $Q_{\alpha\beta}A \subseteq AQ_{\alpha\beta}$,

(d) $Q_{\alpha\beta}$ commutes with $\mathcal{P}M$, G, T.

For reasons of space we omit the proof which is very similar to that of Lemma 3.2. By Lemma 3.3 the space $[\alpha, \beta]_4$ is invariant under the linear operators appearing in (3.11). But $[\alpha, \beta]_4$ is also invariant under the nonlinearity $\mathcal{N}(\ ,\ )$ as asserted by

Lemma 3.4:

(a) if $f \in H^2_{\text{per}}(Q)^3 \cap [\alpha, \beta]_3$, $g \in H^2_{\text{per}}(Q) \cap [\alpha, \beta]_1$ then $(f\nabla)g \in H^1_{\text{per}}(Q) \cap [\alpha, \beta]_1$,

(b) if $f, g \in \text{dom}(A) \cap [\alpha, \beta]_4$ then $\mathcal{N}(f, g) \in \mathcal{E} \cap [\alpha, \beta]_4$.

Proof. Let first $f = e_{\alpha\beta}\widetilde{f}$, $g = e_{\varepsilon\delta}\widetilde{g}$, $\widetilde{f}, \widetilde{g} \in H^2(I)$. We then have

$$f\partial_x g, f\partial_y g, f\partial_z g \in L_{\alpha+\varepsilon,\beta+\delta} \cap H^1_{\text{per}}(Q). \tag{36}$$

Next let $f_N \in H^2_{\text{per}}(Q)^3 \cap [\alpha, \beta]_3$, $g_N \in H^2_{\text{per}}(Q) \cap [\alpha, \beta]_1$ be finite sums:

$$f_N = \sum_{|n| \le N} e_{n(\alpha,\beta)}\widetilde{f}_n, \quad g_N = \sum_{|n| \le N} e_{n(\alpha,\beta)}\widetilde{g}_n, \quad \widetilde{f}_n \in H^2(I)^3, \quad \widetilde{g}_n \in H^2(I).$$

From (3.36) we infer

$$(f_N\nabla)g_N \in H^1_{\text{per}}(Q) \cap [\alpha, \beta]_1. \tag{37}$$

Now assume $f \in H^2_{\text{per}}(Q)^3 \cap [\alpha, \beta]_3$, $g \in H^2_{\text{per}}(Q) \cap [\alpha, \beta]_1$, with Fourier series

$$f = \sum_n e_{n(\alpha,\beta)}\widetilde{f}_n, \quad g_N = \sum_n e_{n(\alpha,\beta)}\widetilde{g}_n, \quad \widetilde{f}_n \in H^2(I)^3, \quad \widetilde{g}_n \in H^2(I)$$

and set also

$$f_N = \sum_{|n| \le N} e_{n(\alpha,\beta)}\widetilde{f}_n, \quad g_N = \sum_{n \le N} e_{n(\alpha,\beta)}\widetilde{g}_n.$$

From standard Fourier theory we infer

$$\|f - f_N\|_{H_2} \to 0, \quad \|g - g_N\|_{H^2} \to 0 \quad \text{as } N \to \infty. \tag{38}$$

We also note: if $u \in H^2_{\text{per}}(Q)^3$, $v \in H^2_{\text{per}}(Q)$ then $(u\nabla)v \in H^1_{\text{per}}(Q)$ and

$$\|(u\nabla)v\|_{H^1} \le c\|u\|_{H^2}\|v\|_{H^2}, \quad \text{some } c; \tag{39}$$

this follows from the continuous embedding $H^2_{\text{per}}(Q) \hookrightarrow C_{\text{per}}(Q)$.

From (3.38), (3.39) we infer:

$$\|(f\nabla)g - (f_N\nabla)g_N\|_{H^1} \to 0 \text{ as N} \to \infty. \tag{40}$$

From (3.37), (3.40) and since $[\alpha,\beta]_1$ is $\mathcal{L}^2$-closed we obtain clause (a) of the Lemma. From (a) we immediately obtain the vector version

$$\text{if } f \in H^2_{\text{per}}(Q)^3 \cap [\alpha,\beta]_3, \quad g \in H^2_{\text{per}}(Q)^4 \cap [\alpha,\beta]_4 \text{ then} \\ (f\nabla)g \in H^1_{\text{per}}(Q)^4 \cap [\alpha,\beta]_4. \tag{41}$$

Clause (b) of the Lemma now follows from (b) of Lemma 3.3 and from (3.41). □

Prior to prove Thm. 1 we list some consequences of Prop. 3.3. With L_0 given by (3.11) we set

$$\mathcal{L}_j = \{\varphi / L_0^j \varphi = 0\}, \quad \mathcal{N}_j = \text{rg}(L_0^j), \quad j \geq 1.$$

Obviously $\mathcal{L}_1$ and $\mathcal{N}_1$ are eigenspace and range of L_0. By Prop. 3.3 and Riesz-Schauder theory there is $\nu \geq 1$ such that

$$\mathcal{L}_1 \subseteq \mathcal{L}_2 \cdots \subseteq \mathcal{L}_\nu = \mathcal{L}_{\nu+1} = \ldots, \qquad \mathcal{N}_1 \supseteq \mathcal{N}_2 \supseteq \cdots \supseteq \mathcal{N}_\nu = \mathcal{N}_{\nu+1} \ldots$$

we set $\mathcal{L}_\nu = \mathcal{L}$, $\mathcal{N}_\nu = \mathcal{N}$. The properties of $\mathcal{L}$, $\mathcal{N}$ are summarised by

Proposition 3.6:

(A) There are bounded projections S, Q from $\mathcal{E}$ onto $\mathcal{L}$, $\mathcal{N}$ respectively such that $S + Q = 1$ and in addition:
 (a) $QL_0 \subseteq L_0 Q$,
 (b) the mapping

$$QL_0 : \mathcal{N} \cap \text{dom}(L_0), \| \, \|_{H^2} \to \mathcal{N}, \| \, \|_{L^2}$$

 is onto, 1-1 and boundedly invertible.

(B) There is a basis $e_1, \ldots, e_N$ of $\mathcal{L}$ and a dual basis $e_1^*, \ldots, e_N^*$ such that $\langle e_j, e_k^* \rangle = \delta_{jk}$ and $f \in \mathcal{N}$ iff $\langle f, e_j^* \rangle = 0$, $j, k \leq N$.

Remarks: Riesz-Schauder theory as summarised above follows readily from Riesz-Schauder theory for compact operators ([3], Chpt. 5) and from Prop. 3.3. With $\mathcal{L}_j, \mathcal{L}, Q, S$ as above and $Q_{\alpha\beta}$ as in Lemma 3.3 we have

Lemma 3.5:

(A) There are $(\alpha_k, \beta_k) \in Z^2$, $k \leq p$ such that $\mathcal{L}_1 \cap \mathcal{E}_{\alpha_k\beta_k} \neq 0$, $k \leq p$ and $\mathcal{L}_j \subseteq \sum_1^p \oplus \mathcal{E}_{\alpha_k\beta_k}$, $j \leq \nu$,

(B) $QQ_{\alpha\beta} = Q_{\alpha\beta}Q$.

Proof. As to (A), let $m \subseteq Z^2$ be the set of (α,β) with $\mathcal{L}_1 \cap \mathcal{E}_{\alpha\beta} \neq 0$. Then m is finite. Otherwise one could find a sequence (α_k, β_k), $k < \infty$ and elements $g_k \in \mathcal{L}_1 \cap \mathcal{E}_{\alpha_k\beta_k}$, $g_k \neq 0$ with $L_0 g_k = 0$. Since the g_k are pairwise orthogonal, $\text{dom}(\mathcal{L}_1) = \infty$ would follow, contradicting Prop. 3.6. Thus $m = \{(\alpha_k, \beta_k)/k \leq p\}$, some p; we set

$$\mathcal{E}' = \sum_{k \leq p} \oplus \mathcal{E}_{\alpha_k\beta_k}. \tag{42}$$

If $\mathcal{L}_1 \not\subseteq \mathcal{E}'$ an $e \in \mathcal{L}^1$, $e \notin \mathcal{E}'$ could be found. Now

$$e = \sum p_{\alpha\beta} e, \quad \langle p_{\alpha\beta} e, p_{\varepsilon\delta} e \rangle = 0 \text{ if } (\alpha,\beta) \neq (\varepsilon,\delta)$$

by virtue of (3.23). Since $e \notin \mathcal{E}'$, $p_{\alpha\beta}e \neq 0$ would hold for some $(\alpha,\beta) \notin m$. By Lemma 3.2 we have

$$L_0 p_{\alpha\beta} e = p_{\alpha\beta} L_0 e = 0, \text{ i.e., } p_{\alpha\beta} e \in \mathcal{L}_1$$

whence $(\alpha,\beta) \in m$, contradicting the definition of m. Thus $\mathcal{L}_1 \subseteq \mathcal{E}'$. Now let $\mathcal{L}_j \subseteq \mathcal{E}'$ be true; let $f \in \mathcal{L}_{j+1}$. Then $L_0 f = \varphi$ for some $\varphi \in \mathcal{L}_j$. Now pick $(\alpha,\beta) \notin m$. Since $\mathcal{L}_j \subseteq \mathcal{E}'$ we have $p_{\alpha\beta}\varphi = 0$ whence

$$L_0 p_{\alpha\beta} f = p_{\alpha\beta} L_0 f = p_{\alpha\beta}\varphi = 0, \quad \text{i.e., } p_{\alpha\beta} f \in \mathcal{L}_1 \cap \mathcal{E}_{\alpha\beta},$$

thus $p_{\alpha\beta} f = 0$ by definition of m. That is, $f \in \mathcal{E}'$ whence $\mathcal{L}_{j+1} \subseteq \mathcal{E}'$, completing the induction and thus the proof of (A).
As to (B) we recall [4], Thm. 6.17 according to which

$$\mathcal{P}' = -(2\pi i)^{-1} \int_\Gamma (L_0 - \zeta)^{-1} d\zeta, \tag{43}$$

Γ a small circle centered at ζ, is a bounded projection onto the generalised eigenspace of L_0 at $\zeta = 0$, i.e., onto $\mathcal{L} = \mathcal{L}_\nu$. Since S in Prop. 3.5 is also a bounded projection onto $\mathcal{L}$, we have

$$\mathcal{P}'S = S, \quad S\mathcal{P}' = \mathcal{P}'. \tag{44}$$

Now S commutes with $(L_0 - \zeta)^{-1}$ by Prop. 3.5 and hence with $\mathcal{P}'$ in (3.44), i.e., $\mathcal{P}'S = S\mathcal{P}'$. By (3.43), (3.44) we get:

$$S = \mathcal{P}' = -(2\pi i)^{-1} \int_\Gamma (L_0 - \zeta)^{-1}\, d\zeta. \tag{45}$$

By Lemma 3.3, $Q_{\alpha\beta}$ commutes with L_0, hence with $(L_0 - \zeta)^{-1}$, finally with S and thus with Q, what proves (B). □

Proof of Theorem 1. By (3.11), we can rewrite (3.10) as follows

$$L_0 w + \varepsilon \mathcal{P} M w - \delta G w + \chi T w + \mathcal{N}(w,w) = 0, \quad w \in \operatorname{dom}(L_0). \tag{46}$$

Based on the spaces $\mathcal{L}, \mathcal{N}$ and on Prop. 3.6 we apply the first half of the Lyapunov-Schmidt method to (3.46), i.e., we make the "Ansatz"

$$w = \zeta e + h, \quad h \in \mathcal{N}, \quad \zeta e = \sum_1^N \zeta_j e_j. \tag{47}$$

We insert (3.47) into (3.46) and apply the projection Q in Prop. 3.5 to the resulting equation so as to get:

$$\begin{aligned} QL_0 h \;&+\; \varepsilon Q\mathcal{P}M(\zeta e + h) - \delta QG(\zeta e + h) + \\ &+\; \chi QT(\zeta e + h) + Q\mathcal{N}(\zeta e + h, \zeta e + h) = 0. \end{aligned} \tag{48}$$

By Prop. 3.5, (A) we have that

$$K = (QL_0)^{-1} : \mathcal{N}, \|\ \|_{L^2} \to \mathcal{N} \cap \operatorname{dom}(L_0),\ \|\ \|_{H^2}$$

is a bounded operator. Application of K to (3.48) gives:

$$h + \varepsilon KQ\mathcal{P}M(\zeta e+h) - \delta KQG(\zeta e+h) + \chi KQT(\zeta e+h) \qquad (49)$$
$$+ \; KQ\mathcal{N}\zeta e+h, \zeta e+h) = 0.$$

We define the space

$$X = \mathcal{N} \cap \mathrm{dom}(L_0), \; \| \; \|_{H_2}, \text{ i.e., } \| \; \|_X = \| \; \|_{H^2}$$

and denote the left-hand side of (3.49) by $\mathcal{F}(h,\zeta,\varepsilon,\delta,\chi)$. Evidently ((2.8), (2.9)) we have

$$\mathcal{F} \text{ maps } X \times R^{N+3} \text{ holomorphically into } X. \qquad (50)$$

Moreover: (1) $\mathcal{F}(0,\dots,0)=0$, (2) $(\partial_h\mathcal{F})(0,\dots,0)=I$. By the the implicit function theorem there are $\zeta_1', \varepsilon_1, \delta_1, \chi_1, d_1 > 0$ and a φ_1 mapping the neighborhood

$$\mathcal{U}_1 = \{(\zeta,\varepsilon,\delta,\chi)/|\zeta| \le \zeta_1', \; |\varepsilon| \le \varepsilon_1, \; |\delta| \le \delta_1, \; |\chi| \le \chi_1\} \subseteq R^{N+3}$$

holomorphically into X such that:
(a) $\mathcal{F}(\varphi_1(\zeta,\varepsilon,\delta,\chi),\zeta,\varepsilon,\delta,\chi) = 0$ on $\mathcal{U}_1$, (b) $\varphi_1(0,\dots,0) = 0$, (c) if $h \in X$ and $(\zeta,\varepsilon,\delta,\chi) \in \mathcal{U}_1$ satisfy $\mathcal{F}(h,\zeta,\varepsilon,\delta,\chi)=0$ and $\|h\|_{H^2} \le d_1$ then $h = \varphi_1(\zeta,\varepsilon,\delta,\chi)$.

Next we define a smaller space $Y \subseteq X$. With $(\alpha_j,\beta_j) \in Z^2$, $j \le p$ as in Lemma 3.5, (A) we claim:

$$a\widehat{\alpha}_j + b\widehat{\beta}_j = -D(\alpha_j,\beta_j) = 0, \quad j \le p. \qquad (51)$$

In fact, by the definition of the set (α_j,β_j), $j \le p$ in the proof of Lemma 3.5, (A) and by Prop. 3.4 we find $\varphi_j \ne 0$ in $\mathcal{L}_1 \cap \mathcal{E}_{\alpha_j\beta_j}$ such that:

$$L_0\varphi_j = (\widetilde{B}+T)\varphi_j = \widetilde{B}\varphi_j - iD(\alpha_j,\beta_j)\varphi_j = 0 \qquad (52)$$

(3.51) now follows from (3.52) and assumption (b) of Thm. 1. By Prop. 3.4 there is (α_0,β_0) in Z^2 with

$$\{(n\alpha_0, n\beta_0)/n \in Z\} = \{(\alpha,\beta)/D(\alpha,\beta)=0, \qquad (53)$$
$$\alpha,\beta \in Z\} = m(\alpha_0,\beta_0).$$

Thus $(\alpha_j,\beta_j) \in m(\alpha_0,\beta_0)$, $j \le p$ by (3.51) whence

$$\text{(a)} \;\; e_j \in \mathrm{dom}(L_0) \cap [\alpha_0,\beta_0]_4, \quad \text{(b)} \; Th = 0, \; h \in \mathrm{dom}(L_0) \cap [\alpha_0,\beta_0]_4. \qquad (54)$$

The space Y is now defined via

$$Y = \mathcal{N} \cap \mathrm{dom}(L_0) \cap [\alpha_0,\beta_0]_4, \quad \| \; \|_Y = \| \; \|_{H^2}. \qquad (55)$$

Y is an $\| \; \|_{H^2}$-closed subspace of X. By Lemmas 3.2–3.4, and (3.54) the mapping $\mathcal{F}$ in (3.52) leaves Y invariant. We now apply the implicit function theorem to $\mathcal{F}$ as a holomorphic mapping from $Y \times R^{N+3}$ to Y so as to find $\zeta_2', \varepsilon_2, \delta_2, \chi_2, d_2 > 0$ and a φ_2 mapping the neighborhood

$$\mathcal{U}_2 = \{(\zeta,\varepsilon,\delta,\chi)/|\zeta| \le \zeta_2', \quad |\varepsilon| \le \varepsilon_2, \quad |\delta| \le \delta_2, \quad |\chi| \le \chi_2\} \subseteq R^{N+3}$$

holomorphically into Y such that:
(α) $\mathcal{F}(\varphi_2(\zeta,\varepsilon,\delta,\chi),\zeta,\varepsilon,\delta,\chi) = 0$ on $\mathcal{U}_2$, (β) $\varphi_2(0,\dots,0) = 0$, (γ) if $(\zeta,\varepsilon,\delta,\chi) \in \mathcal{U}_2$, $h \in Y$ satisfy $\|h\|_{H^2} \le d_2$ and $\mathcal{F}(h,\zeta,\varepsilon,\delta,\chi) = 0$ then $\varphi_2(\zeta,\varepsilon,\delta,\chi) = h$.

By the unicity properties (c) of φ_1, (γ) of φ_2 we find $\varepsilon_0 \leq \min(\varepsilon_1, \varepsilon_2)$, $\delta_0 \leq \min(\delta_1, \delta_2)$, $\chi_0 \leq \min(\chi_1, \chi_2)$, $\zeta_0' \leq \min(\zeta_1', \zeta_2')$ such that:

$$\varphi_1(\zeta, \varepsilon, \delta, \chi) = \varphi_2(\zeta, \varepsilon, \delta, \chi), \quad |\zeta| \leq \zeta_0', \quad |\varepsilon| \leq \varepsilon_0, \quad |\delta| \leq \delta_0, \ |\chi| \leq \chi_0. \tag{56}$$

To conclude the proof, pick $d_0 > 0$ so small that if $w \in \operatorname{dom}(L_0)$, $\|w\|_{H_2} \leq d_0$ then

$$w = \zeta e + h \text{ with } |\zeta| \leq \zeta_0', \quad \|h\|_{H^2} \leq d_1 \text{ where } h \in X. \tag{57}$$

Now let $|\varepsilon| \leq \varepsilon_0$, $|\delta| \leq \delta_0$, $|\chi| \leq \chi_0$ and $w \in \operatorname{dom}(L_0)$ a solution of (3.46) such that $\|w\|_{H^2} \leq d_0$. By (3.57), $w = \zeta e + h$ holds with $|\zeta| \leq \zeta_0$, $\|h\|_{H^2} \leq d_1$ and $\mathcal{F}(h, \zeta, \varepsilon, \delta, \chi) = 0$. Since $|\zeta| \leq \zeta_1', \ldots, |\chi| \leq \chi_1$ and $\|h\|_{H^2} \leq d_1$ we have $h = \varphi_1(\zeta, \varepsilon, \delta, z)$ and since also $|\zeta| \leq \zeta_0', \ldots, |\chi| \leq \chi_0$ we have by (3.56)

$$h = \varphi_1(\zeta, \varepsilon, \delta, \chi) = \varphi_2(\zeta, \varepsilon, \delta, \chi) \in \operatorname{dom}(L_0) \cap [\alpha, \beta]_4$$

whence $Tw = \sum \zeta_j T e_j + Th = 0$ by (3.52). □

Corollary to Theorem 1: If $\tau_0 = 0$ in (3.10) then the conclusion of Thm. 1 holds.

Proof. In this case, $\widetilde{B} = -A + \lambda_0 \mathcal{P} M$ is selfadjoint and has only real eigenvalues. □

Theorem 1 admits a minor generalization. We retain $\widetilde{B}$ in (3.11) but stipulate

$$L_0 = \widetilde{B} + \chi_0 T, \quad \chi_0 \in R, \quad \chi_0 \neq 0 \tag{58}$$

and consider instead of (3.10) the equation

$$L_0 w + \varepsilon \mathcal{P} M w - \delta G w + \chi T + \mathcal{N}(w, w) = 0, \quad L_0 \text{ via (3.58)}. \tag{59}$$

Theorem $1'$**:** With L_0 as in (3.58) and under the assumptions of Thm. 1 there are ε_0, δ_0, χ_0', $d_0 > 0$ as follows: if $|\varepsilon| \leq \varepsilon_0$, $|\chi| \leq \chi_0'$ and if $w \in \operatorname{dom}(L_0)$ is a solution of (3.10) with $\|w\|_{H^2} \leq d_0$ then $Tw = 0$.

Remarks: Theorem $1'$ follows from Theorem 1 via the substitutions $\chi_0 T \to T$, $\chi \chi_0^{-1} \to \chi$.

4. Existence of traveling waves

Theorem 1, $1'$ assert nonexistence of small traveling wave solutions $w \neq 0$ of (3.10) resp. (3.59) under assumptions (a), (b) which in turn imply existence of $(\alpha, \beta) \in Z^2$, $(\alpha, \beta) \neq 0$ with $D(\alpha, \beta) = 0$ what means that aL_1^{-1}, bL_2^{-1} are rationally dependent. In order to remedy the situation we drop (b) while (a) has to be retained in view of Lemma 1. With $\widetilde{B}$, L_0 as in (3.11) we assume: $\underline{H}$: (α) There is precisely one $\sigma > 0$ such that $\sigma(\widetilde{B}) \cap (iR - \{0\}) = \{\pm i\sigma\}$, and $i\sigma$ (and hence $-i\sigma$) has algebraic multiplicity 1, (β) $0 \in \sigma(L_0)$, (γ) aL_1^{-1} and bL_2^{-1} are rationally independent.

Remarks: $H(\alpha)$, which implies $\tau_0 \neq 0$ in (3.10), is reminiscent of Hopf bifurcation ([9], sect. V, [2]). In order to exploit this similarity we need some preparatory steps. By $H(\alpha)$ there is a unique pair $(\alpha, \beta) \in Z^2$ and a $w \in (H^2(I) \cap H_0^1(I))^4$ with

$$\widetilde{B}(e_{\alpha\beta}w) = i\sigma e_{\alpha\beta}w,\ i\widehat{\alpha}w_1 + i\widehat{\beta}w_2 + \partial_z w_3 = 0,\ w = (w_1, \dots, w_4). \tag{1}$$

One easily shows that $\alpha^2 + \beta^2 > 0$ must hold. We assume

$$\alpha\beta \neq 0 \tag{2}$$

but the arguments below apply apart from minor changes also to the case $\alpha\beta = 0$. We note:

Proposition 4.1: $D(\alpha, \beta) = \sigma$ or $D(\alpha, \beta) = -\sigma$.

Proof. Since $0 \in \sigma(\widetilde{B} + T)$ by $H(\beta)$ there is by Lemma 3.2 a pair $(\gamma, \delta) \in Z^2$ and $u \in (H^2(I) \cap H_0^1(I))^4$ with $u \neq 0$ and

$$\widetilde{B}(e_{\gamma\delta}u) + iD(\gamma, \delta)e_{\gamma\delta}u = 0, \quad i\widehat{\gamma}u_1 + \widehat{\delta}u_2 + \partial_z u_3 = 0. \tag{3}$$

Now $\gamma = \delta = 0$ is excluded on elementary grounds. But $D(\gamma, \delta) = 0$, $(\gamma, \delta) \neq 0$ cannot hold either in view of $H(\gamma)$. Thus $D(\gamma, \delta) \neq 0$ whence $D(\gamma, \delta) = \pm\sigma$ by $H(\alpha)$; let, e.g., $D(\gamma, \delta) = \sigma$. Thus $e_{\gamma\delta}u$ is an eigenfunction of $-i\sigma \in \sigma(\widetilde{B})$ whence

$$e_{\gamma\delta}u = ce^{-i(\widehat{\alpha}x + \widehat{\beta}y)}w, \text{ some } c \neq 0.$$

This implies $\gamma = -\alpha$, $\delta = -\beta$ and thus $D(\alpha, \beta) = -\sigma$.

Likewise in case $D(\gamma, \delta) = -\sigma$. □

Remark: Without loss of generality we assume

$$D(\alpha, \beta) = -\sigma, \quad (\alpha, \beta \text{ as in } 4.1). \tag{4}$$

Below $\widetilde{B}^*$, L_0^* are the adjoints of $\widetilde{B}$, L_0 resp.; recalling $e_{\alpha\beta}$, $\mathcal{E}_{\alpha\beta}$ in (3.15), (3.21) we use the shorthand $e_{-\alpha\beta} = e_{-\alpha,-\beta}$, $\mathcal{E}_{-\alpha\beta} = \mathcal{E}_{-\alpha,-\beta}$ and likewise with the projections $\varrho_{\alpha\beta}$, $Q_{\alpha\beta}$.

Proposition 4.2: $\pm i\sigma$ are algebraically simple eigenvalues of $\widetilde{B}$, $\widetilde{B}^*$ respectively and there are φ, $\varphi^* \in \mathrm{dom}(\widetilde{B})$ with:

$$\begin{aligned} &\widetilde{B}\varphi = i\sigma\varphi,\ \widetilde{B}\overline{\varphi} = -i\sigma\overline{\varphi},\ \widetilde{B}^*\varphi^* = -i\sigma\varphi^*,\ \widetilde{B}^*\overline{\varphi}^* = i\sigma\overline{\varphi}^*, \\ &\langle\varphi, \varphi^*\rangle = 1,\ \langle\overline{\varphi}, \overline{\varphi}^*\rangle = 1,\ \langle\varphi, \overline{\varphi}^*\rangle = 0,\ \langle\overline{\varphi}, \varphi^*\rangle = 0, \\ &\varphi, \varphi^* \in \mathcal{E}_{\alpha\beta},\ \overline{\varphi}, \overline{\varphi}^* \subset \mathcal{E}_{-\alpha\beta}. \end{aligned} \tag{5}$$

Proof. With $\varphi = e_{\alpha\beta}w$ as in (4.1) the claim follows from $H(\alpha)$ by standard functional analysis and the fact that the spaces $\mathcal{E}_{\alpha\beta}$, $\mathcal{E}_{-\alpha\beta}$ are invariant under $\widetilde{B}$, $\widetilde{B}^*$ (Lemma 3.3).

Proposition 4.3:

(a) $0 \in \sigma(L_0)$ has algebraic and geometric multiplicity $m = 2$; its eigenspace is spanned by $\varphi, \overline{\varphi}$,

(b) $0 \in \sigma(L_0^*)$ has algebraic and geometric multiplicity $m = 2$; its eigenspace is spanned by φ^*, φ^*.

Proof. We first claim

$$L_0\varphi = L_0\overline{\varphi} = L_0^*\varphi^* = L_0^*\overline{\varphi}^* = 0. \tag{6}$$

This follows from Prop. 3.3, 3.4 and (4.4); e.g.,

$$L_0^*\overline{\varphi}^* = \widetilde{B}^*\overline{\varphi}^* - T\overline{\varphi}^* = i\sigma\overline{\varphi}^* - iD(-\alpha,-\beta)\overline{\varphi}^* = i\sigma\overline{\varphi}^* - i\sigma\overline{\varphi}^* = 0.$$

The other cases in (4.6) follows in the same way.

Next assume

$$L_0\psi = 0, \text{ some } \psi \in \operatorname{dom}(L_0), \quad \psi \neq 0. \tag{7}$$

Now pick $(\gamma,\delta) \in Z^2$ such that

$$(\gamma,\delta) \notin \{(\alpha,\beta),(-\alpha,-\beta)\}. \tag{8}$$

We apply $p_{\gamma\delta}$ to (4.7); by Lemma 3.2 we get

$$\widetilde{B}p_{\gamma\delta}\psi + Tp_{\gamma\delta}\psi = \widetilde{B}p_{\gamma\delta}\psi + iD(\gamma,\delta)p_{\gamma\delta}\psi = 0. \tag{9}$$

Since $\widetilde{B}w \neq 0$ if $w \in (H^2(I) \cap H_0^1(I))^4$, $\operatorname{div} w = 0$, $w \neq 0$, clause (4.9) implies $p_{\gamma\delta}\psi = 0$ if $(\gamma,\delta) = 0$. If $(\gamma,\delta) \neq 0$ then $D(\gamma,\delta) \neq 0$ by $H(\gamma)$. If now $p_{\gamma\delta}\psi \neq 0$ then $D(\gamma,\delta) \in \{\pm\sigma\}$ by $H(\alpha)$ and (4.9) whence

$$D(\alpha+\gamma,\beta+\delta) = 0 \text{ or } D(\gamma-\alpha,\delta-\beta) = 0 \tag{10}$$

by Prop. 4.1. By $H(\gamma)$ and (4.10) we thus have $(\gamma,\delta) = (\alpha,\beta)$ or $(\gamma,\delta) = (-\alpha,-\beta)$, contradicting (4.8).

We thus have

$$p_{\gamma\delta}\psi = 0 \text{ if } (\gamma,\delta) \notin \{(\alpha,\beta),(-\alpha,-\beta)\}. \tag{11}$$

From (4.11) we infer:

$$\psi = p_{\alpha\beta}\psi + p_{-\alpha\beta}\psi. \tag{12}$$

By setting $(\gamma,\delta) = (\alpha,\beta)$ and $(\gamma,\delta) = (-\alpha,-\beta)$ respectively in (4.9) we obtain:

$$p_{\alpha\beta}\psi = c\varphi, \ p_{-\alpha\beta}\psi = c'\overline{\varphi}, \text{ some } c,\ c'. \tag{13}$$

From (4.7), (4.12), (4.13) it now follows that span $(\varphi,\overline{\varphi})$ is the eigenspace of $0 \in \sigma(L_0)$. In a last step we assume

$$L_0 f = a\varphi + b\overline{\varphi}, \text{ some } f \in \operatorname{dom}(L_0),\ a,\ b. \tag{14}$$

We multiply (4.14) scalarly with φ^*, and $\overline{\varphi}^*$ respectively. From (4.5) we infer $a = b = 0$, i.e., $L_0 f = 0$, what concludes the proof. □

Remark: By Prop. 4.3, $0 \in \sigma(L_0)$ has a generalised eigenspace of dimension $m = 2$. Due to the lack of a suitable symmetry, the method of bifurcation from a simple eigenvalue is not applicable. As in case of Hopf bifurcation this difficulty is overcome by the introduction of an amplitude parameter μ, i.e., instead of (3.10) we consider:

$$L_0 w + \varepsilon\mathcal{P}Mw - \delta Gw + \chi Tw + \mu\mathcal{N}(w,w) = 0; \tag{15}$$

if $w \in \mathrm{dom}(L_0)$ is a solution of (4.15) then μw is a solution of (3.10). Since we seek real solutions we introduce real eigenvectors:

$$\begin{aligned} \phi_1 &= \tfrac{1}{\sqrt{2}}(\varphi + \overline{\varphi}), \qquad \phi_2 = \tfrac{i}{\sqrt{2}}(\overline{\varphi} - \varphi), \\ \phi_1^* &= \tfrac{1}{\sqrt{2}}(\varphi^* + \overline{\varphi}^*), \ \phi_2^* = \tfrac{i}{\sqrt{2}}(\overline{\varphi}^* - \varphi^*). \end{aligned} \tag{16}$$

Proposition 4.4: (a) $\mathrm{span}(\varphi, \overline{\varphi}) = \mathrm{span}(\phi_1, \phi_2)$, $\mathrm{span}(\varphi^*, \overline{\varphi}^*) = \mathrm{span}(\phi_1^*, \phi_2^*)$ and $\langle \phi_j, \phi_k^* \rangle = \delta_{jk}$, (b) $f \in \mathrm{rg}(L_0)$ iff $\langle f, \phi_j^* \rangle = 0$, $j = 1, 2$.

Proof. (a) follows from Prop. 4.1 and (4.16). (b) follows from the fact that $0 \in \sigma(L_0)$ has $\mathrm{span}(\phi_1, \phi_2)$ as generalised eigenspace, while $0 \in \sigma(L_0^*)$ has $\mathrm{span}(\phi_1^*, \phi_2^*)$ as generalised eigenspace. □

We add some observations. From Prop. 3.4, 4.2 and from (4.4), (4.16) we infer

$$T\phi_1 = \sigma\phi_2, \ T\phi_2 = -\sigma\phi_1, \ T\phi_1^* = \sigma\phi_2^*, \ T\phi_2^* = -\sigma\phi_1^*. \tag{17}$$

In order to apply Prop. 3.6 to L_0 let $\mathcal{L}_j$, $\mathcal{N}_j$, $\mathcal{L} = \mathcal{L}_\nu$, $\mathcal{N} = \mathcal{N}_\nu$ be the spaces prior to Prop. 3.6. By Prop. 3.5, 4.3, 4.4 we have that

$$\mathcal{L} = \mathcal{L}_1 = \mathcal{L}_\nu = \mathrm{span}(\phi_1, \phi_2), \ \mathcal{N} = \mathcal{N}_1 = \mathcal{N}_\nu = \mathrm{rg}(L_0). \tag{18}$$

The projections S, Q in Prop. 3.5 are now given by

$$Sf = \langle f, \phi_1^* \rangle \phi_1 + \langle f, \phi_2^* \rangle \phi_2, \ Qf = f - Sf, \ f \in \mathcal{E}. \tag{19}$$

Based on (3.5), (4.17), (4.19) a simple computation shows that S, Q commute with T, i.e.:

$$ST \subseteq TS, \quad QT \subseteq TQ. \tag{20}$$

Finally, based on Prop. 3.6 and (4.18) we define the bounded operator K via

$$K = (QL_0)^{-1} : \mathcal{N} \| \ \|_{L^2} \to \mathcal{N} \cap \mathrm{dom}(L_0), \ \| \ \|_{H^2} \tag{21}$$

and introduce the space

$$X = \mathrm{dom}(L_0) \cap \mathrm{rg}(L_0), \ \| \ \|_{H^2} = \mathcal{N} \cap \mathrm{dom}(L_0), \ \| \ \|_{H^2}. \tag{22}$$

We now apply the Ljapounov-Schmidt (L-S) method to (4.15). To this effect we insert the "Ansatz"

$$w = \zeta\phi + h, \quad \zeta\phi = \zeta_1\phi_1 + \zeta_2\phi_2, \quad h \in X$$

into (4.15), multiply the resulting equation with KQ from the left, obtaining the first L-S equation. The other two are the orthogonality conditions which arise from the fact that $\langle L_0 h, \phi_j^* \rangle = 0$, $j = 1, 2$. The L-S equations so obtained are:

$$\begin{aligned} & h + \varepsilon KQ\mathcal{P}M(\zeta\phi + h) - \delta KQG(\zeta\phi + h) + \chi Th \\ & \quad + \mu KQ\mathcal{N}(\zeta\phi + h, \zeta\phi + h) = 0 \\ & \varepsilon\langle \mathcal{P}M(\zeta\phi + h), \phi_j^* \rangle - \delta\langle G(\zeta\phi + h), \phi_j^* \rangle + \chi\langle T(\zeta\phi), \phi_j^* \rangle \\ & \quad + \mu\langle \mathcal{N}(\zeta\phi, h) + \mathcal{N}(h, \zeta\phi) + \mathcal{N}(h, h), \phi_j^* \rangle = 0, \ j = 1, 2. \end{aligned} \tag{23}$$

In (4.23) some simplifications have been taken care of which follow from the relations:

$$QTh = Th \in \mathcal{N}, \ \ QT\zeta\phi = 0, \ \ \langle \mathcal{N}(\zeta\phi, \zeta\phi), \phi_j^* \rangle = 0, \ j = 1, 2. \tag{24}$$

The first two follow from (4.17), (4.20), the third is proved in the appendix. We denote the left-hand sides of (4.23) one after the other by $\mathcal{F}(h, \zeta, \varepsilon, \delta, \chi, \mu)$ and $G_j(h, \zeta, \varepsilon, \delta, \chi, \mu)$, $j = 1, 2$ ($\zeta = (\zeta_1, \zeta_2)$). Clearly $\mathcal{F}$ is a holomorphic mapping from $X \times R^6$ to X while the G_j are holomorphic mappings from $X \times R^6$ to R. Since $\mathcal{F}(0, \ldots, 0) = 0$, $\delta_h \mathcal{F}(0, \ldots, 0) = I$ we have by the implicit function theorem that there are $\zeta_0, \varepsilon_0, \delta_0, \chi_0, \mu_0 > 0$ and a mapping h which maps the neighborhood

$$\mathcal{U} = \{(\zeta, \ldots, \mu) \Big| \, |\zeta| \le \zeta_0, \ |\varepsilon| \le \varepsilon_0, \ldots, |\mu| \le \mu_0\} \subseteq R^6 \tag{25}$$

holomorphically into X such that

$$\mathcal{F}(h(\zeta, \varepsilon, \delta, \chi, \mu), \ \zeta, \varepsilon, \delta, z, \mu) = 0 \text{ on } \mathcal{U}. \tag{26}$$

Since $\mathcal{F}(0, \zeta, 0, \ldots, 0) = 0$ we have by the unicity part of the implicit function theorem:

$$h(\zeta, 0, \ldots, 0) = 0, \quad |\zeta| \le \zeta_0. \tag{27}$$

We now stipulate:

$$\begin{aligned} &H_j(\zeta, \varepsilon, \delta, \chi, \mu) = G_j(h(\zeta, \varepsilon, \delta, \chi, \mu)\zeta, \varepsilon, \delta, \chi, \mu), \ j = 1, 2 \\ &H(\zeta, \ldots, \mu) = (H_1(\zeta, \varepsilon, \ldots, \mu), \ H_2(\zeta, \varepsilon, \ldots, \mu)). \end{aligned} \tag{28}$$

Clearly, H is a holomorphic mapping from $\mathcal{U}$ to R^2 which by (4.27) satisfies

$$H(\zeta, 0, \ldots, 0) = 0, \ |\zeta| \le \zeta_0. \tag{29}$$

Based on (4.27) we would like to solve

$$H(\zeta, \varepsilon, \delta, \chi, \mu) = 0, \quad \zeta \ne 0 \text{ fixed}, \varepsilon, \delta, \chi, \mu \text{ small}. \tag{30}$$

This amounts to investigate the matrices

$$D(\gamma_1, \gamma_2, \zeta) = \left(\frac{\partial H_j}{\partial \gamma_k} (\zeta, 0, 0, 0, 0) \right), \ j, k = 1, 2, \ \zeta \ne 0 \tag{31}$$

for pairs (γ_1, γ_2), $\gamma_j \in \{\varepsilon, \delta, \chi, \mu\}$, $\gamma_1 \ne \gamma_2$ what leads to the study of the determinants

$$d(\gamma_1, \gamma_2, \zeta) = \det(D(\gamma_1, \gamma_2, \zeta)), \ \zeta \ne 0, \ \gamma_j \in \{\varepsilon, \delta, \chi, \mu\} \tag{32}$$

for $\gamma_1 \ne \gamma_2$. Since $d(\gamma_1, \gamma_2, \zeta) = 0$ if $\gamma_1 = \mu$ or $\gamma_2 = \mu$, it remains to investigate three cases:

$$d(\varepsilon, \chi, \zeta), d(\varepsilon, \delta, \zeta), \ d(\delta, \chi, \zeta), \ \zeta = (\zeta_1, \zeta_2) \ne 0. \tag{33}$$

The computation of these determinants is completely straightforward. We content us to summarize the result of this computations. We set

$$\begin{aligned} &a_{jk} = \langle \mathcal{P}M\phi_j, \phi_k^* \rangle = \langle M\phi_j, \phi_k^* \rangle, \ b_{jk} = \langle G\phi_j, \phi_k^* \rangle, \\ &d_{jk} = a_{j1}b_{k2} - a_{j2}b_{k1} \end{aligned} \tag{34}$$

The determinants in (4.34) now assume the form:

$$\begin{aligned} d(\varepsilon,\chi,\zeta) &= \sigma\zeta_1^2 a_{11} - \sigma\zeta_1\zeta_2(a_{12}+a_{21}) + \sigma\zeta_2^2 a_{22} \\ d(\delta,\chi,\zeta) &= -\sigma\zeta_1^2 b_{11} + \sigma\zeta_1\zeta_2(b_{12}+b_{21}) - \sigma\zeta_2^2 b_{22} \\ d(\varepsilon,\delta,\zeta) &= -\zeta_1^2 d_{11} + \zeta_1\zeta_2(d_{12}+d_{21}) - \zeta_2^2 d_{22}. \end{aligned} \tag{35}$$

Theorem 2: (A) If one of a_{11}, $a_{12}+a_{21}$, a_{22} is $\neq 0$ then for almost all $\zeta \in R$, $|\zeta| \le \zeta_0$, $\zeta \neq 0$ there is a neighborhood $\mathcal{U}_\zeta$ of $0 \in R^2$ and holomorphic functions $\varepsilon(\delta,\mu)$, $\chi(\delta,\mu)$ on $\mathcal{U}_\zeta$ such that $\varepsilon(0,0) = \chi(0,0) = 0$ and

$$H(\zeta,\varepsilon(\delta,\mu),\delta,\chi(\delta,\mu),\mu) = 0, \ \ (\delta,\mu) \in \mathcal{U}_\zeta. \tag{36}$$

(B) Likewise with δ, χ and b_{11}, $b_{12}+b_{21}$, b_{22},
(C) Likewise with ε, δ and d_{11}, $d_{12}+d_{21}$, d_{22}.

Proof. Consider (A). If one of a_{11}, $a_{12}+a_{21}$, a_{22} is $\neq 0$ then for almost all $\zeta \in R^2$, $|\zeta| \le \zeta_0, \zeta \neq 0$ the determinant $d(\varepsilon,\chi,\zeta)$ in (4.35) is $\neq 0$. For such ζ the 2×2-matrix $D(\varepsilon,\chi,\zeta)$ in (4.31) is invertible by (4.32), $\gamma_1 = \varepsilon$, $\gamma_2 = \chi$. Since $D(\varepsilon,\chi,\zeta)$ is the Fréchet derivative of $H(\zeta,\ldots,\mu)$ with respect to (ε,χ) at $\varepsilon = \chi = \delta = \mu = 0$ and by (4.29) we can apply the implicit function theorem so as to find a neighborhood $\mathcal{U}_5$ and holomorphic functions $\mathcal{E}(\delta,\mu)$, $\chi(\delta,\mu)$ with the required properties. (B), (C) follows likewise. □

Remarks: Since (4.15) is translation invariant in the x,y-direction, $\varepsilon(\delta,\mu)$, $\chi(\delta,\mu)$ in (A) depend on ζ only via $|\zeta|$; however the smaller $|\zeta| \neq 0$ the smaller the neighborhood $\mathcal{U}_\rho$. Likewise in cases (B), (C).

Corollary: Under the assumptions $H(\alpha)$–(γ) and one of the assumptions in (A), (B) or (C) of Thm. 2 there are arbitrary small values $\varepsilon,\delta,\chi,\mu$, not all $= 0$, and an associated solution $w = w(\varepsilon,\delta,\chi,\mu) \in \mathrm{dom}(\widetilde{B})$ of (4.15) such that $Tw \neq 0$.

Proof. Under the assumption in (A), e.g., $\varepsilon(\delta,\mu)$, $\chi(\delta,\mu)$ and $h = h(\zeta,\varepsilon(\delta,\mu),$ $\delta,\chi(\delta,\mu),\mu)$ satisfy the L-S equations (4.24). Hence $w = \zeta\phi + h$ is a solution of (4.15). Since we are free to pick $\mu \neq 0$, $\delta \neq 0$, not all of $\varepsilon,\chi,\delta,\mu$ are $= 0$. Moreover

$$STw = ST(\zeta\phi + h) = TS(\zeta\phi + h) = T\zeta\phi = \zeta_1\sigma\phi_2 - \zeta_2\sigma\phi_1 \neq 0. \qquad \square$$

Theorem 3: The solution $w \neq 0$ of (4.15) associated with $\varepsilon,\delta,\chi,\mu$ under the assumptions of Thm. 2 give rise to a time periodic traveling wave, i.e., $\widetilde{w}(t,x,y,z) = w(x-at, y-bt, z)$ has time period $d = (\frac{a\alpha}{L_1} + \frac{b\beta}{L_2})^{-1}$ where $(\alpha,\beta) \in Z^2$ is given by (4.1).

Proof. By $H(\alpha)$, a pair $(\alpha,\beta) \in Z^2$, unique up to sign, is determined via (4.1). Recalling $[\alpha,\beta]_4$ in (3.34) we define the space

$$Y = \mathrm{dom}(L_0) \cap \mathcal{N} \cap [\alpha,\beta]_4, \ \ \| \ \|_{H^2} \tag{37}$$

which is an $\| \ \|_{H^2}$-closed subspace of X. Based on Lemmas 3.2–3.4 we conclude as in the proof of Thm. 1 that the mapping $\mathcal{F}$ given by the first equation in (4.23) is a holomorphic mapping from $Y \times R^6$ to Y. As in the proof of Thm. 1 we may replace X by Y in the proof of Thm. 2 in order to infer from the unicity part of the implicit function theorem that the solution $h(\zeta,\varepsilon,\dots,\mu)$ of (4.26) is already in Y; an eventual shrinking of the neighborhood $\mathcal{U}_\zeta$ in (4.25) is thereby taken into account. That is, $w=\zeta\phi+h$ is in Y and has the representation

$$w=\sum_n e^{in(\widehat{\alpha}x+\widehat{\beta}y)}\,w_n(z),\;\; w_n\in(H^2(I)\cap H_0^1(I))^4. \tag{38}$$

This entails

$$\widetilde{w}=\sum_n e^{in(\widehat{\alpha}x+\widehat{\beta}y)}\,w_n(z)e^{-in(\widehat{\alpha}a+\widehat{\beta}b)t}. \tag{39}$$

Recalling $\widehat{\alpha},\widehat{\beta}$ in (3.15), it follows from (4.39) that $d=(\frac{a\alpha}{L_1}+\frac{b\beta}{L_2})^{-1}$ is a time period for $\widetilde{w}$. □

Remarks: (1) By Thm. 3, our bifurcating traveling waves are necessarily time periodic, a fact not immediately evident from the outset.
(2) While it may be difficult to find conditions on the v_j in (3.2) which secure that $H(\alpha)$ and at least one of the assumptions in (A)–(C) are satisfied, there is a case, which although not subject to $H(\alpha)$, can nevertheless be handled by the arguments in the proof of Thm. 2. This case arises if $\chi_0=\tau_0=0$ in (3.11), (3.58), that is (4.15) now assumes the form:

$$\begin{aligned}&\widetilde{B}w+\varepsilon\mathcal{P}Mw-\delta Gw+\chi T+\mu\mathcal{N}(w,w)=0,\;\; w\in\operatorname{dom}(\widetilde{B}),\\ &L_0=\widetilde{B}=-A+\lambda_0\mathcal{P}M.\end{aligned} \tag{40}$$

Since $\widetilde{B}$ is now selfadjoint, condition $H(\alpha)$ cannot hold and is thus omitted. Condition $H(\gamma)$ is retained while condition $H(\beta)$ is replaced by the stronger one:

$\underline{H(\delta):}$ $\quad 0\in\sigma(-A+\lambda_0\mathcal{P}M)$ and $-A+\lambda\mathcal{P}M<0$ if $\lambda<\lambda_0$

By $H(\delta)$, λ_0 is the critical Rayleigh parameter of the Bénard problem ([5],[11]). As to the next assumption we invoke a result of V.I. Judovitch ([5],[6]) which states that there is a set $J\subseteq R^2$ with R^2-J of Lebesgue measure zero such that $(L_1,L_2)\in J$ (L_j the periods) implies that the eigenspace $\mathcal{L}$ of $0\in\sigma(\widetilde{B})$ is as follows. There is a pair $(\alpha,\beta)\in Z^2$, $(\alpha,\beta)\neq 0$ and smooth functions $w,\vartheta\neq 0$ on $[-\frac12,\frac12]$ such that

$$\begin{aligned}&\nu(\partial_z^2-\gamma^2)^2w=\lambda_0\gamma^2\theta,\;\; \mu(\partial_z^2-\gamma^2)\vartheta+\lambda_0 w=0,\;\; \vartheta=w=\partial_z w=0\\ &\text{at } z=\pm\tfrac12\end{aligned} \tag{41}$$

and such that $\mathcal{L}$ is spanned by the four vectors:

$$\varphi_{\varepsilon\delta}=e_{\varepsilon\delta}\left(\frac{i\varepsilon}{\gamma^2}\,\partial_z w,\;\frac{i\delta}{\gamma^2}\,\partial_z w, w,\vartheta\right),\;\; |\varepsilon|=|\alpha|,\;\; |\delta|=|\beta| \tag{42}$$

where $(\varepsilon,\delta)\in Z^2$, $\gamma^2=\alpha^2+\beta^2$. We now assume

$\underline{H(\varepsilon)}$ $\quad (L_1,L_2)\in J$ and $\alpha\beta\neq 0$.

The cases $\alpha=0$ or $\beta=0$ subsume under the arguments below. Our basic assumptions are now $H(\gamma)$, (δ), (ε). By $H(\delta)$, (ε) the eigenspace $\mathcal{L}$ of $0\in\sigma(\widetilde{B})$ has dimension 4. Since one has only four parameters $\varepsilon, \delta, \chi, \mu$, the L-S method cannot be applied straightforwardly. To avoid this difficulty we fix $(\alpha, \beta) \in Z^2$ given by $H(\delta), (\varepsilon)$ and recall the space $[\alpha, \beta]_4$ in (3.34). By Lemmas 3.2–3.4, the left-hand side of (4.40) leaves the space $\mathcal{E} \cap [\alpha, \beta]_4$ invariant. We thus can apply L-S to (4.40) restricted to $\mathcal{E} \cap [\alpha, \beta]_4$. The eigenspace $\mathcal{L}'$ of $0s \in \sigma(\widetilde{B})$, $\widetilde{B}$ restricted to $\mathcal{E} \cap [\alpha, \beta]_4$, has dimension $d = 2$ and is given by spaces $(\varphi_{\alpha\beta}, \varphi_{-\alpha\beta})$, with $\varphi_{\alpha\beta}$ and $\varphi_{-\alpha\beta}$ as in (4.42). We set $\varphi = \varphi_{\alpha\beta}$, $\overline{\varphi} = \varphi_{-\alpha\beta}$. Since $\widetilde{B}$ and hence its restriction to $\mathcal{E} \cap [\alpha, \beta]_4$ is selfadjoint, we have $\varphi^* = \varphi$. A real basis of $\mathcal{L}'$ is again given by (4.16); the orthogonality relations in Prop. 4.4 remain of course valid in the present setting. The projection Q associated by Prop. 3.6 with $\widetilde{B}$ restricted to $\mathcal{E} \cap [\alpha, \beta]_4$ is now given by

$$Qf = f - \langle f, \phi_1 \rangle \phi_1 - \langle f, \phi_2 \rangle \phi_2, \quad f \in \mathcal{E} \cap [\alpha, \beta]_4; \tag{43}$$

it maps $\mathcal{E} \cap [\alpha, \beta]_4$ onto $\mathrm{rg}(\widetilde{B}) \cap [\alpha, \beta]_4$. By Prop. 3.6 the bounded operator $K = (Q\widetilde{B})^{-1}$ is associated with the restriction of $\widetilde{B}$ to $\mathcal{E} \cap [\alpha, \beta]_4$:

$$K : \ \mathrm{rg}(\widetilde{B}) \cap [\alpha, \beta]_4, \ \| \ \|_{L^2} \to \mathrm{dom}(\widetilde{B}) \cap \mathrm{rg}(\widetilde{B}) \cap [\alpha, \beta]_4, \ \| \ \|_{H^2}. \tag{44}$$

With the L-S equations given by (4.23) and reinterpreted as above, we let as before $\mathcal{F}(h, \zeta, \dots, \mu)$, $G_j(h, \zeta, \dots, \mu)$, $j = 1, 2$ be the left-hand sides of (4.23). $\mathcal{F}$, G_j now map $X \times R^6$ holomorphically into X and R respectively, where X is the space

$$X = \mathrm{dom}(\widetilde{B}) \cap \mathrm{rg}(\widetilde{B}) \cap [\alpha, \beta]_4, \ \| \ \|_{H^2}. \tag{45}$$

Since $\mathcal{F}(0) = 0$, $\partial_h \mathcal{F}(0) = I$, there is by the implicit function theorem a neighborhood $\mathcal{U}$ of type (4.26) (some $\zeta_0, \dots, \mu_0$) and a holomorphic mapping h from $\mathcal{U}$ to X which satisfies (4.26), (4.27). In terms of h, G_j, $j = 1, 2$ we define mappings H_j, $j = 1, 2$, and H via (4.28). H is now a holomorphic mapping from $\mathcal{U}$ to R^2 which satisfies (4.25). Since solving the L-S equations (4.23) is reduced to solve (4.30) via the implicit function theorem we are again led to study the regularity of the matrices $D(\gamma_1, \gamma_2, \zeta)$ in (4.32) and hence the determinants in (4.33). Since the eigenfunctions ϕ_j $(= \phi_j^*)$ are known via (4.16), (4.41), (4.42) these determinants can be computed in some cases.

Theorem 4:

(A) For $\zeta \subset R^2$, $\zeta \neq 0$ sufficiently small there are a neighborhood $\mathcal{U}_\zeta$ of $0 \in R^2$ and holomorphic functions $\varepsilon(\delta, \mu)$, $\chi(\delta, \mu)$ on $\mathcal{U}_\zeta$ with;
   (a) $\varepsilon(0, 0) = \chi(0, 0) = 0$,
   (b) $H(\zeta, \varepsilon(\delta, \mu), \delta, \chi(\delta, \mu), \mu) = 0$, $(\delta, \mu) \in \mathcal{U}_\zeta$.

(B) Let v_4 in (3.2) satisfy $\int_I v_4 \partial_z (w\mathcal{V}) dz \neq 0$; for $\zeta \in R^2$, $\zeta \neq 0$ sufficiently small there is a neighborhood $\mathcal{U}_\zeta$ of $v \in R^2$ and holomorphic functions $\delta(\varepsilon, \mu)$, $\chi(\varepsilon, \mu)$ on $\mathcal{U}_\zeta$ with:
   (a) $\delta(0, 0) = \chi(0, 0) = 0$,
   (b) $H(\zeta, \varepsilon, \delta(\varepsilon, \mu), \chi(\varepsilon, \mu), \mu) = 0$, $(\varepsilon, \mu) \in \mathcal{U}_\zeta$.

Proof. As to (A) we evaluate $d(\varepsilon, \chi, \zeta)$ via (4.34), (4.35).

By $H(\gamma)$ we have

$$D(\alpha,\beta) = a\widehat{\alpha} + b\widehat{\beta} = -\sigma \neq 0. \tag{46}$$

Since $\phi_j = \phi_j^*$ we have

$$a_{jj} = \langle \mathcal{P}M\phi_j, \phi_j\rangle,\ a_{21} = a_{12} = \langle \mathcal{P}M\phi_1, \phi_2\rangle. \tag{47}$$

Since ϕ_j, $j = 1,2$ are eigenfunctions of $0 \in \sigma(-A + \lambda_0 \mathcal{P}M)$ we have:

$$\lambda_0 \langle \mathcal{P}M\phi_j\phi_j\rangle = \langle A\phi_j, \phi_j\rangle > 0, \quad \langle \mathcal{P}M\phi_j, \phi_j\rangle > 0. \tag{48}$$

As shown in the appendix it holds that

$$\langle \mathcal{P}M\phi_j, \phi_k\rangle = 0,\ j \neq k, \quad \langle \mathcal{P}M\phi_1, \phi_1\rangle = \langle \mathcal{P}M\phi_2, \phi_2\rangle. \tag{49}$$

By (4.34), (4.35) and (4.46)–(4.49) we thus get:

$$d(\varepsilon,\chi,\zeta) = \sigma(\zeta_1^2 + \zeta_2^2)\langle \mathcal{P}M\phi_1, \phi_1\rangle, \quad \zeta = (\zeta_1,\zeta_2) \neq 0, \tag{50}$$

hence $D(\varepsilon,\chi,\zeta)$ is invertible whence (A) follows by the implicit function theorem.

As to (B) we compute $d(\delta,\chi,\zeta)$ via (4.35) where $b_{jk} = \langle G\phi_j, \phi_k\rangle$. Now we need some results proved in the appendix, namely

$$\begin{aligned} &\langle G\phi_1, \phi_2\rangle = -\langle G\phi_2, \phi_1\rangle,\ \langle G\phi_1, \phi_1\rangle = \langle G\phi_2, \phi_2\rangle, \\ &\langle G\phi_1, \phi_1\rangle = -A\int_I v_4 \partial_z(w\mathcal{V})\,dz, \text{ some } A > 0 \end{aligned} \tag{51}$$

with v_4 given by (3.2). Based on (4.51), the computation of $d(\delta,\chi,\zeta)$ via (4.35) gives

$$d(\delta,\chi,\zeta) = A\sigma(\zeta_1^2 + \zeta_2^2)\int_I v_4 \partial_z(w\mathcal{V})\,dz,\ I = \left[-\tfrac{1}{2}, \tfrac{1}{2}\right]. \tag{52}$$

By the assumption of (B) the claim now follows from (4.31), (4.32), (4.46) and (4.52) via the implicit function theorem □

Remarks: As to (A), (B) it would be desirable to know, e.g., at least the linear terms in the expansion of $\varepsilon(\delta,\mu)$, $\chi(\delta,\mu)$. Below, $(4.23)_j$; $j = 1,2,3$ denotes the jth equation of (4.23). We first consider the expression

$$h(\zeta,\varepsilon,\delta,\chi,\mu) = h_1\varepsilon + h_2\delta + h_3\chi_h + h_4\mu + \ldots, \quad h_j = h_j(\zeta).$$

Inserting this into $(4.23)_1$ a comparison of coefficients yields:

$$\begin{aligned} &h_1 = -KQ\mathcal{P}M\zeta\phi,\ h_2 = KQG\zeta\phi,\ h_4 = -KQ\mathcal{N}(\zeta\phi,\zeta\phi), \\ &h_3 = 0. \end{aligned} \tag{53}$$

We now multiply $(4.23)_2$ with ζ_1, and add $(4.23)_3$ multiplied with ζ_2. By taking care of (4.46), of Prop. 4.4 and (4.49) we get

$$\begin{aligned} &\varepsilon|\zeta|^2\langle \mathcal{P}M\phi_1, \phi_1\rangle - \delta\langle G(\zeta\phi + h), \zeta\phi\rangle + \varepsilon\langle \mathcal{P}Mh, \zeta\phi\rangle \\ &\quad + \mu\langle \mathcal{N}(\zeta\phi + h, \zeta\phi + h), \zeta\phi\rangle = 0, \quad |\zeta|^2 = \zeta_1^2 + \zeta_2^2. \end{aligned} \tag{54}$$

We insert the expansion $\varepsilon = \varepsilon_0\mu + \varepsilon_1\delta + \cdots$ into (4.54).

Comparison of coefficients yields:

$$\varepsilon_0 = 0, \quad \varepsilon_1|\zeta|^2\langle \mathcal{P}M\phi_1, \phi_1\rangle = \langle G\zeta\phi, \zeta\phi\rangle. \tag{55}$$

Next we multiply $(4.23)_3$ with ζ_1 and subtract $(4.23)_2$ multiplied with ζ_2 so as to obtain:

$$\begin{aligned}\chi|\zeta|^2\sigma - \langle G(\zeta\phi + h),\ \zeta_1\phi_2 - \zeta_2\phi_1\rangle\delta + \varepsilon\langle \mathcal{P}Mh, \zeta_1\phi_2 - \zeta_2\phi_1\rangle \\ +\mu\langle \mathcal{N}(\zeta\phi + h, \zeta\phi + h),\ \zeta_1\phi_2 - \zeta_2\phi_1\rangle = 0.\end{aligned} \tag{56}$$

We insert the expansion $\chi = \chi_0\mu + \chi_1\delta + \cdots$ into (4.51); a comparison of coefficients yields:

$$\chi_0 = 0, \quad \chi_1|\zeta|^2\sigma = \langle G\zeta\phi, \zeta_1\phi_2 - \zeta_2\phi_1\rangle. \tag{57}$$

Crucial for the computation of ε_1, χ_1 is Prop. 4.5, to be proved in the appendix:

Proposition 4.5: There is a holomorphic function $F(z)$, $z \in I$, not $\equiv 0$, depending on w, $\mathcal{V}$, α, β in (4.41), (4.42), such that

$$\langle G\phi_1, \phi_2\rangle = \int_I F(z)(\widehat{\alpha}v_1 + \widehat{\beta}v_2)dz \quad (v_j \text{ via } (3.2)). \tag{58}$$

Theorem 5: Assume

$$\text{(a)} \quad \int_I \partial_z(w\mathcal{V})v_4\, dz \neq 0, \quad \text{(b)} \quad \int_I F(z)(\widehat{\alpha}v_1 + \widehat{\beta}v_2)\, dz \neq 0. \tag{59}$$

Then: (A) $\varepsilon_1 \neq 0$, (B) $\chi_1 \neq 0$, (C) for $\zeta \in R^2$, $\zeta \neq 0$ sufficiently small there is a neighborhood $\mathcal{U}_5$ of $O \in R^2$ and holomorphic functions $\varepsilon(\chi,\mu)$, $\delta(\chi,\mu)$ on $\mathcal{U}_\zeta$ with

(a) $\varepsilon(0,0) = \delta(0,0) = 0$,

(b) $H(\zeta, \varepsilon(\chi,\mu), \delta(\chi,\mu), \chi, \mu) = 0 \quad (H \text{ via } (4.28))$.

Proof. From (4.55) we infer

$$\begin{aligned}\delta\varepsilon_1|\zeta|^2\langle \mathcal{P}M\phi_1\phi_1\rangle &= \zeta_1^2\langle G\phi_1\phi_1\rangle + \zeta_2^2\langle G\phi_2, \phi_2\rangle \\ &+ \zeta_1\zeta_2(\langle G\phi_1, \phi_2\rangle + \langle G\phi_2, \phi_1\rangle).\end{aligned} \tag{60}$$

Claim (A) near follows from (4.46), (4.51), from Ass. (4.59)(a) and from (4.60). In order to prove (B) we recall (4.57); computation yields

$$\begin{aligned}\langle G\zeta\phi, \zeta_1\phi_2 - \zeta_2\phi_2\rangle &= \zeta_1^2\langle G\phi_1, \phi_2\rangle - \zeta_2^2\langle G\phi_2, \phi_1\rangle \\ &+ \zeta_1\zeta_2(\langle G\phi_2, \phi_2\rangle - \langle G\phi_1, \phi_1\rangle).\end{aligned}$$

By (4.51) we thus obtain

$$\langle G\zeta\phi, \zeta_1\phi_2 - \zeta_2\phi_1\rangle = (\zeta_1^2 + \zeta_2^2)\langle G\phi_1, \phi_2\rangle. \tag{61}$$

(B) now follows from (4.57), (4.61), Ass. (4.59)(b) and Prop. 4.5.

In order to prove (C) we compute the the determined $d(\varepsilon, \delta, \zeta)$ via (4.34), (4.35). By a straightforward computation based on (4.34), (4.35), (4.49), (4.51) and by using Prop. 4.5 and Ass. (4.59)(b) we get

$$d(\varepsilon, \delta, \zeta) = |\zeta|^2\langle \mathcal{P}M\phi_1, \phi_1\rangle\langle G\phi_1, \phi_2\rangle \neq 0. \tag{62}$$

Claim (C) now follows from (4.62) by the implicit function theorem. □

Remarks: As with Thm. 2 we can rephrase Thms. 4, 5 so as to apply to equation (4.40) and ultimately to equation (0) in the introduction. We content us to summarize the facts for (4.40).

Corollary 1: Let v_j, $j \leq 4$ in (3.2) satisfy assumptions (4.59). Then there are $\varepsilon, \delta, \chi, \mu$ arbitrarily small, all $\neq 0$, and an associated solution $w = w(\zeta, \varepsilon, \delta, \chi, \mu)$ of (4.40) such that $Tw \neq 0$.

Proof. With the assumptions satisfied, we have by Thm. 5 that $\varepsilon(\delta, \mu)$, $\chi(\delta, \mu)$ in Thm. 4 have the form:

$$\varepsilon = \varepsilon_1 \delta + O(\delta^2 + \mu^2), \quad \chi = \chi_1 \delta + O(\delta^2 + \mu^2), \quad \varepsilon_1 \neq 0, \chi_1 \neq 0.$$

Thus for μ, δ both $\neq 0$ and small, ε, χ are both $\neq 0$ and by Thm. 4, (A) we have that

$$w(\zeta, \delta, \mu) = \zeta\phi + h(\zeta, \varepsilon(\delta, \mu), \delta, \chi(\delta, \mu), \mu) = \zeta\phi + h, \ \zeta \neq 0 \qquad (63)$$

is a solution of (4.40). As in the proof of the corollary to Thm. 2 we conclude that $Tw \neq 0$. □

Thus with w as in (4.63) we have that $\widetilde{w} = \delta v + \mu w$ is a genuine traveling wave solution of

$$-A\widetilde{w} + (\lambda_0 + \varepsilon)\mathcal{P}M\widetilde{w} + \chi\mathcal{P}T\widetilde{w} + \mathcal{N}(\widetilde{w}, \widetilde{w}) + \delta f = 0, \qquad (64)$$

i.e., such that

$$\partial_t \widetilde{w}(x - \chi a t, \ y - \chi b t, z) = -\chi(T\widetilde{w})(x - \chi a t, y - \chi b t, z) \neq 0.$$

Since the construction of solutions in the proof of Thm. 4 is in the space $\mathcal{E} \cap [\alpha, \beta]_4$, Thm. 3 retains its validity here, that is, the traveling waves constructed via Thm. 4.5, Cor. 1 are time periodic with period given by Thm. 3. As for the functions $\varepsilon(\chi, \mu)$, $\delta(\chi, \mu)$ it is easy to see that they admit an expansion of the form:

$$\varepsilon(\chi, \mu) = \varepsilon_1 \chi + O(\chi^2 + \mu^2), \quad \delta(\chi, \mu) = \delta_1 \chi + O(\chi^2 + \mu^2). \qquad (65)$$

Based on the assumptions of Thm. 5 one can perform computations similar to the above in order to compute ε_1, δ_1. The result, stated without proof, is

$$\delta_1 = \sigma \langle G\phi_1, \phi_2 \rangle^{-1}, \varepsilon_1 = \sigma \langle \mathcal{P}M\phi_1, \phi_1 \rangle^{-1} \langle G\phi_1, \phi_2 \rangle^{-1} \langle G\phi_1, \phi_1 \rangle, \qquad (66)$$

with $\langle G\phi_1, \phi_2 \rangle$, $\langle G\phi_1, \phi_1 \rangle$ given by (4.51), (4.58) resp.

Remarks: Our construction of traveling wave solutions to (4.40) resp. (0) gives rise to two problems. The first is whether the construction extends to other translation invariant domains such as channels. The second is of course the question whether this waves are stable in some sense, and whether they can be seen either numerically or experimentally. These topics will be treated elsewhere.

Appendix

Below we prove some relations which appear in the proofs of Thm. 4, 5. We recall (4.41), (4.42):

$$\varphi = e_{\alpha\beta}\left(\frac{i\widehat{\alpha}}{\gamma^2}\,\partial_z w, \frac{i\widehat{\beta}}{\gamma^2}\,\partial_z w, w, \mathcal{V}\right), \quad e_{\alpha\beta} = e^{i\tau},\ \tau = \widehat{\alpha}x + \widehat{\beta}x, \tag{A.1}$$

$$\widehat{\alpha} = \frac{2\pi\alpha}{L_1},\ \widehat{\beta} = \frac{2\pi\beta}{L_2},\ \alpha, \beta \in Z, \alpha\cdot\beta \neq 0,\ \gamma^2 = \widehat{\alpha}^2 + \widehat{\beta}^2,$$

α, β given by $H(\delta), (\varepsilon)$. We have $\varphi = a + ib$, a, b real, with

$$a = \left(-\frac{\widehat{\alpha}}{\gamma^2}\,\partial_z w \sin\tau,\ -\frac{\widehat{\beta}}{\gamma^2}\,\partial_z w \sin\tau,\ w\cos\tau,\ \mathcal{V}\cos\tau\right) \tag{A.2}$$

$$b = \left(-\frac{\widehat{\alpha}}{\gamma^2}\,\partial_z w \cos\tau,\ \frac{\widehat{\beta}}{\gamma^2}\,\partial_z w \cos\tau,\ w\sin\tau,\ \mathcal{V}\sin\tau\right).$$

For computational reasons we replace (4.16) by a rescaled definition ϕ_1, ϕ_2, without affecting the assertions to be proved:

$$\phi_1 = \varphi + \overline{\varphi} = 2a, \quad \phi_2 = i(\overline{\varphi} - \varphi) = 2b. \tag{A.3}$$

Let $d_1 = L_1(4\alpha)^{-1}$, $d_2 = L_2(4\beta)^{-1}$; one easily verifies

$$\phi_1(x + d_1, y, z) = -\phi_2(x, y, z), \quad \phi_2(x + d_1, y, z) = \phi_1(x, y, z) \tag{A.4}$$

likewise with d_2 and y. We also need the Euclidean scalar product

$$u\cdot v = \sum_1^4 u_j v_j, \quad u = (u_1, \ldots, u_4),\ v = (v_1, \ldots, v_4)$$

and the real scalar product

$$[u, v] = \langle u, \overline{v}\rangle = \int_Q u\cdot v\,dx^3, \quad dx^3 = dx\,dy\,dz.$$

Proposition A1:

(a) $\langle \mathcal{P}M\phi_1, \phi_1\rangle = \langle \mathcal{P}M\phi_2, \phi_2\rangle$,

(b) $\langle \mathcal{P}M\phi_j, \phi_k\rangle = 0$, $j \neq k$.

Proof. As to (a) we have by periodicity

$$\langle \mathcal{P}M\phi_1, \phi_1\rangle = \int_0^{L_1}\int_0^{L_2}\int_I M\phi_1\cdot\phi_1\,dx^3 = \int_{d_1}^{L_1+d_1}\int_0^{L_2}\int_I M\phi_1\cdot\phi_1\,dx^3.$$

On the last integral we perform the substitution $x + d_1 \to x$; by (A.4) we get

$$\langle \mathcal{P}M\phi_1, \phi_1\rangle = [M(-\phi_2), (-\phi_2)] = [M\phi_2, \phi_2] = \langle \mathcal{P}M\phi_2, \phi_2\rangle.$$

In case of (b) the same substitution yields:

$$\langle \mathcal{P}M\phi_1, \phi_2\rangle = [M\phi_1, \phi_2] = [M(-\phi_2), \phi_1] = -\langle \mathcal{P}M\phi_1, \phi_2\rangle,$$

whence $\langle \mathcal{P}M\phi_1, \phi_2\rangle = \langle \mathcal{P}M\phi_2, \phi_1\rangle = 0$. □

Proposition A2: $\langle \mathcal{N}(\phi_j, \phi_k), \phi_l\rangle = 0$, $j, k, l = 1, 2$.

Proof. Since the ϕ_j are real we have

$$\tau = \langle \mathcal{N}(\phi_j, \phi_k), \phi_l\rangle = -[(\phi_j\nabla)\phi_k, \phi_l].$$

Recalling (A.1)–(A.3) we have that I is a sum

$$I = c\sum[(\psi_p\nabla)\psi_q, \psi_r], \quad \psi_p, \psi_q, \psi_r \in \{\varphi, \overline{\varphi}\}$$

with φ, $\overline{\varphi}$ via (1) and c a constant; due to the structure of φ, $\overline{\varphi}$, each term $[(\psi_p\nabla)\psi_q, \psi_r]$ is a product of two integrals π_1, π_2 say, which have the form:

$$\pi_1 = \int_0^{L_1}\int_0^{L_1} e_{\alpha_1\beta_1}e_{\alpha_2\beta_2}e_{\alpha_3\beta_3}\,dx\,dy, \quad \pi_2 = \int_I \mathcal{P}(z)\,dz, \tag{A.5}$$

and where

$$(\alpha_j, \beta_j) \in \{(\alpha, \beta), (-\alpha, -\beta)\}, \quad \alpha, \beta \in Z,\ \alpha\cdot\beta \neq 0 \tag{A.6}$$

with α, β fixed by $H(\delta)$, (ε) in Sec. 4. With the exponential factors $e_{\alpha_j\beta_j}$ given by (1), and by (A.6), the factors π_1 in (A.5) vanish, whence the claim follows. □

We now come to the claims regarding G defined via (3.2), (3.9).

Proposition A3:

(a) $\langle G\phi_1\phi_1\rangle = \langle G\phi_2, \phi_2\rangle$,

(b) $\langle G\phi_1, \phi_2\rangle = -\langle G\phi_2, \phi_1\rangle$.

Proof. As to (b) we recall (2.8), (3.9) according to which

$$\begin{aligned}\langle G\phi_1, \phi_2\rangle &= -\langle \mathcal{N}(v, \phi_1, \phi_2\rangle - \langle \mathcal{N}(\phi_1, v), \phi_2\rangle = \\ &= [(v\nabla)\phi_1, \phi_2] + [(\phi_1\nabla)v, \phi_2] = I_1 + I_2.\end{aligned}$$

By periodicity we have

$$I_1 = \int_{d_1}^{L_1+d_1}\int_0^{L_2}\int_I (v\nabla)\phi_1\cdot\phi_2\,dx^3.$$

Since the substitution $x + d_1 \to x$ commutes with differentiation we have by (A.4):

$$I_1 = \int_0^{L_1}\int_0^{L_2}\int_I ((v\nabla)(-\phi_2))\cdot\phi_1\,dx^3 = [\mathcal{N}(v, \phi_2), \phi_1].$$

Likewise we find $I_2 = [\mathcal{N}(\phi_2, v), \phi_1]$ whence

$$\langle G\phi_1, \phi_2\rangle = [\mathcal{N}(v, \phi_2), \phi_1] + [\mathcal{N}(\phi_2, v), \phi_1] = -\langle G\phi_2, \phi_1\rangle,$$

proving (b); the proof of (a) is similar and omitted. □

We now come to the proof of Prop. 4.5 which requires some preparatory steps, labeled by S1, S2,...

S1: Based on (3.9) and (A.1)–(A.3), we have

$$\begin{aligned}\langle G\phi_1, \phi_2\rangle &= [(v\nabla)]\phi_1, \phi_2] + [(\phi_1\nabla)v, \phi_2] \\ &= 4[(v\nabla)a, b] + 4[(a\nabla)v, b] = \mathrm{I} + \mathrm{II}.\end{aligned} \tag{A.7}$$

In order to evaluation I, II we note that they depend linearly on $v = (v_1, v_2, 0, v_4)$ in (3.2). This allows us to split the evaluation of I, II into three cases:

$$\text{(a) } v = (r, 0, 0, 0), \quad \text{(b) } v = (0, r, 0, 0), \quad \text{(c) } v = (0, 0, 0, r), \tag{A.8}$$

with $r \in H^2(I) \cap H_0^1(\mathrm{I})$. We first compute I under Ass. (A.8)(a).
We first note:

$$\int_0^{L_1}\int_0^{L_2} \sin^2\tau\, dx\, dy = \int_0^{L_1}\int_0^{L_2} \cos^2\tau\, dx\, dy = \frac{L_1L_2}{2}, \quad \tau = \widehat{\alpha}x + \widehat{\beta}y. \tag{A.9}$$

Based on (2) we compute the term $(v\nabla)a$ under the assumption (A.8)(a) and obtain that $(v\nabla)a$ is equal to

$$\left(-\frac{\widehat{\alpha}^2}{\gamma^2}\, r\partial_z w \cos\tau, -\frac{\widehat{\alpha}\widehat{\beta}}{\gamma^2}\, r\partial_z w\cos\tau, -wr\widehat{\alpha}\sin\tau, -\mathcal{V}r\widehat{\alpha}\sin\tau\right). \tag{A.10}$$

The Euclidean scalar product of (A.10) with b gives for $(v\nabla)a \cdot b$ the expression:

$$\begin{aligned}-\frac{\widehat{\alpha}^3}{\gamma^4}\, r(\partial_z w)^2\cos^2\tau - \frac{\widehat{\alpha}\widehat{\beta}^2}{\gamma^4}\, r(\partial_z w)^2\cos^2\tau \\ -rw^2\widehat{\alpha}\sin^2\tau - r\mathcal{V}^2\widehat{\alpha}\sin^2\tau.\end{aligned} \tag{A.11}$$

After a rearrangement of terms in (A.11), integration over Q based on (A.9) yields:

$$I = 4[(v\nabla)a, b] = -2L_1L_2\int_I\left(\frac{(\partial_z w)^2}{\gamma^2} + w^2 + \mathcal{V}^2\right)\widehat{\alpha}r\, dz. \tag{A.12}$$

In order to compute II in (A.7) under Ass. (A.8)(a) we note that $\mathrm{II} = -4[(a\nabla)b, v]$; by (A.8)(a) it suffices to compute only the first component of $(a\nabla)b$, denoted by $(a\nabla)b_1$. Computation based on (A.2) yields:

$$(a\nabla)b_1 = \frac{\widehat{\alpha}^3}{\gamma^4}(\partial_z w)^2\sin^2\tau + \frac{\widehat{\alpha}\widehat{\beta}^2}{\gamma^4}(\partial_z w)^2\sin^2\tau + \frac{\widehat{\alpha}}{\gamma^2}\, w\partial_z^2 w\cos^2\tau$$

which, after rearrangements of terms, reduces to

$$(a\nabla)b_1 = \frac{\widehat{\alpha}}{\gamma^2}\left((\partial_z w)^2\sin^2\tau + w\partial_z^2 w\cos^2\tau\right). \tag{A.13}$$

Based on (A.9), (A.13), computation yields

$$\mathrm{II} = -4\int_Q (a\nabla)b_1 r\, dx^3 = -2L_1L_2\int_I ((\partial_z w)^2 + w\partial_z^2 w)\frac{\widehat{\alpha}}{\gamma^2}\, r\, dz. \tag{A.14}$$

By combining (A.7), (A.12), (A.14) we infer that under the assumption of case (A.8)(a), $\langle G\phi_1, \phi_2\rangle$ has the form

$$\langle G\phi_1, \phi_2\rangle = -2L_1L_2 \int_I (P_1(z) + P_2(z))\widehat{\alpha} r\, dz, \tag{A.15}$$

$$\mathcal{P}_1(z) = \frac{1}{\gamma^2}((\partial_z w)^2 + w\partial_z^2 w), \quad \mathcal{P}_2(z) = \frac{\partial_z w}{\gamma^2} + w^2 + \mathcal{V}^2.$$

S2: While (A.15) gives an expression of $\langle G\phi_1, \phi_2\rangle$ under Ass. (A.8)(a), a completely analogous computation describes $\langle G\phi_1, \phi_2\rangle$ under Ass. (A.8)(b). Thus with v as in (A.8)(b) we have

$$\langle G\phi_1, \phi_2\rangle = -2L_1L_2 \int_I (\mathcal{P}_1(z) + \mathcal{P}_2(z))\widehat{\beta} r\, dz. \tag{A.16}$$

S3: It remains to evaluate $\langle G\phi_1, \phi_2\rangle$ under Ass. (A.8)(c). Since $\widetilde{v} = (v_1, v_2, v_3) = 0$ under Ass. (A.8)(c) and according (1.4) we have that I defined via (A.7) vanishes: $I = 0$. By (A.7) we thus have $\langle G\phi_1, \phi_2\rangle = 4[(a\nabla)v, b]$ under Ass. (A.8)(c) which also implies that the first three components $(a\nabla)v_j$, $j = 1, 2, 3$ of $(a\nabla)v$ vanish. As to the fourth component $(a\nabla)v_4$ of $(a\nabla)v$, computation based on (A.2), (A.8)(c) yields $(a\nabla)v_4 = w\partial_z r \cos\tau$ whence

$$[(a\nabla)v, b] = \int_I w\mathcal{V}\partial_z r\, dz \cdot \int_0^{L_1}\int_0^{L_2} \sin\tau\cos\tau\, dx\, dy. \tag{A.17}$$

Since the trigonometric integral in (A.17) vanishes, we have

$$\langle G\phi_1, \phi_2\rangle = 0 \text{ under Assumption (A.8)(c)}. \tag{A.18}$$

S4: It remains to compute $\langle G\phi_1, \phi_1\rangle$ for $v = (v_1, v_2, 0, v_4)$ in (3.2), (3.9) arbitrary. By (3.9) and (A.3) we have

$$\langle G\phi_1, \phi_1\rangle = -\langle \mathcal{N}(\phi_1, v), \phi_1\rangle = [(\phi_1\nabla)v, \phi_1] = 4[(a\nabla)v, a].$$

According to (A.2) we have that

$$(a\nabla)v = (w\partial_z v_1 \cos\tau, w\partial_z v_2\cos\tau, 0, w\partial_z v_4\cos\tau). \tag{A.19}$$

The Euclidean scalar product of (A.19) with a in (A.2) gives:

$$\begin{aligned}(a\nabla)v\cdot a = -\frac{w}{\gamma^2}\,\partial_z w(\widehat{\alpha}\partial_z v_1 + \widehat{\beta}\partial_z v_2)\sin\tau\cos\tau \\ + w\mathcal{V} + w\mathcal{V}\partial_z v_4\cos\tau.\end{aligned} \tag{A.20}$$

Since $\int_0^{L_1}\int_0^{L_2} \sin\tau\cos\tau\, dxdy = 0$, we infer from (A.20):

$$\begin{aligned}\langle G\phi_1, G\phi_1\rangle = 4\int_0^{L_1}\int_0^{L_2}(a\nabla)v\cdot a\, dx^3 &= 2L_1L_2\int_I w\mathcal{V}\partial_z v_4\, dz \\ &= -2L_1L_2\int_I \partial_z(w\mathcal{V})v_4\, dz\end{aligned} \tag{A.21}$$

S5: *Proof of Prop.* 4.5. We invoke (A.15), (A.16), (A.18) in S1, S2, S3 respectively and recall that G in (3.9) depends linearly on v. Since

$$v = (v_1, 0, 0, 0) + (0, v_2, 0, 0) + (0, 0, 0, v_4)$$

we infer from (A.15), (A.16), (A.18):

$$\langle G\phi_1, \phi_2 \rangle = -2L_1 L_2 \int_I (P_1(z) + P_2(z))(\widehat{\alpha} v_0 + \widehat{\beta} v_2) dz \qquad \text{(A.22)}$$

with $P_j(z)$, $j = 1, 2$ given by (A.15). We claim:

$$\mathcal{P}_1(z) + \mathcal{P}_2(z) \not\equiv 0. \qquad \text{(A.23)}$$

In fact, in case of the contrary we would have:

$$\int_I (P_1(z) + P_2(z))\, dz = 0. \qquad \text{(A.24)}$$

But according to the definition of the P_j in (15) and since $w(\pm\frac{1}{2}) = 0$ it holds that

$$\int_I P_2(z)\, dz > 0, \quad \int_I P_1(z)\, dz = \tfrac{1}{\gamma^2} \int_I \partial_z (w \partial_z w)\, dz = 0. \qquad \text{(A.25)}$$

Thus (A.24) cannot hold, whence (A.23) follows. Prop. 4.5 now follows by setting $F(z) = c(P_1(z) + P_2(z))$, $c \neq 0$ a suitable constant. □

References

[1] S. Chandrasekhar, Hydrodynamic and hydromagnetic stability. Dover Publ., New York 1961.

[2] G. Crandall, P. Rabinowitz: The Hopf bifurcation theorem in infinite dimensions. Arch. Rat. Mech. Analysis 67, no. 1, 53–72, 1977.

[3] A. Friedman, Foundations of modern analysis. Dover Publ., New York, 1982

[4] T. Kato, Perturbation theory for linear operators. Springer, Berlin, New York, 1980.

[5] K. Kirchgässner, H. Kielhöfer, Stability and bifurcation in fluid dynamics. Rocky Mountains J. of Math., vol. 3, no. 2, 275–318, 1973.

[6] K. Kirchgässner, Bifurcation in nonlinear hydrodynamic stability. SIAM Review, vol. 17, no. 4, 452–483 (1975).

[7] A. Pazy, Semigroups of linear operators and applications to PDE's. Appl. Math. Sc. 44, Springer, New York, 1983.

[8] P. Roberts, An introduction to magneto-hydrodynamics, Longmans, 1974.

[9] D. Sattinger, Topics in stability and bifurcation theory. Lecture Notes in Math. 309, Springer, Berlin, New York, 1973.

[10] B. Scarpellini, Equilibrium solutions of the Bénard equations with an exterior force. J. Diff. & Integral Eq. 16, no. 2, 129–1958, 2003.

[11] B. Scarpellini, W. von Wahl, Stability properties of the Boussinesq equations. ZAMP 49, 294–321, 1998.

[12] B. Scarpellini, Stability, instability and direct integrals. Chapman & Hall/CRC, Boca Raton, 1999.

[13] W. von Wahl, The Boussinesq equations in terms of poloidal and toroidal fields and the mean flow. Bayreuther Math. Schriften 40, 203–290, 1992.

[14] B. Sandstede: Stability of traveling waves. Handbook of dynamical systems, Vol. 2, (2002) 983–1055, Elsevier.

[15] D. Chae, P. Dubovski: Traveling wave-like solutions of the Navier-Stokes and related equations. J. Math. Anal. Appl. 204 (1996), 930–939.

B. Scarpellini
Mathematisches Institut
Universität Basel
Rheinsprung 21
CH-4051 Basel, Switzerland

Progress in Nonlinear Differential Equations
and Their Applications, Vol. 64, 463–472

Vector-valued Sobolev Spaces and Gagliardo-Nirenberg Inequalities

Hans-Jürgen Schmeißer and Winfried Sickel

Abstract. We prove Gagliardo-Nirenberg inequalities for vector-valued Lizorkin-Triebel spaces. Therefrom we derive the corresponding assertions for the vector-valued Sobolev spaces $W_p^m(\mathbb{R}^n, E)$. Here we do not assume the UMD property for E.

Mathematics Subject Classification (2000). 46E35.

Keywords. Vector-valued Sobolev spaces, vector-valued Lizorkin-Triebel spaces, Littlewood-Paley theory, Gagliardo-Nirenberg inequalities, embeddings.

1. Introduction

Vector-valued Sobolev and Besov spaces are now widely used in abstract evolution equations, cf., e.g., Amann [Am 1, Am 2] and Prüß [Pr], and in the theory of integral operators, cf., e.g., König [Ko] and Pietsch [Pi]. For certain applications it is quite inconvenient to assume that the Banach space E has the UMD property which is equivalent to the boundedness of the Hilbert transform in $L_2(\mathbb{R}, E)$ by Bourgain [Bo]. There are simple examples of Banach spaces which do not have this property. The UMD property implies reflexivity of E. So it rules out Hölder spaces and L_1 and variants of those, cf. Amann [Am 1]. On the other hand the UMD property is essential for the validity of Michlin-Hörmander type Fourier multiplier assertions in corresponding L_p-spaces, cf. McConnell [Mc] and Zimmermann [Zi]. So, the question is which properties of distribution spaces really depend on the UMD property and which do not. Here we are concerned with Gagliardo-Nirenberg inequalities which state that (in the scalar-valued case)

$$\| D^\beta f \,|L_r\| \le C \,\| f \,|L_q\|^{1-|\beta|/m} \Big(\sum_{|\alpha|=m} \| D^\alpha f \,|L_p\| \Big)^{|\beta|/m}$$

holds for all $f \in L_p(\mathbb{R}^n)$ belonging to the Sobolev spaces $W_p^m(\mathbb{R}^n)$, provided that

$$1 \le p, q \le \infty, \quad 0 < |\beta| < m, \quad \text{and} \quad \frac{1}{r} = \frac{1 - |\beta|/m}{q} + \frac{|\beta|/m}{p}.$$

Further, the constant C does not depend on f. To prove its vector-valued variant one has to avoid techniques which require the UMD property of the Banach space under consideration.

Hence one has to avoid techniques which are based on such multipliers. A way out is the use of vector-valued spaces of Lizorkin-Triebel type, where multiplier theorems are available and their Fourier-analytic definition uses Littlewood-Paley type decompositions. This scale of spaces seems to be of interest also for themselves. However, in contrast to the scalar case they do not contain (fractional) Sobolev spaces as special cases. Nevertheless, it turns out that some assertions as the Gagliardo-Nirenberg inequalities, can be derived via embeddings.
Agreement: Throughout the paper E denotes a Banach space and $\| \cdot \|_E$ its norm without any further restriction.

2. Vector-valued function spaces

2.1. Preliminaries

Our main tool will be Fourier analysis. There is not much information about the Fourier transform in the literature in the context of vector-valued functions. We refer to Amann [Am 1, Am 2, Am 3] and [Sc, SS] for more detailed reports.
A linear mapping $f : \mathcal{S}(\mathbb{R}^n) \mapsto E$ is called a tempered distribution with values in E if there exist a constant c and nonnegative integers k and ℓ such that

$$\| f(\varphi) \|_E \le c \sup_{x \in \mathbb{R}^n} \sup_{|\alpha| \le \ell} (1 + |x|)^k \, |D^\alpha \varphi(x)|$$

holds for all $\varphi \in \mathcal{S}(\mathbb{R}^n)$. The collection of all such mappings f is denoted by $\mathcal{S}'(\mathbb{R}^n, E)$. In this paper all derivatives have to be understood in the sense of distributions (either in $\mathcal{S}'(\mathbb{R}^n, E)$ or in $\mathcal{S}'(\mathbb{R}^n)$). As usual we define the vector-valued Fourier transform for $f \in \mathcal{S}'(\mathbb{R}^n, E)$ and $\varphi \in \mathcal{S}(\mathbb{R}^n)$ by

$$\mathcal{F} f\,(\varphi) := f(\mathcal{F}\varphi)\,,$$

where

$$\mathcal{F}\varphi(\xi) = (2\pi)^{-n/2} \int_{\mathbb{R}^n} e^{-ix\xi}\, \varphi(x)\, dx\,, \qquad \Big(x\,\xi = \sum_{i=1}^n x_i\, \xi_i \Big)\,,$$

denotes the scalar-valued Fourier transform on $\mathbb{R}^n$. As in the preceding two lines we use the same symbol for both versions, hoping that the meaning will be clear from the context. Finally, let us mention that, as in the scalar case, the Fourier transform and its inverse transform are continuous one-to-one mappings of $\mathcal{S}(\mathbb{R}^n, E)$ onto itself and of $\mathcal{S}'(\mathbb{R}^n, E)$ onto itself.
We shall introduce the scale of Lizorkin-Triebel spaces by Fourier-analytic means. To this end we use smooth dyadic decompositions of unity. Let ψ be an infinitely

differentiable function such that $0 \le \psi(x) \le 1$, $\psi(x) = 1$ if $|x| \le A$, and $\psi(x) = 0$ if $|x| > B$ for some numbers $0 < A < B < \infty$. Then we put

$$\varphi_0(x) = \psi(x)\,, \quad \varphi_1(x) = \varphi_0(x/2) - \varphi_0(x)\,, \quad \varphi_j(x) = \varphi_1(2^{-j+1}x)\,, j = 2, 3, \ldots\,. \tag{1}$$

Obviously

$$\sum_{j=0}^{\infty} \varphi_j(x) = 1\,, \qquad x \in \mathbb{R}^n\,.$$

We denote the set of all such smooth dyadic decompositions of the unity by Φ. For any $f \in \mathcal{S}'(\mathbb{R}^n, E)$ it holds

$$f = \sum_{j=0}^{\infty} \mathcal{F}^{-1}[\varphi_j(\xi)\, \mathcal{F}f(\xi)](x)$$

(convergence in $\mathcal{S}'(\mathbb{R}^n, E)$), see, e.g., [Sc] and [Am 3]. In what follows we work quite often with such decompositions. For the sake of brevity we put

$$f_j(x) = \mathcal{F}^{-1}[\varphi_j(\xi)\, \mathcal{F}f(\xi)](x), \qquad j = 0, 1, \ldots\,, \quad f \in \mathcal{S}'(\mathbb{R}^n, E). \tag{2}$$

2.2. Definition and basic properties

For us it is most convenient to introduce Lizorkin-Triebel classes by means of the so-called decomposition method. It has the disadvantage of some non-transparency but it is nearest to Fourier analysis.

Definition 1. *Let* $\{\varphi_j\}_{j=0}^{\infty} \in \Phi$. *Let* $-\infty < s < \infty$ *and* $1 \le q \le \infty$.

(i) *Let* $1 \le p < \infty$. *The Lizorkin-Triebel space* $F^s_{p,q}(\mathbb{R}^n, E)$ *is the set of all* $f \in \mathcal{S}'(\mathbb{R}^n, E)$ *such that*

$$\| f \,|F^s_{p,q}(\mathbb{R}^n, E) \| = \left\| \left(\sum_{j=0}^{\infty} 2^{jsq} \, \| \mathcal{F}^{-1}\big[\varphi_j(\xi)\, \mathcal{F}f(\xi)\big](\cdot) \, \|_E^q \right)^{1/q} \Big| L_p(\mathbb{R}^n) \right\| < \infty\,,$$

if $q < \infty$ *and*

$$\| f \,|F^s_{p,\infty}(\mathbb{R}^n, E)\| = \left\| \sup_{j=0,1,\ldots} 2^{jsq} \, \| \mathcal{F}^{-1}\big[\varphi_j(\xi)\, \mathcal{F}f(\xi)\big](\cdot) \, \|_E \Big| L_p(\mathbb{R}^n) \right\| < \infty\,,$$

if $q = \infty$.

(ii) *Let* $1 \le p \le \infty$. *The Besov-Nikol'skij space* $B^s_{p,\infty}(\mathbb{R}^n, E)$ *is the set of all* $f \in \mathcal{S}'(\mathbb{R}^n, E)$ *such that*

$$\| f \,|B^s_{p,\infty}(\mathbb{R}^n, E)\| = \sup_{j=0,1,\ldots} 2^{js} \, \| \mathcal{F}^{-1}\big[\varphi_j(\xi)\, \mathcal{F}f(\xi)\big](\cdot) \,|L_p(\mathbb{R}^n, E)\| < \infty\,.$$

Remark 1. *In the scalar case the above definitions go back to papers by Peetre, Nikol'skij, Lizorkin, and Triebel, for the vector-valued case we refer to* [Tr 2]. *Vector-valued Besov spaces have been introduced in this way in Sparr* [Sp], *see also Schmeisser* [Sc] *and Amann* [Am 1, Am 2, Am 3]. *The classical approach to*

Besov spaces uses moduli of continuity, differences and derivatives or interpolation of Sobolev spaces. We refer to Amann [Am 1], *Grisvard* [Gr], *Muramatu* [Mu], *König* [Ko], *Pietsch* [Pi] *and Prüß* [Pr].

Proposition 1. *The spaces defined above are Banach spaces. They are independent of the chosen system* $\{\varphi_j\}_{j=0}^{\infty} \in \Phi$ *in the sense of equivalent norms. We have the chain of continuous embeddings*

$$\mathcal{S}(\mathbb{R}^n, E) \hookrightarrow F^s_{p,q}(\mathbb{R}^n, E) \hookrightarrow B^s_{p,\infty}(\mathbb{R}^n, E) \hookrightarrow \mathcal{S}'(\mathbb{R}^n, E)\,.$$

Remark 2. *Proposition* 1 *is a consequence of a Fourier multiplier assertion for entire analytic E-valued functions, cf. Triebel* [Tr 2, 15.4]. *Based on this tool the proof in the vector-valued situation is the same as in the scalar case, cf., e.g.,* [Tr 1, 2.3.2]. *For the embeddings see* [Tr 1, 2.3.3] *and* [Sc].

Remark 3. *We do not distinguish between spaces equipped with equivalent norms.*

Definition 2.

(i) *Let* $1 \le p \le \infty$ *and let m be a natural number. Then we define*

$$W^m_p(\mathbb{R}^n, E) = \left\{ f \in L_p(\mathbb{R}^n, E) : \, D^\alpha f \in L_p(\mathbb{R}^n, E) \quad \textit{for all} \quad |\alpha| \le m \right\} \tag{3}$$

and put

$$\| \, f \, | W^m_p(\mathbb{R}^n, E) \, \| = \sum_{|\alpha| \le m} \| \, D^\alpha f \, | L_p(\mathbb{R}^n, E) \| \,.$$

(ii) *Let* $1 < p < \infty$ *and let* $s \in \mathbb{R}$. *Then we define*

$$H^s_p(\mathbb{R}^n, E) = \left\{ f \in \mathcal{S}'(\mathbb{R}^n, E) : \, \mathcal{F}^{-1}[(1+|\xi|^2)^{s/2} \, \mathcal{F}f(\xi)](x) \in L_p(\mathbb{R}^n, E) \right\} \tag{4}$$

and put

$$\| \, f \, | H^s_p(\mathbb{R}^n, E) \, \| = \| \, \mathcal{F}^{-1}[(1+|\xi|^2)^{s/2} \, \mathcal{F}f(\xi)](x) \, | L_p(\mathbb{R}^n, E) \| \,.$$

Remark 4. *(UMD spaces). It is well known (cf. McConnell* [Mc], *Zimmermann* [Zi]*), that*

$$H^m_p(\mathbb{R}^n, E) = W^m_p(\mathbb{R}^n, E)\,, \qquad 1 < p < \infty\,,$$

if the Banach space E has the UMD property (unconditionality of martingale differences). A survey of characterizations and properties of UMD spaces can be found in Amann [Am 1] *and in Pietsch and Wenzel* [PW]. *In general we cannot expect the above coincidence. For example, it is shown in Bukhvalov* [Bu] *that in the case* $n = 1$ *and if m is odd the UMD property of E is necessary.*

Remark 5. *(Littlewood-Paley theory). In the scalar case we have (equivalent norms)*

$$H^s_p(\mathbb{R}^n) = F^s_{p,2}(\mathbb{R}^n)\,, \qquad 1 < p < \infty\,. \tag{5}$$

This is based on the Littlewood-Paley-Theorem

$$\| \, f \, | L_p(\mathbb{R}^n) \| \quad \sim \quad \left\| \left(\sum_{j=0}^{\infty} | \, \mathcal{F}^{-1}[\varphi_j \, \mathcal{F}f](x) \, |^2 \right)^{1/2} \middle| L_p(\mathbb{R}^n) \right\|, \qquad 1 < p < \infty\,,$$

and the lift property of the spaces $H^s_p(\mathbb{R}^n)$ *and* $F^s_{p,2}(\mathbb{R}^n)$*, see, e.g., Lizorkin* [Li] *and Triebel* [Tr 1]*. The arguments in Rubio de Francia and Torrea* [RT, p. 283] *(due to Pisier* [Pis]*) show in the E-valued case*

$$\| f \,|L_p(\mathbb{R}^n, E)\| \sim \left\| \Big(\sum_{j=0}^{\infty} \| \mathcal{F}^{-1}[\varphi_j \, \mathcal{F} f](x) \|_E^2 \Big)^{1/2} \Big| L_p(\mathbb{R}^n) \right\|, \quad 1 < p < \infty ,$$

(or equivalently $L_p(\mathbb{R}^n, E) = F^0_{p,2}(\mathbb{R}^n, E)$*) if and only if E can be renormed as a Hilbert space. This extends to the E-valued version of (5) by the lift property. Analogously, it holds* $F^m_{p,2}(\mathbb{R}^n, E) = W^m_p(\mathbb{R}^n, E)$*,* $1 < p < \infty$*, if and only if E is Hilbertian.*

2.3. Some embeddings

Here we collect a few embedding relations between these different types of spaces.

Proposition 2.

(i) *Suppose* $1 < p < \infty$ *and* $m \in \mathbb{N}_0$*. Then we have the chain of continuous embeddings*

$$F^m_{p,1}(\mathbb{R}^n, E) \hookrightarrow W^m_p(\mathbb{R}^n, E) \hookrightarrow F^m_{p,\infty}(\mathbb{R}^n, E) . \tag{6}$$

(ii) *Suppose* $1 < p < \infty$ *and* $s \in \mathbb{R}$*. Then we have the continuous embeddings*

$$F^s_{p,1}(\mathbb{R}^n, E) \hookrightarrow H^s_p(\mathbb{R}^n, E) \hookrightarrow F^s_{p,\infty}(\mathbb{R}^n, E) . \tag{7}$$

(iii) *Suppose* $m \in \mathbb{N}_0$*. Then we have*

$$F^m_{1,1}(\mathbb{R}^n, E) \hookrightarrow W^m_1(\mathbb{R}^n, E) \hookrightarrow B^m_{1,\infty}(\mathbb{R}^n, E) . \tag{8}$$

(iv) *Suppose* $m \in \mathbb{N}_0$*. Then we have*

$$W^m_\infty(\mathbb{R}^n, E) \hookrightarrow B^m_{\infty,\infty}(\mathbb{R}^n, E) . \tag{9}$$

Proof. Observe that in [Tr 2, Sect. 15] one can find the tools needed to reduce the proof of all these embeddings to the case $m = 0$ (or $s = 0$) as in the scalar case. For the latter one see, e.g., [Tr 1, 2.5]. Let $1 \le p < \infty$. Then

$$\| f \,|L_p(\mathbb{R}^n, E)\| \le \left\| \sum_{j=0}^{\infty} \| f_j \|_E \Big| L_p(\mathbb{R}^n) \right\|$$

is a consequence of the triangle inequality. This proves the first embeddings in (6) and (8).

It follows from the construction of our smooth decomposition of unity that

$$\| f_j(x) \|_E \le 2 \sup_{j=0,1,\ldots} \int |\mathcal{F}^{-1}[\varphi_0(2^{-j}\xi)](z)| \, \| f(x-z) \|_E \, dz .$$

The right-hand side can be estimated by $c\,(M \| f(\cdot) \|_E)(x)$, see Stein [St, p. 56/57]. Applying the Hardy-Littlewood maximal inequality we get

$$\| f \,|F^0_{p,\infty}(\mathbb{R}^n, E)\| \le c \left\| \Big(M \| f(\cdot) \|_E \Big)(x) \Big| L_p(\mathbb{R}^n) \right\|$$

for all $f \in L_p(\mathbb{R}^n, E)$, where $1 < p < \infty$. If $p = 1$ or if $p = \infty$ we argue that

$$\begin{aligned} \| f_j \,|L_p(\mathbb{R}^n, E)\| &\leq c \Big\| \int |\mathcal{F}^{-1}\varphi_j(z)| \, \| f(x-z) \|_E \, dz \Big| L_p(\mathbb{R}^n) \Big\| \\ &\leq c \|\mathcal{F}^{-1}\varphi_0 | L_1(\mathbb{R}^n)\| \; \| f \,| L_p(\mathbb{R}^n, E)\| \,. \end{aligned}$$

This proves

$$\| f \,| B^0_{p,\infty}(\mathbb{R}^n, E)\| \leq c \, \| f \,| L_p(\mathbb{R}^n, E)\| \,.$$

The proof is complete. □

3. Gagliardo-Nirenberg inequalities

The aim of this section consists in determining upper bounds for the $L_p(\mathbb{R}^n, E)$-norms of intermediate derivatives $D^\beta f$, $|\beta| < m$ in terms of the $L_p(\mathbb{R}^n, E)$-norms of f and its derivatives $D^\alpha f$, $|\alpha| = m$. Such interpolation inequalities have been obtained by many authors, e.g., Ehrling, Il'yin, Nirenberg, Browder, and Gagliardo. A proof which does not make use of Fourier analytical techniques may be found in Adams [Ad, IV] (there also exact references are given, but see also Besov, Kudryavtsev, Lizorkin and Nikol'skij [BKLN] for this). Recently, Brezis and Mironescu [BM] published a version based on Littlewood-Paley theory. Surprisingly these inequalities are very simple to prove (even in the vector-valued situation) for the most complicated spaces, namely the Lizorkin-Triebel classes.

3.1. Gagliardo-Nirenberg inequalities for Lizorkin-Triebel classes

We follow Brezis and Mironescu [BM].

Proposition 3. *Let $1 \leq q, q_0, q_1 \leq \infty$, $-\infty < s_0 < s_1 < \infty$, and $0 < \Theta < 1$.*

(i) *Let $1 \leq p_0, p_1 < \infty$. We put*

$$\frac{1}{p} = \frac{1-\Theta}{p_0} + \frac{\Theta}{p_1} \qquad \text{and} \qquad s = (1-\Theta)\, s_0 + \Theta\, s_1 \,.$$

Then there exists a positive constant C such that

$$\| f \,| F^s_{p,q}(\mathbb{R}^n, E) \,\| \leq C \, \| f \,| F^{s_0}_{p_0,q_0}(\mathbb{R}^n, E)\|^{1-\Theta} \| f \,| F^{s_1}_{p_1,q_1}(\mathbb{R}^n, E)\|^{\Theta} \tag{10}$$

holds for all $f \in \mathcal{S}'(\mathbb{R}^n, E)$.

(ii) *Let $1 \leq p_0 < \infty$. We put*

$$\frac{1}{p} = \frac{1-\Theta}{p_0} \qquad \text{and} \qquad s = (1-\Theta)\, s_0 + \Theta\, s_1 \,.$$

Then there exists a positive constant C such that

$$\| f \,| F^s_{p,q}(\mathbb{R}^n, E) \,\| \leq C \, \| f \,| F^{s_0}_{p_0,q_0}(\mathbb{R}^n, E)\|^{1-\Theta} \| f \,| B^{s_1}_{\infty,\infty}(\mathbb{R}^n, E)\|^{\Theta} \tag{11}$$

holds for all $f \in \mathcal{S}'(\mathbb{R}^n, E)$.

Proof. The proof is based on the following elementary inequality, cf. [BM]. For any sequence $\{a_\ell\}_\ell$ of complex numbers it holds

$$\| \, 2^{sj} \, a_j \, |\ell_q\| \le C \, \| \, 2^{s_0 j} \, a_j \, |\ell_\infty\|^{1-\Theta} \, \| \, 2^{s_1 j} \, a_j \, |\ell_\infty\|^{\Theta} \, ,$$

where $s = (1-\Theta)\, s_0 + \Theta\, s_1$ and $1 \le q \le \infty$. This can be applied with $a_j = \| \, f_j(x) \, \|_E$. Then Hölder's inequality yields

$$\| \, f \, |F^s_{p,q}(\mathbb{R}^n, E) \, \| \le C \, \| \, f \, |F^{s_0}_{p_0,\infty}(\mathbb{R}^n, E)\|^{1-\Theta} \| \, f \, |F^{s_1}_{p_1,\infty}(\mathbb{R}^n, E)\|^{\Theta} \, ,$$

as long as $p_1 < \infty$. In the case $p_1 = \infty$ we obtain

$$\| \, f \, |F^s_{p,q}(\mathbb{R}^n, E) \, \| \le C \, \| \, f \, |F^{s_0}_{p_0,\infty}(\mathbb{R}^n, E)\|^{1-\Theta} \sup_{x \in \mathbb{R}^n} \Big(\sup_{j=0,1,\ldots} \| \, 2^{s_1 j} \, f_j(x) \, \|_E \Big)^{\Theta} .$$

The monotonicity with respect to the third parameter complements the proof. □

Remark 6. *The proof given above is due to Brezis and Mironescu* [BM] *for* $E = \mathbb{C}$. *The surprising fact is the independence of the microscopic parameters* q, q_0, q_1.

3.2. The vector-valued version of the Gagliardo-Nirenberg inequalities

As a consequence of the preceding proposition and Proposition 2 we now obtain the Gagliardo-Nirenberg inequalities in the vector-valued setting.

Theorem 1. *Let* $m \in \mathbb{N}$, $1 < r \le \infty$, $1 < p, q \le \infty$, *and* $0 < |\beta| < m$. *For*

$$\Theta = |\beta|/m \qquad \text{and} \qquad \frac{1}{r} = \frac{1-\Theta}{q} + \frac{\Theta}{p}$$

there exists a constant c *such that*

$$\| \, D^\beta f \, |L_r(\mathbb{R}^n, E)\| \le c \, \| \, f \, |L_q(\mathbb{R}^n, E)\|^{1-\Theta} \Big(\sum_{|\alpha|=m} \| \, D^\alpha f \, |L_p(\mathbb{R}^n, E)\| \Big)^{\Theta} \qquad (12)$$

holds for all $f \in W^m_p(\mathbb{R}^n, E) \cap L_q(\mathbb{R}^n, E)$.

Proof. We choose $s_0 = 0$, $s_1 = m$ and $\Theta = |\beta|/m$.

Step 1. We assume $1 < p, q < \infty$. Applying Proposition 3 with $q = 1$ and $q_0 = q_1 = \infty$ it follows

$$\| \, f \, |F^{|\beta|}_{r,1}(\mathbb{R}^n, E)\| \le C \, \| \, f \, |F^0_{q,\infty}(\mathbb{R}^n, E)\|^{1-\Theta} \| \, f \, |F^m_{p,\infty}(\mathbb{R}^n, E)\|^{\Theta}$$

Now we use Proposition 2 and obtain

$$\| \, f \, |W^{|\beta|}_r(\mathbb{R}^n, E)\| \le c \, \| \, f \, |L_q(\mathbb{R}^n, E)\|^{1-\Theta} \| \, f \, |W^m_p(\mathbb{R}^n, E)\|^{\Theta} . \qquad (13)$$

To achieve at (12) we replace the complete Sobolev norm on the left-hand side by $\| \, D^\beta f \, |L_p(\mathbb{R}^n, E)\|$ and consider the family $f(\lambda x)$, $\lambda > 0$. Then (13) yields

$$\lambda^{|\beta| - n/r} \, \| \, D^\beta f |L_r(\mathbb{R}^n, E)\| \qquad (14)$$

$$\le c \, \lambda^{-n(1-\Theta)/q} \, \| \, f \, |L_q(\mathbb{R}^n, E)\|^{1-\Theta} \Big(\sum_{|\alpha| \le m} \lambda^{|\alpha| - n/p} \, \| \, D^\alpha f \, |L_p(\mathbb{R}^n, E)\| \Big)^{\Theta}$$

with some c independent of λ. Dividing both sides by $\lambda^{|\beta|-n/r}$ we end up with

$$\| D^\beta f \,|L_r(\mathbb{R}^n, E)\| \le c \,\| f \,|L_q(\mathbb{R}^n, E)\|^{1-\Theta} \Big(\sum_{|\alpha|\le m} \lambda^{|\alpha|-m} \,\| D^\alpha f \,|L_p(\mathbb{R}^n, E)\| \Big)^\Theta .$$

For $\lambda \to \infty$ the desired result follows.

Step 2. We consider $p = \infty$ and $1 < q < \infty$. We proceed as above. It follows from (11) that

$$\| f \,|F^{|\beta|}_{r,1}(\mathbb{R}^n, E)\| \le C \,\| f \,|F^0_{q,\infty}(\mathbb{R}^n, E)\|^{1-\Theta} \| f \,|B^m_{\infty,\infty}(\mathbb{R}^n, E)\|^\Theta$$

Now we use Proposition 2 and obtain

$$\| f \,|W^{|\beta|}_r(\mathbb{R}^n, E)\| \le c \,\| f \,|L_q(\mathbb{R}^n, E)\|^{1-\Theta} \| f \,|W^m_\infty(\mathbb{R}^n, E)\|^\Theta .$$

The rest is the same as in Step 1.

Step 3. Let $q = \infty$ and $1 < p < \infty$. It follows

$$\| f \,|F^{|\beta|}_{r,1}(\mathbb{R}^n, E)\| \le C \,\| f \,|B^0_{\infty,\infty}(\mathbb{R}^n, E)\|^{1-\Theta} \| f \,|F^m_{p,\infty}(\mathbb{R}^n, E)\|^\Theta$$

Now we use Proposition 2 and obtain

$$\| f \,|W^{|\beta|}_r(\mathbb{R}^n, E)\| \le c \,\| f \,|L_\infty(\mathbb{R}^n, E)\|^{1-\Theta} \| f \,|W^m_p(\mathbb{R}^n, E)\|^\Theta .$$

The rest is the same as in Step 1.

Step 4. It remains to consider the case $r = p = q = \infty$. Let $m \in \mathbb{N}$ and $0 < \Theta < 1$. In [SS] we proved the following continuous embedding for the real interpolation of Nikol'skij-Besov spaces

$$\Big(B^0_{\infty,\infty}(\mathbb{R}^n, E), B^m_{\infty,\infty}(\mathbb{R}^n, E)\Big)_{\Theta,1} \hookrightarrow W^{m\Theta}_\infty(\mathbb{R}^n, E) .$$

By standard properties of the real interpolation method it follows

$$\| f \,|W^{m\Theta}_\infty(\mathbb{R}^n, E)\| \le c \,\| f \,|B^0_{\infty,\infty}(\mathbb{R}^n, E)\|^{1-\Theta} \;\| f \,|B^m_{\infty,\infty}(\mathbb{R}^n, E)\|^\Theta$$

with some constant c independent of f. Applying Proposition 2 and a scaling argument as above we have finished. □

Remark 7. *In case $p = 1$ and/or $q = 1$ our knowledge is less complete. Only in case $p = q = 1$ we have a definite answer. Observe, if $m \in \mathbb{N}$ and $0 < \Theta < 1$ then*

$$\Big(B^0_{1,\infty}(\mathbb{R}^n, E), B^m_{1,\infty}(\mathbb{R}^n, E)\Big)_{\Theta,1} \hookrightarrow F^{m\Theta}_{1,1}(\mathbb{R}^n, E) ,$$

cf. [SS]. *The same type of arguments as used in Step 4 of the proof of the preceding theorem leads to the inequality*

$$\| D^\beta f \,|L_1(\mathbb{R}^n, E)\| \le c \,\| f \,|L_1(\mathbb{R}^n, E)\|^{1-\Theta} \Big(\sum_{|\alpha|=m} \| D^\alpha f \,|L_1(\mathbb{R}^n, E)\| \Big)^\Theta ,$$

where $0 < \Theta < 1$, $|\beta| = m\Theta$, and $f \in W^m_1(\mathbb{R}^n, E)$.

By using simply the sandwich argument, cf. Proposition 2(ii), one obtains the following result with respect to the classes $H^s_p(\mathbb{R}^n, E)$.

Theorem 2. *Let* $-\infty < s_0 < s_1 < \infty$, *and* $0 < \Theta < 1$.

(i) *Let* $1 < p_0, p_1 < \infty$. *We put*

$$\frac{1}{p} = \frac{1-\Theta}{p_0} + \frac{\Theta}{p_1} \qquad \text{and} \qquad s = (1-\Theta)\, s_0 + \Theta\, s_1 \,.$$

Then there exists a positive constant c *such that*

$$\| \, f \, | H^s_p(\mathbb{R}^n, E) \, \| \le c \, \| \, f \, | H^{s_0}_{p_0}(\mathbb{R}^n, E) \|^{1-\Theta} \| \, f \, | H^{s_1}_{p_1}(\mathbb{R}^n, E) \|^{\Theta} \tag{15}$$

holds for all $f \in \mathcal{S}'(\mathbb{R}^n, E)$.

(ii) *Let* $1 < p_0 < \infty$. *We put*

$$\frac{1}{p} = \frac{1-\Theta}{p_0} \qquad \text{and} \qquad s = (1-\Theta)\, s_0 + \Theta\, s_1 \,.$$

Then there exists a positive constant c *such that*

$$\| \, f \, | H^s_p(\mathbb{R}^n, E) \, \| \le c \, \| \, f \, | H^{s_0}_{p_0}(\mathbb{R}^n, E) \|^{1-\Theta} \| \, f \, | F^{s_1}_{\infty,\infty}(\mathbb{R}^n, E) \|^{\Theta} \tag{16}$$

holds for all $f \in \mathcal{S}'(\mathbb{R}^n, E)$.

Remark 8. *Based on the above strategy one can deal with the Sobolev embeddings and its refined versions, the Jawerth-Franke embeddings (comparing Lizorkin-Triebel classes with Besov classes). We refer to* [SS] *for details. Let us mention in this context also the paper of Pelczynski and Wojciechowski* [PeWo], *where this problem is treated for* $W^1_1(\mathbb{R}^n, E)$. *Moreover, the sandwich strategy to prove a result for the more complicated Lizorkin-Triebel spaces and afterwards to use the independence of the microlocal parameter* q *to deduce results for the Sobolev spaces is also applicable for the trace problem. We again refer to* [SS].

References

[Ad] R.A. Adams, *Sobolev spaces*, Academic Press, New York, 1975.

[Am 1] H. Amann, *Linear and quasilinear parabolic problems*, Vol. **1**, Birkhäuser, Basel, 1995.

[Am 2] H. Amann, *Operator-valued Fourier multipliers, vector-valued Besov spaces, and applications*, Math. Nachr. **186** (1997), 5–56.

[Am 3] H. Amann, *Vector-valued distributions and Fourier multipliers*, Preprint, Zürich, 1997.

[BKLN] O.V. Besov, L.D. Kudrayavtsev, P.I. Lizorkin and S.M. Nikol'skij, *Investigations in the theory of spaces of differentiable functions of several variables*, Proc. Steklov Inst. **182** (1990), 73–139.

[Bo] J. Bourgain, *Some remarks on Banach spaces in which martingale difference sequences are unconditional*, Ark. Mat. **21** (1983), 163–168.

[BM] H. Brezis and P. Mironescu, *Gagliardo-Nirenberg, composition and products in fractional Sobolev spaces*, J. Evol. Equ. **1** (2001), 387–404.

[Bu] A.V. Bukhvalov, *Sobolev spaces of vector-valued functions*, J. Math. Sciences **71** (1994), 2173–2179.

[Gr] P. Grisvard, *Commutativité de deux foncteurs d'interpolation at applications*, J. Math. Pures Appl. **45** (1966), 143–290.

[Ko] H. König, *Eigenvalue distribution of compact operators*, Birkhäuser, Basel, 1986.

[Li] P.I. Lizorkin, *On multipliers of Fourier integrals in the spaces* $L_{p\Theta}$, Trudy Math. Inst. Steklov **89** (1967), 231–248.

[Mc] T.R. McConnell, *On Fourier multiplier transforms of Banach-valued functions*, Trans. Amer. Math. Soc. **285** (1984), 739–757.

[Mu] T. Muramatu, *Besov spaces and Sobolev spaces of generalized functions defined on a general region*, Publ. R.I.M.S. Kyoto Univ. **9** (1974), 325–396.

[Pe] J. Peetre, *New thoughts on Besov spaces*, Duke Univ. Press, Durham, 1976.

[PeWo] A. Pelczynski and M. Wojciechowski, *Molecular decompositions and embedding theorems for vector-valued Sobolev spaces with gradient norm*, Studia Math **107** (1993), 61–100.

[Pi] A. Pietsch, *Eigenvalues and s-numbers*. Cambridge Univ. Press, Cambridge, 1987.

[PW] A. Pietsch and J. Wenzel, *Orthonormal systems and Banach space geometry*. Cambridge Univ. Press, Cambridge, 1998.

[Pis] G. Pisier, *Type des espaces normés*. Seminaire Maurey-Schwartz 1973–1974, Ecole Polytechnique, Palaiseau.

[Pr] J. Prüss, *Evolutionary integral equations and applications*, Birkhäuser, Basel, 1993.

[RT] J.L. Rubio de Francia and J.L. Torrea, *Some Banach techniques in vector-valued Fourier analysis*, Coll. Math. **54** (1987), 273–284.

[Sc] H.-J. Schmeisser, *Vector-valued Sobolev and Besov spaces*, In: Seminar Analysis of the Karl-Weierstraß-Institute, 4-44, Teubner-Texte Math. **96**, Teubner, Leipzig, 1987.

[SS] H.-J. Schmeißer and W. Sickel, *Traces, Gagliardo-Nirenberg inequalities and Sobolev type embeddings for vector-valued function spaces*, Jenaer Schriften zur Mathematik und Informatik Math/Inf/24/01, Jena, 2001.

[Sp] G. Sparr, *Interpolation of several Banach spaces*, Ann. Mat. Pura Appl. **99** (1974), 247–316.

[St] E.M. Stein, *Harmonic analysis. Real-variable methods, orthogonality, and oscillatory integrals*. Princeton Univ. Press, Princeton, 1993.

[Tr 1] H. Triebel, *Theory of function spaces*, Birkhäuser, Basel, 1983.

[Tr 2] H. Triebel, *Fractals and spectra related to Fourier analysis and function spaces*, Birkhäuser, Basel, 1997.

[Zi] F. Zimmermann, *On vector-valued Fourier multiplier theorems*, Stud. Math. **93** (1989), 201–222.

Hans-Jürgen Schmeißer and Winfried Sickel
Mathematisches Institut
Friedrich-Schiller-Universität Jena
Ernst-Abbe-Platz 1–2
D-07743 Jena, Germany

Progress in Nonlinear Differential Equations
and Their Applications, Vol. 64, 473–495

The Influence of Gradient Perturbations on Blow-up Asymptotics in Semilinear Parabolic Problems: A Survey

Philippe Souplet

Dedicated to Professor Herbert Amann

Abstract. In this paper, we give a survey of results concerning semilinear parabolic problems of the form $u_t - \Delta u = u^p + g(x, t, u, \nabla u)$. Our goal is to examine the effect of the (gradient) perturbation term g on the asymptotic behavior of blow-up solutions. It turns out that, if the perturbation becomes critical or supercritical in a scaling sense, then the blow-up rate as well as the blow-up profiles may become notably different from those known in the unperturbed case. In some cases, we give precise asymptotic estimates on blow-up solutions.

Mathematics Subject Classification (2000). 35K55, 35B40, 35B33.

1. Introduction

Consider the semilinear heat equation

$$\begin{cases} u_t - \Delta u &= |u|^{p-1}u, \qquad x \in \Omega, \quad t > 0, \\ u &= 0, \qquad x \in \partial\Omega, \quad t > 0, \\ u(x,0) &= u_0(x), \qquad x \in \Omega, \end{cases} \tag{1.1}$$

where $p > 1$, Ω is a regular domain of $\mathbb{R}^N$ and $u_0 \in C_0(\Omega)$. It is well known that for large initial data (in a suitable sense), solutions of (1.1) do not exist globally. Namely, they blow up in finite time $T = T(u_0)$ in sup norm:

$$\lim_{t \to T} \|u(t)\|_\infty = \infty.$$

The description of the asymptotic behavior of blowing-up solutions of (1.1) has attracted considerable interest in the past two decades (see, e.g., [24, 14] for recent surveys). Main elements in the description are the blow-up rate, set, profiles and the possible continuation after blow-up.

First of all, it is well known that any blow-up solution of (1.1) satisfies the lower blow-up rate estimate

$$\|u(t)\|_\infty \geq \kappa(T-t)^{-1/(p-1)}, \quad 0 \leq t < T, \quad \kappa = (p-1)^{-1/(p-1)}. \tag{1.2}$$

On the other hand, under certain assumptions, the upper blow-up rate estimate

$$\|u(t)\|_\infty \leq C(T-t)^{-1/(p-1)}, \quad 0 < t < T, \tag{1.3}$$

is also satisfied for some $C > 0$. In this case, the order of growth is the same as for the related ODE $y' = y^p$, and (1.3) is referred to as *type I blow-up*. This holds for instance if $u, u_t \geq 0$, with Ω bounded [20], or if

$$p < p_S := \begin{cases} \frac{N+2}{N-2} & \text{if } N \geq 3 \\ \infty & \text{if } N \leq 2, \end{cases}$$

and $\Omega = \mathbb{R}^N$ or Ω is convex bounded [27, 29, 30]. Actually, for radial positive solutions (when Ω is the ball B_R or $\Omega = \mathbb{R}^N$), the critical exponent for the blow-up rate is known to be

$$p_* := \begin{cases} 1 + \frac{4}{N-4-2\sqrt{N-1}} & \text{if } N \geq 11 \\ \infty & \text{if } N \leq 10. \end{cases}$$

More precisely, if $p < p_*$ and $u_0 \geq 0$ is radial (and satisfies some additional assumptions when $\Omega = \mathbb{R}^N$), then (1.3) is satisfied [39], whereas if $p > p_*$ and $\Omega = \mathbb{R}^N$, then there are examples of radial positive solutions such that

$$\lim_{t \to T} (T-t)^{1/(p-1)} \|u(t)\|_\infty = \infty \tag{1.4}$$

(see [32]). Property (1.4) is referred to as *fast* or *type II blow-up*. On the other hand, for radial sign-changing solutions, formal arguments suggest that type II blow-up can occur when $p = p_S$ [18]. In the nonradial case, the question is still open for $p \geq p_S$. Finally, let us mention that less is known about the blow-up rate for the Neumann problem (see [42] and the references therein).

As for the blow-up set and profiles, let us mention a few of the many known results. First, if Ω is convex bounded and $u_0 \geq 0$, then the blow-up set is a compact subset of Ω, i.e., u remains bounded near $\partial\Omega$ [20]. Next, if $\Omega = B_R$ and $u_0 \geq 0$ is radial, with $u_{0,r} \leq 0$, $r = |x|$, then u blows up only at the origin and satisfies

$$u(x,t) \leq C_\varepsilon |x|^{-2/(p-1)-\varepsilon}, \qquad 0 < |x| < R, \quad 0 < t < T, \tag{1.5}$$

for each $\varepsilon > 0$ (see [20] and cf. also [43, 21]). Actually, if in addition $p < p_S$, then the final space profile satisfies the precise estimate

$$u(x,T) \sim c_p |x|^{-2/(p-1)} \big|\log|x|\big|^{1/(p-1)}, \quad |x| \to 0 \tag{1.6}$$

(see [56, 40]). On the other hand, the asymptotic behavior can be described by a different notion of profile, based on self-similar variables. Namely, if $p < p_S$, and

$\Omega = \mathbb{R}^N$ or Ω is convex bounded, then for any blow-up point $a \in \Omega$, it holds

$$\lim_{t\to T} (T-t)^{1/(p-1)} u(a + y\sqrt{T-t}, t) = \pm\,\kappa, \tag{1.7}$$

uniformly for y bounded (see [26–30] and cf. also [21]).

Finally, it is known [3] that if, e.g., $p < p_S$ or $u_t \geq 0$, then solutions cannot be continued after $t = T$, even in a weak sense. On the contrary, for $p_S < p < p_*$, there are examples of blow-up solutions which can be continued and become immediately regular again after T [15]. We refer to [14, 47] for more details.

The aim of this paper is to examine the effect of a perturbation, especially by a gradient term, on the asymptotic behavior of blow-up solutions. For simplicity, we shall restrict our attention to positive solutions. Thus we shall consider perturbed problems of the form

$$\begin{cases} u_t - \Delta u &= u^p + g(x,t,u,\nabla u), & x \in \Omega, \quad t > 0, \\ u &= 0, & x \in \partial\Omega, \quad t > 0, \\ u(x,0) &= u_0(x) \geq 0, & x \in \Omega. \end{cases} \tag{1.8}$$

It will turn out that whether or not estimates such as (1.3), (1.5)–(1.7) are preserved for problem (1.8) strongly depends on the size of the perturbation term.

Unless otherwise stated, we shall assume that g is a C^1 function, with $g(x,t,0,0) \geq 0$, and that $u_0 \in X$, where

$$X := \big\{0 \leq w \in C^1(\overline{\Omega}) : w, \nabla w \in L^\infty(\Omega),\ w = 0 \text{ on } \partial\Omega\big\},$$
$$\|w\|_X := \|w\|_\infty + \|\nabla w\|_\infty.$$

Problem (1.8) then admits a unique, maximal solution $u \geq 0$, classical for $t > 0$ (see, e.g., [48] and the references therein). Its maximal existence time will be denoted by $T = T(u_0)$. Also, when $\Omega = B_R$ or $\mathbb{R}^N$, a solution will be said to be radial (or symmetric when $N=1$) if $u(x,t) = u(|x|,t)$, and we shall always denote $r = |x|$.

In some respects, the present article constitutes a sequel to our survey paper [49], incorporating some more recent developments and focusing on the special question of blow-up asymptotics. The outline of the rest of the paper is as follows. In Section 2, we introduce a notion of sub-/supercritical perturbation, based on scaling considerations, which will turn out to be useful for a heuristic interpretation of many results. In Section 3, we present results which show that blow-up remains type I if the perturbation satisfies a suitable subcriticality condition. A closely related property, concerning bounds of global solutions, will be also discussed. In Section 4, we report on some recent results that go in the opposite direction: for certain (critical) perturbations, even in one space dimension, fast or type II blow-up occurs. Then in Section 5, we present various results concerning blow-up profiles in space. In particular, the blow-up profile estimate (1.5) ceases to be true if the perturbation becomes supercritical and we give precise estimates of the different profiles that occur. In Section 6, we treat similar questions concerning blow-up profiles in self-similar variables. Finally in Section 7, we briefly discuss the questions of blow-up sets and continuation after blow-up. In each section, we have intended to present proofs for some of the main results (or at least the principal ideas thereof).

2. Sub-/supercritical perturbations and examples

It is a well-known fact that scaling and self-similarity play an important role in the analysis of nonlinear PDE's. For our problem (1.8), the optimal or critical conditions that will appear below will be often interpretable heuristically via scaling considerations. Namely, assume for simplicity that $g = g(u, \nabla u)$ and consider the dilations

$$u_\alpha(x,t) = \alpha^{-1} u(\alpha^{(1-p)/2}x, \alpha^{1-p}t), \qquad g_\alpha(z,\xi) = \alpha^{-p} g(\alpha z, \alpha^{(p+1)/2}\xi), \qquad \alpha > 0.$$

Observe that for any solution of $(1.8)_1$, the function u_α solves the same equation with g replaced by g_α. Let us agree to say roughly that the perturbation g is *subcritical* (resp., *critical or supercritical*), if

$$g(\alpha z, \alpha^{(p+1)/2}\xi) = o(\alpha^p) \quad (\text{resp.}, \sim \alpha^p \text{ or } \gg \alpha^p), \quad \text{as } \alpha \to \infty$$

(for fixed z, ξ, say). Then it is reasonable to expect that for subcritical perturbations, asymptotic properties of blow-up solutions to (1.8) should be close to those of the unperturbed equation (1.1). Likewise, one can expect that the behavior might be different in some respect for supercritical (or even critical) perturbations. Although certain open questions remain, such expectations will turn out to be true.

Let us give some typical examples of problem (1.8) which have been studied in the literature. In the following equation:

$$u_t - \Delta u = u^p - \mu|\nabla u|^q, \tag{2.1}$$

with homogeneous Dirichlet boundary conditions, where $p, q > 1$, $\mu > 0$, the gradient term has a dissipative effect which may prevent blow-up. Equation (2.1) has been first considered in [9] and has next been studied by several authors (see [49] for a survey and list of references). As a second example, let us mention the equation

$$u_t - \Delta u = u^p + a \cdot \nabla(u^q), \tag{2.2}$$

where $p > 1$, $q \geq 1$, $0 \neq a \in \mathbb{R}^N$, which contains a convective gradient term. For results on blow-up for equation (2.2), we refer to, e.g., [1, 38]. Note that equation (2.1) (resp., (2.2)) is (sub-, super-) critical in the above sense if $q\ (<,>) = 2p/(p+1)$ (resp., $(p+1)/2$).

It is known that, in a bounded domain for instance, finite time blow-up for (2.1) with $\mu > 0$ may happen if and only if $p > q$, and that it occurs in L^∞ norm (see [54, 13] and cf. also the references in [49]). The same is true for (2.2). Observe that the critical blow-up exponent $q = p$ for (2.1) does not coincide with the scaling exponent $q = 2p/(p+1)$ of the perturbation, but is rather related to its "size" as compared with that of the reaction term. The situation will be different for blow-up asymptotics.

As a generalization of (2.1), one may also consider the equation

$$u_t - \Delta u = u^p - \lambda u^s |\nabla u|^q, \tag{2.3}$$

with $q > 1$ and $s, \lambda \in \mathbb{R}$. In the case $s < 0$, the homogeneous boundary conditions will be replaced by $u = b > 0$ on $\partial\Omega$ and initial conditions will be assumed to verify $u_0(x) \geq b$ in Ω (and $u_0(x) = b$ on $\partial\Omega$). Note that these assumptions guarantee $u \geq b$ for $t > 0$, so that the gradient term is nonsingular. (Although the boundary conditions are different from those in (1.8), this change is unessential: (2.3) can be recast in the form (1.8) by setting $v = u - b$ and taking $g(x, t, v, \nabla v) = (v + b)^p - v^p - \lambda(v + b)^s |\nabla v|^q$.)

If $\lambda < 0$, then blow-up for (2.3) is of course always possible, but when, e.g., $q > \max(2, p)$ and $s = 0$, gradient blow-up, instead of L^∞ blow-up, may occur (see [50]), and this is out of the scope of the present paper. Let us thus assume $\lambda > 0$. If $p > q + s$, then we have $T < \infty$ for large initial data (this follows from [54, Theorem 1] if $s > 0$ and from simple modifications of its proof if $s < 0$). Moreover, blow-up has to occur in sup norm whenever $q \leq 2$ [36] or $s \geq 0$ [5]. Finally, if $p \leq q + s$ then solutions cannot blow up, at least in L^∞ (this follows from the proofs in [13, 4]).

We shall devote particular attention to the following special, one-dimensional case of (2.3):

$$\begin{cases} u_t - u_{xx} = u^p - \lambda \dfrac{|u_x|^2}{u^m}, & -1 < x < 1, \quad t > 0, \\ u(\pm 1, t) = b, & t > 0, \\ u(x, 0) = u_0(x) \geq b, & -1 < x < 1, \end{cases} \tag{2.4}$$

with $m, \lambda, b > 0$, on which several interesting phenomena can be observed. Let us note that (2.3) is (sub-, super-) critical in the above sense for $s + q(p+1)/2$ $(<, >) = p$. In particular for (2.4), the conditions become just m $(>, <) = 1$.

3. Type I blow-up rates

In this section, we shall see that if the perturbation term in (1.8) satisfies a suitable subcriticality condition and p is not too large, then blow-up is type I.

A property closely related to type I blow-up is the boundedness of global solutions, i.e.:

$$\sup_{t \geq 0} \|u(t)\|_\infty \leq C(u_0) < \infty, \quad \text{whenever } T(u_0) = \infty. \tag{3.1}$$

This property is true for (1.1) when $p < p_S$, actually in the stronger version (usually called *a priori* bound) where $C(u_0) = C(\|u_0\|_\infty)$ (see [25, 46] and the references therein). Likewise, it will be preserved under subcritical perturbation.

Although the two properties concern different classes of solutions, there are amenable to a unified approach, based on scaling and Fujita- or Liouville-type theorems (see Lemma 3.1) and on a doubling property (see Lemma 3.2); they will be given a single proof. Note that in the classical proofs of (1.3) or (3.1) for problem (1.1) with $p < p_S$, essential use is made of the energy functional E associated to the problem. Here (at the expense of further restriction on p, though), we do not use any energy functional (and indeed, problem (1.8) is not known to possess a variational structure in general).

The idea that we use was first introduced in [33] to establish blow-up rates for problems with nonlinear boundary conditions, using Fujita theorems. It was adapted in [10, 16] to other problems, e.g., equations of the form (1.8) or parabolic systems, and it was modified in [48] to prove a priori bounds for global solutions. Next, in [41], parabolic Liouville theorems were used instead of Fujita results, allowing to increase the range of p, and a unified proof, covering blow-up rates and a priori bounds at the same time, was given. The present result is an extension of [41, Theorem 3.2].

Theorem 3.1. *Let $p > 1$ satisfy*

$$\begin{cases} p < p_B := N(N+2)/(N-1)^2 & \text{if } \Omega = \mathbb{R}^N, \\ p \le (N+3)/(N+1) & \text{otherwise.} \end{cases} \tag{3.2}$$

Assume that g satisfies the growth assumption

$$|g(x,t,u,\xi)| \le C_0(1 + |u|^{p_1} + |\xi|^q), \text{ for some } 1 \le p_1 < p \text{ and } 1 < q < \tfrac{2p}{p+1}. \tag{3.3}$$

There exists $C = C(\|u_0\|_X) > 0$ such that that all solutions u of (1.8) verify the following.

(i) *if $T < \infty$, then*

$$\|u(t)\|_\infty + \|\nabla u(t)\|_\infty^{2/(p+1)} \le C\big(1 + (T-t)^{-\frac{1}{p-1}}\big), \quad 0 < t < T; \tag{3.4}$$

(ii) *if $T = \infty$, then*

$$\|u(t)\|_\infty + \|\nabla u(t)\|_\infty \le C, \quad t \ge 0. \tag{3.5}$$

Remarks 3.1. (i) Theorem 3.1 applies for instance to (2.1) and (2.2) in the subcritical case ($q < 2p/(p+1)$ or $q < (p+1)/2$, respectively). Although condition (3.3) is relatively general, it is however not satisfied by all subcritical perturbations in the sense defined in Section 2. In particular, it does not hold for (2.3) when $q \ge 2p/(p+1)$, although the perturbation may be subcritical for certain $s < 0$. For a type I blow-up result under a different subcriticality condition, see Theorem 4.2 below.

(ii) If we assume $g(x,t,u,0) \le 0$, then any blow-up solution of (1.8) satisfies the lower estimate (1.2). This follows immediately from the proof of [20, Theorem 4.5].

(iii) In some cases, the assumption (3.2) on p can be weakened. For instance, if $\Omega = \mathbb{R}^N$ or B_R, and u is radial, then the conclusions of Theorem 3.1 are true for all $p < p_S$.

(iv) If we allow the constants in (3.4), (3.5) to depend on the solution u, then Theorem 3.1 remains true for all $p < p_B$ if Ω is convex and bounded (in case $T = \infty$, we thus have boundedness of global solutions).

(v) In the case of equation (2.1) with Ω convex and bounded, it was proved in [11] that blow-up is type I for all $p > 1$ and $1 < q < 2p/(p+1)$, under the assumption $u_t \geq 0$. The proof is based on a modification of maximum principle arguments from [20].

Let us turn to the proof of Theorem 3.1. For a given solution u of (1.8), let us denote

$$M_u(t) := \sup_{s\in[0,t]} \left(\|u(s)\|_\infty + \|\nabla u(s)\|_\infty^{2/(p+1)}\right), \qquad t \in [0, T(u_0)).$$

Clearly, M_u is a nonnegative, continuous and nondecreasing function on $[0, T(u_0))$. For any given $t \in [0, T(u_0))$ we can define t_u^+ by

$$t_u^+ = \sup \left\{s \in [t, T(u_0));\ M_u(s) \leq 2M_u(t)\right\} \leq T(u_0) \leq \infty.$$

We shall first prove the following Lemma.

Lemma 3.1. *Assume* (3.2), (3.3) *and let* $K > 0$. *There exists* $C_K > 0$ *such that for all solutions of* (1.8) *with* $\|u_0\|_X \leq K$ *and all* $t \in [0, T(u_0))$, *it holds either*

$$M_u(t) \leq C_K, \tag{3.6}$$

or

$$(t_u^+ - t)M_u^{p-1}(t) \leq C_K. \tag{3.7}$$

Proof. Suppose that the Lemma is false. Then there exists a sequence (u_j) of solutions and $t_j \in [0, T_j)$, with $T_j := T(u_{j,0})$, such that for all j,

$$\|u_{j,0}\|_X \leq K, \tag{3.8}$$

$$M_j \geq j, \tag{3.9}$$

$$(t_j^+ - t_j)M_j^{p-1} \geq j, \tag{3.10}$$

where $M_j := M_{u_j}(t_j)$ and $t_j^+ := t_{j,u_j}^+$.

By the well-posedness of (1.8) in X (see, e.g., [48]), (3.8) implies the existence of $\delta > 0$ such that, for all j,

$$T_j > \delta, \qquad \|u_j(t)\|_X \leq 2K, \quad 0 \leq t \leq \delta.$$

For each j, choose $\hat{t}_j \in (0, t_j]$ and $x_j \in \overline{\Omega}$ such that

$$\tfrac{1}{2}M_j \leq u(x_j, \hat{t}_j) + |\nabla u(x_j, \hat{t}_j)|^{2/(p+1)} \leq M_j. \tag{3.11}$$

Then we have

$$\hat{t}_j \geq \delta \quad \text{for large } j. \tag{3.12}$$

Put $\lambda_j := M_j^{-(p-1)/2}$. By (3.9) and (3.10), we have

$$\lambda_j \to 0 \quad \text{as } j \to \infty, \tag{3.13}$$

$$\lambda_j^{-2}(t_j^+ - \hat{t}_j) \to \infty \quad \text{as } j \to \infty. \tag{3.14}$$

Next we rescale the solution u about $(x_j, \hat{t}_j)$ with the scaling factor λ_j as follows:

$$\varphi_j(y,s) := \lambda_j^{2/(p-1)} u_j(\lambda_j y + x_j, \lambda_j^2 s + \hat{t}_j), \quad (y,s) \in D_j$$

where $D_j := \overline{\Omega}_j \times [-\lambda_j^{-2}\hat{t}_j, \lambda_j^{-2}(T_j - \hat{t}_j))$, $\Omega_j = \{y \in \mathbb{R}^N;\ \lambda_j y + x_j \in \Omega\}$. Then φ_j is a solution of the equation

$$\varphi_s = \Delta\varphi + \varphi^p + g_j \quad \text{in } D_j$$

with $\varphi_j(0,0) + |\nabla\varphi_j(0,0)|^{2/(p+1)} \geq 1/2$, where

$$g_j(y,s) := \lambda_j^{2p/(p-1)} g\big(\lambda_j y + x_j, \lambda_j^2 s + \hat{t}_j, \lambda_j^{-2/(p-1)}\varphi_j(y,s), \lambda_j^{-(p+1)/(p-1)}\nabla\varphi_j(y,s)\big).$$

As a consequence of the definition of t_j^+ and (3.12), we have

$$0 \leq \varphi_j + |\nabla\varphi_j|^{2/(p+1)} \leq 2, \quad \text{in } \widetilde{D}_j := \Omega_j \times (-\delta\lambda_j^{-2}, \lambda_j^{-2}(t_j^+ - \hat{t}_j)).$$

The growth assumption (3.3) implies that $|g_j| \leq C\lambda_j^\varepsilon\big(1 + |\varphi_j|^{p_1} + |\nabla\varphi_j|^q\big)$, where $\varepsilon := \min\big\{\frac{2(p-p_1)}{p-1}, \frac{p+1}{p-1}\big(\frac{2p}{p+1} - q\big)\big\} > 0$. Consequently,

$$|g_j| \leq C\lambda_j^\varepsilon \quad \text{in } \widetilde{D}_j. \tag{3.15}$$

Now, we distinguish the cases (i) $d_j/\lambda_j \to \infty$ and (ii) $d_j/\lambda_j \to c \geq 0$, where $d_j = \mathrm{dist}(x_j, \partial\Omega)$. Using parabolic estimates similarly as in [25] (see also [48]), we obtain a subsequence converging to a (bounded) solution of either

$$\varphi_s = \Delta\varphi + \varphi^p, \qquad y \in \mathbb{R}^N, \quad s \in \mathbb{R}, \tag{3.16}$$

or

$$\begin{cases} \varphi_s = \Delta\varphi + \varphi^p, & y \in H_c, \quad s \in \mathbb{R}, \\ \varphi = 0, & y \in \partial H_c, \quad s \in \mathbb{R}, \end{cases} \tag{3.17}$$

for some $c \in \mathbb{R}$, where $H_c = \{y \in \mathbb{R}^N; y_N > c\}$, and such that $\varphi(0,0) + |\nabla\varphi(0,0)|^{2/(p+1)} \geq 1/2$. But, as a consequence of [7, Théorème 0.1], (3.16) does not admit any nontrivial solution if $p < p_B$. As for (3.17), it does not admit any nontrivial solution (even just for $s > 0$) whenever $p \leq (N+3)/(N+1)$ (see, e.g., [37]). The Lemma follows. □

Since $M_u(t)$ is a nondecreasing function, then either:

(i) (3.6) holds on the whole $[0, T(u_0))$, hence in particular $T(u_0) = \infty$,

or:

(ii) there exists $\tau \in [0, T(u_0))$ such that we have (3.6) on $[0,\tau]$ and (3.7) on $[\tau, T(u_0))$.

Theorem 3.1 is then an immediate consequence of the following Lemma, which is essentially contained in [33].

Lemma 3.2. *Assume that (ii) above occurs. Then $T = T(u_0) < \infty$ and*

$$\|u(t)\|_\infty \leq C_1\big(1 + (T-t)^{-\frac{1}{p-1}}\big), \quad 0 < t < T,$$

where $C_1 = C_1(C_K) > 0$.

Proof. Fix $t \in [\tau, T)$ and define an increasing sequence by $t_0 := t$ and $t_{i+1} := t^+_{i,u}$, $i = 0, 1, 2, \ldots$ Then

$$M_i := M_u(t_i) = 2^i M(t) \quad \text{and} \quad t_{i+1} - t_i \le C_K M_i^{1-p} = \left(2^{1-p}\right)^i C_K M^{1-p}(t). \tag{3.18}$$

It follows that t_i converges to a limit $t_\infty < \infty$, necessarily equal to T, since $M_u(t_i) \to \infty$. Formula (3.18) then implies that

$$T - t \le C_0 M^{1-p}(t), \quad \tau \le t < T,$$

where $C_0 := C_K \sum_{i=0}^{\infty} \left(2^{1-p}\right)^i$. Since $\|u(t)\|_\infty \le C_K$ on the remaining interval $[0, \tau]$, the Lemma follows. □

Remarks 3.2. (i) For $p < p_S$, (3.16) does not admit any bounded, positive radial solution (see [58]). Suitable modifications of the proof of Lemma 3.1 then yield the result mentioned in Remark 3.1 (iii). It seems to be open whether problem (3.16) can possess positive (bounded) solutions if $p_B \le p < p_S$. The analogous question for (3.17) is also open for $p > (N+3)/(N+1)$. For a given p, negative answers to those questions would imply the conclusions of Theorem 3.1 for any domain Ω. Let us note that even for equation (1.1), type I blow-up is still an open question for $p > (N+3)/(N+1)$ if Ω is nonconvex (and u does not satisfy $u, u_t \ge 0$).

(ii) The result mentioned in Remark 3.1 (iv) follows from the proof of Lemma 3.1 and the fact that case (ii) in that proof does not occur, due to a moving plane argument from [11, Theorem 3.1].

4. Fast blow-up rates

In this section, we shall show that for certain *critical* perturbations, namely problem (2.4) with $m = 1$, blow-up may become type II.

Theorem 4.1. [31] *Let $m = 1$ and $\lambda > p$. Assume that u_0 is symmetric, with $u_{0,r} \le 0$, and let the solution u of (2.4) satisfy $T < \infty$. Then*

$$\lim_{t \to T} (T - t)^{1/(p-1)} \|u(t)\|_\infty = \infty. \tag{4.1}$$

It was moreover proved in [31] that the assumption on λ in Theorem 4.1 is optimal: if $m = 1$ and $0 < \lambda \le p$, then blow-up is type I (at least if $u_t \ge 0$). This shows a phenomenon of strong sensitivity to gradient perturbations. By using results in [52], it is actually possible to confirm the optimality of Theorem 4.1 by extending the type I blow-up result to a class of parabolic *inequalities* which contains the sub-critical case $m > 1$.

Theorem 4.2. *Assume that u is symmetric, with $u_r \le 0$ and $u_t \ge 0$ in $D = (-1,1)\times(0,T)$. Let $\lambda < 1$, $K > 0$ and assume that*

$$u_t - u_{xx} \ge u^p - \lambda \frac{|u_x|^2}{u}, \qquad \text{for all } (x,t) \in D \text{ such that } u(x,t) \ge K,$$

Then (1.3) *holds. In particular, this applies to problem* (2.4) *for any $m > 1$ and $\lambda > 0$.*

Remarks 4.1. (i) The exact type II blow-up rate in Theorem 4.1 is presently unknown. However, we conjecture that there should not exist a single rate for all solutions. Such a situation occurs for problem (1.1) with $p > p_*$ [32].

(ii) We have mentioned in Section 3 that there is a connection between type I blow-up and property (3.1) for global solutions. Actually, it is known that for stronger perturbations, property (3.1) may fail, too. In case of equation (2.1), an example of unbounded global solution was constructed in [12] for any $p > q = 2$, with $N = 1$, $\Omega = (-1,1)$ and some $\mu = \mu(p) > 0$. In that example, the solution stabilizes (monotonically) in infinite time to a stationary solution singular at $x = 0$. On the other hand, for any $q \ge p$ and $\mu > 0$, and, e.g. $\Omega = \mathbb{R}^N$, it was shown in [55] that all sufficiently large, compactly supported initial data give rise to a growing-up global solution, such that $u(x,t) \to \infty$ as $t \to \infty$ for each $x \in \mathbb{R}^N$.

(iii) The blow-up rate for problem (2.4) with $m = 1$ is actually quite unstable. Consider the following mild modification:

$$\begin{cases} u_t - u_{xx} = a(x,t)u^p - \lambda \dfrac{|u_x|^2}{u}, & -1 < x < 1, \quad t > 0, \\ u(\pm 1, t) = b(t), & t > 0, \\ u(x,0) = u_0(x) \ge 1, & -1 < x < 1, \end{cases} \tag{4.2}$$

with $p > 1$ and any $\lambda \in \mathbb{R}$. Then there exist continuous functions $a = a(|x|,t)$ on $[-1,1]\times[0,T)$, $b(t)$ on $[0,T)$ satisfying

$$0 < c_1 \le a(|x|,t), b(t) \le c_2, \qquad |x| < 1, \quad 0 < t < T,$$

and such that for some symmetric u_0, with $u_{0,r} \le 0$, problem (4.2) admits a solution with type I blow-up. An explicit example can be easily found under the form $u(x,t) = [x^2 + C(T-t)]^{-1/(p-1)}$ (cf. [45, 47] and [31] for similar examples related to complete blow-up and dead-core rates, respectively).

The rest of this section will be essentially devoted to present (the sketch of) a proof of Theorem 4.1 (the short proof of Theorem 4.2 will be given afterwards). Actually, Theorem 4.1 is a consequence of a recent result of [31] on *dead-cores* for the absorption problem

$$\begin{cases} w_t - w_{xx} = -w^\gamma, & -1 < x < 1, \quad t > 0, \\ w(\pm 1, t) = k, & t > 0, \\ w(x,0) = w_0(x), & -1 < x < 1, \end{cases} \tag{4.3}$$

where $0 < \gamma < 1$ and $k > 0$. Indeed, similarly as in [34], we notice that (2.4) with $m = 1$ is transformed into (4.3) by the change of unknown

$$w = au^{1-\lambda}, \qquad \text{with} \quad a = (\lambda - 1)^{(1-\lambda)/\lambda}, \quad \gamma = \frac{\lambda - p}{\lambda - 1}, \quad k = ab^{1-\lambda}. \tag{4.4}$$

Now w is nondecreasing in $|x|$ and blow-up of u at $t = T$ is equivalent to appearance of a dead-core for w, i.e., $w(0, T) = 0$. Note that $w \geq 0$ exists for all times $t > 0$. The fast blow-up estimate (4.1) becomes equivalent to

$$\lim_{t \to T} (T - t)^{-\alpha} w(0, t) = 0, \qquad \alpha = \frac{1}{1 - \gamma}. \tag{4.5}$$

The proof of (4.5) relies on the use of self-similar variables (and will actually provide additional information – cf. Theorems 5.4 and 6.3 below). Namely, following [26, 21] and [19], we set

$$T - t = e^{-s}, \qquad y = x/\sqrt{T-t} \qquad \text{and} \qquad w(x, t) = (T - t)^{\alpha} v(y, s).$$

Then v satisfies the equation

$$v_s = v_{yy} - \frac{y}{2} v_y + \alpha v - v^{\gamma} \quad \text{in } D, \tag{4.6}$$

where $D := \{(y, s);\ -\log T < s < \infty,\ |y| < e^{s/2}\}$. Under the assumptions of Theorem 4.1, we shall actually show the more precise convergence statement

$$\lim_{s \to \infty} v(y, s) = V_1(y) := k_{\gamma} |y|^{2\alpha}, \qquad k_{\gamma} = \left[\tfrac{(1-\gamma)^2}{2(1+\gamma)}\right]^{\alpha}, \tag{4.7}$$

uniformly on $\{|y| < R\}$ for each $R > 0$, from which (4.5) (and hence Theorem 4.1) readily follows.

A quick check reveals that the right-hand side $V_1(y)$ of (4.7) provides an (unbounded) stationary solution of (4.6), more precisely a solution of

$$V_{yy} - \frac{y}{2} V_y + \alpha V - V^{\gamma} = 0, \quad y \in \mathbb{R}. \tag{4.8}$$

Note that each solution of (4.8) corresponds to a self-similar solution of $w_t = w_{xx} - w^{\gamma}$ in $(-\infty, T) \times \mathbb{R}$ given by $w(x, t) = (T - t)^{1/(1-\gamma)} V(x/\sqrt{T-t})$. On the other hand $V_1(x)$, restricted to $[-1, 1]$, is also a stationary solution of (4.3) with $k = k_{\gamma}$.

The proof of (4.7) will then be carried out in 3 steps.

(i) identify the stationary solutions of (4.6) (in a suitable set);

(ii) prove that all the global solutions of (4.6) are attracted by the set of stationary solutions of (4.6) (in the locally compact topology);

(iii) discard all the possible limits other than the stationary solution V_1.

We shall only briefly outline the main ideas of each step. We refer to [31] for details. We need three lemmas. We start with a lower estimate which is the key ingredient in step (iii).

Lemma 4.1. *There exist $c_1 > 0$ (depending on w) such that*

$$w(x, t) \geq \left[w^{1-\gamma}(0, t) + c_1 x^2\right]^{\alpha}, \qquad |x| \leq 1, \quad T/2 \leq t \leq T. \tag{4.9}$$

Sketch of proof. We consider the auxiliary function

$$J = w_x - \varepsilon x w^\gamma.$$

By a suitable modification of an argument of [20], and using a nonlocal version of the maximum principle, it can be proved that for $\varepsilon > 0$ small enough, it holds

$$J \geq 0 \quad \text{in } [0,1] \times (T/2, T). \tag{4.10}$$

This yields $(w^{1-\gamma})_x = (1-\gamma)w^{-\gamma}w_x \geq \varepsilon(1-\gamma)x$ in $[0,1] \times (T/2, T)$ and the estimate then immediately follows by integrating in space between 0 and x. □

We next derive similar upper estimates on w which yield polynomial upper bounds for v.

Lemma 4.2. *There exist $c_2 > 0$ (depending on w) such that*

$$w(x,t) \leq \left[w^{\frac{1-\gamma}{2}}(0,t) + c_2|x|\right]^{2\alpha}, \qquad |x| \leq 1, \quad T/2 \leq t \leq T. \tag{4.11}$$

Moreover, the corresponding global solution v of (4.6) satisfies

$$v(y,s) \leq C(1+|y|)^{\frac{2}{1-\gamma}} \quad \text{and} \quad |v_y(y,s)| \leq C(1+|y|)^{\frac{1+\gamma}{1-\gamma}} \tag{4.12}$$

for all $-\log(T/2) =: s_0 < s < \infty$, $|y| < e^{s/2}$.

Sketch of proof. We consider the auxiliary function $J(x,t) := \frac{1}{2}w_x^2 - Cw^{\gamma+1}$, where $C > 0$ is a suitably large constant. By using the maximum principle, it can be proved that $J \leq 0$ in $Q := [0,1] \times (T/2, T)$. This implies (4.11) and we get in turn the estimate

$$|w_x| \leq C w^{(\gamma+1)/2}, \qquad |x| \leq 1, \quad T/2 \leq t \leq T. \tag{4.13}$$

For the proof of (4.12), we first notice that, since $w_{xx}(0,t) \geq 0$, we have $w_t(0,t) \geq -w^\gamma(0,t)$, and by integrating between t and T we get $w(0,t) \leq C(T-t)^\alpha$. By combining this with (4.11), we obtain

$$\begin{aligned} v(y,s) &= (T-t)^{-\alpha} w(y\sqrt{T-t}, t) \leq (T-t)^{-\alpha}\left[w^{\frac{1-\gamma}{2}}(0,t) + c_2|y|\sqrt{T-t}\right]^{\frac{2}{1-\gamma}} \\ &\leq C(T-t)^{-\alpha}\left[\sqrt{T-t} + |y|\sqrt{T-t}\right]^{\frac{2}{1-\gamma}} = C(1+|y|)^{\frac{2}{1-\gamma}}. \end{aligned}$$

The estimate of v_y then follows from (4.13). □

Next, for step (i), we have the following.

Lemma 4.3. *Let $V \in C^2(\mathbb{R})$ by a solution of (4.8) such that*

$$V = V(|y|), \quad \text{with} \quad V' \geq 0, \quad V > 0 \quad \text{for all } y > 0,$$

and such that V is polynomially bounded. Then

$$V = V_1 = k_\gamma |y|^{2\alpha} \quad \text{or} \quad V = V_2 := \kappa_\gamma := (1-\gamma)^{1/(1-\gamma)}.$$

Sketch of proof. We make use of the auxiliary function

$$Z := |W - \tfrac{y}{2}W'|^{\frac{1}{1-\gamma}}, \quad W = V^{1-\gamma}.$$

By some tedious calculation, one can show that at each y such that $Z(y) > 0$, it holds $\frac{y}{2}Z' - Z'' \le 0$, i.e.,

$$(e^{-y^2/4}Z')' \ge 0. \tag{4.14}$$

It then follows that the function $Z \ge 0$ is nonincreasing in $(0, \infty)$. Indeed, otherwise, there would exist y_0 such that $Z(y_0) > 0$ and $Z'(y_0) > 0$, hence $Z' \ge Ce^{y^2/4}$ for $y \ge y_0$ by (4.14). Due to $|(y^{-2}W)'| = 2y^{-3}Z^{1-\gamma}$, we would get $W \ge e^{\eta y^2}$ as $y \to \infty$, for some $\eta > 0$, contradicting the polynomial bound assumed on V.

Now assume for contradiction that Z is nonconstant on $(0, \infty)$. Then there is $R > 0$ such that $Z(0) > Z(R)$ and we may choose $\varepsilon > 0$ so small that $f := Z + \varepsilon e^{y^2/2}$ satisfies $f(0) > Z(0) > f(R)$. It follows that f has a local maximum at some $y_1 \in (-R, R)$ and that $Z(y_1) > 0$. Therefore, we get $0 \le (y/2)f' - f'' \le \varepsilon((y^2/2) - 1 - y^2)e^{y^2/2} < 0$, a contradiction.

We deduce that $W - (y/2)W' = C$ on $(0, \infty)$. By integration, we finally get $W = A + By^2$ and the conclusion follows easily by substituting into the equation (4.8). □

Now, the rest of the proof of (4.7), in particular Step (ii), will be a consequence of energy arguments close to those from Giga and Kohn [26] for blow-up problems. A difference with [26] is that here v is *not* uniformly bounded; and indeed it will be proved that, unlike in those works, v converges to an unbounded self-similar profile.

End of proof of Theorem 4.1 *(sketch).* By using the energy functional

$$E(s) = \int_0^{R(s)} \Big(\frac{v_y^2}{2} + \frac{v^{\gamma+1}}{\gamma+1} - \frac{\alpha v^2}{2}\Big)(y, s)\rho(y)\, dy,$$

where $\rho(y) = e^{-\frac{y^2}{4}}$ and $R(s) = e^{\frac{s}{2}}$, along with estimate (4.12), one can show that $\int_{s_0}^{\infty}\int_0^{R(s)} \rho v_s^2\, dy\, ds < \infty$. Therefore, by standard arguments, using (4.12) and parabolic L^p estimates, we deduce that, for each sequence $s_n \to \infty$, there exists a subsequence s_n' such that $v(\cdot, s_n')$ converges to a solution V of (4.8), uniformly on $\{|y| < R\}$ for each $R > 0$.

But on the other hand, by the lower bound in Lemma 4.1, for each $y \in \mathbb{R}$ and $s > 2\log|y|$, we have

$$v(y, s) = (T - t)^\alpha w(y\sqrt{T-t}, t) \ge (T-t)^\alpha (c_1|y\sqrt{T-t}|^2)^\alpha = c_1^\alpha |y|^{2\alpha}.$$

In view of Lemma 4.3 and the polynomial bound (4.12), this shows that necessarily $V = V_1$. The convergence assertion (4.7) follows, hence Theorem 4.1. □

Proof of Theorem 4.2. Set

$$v = u^{1-\lambda}, \qquad \gamma = \frac{p - \lambda}{1 - \lambda} > 1. \tag{4.15}$$

Then v satisfies $v_t - v_{xx} \ge (1-\lambda)v^\gamma$ at each $(x, t) \in D$ such that $v \ge K^{1-\lambda}$. Since moreover $v_r \le 0$ and $v_t \ge 0$, we may apply [52, Theorem 5] to deduce that $v(x, t) \le C(T-t)^{-1/(\gamma-1)}$ in D, that is, $u(x, t) \le C(T-t)^{-1/(p-1)}$. □

5. Blow-up profiles in space

The results of this section show that the influence of gradient perturbations on spatial blow-up profiles is connected in a precise way with the sub- or super-critical nature of the perturbation (in the sense defined in Section 2). In particular, for equation (2.1), the absorbing gradient term has no significant effect on the spatial blow-up profile if $q < q_0 := 2p/(p+1)$, whereas its influence becomes quite strong for $q > q_0$, making it more and more singular as q increases to p (recall that blow-up for (2.1) occurs if and only if $p > q$).

Theorem 5.1. [11] *Assume* $\Omega = B_R$, $1 < q < p$, $\mu > 0$. *Let* u_0 *be radial, with* $u_{0,r} \leq 0$, *and let the solution of* (2.1) *satisfy* $T < \infty$. *Then* 0 *is the only blow-up point. Moreover, for all* $\alpha > \alpha_0$, *it holds*

$$u(x,t) \leq C_\alpha |x|^{-\alpha}, \quad 0 < |x| < R, \quad 0 < t < T, \tag{5.1}$$

with

$$\alpha_0 = \begin{cases} 2/(p-1) & \text{if } 1 < q \leq 2p/(p+1) \\ q/(p-q) & \text{if } 2p/(p+1) < q < p \end{cases}$$

(observe that $q/(p-q) > 2/(p-1)$ *if* $2p/(p+1) < q < p$*).*

As an indication of the optimality of α_0 in Theorem 5.1, it was shown in [11] that (5.1) can hold for no $\alpha < \alpha_0$ if in addition $N = 1$ and $u_t \geq 0$. Actually, we have the following lower bound on the profile, which shows that Theorem 5.1 is sharp.

Theorem 5.2. [51] *Under the hypotheses of Theorem* 5.1, *assume in addition that* $u_t \geq 0$. *Then it holds*

$$u(x,T) \geq C|x|^{-\alpha_0}, \qquad 0 < |x| < \eta, \qquad \text{for some } C, \eta > 0. \tag{5.2}$$

Here and in what follows, we have denoted $u(x,T) = \lim_{t\to T} u(x,t)$. Recall that if 0 is the only blow-up point (without even assuming $u_t \geq 0$), then this limit exists and finite for any $x \neq 0$, due to standard parabolic estimates.

Remarks 5.1. (i) The proof of Theorem 5.2 actually shows that the lower blow-up profile estimate (5.2) with $\alpha_0 = 2/(p-1)$ is true for the general perturbed problem (1.8) in $\Omega = B_R$ under the assumption $g \leq 0$, whenever u is radial, $u_r < 0$ for $r > 0$, $u_t \geq 0$, and u blows up in sup norm. Moreover, no boundary conditions are actually required.

(ii) Consider equation $(2.3)_1$ with $q \geq 1, \lambda > 0$, $s \in \mathbb{R}$ and $q + s < p$. Assume $\Omega = B_R$, u radial, $u_r \leq 0$, $u_t \geq 0$ (no boundary conditions are required). Then the proof of Theorem 5.2 yields the lower blow-up profile estimate (5.2) with $\alpha_0 = q/(p-q-s)$, whenever blow-up occurs in sup norm. We observe that $\alpha_0 > 2/(p-1)$ when $q > 2(p-s)/(p+1)$. The argument of the proof can be easily extended to obtain estimates of this kind for more general g's and/or nonlinearities other than u^p, with nonlinear diffusion as well (see [51]).

(iii) In Theorem 5.2, we may take $C = C(p)$ if $q \leq 2p/(p+1)$ and $C = C(p, q, \mu)$ if $2p/(p+1) < q < p$.

The proof of Theorem 5.1 was based on nontrivial modifications of the argument of [20] (which consists in applying the maximum principle to a function of the form $J = u_r + c_\varepsilon(r)F(u)$). Moreover, in order to reach the exponent α_0, the authors had to incorporate an additional bootstrap argument (on the value of γ for $F(u) = u^\gamma$). We shall not reproduce this proof here for reasons of space.

Proof of Theorem 5.2. First assume $1 < q \leq 2p/(p+1)$. Since $u_t \geq 0$ and $u_r \leq 0$, we have

$$\left(\tfrac{1}{2}u_r^2 + \tfrac{1}{p+1}u^{p+1}\right)_r = (u_{rr} + u^p)u_r = \left(u_t + \mu|u_r|^q - \tfrac{N-1}{r}u_r\right)u_r \leq 0,$$

hence $\left(\frac{1}{2}u_r^2 + \frac{1}{p+1}u^{p+1}\right)(r,t) \leq \frac{1}{p+1}u^{p+1}(0,t)$. Therefore, we get

$$\|u_r(t)\|_\infty \leq C_1 u^{(p+1)/2}(0,t). \tag{5.3}$$

For $0 < t < T$, let $r_0(t)$ be such that $u(r_0(t),t) = \frac{1}{2}u(0,t)$. Note that, since $u_r < 0$ for $0 < t < T$ and $0 < r \leq R$, the implicit function theorem implies that $r_0(t)$ is unique and is a continuous function of t. Since u is nonglobal, it is known that u cannot remain bounded in sup norm (see Section 2). Consequently, $u(0,t) \to \infty$ as $t \to T$, due to $u_r \leq 0$ and $u_t \geq 0$. We may also assume that 0 is the only blow-up point, since otherwise the result is trivial. This implies that $r_0(t) \to 0$ as $t \to T$. Now we have

$$-u_r \leq C_2 u^{(p+1)/2}, \qquad 0 \leq r \leq r_0(t).$$

Integrating, we get

$$\begin{aligned} u^{-(p-1)/2}(r_0(t),t) &\leq u^{-(p-1)/2}(0,t) + C_3 r_0(t) \\ &= 2^{-(p-1)/2}u^{-(p-1)/2}(r_0(t),t) + C_3 r_0(t) \end{aligned}$$

hence $u(r_0(t),t) \geq C_4(r_0(t))^{-2/(p-1)}$. Using $u_t \geq 0$, it follows that

$$u(r_0(t),T) \geq C_4(r_0(t))^{-2/(p-1)}, \quad 0 < t < T.$$

Since r_0 is continuous and $r_0(t) \to 0$ as $t \to T$, we deduce that the range $r_0((0,T))$ contains an interval of the form $(0,\eta)$ and the conclusion follows.

Let us now consider $2p/(p+1) < q < p$ and put $\gamma = p/q > 1$. For each $t \in (0,T)$, at a point $r \in (0,R]$ where $|u_r(\cdot,t)|$ achieves its maximum, we have

$$\mu|u_r|^q = u^p + u_{rr} - u_t + \tfrac{N-1}{r}u_r \leq u^p,$$

due to $u_t \geq 0$, $u_r \leq 0$ and $u_{rr}(r,t) \leq 0$. It follows that

$$\|u_r(t)\|_\infty \leq C_1 u^\gamma(0,t). \tag{5.4}$$

Using (5.4) in place of (5.3), the conclusion follows exactly as in the case $q \leq 2p/(p+1)$. □

In some special cases, it is possible to obtain sharper estimates or even exact profiles. Let us first consider the critical case $q = 2p/(p+1)$ of equation (2.1).

Theorem 5.3. [53] *Let* $\Omega = \mathbb{R}^N$, $q = 2p/(p+1)$ *and* $\mu \in (0,2)$. *There exists* $p_0(N,\mu) > 1$ *such that for all* $p \in (1,p_0)$ *and some radial initial data* u_0, $u_{0,r} \le 0$, *the equation* (2.1) *admits a solution with single-point blow-up and the final profile*

$$u(x,T) = C|x|^{-2/(p-1)}, \quad x \in \mathbb{R}^N, \quad \textit{for some } C > 0.$$

Theorem 5.3 is a consequence of Theorem 6.2 below. As for the critical problem (2.4) with $m = 1$, considered in Section 4, we have the following result:

Theorem 5.4. [31] *Let* $m = 1$ *and* $\lambda > p$. *Assume that* u_0 *is symmetric, with* $u_{0,r} \le 0$, *and let the solution* u *of* (2.4) *satisfy* $T < \infty$. *Then we have single-point blow-up and the final profile satisfies*

$$C_1 \le |x|^{2/(p-1)} u(x,T) \le C_2, \quad 0 < |x| < 1,\ x \ne 0, \qquad \textit{for some } C_1, C_2 > 0.$$

Proof of Theorem 5.4. The (globally defined) solution $w \ge 0$ of the transformed problem (4.3) (cf. (4.4)) satisfies estimates (4.9) and (4.11). Since $w(0,T) = 0$, we have in particular $(c_1|x|^2)^\alpha \le w(x,T) \le (c_2|x|)^{2\alpha}$, for $|x| < 1$, and the assertion concerning $u(x,T)$ follows immediately. □

Remarks 5.2. (i) Let us note that for the critical problems in Theorems 5.3, 5.4, we have the same order of growth for the space profiles near $x = 0$ (slightly less singular than the usual profile (1.5) for the model problem (1.1)), but the time rates are different: type I or type II, respectively. On the other hand, comparison of these results with Theorems 5.1, 5.2 yields the interesting observation that the dissipative gradient term can have opposite effects on the blow-up profile: slightly less singular or strongly more singular profile.

(ii) In view of the results of Sections 3–5, it is an interesting open question whether blow-up is type I or type II for equation (2.1) in the supercritical case $2p/(p+1) < q < p$. For the critical case $q = 2p/(p+1)$, see Remark 6.1 (ii).

6. Blow-up profiles in self-similar variables

This section concerns the description of blow-up in self-similar variables for problem (1.8). There again, the results can be interpreted in terms of criticality of the perturbations. For instance, as long as $q < q_0 = 2p/(p+1)$ in equation (2.1), the classical Giga-Kohn profile (1.7) is preserved (under suitable assumptions), while a new, nonflat profile appears for $q = q_0$ (compare formulas (6.1) and (6.2)).

Theorem 6.1. *Let* $1 < p \le p_S$, $\Omega = B_R$ *or* $\Omega = \mathbb{R}^N$. *Assume that* $g = g(|x|, t, u, |\nabla u|)$ *satisfies* $g(0,t,u,0) \le 0$, *along with the subcritical growth assumption* (3.3). *Moreover, assume that the solution* u *of* (1.8) *is radial, with* $u_r \le 0$, *that it blows up in* L^∞, *and satisfies the type I estimate* (1.3). *Then it holds,*

$$\lim_{t\to T} (T-t)^{1/(p-1)} u(y\sqrt{T-t}, t) = \kappa = (p-1)^{-1/(p-1)}, \tag{6.1}$$

uniformly for y *bounded.*

For $q = 2p/(p+1)$, equation (2.1) is actually scale-invariant, which allows for the possibility of existence of self-similar solutions.

Theorem 6.2. [53] *Let* $\Omega = \mathbb{R}^N$, $q = 2p/(p+1)$ *and* $\mu \in (0,2)$. *There exists* $p_0(N,\mu) > 1$ *such that for all* $p \in (1,p_0)$, (2.1) *admits a positive self-similar solution of the form*

$$u(x,t) = (T-t)^{-1/(p-1)} W(x/\sqrt{T-t}), \quad x \in \mathbb{R}^N, \quad -\infty < t < T,$$

equivalently

$$(T-t)^{1/(p-1)} u(y\sqrt{T-t}, t) = W(y), \tag{6.2}$$

where $W \in C^2(\mathbb{R}^N)$ *is radial, with* $W_r \le 0$. *Moreover, for all such solution,* W *satisfies*

$$\lim_{|y|\to\infty} |y|^{2/(p-1)} W(y) = C > 0.$$

As for the critical problem (2.4) with $m = 1$, it leads to a still distinctly different self-similar profile, namely a singular one.

Theorem 6.3. [31] *Let* $m = 1$ *and* $\lambda > p$. *Assume that* u_0 *is symmetric, with* $u_{0,r} \le 0$, *and let the solution* u *of* (2.4) *satisfy* $T < \infty$. *Then it holds*

$$\lim_{t\to T} (T-t)^{1/(p-1)} u(y\sqrt{T-t}, t) = c_p |y|^{-2/(p-1)},$$

uniformly for y *bounded and bounded away from* 0.

Theorem 6.2 was proved in [53] by using ODE methods. As for Theorem 6.3, its proof was given in Section 4 above (cf. Formula (4.7)). Theorem 6.1 seems to be new and we now give a proof.

Proof of Theorem 6.1. Following [26, 21], we rescale equation (1.8) by similarity variables, by setting

$$T - t = e^{-s}, \qquad y = x/\sqrt{T-t} \qquad \text{and} \qquad v(y,s) = (T-t)^{\beta} u(x,t),$$

with $\beta = 1/(p-1)$. Then v satisfies the equation

$$v_s = \Delta v - \tfrac{y}{2}.\nabla v + v^p - \beta v + h(y,s) \quad \text{in } D, \tag{6.3}$$

where

$$h(y,s) = e^{-p\beta s} g(x,t,u(x,t),|\nabla u(x,t)|)$$

and $D = \{(y,s) \in \mathbb{R}^N \times \mathbb{R},\ s > s_0 := -\log T,\ |y| \le Re^{s/2}\}$. It was proved in [41, Lemma 5.1] that under assumption (3.3) (even with $q \le 2p/(p+1)$), it holds

$$\|\nabla u(t)\|_\infty \le C \sup_{\tau\in[0,\tau]} \|u(\tau)\|_\infty^{(p+1)/2}$$

for t close enough to T. Therefore, the assumption (1.3) implies that

$$\|u(t)\|_\infty \le C(T-t)^{-\beta} = Ce^{\beta s}, \qquad \|\nabla u(t)\|_\infty \le C(T-t)^{-(p+1)\beta/2} = Ce^{\frac{p+1}{2}\beta s}.$$

Letting $\gamma = \beta \min(p - p_1, p - q(p+1)/2) > 0$, it follows from (3.3) that

$$\|h(s)\|_\infty \le Ce^{-p\beta s}(1 + Ce^{\beta p_1 s} + Ce^{\frac{(p+1)q}{2}\beta s}) \le Ce^{-\gamma s}.$$

Moreover, (1.3) implies that v is bounded in D. Since the perturbation in equation (6.3) is decaying and integrable in time, the energy arguments in [26] (derived there for $h = 0$) carry over, and we deduce that $v(y,s)$ converges to 0 or κ, as $s \to \infty$, uniformly for y bounded. Finally, since $g(0,t,u,0) \le 0$ and u blows up in L^∞, we easily obtain $(T-t)^\beta u(0,t) \ge C_1 > 0$, hence $v(0,s) \ge C_1$. Consequently, the 0 limit cannot occur and the result is proved. □

Remarks 6.1. (i) On a general domain, under the assumptions of Theorem 6.1 (but for radial symmetry), the proof shows that, for any blow-up point $a \in \Omega$ (i.e., a such that u is not locally bounded nearby (a,T)), it still holds

$$\lim_{t\to T}(T-t)^{1/(p-1)}u(a+y\sqrt{T-t},t) = \kappa \text{ or } 0,$$

uniformly for y bounded. In the case $g = 0$, the nondegeneracy results in [28] (based, among other things, on weighted energy arguments) show that the 0 limit cannot occur. With more work, it might be possible to extend this to our case, but this extension does not seem immediate.

(ii) Consider equation (2.1) in the critical case $q = 2p/(p+1)$. It is known [52] that blow-up is type I if μ is small enough (under suitable assumptions), but the question is open for $\mu > 0$ large. However, the gradient term may already have a (mild) influence on the blow-up rate for $\mu > 0$ small. Namely, the solution u from Theorem 6.2 satisfies $\lim_{t\to T}(T-t)^{1/(p-1)}\|u(t)\|_\infty = W(0) > \kappa$ (see [53]), whereas, for $q < 2p/(p+1)$ and, e.g., $p < p_B$, any radially decreasing blow-up solution in $\mathbb{R}^N$ satisfies $\lim_{t\to T}(T-t)^{1/(p-1)}\|u(t)\|_\infty = \kappa$ by Theorems 6.1 and 3.1.

(iii) As in Theorem 4.1, the assumption $\lambda > p$ in Theorems 5.2 and 6.3 is optimal, for there is a bifurcation in the spatial and self-similar profiles at $\lambda = p$. Indeed, for $\lambda \in (1,p)$, the transformation (4.4) converts equation (2.4) with $m = 1$ into the *quenching* problem $u_t - u_{xx} = -u^{-\gamma}$ for some $\gamma > 0$. Assume for instance that u is symmetric with $u_r \le 0$, $u_t \ge 0$. By using the results in [17] (and references therein) on the quenching problem, and transforming back to (2.4), we find that the space profile at $a = 0$ is given by (1.6) and the self-similar profile by (1.7).

(iv) For the model equation (1.1), more precise self-similar profiles based on the larger time-scale $\sqrt{(T-t)|\log(T-t)|}$ have been studied by several authors, see [56, 8] and the references therein. Some of the results remain valid for equation (1.8) under assumption (3.3) (see [8, Remark 3 p. 541]).

(v) For equation (2.1) but now with $\mu < 0$, self-similar profiles have been studied in [23] for $q = 2$ and $\Omega = \mathbb{R}^N$. It was proved that radial blow-up solutions behave asymptotically like a self-similar solution w of the Hamilton-Jacobi equation $w_t = w^p + |\nabla w|^2$, with w having the form $w(x,t) = (T-t)^{-1/(p-1)}\theta(x(T-t)^{-\sigma})$ and $\sigma = (2-p)/2(p-1)$. Note that this kind of self-similar behavior is quite different from that in Theorem 6.2; indeed, σ describes the range $(-\infty, 1/2)$ for $p \in (1,\infty)$.

7. Blow-up sets and continuation after blow-up

We have already seen several results which ensure single point blow-up for equation (1.8) in the radial case (cf. Theorems 5.1, 5.3 and 5.4). In particular, unlike blow-up rates and profiles, the property of single-point blow-up in the radial case does not seem to be affected by the presence of a dissipative gradient term, even if it is critical or supercritical. On the other hand, it was proved in [11] that if Ω is convex bounded then, as for (1.1) with $u_0 \geq 0$, the blow-up set of any solution of (2.1) is a compact subset of Ω when $\mu > 0$ and $q < 2p/(p+1)$.

The situation is quite different when $\mu < 0$ (see [35, 34, 23]). Namely, one has global blow-up if $1 < p < q = 2$ and regional blow-up if $p = q = 2$. However, if $p > q = 2$, $\Omega = B_R$ and u_0 is radial, with $u_{0,r} \leq 0$, then there is still single point blow-up. The proof involves the transformation $v = e^u - 1$, which changes (2.1) into the equation with mildly superlinear source $v_t - \Delta v = (1+v)\log^p(1+v)$. The authors of [34] interpret this result in the following way. While the term u^p alone would force the solution to develop a spike at the maximum point, hence causing single point blow-up, the gradient term now has a positive sign and tends to push up the steeper parts of the graph of $u(.,t)$. This enhances regional or even global blow-up, the influence of the gradient term becoming more important as the value of p decreases.

We note that the change of behavior is here no longer related to scaling properties of the perturbation, but rather to its size as compared with that of the reaction term (see the end of paragraph 2 of Section 2 for a related phenomenon).

As for continuation after blow-up, it can be seen that some solutions of the critical problem (2.4) for $m = 1$ and $\lambda > p$ admit a *nontrivial continuation after blow-up*, in sharp contrast with the unperturbed problem (1.1) in the one-dimensional case. Moreover, this continuation is unique can be described in a precise way. It is to be noted that the continuation does not enjoy the property of immediate regularization mentioned in Introduction for (1.1) with $p_S < p < p_*$.

For simplicity, consider for instance the initial data $u_0 \equiv b$, with $b > b_0(\lambda, p)$ large enough (but this may be extended to more general initial data). Then we have $T < \infty$ and, by using the transformation (4.4) and known results [2, 57] on the dead-core problem (4.3), one obtains the following. There exists a *blow-up interface* $\sigma : [T, \infty) \to [0, 1)$, monotonically increasing, locally Lipschitz, with $\sigma(T) = 0 < L = \sigma(\infty) < 1$, and an extension $\overline{u}$ of u, with $\overline{u} \in C(D; (0, \infty])$, $D = [-1, 1] \times [0, \infty)$, such that

$$\begin{cases} \overline{u} = \infty \text{ in } D_\infty := \big\{(x,t);\ T \leq t < \infty,\ |x| \leq \sigma(t)\big\}, \\ \overline{u} \text{ is a classical solution to } (2.4)_1 \text{ in } D \setminus D_\infty. \end{cases}$$

Moreover, for $x \in (L, 1)$, $\overline{u}(x,t)$ converges monotonically as $t \to \infty$ to the unique stationary solution v of $(2.4)_1$ in $(L, 1)$ such that $v(1) = b$ and $v(L) = \infty$. We refer to [6] for similar results on a different blow-up problem with gradient perturbation.

Remark 7.1. In the opposite direction, note that if $\lambda < 1$ in problem (2.4) for $m = 1$, then blow-up is complete, due to the transformation (4.15) and the known results on (1.1). In fact, the threshold is given by $\lambda = (p+1)/2$. Indeed, for $\lambda \in (1,p)$, one can use the results in [22, 44] on the quenching problem (cf. Remark 6.1 (iii)) to deduce that incomplete blow-up may occur if and only if $\lambda > (p+1)/2$.

References

[1] J. Aguirre, M. Escobedo, On the blow up of solutions for a convective reaction diffusion equation, *Proc. Roy. Soc. Edinburgh.*, 123A, 3 (1993), 433–460.

[2] C. Bandle, I. Stakgold, The formation of the dead core in parabolic reaction-diffusion problems, *Trans. Amer. Math. Soc.*, 286 (1984), 275–293.

[3] P. Baras, L. Cohen, Complete blow-up after $T_{\max}$ for the solution of a semilinear heat equation, *J. Funct. Anal.*, 71 (1987), 142–174.

[4] J.-Ph. Bartier, Global behavior of solutions of a reaction-diffusion equation with gradient absorption in unbounded domains, to appear.

[5] J.-Ph. Bartier, Ph. Souplet, Gradient bounds for solutions of semilinear parabolic equations without Bernstein's quadratic condition, *C.R. Acad. Sc. Paris*, 338 (2004), 533–538.

[6] J. Bebernes, D. Eberly, Characterization of blow-up for a semilinear heat equation with a convection term, *Quart. J. Mech. Appl. Mech.*, 42 (1989) 447–456.

[7] M.-F. Bidaut-Véron, Initial blow-up for the solutions of a semilinear parabolic equation with source term, *in:* Equations aux dérivées partielles et applications, articles dédiés à Jacques-Louis Lions. Gauthier-Villars, Paris, 1998, pp. 189–198.

[8] J. Bricmont, A. Kupiainen, Universality in blow-up for nonlinear heat equations, *Nonlinearity*, 7 (1994), 539–575.

[9] M. Chipot, F.B. Weissler, Some blow up results for a nonlinear parabolic problem with a gradient term, *SIAM J. Math. Anal.*, 20 (1989), 886–907.

[10] M. Chlebik, M. Fila, From critical exponents to blowup rates for parabolic problems, *Rend. Mat. Appl., Ser. VII*, 19 (1999), 449–470.

[11] M. Chlebik, M. Fila, P. Quittner, Blowup of positive solutions of a semilinear parabolic equation with a gradient term, *Dyn. Contin. Discrete Impulsive Syst. Ser. A*, 10 (2003), 525–537.

[12] K. Deng, Stabilization of solutions of a nonlinear parabolic equation with a gradient term, *Math. Z.*, 216 (1994), 147–155.

[13] M. Fila, Remarks on blow up for a nonlinear parabolic equation with a gradient term, *Proc. Amer. Math. Soc.*, 111 (1991), 795–801.

[14] M. Fila, Blow-up of solutions of supercritical parabolic problems, preprint (2004).

[15] M. Fila, H. Matano, P. Polacik, Immediate regularization after blow-up, *SIAM J. Math. Anal.*, to appear.

[16] M. Fila, Ph. Souplet, The blow-up rate for semilinear parabolic problems on general domains, *Nonlinear Differ. Eq. and Appl. NoDEA*, 8 (2001), 473–480.

[17] S. Filippas, J.-S. Guo, Quenching profiles for one-dimensional semilinear heat equations, *Quart. Appl. Math.*, 51 (1993), 713–729.

[18] S. Filippas, M.A. Herrero, J.J.L. Velázquez, Fast blow-up mechanisms for sign-changing solutions of a semilinear parabolic equation with critical nonlinearity, *Proc. R. Soc. Lond. A*, 456 (2000), 2957–2982.

[19] A. Friedman, M.A. Herrero, Extinction properties of semilinear heat equations with strong absorption, *J. Math. Anal. Appl.*, 124 (1987), 530–546.

[20] A. Friedman, J.B. McLeod, Blowup of positive solutions of semilinear heat equations, *Indiana Univ. Math. J.*, 34 (1985), 425–447.

[21] V.A. Galaktionov, S. Posashkov, Application of new comparison theorems to the investigation of unbounded solutions of nonlinear parabolic equations (Russian), *Differ. Uravn.*, 22 (1986), 1165–1173. (Engl. transl.: *Differ. Equations*, 22 (1986), 809–815.)

[22] V.A. Galaktionov, J.L. Vázquez, Necessary and sufficient conditions for complete blow-up and extinction for one-dimensional quasilinear heat equations, *Arch. Rational Mech. Anal.*, 129 (1995), 225–244.

[23] V.A. Galaktionov, J.L. Vázquez, Blowup for quasilinear heat equations described by means of nonlinear Hamilton-Jacobi equations, *J. Differ. Equations*, 127 (1996), 1–40.

[24] V.A. Galaktionov, J.L. Vázquez, The problem of blow-up in nonlinear parabolic equations, *Discrete Contin. Dyn. Syst.*, 8 (2002), 399–433.

[25] Y. Giga, A bound for global solutions of semilinear heat equations, *Comm. Math. Phys.*, 103 (1986), 415–421.

[26] Y. Giga, R.V. Kohn, Asymptotically self-similar blow-up of semilinear heat equations, *Comm. Pure Appl. Math.*, 38 (1985), 297–319.

[27] Y. Giga, R.V. Kohn, Characterizing blowup using similarity variables, *Indiana Univ. Math. J.*, 36 (1987), 1–40.

[28] Y. Giga, R.V. Kohn, Nondegeneracy of blowup for semilinear heat equations, *Comm. Pure Appl. Math.*, 42 (1989), 845–884.

[29] Y. Giga, S. Matsui, S. Sasayama, Blow up rate for semilinear heat equations with subcritical nonlinearity, *Indiana Univ. Math. J.*, 53 (2004), 483–514.

[30] Y. Giga, S. Matsui, S. Sasayama, On blow-up rate for sign-changing solutions in a convex domain, *Math. Methods Appl. Sci.*, 27 (2004), 1771–1782.

[31] J.-S. Guo, Ph. Souplet, Fast rate of formation of dead-core for the heat equation with strong absorption and applications to fast blow-up, *Math. Ann.*, to appear.

[32] M.A. Herrero, J.J.L. Velazquez, Explosion de solutions des équations paraboliques semilinéaires supercritiques, *C. R. Acad. Sc. Paris Série I*, 319 (1994), 141–145.

[33] B. Hu, Remarks on the blowup estimate for solution of the heat equation with a nonlinear boundary condition, *Differ. Integral Equations*, 9 (1996), 891–901.

[34] B. Kawohl, L. Peletier, Remarks on blowup and dead cores for nonlinear parabolic equations, *Math. Z.*, 202 (1989), 207–217.

[35] A.A. Lacey, Global blowup of a nonlinear heat equation, *Proc. Royal Soc. Edinburgh*, 104 (1986), 161–167.

[36] O. Ladyzenskaya, V.A. Solonnikov, N.N. Uralceva, Linear and Quasilinear Equations of Parabolic Type, Amer. Math. Soc. Translations, Providence, RI, 1967.

[37] H.A. Levine, P. Meier, A blowup result for the critical exponent in cones, *Israel J. Math.*, 67 (1989), 129–136.

[38] H.A. Levine, L.N. Payne, P.E. Sacks, B. Straughan, Analysis of convective reaction-diffusion equation (II), *SIAM J. Math. Anal.*, 20 (1989), 133–147.

[39] H. Matano, F. Merle, On nonexistence of type II blow-up for a supercritical nonlinear heat equation, *Commun. Pure Appl. Math.*, 57 (2004), 1494–1541.

[40] J. Matos, Self-similar blow up patterns in supercritical semilinear heat equations, *Commun. Appl. Anal.*, 5 (2001), 455–483.

[41] J. Matos, Ph. Souplet, Universal blow-up rates for a semilinear heat equation and applications, *Advances Differ. Equations*, 8 (2003), 615–639.

[42] N. Mizoguchi, Blowup rate of solutions for a semilinear heat equation with the Neumann boundary condition, *J. Differ. Equations*, 193 (2003), 212–238.

[43] C. Mueller, F.B. Weissler, Single point blow-up for a general semilinear heat equation, *Indiana Univ. Math. J.*, 34 (1985), 881–913.

[44] D. Phillips, Existence of solutions of quenching problems, *Appl. Anal.*, 24 (1987), 253–264.

[45] M. Pierre, unpublished manuscript (2003).

[46] P. Quittner, Continuity of the blow-up time and a priori bounds for solutions in superlinear parabolic problems, *Houston J. Math.*, 29 (2003), 757–799.

[47] P. Quittner, Complete and energy blow-up in superlinear parabolic problems, Proc. NCTS Intern. Conf. on Elliptic and Parabolic Problems, Hsinchu 2004, World Scientific 2005.

[48] P. Quittner, Ph. Souplet, A priori estimates of global solutions of superlinear parabolic problems without variational structure, *Discrete Contin. Dyn. Syst.*, 9 (2003), 1277–1292.

[49] Ph. Souplet, Recent results and open problems on parabolic equations with gradient nonlinearities, *Electronic J. Differ. Equations*, Vol. 2001 (2001), No. 20, pp. 1–19.

[50] Ph. Souplet, Gradient blow-up for multidimensional nonlinear parabolic equations with general boundary conditions, *Differ. Integral Equations*, 15 (2002), 237–256.

[51] Ph. Souplet, Blow-up profile estimates for nonlinear parabolic equations, in preparation.

[52] Ph. Souplet, S. Tayachi, Blow-up rates for nonlinear heat equations with gradient terms and for parabolic inequalities, *Colloq. Math.*, 88 (2001), 135–154.

[53] Ph. Souplet, S. Tayachi, F.B. Weissler, Exact self-similar blow-up of solutions of a semilinear parabolic equation with a nonlinear gradient term, *Indiana Univ. Math. J.*, 48 (1996), 655–682.

[54] Ph. Souplet, F.B. Weissler, Self-similar sub-solutions and blow-up for nonlinear parabolic equations, *J. Math. Anal. Appl.*, 212 (1997), 60–74.

[55] Ph. Souplet, F.B. Weissler, Poincaré's inequality and global solutions of a nonlinear parabolic equation, *Ann. Inst. H. Poincaré, Analyse non linéaire*, 16 (1999), 337–373.

[56] J.J.L. Velázquez, Blow up for semilinear parabolic equations, *in:* Research in applied mathematics, Recent advances in partial differential equations, M.A. Herrero and E. Zuazua (eds.), John Wiley & Sons, 1994, pp. 131–145.

[57] H.-M. Yin, The Lipschitz continuity of the interface in the heat equation with strong absorption, *Nonlinear Anal.*, 20 (1993), 413–416.

[58] P. Poláčik, P. Quittner, A Liouville-type theorem and the decay of radial solutions of a semilinear heat equation, preprint (2005).

Philippe Souplet
Département de Mathématiques INSSET
Université de Picardie
F-02109 St-Quentin, France

and

Laboratoire de Mathématiques Appliquées
UMR CNRS 7641
Université de Versailles
45 avenue des Etats-Unis
F-78035 Versailles, France
e-mail: `souplet@math.uvsq.fr`

Progress in Nonlinear Differential Equations
and Their Applications, Vol. 64, 497–507

Non-existence of Positive Solutions for Diffusive Logistic Equations with Nonlinear Boundary Conditions

Kenichiro Umezu

Dedicated to Professor Herbert Amann on the occasion of his 65th birthday

Abstract. In this paper a diffusive logistic equation with large diffusion is considered under nonlinear boundary conditions. Non-existence of the corresponding stationary positive solutions is discussed by use of variational techniques.

1. Introduction

This paper is devoted to the study of the following semilinear elliptic problem with a nonlinear boundary condition arising in population dynamics (cf. [3], [4], [5], [6], [8], [10]):

$$\begin{cases} -\Delta u = \lambda(m(x) - u)u & \text{in } \Omega, \\ \dfrac{\partial u}{\partial \boldsymbol{n}} = \lambda b(x) u^p & \text{on } \partial\Omega. \end{cases} \tag{1.1}$$

Here Ω is a bounded domain of $\mathbb{R}^N$, $N \geq 2$, with smooth boundary $\partial\Omega$, $\lambda > 0$ is a parameter, $p > 1$ is a constant, $m \in C^\theta(\overline{\Omega})$ satisfies that $m(x_0) > 0$ for some $x_0 \in \Omega$ and may change sign, $b \in C^{1+\theta}(\partial\Omega)$ is non-negative, and $\boldsymbol{n}$ is the unit exterior normal to $\partial\Omega$.

Problem (1.1) denotes steady states of solutions of the parabolic problem

$$\begin{cases} \dfrac{\partial u}{\partial t}(t,x) = \nabla \cdot (1/\lambda)\nabla u(t,x) + (m(x) - u(t,x))u(t,x) & \text{in } (0,\infty)\times\Omega, \\ u(0,x) = u_0(x) & \text{in } \Omega, \\ (1/\lambda)\nabla u(t,x)\cdot \boldsymbol{n} = b(x)u(t,x)^p & \text{on } (0,\infty)\times\partial\Omega. \end{cases}$$

From the viewpoint of population dynamics, the unknown function u denotes the population density of some species with diffusion rate $1/\lambda$ where m represents the birth or decay rate, and the nonlinear boundary condition means that the law of the population flux $(1/\lambda)\nabla u \cdot \boldsymbol{n}$ to the direction $\boldsymbol{n}$ on the boundary $\partial\Omega$ is determined nonlinearly on u, having the flux rate b.

If $u \in C^2(\overline{\Omega})$ is a solution of (1.1) for some $\lambda > 0$, then we call (λ, u) a *solution* of (1.1). A solution (λ, u) of (1.1) is called *positive* if $u > 0$ in Ω.

The existence and multiplicity of positive solutions of (1.1) have been studied in the previous work [10], in which, if $p = 2$, $\int_\Omega m\,dx \geq 0$ and $\int_{\partial\Omega} b\,d\sigma < |\Omega|$, then it is proved that there exists a positive solution of (1.1) for any $\lambda > 0$ small, converging to the constant $c^*(b) = \int_\Omega m\,dx/(|\Omega| - \int_{\partial\Omega} b\,d\sigma)$ in the topology of $C^2(\overline{\Omega})$ as $\lambda \downarrow 0$, see [10, Theorems 1.1 and 1.6]. Here $d\sigma$ denotes the surface element of $\partial\Omega$ and $|\Omega|$ the volume of Ω. Under the same assumption a multiplicity result can be obtained for such $\lambda > 0$ small by use of a variational technique, if additionally the space dimension N is equal to two or three, see [10, Theorem 1.8], where the second positive solution tends to $+\infty$ in $C(\overline{\Omega})$, equivalently in the usual Sobolev space $W_2^1(\Omega)$, as $\lambda \downarrow 0$ (in fact, only the case $\int_\Omega m\,dx = 0$ is mentioned in [10, Theorem 1.8], but one can check easily that the assertion of the theorem is also true for the case $\int_\Omega m\,dx > 0$). The purpose of this paper is to discuss sufficient conditions for the non-existence of positive solutions in the case when $p = 2$ and $\int_\Omega m\,dx \geq 0$. It should be remarked that Hess [8] treats the linear case $b = 0$ and proves that a unique positive solution for each $\lambda > 0$ converges to $c^*(0)$ as $\lambda \downarrow 0$, see [8, Example 28.6].

Now our main result is the following.

Theorem 1.1. *Let the space dimension N equals two or three. Assume that $p = 2$ and $\int_\Omega m\,dx \geq 0$. If $\int_{\partial\Omega} b\,d\sigma > |\Omega|$, then problem* (1.1) *has no positive solutions for any $\lambda > 0$ small (simply for $0 < \lambda \ll 1$).*

For constants $a > 0$ and $p, q > 1$, Chipot, Fila and Quittner [7] consider a semilinear elliptic problem with a nonlinear boundary condition of the form

$$\begin{cases} -\Delta u = -au^p & \text{in } \Omega, \\ \dfrac{\partial u}{\partial \boldsymbol{n}} = u^q & \text{on } \partial\Omega. \end{cases} \tag{1.2}$$

The existence and multiplicity of positive solutions and also their stability are studied there. Our argument is variational, inspired by them. It will be noted that the smallness of parameter $\lambda > 0$ is crucial for applying techniques in [7] in our setting.

At the end of Introduction it should be mentioned that the space dimension N is assumed to be two or three throughout this paper. This implies that the following three compact embeddings hold true from Sobolev's embedding theorem:

$$W_2^1(\Omega) \subset L^2(\Omega),\ W_2^1(\Omega) \subset L^3(\Omega) \text{ and } W_2^1(\Omega) \subset L^3(\partial\Omega). \tag{1.3}$$

The next section is devoted to the proof of Theorem 1.1.

2. Proof of Theorem 1.1

2.1. Weak super- and subsolutions

In this section we consider (1.1) with $p = 2$:

$$\begin{cases} -\Delta u = \lambda(m(x) - u)u & \text{in } \Omega, \\ \dfrac{\partial u}{\partial \boldsymbol{n}} = \lambda b(x)u^2 & \text{on } \partial\Omega, \end{cases} \tag{2.1}$$

under the assumption that $N = 2$ or 3, $\int_\Omega m\, dx \geq 0$ and $\int_{\partial\Omega} b\, d\sigma > |\Omega|$. It has been discussed in [7] that the weak super- and subsolution method constructs a weak solution of (1.2) to the Sobolev space $X := W_2^1(\Omega)$ such that it is a local minimizer of the corresponding energy functional. Our proof starts with its formulation in our setting.

A function $u \in X$ is called a *weak subsolution* of (2.1) if we have

$$\int_\Omega \nabla u \nabla \varphi\, dx - \lambda \int_\Omega m u \varphi\, dx + \lambda \int_\Omega u^2 \varphi\, dx - \int_{\partial\Omega} b u^2 \varphi\, d\sigma \leq 0$$

for any non-negative $\varphi \in X$. Similarly a *weak supersolution* of (2.1) is defined by the reverse of the inequality. Usual pointwise sub- and supersolutions of (2.1) to class $C^2(\overline{\Omega})$ are called *strong* (see [2]). Strong sub- and supersolutions of (2.1) are called *strict* if they are not solutions of (2.1). It is easy to check that strong sub- and supersolutions imply weak ones, respectively.

Now we prove the following (cf. [7, Lemma 2.4]).

Lemma 2.1. *Let $\underline{u}, \overline{u} \in C^2(\overline{\Omega})$ be strict, strong sub- and supersolutions of* (2.1), *respectively, such that $0 \leq \underline{u} \leq \overline{u}$ in $\overline{\Omega}$. Then there exists a positive solution $u \in C^2(\overline{\Omega})$ of* (2.1) *such that*

$$\underline{u} < u < \overline{u} \quad \text{in } \overline{\Omega}. \tag{2.2}$$

Moreover, the positive solution u is a local minimizer of the corresponding energy functional

$$E_\lambda(v) = \frac{1}{2}\int_\Omega |\nabla v|^2\, dx - \frac{\lambda}{2}\int_\Omega m v^2\, dx + \frac{\lambda}{3}\int_\Omega v^3\, dx - \frac{\lambda}{3}\int_{\partial\Omega} b v^3\, d\sigma, \quad v \in X,$$

meaning that there exists a constant $\varepsilon > 0$ such that $\|v - u\|_X \leq \varepsilon$ implies $E_\lambda(u) \leq E_\lambda(v)$.

Proof. The assertion (2.2) is a direct consequence of applying the strong maximum principle and boundary point lemma (cf. [9]). The remainder assertion will be verified in accordance with the argument in [7, Lemma 2.4].

(I) First let us show how to use the weak sub- and supersolution method to provide us with the existence of a positive solution u. Let

$$M = \{u \in X : \underline{u} \leq u \leq \overline{u} \text{ a.e. both in } \Omega \text{ and on } \partial\Omega\}.$$

One can check easily that E_λ is coercive and weakly lower semi-continuous on M. Hence, by the standard argument, $\min_{v \in M} E_\lambda(v)$ is attained by some $u \in M$. We claim here that the function u is a weak solution, and consequently a solution of

(2.1). Indeed, the assertion that u is a solution is obtained by elliptic regularity with the aid of (1.3). Let $\varphi \in C^1(\overline{\Omega})$ and $\varepsilon > 0$, and we put

$$u_\varepsilon(x) = \min\{\overline{u}(x),\ \max\{\underline{u}(x),\ u(x) + \varepsilon\varphi(x)\}\}, \quad x \in \overline{\Omega}.$$

It follows that $u_\varepsilon \in M$. Since M is convex, we have $u + \tau(u_\varepsilon - u) \in M$ for all $\tau \in [0,1]$. Since $u = \min_{v\in M} E_\lambda(v)$, it follows that

$$0 \leq \frac{d}{d\tau} E_\lambda(u + \tau(u_\varepsilon - u))|_{\tau=0}. \tag{2.3}$$

Letting $u_\varepsilon = (u + \varepsilon\varphi) - \varphi^\varepsilon + \varphi_\varepsilon$, where

$$\varphi^\varepsilon = \max\{0,\, u + \varepsilon\varphi - \overline{u}\} \geq 0, \quad \varphi_\varepsilon := -\min\{0,\, u + \varepsilon\varphi - \underline{u}\} \geq 0,$$

we have by a direct computation

$$\begin{aligned}
&\frac{d}{d\tau} E_\lambda(u + \tau(u_\varepsilon - u))|_{\tau=0} \\
&= \varepsilon \left(\int_\Omega \nabla u \nabla \varphi \, dx - \lambda \int_\Omega m u \varphi \, dx + \lambda \int_\Omega u^2 \varphi \, dx - \lambda \int_{\partial\Omega} b u^2 \varphi \, d\sigma \right) \\
&\quad - \left(\int_\Omega \nabla u \nabla \varphi^\varepsilon \, dx - \lambda \int_\Omega m u \varphi^\varepsilon \, dx + \lambda \int_\Omega u^2 \varphi^\varepsilon \, dx - \lambda \int_{\partial\Omega} b u^2 \varphi^\varepsilon \, d\sigma \right) \\
&\quad + \left(\int_\Omega \nabla u \nabla \varphi_\varepsilon \, dx - \lambda \int_\Omega m u \varphi_\varepsilon \, dx + \lambda \int_\Omega u^2 \varphi_\varepsilon \, dx - \lambda \int_{\partial\Omega} b u^2 \varphi_\varepsilon \, d\sigma \right).
\end{aligned}$$

It follows from (2.3) that

$$\int_\Omega \nabla u \nabla \varphi \, dx - \lambda \int_\Omega m u \varphi \, dx + \lambda \int_\Omega u^2 \varphi \, dx - \lambda \int_{\partial\Omega} b u^2 \varphi \, d\sigma \geq \frac{1}{\varepsilon}(I_{1,\varepsilon} - I_{2,\varepsilon}), \tag{2.4}$$

where

$$\begin{aligned}
I_{1,\varepsilon} &= \int_\Omega \nabla u \nabla \varphi^\varepsilon \, dx - \lambda \int_\Omega m u \varphi^\varepsilon \, dx + \lambda \int_\Omega u^2 \varphi^\varepsilon \, dx - \lambda \int_{\partial\Omega} b u^2 \varphi^\varepsilon \, d\sigma, \\
I_{2,\varepsilon} &= \int_\Omega \nabla u \nabla \varphi_\varepsilon \, dx - \lambda \int_\Omega m u \varphi_\varepsilon \, dx + \lambda \int_\Omega u^2 \varphi_\varepsilon \, dx - \lambda \int_{\partial\Omega} b u^2 \varphi_\varepsilon \, d\sigma.
\end{aligned}$$

Since $\overline{u}$ is a weak supersolution, it follows that

$$\begin{aligned}
I_{1,\varepsilon} &= \int_\Omega \nabla \overline{u} \nabla \varphi^\varepsilon \, dx - \lambda \int_\Omega m \overline{u} \varphi^\varepsilon \, dx + \lambda \int_\Omega \overline{u}^2 \varphi^\varepsilon \, dx - \lambda \int_{\partial\Omega} b \overline{u}^2 \varphi^\varepsilon \, d\sigma \\
&\quad + \int_\Omega \nabla(u - \overline{u}) \nabla \varphi^\varepsilon \, dx - \lambda \int_\Omega m(u - \overline{u}) \varphi^\varepsilon \, dx + \lambda \int_\Omega (u^2 - \overline{u}^2) \varphi^\varepsilon \, dx \\
&\quad - \lambda \int_{\partial\Omega} b(u^2 - \overline{u}^2) \varphi^\varepsilon \, d\sigma \\
&\geq \int_\Omega \nabla(u - \overline{u}) \nabla \varphi^\varepsilon \, dx - \lambda \int_\Omega m(u - \overline{u}) \varphi^\varepsilon \, dx + \lambda \int_\Omega (u^2 - \overline{u}^2) \varphi^\varepsilon \, dx \\
&\quad - \lambda \int_{\partial\Omega} b(u^2 - \overline{u}^2) \varphi^\varepsilon \, d\sigma.
\end{aligned} \tag{2.5}$$

Now introduce

$$\Omega_\varepsilon = \{x \in \Omega : u + \varepsilon\varphi \geq \overline{u} > u\}, \quad \Gamma_\varepsilon := \{x \in \partial\Omega : u + \varepsilon\varphi \geq \overline{u} > u\},$$

and then $\varphi^\varepsilon = u + \varepsilon\varphi - \overline{u}$ for $x \in \Omega_\varepsilon$ and for $x \in \Gamma_\varepsilon$. Meanwhile, $\varphi^\varepsilon = 0$ for $x \notin \Omega_\varepsilon$ and for $x \notin \Gamma_\varepsilon$. It follows from (2.5) that $I_{1,\varepsilon} \geq J_{1,\varepsilon} + \varepsilon J_{2,\varepsilon}$, where

$$\begin{aligned} J_{1,\varepsilon} = & \int_{\Omega_\varepsilon} |\nabla(u - \overline{u})|^2 \, dx - \lambda \int_{\Omega_\varepsilon} m(u - \overline{u})^2 \, dx + \lambda \int_{\Omega_\varepsilon} (u^2 - \overline{u}^2)(u - \overline{u}) \, dx \\ & - \lambda \int_{\Gamma_\varepsilon} b(u^2 - \overline{u}^2)(u - \overline{u}) \, d\sigma \\ J_{2,\varepsilon} = & \int_{\Omega_\varepsilon} \nabla(u - \overline{u}) \nabla\varphi \, dx - \lambda \int_{\Omega_\varepsilon} m(u - \overline{u})\varphi \, dx + \lambda \int_{\Omega_\varepsilon} (u^2 - \overline{u}^2)\varphi \, dx \\ & - \lambda \int_{\Gamma_\varepsilon} b(u^2 - \overline{u}^2)\varphi \, d\sigma. \end{aligned}$$

Since $|\Omega_\varepsilon|, |\Gamma_\varepsilon| \to 0$ as $\varepsilon \downarrow 0$, $J_{2,\varepsilon}$ converges to zero as $\varepsilon \downarrow 0$ by use of Lebesgue's convergence theorem. On the other hand, either if $x \in \Omega_\varepsilon$, or if $x \in \Gamma_\varepsilon$, then $|u - \overline{u}| \leq \varepsilon \|\varphi\|_{C(\overline{\Omega})}$, so that $J_{1,\varepsilon}$ is estimated below by $-C_1\varepsilon^2$ for some constant $C_1 > 0$. The following assertion has been deduced.

$$I_{1,\varepsilon} \geq -C_1\varepsilon^2 + \varepsilon\, \mathrm{o}(1), \quad \varepsilon \downarrow 0. \tag{2.6}$$

Similarly, since $\underline{u}$ is a weak subsolution, there exists a constant $C_2 > 0$ such that

$$I_{2,\varepsilon} \leq C_2\varepsilon^2 + \varepsilon \mathrm{o}(1), \quad \varepsilon \downarrow 0. \tag{2.7}$$

Combining (2.4) with (2.6) and (2.7), it follows that

$$\begin{aligned} & \int_\Omega \nabla u \nabla\varphi \, dx - \lambda \int_\Omega m u\varphi \, dx + \lambda \int_\Omega u^2\varphi \, dx - \lambda \int_{\partial\Omega} bu^2\varphi \, d\sigma \\ & \geq -(C_1 + C_2)\varepsilon + \mathrm{o}(1), \quad \varepsilon \downarrow 0, \end{aligned}$$

which implies

$$\int_\Omega \nabla u \nabla\varphi \, dx - \lambda \int_\Omega m u\varphi \, dx + \lambda \int_\Omega u^2\varphi \, dx - \lambda \int_{\partial\Omega} bu^2\varphi \, d\sigma \geq 0.$$

By replacing φ by $\psi = -\varphi$ and carrying out the same argument, it is proved that the reverse inequality holds. The claim has been now verified.

(II) Next let us show how to argue by contradiction to obtain that the solution (λ, u) is a local minimizer of E_λ for each $\lambda > 0$. Assume to the contrary that there exist $u_k \in X$, $k = 1, 2, 3, \ldots$, satisfying

$$\|u_k - u\|_X \leq \frac{1}{k} \quad \text{and} \quad E_\lambda(u) > E_\lambda(u_k). \tag{2.8}$$

Since E_λ is weakly lower semi-continuous in X, there exists $w_k \in X$ such that $E_\lambda(w_k) = \min_{\|v-u\|_X \leq 1/k} E_\lambda(v)$. Hence we may assume that

$$E_\lambda(u) > E_\lambda(u_k) = \min_{\|v-u\|_X \leq 1/k} E_\lambda(v).$$

Then there exists a Lagrange multiplier $\Lambda_k \in \mathbb{R}$ such that for any $\varphi \in X$

$$E'_\lambda(u_k)\varphi + \Lambda_k \left(\int_\Omega \nabla(u_k - u)\nabla\varphi \, dx + \int_\Omega (u_k - u)\varphi \, dx \right) = 0. \tag{2.9}$$

We assert here that

$$\liminf_{k\to\infty} \Lambda_k \geq 0. \tag{2.10}$$

To verify (2.10), we let $\varphi = u_k - u$ and then

$$\Lambda_k \|u_k - u\|_X^2 = -\left\{ \int_\Omega \nabla u_k \nabla(u_k - u) \, dx - \lambda \int_\Omega m u_k (u_k - u) \, dx \right.$$
$$\left. + \lambda \int_\Omega u_k^2 (u_k - u) \, dx - \lambda \int_{\partial\Omega} b u_k^2 (u_k - u) \, d\sigma \right\}.$$

Since u is a weak solution of (2.1), it follows that

$$\Lambda_k \|u_k - u\|_X^2 = -\left\{ \int_\Omega |\nabla(u_k - u)|^2 \, dx - \lambda \int_\Omega m (u_k - u)^2 \, dx \right.$$
$$\left. + \lambda \int_\Omega (u_k + u)(u_k - u)^2 \, dx - \lambda \int_{\partial\Omega} b(u_k + u)(u_k - u)^2 \, d\sigma \right\}. \tag{2.11}$$

Taylor's formula (cf. [11]) gives us that

$$E_\lambda(u_k) - E_\lambda(u) = E'_\lambda(u)(u_k - u) + \frac{1}{2} E''_\lambda(u)(u_k - u)^2 + \mathrm{o}(\|u_k - u\|_X^2), \quad k \to \infty \tag{2.12}$$

By definition,

$$E''_\lambda(u)(u_k - u)^2 = \int_\Omega |\nabla(u_k - u)|^2 \, dx - \lambda \int_\Omega m (u_k - u)^2 \, dx$$
$$+ 2\lambda \int_\Omega u (u_k - u)^2 \, dx - 2\lambda \int_{\partial\Omega} u (u_k - u)^2 \, d\sigma. \tag{2.13}$$

Substituting (2.12) and (2.13) into (2.11), and having in mind u is a weak solution of (2.1), it follows that

$$\Lambda_k \|u_k - u\|_X^2 = -2(E_\lambda(u_k) - E_\lambda(u)) - \lambda \int_\Omega (u_k - u)^3 \, dx$$
$$+ \lambda \int_{\partial\Omega} b (u_k - u)^3 \, d\sigma + \mathrm{o}(\|u_k - u\|_X^2), \quad k \to \infty.$$

This implies (2.10), in view of (2.8) and (1.3).

By the standard argument, it follows from (2.9), (2.10) and (1.3) that u_k is a solution to $C^2(\overline{\Omega})$ of the problem

$$\begin{cases} -\Delta\left(u_k - \dfrac{\Lambda_k}{1+\Lambda_k}u\right) = \dfrac{\Lambda_k+\lambda m}{1+\Lambda_k}u_k - \dfrac{\lambda}{1+\Lambda}u_k^2 + \dfrac{\Lambda_k}{1+\Lambda_k}u & \text{in } \Omega, \\ \dfrac{\partial}{\partial \boldsymbol{n}}\left(u_k - \dfrac{\Lambda_k}{1+\Lambda_k}u\right) = \dfrac{\lambda}{1+\Lambda_k}bu_k^2 & \text{on } \partial\Omega. \end{cases}$$

By Sobolev's embedding theorem and elliptic regularity we have

$$\sup_{k\geq 1}\|u_k - u\|_{W_r^1(\Omega)} < \infty \quad \text{for any } r > N.$$

By compactness arguments there exists $v \in C(\overline{\Omega})$ such that some sub-sequence of $\{u_k - u\}$, still denoted by $\{u_k - u\}$, converges to v in $C(\overline{\Omega})$, and then we note

$$\begin{aligned} \|v\|_{L^2(\Omega)} &\leq \|v - (u_k - u)\|_{L^2(\Omega)} + \|u_k - u\|_{L^2(\Omega)} \\ &\leq |\Omega|^{1/2}\|v - (u_k - u)\|_{C(\overline{\Omega})} + \|u_k - u\|_{L^2(\Omega)}. \end{aligned}$$

Hence we get $v = 0$ in $L^2(\Omega)$, since $u_k \to u$ in X. By the continuity of v it follows that $v = 0$ in $\overline{\Omega}$, so that $u_k \to u$ in $C(\overline{\Omega})$. In view of (2.2) this implies that $\underline{u} < u_k < \overline{u}$ in $\overline{\Omega}$ for $k \geq 1$ large, and thus that $u_k \in M$ for such k. Since $E_\lambda(u) = \min_{v\in M} E_\lambda(v)$, it follows that $E_\lambda(u) \leq E_\lambda(u_k)$, contradictory for (2.8). The proof of Lemma 2.1 is now complete. □

2.2. Proof of Theorem 1.1

We show here how to use Lemma 2.1 to prove Theorem 1.1. The proof is due to a contradiction argument. Assume to the contrary that there exists a sequence (λ_j, u_j), $\lambda_j \downarrow 0$, of positive solutions of (2.1). Since $\int_{\partial\Omega} b\, d\sigma > |\Omega|$, we can choose a function $b_1 \in C^{1+\theta}(\partial\Omega)$ independent of (λ_j, u_j) such that $0 \leq b_1 \leq b$, $b_1 \not\equiv b$, $\int_{\partial\Omega} b_1\, d\sigma > |\Omega|$ and (λ_j, u_j) is a strict strong supersolution of (2.1) with $b = b_1$. On the other hand, Since $\int_\Omega m\, dx \geq 0$, the eigenvalue problem

$$\begin{cases} -\Delta\phi - \lambda m\phi = \mu\phi & \text{in } \Omega, \\ \dfrac{\partial\phi}{\partial \boldsymbol{n}} = 0 & \text{in } \partial\Omega \end{cases}$$

has a unique principal eigenvalue $\mu_1 < 0$ for $\lambda > 0$ (cf. [1]). Let $\phi_1 \in C^2(\overline{\Omega})$ be the corresponding eigenfunction, which can be chosen as strictly positive in $\overline{\Omega}$. It is easy to check that $(\lambda_j, \varepsilon_j\phi)$ is a strict, strong subsolution of (2.1) with $b = b_1$ and satisfies $\varepsilon_j\phi \leq u_j$ in $\overline{\Omega}$ if constant $\varepsilon_j > 0$ is so small. Lemma 2.1 provides us with the existence of a positive solution (λ_j, v_j) of (2.1) with $b = b_1$ such that $\varepsilon_j\phi < v_j < u_j$ in $\overline{\Omega}$ and v_j is a local minimizer of the corresponding energy functional. So, without loss of generality we may assume that there exists a sequence (λ_j, u_j), $\lambda_j \downarrow 0$, of positive solutions of (2.1) which are local minimizers of E_{λ_j}. Then we prove the following.

Proposition 2.2. *For some $j_0 \geq 1$ there exists a constant $\eta_{j_0} > 0$ such that*

$$E_{\lambda_{j_0}}(u_{j_0} + \eta u_{j_0}) - E_{\lambda_{j_0}}(u_{j_0}) < 0 \quad \text{for any } 0 < \eta < \eta_{j_0}. \tag{2.14}$$

Proof. Denoting the positive solution (λ_j, u_j) simply by (λ, u), it follows that

$$E''_\lambda(u)u^2 = \int_\Omega |\nabla u|^2\,dx - \lambda \int_\Omega mu^2 + 2\lambda \int_\Omega u^3\,dx - 2\lambda \int_{\partial\Omega} bu^3\,dx.$$

Since (λ, u) is a solution of (2.1), we have

$$E''_\lambda(u)u^2 = \lambda\left(\int_\Omega u^3\,dx - \int_{\partial\Omega} bu^3\,d\sigma\right) \tag{2.15}$$

$$= -\left(\int_\Omega |\nabla u|^2\,dx - \lambda \int_\Omega mu^2\,dx\right). \tag{2.16}$$

For $u \in X$ we set $u = \xi + w$ where $\xi = |\Omega|^{-1}\int_\Omega u\,dx$. The function w is an orthogonal complement to constants in X, or $\int_\Omega w\,dx = 0$. From Poincaré's inequality (see (2.21) below), $\|w\|_X$ is equivalent to $(\int_\Omega |\nabla w|^2\,dx)^{1/2}$. It is well known that $\|u\|_X$ is equivalent to $|\xi| + \|w\|_X$. We note that if $u \in C^1(\overline{\Omega})$ is positive in D, then $\xi > 0$.

Let $\delta_0 > 0$ be a constant. For the positive solution (λ, u) with $u = \xi + w$ as above, if we consider the case $\|w/\xi\|_X \leq \delta_0$, then assertion (2.15) gives us that

$$E''_\lambda(u)u^2 = \xi^3\lambda\left(\int_\Omega \left(1 + \frac{w}{\xi}\right)^3 dx - \int_{\partial\Omega} b\left(1 + \frac{w}{\xi}\right)^3 d\sigma\right). \tag{2.17}$$

Since u is a weak solution of (2.1), we have by Taylor's formula

$$E_\lambda(u + \eta u) - E_\lambda(u) = \frac{\eta^2}{2}E''_\lambda(u)u^2 + R_3,$$

where

$$R_3 = \int_0^1 \frac{(1-s)^2}{2} E'''_\lambda(u + s\eta u)(\eta u)^3\,ds.$$

By definition,

$$E'''_\lambda(u)[\varphi, \psi, \phi] = 2\lambda\left(\int_\Omega \varphi\psi\phi\,dx - \int_{\partial\Omega} b\varphi\psi\phi\,d\sigma\right),$$

so that condition (1.3) gives us that there exists a constant $C_3 > 0$ independent of (λ, u) and $0 < \eta < 1$ both, satisfying

$$E_\lambda(u + \eta u) - E_\lambda(u) \leq \frac{\eta^2}{2}E''_\lambda(u)u^2 + C_3\lambda\eta^3\|u\|_X^3. \tag{2.18}$$

From (2.17) it follows that

$$E_\lambda(u + \eta u) - E_\lambda(u) \leq \frac{\lambda\eta^2\xi^3}{2}\left\{\int_\Omega \left(1 + \frac{w}{\xi}\right)^3 dx - \int_{\partial\Omega} b\left(1 + \frac{w}{\xi}\right)^3 d\sigma + C_4\eta\right\} \tag{2.19}$$

for some constant $C_4 > 0$ depending on δ_0. The compact embeddings (1.3) enable us to have

$$\int_\Omega \left(1+\frac{w}{\xi}\right)^3 dx - \int_{\partial\Omega} b\left(1+\frac{w}{\xi}\right)^3 d\sigma < \frac{3}{4}\left(|\Omega| - \int_{\partial\Omega} b\, d\sigma\right) < 0,$$

if $\delta_0 > 0$ is chosen small enough. So, from (2.19) it follows that if $\|w/\xi\|_X \le \delta_0$, then

$$E_\lambda(u+\eta u) - E_\lambda(u) < 0 \quad \text{for any } 0 < \eta < \frac{3(\int_{\partial\Omega} b\, d\sigma - |\Omega|)}{4C_4} \tag{2.20}$$

Next we consider the case $\|w/\xi\|_X \ge \delta_0$, where δ_0 has been determined in (2.20). Let $\alpha_0 > 0$ be a constant from Poincaré's inequality, that is,

$$\alpha_0 \|w\|_X^2 \le \int_\Omega |\nabla w|^2\, dx, \quad \forall w \in \{w \in X : \int_\Omega w\, dx = 0\}. \tag{2.21}$$

Then Schwarz's inequality shows

$$\begin{aligned}&\int_\Omega |\nabla u|^2\, dx - \lambda \int_\Omega m u^2\, dx \\ &\ge \lambda \xi^2 \left\{ \left(\frac{\alpha_0}{\lambda} - \|m_+\|_{C(\overline{\Omega})}\right) \left\|\frac{w}{\xi}\right\|_X^2 - 2\|m\|_{L^2(\Omega)} \left\|\frac{w}{\xi}\right\|_X - \int_\Omega m\, dx \right\},\end{aligned} \tag{2.22}$$

where $m_+ = \max\{m, 0\}$. Let $K > 4$ be chosen so large that

$$\delta_0 > \frac{2\displaystyle\int_\Omega m\, dx}{(K-4)\|m\|_{L^2(\Omega)}}. \tag{2.23}$$

If we assume that

$$0 < \lambda \le \frac{\alpha_0}{\|m_+\|_{C(\overline{\Omega})} + (K\|m\|_{L^2(\Omega)})/(2\delta_0)}, \tag{2.24}$$

then we have

$$\frac{\alpha_0}{\lambda} - \|m_+\|_{C(\overline{\Omega})} \ge \frac{K\|m\|_{L^2(\Omega)}}{2\delta_0},$$

by which (2.22) implies

$$\begin{aligned}&\int_\Omega |\nabla u|^2\, dx - \lambda \int_\Omega m u^2\, dx \\ &\ge \lambda \xi^2 \left\{ \frac{K\|m\|_{L^2(\Omega)}}{2\delta_0} \left\|\frac{w}{\xi}\right\|_X^2 - 2\|m\|_{L^2(\Omega)} \left\|\frac{w}{\xi}\right\|_X - \int_\Omega m\, dx \right\}.\end{aligned}$$

Hence it follows from (2.23) that

$$\int_\Omega |\nabla u|^2\, dx - \lambda \int_\Omega m u^2\, dx \ge C_5 \lambda \xi^2, \tag{2.25}$$

where C_5 is a constant given as

$$C_5 = \left(\frac{K-4}{2}\right) \|m\|_{L^2(\Omega)}\delta_0 - \int_\Omega m\,dx > 0.$$

From (2.16) and (2.25) we see that $E''_\lambda(u)u^2 \leq -C_5\lambda\xi^2$. From (2.18) it follows that

$$E_\lambda(u+\eta u) - E_\lambda(u) \leq \lambda\eta^2\left(-\frac{C_5\xi^2}{2} + C_3\eta\|u\|_X^3\right) < 0 \quad \text{if } 0 < \eta < \frac{C_5\xi^2}{2C_3\|u\|_X^3}. \tag{2.26}$$

In view of (2.20), (2.24) and (2.26), such λ_j as in (2.24) verifies (2.14). The proposition has been proved. □

Now assertion (2.14) is contradictory for u_{j_0} being a local minimizer of $E_{\lambda_{j_0}}$. The proof of Theorem 1.1 is now complete. □

Acknowledgements

This research was partly supported by the Grant-in-Aid for Scientific Research (C)(No. 16540165), Japan Society for the Promotion of Science.

References

[1] G.A. Afrouzi and K.J. Brown, On principal eigenvalues for boundary value problems with indefinite weight and Robin boundary conditions, Proc. Amer. Math. Soc. **127** (1999), 125–130.

[2] H. Amann, Nonlinear elliptic equations with nonlinear boundary conditions, In : New Developments in differential equations (W. Eckhaus ed.), Math. Studies Vol. 21, North–Holland, Amsterdam, 1976, pp. 43–63.

[3] R.S. Cantrell and C. Cosner, Diffusive logistic equations with indefinite weights: population models in disrupted environments, Proc. Roy. Soc. Edinburgh **112A** (1989), 293–318.

[4] R.S. Cantrell and C. Cosner, Diffusive logistic equations with indefinite weights: population models in disrupted environments II, SIAM J. Math. Anal. **22** (1991), 1043–1064.

[5] R.S. Cantrell and C. Cosner, The effects of spatial heterogeneity in population dynamics, J. Math. Biology **29** (1991), 315–338.

[6] R.S. Cantrell, C. Cosner and V. Hutson, Ecological models, permanence and spatial heterogeneity, Rocky Mountain J. Math. **26** (1996), 1–35.

[7] M. Chipot, M. Fila and P. Quittner, Stationary solutions, blow up and convergence to stationary solutions for semilinear parabolic equations with nonlinear boundary conditions, Acta. Math. Univ. Comenianae **60** (1991), 35–103.

[8] P. Hess, Periodic-parabolic boundary value problems and positivity, Pitman Research Notes in Math. Series Vol. 247, Longman Scientific & Technical, Harlow, Essex, 1991.

[9] M.H. Protter and H.F. Weinberger, Maximum principles in differential equations, Prentice–Hall, Englewood Cliffs New Jersey, 1967.

[10] K. Umezu, Multiplicity of positive solutions under nonlinear boundary conditions for diffusive logistic equations, Proc. Edinburgh Math. Soc. **47** (2004), 495-512.
[11] E. Zeidler, Nonlinear functional analysis and its applications I: Fixed-point theorems, Springer-Verlag, New York Berlin Heidelberg Tokyo, 1993.

Kenichiro Umezu
Faculty of Engineering
Maebashi Institute of Technology
Maebashi 371-0816, Japan

Progress in Nonlinear Differential Equations
and Their Applications, Vol. 64, 509–516

Extremal Equilibria and Asymptotic Behavior of Parabolic Nonlinear Reaction-diffusion Equations

Aníbal Rodríguez-Bernal and Alejandro Vidal-López

"Nonlinear Elliptic and Parabolic Problems: A Special Tribute to the Work of Herbert Amann", Zürich, June 2004

1. Introduction

We give some conditions on the nonlinear term that ensure some "nice" properties for the asymptotic behavior of parabolic nonlinear reaction diffusion equations. Namely, the existence of two ordered extremal equilibria such that the order interval they define attracts the dynamics of the problem uniformly in space for bounded sets of initial data. The proof relies in the use of a parabolic version of the sub-supersolution method as well as the comparison principle that parabolic equations of this type satisfy.

As a consequence of our results we can recover, among others, some results of a very well-known paper of H. Amann "Fixed points equations in ordered Banach spaces and nonlinear eigenvalue problem" where he studies elliptic equations by mean of topological techniques such as the topological degree or Leray-Schauder index. It is also in this paper that H. Amann introduces the sub-supersolution method for elliptic equations.

2. Our main result

We consider the following problem

$$\left\{\begin{array}{rcll} u_t - \Delta u &=& f(x,u) & \text{in} \quad \Omega \\ u(0) &=& u_0 & \\ Bu &=& 0 & \text{on} \quad \partial\Omega \end{array}\right. \tag{1}$$

where $\Omega \subset \mathbf{R}^N$ is a bounded domain, $u_0 \in X$ (e.g., $X = C(\overline{\Omega})$), B is a boundary operator (Dirichlet, Robin or Neumann) and $f : \Omega \times \mathbf{R} \to \mathbf{R}$ is a continuous function, locally Lipschitz in u and satisfies the structure condition

$$s\ f(x,s) \le C(x)s^2 + D(x)|s| \quad \forall s \in \mathbf{R} \tag{2}$$

with $C \in L^p(\Omega)$, $p > N/2$, and $0 \le D \in L^q(\Omega)$, $q > N/2$.

Remark 2.1. *The above condition on f means that is bounded above by a linear function for s positive and bounded below by another for s negative.*

Remark 2.2. *For the sake of simplicity we deal with the Laplacian operator but due to the methods we will use the results remains true for more general operators.*

Under the above conditions, problem (1) has a unique solution for all time $t \ge 0$ that we denote by $u_f(t,x;u_0)$ (see, e.g., Henry [5]). Therefore, the solutions of the problem (1) define a dynamical system, $S(t)$, that maps every initial data u_0 to the solution at time t of problem (1) starting at u_0.

As a consequence of the parabolic maximum principle, the following comparison principle holds for (1) (see Lunardi [7]).

Theorem 2.3. *Let $f, g(x,u) : \Omega \times \mathbf{R} \to \mathbf{R}$ be continuous functions, locally Lipschitz in u. Suppose that for all $x \in \Omega$ and $u \in \mathbf{R}$ we have*

$$f(x,u) \le g(x,u).$$

Then,

$$u_f(t,x;u_0) \le u_g(t,x;u_0)$$

while both solutions exist, where we have denoted by $u_f(t,x;u_0)$ the solution at time t of problem (1) starting at u_0.

Moreover, if $u_0 \le v_0$ are two ordered initial conditions then

$$u_f(t,x;u_0) \le u_f(t,x;v_0)$$

while both solutions exist.

Finally, the solutions of (1) are regular because of the smoothing property of the equation (see, e.g., Lunardi [7]).

We state now our main result.

Theorem 2.4. *Assume f satisfies (2). Suppose that $S_{\Delta+C}(t)$ has exponential decay, i.e., solutions of problem*

$$\begin{cases} w_t - \Delta w &= C(x)w \\ Bw &= 0. \end{cases}$$

decay exponentially to zero as t goes to ∞.

Then, there exist maximal and minimal equilibrium points of (1), $\varphi_m \le \varphi_M$, such that

$$\varphi_m \le \liminf_{t\to\infty} u(t,x;u_0) \le \limsup_{t\to\infty} u(t,x;u_0) \le \varphi_M \tag{3}$$

uniformly for $x \in \Omega$ and for u_0 in bounded sets of X.

In particular, the attractor for the problem (1) exists and is contained between the extremal equilibria and, of course, these equilibria belong to the attractor.

Moreover, the maximal equilibrium is stable from above and the minimal one from below.

Remark 2.5. *Notice that the extremal equilibria do not have to be sign-constant.*

Remark 2.6. *The property* (3) *implies, in particular, that any equilibrium* ψ *satisfies*

$$\varphi_m \leq \psi \leq \varphi_M.$$

For the proof of Theorem 2.4 we need the following lemma that give us an estimate concerning the asymptotic behavior of the solutions of the nonlinear problem in terms of the linear one coming from the structure condition (2).

Lemma 2.7. *Let* ϕ *be the unique solution of*

$$\begin{cases} -\Delta\phi & = & C(x)\phi + D(x) & in & \Omega \\ B\phi & = & 0 & on & \partial\Omega. \end{cases}$$

Then

$$\limsup_{t\to\infty} |u(t,x;u_0)| \leq \phi(x)$$

uniformly for $x \in \Omega$ *and for* u_0 *in bounded sets of* X.

Remark 2.8. *The key in the proof of this lemma is the exponential decay of the semigroup associated to* $\Delta + C$.

Proof of Theorem 2.4 *(sketch).* First, we set $\eta = \phi + \delta$ with δ a positive constant. Now, from Lemma 2.7, we obtain that $S(T)\eta \leq \eta$ at a certain time T large enough. Making the semigroup act on both sides of this inequality, we have that the sequence $S(nT)\eta$, is decreasing by the monotonicity property of the solutions. Moreover, it is bounded from below by $-\eta$. So it converges to a function φ_M pointwise. It turns out that by the Ascoli-Arzelà theorem, this convergence is uniform. In fact, the solution starting at η converges to φ_M as t goes to ∞.

Furthermore, by the continuity of $S(t)$, φ_M is an equilibrium point. Finally, we prove that the solutions of (1) satisfies (3) uniformly for $x \in \Omega$ and for u_0 in bounded sets of initial data. In particular, this implies that φ_M is maximal.

The rest of the statements of the theorem follow in a straightforward way from the above arguments. □

3. Some consequences for positive solutions

We state now some consequences of Theorem 2.4 dealing with nonnegative and positive equilibria of problem (1). The first result gives a condition for the existence of a minimal nonnegative equilibrium which is easy to check.

Corollary 3.1. *Suppose that* $f(\cdot,0) \geq 0$, $f(\cdot,0) \not\equiv 0$, *and there exist* $C \in L^p(\Omega)$, $p > N/2$ *and* $0 \leq D \in L^q(\Omega)$ *such that*

$$f(x,s) \leq C(x)s + D(x) \tag{4}$$

for almost $x \in \Omega$ and for all $s \geq 0$. Then (1) *has two ordered extremal nonnegative equilibria provided that the semigroup associated to $\Delta + C$, $S_{\Delta+C}$, has exponential decay.*

Proof. The existence of the maximal equilibrium φ follows from Theorem 2.4 (notice that f satisfies (2) and $S_{\Delta+C}$ has exponential decay). Moreover it is nonnegative since $f(\cdot, 0) \geq 0$.

Now $f(\cdot, 0) \geq 0$, $f(\cdot, 0) \not\equiv 0$, implies that 0 is a strict subsolution. So, $S(t)0$ is increasing. Furthermore, $S(t)0$ is bounded by the maximal solution φ_M whose existence follows from Theorem 2.4 . Thus, $S(t)0$ converges to a certain function φ_m^+ that is positive (by the strong maximum principle). By continuity of $S(t)$, φ_m^+ is an equilibrium.

Finally, notice that from the comparison principle, if ψ is a nonnegative equilibrium then $\psi = S(t)\psi \geq S(t)0 \geq 0$. Thus, taking limits as t goes to ∞ we have that φ_m^+ is the minimal nonnegative equilibrium for (1). □

Remark 3.2. *If* 0 *is solution, that is, $f(\cdot, 0) \equiv 0$, then it is the minimal nonnegative equilibrium.*

Another interesting question is when the problem (1) has positive equilibria even with $f(x, 0) = 0$. The next result goes in that direction when the former assumption holds (i.e., problem (1) has zero as an equilibrium solution). Essentially, what is needed for the existence of positive equilibria is the instability of 0 and the global boundedness of solutions starting near of 0. This holds, in particular, if all the solutions are globally bounded.

Corollary 3.3. *Let $f(\cdot, 0) \equiv 0$. Suppose that there exist $C \in L^p(\Omega)$, $p > N/2$ and $0 \leq D \in L^q(\Omega)$ such that*

$$f(x, s) \leq C(x)s + D(x) \tag{5}$$

for almost $x \in \Omega$ and for all $s \geq 0$.

Then (1) *has at least one positive solution provided that* 0 *is unstable and the semigroup generated by $\Delta + C$ has exponential decay.*

Proof. From Theorem 2.4 there exists a maximal equilibrium φ_M which is stable from above. We recall that we obtain the existence of the maximum equilibrium as limits of the solution of (1) starting at $\phi \geq 0$. Since $f(\cdot, 0) \equiv 0$ we have that φ_M is nonnegative, by the comparison principle. Now, since 0 is unstable it follows that φ_M is positive. Otherwise, $\varphi_M \equiv 0$ (by the maximum principle φ_M is either zero or positive). Thus, 0 would be stable that is a contradiction. □

Remark 3.4. *Notice that the positive equilibrium which existence we have obtained is the maximal equilibrium of* (1).

Remark 3.5. *In the last case, there exists an attractor of nonnegative solutions for problem* (1) *which is contained in the order interval $[0, \varphi_M]$. We recall that given two ordered elements of X, $u \leq v$, the order interval is defined by $[u, v] = \{x \in X : u \leq x \leq v\}$.*

4. Amann's result revisited

In [1], H. Amann studies the existence of positive solutions of elliptic problems of the form

$$\begin{cases} Lu = f(x,u) & \text{in} \quad \Omega \\ Bu = 0 & \text{on} \quad \partial\Omega \end{cases} \tag{6}$$

where $\Omega \subset \mathbf{R}^N$ is a bounded domain of $\mathbf{R}^N$; L is a uniformly elliptic operator of the form

$$Lu = -\sum_{i,j=1}^{N} a_{ij} D_i D_k u + \sum_{i=1}^{N} a_i D_i u + au$$

with $a_{ij}, a_i, a \in C^{\mu}(\overline{\Omega})$, (a_{ij}) is symmetric and a is nonnegative; B is a boundary operator of the form

$$Bu = b(x)u + \delta\frac{\partial u}{\partial \nu}, \qquad \delta = 0, 1$$

with $0 \le b \in C^{1+\mu}(\partial\Omega)$; and $f \in C^{\mu}(\overline{\Omega} \times \mathbf{R}_+)$ such that there exists $\omega \ge 0$ for which

$$f(x,\xi) - f(x,\eta) > -\omega(\xi - \eta) \tag{7}$$

for all $x \in \overline{\Omega}$, $\xi > \eta \ge 0$.

Remark 4.1. *We consider more general functions f. Notice that we do not require (7) holds for f nor f to be Hölder continuous. We just need f continuous and locally Lipschitz in u.*

Theorem 4.2 (Theorem 9.6 in Amann [1], p. 649). *Assume that $f(\cdot, 0) \ge 0$ and there exist $\hat{f} \in C(\overline{\Omega})$ and $0 < m \in C^{\mu}(\overline{\Omega})$ such that*

$$f(x,\xi) \le \hat{f}(x) + \lambda m(x)\xi \quad (x,\xi) \in \overline{\Omega} \times \mathbf{R}_+ \tag{8}$$

with $\lambda > 0$.

Then (6) has a minimal nonnegative solution provided that $\lambda < \lambda_0(m)$, where $\lambda_0(m)$ is the principal eigenvalue associated to m.

This result can be obtained as a consequence of Corollary 3.1. Namely, notice that (8) implies that (2) holds with $C(x) = \lambda m(x)$ and $D(x) = \hat{f}(x)$ and $\lambda < \lambda_0(m)$ implies the exponential decay of the semigroup generated by $\Delta + \lambda m(x)$. The result follows now from Theorem 3.1.

Remark 4.3. *Note that in our proof we do not need to assume the positivity of m or λ, nor such regularity in the coefficients. In fact, what we need is that $C(x) = \lambda m(x)$ belongs to some $L^q(\Omega)$ space with $q > N/2$ without restriction over its sign.*

Theorem 4.4 (Theorem 9.8 in Amann [1], p. 650). *Assume that $f(\cdot, 0) \equiv 0$. Suppose that $\partial_u f$ exists, is continuous in a neighborhood of zero, and satisfies $0 < \partial_u f(\cdot, 0) \in C^{\mu}(\overline{\Omega})$.*

If $\psi > 0$ is a supersolution for (6), then there exists a maximal positive solution in the order interval $[0, \psi]$, provided that

$$\lambda_0(\partial_u f(\cdot, 0)) < 1. \tag{9}$$

This result is easy to obtain following the argument in the proof of Corollary 3.3 since condition (9) implies the instability of the zero solution. Namely, since ψ is a positive supersolution, the solution of (1) starting at ψ is nonincreasing. Furthermore, it is bounded from below. Thus, it converges to some equilibrium $\psi_M \geq 0$. Now, the proof follows from the proof of Corollary 3.3 with ψ_M instead of φ_M.

Notice that to obtain the result we do not need as much regularity for the coefficients nor the positivity of $\partial_u f(\cdot, 0)$. In fact, it is enough for f to be continuous and locally Lipschitz in u, and such that there exists $a \in L^q(\Omega)$ for some $q > N/2$ with $a(x) \leq \liminf_{s \to 0^+} f(x, u)/u$ and 0 is unstable for the semigroup generated by $\Delta + a$ (i.e., the first eigenvalue of $\Delta + a$ is negative).

Theorem 4.5 (Theorem 9.9 in Amann [1], p. 650). *Assume that $f(\cdot, 0) \equiv 0$. Suppose that $\partial_u f$ exists, is continuous in a neighborhood of zero, and satisfies $0 < \partial_u f(\cdot, 0) \in C^\mu(\overline{\Omega})$. Moreover, suppose that there exist $\hat{f} \in C(\overline{\Omega})$ and $0 < m \in C^\mu(\overline{\Omega})$ such that*

$$f(x, \xi) \leq \hat{f}(x) + \lambda m(x)\xi \quad (x, \xi) \in \overline{\Omega} \times \mathbf{R}_+$$

with $\lambda \geq 0$.

Then (6) has at least one positive solution provided that

$$\lambda_0(\partial_u f(\cdot, 0)) < 1 \quad \text{and} \quad \lambda < \lambda_0(m).$$

Note again that with our techniques we can prove the result above with no sign and weaker regularity assumptions on $\partial_u f$, $m(x)$ and λ, the following result

Remark 4.6. *Some other related results that can be obtained as a consequence of our results can be found in Berestycki and P.L. Lions [2], P.L. Lions [6] or de Figueiredo [4], for instance.*

5. An example: the logistic equation

We consider now the logistic equation

$$\left\{ \begin{array}{rcll} u_t - \Delta u & = & a(x)u - b(x)u^3 & \text{in} \quad \Omega \\ u(0) & = & u_0 & \\ u & = & 0 & \text{on} \quad \partial\Omega. \end{array} \right. \tag{10}$$

in X^+. Suppose that $a(x) = a_0(x) + a_1(x)$ and $S_{\Delta + a_0}$ has exponential decay. Using Young's Inequality, for $u \geq 0$, we have

$$f(x, u) = a(x)u - b(x)u^3 \leq a_0(x)u + M\left(\frac{a_1^3(x)}{b(x)}\right)^{1/2}$$

with a similar computation for $u \leq 0$. Thus, setting $C(x) = a_0(x)$, property (2) is satisfied provided that

$$\frac{a_1^3}{b} \in L^r(\Omega) \tag{11}$$

for some $r > N/4$.

From the results on Section 3 it is possible to obtain the following result concerning the existence of a positive solution of problem (10) and its stability.

Theorem 5.1. *Suppose that there exists a nontrivial decomposition of $a(x) = a_0(x) + a_1(x)$ (i.e., a_0 and a_1 are not identically zero) such that $S_{\Delta + a_0}$ has exponential decay and* (11) *holds. Moreover, suppose that* 0 *is unstable for the problem*

$$\begin{cases} u_t - \Delta u &= a(x)u \quad \text{in} \quad \Omega \\ u(0) &= u_0 \\ u &= 0 \qquad\quad \text{on} \quad \partial\Omega. \end{cases} \tag{12}$$

Then, there exist a unique positive solution for the problem (10). *Moreover, this solution is globally asymptotically stable for positive solutions.*

Remark 5.2. *Notice that if* 0 *is stable for the problem* (12) *then it is the unique solution of the problem* (10) *and it is asymptotically globally stable.*

Proof. The existence is an straightforward application of Corollary 3.3. Actually, what we prove is the existence of the maximal positive equilibrium φ (see Remark 3.4). For the uniqueness, let $\psi \leq \varphi$ another equilibrium. Then,

$$-\Delta\varphi = f(x, \varphi) \qquad \text{and} \qquad -\Delta\psi = f(x, \psi)$$

where $f(x, u) = a(x)u - b(x)u^3$. Multiplying the first equation by ψ, the second one by φ, subtracting and integrating in Ω, we have

$$0 = \int_\Omega \frac{f(x,\varphi)}{\varphi}\varphi\psi - \int_\Omega \frac{f(x,\psi)}{\psi}\varphi\psi = \int_\Omega \left(\frac{f(x,\varphi)}{\varphi} - \frac{f(x,\psi)}{\psi}\right)\varphi\psi$$

Now, notice that f is such that $f(x, s)/s$ is strictly decreasing. Thus,

$$\frac{f(x,\varphi)}{\varphi} - \frac{f(x,\psi)}{\psi} \neq 0.$$

Therefore, must be $\psi \equiv 0$. The stability of the positive solution is clear. □

Remark 5.3. *For the uniqueness argument in the proof above we follow the argument in Stackgold and Payne* [8] *(see also Ex. 7, p. 127, in Henry* [5]*).*

References

[1] H. Amann. Fixed point equations and nonlinear eigenvalue problems in ordered Banach spaces. *SIAM Review*, 18(4):620–709, 1976.

[2] H. Berestycki and P.L. Lions. Some applications of the method of super and subsolutions. In *Bifurcation and nonlinear eigenvalue problems (Proc., Session, Univ. Paris XIII, Villetaneuse, 1978)*, volume 782 of *Lecture Notes in Math.*, pages 16–41. Springer, Berlin, 1980.

[3] H. Brézis and Luc Oswald. Remarks on sublinear elliptic equations. *Nonlinear Anal., Theory Methods Appl.*, 10:55–64, 1986.

[4] D.G. de Figueiredo. Positive solutions of semilinear elliptic problems. In *Differential equations (São Paulo, 1981)*, volume 957 of *Lecture Notes in Math.*, pages 34–87. Springer, Berlin, 1982.

[5] D. Henry. *Geometric Theory of Semilinear Parabolic Equations.* Number 840 in Lecture Notes in Mathematics. Springer-Verlag, 1981.

[6] P.L. Lions. On the existence of positive solutions of semilinear elliptic equations. *SIAM Review*, 24(4):441–467, 1982.

[7] A. Lunardi. *Analytic semigroups and optimal regularity in parabolic problems.* Progress in Nonlinear Differential Equations and their Applications, 16. Birkhäuser Verlag, Basel, 1995.

[8] I. Stakgold and L.E. Payne. Nonlinear problems in nuclear reactor analysis. In *Nonlinear problems in the physical sciences and biology* (*Proc. Battelle Summer Inst., Seattle, Wash.*, 1972), volume 322 of *Lecture Notes in Math.*, pages 298–307. Springer, Berlin, 1979.

Aníbal Rodríguez-Bernal and Alejandro Vidal-López
Departamento de Matemática Aplicada
Universidad Complutense de Madrid
E-28040 Madrid, Spain

Progress in Nonlinear Differential Equations
and Their Applications, Vol. 64, 517–528

A Remark on Continuous Coagulation-Fragmentation Equations with Unbounded Diffusion Coefficients

Christoph Walker

Abstract. Continuous coagulation-fragmentation processes with diffusion are studied. It is shown that the parameter dependent diffusion term $d(y)\Delta$ generates an analytic semigroup in suitable state spaces even for unbounded diffusion coefficients $d(y)$. This yields existence and uniqueness of local-in-time smooth solutions that are global for small initial values in the absence of fragmentation.

1. Introduction

In the present paper we demonstrate how to extend a recent result of H. Amann and the author on diffusive continuous coagulation-fragmentation equations [4] to the case of unbounded diffusion coefficients. Recall that these equations describe the time evolution of a system of a large number of particles that may change size due to coalescence or breakage. The mechanism leading to aggregation is assumed to be governed merely by Brownian motion. Applications of these processes can be found in various scientific and industrial disciplines, such as biology, physics, chemistry, or oil industry (e.g., see [9] and the references therein).

More precisely, denoting by y the particle size and by $u = u(y) = u(t, y, x)$ the particle size distribution function at time t and position x, the continuous version of the diffusive coagulation-fragmentation equations reads as

$$\begin{aligned} \partial_t u(y) - d(y)\Delta_x u(y) &= L[u](y) && \text{in } \Omega\,, && t>0\,, && y\in(0,y_0)\,,\\ \partial_\nu u(y) &= 0 && \text{on } \partial\Omega\,, && t>0\,, && y\in(0,y_0)\,,\\ u(0,y,\cdot) &= u^0(y) && \text{in } \Omega\,, && y\in(0,y_0)\,. \end{aligned} \tag{1}$$

Here $\Omega \subset \mathbb{R}^n$ is a bounded domain with smooth boundary $\partial\Omega$ and $u^0 = u^0(y, x)$ is a given initial distribution. The right-hand side

$$L[u] := L_{\mathsf{b}}[u] + L_{\mathsf{c}}[u, u] + L_{\mathsf{s}}[u, u]$$

consists of the integral operators

$$L_{\mathsf{b}}[u](y) := \int_y^{y_0} \gamma(y',y)\, u(y')\, \mathrm{d}y' - u(y) \int_0^y \frac{y'}{y}\, \gamma(y,y')\, \mathrm{d}y' ,$$

$$\begin{aligned} L_{\mathsf{c}}[u,v](y) := \frac{1}{2} &\int_0^y K(y',y-y')\, P(y',y-y')\, u(y-y')\, v(y')\, \mathrm{d}y' \\ &+ \frac{1}{2} \int_y^{y_0} \int_0^{y'} K(y'',y'-y'')\, Q(y'',y'-y'') \\ &\qquad \times \beta_{\mathsf{c}}(y',y)\, u(y'')\, v(y'-y'')\, \mathrm{d}y''\mathrm{d}y' \\ &- u(y) \int_0^{y_0-y} K(y,y')\, \{P(y,y')+Q(y,y')\}\, v(y')\, \mathrm{d}y' , \\ L_{\mathsf{s}}[u,v](y) := \frac{1}{2} &\int_{y_0}^{2y_0} \int_{y'-y_0}^{y_0} K(y'',y'-y'')\, \beta_{\mathsf{s}}(y',y)\, u(y'')\, v(y'-y'')\, \mathrm{d}y''\mathrm{d}y' \\ &- u(y) \int_{y_0-y}^{y_0} K(y,y')\, v(y')\, \mathrm{d}y' , \end{aligned}$$

for $y \in (0, y_0)$, where L_{b}, L_{c}, and L_{s} account for the formation and depletion of particles due to spontaneous breakage, coalescence and collisional breakage, and due to scattering, respectively. We refer to [4], [13] for a precise definition of the kernels and a more detailed interpretation of the above terms. By $Y := (0, y_0)$ we denote the admissible range for particle sizes, which is either unbounded if particles are allowed to become arbitrarily large, that is, if $y_0 = \infty$, or bounded if $y_0 \in (0,\infty)$. In the former case of the classical coagulation-fragmentation model, the scattering operator L_{s} is identically zero. As in [4] we treat both cases simultaneously.

We refrain from recalling the present state of research on continuous coagulation-fragmentation processes with diffusion, but refer instead to [2], [4], [5], [8], [13] and the references therein. It has been shown in [4] that the above system of equations possesses a unique smooth solution locally in time, which preserves the total mass. Moreover, this solution exists globally for small initial values provided that the linear fragmentation terms are neglected. These results rely on the fact that the analytic semigroup generated by the size-dependent diffusion term $d(y)\Delta_x$ on the state spaces $L_1(Y, L_p(\Omega))$, $1 \le p < \infty$, has smoothing properties. The basic assumption in [4] for the generation result and the regularizing effects is that the diffusion coefficients satisfy a bound of the form

$$0 < d_\star \le d(y) \le d^\star < \infty , \quad y \in Y .$$

The main motivation for the present paper is to relax this restriction in order to include unbounded coefficients. We will show that neither the upper nor the lower bound for the diffusion coefficients is necessary to obtain the generation result. Nevertheless, we point out that the lower bound $d_\star > 0$ is crucial in our analysis to guarantee existence of smooth solutions. This bound provides a suitable control for the time singularities arising from the regularizing effects.

2. The diffusion semigroup

We briefly recall the most important notations and abbreviations already used in [4] and refer to [4] for more details. The abbreviation $L_p := L_p(\Omega)$ stands for the Lebesgue spaces and $H^\mu_{p,\mathcal{B}} := H^\mu_{p,\mathcal{B}}(\Omega)$ for the Bessel potential spaces including Neumann boundary conditions (if meaningful). For $p \in [1,\infty)$ we denote by Δ_p the (well-defined) closure in L_p of the linear operator $\Delta\mid_{C^2(\bar{\Omega})}$ subject to Neumann boundary conditions, which generates a positive analytic semigroup $\{e^{t\Delta_p}\,;\, t \geq 0\}$ of contractions on L_p. The domain of Δ_p equals $H^2_{p,\mathcal{B}}$ provided $p > 1$. Since the restriction of $e^{t\Delta_1}$ to L_p coincides with $e^{t\Delta_p}$, we may put $\Delta := \Delta_1$ and obtain the estimates

$$\|e^{t\Delta}\|_{\mathcal{L}(L_p, H^\alpha_{q,\mathcal{B}})} \leq c\,(1 \wedge t)^{-\frac{n}{2}(\frac{1}{p}-\frac{1}{q})-\frac{\alpha}{2}}\ , \quad t > 0\ , \tag{2}$$

for $\alpha \in [0,2] \setminus \{1+1/q\}$ and $1 \leq p \leq q \leq \infty$, where $q \in (1,\infty)$ if $\alpha > 0$, and also

$$\|e^{t\Delta}\|_{\mathcal{L}(H^\alpha_{p,\mathcal{B}}, H^\mu_{p,\mathcal{B}})} \leq c\,(1 \wedge t)^{-\frac{\mu-\alpha}{2}}\ , \quad t > 0\ , \tag{3}$$

for $1 < p < \infty$ and $0 \leq \alpha \leq \mu \leq 2$ with $\alpha, \mu \neq 1 + 1/p$.

We then set

$$\mathbb{L}_p := L_1\big(Y, L_p, (1+y)\mathrm{d}y\big)\ , \quad 1 \leq p \leq \infty\ ,$$

and

$$\mathbb{H}^\alpha_{p,\mathcal{B}} := L_1\big(Y, H^\alpha_{p,\mathcal{B}}, (1+y)\mathrm{d}y\big)\ , \quad 1 < p < \infty\ , \quad \alpha \in [0,2] \setminus \{1+1/p\}\ ,$$

with the convention $\mathbb{H}^0_{p,\mathcal{B}} := \mathbb{L}_p$ for $p \in \{1,\infty\}$. By $\mathbb{L}^+_p$ we denote the positive cone of $\mathbb{L}_p$.

Throughout this paper we assume for the diffusion coefficients that

$$d \in L_{1,\mathrm{loc}}(Y, (0,\infty)) \tag{4}$$

and that there exists $d_\star$ with

$$d(y) \geq d_\star > 0\ , \quad y \in Y\ . \tag{5}$$

If d satisfies (4), we define for each $p \in [1,\infty)$ an operator $\mathbb{A}_p$ by virtue of

$$(\mathbb{A}_p u)(y) := -d(y)\Delta_p u(y)\ , \quad \text{a.e. } y \in Y\ ,$$

for

$$u \in \mathsf{D}(\mathbb{A}_p) := \{u \in \mathbb{L}_p\,;\, u(y) \in \mathsf{D}(\Delta_p) \text{ for a.e. } y \in Y\,,\, d(\cdot)\Delta_p u \in \mathbb{L}_p\}\ .$$

Observe then that, provided d additionally satisfies (5), the continuous injections

$$L_1\big(Y, H^2_{p,\mathcal{B}}, d(y)(1+y)\mathrm{d}y\big) \hookrightarrow \mathsf{D}(\mathbb{A}_p) \hookrightarrow \mathbb{H}^2_{p,\mathcal{B}}\ , \quad p \in (1,\infty)\ , \tag{6}$$

hold. Finally, let the projection $\mathbb{P} \in \mathcal{L}(\mathbb{L}_p)$ be defined by

$$\mathbb{P}u := \frac{1}{|\Omega|}\int_\Omega u(\cdot, x)\,\mathrm{d}x\ , \quad u \in \mathbb{L}_p\ , \quad 1 \leq p < \infty\ ,$$

so that the space $\mathbb{L}_p$ has the direct sum decomposition

$$\mathbb{L}_p = \mathbb{P}(\mathbb{L}_p) \oplus (1-\mathbb{P})(\mathbb{L}_p)\ . \tag{7}$$

Note that $\mathbb{P}(\mathbb{L}_p) = L_1(Y,(1+y)\mathrm{d}y) \subset \mathbb{L}_p$. We put $\mathbb{L}_p^\bullet := (1-\mathbb{P})(\mathbb{L}_p)$ to shorten notation.

Now we can prove analogous statements to [4, Thm. 2, Prop. 3] for diffusion coefficients obeying (4) and (5).

Theorem 1. Suppose d satisfies (4) and (5). Then $-\mathbb{A}_p$ generates a positive strongly continuous analytic semigroup of contractions on $\mathbb{L}_p$ for each $p \in [1,\infty)$. It is given by

$$(e^{-t\mathbb{A}_p}u)(y) = e^{td(y)\Delta_p}u(y)\ , \quad \text{a.e. } y \in Y\ , \quad t \geq 0\ , \quad u \in \mathbb{L}_p\ , \tag{8}$$

and it holds that $\mathbb{A}_p \supset \mathbb{A}_q$ for $1 \leq p < q < \infty$. Furthermore, the estimates

$$\|e^{-t\mathbb{A}_p}\|_{\mathcal{L}(\mathbb{L}_p,\mathbb{H}^\alpha_{q,\mathcal{B}})} \leq c(T)\, t^{-\frac{n}{2}(\frac{1}{p}-\frac{1}{q})-\frac{\alpha}{2}}\ , \quad 0 < t \leq T\ , \tag{9}$$

for $\alpha \in [0,2]\setminus\{1+1/q\}$ and $1 \leq p \leq q \leq \infty$, where $q \in (1,\infty)$ if $\alpha > 0$, and

$$\|e^{-t\mathbb{A}_p}\|_{\mathcal{L}(\mathbb{H}^\alpha_{p,\mathcal{B}},\mathbb{H}^\mu_{p,\mathcal{B}})} \leq c(T)\, t^{-\frac{\mu-\alpha}{2}}\ , \quad 0 < t \leq T\ , \tag{10}$$

for $1 < p < \infty$ and $0 \leq \alpha \leq \mu \leq 2$ with $\alpha, \mu \neq 1 + 1/p$ are valid. Moreover, for $1 \leq p < \infty$, (7) decomposes $e^{-t\mathbb{A}_p}$ into

$$e^{-t\mathbb{A}_p} = 1 \oplus \left(e^{-t\mathbb{A}_p}\,|_{\mathbb{L}_p^\bullet}\right)\ , \quad t \geq 0\ ,$$

and there exists $\omega_0 > 0$ such that, for $1 < p < q \leq \infty$ and some $M := M(p,q) > 0$,

$$\|e^{-t\mathbb{A}_p}\,|_{\mathbb{L}_p^\bullet}\|_{\mathcal{L}(\mathbb{L}_p^\bullet,\mathbb{L}_q^\bullet)} \leq M\, e^{-\omega_0 t}\, t^{-\frac{n}{2}(\frac{1}{p}-\frac{1}{q})}\ , \quad t > 0\ . \tag{11}$$

Proof. Fix $1 \leq p < \infty$. Analogously to [4, Lem. 9] one shows that the tensor product $\mathcal{D}(Y)\otimes\mathcal{D}(\Omega)$ is dense in $\mathbb{L}_p$, where $\mathcal{D}(X)$ denotes the test functions on an open subset $X \subset \mathbb{R}^m$. Due to (4), $\mathcal{D}(Y)\otimes\mathcal{D}(\Omega)$ is contained in the domain of $\mathbb{A}_p$. Therefore, $\mathbb{A}_p$ is densely defined. That it is a closed operator in $\mathbb{L}_p$ follows from the fact that $d(y)\Delta_p$ is closed in L_p for a.e. $y \in Y$. Next we write $L_p = \mathbb{R}\cdot 1 \oplus L_p^\bullet$ with $L_p^\bullet := (1-\mathbb{P})(L_p)$ as in the proof of [4, Prop. 3], which then decomposes Δ_p into $\Delta_p = 0 \oplus \Delta_p^\bullet$. Hereby, $\Delta_p^\bullet := \Delta_p\,|_{\mathsf{D}(\Delta_p)\cap L_p^\bullet}$ generates an analytic semigroup

$$\{e^{t\Delta_p^\bullet} = e^{t\Delta_p}\,|_{L_p^\bullet}\,;\, t \geq 0\}$$

on $L_p^\bullet$ and

$$[\mathsf{Re}\, z \geq -\omega] \subset \varrho(\Delta_p^\bullet) = \varrho(\Delta_p)\cup\{0\}$$

for some $\omega > 0$, where $\varrho(\Delta_p^\bullet)$ and $\varrho(\Delta_p)$ denote the resolvent sets of the operators $\Delta_p^\bullet$ and Δ_p, respectively. Therefore, there exist $N \geq 1$ and a number $\alpha \in (\pi/2,\pi)$ such that

$$\Sigma_\alpha := \{z \in \mathbb{C}\setminus\{0\}\,;\, |\arg z| < \alpha\} \subset \varrho(\Delta_p^\bullet)$$

and

$$\|(\lambda - \Delta_p^\bullet)^{-1}\|_{\mathcal{L}(L_p^\bullet)} \leq \frac{N}{|\lambda|}\ , \quad \lambda \in \Sigma_\alpha\ .$$

Observing the resolvent decomposition

$$(\lambda - d(y)\Delta_p)^{-1} = d(y)^{-1}\left(\frac{\lambda}{d(y)} - \Delta_p\right)^{-1} = d(y)^{-1}\left[\frac{d(y)}{\lambda} \oplus \left(\frac{\lambda}{d(y)} - \Delta_p^\bullet\right)^{-1}\right]$$

for $\lambda \in \Sigma_\alpha$ and a.e. $y \in Y$, we derive

$$\|(\lambda - d(y)\Delta_p)^{-1}\|_{\mathcal{L}(L_p)} \leq \frac{N'}{|\lambda|}\ , \quad \lambda \in \Sigma_\alpha\ , \quad \text{a.e. } y \in Y,$$

for some $N' \geq 1$. Clearly, this implies that Σ_α belongs to the resolvent set of the operator $-\mathbb{A}_p$ and that the resolvent is given by

$$\big((\lambda + \mathbb{A}_p)^{-1}u\big)(y) = (\lambda - d(y)\Delta_p)^{-1}u(y)\ , \quad \text{a.e. } y \in Y\ , \quad u \in \mathbb{L}_p\ , \quad \lambda \in \Sigma_\alpha\ .$$

Hence we deduce that

$$\|(\lambda + \mathbb{A}_p)^{-1}\|_{\mathcal{L}(\mathbb{L}_p)} \leq \frac{N'}{|\lambda|}\ , \quad \lambda \in \Sigma_\alpha\ ,$$

so that well-known generation results (e.g., [6, Thm. 4.2.1]) then ensure that $-\mathbb{A}_p$ is the generator of a strongly continuous analytic semigroup on $\mathbb{L}_p$ for $1 \leq p < \infty$. Moreover, from [7, Thm. 11.6.6] we infer (8). Consequently, the semigroup generated by $-\mathbb{A}_p$ is a positive semigroup of contractions and satisfies the estimates (9) and (10) in view of (2), (3) and assumption (5). Finally, that $e^{-t\mathbb{A}_p}$ leaves both of the the spaces $\mathbb{P}(\mathbb{L}_p)$ and $\mathbb{L}_p^\bullet$ invariant and that the estimate (11) holds can be shown exactly as in the proof of [4, Prop. 3]. □

Remark 2. *Observe that* (5) *is required for the estimates* (9)–(11) *but not for the analyticity of the semigroup* $\{e^{-t\mathbb{A}_p}\ ;\ t \geq 0\}$.

3. Well-posedness

In the sequel, for $\vartheta \geq 0$ given, we say that *hypothesis* $H(\vartheta)$ *is satisfied* provided that

(H_1) K is a non-negative symmetric function defined on $Y \times Y$ and there is $k > 0$ such that

$$K(y,y')\big(d(y+y')^\vartheta + d(y)^\vartheta\big) \leq k\ , \quad (y,y') \in Y \times Y\ .$$

P and Q are non-negative and symmetric functions belonging to $L_\infty(\Xi)$, where $\Xi := \{(y,y') \in Y \times Y\ ;\ y + y' \in Y\}$, such that

$$0 \leq P(y,y') + Q(y,y') \leq 1 \quad \text{for a.e. } (y,y') \in \Xi\ .$$

(H_2) γ is a measurable function from $\{(y,y')\ ;\ 0 < y' < y < y_0\}$ into $\mathbb{R}^+$ such that there exists $m_\gamma > 0$ with

$$\int_0^y d(y')^\vartheta \gamma(y,y')\ \mathrm{d}y' + d(y)^\vartheta \int_0^y \frac{y'}{y}\gamma(y,y')\ \mathrm{d}y' \leq m_\gamma \quad \text{for a.e. } y \in Y\ .$$

(H_3) β_{c} is a non-negative measurable function on $\{(y,y')\ ;\ 0 < y' < y < y_0\}$ such that

$$Q(y,y')\left(\int_0^{y+y'} y''\beta_{\mathsf{c}}(y+y',y'')\ \mathrm{d}y'' - y - y'\right) = 0 \quad \text{for a.e. } (y,y') \in \Xi\ ,$$

and there exists $m_{\mathsf{c}} > 0$ with

$$Q(y,y') \int_0^{y+y'} d(y'')^{\vartheta} \beta_{\mathsf{c}}(y+y',y'')\,\mathrm{d}y'' \le m_{\mathsf{c}} \quad \text{for a.e. } (y,y') \in \Xi\ .$$

(H_4) β_{s} is a measurable function from $(y_0, 2y_0) \times (0, y_0)$ into $\mathbb{R}^+$ such that

$$\int_0^{y_0} y'' \beta_{\mathsf{s}}(y+y',y'')\,\mathrm{d}y'' = y+y' \quad \text{for a.e. } y+y' \in (y_0, 2y_0)\ ,$$

and there exists $m_{\mathsf{s}} > 0$ with

$$\int_0^{y_0} d(y'')^{\vartheta} \beta_{\mathsf{s}}(y+y',y'')\,\mathrm{d}y'' \le m_{\mathsf{s}} \quad \text{for a.e. } y+y' \in (y_0, 2y_0)\ .$$

Clearly, hypothesis $H(\vartheta)$ with $\vartheta \ge 0$ implies hypothesis $H(0)$ in view of (5). Let us mention right now that $H(\vartheta)$ ensures 'the gain of integrability' (see [12, Lem. 2.6])

$$L[u] \in L_1\big(Y, d(y)^{\vartheta}(1+y)\mathrm{d}y\big) \quad \text{for} \quad u \in L_1(Y,(1+y)\mathrm{d}y)\ . \tag{12}$$

Also note that hypothesis $H(0)$ coincides with the assumptions made in [4]. We therefore refer to [4, Ex. 1] for kernels obeying $H(0)$. If, in the case $y_0 < \infty$, the kernels are of the form

$$\begin{array}{ll} \gamma(y,y') \propto y^{\alpha-\xi-1}\,(y')^{\xi}\ , & 0 < y' < y < y_0\ , \\ \beta_{\mathsf{c}}(y,y') := (\zeta+2)\,y^{-1-\zeta}\,(y')^{\zeta}\ , & 0 < y' < y < y_0\ , \\ \beta_{\mathsf{s}}(y,y') := (\nu+2)\,y_0^{-2-\nu}\,y\,(y')^{\nu}\ , & 0 < y' < y_0 \le y < 2y_0\ , \\ Q(y,y') \propto (y+y')^{\tau}\ , & (y,y') \in \Xi\ , \\ K(y,y') \propto (yy')^{\mu}\ , & y,y' \in Y\ , \end{array}$$

with $0 \ge \xi, \zeta, \nu > -1$ and $\alpha, \mu, \tau > 0$, we can choose $\vartheta > 0$ sufficiently small such that hypothesis $H(\vartheta)$ is satisfied for diffusion coefficients $d(y) \sim y^{-\lambda}$, $\lambda > 0$.

In order to show that problem (1) is well posed, we assume in the sequel that (4), (5), and at least hypothesis $H(0)$ are satisfied. Then it follows, for $1 \le p < \infty$, that

$$L_{\mathsf{b}} \in \mathcal{L}(\mathbb{L}_p) \quad \text{and} \quad G := L_{\mathsf{c}} + L_{\mathsf{s}} \in \mathcal{L}^2(\mathbb{L}_{2p}, \mathbb{L}_p) \tag{13}$$

since the pointwise product $L_{2p} \times L_{2p} \to L_p$ is a multiplication. If J denotes an interval in $\mathbb{R}^+$ containing 0, we put $\dot{J} := J \setminus \{0\}$. We then call $u \in C(J, \mathbb{L}_p)$ a *mild* $\mathbb{L}_p$-solution to the re-written problem

$$\dot{u} + \mathbb{A}_p u = L[u]\ , \quad t > 0\ , \quad u(0) = u^0\ , \tag{14}$$

provided it solves the fixed point equation

$$u(t) = U(t)u^0 + U \star L[u](t) \quad \text{in} \quad \mathbb{L}_p\ , \quad t \in J\ , \tag{15}$$

where $U(t) := e^{-t\mathbb{A}}$, $t \ge 0$, with $\mathbb{A} := \mathbb{A}_1$ and

$$U \star v(t) := \int_0^t U(t-s)\,v(s)\,\mathrm{d}s\ , \quad t \in J\ .$$

A mild solution is a *strong* $\mathbb{L}_p$-solution if $u \in C^1(\dot{J}, \mathbb{L}_p) \cap C(\dot{J}, \mathsf{D}(\mathbb{A}_p))$.

Given a Banach space E and $\mu \in \mathbb{R}$, we denote by $BC_\mu(\dot{J}, E)$ the Banach space of all functions $u : \dot{J} \to E$ such that $\big(t \mapsto t^\mu u(t)\big)$ is bounded and continuous from $\dot{J}$ into E, equipped with the norm

$$u \mapsto \|u\|_{BC_\mu(\dot{J},E)} := \sup_{t\in\dot{J}} t^\mu \, \|u(t)\|_E \ .$$

We write $C_\mu(\dot{J}, E)$ for the closed linear subspace thereof consisting of all u satisfying $t^\mu u(t) \to 0$ in E as $t \to 0$.

Due to Theorem 1 we have (see [4, Prop. 4])

Proposition 3. Let $1 \le p \le q \le \infty$ and $\alpha \in [0,2) \setminus \{1+1/q\}$ be such that $n(1/p - 1/q)/2 + \alpha/2 < 1$ and either $q \in (1,\infty)$ or $\alpha = 0$. Then, for $\mu < 1$,

$$\big(u \mapsto U \star L_{\mathsf{b}}[u]\big) \in \mathcal{L}\big(C_\mu(\dot{J}, \mathbb{L}_p), C_{\mu+\frac{n}{2}(\frac{1}{p}-\frac{1}{q})+\frac{\alpha}{2}-1}(\dot{J}, \mathbb{H}^\alpha_{q,\mathcal{B}})\big) \tag{16}$$

and

$$\big(u \mapsto U \star G[u,u]\big) \in \mathcal{L}^2\big(C_{\mu/2}(\dot{J}, \mathbb{L}_{2p}), C_{\mu+\frac{n}{2}(\frac{1}{p}-\frac{1}{q})+\frac{\alpha}{2}-1}(\dot{J}, \mathbb{H}^\alpha_{q,\mathcal{B}})\big) \ . \tag{17}$$

Moreover, the analogue of [4, Prop. 6]) is still valid:

Proposition 4. Let $1 < p \le q < \infty$, $\alpha \in [0,2) \setminus \{1+1/q\}$ and assume that $n(1/p - 1/q)/2 + \alpha/2 < 1$, where either $\alpha = 0$ and $p < q$ or $\alpha > 0$ and $q \in (1,\infty)$. Then, for $u^0 \in \mathbb{L}_p$,

$$Uu^0 := \big(t \mapsto U(t)u^0\big) \in C_{\frac{n}{2}(\frac{1}{p}-\frac{1}{q})+\frac{\alpha}{2}}(\dot{J}, \mathbb{H}^\alpha_{q,\mathcal{B}}) \ .$$

Proof. Owing to (6), $n(1/p - 1/q)/2 + \alpha/2 < 1$, and classical embedding arguments we have

$$\mathsf{D}(\mathbb{A}_p) \hookrightarrow \mathbb{H}^2_{p,\mathcal{B}} \hookrightarrow \mathbb{H}^\alpha_{q,\mathcal{B}} \ .$$

Hence, since $U(t)u^0 \in \mathsf{D}(\mathbb{A}_p)$, $t > 0$, due to Theorem 1, we deduce

$$\|U(t+h)u^0 - U(t)u^0\|_{\mathbb{H}^\alpha_{q,\mathcal{B}}} \le c \int_Y \|e^{(t+h)d(y)\Delta} u^0(y) - e^{td(y)\Delta} u^0(y)\|_{H^2_{p,\mathcal{B}}} \, (1+y) \, dy$$

for $0 \le |h| < t$. Taking (2) and (5) into account and that $\{e^{\tau\Delta} \, ; \, \tau \ge 0\}$ is a bounded and strongly continuous semigroup on $H^2_{p,\mathcal{B}}$ in view of [1, V. Thm. 2.1.3], we infer from Lebesgue's theorem that the right-hand side of the above inequality tends to zero as $h \to 0$, and thus $Uu^0 \in C(\dot{J}, \mathbb{H}^\alpha_{q,\mathcal{B}})$. Theorem 1 then yields that the map $t \mapsto t^\zeta \|U(t)u^0\|_{\mathbb{H}^\alpha_{q,\mathcal{B}}}$ remains bounded on $\dot{J}$ for $\zeta := n(1/p - 1/q)/2 + \alpha/2 < 1$. The fact that $\mathbb{H}^\alpha_{q,\mathcal{B}}$ is dense in $\mathbb{L}_p$ implies $Uu^0 \in C_\zeta(\dot{J}, \mathbb{H}^\alpha_{q,\mathcal{B}})$ as in the proof of [4, Prop. 6]. □

Under hypothesis $H(0)$ we can prove now existence and uniqueness of mild solutions to problem (14). Subsequently, we will derive more regularity (with respect to time) for these solutions.

Our global-in-time statement requires that collisional breakage is dominated by coalescence, i.e., that

$$Q(y,y')\left(\int_0^{y+y'} \beta_{\mathsf{c}}(y+y',y'')\,\mathrm{d}y'' - 2\right) \le P(y,y')\ ,\quad y+y'\in Y\ , \tag{18}$$

and that scattering is a binary processes, meaning that

$$\beta_{\mathsf{s}}(y,y') = \beta_{\mathsf{s}}(y,y-y')\ ,\quad (y,y')\in\Xi\ , \tag{19}$$

and

$$\beta_{\mathsf{s}}(y,y') = 0\ ,\quad 0<y'<y-y_0<y_0\ . \tag{20}$$

Then we have the following existence and uniqueness result.

Theorem 5. Let (4), (5), and hypothesis $H(0)$ be satisfied and assume that $(n/2\vee 1)<p<\infty$. Then, given any non-negative initial value $u^0\in\mathbb{L}_p$, problem (14) possesses a unique maximal non-negative mild $\mathbb{L}_p$-solution $u:=u(\cdot;u^0)$ on an open interval $J(u^0)\subset\mathbb{R}^+$ such that

$$t^{n/4p}\,\|u(t)\|_{\mathbb{L}_{2p}}\to 0\quad\text{as}\quad t\to 0^+\ .$$

In addition, u has the regularity

$$u\in C(\dot{J}(u^0),\mathbb{H}^2_{q,\mathcal{B}})\ ,\quad q\in(1,\infty)\ . \tag{21}$$

If $t^+ := \sup J(u^0)<\infty$, then

$$\sup_{t^+/2<t<t^+}\|u(t)\|_{\mathbb{L}_q}=\infty\ ,\quad (n/2\vee 1)<q<\infty\ . \tag{22}$$

The solution u preserves the total mass, that is,

$$\int_\Omega\int_Y y\,u(t,y,x)\,\mathrm{d}y\mathrm{d}x = \int_\Omega\int_Y y\,u^0(y,x)\,\mathrm{d}y\mathrm{d}x\ ,\quad t\in J(u^0)\ . \tag{23}$$

Finally, if $\gamma\equiv 0$ and (18)–(20) are satisfied, then, for $p\in(n/2,\infty)$ and $p\ge 2$, the solution remains bounded in $\mathbb{L}_p$ whenever $\|u^0\|_{\mathbb{L}_p}$ is small. In particular, u exists globally in time in this case.

Proof. The proofs of the existence and uniqueness parts of this theorem including the blow-up behavior (22) are, verbatim, the same as the ones of [4, Thm. 7] (see step (i) and (iii) of the proof of [4, Thm. 7] and Propositions 3 and 4). For the regularity result (21) we may focus on $q>p$. We choose $T\in\dot{J}(u^0)$ and $\varepsilon>0$ sufficiently small and derive

$$L[u(\cdot+\varepsilon)]\in C([0,T],\mathbb{H}^\nu_{q,\mathcal{B}})\quad\text{and}\quad u\in C((0,T],\mathbb{H}^\nu_{q,\mathcal{B}})$$

for some $\nu>0$. This can be done as in the second step of the proof of [4, Thm. 7] in view of Propositions 3 and 4. Since U is strongly continuous on $\mathbb{H}^\nu_{q,\mathcal{B}}$ due to Lebesgue's theorem (see the proof of Proposition 4), estimate (10) implies

$$U\star L[u(\cdot+\varepsilon)]\in C([0,T],\mathbb{H}^2_{q,\mathcal{B}})\quad\text{and}\quad Uu(\varepsilon)\in C((0,T],\mathbb{H}^2_{q,\mathcal{B}})\ .$$

Due to

$$u(t+\varepsilon) = U(t)u(\varepsilon) + U \star L[u(\cdot+\varepsilon)](t) \ , \quad t \in [0,T] \ , \tag{24}$$

assertion (21) follows since $\varepsilon > 0$ was arbitrary.
Next observe that, for $T \in \dot{J}(u^0)$, the mild solution $u = u(\cdot;u^0)$ is bounded on $[0,T]$ with values in $\mathbb{H}^\alpha_{q,\mathcal{B}}$ provided $u^0 \in \mathbb{H}^\alpha_{q,\mathcal{B}}$. Since $\mathbb{H}^\alpha_{q,\mathcal{B}}$ embeds continuously in $\mathbb{L}_\infty$ for q sufficiently large and since u solves the fixed point equation

$$u(t) = U_\omega(t)u^0 + U_\omega \star (L[u] + \omega u)(t) \ , \quad t \in J(u^0) \ ,$$

we deduce that u is non-negative for non-negative initial values $u^0 \in \mathbb{H}^\alpha_{q,\mathcal{B}}$ as in [4, Thm. 10]. Here, the constant $\omega > 0$ suitably depends on $\sup_{0\le t\le T} \|u(t)\|_{\mathbb{L}_\infty}$ and U_ω is given by $U_\omega(t) := e^{-\omega t}U(t)$. The continuous dependence of $u(\cdot;u^0)$ on u^0 and the density of $\mathbb{H}^\alpha_{q,\mathcal{B}} \cap \mathbb{L}_p^+$ in $\mathbb{L}_p^+$ (see [4, Lem. 9]) yields then the positivity assertion.
The conservation of mass formula (23) follows from (15) by taking into account the equality

$$\begin{aligned}\int_Y y \int_\Omega [U(t)v](y) \, \mathrm{d}x \, \mathrm{d}y &= \int_Y y \int_\Omega e^{td(y)\Delta} v(x,y) \, \mathrm{d}x \, \mathrm{d}y \\ &= \int_Y y \int_\Omega v(x,y) \, \mathrm{d}x \, \mathrm{d}y \ , \quad t > 0 \ , \quad v \in \mathbb{L}_p \ ,\end{aligned}$$

which is due to the Neumann boundary conditions, and observing that

$$\int_Y y \, L[v](y) \, \mathrm{d}y = 0 \ , \quad v \in L_1(Y, y\mathrm{d}y) \ ,$$

(see hypotheses (H_1)–(H_4) and [12, Lem. 2.6]). Finally, based on estimate (11), the global-in-time existence result is obtained as in [4, Thm. 16] noticing that the proof of the latter merely requires mild solutions. This proves the assertion. □

Before we prove that hypothesis $H(\vartheta)$ with $\vartheta > 0$ provides strong solutions, let us show that also the weaker assumption $H(0)$ implies that the original equation (1) is satisfied pointwise almost everywhere. More precisely, defining

$$\mathcal{L}_q := L_1(Y, L_q, \mathrm{d}\mu(y)) \quad \text{with} \quad \mathrm{d}\mu(y) := d(y)^{-1}(1+y)\mathrm{d}y \ ,$$

we have

Theorem 6. The mild solution $u = u(\cdot;u^0)$ provided by Theorem 5 belongs to $C^1(\dot{J}(u^0), \mathcal{L}_q)$ for each $q \in [1,\infty)$ and it holds that

$$\dot{u}(t) = d\Delta u(t) + L[u(t)] \quad \text{in} \quad \mathcal{L}_q \ , \quad t \in \dot{J}(u^0) \ .$$

In particular, for almost every $y \in Y$, it holds that

$$\dot{u}(t,y) - d(y)\Delta u(t,y) = L[u(t)](y) \quad \text{in} \quad L_q \ , \quad t \in \dot{J}(u^0) \ .$$

Proof. Fix $T \in \dot{J}(u^0)$ and $\varepsilon > 0$ sufficiently small and set

$$u_\varepsilon := u(\cdot+\varepsilon) \in C([0,T], \mathbb{H}^2_{q,\mathcal{B}}) \ , \quad q \in (1,\infty) \ .$$

Then there is $\nu > 0$ such that $f_\varepsilon := L[u_\varepsilon] \in C([0,T], \mathbb{H}^\nu_{q,\mathcal{B}})$ (see [10, Cor. 4.5.2]). Given $t \in [0,T]$ and $h > 0$ sufficiently small, we deduce

$$\begin{aligned}
&\big\| h^{-1}\big(U(t+h)u(\varepsilon) - U(t)u(\varepsilon)\big) + \mathbb{A}\, U(t)u(\varepsilon)\big\|_{\mathcal{L}_q} \\
&\qquad\qquad \le h^{-1} \int_Y d(y) \int_t^{t+h} \big\|\Delta \left(e^{\tau d(y)\Delta} - e^{td(y)\Delta}\right) u_\varepsilon(y)\big\|_{L_q} \,\mathrm{d}\tau\, \mathrm{d}\mu(y) \\
&\qquad\qquad \le h^{-1} \int_Y \int_t^{t+h} \big\|\left(e^{\tau d(y)\Delta} - e^{td(y)\Delta}\right) u_\varepsilon(y)\big\|_{H^2_{q,\mathcal{B}}} \,\mathrm{d}\tau\, (1+y)\mathrm{d}y\ .
\end{aligned}$$

Since $\{e^{\tau d(y)\Delta}\, ;\, \tau \ge 0\}$ is for a.e. $y \in Y$ a bounded and strongly continuous semigroup on $H^2_{q,\mathcal{B}}$, we may apply Lebesgue's theorem to deduce that the right-hand side of the above inequality tends to zero as h does. Hence

$$\partial_+ U(t)u(\varepsilon) = -\mathbb{A}\, U(t)u(\varepsilon) \quad \text{in} \quad \mathcal{L}_q\ , \quad t \in [0,T]\ , \tag{25}$$

and, since by the same arguments $\partial_+ U u(\varepsilon) \in C([0,T], \mathcal{L}_q)$, we obtain that $Uu(\varepsilon)$ belongs to $C^1([0,T], \mathcal{L}_q)$ with derivative given by (25). Next, for $h > 0$ small and $t \in (0,T]$, we write

$$\begin{aligned}
&h^{-1}\big(U \star f_\varepsilon(t+h) - U \star f_\varepsilon(t)\big) \\
&= h^{-1} \int_0^t \big(U(t+h-s) - U(t-s)\big) f_\varepsilon(s)\, \mathrm{d}s + h^{-1} \int_t^{t+h} U(t+h-s) f_\varepsilon(s)\, \mathrm{d}s \\
&=: M_h + N_h\ .
\end{aligned}$$

Then, as above, we compute

$$\begin{aligned}
&\|M_h + \mathbb{A}\,(U \star f_\varepsilon)(t)\|_{\mathcal{L}_q} \\
&\le c h^{-1} \int_Y \int_0^t \int_t^{t+h} \big\|\big(e^{(\tau - s)d(y)\Delta} - e^{(t-s)d(y)\Delta}\big) f_\varepsilon(s,y)\big\|_{H^2_{q,\mathcal{B}}} \,\mathrm{d}\tau\, \mathrm{d}s\, (1+y)\mathrm{d}y\ .
\end{aligned}$$

According to (3) and (5), for $h \le 1$ and $0 < s < t \le T$, it holds that

$$\begin{aligned}
h^{-1} &\int_t^{t+h} \big\|\big(e^{(\tau - s)d(y)\Delta} - e^{(t-s)d(y)\Delta}\big) f_\varepsilon(s,y)\big\|_{H^2_{q,\mathcal{B}}} \,\mathrm{d}\tau \\
&\qquad\qquad \le c \sup_{t \le \tau \le t+1} \big(1 \wedge (\tau - s)d(y)\big)^{-1+\nu/2} \|f_\varepsilon(s,y)\|_{H^\nu_{q,\mathcal{B}}} \\
&\qquad\qquad \le c(T)\,(t-s)^{-1+\nu/2} \|f_\varepsilon(s,y)\|_{H^\nu_{q,\mathcal{B}}}
\end{aligned}$$

so that we may again apply Lebesgue's theorem to deduce

$$M_h \longrightarrow -\mathbb{A}\,(U \star f_\varepsilon)(t) \quad \text{in} \quad \mathcal{L}_q\ , \quad h \to 0^+\ . \tag{26}$$

On the other hand, we have

$$\begin{aligned}
&\|N_h - f_\varepsilon(t)\|_{\mathcal{L}_q}\\
&\le h^{-1}\int_Y\int_t^{t+h} \|e^{(t+h-s)d(y)\Delta}\|_{\mathcal{L}(L_q)}\|f_\varepsilon(s,y)-f_\varepsilon(t,y)\|_{L_q}\,\mathrm{d}s\,\mathrm{d}\mu(y)\\
&\quad + h^{-1}\int_Y\int_t^{t+h} \big\|e^{(t+h-s)d(y)\Delta}f_\varepsilon(t,y)-f_\varepsilon(t,y)\big\|_{L_q}\,\mathrm{d}s\,\mathrm{d}\mu(y)\\
&\le d_\star^{-1}\sup_{t\le s\le t+h}\|f_\varepsilon(s)-f_\varepsilon(t)\|_{\mathbb{L}_q}\\
&\quad + d_\star^{-1}\int_Y \sup_{0<\tau<hd(y)}\big\|e^{\tau\Delta}f_\varepsilon(t,y)-f_\varepsilon(t,y)\big\|_{L_q}\,(1+y)\,\mathrm{d}y\ .
\end{aligned}$$

Lebesgue's theorem shows that the right side converges towards zero as $h\to 0^+$, whence $N_h \to f_\varepsilon(t)$ in $\mathcal{L}_q$. Taking (26) into account we obtain

$$\partial_+(U\star f_\varepsilon)(t) = -\mathbb{A}\,U\star f_\varepsilon(t) + f_\varepsilon(t) \quad \text{in} \quad \mathcal{L}_q\ , \quad t\in(0,T]\ . \tag{27}$$

Due to (3), (5), and the strong continuity of U on $\mathbb{H}^\nu_{q,\mathcal{B}}$, we derive

$$\partial_+(U\star f_\varepsilon)\in C((0,T],\mathcal{L}_q) \quad \text{and} \quad U\star f_\varepsilon \in C((0,T],\mathcal{L}_q)\ .$$

Thus $U\star f_\varepsilon \in C^1((0,T],\mathcal{L}_q)$. Recalling (25), we infer from equation (24) that $u_\varepsilon\in C^1((0,T],\mathcal{L}_q)$. Since $\varepsilon>0$ was arbitrary, the assertion follows. □

Stronger assumptions on the kernels ensure that the mild solutions actually are strong solutions as the concluding theorem shows.

Theorem 7. Suppose that $H(\vartheta)$ holds with $\vartheta>0$. Then, given the assumptions of Theorem 5, the mild solution $u=u(\cdot;u^0)$ is a strong $\mathbb{L}_p$-solution to problem (14) and has the additional regularity

$$u\in C^1(\dot J(u^0),\mathbb{L}_q)\cap C(\dot J(u^0),\mathsf{D}(\mathbb{A}_q))\ , \quad q\in(1,\infty)\ .$$

Proof. We may assume that $q\in(1,\infty)$ is large. By $[\cdot,\cdot]_\theta$ we denote the complex interpolation functor of exponent $\theta\in(0,1)$. Then, in view of [11, Thm. 1.18.4], we may follow the lines of the second step of the proof of [11, Thm. 1.18.5] to show that

$$\big[L_1(Y,L_q,(1+y)\mathrm{d}y),L_1(Y,H^2_{q,\mathcal{B}},d(y)(1+y)\mathrm{d}y)\big]_\theta = L_1\big(Y,H^{2\theta}_{q,\mathcal{B}},d(y)^\theta(1+y)\mathrm{d}y\big)$$

provided $2\theta\in(0,1)\setminus\{1+1/q\}$. Consequently, (6) implies, for $2\theta\ne 1+1/q$,

$$L_1\big(Y,H^{2\theta}_{q,\mathcal{B}},d(y)^\theta(1+y)\mathrm{d}y\big)\hookrightarrow\big[\mathbb{L}_q,\mathsf{D}(\mathbb{A}_q)\big]_\theta =: \mathbb{E}_\theta\ . \tag{28}$$

Let now $u\in C(J(u^0),\mathbb{L}_p)\cap C(\dot J(u^0),\mathbb{H}^2_{q,\mathcal{B}})$ be the mild solution provided by Theorem 5. Since q is large, we infer from [10, Cor. 4.5.2] that there is some $\nu>0$ small such that the pointwise product $H^2_{q,\mathcal{B}}\times H^2_{q,\mathcal{B}}\hookrightarrow H^{2\nu}_{q,\mathcal{B}}$ is continuous. By making ϑ smaller if necessary, we may assume that $\nu=\vartheta$. Therefore, hypothesis $H(\vartheta)$ guarantees that the map $t\mapsto L[u(t)]$ is continuous on $\dot J(u^0)$ with values in $L_1(Y,H^{2\vartheta}_{q,\mathcal{B}},d(y)^\vartheta(1+y)\mathrm{d}y)$ (see (12)). Taking (28) into account we thus obtain

$L[u] \in C(\dot{J}(u^0), \mathbb{E}_\vartheta)$. As in part (ii) of the proof of [4, Thm. 7] we can now shift the equation (14) and apply [1, Thm. IV.1.5.1] in order to deduce that $u(\cdot + \varepsilon)$ is a strong $\mathbb{L}_q$-solution for each $\varepsilon > 0$ small. This proves the claim. □

Acknowledgement

I would like to thank G. Simonett for fruitful discussions on the subject.

References

[1] H. Amann. *Linear and quasilinear parabolic problems. Volume I: Abstract linear theory.* Birkhäuser. Basel, Boston, Berlin 1995.

[2] H. Amann. *Coagulation-fragmentation processes.* Arch. Rat. Mech. Anal. **151** (2000), 339–366.

[3] H. Amann, J. Escher. *Strongly continuous dual semigroups.* Ann. Mat. Pura Appl. **CLXXI** (1996), 41–62. **151** (2000), 339-366.

[4] H. Amann, C. Walker. *Local and global strong solutions to continuous coagulation-fragmentation equations with diffusion.* To appear in: J. Diff. Eq.

[5] H. Amann, F. Weber. *On a quasilinear coagulation-fragmentation model with diffusion.* Adv. Math. Sci. Appl. **11** (2001), 227–263.

[6] H.O. Fattorini. *The Cauchy problem.* Addison-Wesley Publishing Company. London, Amsterdam 1983.

[7] E. Hille, R.S. Phillips. *Functional Analysis and Semigroups.* American Mathematical Society Colloquium Publications. Volume XXXI (Revised edition). Providence, Rhode Island, 1957.

[8] P. Laurençot, S. Mischler. *The continuous coagulation-fragmentation equations with diffusion.* Arch. Rat. Mech. Anal. **162**, No. 1 (2002), 45–99.

[9] P. Laurençot, S. Mischler. *On coalescence equations and related models.* In: P. Degond, L. Pareschi, G. Russo (eds.), *Modeling and computational methods for kinetic equations.* Birkhäuser. Boston 2004.

[10] T. Runst, W. Sickel. *Sobolev spaces of fractional order, Nemytskij operators, and nonlinear partial differential equations.* De Gruyter Series in Nonlinear Analysis and Applications **3**. De Gruyter. Berlin, New York 1996.

[11] H. Triebel. *Interpolation theory, function spaces, differential operators.* Second edition. Johann Ambrosius Barth. Heidelberg, Leipzig 1995.

[12] C. Walker. *Coalescence and breakage processes.* Math. Meth. Appl. Sci. **25** (2002), 729–748.

[13] C. Walker. *On a new model for continuous coalescence and breakage processes with diffusion.* Adv. Diff. Equ. **10**, no. 2 (2005), 121–152.

Christoph Walker
Vanderbilt University
Department of Mathematics
1326 Stevenson Center
Nashville, TN 37240, USA
e-mail: cwalker@math.vanderbilt.edu

Progress in Nonlinear Differential Equations
and Their Applications, Vol. 64, 529–536

On L_p-Estimates of Optimal Type for the Parabolic Oblique Derivative Problem with VMO-Coefficients – A Refined Version

Peter Weidemaier

Dedicated to Prof. Dr. Herbert Amann

Abstract. We prove $W_p^{2,1}(\Omega_T)$-estimates ($1 < p < \infty$) for parabolic operators with a second-order elliptic part in non-divergence form with essentially bounded VMO-coefficients. The boundary condition $\sum_{i=1}^n b_i(\xi,t)\partial_i u(\xi,t) = g(\xi,t)$ on $\partial\Omega_T$ is considered in the non-degenerate case, and the b_i are only assumed to be in space-time Sobolev-spaces (see condition (B)).

1. Introduction

As in [7] we consider the initial-boundary value problem (IBVP)

$$u_t - a_{ij}(x,t)\partial_i\partial_j u = f(x,t) \quad \text{in } \Omega_T \tag{1}$$

$$u(x,0) = u_0(x) \quad \text{in } \Omega \tag{2}$$

$$\langle b(\xi,t), \nabla_x u \rangle = g(\xi,t) \quad \text{on } \Gamma_T, \tag{3}$$

where $\langle a, b \rangle := \sum_{i=1}^n a_i b_i$ and $\Omega \subset \mathbb{R}^n$. The focus is on weak assumptions for the coefficients, which still allow estimates of the solution in $W_p^{2,1}$ spaces. As in [7] we assume that $a_{ij} \in L_\infty \cap VMO(\Omega_T)$. While in [7] we assumed $b_i \in C^{\alpha,\alpha/2}(\partial\Omega_T)$ for some $\alpha > 1 - 1/p$, in the present work the b_i are assumed to be in space-time Sobolev spaces (see condition (B)). The method of proof is similar to that in [7].

Let us introduce some notation : Ω denotes an open subset of $\mathbb{R}^n$ with boundary $\Gamma := \partial\Omega$, while Ω_T stands for $\Omega \times (0,T)$. The conjugate exponent of p is denoted p', $\frac{1}{p} + \frac{1}{p'} = 1$. The Sobolev-Slobodečki spaces are denoted W_p^s, $s > 0$ with norm $\|\cdot\|_{s,p}$. They coincide with the usual Sobolev spaces for integer s and are defined for non-integer s and X-valued functions, X a Banach space, as

$$W_p^s(\Omega,\, X) = \{f \in W_p^{[s]}(\Omega,\, X) \mid |\partial^\alpha f|_{W_p^{s-[s]}(\Omega,X)} < \infty \quad \forall\, |\alpha| = [s]\}\ ,$$

where $[s]$ is the biggest integer smaller s and for $\theta \in (0,1)$

$$|f|_{W_p^\theta(\Omega,X)} := \left(\int_\Omega \int_\Omega \frac{\|f(x) - f(y)\|_X^p}{|x-y|^{n+\theta p}} dxdy \right)^{1/p} .$$

The norm in $W_p^s(\Omega, X)$ is given by $\|f\|_{W_p^{[s]}(\Omega,X)} + \sum_{|\alpha|=[s]} |\partial^\alpha f|_{W_p^{s-[s]}(\Omega,X)}$. The Sobolev-Slobodečki spaces on manifolds are defined by local charts. In particular, if $\mathbb{R}^{n-1} \supset V_i \xrightarrow{\alpha_i} \Gamma, i = 1, \ldots, N$ are local charts whose images cover Γ and $(\phi_i)_{i=1,\ldots,N}$ is a smooth partition of unity on Γ with $\operatorname{supp} \phi_i \subset \alpha_i(V_i)$, then

$$W_p^s(\Gamma) = \{\, f \in L_p(\Gamma) \mid \alpha_i^*(\phi_i f) \in W_p^s(V_i) \text{ all i } \},$$

$$\|f\|_{W_p^s(\Gamma)} := \sum_{i=1}^N \|\alpha_i^*(\phi_i f)\|_{W_p^s(V_i)},$$

where α_i^* is the pull-back defined by α_i, i.e., $\alpha_i^*(g) = g \circ \alpha_i$. The spaces $L_p(\Gamma)$ are also defined by local charts in a similar way. We write $\|\cdot\|_{p,\Gamma}$ for $\|\cdot\|_{L_p(\Gamma)}$. Furthermore, $W_{p,q}^{\alpha,\beta}(\Gamma_T) := L_q(0,T;\, W_p^\alpha(\Gamma)) \cap W_q^\beta(0,T;\, L_p(\Gamma))$ and $W_p^{k,m} = W_{p,p}^{k,m}$. Finally we denote

$$|f|_{W_{p,q}^{\alpha,\beta}(\Gamma_T)} := \left(\int_0^T |f(\cdot,t)|_{W_p^\alpha(\Gamma)}^q dt \right)^{1/q} + |f|_{W_q^\beta(0,T;\, L_p(\Gamma))}.$$

The Hölder spaces are denoted C^α with seminorm $[\cdot]_\alpha$, resp. $C^{\alpha,\beta}$ for space-time Hölder spaces. Furthermore, $Lip(\Gamma_T)$ denotes the globally Lipschitz-continuous functions on Γ_T. Finally, c denotes a constant which may vary from line to line and the notation c^* is used to emphasize the uniform boundedness of the constant $c(T)$ for $T \leq 1$.

2. Main result

We formulate condition (B).

$$\left.\begin{array}{l} b_i \in W_{q_1,q_2}^{1-\frac{1}{p}+\delta,(1-\frac{1}{p}+\delta)/2}(\Gamma_T), \text{ where } q_1,\, q_2 \text{ can be} \\ \text{any of the following numbers parametrized by } \theta: \\ \quad \begin{array}{lll} q_1 = \frac{n-1}{\theta - 1/p} & \text{and } \delta > 0 & \text{if } \theta \in (\frac{1}{p}, \min(1, \frac{n}{p})) \\ q_1 > p & \text{and } \delta > 0 & \text{if } \theta = \frac{n}{p} < 1 \\ q_1 = p & \text{and } \delta = 0 & \text{if } \theta \in (\frac{n}{p}, 1) \end{array} \\ \text{and } q_2 = \frac{2p}{1-\theta} \text{ in each case.} \end{array}\right\} (B)$$

Theorem 2.1. *Assume that Ω is a bounded $C^{1,1}$-domain (see Kufner et al. [3], 6.2.2. Definition) in $\mathbb{R}^n$ with ν denoting the vector field of exterior unit normals on Γ. Assume further that*

(A) $A(x,t) := (a_{ij}(x,t))_{1\le i,j\le n}$ *is symmetric and positive definite uniformly in* $\overline{\Omega} \times [0,T]$, *say* $A(x,t) \ge \lambda I$ *with* $\lambda > 0$, $a_{ij}(x,t) \in L_\infty \cap VMO(\Omega_T)$

(E) $1 < p < \infty,\ p \ne 3$

(B) $\langle\, b(\xi,t), \nu(\xi)\,\rangle > 0$ *on* Γ_T *("nondegeneracy condition") and the* b_i *satisfy assumption* (B).

(F) $f \in L_p(\Omega_T)$

(G) $g \in W_p^{1-1/p,(1-1/p)/2}(\Gamma_T)$

(Iv) $u_0 \in W_p^{2-2/p}(\Omega)$

(C) $\langle\, b(\cdot,0), \nabla u_0(\cdot)\,\rangle = g(\cdot,0)$ *on* Γ *for* $p > 3$.

Then (1) $\to$ (3) *has a unique solution* $u \in W_p^{2,1}(\Omega_T)$ *and there is a constant* $c^*(p,n,\lambda,\partial\Omega,T,\|b\|_{W_{q_1,q_2}^{1-\frac{1}{p}+\delta,(1-\frac{1}{p}+\delta)/2}(\Gamma_T)}, \|a_{ij}\|_{L_\infty\cap VMO})$ *such that*

$$\|u\|_{W_p^{2,1}(\Omega_T)} \le c^* \cdot \left(\|u_0\|_{W_p^{2-2/p}(\Omega)} + \|f\|_{L_p(\Omega_T)} + \|g\|_{W_p^{1-1/p,(1-1/p)/2}(\Gamma_T)}\right). \quad (4)$$

Remark 2.2. i.) The theorem is a refinement of previous results of Softova ([6], [5]) who assumed boundary coefficients $b_i \in \mathrm{Lip}\,(\partial\Omega_T)$, and of the present author ([7]) who assumed $b_i \in C^{\alpha,\alpha/2}(\partial\Omega_T)$ for some $\alpha > 1 - 1/p$.
ii) Condition (B) implicitly implies that the coefficients are Hölder-continuous in space-time: Indeed an easy calculation based on the interpolation inequality

$$\|f\|_{\sigma,q_1,\Omega} \le c \cdot \|f\|_{q_1,\Omega}^{1-\theta}\|f\|_{s_1,q_1,\Omega}^{\theta}, \quad \forall f \in W_{q_1}^{s_1}(\Omega),$$

where $\sigma = \theta s_1 \notin \mathbb{N}$, $\theta \in [0,1]$ and $q_1 > 1$ (cf. [2], 4.1 Proposition), together with Hölder's inequality shows that

$$W_{q_1,q_2}^{s_1,s_2}(\Gamma_T) \hookrightarrow W_{q_2}^{\alpha}(0,T;\, W_{q_1}^{\sigma}(\Gamma)) \quad \text{if} \quad \frac{\alpha}{s_2} + \frac{\sigma}{s_1} < 1,\ \sigma \notin \mathbb{N}.$$

But the last space embeds continuously into $C^{\beta_2}(0,T;\, C^{\beta_1}(\Gamma))$, if $\alpha - \frac{1}{q_2} > \beta_2$ and $\sigma - \frac{n-1}{q_1} > \beta_1$, by standard Sobolev embeddings (cf. [1]). Altogether such positive $\beta's$ can be chosen iff

$$\frac{1}{s_2\, q_2} + \frac{n-1}{s_1\, q_1} < 1.$$

This condition is satisfied in all cases in (B). However, for δ and q_1 small, the Hölder exponents $\beta_1,\ \beta_2$ thus obtained are smaller than $1 - \frac{1}{p}$, $(1 - \frac{1}{p})/2$ respectively.

Proof. The proof is by localizing the problem on small domains as detailed in [7], i.e., we consider the boundary value problems fulfilled by $u_k = u\phi_k$ with ϕ_k $(k \ge 1)$ being a smooth function from a partition of unity on Γ with $supp\,\phi_k \subset B_R(\xi_k), \xi_k \in \Gamma$. We may assume $\|\nabla\phi_k\|_\infty \le C/R$. The parameter $R \in (0,1)$ will be chosen small later. Then u_k satisfies the boundary condition

$$\langle\, b(\xi_k,0),\, \nabla u_k\,\rangle = g_k + \bar{g} \quad \text{with} \quad \bar{g} := \langle\, b(\xi_k,0) - b(\xi,t), \nabla u_k\,\rangle + \langle\, b, \nabla\phi_k\,\rangle u\,.$$

The key estimate in [7] for the perturbation terms in $\bar{g}$ was that for all $\varepsilon > 0$ there exist $R(\varepsilon)$, $T(\varepsilon, R)$ such that

$$\|\bar{g}\|_{W_p^{1-1/p,(1-1/p)/2}(\Gamma_T)} \leq \varepsilon \left(\|u_k\|_{W_p^{2,1}(\Omega_T)} + \|u\|_{W_p^{2,1}(\Omega_T)} \right) + c(\frac{1}{\varepsilon})\|u_0\|_{p,\Omega} \,. \quad (5)$$

This estimate then implies (4) in the sense of an a priori estimate (see [7]), and the rest of the proof is along the lines of [7]. Estimate (5) follows from Lemma 3.1 applied to each of the two terms of the sum in $\bar{g}$. In this lemma we choose ψ such that $\psi \big| B_R(\xi_k) \equiv 1$, $supp\, \psi \subset B_{2R}(\xi_k)$, $\|\nabla\psi\|_\infty \leq C/R$, $[\psi]_\alpha \leq C/R^\alpha$. As for the first term of $\bar{g}$ we observe that, by the uniform continuity of b on Γ_T,

$$\|(b(\xi_k, 0) - b)\psi\|_{\infty, \Gamma_T} \leq \varepsilon$$

provided $R(\varepsilon), T(\varepsilon)$ are sufficiently small. From now on R will be kept fix. As for the second term in $\bar{g}$ we apply Lemma 3.1 with $m_2(\xi, t) := \langle b(\xi, t), \nabla\phi_k(\xi)\rangle$ and $v = u$. Then we estimate

$$\begin{aligned} \|m_2\psi\|_{\infty, \Gamma_T} &\leq c^* \frac{\|b\|_{\infty, \Gamma_T}}{R} \\ \|u\|_{W_p^{1-1/p,(1-1/p)/2}(\Gamma_T)} &\leq c^* \left(\|u\|_{L_p(0,T;\, W_p^1(\Omega)} + T^{1/2}\|u\|_{W_p^{2,1}(\Omega_T)} \right), \end{aligned} \quad (6)$$

where in the second estimate we used $\|u\|_{L_p(0,T;\, W_p^{1-1/p}(\Gamma)} \leq c^*\|u\|_{L_p(0,T;\, W_p^1(\Omega)}$, the estimate

$$|u|_{W_p^{\beta_1}(0,T;\, L_p(\Gamma))} \leq T^{\beta_2 - \beta_1}|u|_{W_p^{\beta_2}(0,T;\, L_p(\Gamma))} \,, \quad (7)$$

for $0 < \beta_1 < \beta_2 < 1$ and the embedding $W_p^{2,1}(\Omega_T) \hookrightarrow W_p^{1-1/(2p)}(0, T;\, L_p(\Gamma))$ (cf. [4], Chapter II, Lemma 3.4.). Estimate (7) follows directly from the definition of the seminorms and is applied with $\beta_1 = (1 - 1/p)/2$, $\beta_2 = 1 - 1/(2p)$. Using in (6) that

$$\|u\|_{L_p(0,T;\, W_p^1(\Omega)} \leq \varepsilon\|u\|_{W_p^{2,1}(\Omega_T)} + T^{1/p}c(1/\varepsilon)\|u_0\|_{p,\Omega}$$

for $T \leq T_0(\varepsilon)$ (cf. Lemma 3.4), we see that $\|m_2\psi\|_{\infty, \Gamma_T}\|u\|_{W_p^{1-1/p,(1-1/p)/2}(\Gamma_T)}$ can be estimated by the right-hand side in (5). □

3. Some tools from calculus

First we prove a result on estimating products of functions in the Sobolev spaces on the boundary.

Lemma 3.1. *Assume that* $u \in W_p^{2,1}(\Omega_T)$, $p \in (1, \infty)$ *and that* $m = m(\xi, t)$ *belongs to the class described in (B). Let* ψ *be a smooth function on* Γ *with* $supp\, \psi \equiv 1$ *on* $supp\, u(\cdot, t)$ *for a.e.* $t \in (0, T)$, $0 \leq \psi \leq 1$. *Then it holds for*

$v \in \{u, \partial_j u \mid j = 1, \ldots, n$:

$$\|mv\|_{W_p^{1-1/p,(1-1/p)/2}(\Gamma_T)} \leq c\Big(\|m\psi\|_{\infty,\Gamma_T}\|v\|_{W_p^{1-1/p,(1-1/p)/2}(\Gamma_T)}$$
$$+|m|_{W_{q_1,q_2}^{1-\frac{1}{p}+\delta,(1-\frac{1}{p}+\delta)/2}(\Gamma_T)}\Big[(\varepsilon + c(\frac{1}{\varepsilon})T^{1/p'})\|u\|_{W_p^{2,1}(\Omega_T)} + c(\frac{1}{\varepsilon})\|u_0\|_{p,\Omega}\Big]\Big)\,,$$

where $\varepsilon > 0$ is arbitrary and $c(\frac{1}{\varepsilon})$ is a constant that increases with decreasing ε.

Proof. The Lemma follows from Lemma 3.2, applied with $s = 1 - 1/p$, $v = \partial_j u$ and $v = u$, and Lemma 3.3. □

Lemma 3.2. *Let ψ be as in the preceding Lemma and let $s \in (0,1), p, r \in (1,\infty)$, $q \in (1,\infty]$. Then it holds*

$$|mv|_{W^{s,\frac{s}{2}}(\Gamma_T)} \leq c\left(\|m\psi\|_{\infty,\Gamma_T}|v|_{W^{s,\frac{s}{2}}(\Gamma_T)} + |m|_{W_{pq',pr'}^{s+\delta,\frac{s+\delta}{2}}(\Gamma_T)}\|v\|_{L_{pr}(0,T;L_{pq}(\Gamma))}\right).$$

Here $\delta > 0$ is arbitrary, if $q < \infty$ and $\delta = 0$, if $q = \infty$.

Proof. We give the proof for $q < \infty$. The case $q = \infty$ follows by obvious modifications. Suppressing the t-argument of v for convenience, we have

$$|\alpha_i^*(\phi_i mv)|_{W_p^s(V_i)} = \left(\int\limits_{V_i}\int\limits_{V_i}\frac{|\alpha_i^*(\phi_i mv)(x) - \alpha_i^*(\phi_i mv)(y)|^p}{|x-y|^{n-1+sp}}dx\,dy\right)^{1/p}.$$

Since $v\psi \equiv v$ on Γ, we replace v by $v\psi$ in the last identity. Using $fg(x) - fg(y) = [f(x) - f(y)]\,g(x) + [g(x) - g(y)]\,f(y)$ with $f = \alpha_i^*(\phi_i v)$, $g = \alpha_i^*(m\,\psi)$ the last expression is majorized by

$$\|m\psi\|_{\infty,\Gamma}|v|_{W_p^s(\Gamma)} + \left(\int\limits_{V_i}\int\limits_{V_i}\frac{|\alpha_i^*(m\,\psi)(x) - \alpha_i^*(m\,\psi)(y)|^p}{|x-y|^{n-1+sp}}|\alpha_i^*(\phi_i v)(y)|^p dx\,dy\right)^{1/p}.$$

We continue to estimate the last integral. Splitting

$$n - 1 + sp = [\frac{n-1}{q'} + (s+\delta)p] + [\frac{n-1}{q} - \delta p]$$
$$\text{with } \frac{1}{q} + \frac{1}{q'} = 1,\ q > 1 \text{ and } \delta > 0$$

and applying Hölder's inequality with q, q' this integral is estimated by

$$\left(\int\limits_{V_i}\int\limits_{V_i}\frac{|\alpha_i^*(m\,\psi)(x) - \alpha_i^*(m\,\psi)(y)|^{pq'}}{|x-y|^{n-1+(s+\delta)pq'}}dx\,dy\right)^{1/q'}\left(\int\limits_{V_i}\int\limits_{V_i}\frac{|\alpha_i^*(\phi_i v)(y)|^{pq}}{|x-y|^{n-1-\delta pq}}dx\,dy\right)^{1/q}.$$

The second factor can be estimated by

$$\left(\int\limits_{V_i} |\alpha_i^*(\phi_i v)(y)|^{pq} dy\right)^{1/q}, \quad \text{since } \sup_{y\in V_i} \int\limits_{V_i} \frac{dx}{|x-y|^{n-1-\delta pq}} < \infty .$$

By the estimates derived so far, we get for functions depending on t the estimate

$$\left(\int_0^T |m(t)v(t)|^p_{W_p^s(\Gamma)} dt\right)^{1/p} \leq \|m\psi\|_{\infty,\Gamma_T} \left(\int_0^T |v(t)|^p_{W_p^s(\Gamma)} dt\right)^{1/p}$$
$$+ c\left(\int_0^T |m(t)\psi(t)|^p_{W^{s+\delta}_{pq'}(\Gamma)} \|v(t)\|^p_{pq,\Gamma} dt\right)^{1/p} .$$

After applying Hölder's inequality with r, r' in the t-integral, the last term can be estimated by

$$c\|m\psi\|_{L_{pr'}(0,T;\, W^{s+\delta}_{pq'}(\Gamma))} \|v\|_{L_{pr}(0,T;\, L_{pq}(\Gamma))} \,.$$

Next we treat the terms corresponding to $|mv|_{W_p^{s/2}(0,T;\, L_p(\Gamma))}$ analogously. We have to consider

$$\left(\int_0^T \int_0^T \frac{\|[m(t)-m(\tau)]v(t)\|^p_{p,\Gamma}}{|t-\tau|^{1+\frac{s}{2}p}} d\tau\, dt\right)^{1/p} . \tag{8}$$

After application of Hölder's inequality with q', q in the integral over Γ, splitting the term with $\frac{1}{|t-\tau|}$ according to

$$1 + \frac{s}{2}p = (\frac{1}{r'} + (\frac{s}{2}+\delta)p) + (\frac{1}{r} - \delta p) \quad \text{with} \quad \delta > 0$$

and applying Hölder's inequality with r', r in the $\tau - t$-integral, the expression in (8) can be estimated by

$$\left(\int_0^T \int_0^T \frac{\|m(t)-m(\tau)\|^{pr'}_{pq',\Gamma}}{|t-\tau|^{1+(\frac{s}{2}+\delta)pr'}} d\tau\, dt\right)^{1/pr'} \left(\int_0^T \int_0^T \frac{\|v(t)\|^{pr}_{pq,\Gamma}}{|t-\tau|^{1-\delta pr}} d\tau\, dt\right)^{1/pr} .$$

Carrying out the integration with respect to τ in the last integral, we arrive at the desired result. □

Lemma 3.3. *For $u \in W_p^{2,1}(\Omega_T)$ the inequality*

$$\|u\|_{L_{pr}(0,T;L_{pq}(\Gamma))} + \|\nabla_x u\|_{L_{pr}(0,T;\, L_{pq}(\Gamma))}$$
$$\leq \varepsilon \|u\|_{W_p^{2,1}(\Omega_T)} + c(\frac{1}{\varepsilon}) \left(\|u_0\|_{p,\Omega} + T^{1/p'}\|\partial_t u\|_{p,\Omega_T}\right)$$

holds, where $\varepsilon > 0$ *is arbitrary,* $c(\frac{1}{\varepsilon})$ *is a constant that increases with decreasing* ε *and* r, q *are as follows:*

$$\begin{array}{ll} q = \frac{n-1}{n-\theta p} & \text{if } \theta \in (\frac{1}{p}, \min(1, \frac{n}{p})) \\ q < \infty & \text{if } \theta = \frac{n}{p} < 1 \\ q = \infty & \text{if } \theta \in (\frac{n}{p}, 1) \end{array}$$

and $r = \frac{2}{1+\theta}$ *in each case.*

Proof. By standard embedding results (cf. [1], 5.14 Lemma, 5.16 Corollary)

$$W_p^\theta(\Omega) \hookrightarrow W_p^{\theta-1/p}(\Gamma) \hookrightarrow L_s(\Gamma)$$
$$\text{for } \frac{1}{s} \geq \frac{1}{s^*} := \frac{1}{p} - \frac{\theta - 1/p}{n-1}, \text{ if } \theta > \frac{1}{p} \text{ and } s^* > 0 \,.$$

The last condition is equivalent with $\theta p < n$. If $s^* = 0$, s arbitrary large is allowed, and if $s^* < 0$, $s = \infty$ is allowed. Putting $s = pq$, we thus have

$$\|u(t)\|_{pq,\,\Gamma} + \|\partial_j u(t)\|_{pq,\,\Gamma} \leq c \cdot \|u(t)\|_{1+\theta,\,p,\,\Omega} \leq c \cdot \|u(t)\|_{2,\,p,\,\Omega}^{\frac{1+\theta}{2}} \|u(t)\|_{p,\,\Omega}^{\frac{1-\theta}{2}}$$

for q as given in the Lemma and $\theta < 1$. The last estimate is the interpolation inequality (cf. [2], 4.1 Proposition). From the last estimate we get for $r = \frac{2}{1+\theta}$ and $v \in \{u, \partial_j u\}$

$$\int_0^T \|v(t)\|_{pq,\,\Gamma}^{pr} d\,t \leq c \int_0^T \|u(t)\|_{2,\,p,\,\Omega}^{p} \|u(t)\|_{p,\,\Omega}^{p(r-1)} d\,t,$$

since

$$\frac{1-\theta}{2} = \frac{1}{r'}, \; \frac{pr}{r'} = p(r-1) \,.$$

Inserting the estimate

$$\sup_{t \in [0,T]} \|u(t)\|_{p,\,\Omega} \leq \|u_0\|_{p,\,\Omega} + T^{1/p'} \|\partial_t u(t)\|_{p,\,\Omega_T},$$

which follows easily by Hölder's inequality from

$$u(\cdot, t) = u_0(\cdot) + \int_0^t \partial_t u(\cdot, s) d\,s \,,$$

we get

$$\|v\|_{L_{pr}(0,T;\,L_{pq}(\Gamma))} \leq \|u\|_{W_p^{2,1}(\Omega_T)}^{1/r} \left(\|u_0\|_{p,\,\Omega} + T^{1/p'} \|\partial_t u\|_{p,\,\Omega_T} \right)^{1/r'} \,.$$

Hölder's inequality "with ε" with r, r' yields the assertion of the Lemma. □

Lemma 3.4. *Let* $1 \leq p < \infty$ *and* $0 < s < 2$*. For every* $\varepsilon > 0$ *there exists* $T_0(\varepsilon)$ *such that for* $T \leq T_0$

$$\|u\|_{L_p(0,T;\,W_p^s(\Omega))} \leq \varepsilon \|u\|_{W_p^{2,1}(\Omega_T)} + T^{1/p} c(1/\varepsilon) \|u_0\|_{p,\,\Omega} \,.$$

Proof. See [7], Lemma 3.4. □

References

[1] Adams, R.A., Sobolev Spaces, Academic Press, New York-San Francisco-London, 1975.

[2] Amann, H., Global existence for semilinear parabolic systems, J. Reine Angew. Math. 360 (1985), 47–83.

[3] Kufner, A., John, O., and Fučik, S., Function Spaces, Noordhoff Int. Publ., Leyden, 1977.

[4] Ladyzhenskaya, O.A., Solonnikov, V.A., and Uralceva, N.N., Linear and Quasi-linear Equations of Parabolic Type, Transl. Math. Monogr., 23, Amer. Math. Soc., Providence, RI, 1968.

[5] Maugeri, A., Palagachev, D.K., and Softova, L.G., Elliptic and Parabolic Equations with Discontinuous Coefficients, Wiley-VCH, Berlin, 2000.

[6] Softova, L.G., Oblique derivative problem for parabolic operators with VMO coefficients, Manuscripta Math. 103 (2000), 203–220.

[7] Weidemaier P., On L_p-estimates of optimal type for the parabolic oblique derivative problem with VMO-coefficients, Differential Integral Equations 18 (2005), 935–946.

Peter Weidemaier
Fraunhofer-Institut Kurzzeitdynamik
Eckerstr. 4
D-79104 Freiburg, Germany
e-mail: weide@emi.fhg.de

Abbas Bahri
Flow Lines and Algebraic Invariants in Contact Form Geometry
Birkhäuser